Weaponeering: An Introduction

Third Edition, Volume 1

Weaponeering: An Introduction

Third Edition, Volume 1

Morris R. Driels
Professor Emeritus, U.S. Naval Postgraduate School
Monterey, California

AIAA EDUCATION SERIES

AIAA
American Institute of Aeronautics and Astronautics, Inc.

American Institute of Aeronautics and Astronautics, Inc., Reston, Virginia

1 2 3 4 5

Library of Congress Cataloging-in-Publication Data

Names: Driels, Morris R., author.
Title: Weaponeering: an introduction / Morris R. Driels, U.S. Naval Postgraduate School, Monterey, California.
Description: Third edition. | Reston : American Institute of Aeronautics and Astronautics, Inc., [2019] | Series: AIAA education series | "The book is written in two volumes: Volume 1: Introduction to Weaponeering ... Volume 2: Advanced Weaponeering."—ECIP galley. | Includes bibliographical references and index. | Contents: v.1. Introduction to Weaponeering—v.2. Volume 2: Advanced Weaponeering.
Identifiers: LCCN 2019050999 | ISBN 9781624105357 (v. 1 ; hardcover)
Subjects: LCSH: United States—Armed Forces—Weapons systems—Evaluation.
Classification: LCC UF503 .D75 2019 | DDC 355.8–dc23
LC record available at https://lccn.loc.gov/2019050999

This book is dedicated to all the men and women who have served in our Armed Forces, past, present, and future. They and their families bear the high cost of the freedom we enjoy to live our lives the way we choose.

CONTENTS

PREFACE

History of the Joint Technical Coordinating Group for Munitions Effectiveness (JTCG/ME)

In 1963, an Army–Air Force panel known as the Close Air Support Board issued a report calling attention to large gaps and gross inaccuracies in data then published on air-to-surface (AS) nonnuclear munitions. To remedy these inconsistencies, the board recommended production of a joint service publication containing a comprehensive list of targets with corresponding data on the effectiveness of aerially delivered munitions suitable for defeating those targets. Responding to this challenge, the Joint Chiefs of Staff requested that a joint service working group correct the data deficiencies and prepare a Joint Munitions Effectiveness Manual (JMEM) for air-to-surface weapons.

The Army was tasked to be the lead service. The chairman of the JTCG/ME established an ad hoc group of military and civilian scientists scattered throughout the Defense Department. The group developed the first standardized methodology for evaluating weapons, and this product was a coordination draft of a multiservice manual entitled *Joint Munitions Effectiveness Manual for Air-Delivered Non-nuclear Weapons.* This manual was accepted not only by the scientists and military professionals, but also by the Secretary of Defense. The latter requested that a similar approach be applied to surface-to-surface weapons.

In the fall of 1965, the original ad hoc group was given formal status as the Joint Technical Coordinating Group for Munitions Effectiveness (JTCG/ME) by the Joint Logistics Commanders. By mid-1966, the JTCG/ME was supported by three working groups: Target Vulnerability, Chemical and Biological, and the original Air-to-Surface (JMEM/AS) Group. These working groups, in turn, controlled the activities of subgroups created and tailored for specific aspects of the parent group's mission. The JMEM/AS had subgroups for weapon characteristics, delivery accuracy, flame and incendiary effects, methodology, and publications. A separate Basic Manual Working Group, chaired by the Defense Intelligence Agency, was established to maintain the *Weapon Effectiveness, Selection, and Requirements—Basic JMEM/AS Manual.*

In January 1967, JTCG/ME brought in a JMEM Production Contractor to provide support for the development and production of technical handbooks, JMEM, special reports, and related publications. By 1967, the JTCG/ME was concerned with deriving or validating data not only through tests, experiments, and mathematical models, but also from direct inputs of data from the battlefield. The JTCG/ME Wound Data and Munitions Effectiveness Team gathered wound data from southeast Asia (SEA). Later, the team expanded the scope of its battlefield collection to include materiel and sent specialized teams to SEA under an effort labeled the Battle Damage Assessment and Reporting Program (BDARP). These data are currently stored in the Survivability/Vulnerability Information Analysis Center (SURVIAC) at Wright-Patterson Air Force Base and are available for study and research by DoD agencies and contractors.

In September 1967, a separate major group, JMEM/SS, was created to examine and produce data on surface-to-surface munitions. Manuals containing data on individual surface-to-surface weapons were published and revised as new targets or munitions were developed. To complete the weapon–target interface, the JTCG/ME established the Anti-Air Working Group in 1976. Additionally, a Red-on-Blue Working Group was formed in 1977 to address the effectiveness of Red munitions on Blue targets. A Special Operations Working Group, initiated in 1983, provided target vulnerability and weapon effectiveness studies for Special Forces.

In 1994, the JTCG/ME was reorganized with four major working groups: Air-to-Surface, Surface-to-Surface, Anti-Air, and Vulnerability (including Special Operations) to cover the spectrum of weapon effects issues. In addition, each working group is supported by a formally chartered Operational Users Working Group (OUWG). The JTCG/ME Program Office is the focal point for all JTCG/ME efforts. It coordinates the efforts of the working groups while the execution of those efforts is the responsibility of the working group chairmen.

In May 1999, the Office of the Secretary of Defense (OSD) revised DoD Directive 5000.2R—Mandatory Procedures for Major Defense Acquisition Programs (MDAPS) and Major Automated Information Systems (MAIS) Acquisition programs—to require the procuring agency to provide weapon effectiveness data for use in JMEM for weapons in the acquisition process prior to their achieving initial operational capability. OSD also required that these data be prepared using methodology coordinated with the JTCG/ME.

The resources to operate the program are distributed to the JTCG/ME directly from the Director of Operational Test and Evaluation, Office of the Secretary of Defense. Chartered by the Joint Logistics Commanders with two-star or civilian-equivalent Offices of Primary Responsibility

(OPR), the JTCG/ME is guided by a steering committee composed of members from each of the services, the Joint Chiefs of Staff (J-8), the Defense Intelligence Agency, and the Defense Threat Reduction Agency. The Army continues as DoD Executive Agent, as it had been with the original ad hoc group, and the Director of the U.S. Army Materiel Systems Analysis Activity at the Aberdeen Proving Ground, Maryland, presides.

The committee is composed of civilian and military weapon/munitions experts throughout the Defense Department to ensure objective, scientific guidance in the development and employment of nonnuclear munitions. The following organizational chart reflects the original structure of the JTCG/ME and indicates the domain stovepipes created for individual weaponeering tools were in the three main areas of application:

1. Air-to-surface (AS)
2. Surface-to-surface (SS)
3. Anti-air (AA)

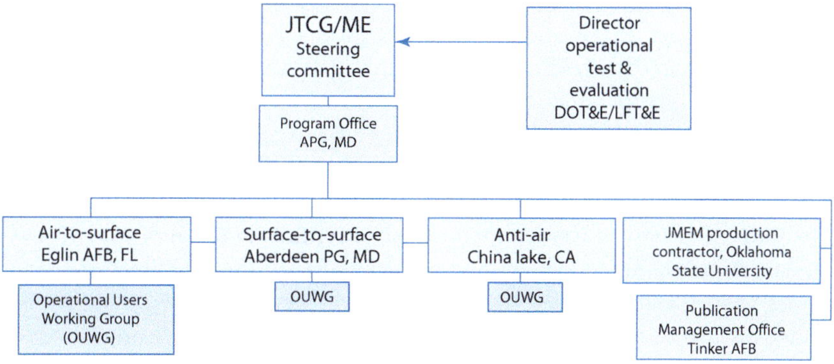

Original organization of the JTCG/ME.

The principal products of the JTCG/ME are known as the Joint Munitions Effectiveness Manuals (JMEM). Given the organizational structure shown, it is no surprise to find that each working group developed its own weaponeering tool to serve its customer and service base. The JMEM were developed to provide a set of data and methodologies that would permit a standardized comparison of weapon effectiveness across all service communities. For example, these data and methodologies should enable a Department of Defense planner in the Pentagon, a Navy weapons officer onboard a ship, and an Air Force targeting specialist in Korea to arrive at similar estimates for the effectiveness of a particular weapon employed under specified conditions. This use of standardized data and methodologies facilitates a wide range of activities beyond the planning and execution of operations, such as weapon development and procurement, management of stockpile levels, and prepositioning of war-loads.

Of greatest importance to the targeting process, JMEM data and methodologies are crucial for the development of force employment options contained in the Courses of Action (COA) nominations developed for theater commanders, and the resulting execution tasking orders for tactical units. The JMEM include detailed data on the physical characteristics and performance of weapons and weapon systems, descriptions of the mathematical methodologies that employ these data to generate effectiveness estimates, software that permit users to calculate effectiveness estimates, and precalculated weapon effectiveness estimates.

Historically, JMEM began with sets of manuals that provided data and methodologies for weaponeering. Over time these manuals were supplemented by computer programs to ease the computational burden on the user, to derive results faster, and to allow more realistic models to be incorporated. In recent years, the JMEM have evolved into CD-ROM products. The first and most widely used was for the air-to-surface community and was known as the Joint Air-to-Surface Weaponeering System (JAWS). Subsequently, the SS Working Group produced a functionally similar product known as the JMEM/SS Weapons Effectiveness System (JWES). The corresponding product for the Anti-Air group is the Joint Anti-Air Combat Effectiveness Model (J-ACE).

Around 2005, several of the Combatant Commanders (COCOM) became involved in guiding the development of JMEM tools and moved them towards a target-centric model. From their perspective, it is more important when investigating an attack on a target to estimate the effectiveness of all weapons that could logically be used irrespective of whether they are air launched or ground launched. Up until this point, the scenario-based tools were separate programs, usually running on different computers. As a result, it was decided to integrate the principle tools embodying the JMEM into a single product, the JMEM Weaponeering System (JWS).

This involved a substantial effort in rationalization between the working groups that involved the following subtasks:

1. Review target vulnerability data for common targets (e.g., T-72 tank) to ensure consistency.
2. Review effectiveness methodologies for common weapon systems (e.g., air and surface machine guns), again to ensure consistency.
3. Provide user transparency regarding the methodologies behind the effectiveness calculations.

This led not only to a different software structure for the single flagship product JWS, but also to a structural reorganization of the JTCG/ME itself in order to ensure a less stovepipe approach to the weapon effectiveness process. This revised organizational structure is shown in the following figure.

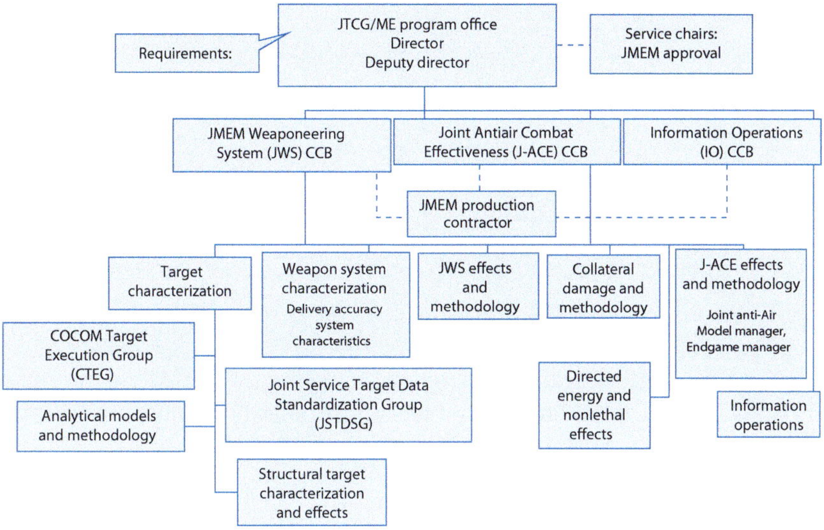

Revised organization of the JTCG/ME.

As a result of this amalgamation, JWS incorporates both JWES and JAWS together with a common database and allows almost any type of conventional weapon to be played against any type of target supported by the JMEM. Incorporation of J-ACE within the JWS framework is also planned for the future.

OBJECTIVES OF THE BOOK

The main objective of this book is to assist those engaged in assessing the effects of conventional weapons to understand the components, limitations, and techniques needed to perform that assessment. The intent is to explain the underlying methodologies behind various weaponeering tools rather than how to use the tools themselves.

This book describes the principle methods used by the JMEM to predict the probability of successfully attacking ground targets using air-launched or ground-launched weapons. Air-launched weapons include guided and unguided bombs, air-to-ground missiles, laser- and GPS-guided weapons, rockets, and guns. Surface engagements will cover both direct and indirect fire weapons. A chapter on anti-air weapons effects is also included, which involves airborne targets attacked with air- or ground-based threats.

The author outlines the various methodologies used in operational products of the JTCG/ME including JWS that are widely used in the U.S. Army, Navy, Air Force, and Marine Corps, as well as military services in some other countries.

Over the years, I have frequently seen research work being undertaken only to discover at a later date previous material dealing with the same subject already in existence but unknown to those involved in the research. As a result, I have tried to include in the book everything I know about weaponeering in the hope it will serve as a reference work to avoid wasting time and other resources "reinventing the wheel." The book is written in two volumes:

1. *Volume 1: Introduction to Weaponeering* covers the basic subjects that need to be understood to compute weapon effects based on the concept of an effectiveness index (EI). These subjects include weapon delivery accuracy, warhead lethality, and target vulnerability. This approach gives reasonable, rapid results and allows an easier understanding of current methods used to estimate weapon effects.

2. *Volume 2: Advanced Weaponeering* describes some of the more complex methodologies used in JWS and includes specific techniques for dealing with particular targets such as personnel, buildings, bunkers, bridges, and collateral damage concerns. In addition, some of the basic science underlying the methods is expanded including topics such as air blast, ground penetration, and projectile trajectory.

ACKNOWLEDGEMENTS

This book been produced with the assistance of many people and organizations, and represents an accumulation of our collective knowledge of this subject. In particular, the considerable support of Bryan Paris and David Chung of the JTCG/ME Program Office and Chantal Roman of AMSAA in helping bring this work to the widest possible audience is greatly appreciated.

I am also very happy to acknowledge the time, effort, and tolerance of the subject matter experts involved in the review process for this third edition. I hope I have represented their various areas of expertise in a way that truly reflects their valuable contributions to the subject. Any errors in the manuscript are mine, not theirs.

NOMENCLATURE

A_{ET}	effective target area
A_L	lethal area
A'_{ET}	single submunition effective target area
A_P	presented area
A_T	target area
A_V	vulnerable area
a	ratio of weapon radii
a_L	single submunition lethal area
C	apparent contrast
C_d	coefficient of drag
C_T	threshold contrast
C_0	intrinsic contrast
CI_X	center of impact, x
CI_Z	center of impact, z
D_0	shaping factor in Carleton damage function
$E(x)$	expected value of x
F_d	drag force
F_R	fractional coverage in range
$F(x)$	cumulative density function *CDF*
$f(x)$	probability density function *PDF*
g	acceleration due to gravity
H	Harp angle
H_A	door or tunnel height
h_{AGL}	height/altitude above ground level
h_{ASL}	height/altitude above mean sea level
h_{AT}	height/altitude above target
I	impact angle, angle of fall
L	submunition pattern length
L_A	length of target area
L_B	length of lethal area with ballistic dispersion
L_{EP}	effective pattern length
L_{ET}	length of effective target area
L_M	minefield length

L_P	stick pattern length
L_S	stick length
L_{SH}	shadow length
L_V	volley length
L'_{ET}	2 x weapon radius in range
M	Mach number
MAE_B	mean area of effectiveness for blast
MAE_{bldg}	mean area of effectiveness for buildings
MAE_F	mean area of effectiveness for fragments
m	mass
N	number of cycles resolved by the sensor on the target
n_b	number of submunitions
n_{od}	degree of stick overlap in deflection
n_{or}	degree of stick overlap in range
n_p	number of intervalometer pulses
n_r	number of weapons released per pulse
P_{CAS}	probability of casualty
P_{CD}	conditional probability of damage
P_{CDS}	PCD in final aggregated stick
P_{CD1}	PCD in lethal area expended for ballistic dispersion
P_{GE}	proportion of gross errors
P_{HIT}	probability of hit; proportion of direct hits
$P_{I/H}$	probability of incapacitation given a hit
P_K	probability of kill
$P_{K/D}$	probability of kill given a detonation
$P_{K/H}$	probability of kill given a hit
$P_{k/sh}$	shotline PK
$P_{K\text{-}BLAST}$	probability of kill due to blast
$P_{K\text{-}FRAG}$	probability of kill due to fragments
P_L	probability of launch
P_{NM}	proportion of near misses
P_S	probability of damage due to a single submunition
$P_{S/H}$	probability of survival given a hit
P_{VA}	probability of visually detecting a mine
R	reliability
R_A	cluster munition edge effect
R_B	bomb fall (ballistic) range
R_D	designation range
R_G	ground range
R_R	range at which weapon is released
R_{RQ}	required range for maneuver
R_{UNMASK}	unmask range of terrain
R_{VIS}	visibility range
R_x	weapon radius

R_y	weapon radius
R_0	range offset
s	Laplace operator
T	minefield threat
t	time
t_f	functioning time for cluster munitions
W	submunition pattern width
W_A	target area width
W_{ACT}	actuation width of a sea mine
W_B	width of lethal area enlarged for ballistic dispersion
W_b	bridge width
W_D	actuation width of a mine
W_{EP}	effective pattern width
W_{ET}	effective target area width
W_M	minefield width
W_P	pattern width
WR_d	weapon radius-deflection
WR_r	weapon radius-range
W_S	stick width
W_V	volley width
W_x	wind speed in range
W_z	crosswind speed
W'_{ET}	2 x weapon radius in deflection
x_{bd}	standard deviation of ballistic dispersion, deflection
x_{br}	standard deviation of ballistic dispersion, range
z	standardized normal random variable

Acronyms

AAM	air-to-air missile
AGM	air-to-ground missile
AGR	air-to-ground ranging
AIS	abbreviated injury scale
AJEM	advanced joint effectiveness model
AOA	angle of attack
AOF	angle of fall
AP	antipersonnel; armor piercing
APAM	antipersonnel and antimateriel
API	armor-piercing incendiary
ARM	antiradiation missile
ASII	adjusted severity of injury scale
ASL	azimuth steering line
ASUGM	air-to-surface guided munition
AT	antitank

ATA	apparent target area
ATACMS	Army tactical missile system
AVMAT	averaging matrix (program)
BAD	behind armor debris
BAS	Bridge Analysis System
BBF	base burn factor
BCS	battery computer system
BEI	bridge effectiveness index
BIM	Bridge Identification Module
BINOMDIST	binomial distribution function in Excel
BLU	bomb live unit
BOC	bombing on coordinates
CALCM	conventional air-launched cruise missile
CBU	cluster bomb unit
CCIP	continuously computer impact point
CCRP	continuously computed release point
CD	collateral damage
CDE	collateral damage estimation
CDF	cumulative density function
CE	casualty estimation
CEB	combined effects bomblet
CEM	combined effects munition
CEP	circular error probable
CER	collateral effects radius
CHA	collateral hazard area
CJCSI	Chairman of the Joint Chiefs of Staff Instruction
CLCL	common layered cratering library
CM	countermeasures
CO	collateral object
COV	center of vulnerability
COVART	Computation Of Vulnerable Area Tool
cd	critical dimension
c.g.	center of gravity
c.p.	center of pressure
DA	minefield area density
DEP	deflection error probable
DEP′	deflection error probable with ballistic dispersion
DF	minefield frontal density
DGPS	differential GPS
DMPI	desired mean point of impact
DOB	depth of burst
DOF	degrees of freedom
DOP	dilution of precision
DPI	desired point of impact

DSMAC	digital scene matching area correlation
DT	rocket motor delay time
DTRA	Defense Threat Reduction Agency
DU	depleted uranium
ECEF	Earth centered, Earth fixed
EEW	equivalent explosive weight
EFC	expected fractional coverage
EFD	expected fractional damage
EFP	explosively formed projectile
EI	effectiveness index
EMD	effective miss distance
EO	electro-optical
ER	extended range
FA	field artillery
FAE	fuel-air explosive
FALT	failure analysis logic tree
FASCAM	family of scatterable mines
FASGEN	Fast Shotline Generator
FATEPEN	Fast Air Target Encounter Penetration Method
FC	fractional coverage
FCC	fire control computer
FCS	fire control system
FD	fractional coverage in deflection
FEM	finite element modeling
FFAR	folding fin aircraft rocket
FFE	fire for effect
FIST	Fast Integrated Structural Tool
FLIR	forward-looking infrared
FMU	fuze module unit
FOR	field of regard
FOV	field of view
FS	fuze setting
FTT	fixed target track
GAM	GPS-aided munition
GATS	GPS-aided targeting system
GAU	gun ammunition unit
GBU	guided bomb unit munition unit
GFSM	General Full Spray Materiel (Program)
GFSP	General Full Spray Personnel (Program)
GP	general purpose (bomb)
GPS	global positioning system
GUI	graphical user interface
GunCAJ	gun computer aided JMEM
GWS	gun weapon system

GWTS	guided weapon trajectory software
HARM	high-speed antiradiation missile
HE	high explosive
HEAT	high explosive antitank
HEI	high explosive incendary
HEP	height error probable
HOB	height of burst
HTGW	hard target guided weapon
HTSF	hard target smart fuze
HUD	head-up display
ICM	improved conventional munition
IMEA	Integrated Munitions Effects Assessment
IMU	inertial measurement unit
INS	inertial navigation system
IR	infrared
ISS	injury severity score
J-ACE	Joint Antiair Combat Effectiveness Model
JAAM	Joint Antiair Model
JASSM	Joint Air-to-Surface Standoff Missile
JBAM	Joint Blast Analysis Methodology
JDAM	Joint Direct Attack Munition
JDAP	Joint Delivery Accuracy Program
JGEM	Joint Gun Engagement Model
JMAE	Joint Mean Area of Effects (Program)
JPF	joint programmable fuze
JS	Joint Staff
JSOW	joint standoff weapon
JWAM	Joint Weapon Accuracy Model
JWS	JMEM Weaponeering System
K–B	Kingery–Bulmash
LAM	lethal area matrix
LANTIRN	low-altitude navigation and targeting infrared night
LAR	launch acceptable region
LB	luminance of the background
LD	low drag
LGB	laser-guided bomb
LLLGB	low-level laser-guided bomb
LLW	low-level winds
LMD	lethal miss distance
LOS	line of sight
Ls	luminance of the stimulus
LVDG	large volley damage grid
MAI	mean area of injury
MC	Monte Carlo

ME	matrix evaluator
MED	mean effective diameter of a mine
MER	multiple ejection rack
MET	meteorological
MEVA	Modular Effectiveness Vulnerability Assessments
MIDB	modernized integrated database
MLRS	multiple launch rocket system
MOAB	massive ordnance air blast
MOP	massive ordnance penetrator
MPI	mean point of impact
MSL	mean sea level
MTS	minimum target size
MUVES	Modular Unix Based Vulnerability Suite
NCV	noncombatant casualty cutoff value
ND	number of target elements damaged; NORMDIST Excel function
NORMDIST	normal distribution function in Excel
NVDG	*N* volley damage grid
ORCA	Operational Requirements-Based Casualty Assessment
PD	probability of damage; point detonating (fuze)
PDF	probability density function
PE	probability of exposure; probability of encountering a mine; probable error
PH	personnel height
PHD	probability of damage given a hit
PID	positive identification
PMPT	Precision Weapon Planning Tool
PSC	prestressed concrete
PT	point track
PTO	prevent takeoff
PVTM	Passive Vehicle Target Model
PW	personnel width
PX	proximity (fuze)
QE	quadrant elevation
RAP	rocket-assisted projectile
RB_1	inner blast radius
RB_2	outer blast radius
RC	reinforced concrete
RCER	raw collateral effects radius
RED	risk estimates distance
REP	range error probable
REP′	range error probable including ballistic dispersion
ROE	rules of engagement

Rp	radius of submunition circular pattern (normal plane)
Rs	submunition reliability
RSS	root sum square
SADARM	sense and destroy armor
SAK-MD	suite of antiair kill chain models and data
SAM	surface-to-air missile
SAR	synthetic aperture radar
SDF	single degree of freedom
SEP	spherical error probable
SFPK	single-fragment probability of kill
SFW	sensor fuzed weapon
SGLR	sky/ground luminance ratio
SLAM	standoff land attack missile
SR	slant range
SSBM	surface-to-surface ballistic munition
SSPD	single sortie probability of damage
$SSPD_i$	SSPD for the *i*th sample sortie
STAR	sensitive target approval and review process
STMG	Smart Target Model Generator
SVDG	single volley damage grid
TAB	Tunnel Air Blast model
TARCOM	target complex
TAS	true airspeed
TD	decision time
TDD	target detection device
TDR	time between designation and release
TER	triple ejection rack
TF	rocket motor thrust factor
TFT	tabular firing tables
TGM	target geometric model
TLAM	Tomahawk land attack missile
TLE	target location error
TOF	time of flight
TOW	tube launched
TSPI	time-space-position information
TST	time-sensitive targets
TTG	time to go
UERE	user-expected range error
UWTS	unguided weapon trajectory software
VT	variable time (fuze)
WCER	warhead collateral effects radius
WCMD	wind corrected munitions dispenser

Chapter 1 Overview of Weaponeering and Weapon Effectiveness

1.1 Definitions

In general terms, *weaponeering* can be defined as the process of determining the quantity of a specific type of weapon required to achieve a defined level of target damage, considering target vulnerability, weapon effects, munitions delivery error, damage criteria, probability of kill, weapon reliability, and so forth. In the operational arena, where planners are striving for the most effective use of limited resources, efficiency is a critical factor that must be considered in the weaponeering process.

Commanders want to use their weapons systems to inflict maximum damage on enemy plans, forces, and facilities. Under these conditions, weaponeering is probably best defined as the process by which the minimum force level and optimum ordnance needed to achieve a desired level of damage to a target or critical component, such that enemy warfighting is affected, are determined. Weaponeering methods may be employed to evaluate air-to-surface (AS) or surface-to-surface (SS) engagements; however, the nature of weaponeering is different for each domain. For AS applications there is usually a wide range of weapon options (e.g., missiles, guided or unguided bombs, one large weapon, several small weapons, rockets, guns, etc.). For SS, however, the choices are more limited (e.g., for an M1A1 tank shooting at a T72 tank, the round chosen is almost always the depleted uranium sabot penetrator). The Joint Munitions Effectiveness Manual (JMEM) will predict the performance of this round against the target at different ranges and target aspects.

1.2 Examples of a Weaponeering Outcome: Air Tasking Order

One product commonly associated with the output of the AS weaponeering process is the air tasking order (ATO). Table 1.1 shows a sample ATO that would be given to aviators prior to flying a mission.

Table 1.1 Sample Air Tasking Order (ATO)

MSN:	8088 **UNIT:** 222FS **FORCE:** 2XF/A-18 **C/S:** REGAL55 **TOTZ:** 0835
ORD:	SCL-855, FUZE FOR MAX CRATERING
PRITGT:	VUL HWY SEG, BV981.335 (GCC REQ FM-668)
DMPI:	RTE SEGMENT CROSSING DIKE
RMKS:	COODR WITH MSN 7345

Although many of these terms will be dealt with in detail later, a brief explanation is in order.

MSN	Mission number
UNIT	Operational unit executing the mission
FORCE	Aircraft to be used (two F/A-18s in this case)
C/S	Call sign for the mission
TOTZ	Time over the target (Zulu time)
ORD	Ordnance to use (standard conventional load 855 with fuze settings set to maximum cratering)
PRITGT	Description and coordinates of target (highway)
DMPI	Aimpoint (highway crossing a dyke)
RMKS	Remarks (coordinate with mission #7345)

Given the availability of aircraft, weapons, and the target to be attacked, the weaponeering process determines which weapons and how many are to be used in the attack in order to maximize the probability of damaging the target while minimizing the number of weapons and aircraft utilized, thereby minimizing the risk to aircrew. Figure 1.1 shows an F/A-18 aircraft with a standard conventional load of eight 500-lb Mk-82 unguided bombs.

Another example of the utility of weaponeering is in the assessment of collateral damage. Consider the image of a planned target area shown in Fig. 1.2. Here the weapon aimpoint is located at the center of the target (building) and a collateral effects radius (CER) circle is centered on it. Weaponeering involves the estimation of lethal effects expected from the warhead and the accuracy with which the weapon is delivered, both of which contribute to the CER. If no strike elements or collateral objects (COs) are identified in the vicinity of the target, the CER helps to support a decision on whether to proceed with the attack. Collateral damage and the calculation of the CER circles is a complex task and is dealt with in detail in Volume 2, Chapter 17.

The daily ATO is the published order that directs all the air missions in a campaign. Weaponeering, together with an affiliated activity known as *targeting,* will provide the planner with information on the selection of weapons, fuzes, aimpoints, and force levels required. Clearly both targeting and weaponeering do not take place in isolation from the strategic planning

Fig. 1.1 F/A-18 with a standard conventional load.

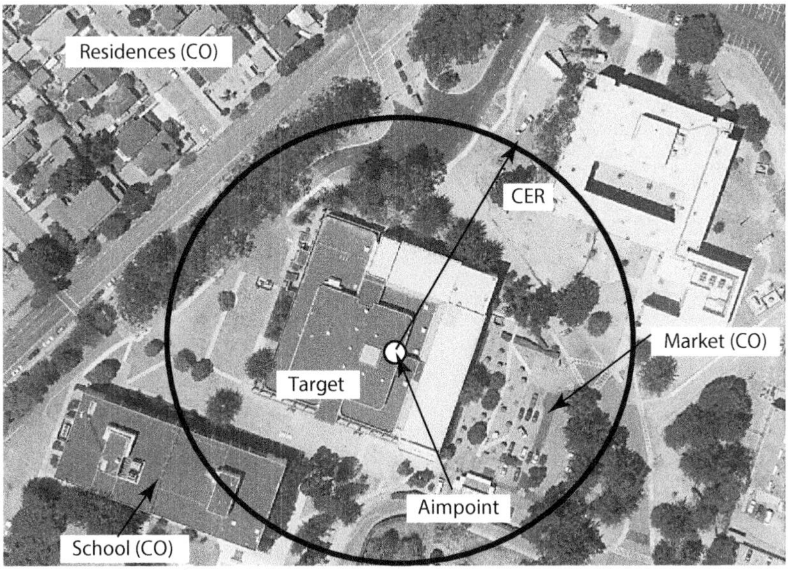

Fig. 1.2 Building target with surrounding collateral objects.

that accompanies warfare, but are an intrinsic part of it. A notional overview of this "bigger picture" is given in the next section.

1.3 Weaponeering: Part of a Larger Planning Cycle

Although weaponeering is an important piece of the overall attack plan, it is only one part. It is important to recognize that weaponeering lies at the midpoint in the recurring force application cycle, as indicated in Fig. 1.3, requiring inputs from previous steps in the cycle while providing a portion of the inputs to subsequent steps in the cycle.

The individual phases may be characterized by the following activities:

- *Objectives:* Outlined by the immediate commander and senior headquarters in their employment guidance, the mission objectives define which facets of enemy activity are to be affected by the mission. An example of a tactical objective would be to deprive the enemy of the use of electrical power within a given geographical region for a period of 14 days.
- *Target development:* Guided by defined tactical objectives and working with the target list developed by the command structure, targeting specialists select suitable target elements to achieve these objectives. These specialists, known as targeteers, have knowledge of the infrastructure of the region and can identify critical elements relating to the tactical objective. Using the previous example, targeteers would have access to the type of information shown in Fig. 1.4 for the power distribution system in the region of interest [1].

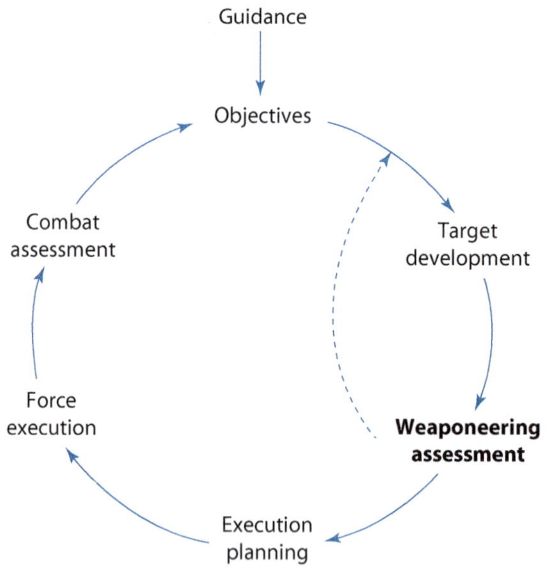

Fig. 1.3 Force application planning cycle.

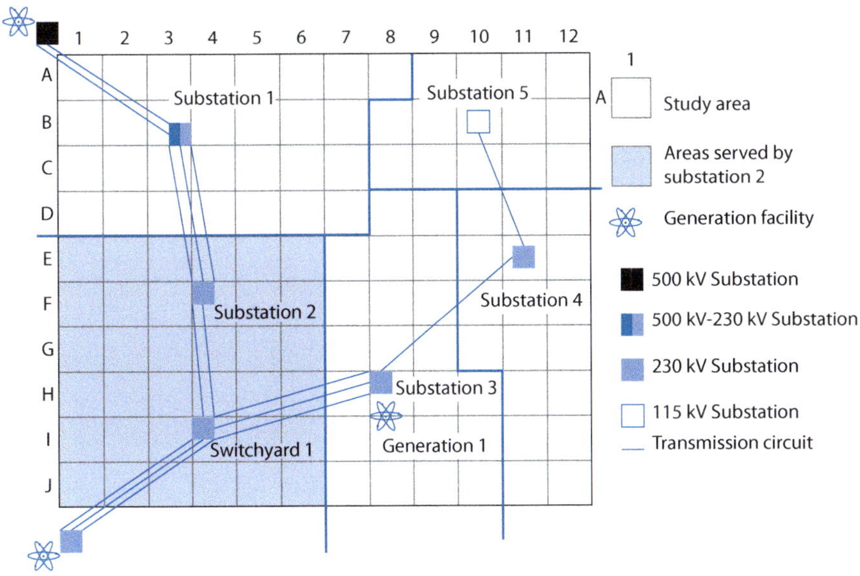

Fig. 1.4 Power distribution system in region of interest.

For example, assuming the substations are the focal points for power distribution in the regions shown, to meet the objective listed, substation 2 would be selected for attack to black out the lower left region of Fig. 1.4, and aimpoints in that station (Fig. 1.5) would be defined to produce the desired effect.

In this example, perhaps replacing the damaged transformers would take about 14 days; however, rebuilding the control building would take much longer, hence the transformers would be targeted. Note also that selection of substation 2 rather than switchyard 1 achieves the tactical objective while minimizing disruption to other regions, an example of mitigating collateral damage.

- *Weaponeering assessment:* Given the aimpoint and target elements to be attacked, the weaponeer selects the type and quantity of weapons needed to produce the required level of damage. This essentially defines the mission in terms of delivery system (aircraft, surface rockets), warhead, sorties, and target, and for air-launched weapons would produce the air tasking order described earlier; for surface-launched rockets it would generate firing orders.
- *Execution planning:* The outputs of the weaponeering process are essential inputs to execution planning, which is conducted on two levels. At the headquarters level, operations staffers assign missions to specific units, perform attack force packaging, task required mission support forces, determine attack timing, and outline communications and coordination

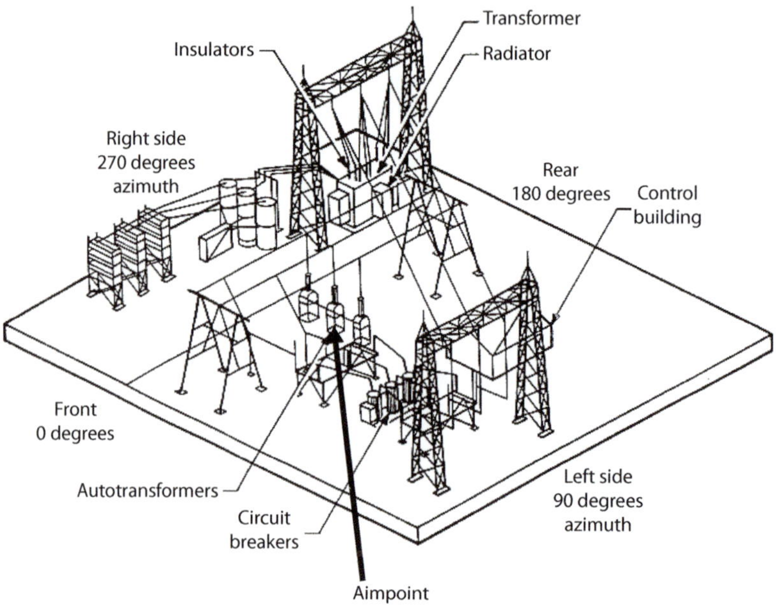

Fig. 1.5 Layout of substation 2 shown in Figure 1.4.

requirements; the products of this effort are also embodied in the daily air operations or air task orders. At the unit level, the execution-planning phase includes target study and detailed mission planning.

- *Force execution and combat assessment:* Once the mission is planned and executed, the results are assessed to determine if the objectives have been met or whether a restrike is required.

Note that this cycle may have internal feedback loops depending upon the operational level at which the weaponeering is done; for example, after execution of initial sorties, pilot feedback may suggest a more effective weapon is needed for the next sortie or restrike.

Table 1.2 outlines some of the key planning questions that should be resolved at each step in the planning cycle.

1.4 An Example of a Weaponeering Tool

Weaponeering has developed into a highly automated process for most operational users, principally through the availability of products such as the JMEM Weaponeering System (JWS). With such programs it is possible to select a target, choose a weapon, specify release conditions of the weapon from the aircraft, and compute the resulting probability of damage due to a single weapon, denoted at PD_1. JWS may also be used to determine if a surface-launched weapon can be used to attack the same target. Figure 1.6

Table 1.2 Key Force Application Planning Questions

Planning Step	Questions to Resolve
Objectives	What enemy activity is to be affected? Where should air effort be focused? What effects are desired? When and for how long?
Target development	Which targets relate to the objectives? Of those, which are most important? Of those, which are vulnerable? What are the critical target elements? Which target defenses must be attacked? What is the target prioritization? What is the optimum timing of attacks?
Weaponeering assessment	What are the appropriate damage criteria? What probability of damage is required? How much force and ordnance are needed? What are the optimum aimpoints? What are the recommended ordnance and fusing?
Execution planning (HQ level)	Which units are best suited to the task? When should the mission be executed?
Execution planning (unit level)	What support is required? What C^3 is required? What are the ingress/egress routes and tactics? What are the weapon delivery tactics?
Force execution/combat assessment	What was destroyed or damaged? Did weapons work as expected? Did attack produce intended effect?

shows a user interface to an unclassified version of an AS weaponeering program to compute PD_1; it indicates the various items of data needed for the computation.

In order to use a tool such as this, a weaponeer must understand the nature of the data that have to be input to the program, where to obtain them, and any constraints associated with them. The main areas of interest (and their associated data) needed to compute damage levels are:

- Weapon/aircraft accuracy
- Weapon trajectory characteristics
- Warhead damage mechanisms
- Target vulnerability
- A methodology to integrate these into a damage measure

More specifically, the main types of input data needed, at least for the sample program shown in Fig. 1.6, include:

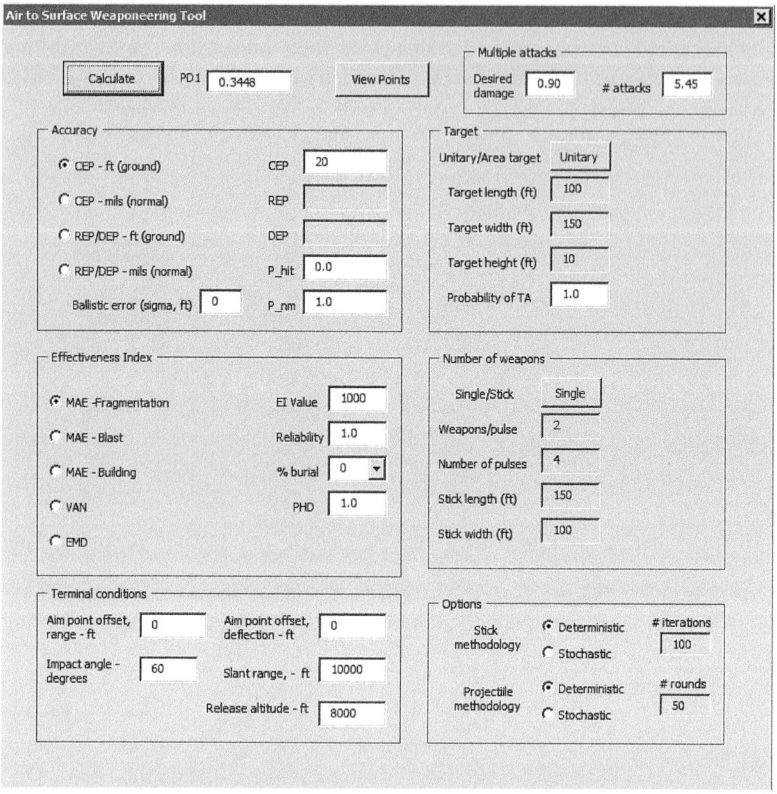

Fig. 1.6 Example of an automated weaponeering tool.

- Damage mechanism of the weapon
- What level of damage the target is to sustain
- Weapon accuracy
- Weapon effectiveness for the particular target
- Type of weapon used
- Release and terminal conditions
- Number of weapons used per pass
- Point at which the weapons are aimed
- Required damage level for the whole mission
- Trajectory of the weapon and its impact, or arrival state

The scenario depicted in the tool is for the release of a single unguided weapon against a single target. The specific data used are as follows:

- The accuracy is in terms of a CEP = 20 ft.
- The weapon lethal area is given by an $MAE_F = 1000$ ft^2
- Release conditions are such that the slant range is 10,000 ft with a weapon impact angle of 60 deg.

At the top of the sheet is the computed value of PD_1 against the target (one pass by one aircraft dropping one bomb) and is 0.3448, or 34.48%. The user has also entered the required total damage to be 0.9 (top right) and the tool has computed that 5.45 (i.e., 6) sorties will be required to achieve (at least) this much damage.

Each of these data requirements will be discussed in detail in the next section and in subsequent chapters of this book. The purpose is to provide users with as much background as possible into all the information needed to compute an estimate of damage to the target as well as the actual computational process itself. This will allow those who generate damage estimates to be aware of the assumptions and limitations inherent in the process, and as such to rely on the result to an appropriate degree.

1.5 Basic Components Needed to Compute Effectiveness

In this section we will outline those topics that have to be understood in order to compute the effectiveness of a particular weapon on a given target; in doing so we will suggest a road map for the rest of the material in the first part of this book. The reader is encouraged to observe the relationship between the topics that follow and the data entry requirements of the example weaponeering tool shown in Fig. 1.6.

1.5.1 Weapon Damage Mechanisms

Weapons may damage targets through a variety of methods; in general, the specification of the target will suggest the appropriate damage mechanism. For example, personnel targets are susceptible to blast and fragments, whereas an armored vehicle is susceptible to shaped charge. Most explosive-filled warheads will produce both blast and fragments, although a few are primarily blast weapons. The assessment of the damage-producing ability of a particular warhead is generally known as the process of determining the weapon or warhead characteristics. The basic damage mechanisms will now be reviewed.

1.5.1.1 Blast

A blast warhead is designed to achieve target damage primarily from overpressure effect. When a high explosive detonates, it is converted almost instantly into a gas at very high pressure and temperature. Under the pressure of the gases generated, the weapon case expands and breaks into fragments. The air surrounding the casing is compressed, and a shock (blast) wave is transmitted into it. Typical initial values for a high-explosive weapon are 200 kilobars of pressure (1 bar = 1 atmosphere) and 5000°C. As the wave passes a fixed point, the blast wave is perceived as a rapid rise from ambient to some peak pressure followed by an exponential decay to a

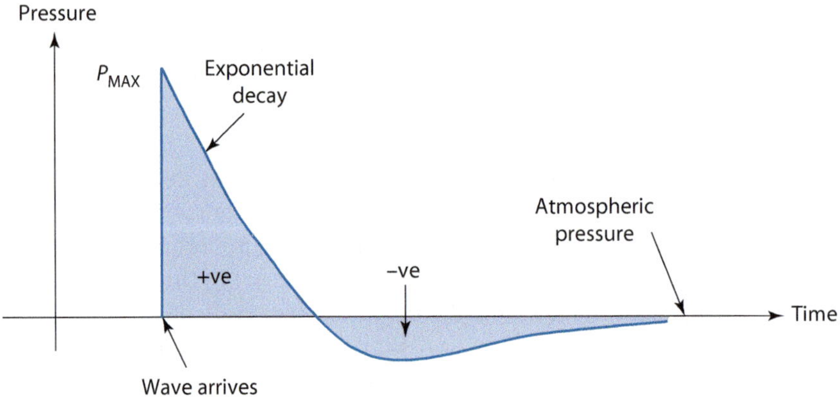

Fig. 1.7 Blast wave pressure—time profile.

value below atmospheric, followed by a return to atmospheric, as shown in Fig. 1.7.

The peak pressure is attenuated as the wave propagates from the detonation point, where the rate of attenuation is proportional to the expansion of the volume of gases behind the blast wave. In other words, the blast pressure is inversely proportional to the cube of the distance from the blast center ($1/R^3$)—see Fig. 1.8.

If the outside of a target is exposed to the blast wave, it will experience an overpressure followed by an underpressure, causing motion initially in the direction of the blast wave and then in the opposite direction. For buildings, damage to a single floor causing it to collapse may produce structural failure

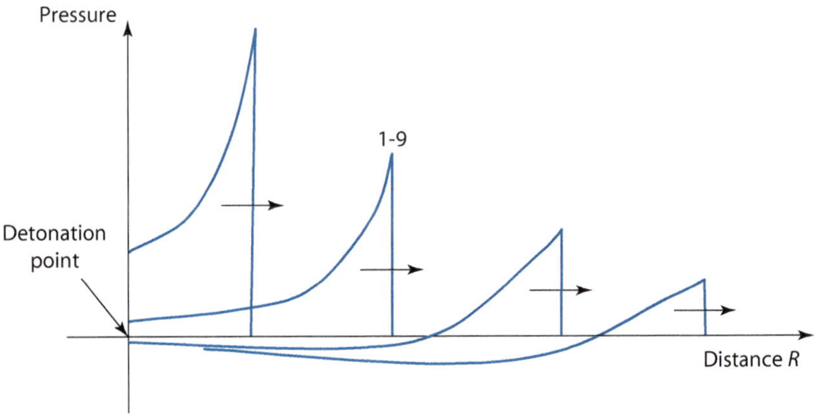

Fig. 1.8 Blast wave attenuation.

Fig. 1.9 Building damage due to external blast.

in the whole building through a process called *progressive collapse,* as shown in Fig. 1.9.

When an explosion occurs within a structure, the peak pressure associated with the initial shock front will be extremely high and, in turn, will be amplified by reflections from walls, floors, and ceilings within the structure. In addition, the expansion of gases from the explosion will exert additional pressures and increase the load duration within the structure. Thus the loading is the combined effect of shock (including reflections) and the quasi-static gas pressure over time. The effects of both pressures eventually may destroy the structure if it is not strengthened sufficiently or adequate venting for the gas and the shock pressure is not provided. Blast is particularly destructive against unreinforced buildings and personnel in open terrain, but may also damage vehicles if the warhead is detonated close enough.

An example of a weapon designed just to damage targets with blast is the Massive Ordnance Air Blast (MOAB) bomb shown in Fig. 1.10.

Although blast seems like a simple damage process—a spherically radiating pressure wave of diminishing overpressure—it is difficult to predict the damage sustained by a particular target type. This is because the interaction of the blast wave with the target produces many reflected and propagated waves that quickly turn the problem into a three-dimensional multiwave computation.

Fig. 1.10 GBU-43 guided bomb with blast warhead.

1.5.1.2 Fragmentation

Approximately 30% of the energy released by the explosive detonation is used to break up the case and impart kinetic energy to the fragments. The balance of available energy is used to create a shock front and blast effects. The fragments are propelled at high velocity, and after a short distance they overtake and pass through the shock wave. The rate at which the velocity of the shock front accompanying the blast decreases is generally much greater than the decrease in velocity of fragments, which occurs due to air friction. Therefore, the advance of the shock front lags behind that of the fragments. The radius of effective fragment damage, although target dependent, thus exceeds considerably the radius of effective blast damage in an air burst.

When a shell or bomb detonates, it expands to two or three times its static dimensions before breaking apart. A general purpose warhead then disintegrates into irregular-shaped fragments having a wide variety of masses and projection velocities (see Fig. 1.11).

In order to tailor the effect of fragments to specific target types, the breakup of the casing may be made more predictable by scoring either the inside or outside prior to filling with explosive (Fig. 1.12).

Whereas the effects of an idealized blast payload are attenuated by a factor roughly equal to $1/R^3$ (where R is measured from the detonation point), the attenuation of idealized fragmentation effects will vary as $1/R^2$ and $1/R$, depending upon the specific design of the payload. Herein lies

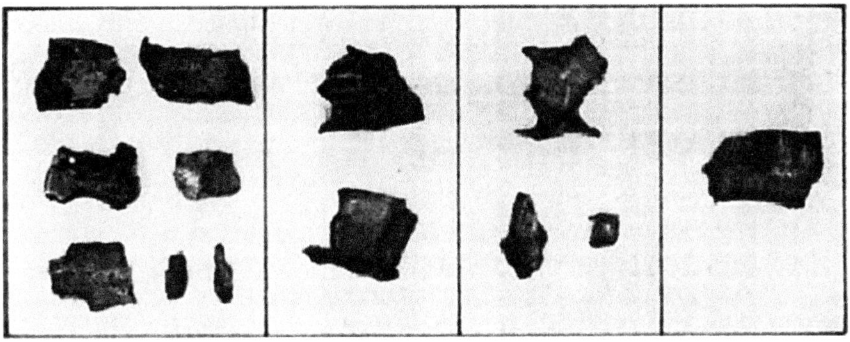

Fig. 1.11 Naturally formed fragments.

the principle advantage of a fragmentation payload: it can afford a greater miss distance and still remain effective because its attenuation is less.

Fragments are known as primary or secondary fragments depending on their origin. *Primary fragments* are formed as a result of the shattering of the casing of conventional munitions. These fragments usually are small and travel initially at velocities of the order of thousands of feet per second. *Secondary fragments* are formed as a result of high blast pressures on structural components and items in close proximity to the explosion. These fragments are somewhat larger in size than primary fragments and travel initially at velocities in the order of hundreds of feet per second. A hazardous fragment is one having an impact energy of 58 ft · lb (79 J) or greater. Figure 1.13 shows fragments projected radially from a detonation.

In the same class of warheads are single projectiles such as armor-piercing (AP) rounds and long rod penetrators, which damage targets

Fig. 1.12 Internal scoring of warhead casing.

Fig. 1.13 Fragments (outer ring) propagating from a detonating warhead.

using the same mechanism. Figure 1.14 shows the M829 subcaliber (120-mm) discarding sabot, depleted uranium (DU) round fired by the M1A1 tank. In this figure the larger diameter sleeve (sabot) ensures the round fits into the bore, but on exiting it peels back to leave the smaller diameter (about 1 in.) DU dart.

Even very small fragments of the order of a few grains (1 lb = 7000 grains) will cause severe injury to personnel; larger ones may damage unprotected vehicles and aircraft, as shown in Fig. 1.15.

We will not be able to specify exactly the number, size, velocity, and direction of individual, naturally formed fragments generated by a particular warhead. Through extensive testing, however, we are able to produce statistical estimates of these quantities.

Fig. 1.14 Kinetic energy penetrator.

Fig. 1.15 Vehicle damage due to fragments.

Another example of a warhead that produces only penetration effects is a 105-mm round that may be fired by either an artillery cannon or a tank. It produces several thousand 1-in. darts, or flechettes, that are effective against personnel and unprotected materiel targets (Fig. 1.16).

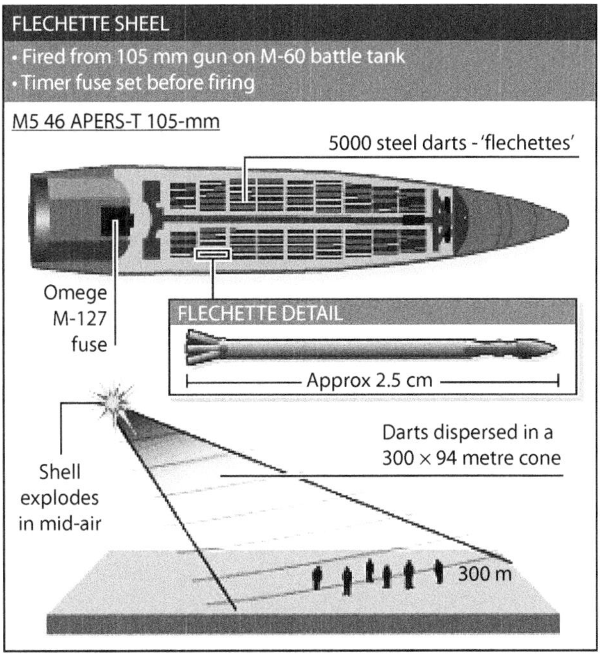

Fig. 1.16 Flechette artillery round.

1.5.1.3 Shaped Charge

A shaped charge warhead consists basically of a hollow liner of metal material, usually copper or aluminum, of conical, hemispherical, or other shape, backed on the convex side by explosive. A container, fuze, and detonating device complete the round. Figure 1.17 shows the M77 81-mm submunition that has a shaped charge in the lower portion and also a steel case for producing fragments.

When this warhead is in close proximity to the target, the fuze detonates the charge from the rear. A detonation wave sweeps forward and begins to collapse the metal cone liner at its apex. The collapse of the cone results in the formation and ejection of a continuous high-velocity solid jet of liner material. At the pressures produced by the explosion, the liner material flows like a fluid, and is therefore referred to as a jet.

The use of copper as a shaped charge jet will be discussed later in Chapter 7, but one important issue is the phase transformation from solid to liquid has to take place very rapidly; therefore, highly conductive materials are required.

Velocity of the tip of the jet is about 8500 m/s, while the trailing end of the jet has a velocity on the order of 1500 m/s. This produces a velocity gradient that tends to stretch out or lengthen the jet. The jet is then followed by a slug that consists of about 80% of the liner mass. The slug has a velocity on the order of 600 m/s.

When the jet strikes a target of armor plate or mild steel, pressures in the range of hundreds of kilobars are produced at the point of contact. This pressure produces stresses far above the yield strength of steel, and the target material also flows like a fluid out of the path of the jet, as shown in Fig. 1.18. This phenomenon is called *hydrodynamic penetration,* and the effect on a thick steel plate is shown in Fig. 1.19. In displacing the steel to produce the hole, the kinetic energy of the jet is expended.

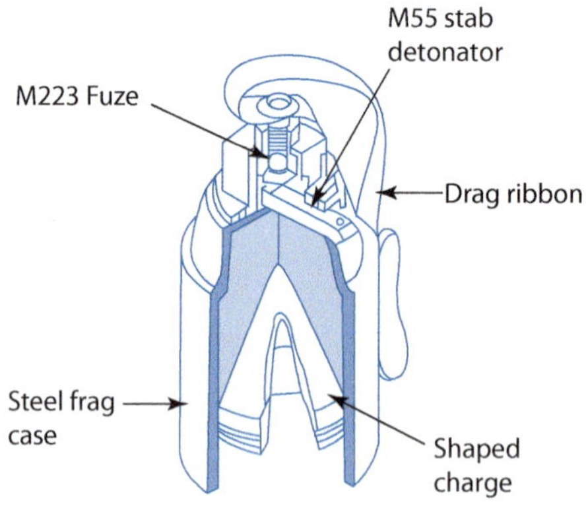

Fig. 1.17 Dual purpose fragment/shaped charge warhead.

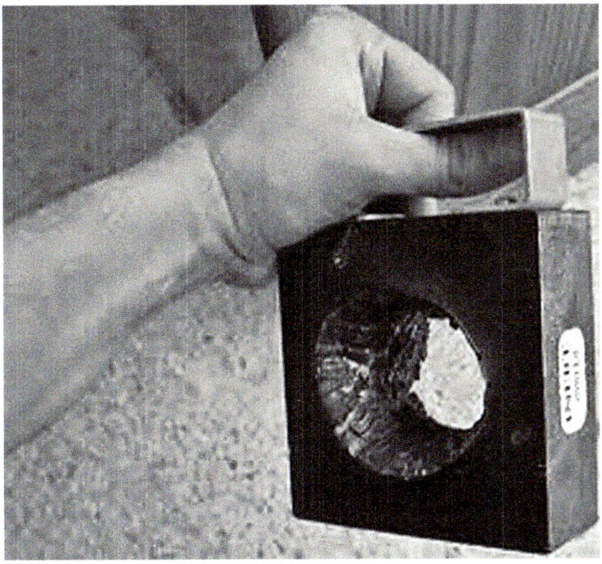

Fig. 1.18 Effect of a shaped charge on a steel plate.

The goal for damage to a target such as an armored vehicle is for the jet to produce a hole such as that shown in Fig. 1.18 while still having sufficient jet material remaining to enter the interior of the target and cause fire, injure the crew, and detonate munitions, as shown in Figure 1.19.

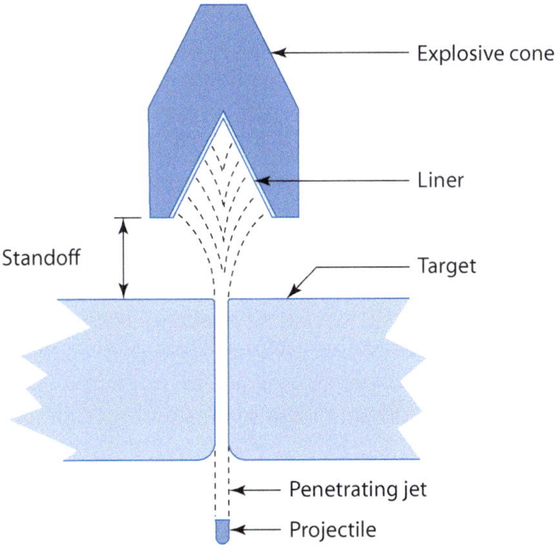

Fig. 1.19 Shaped charge damage mechanism.

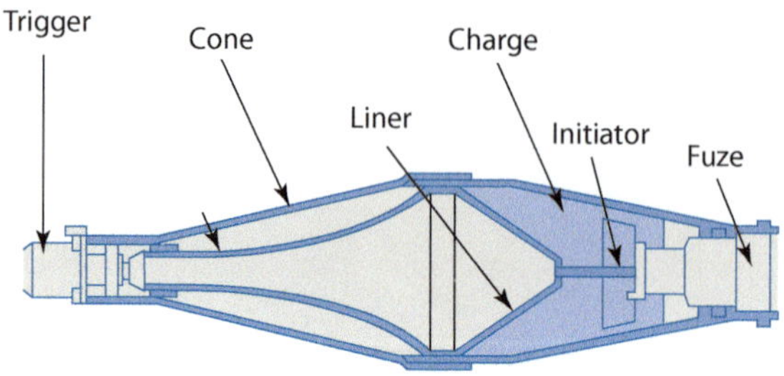

Fig. 1.20 Shaped charge (RPG-7) warhead.

The depth of penetration is maximized for a particular standoff (i.e., the distance between the warhead and target at detonation), and shaped charged warheads are fuzed at this optimum—see Fig. 1.20. In this design, the critical standoff is achieved by mounting a trigger on a hollow cone ahead of the charge with an electrical connection to the fuze located behind.

Unlike blast or fragmentation, a shaped charge weapon has to hit the target in order to damage it; as a result, a simple protective metal chain curtain suspended a few inches from the side of an armored vehicle will often cause the weapon to detonate harmlessly away from the hull.

1.5.1.4 Explosively Formed Projectile (EFP)

An EFP is similar to a shaped charge but is designed to travel over a larger standoff distance and maintains its solid structure instead of a liquid jet. A conventional shaped charge generally has a conical metal liner that projects a hypervelocity jet of metal able to penetrate to great depths into steel armor; however, in traveling over a long distance, the jet breaks up along its length into particles that drift out of alignment, greatly diminishing its effectiveness at a distance.

An EFP, on the other hand, has a liner in the shape of a shallow dish. The force of the blast molds the copper plate into any of a number of configurations, depending on how the plate is formed and how the explosive is detonated. Sophisticated EFP warheads have multiple detonators that can be fired in different arrangements, causing various types of waveforms in the explosive, resulting in a long rod penetrator, an aerodynamic slug projectile, or multiple high-velocity fragments. These generally remain intact and are therefore able to penetrate armor at long range, delivering a wide spray of fragments of liner material and vehicle armor backspall into the vehicle's interior, injuring its crew and damaging other systems.

As a rule of thumb, an EFP will perforate a thickness of armor equal to only about the diameter of its charge, whereas a typical shaped charge will go through six or more diameters.

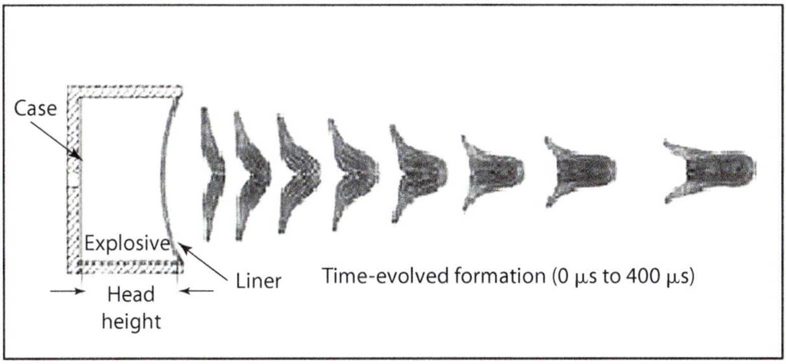

Fig. 1.21 Formation of an EFP.

Figure 1.21 shows the formation of the EFP following detonation of the warhead, and Fig. 1.22 shows the final form of the projectile.

Detonation distances for EFPs are on the order of feet rather than the few inches for a shaped charge.

1.5.2 Fuzing Systems

A fuze initiates warhead detonation when a particular event occurs, such as elapsed time from firing or large negative acceleration (impact), and under the optimal circumstances. Fuzes may be located in the nose or tail of the munitions, or both. Our discussion focuses on bombs dropped from aircraft, but the principles are the same for all fuzed munitions. There are two delays associated with general purpose fuzes:

1. The arming delay is the time after release before the bomb becomes armed.
2. The functioning delay is the time after impact before the bomb detonates.

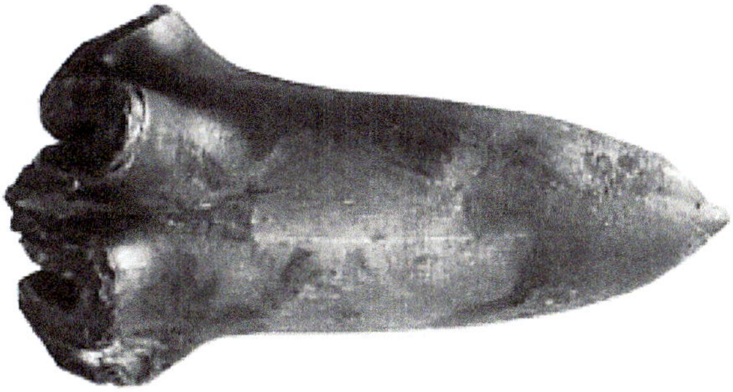

Fig. 1.22 Projectile shape.

Bombs are armed by one or more of the following methods:

- The arming vane, a small propeller, is rotated by airflow after weapon release. A specified number of rotations arm the fuze.
- The arming pin is ejected or withdrawn by a spring action or lanyard, releasing the arming mechanism and allowing the fuze to arm.
- The inertia fuze is armed by abrupt changes in the velocity of the bomb caused by the deployment of fins or ballutes.
- The electric fuze is armed by a time-delay circuit powered by a thermal battery activated by extraction of the arming lanyard upon bomb release.

Fuzes may be mechanical, electrical, or electronic in design.

- A *mechanical fuze* operates in a manner similar to the hammer firing a cartridge in a gun. When the bomb impacts a target, it drives a striker into the detonator setting off a train of explosives. More sophisticated mechanical components in the fuze produce functioning delays.
- An *electrical fuze* is similar to a mechanical fuze, but a series of resistor/capacitor networks discharge over set times to produce both arming and functioning delays.
- An *electronic fuze* is a solid state electrical fuze with some delays selectable from the aircraft.

Different options are obtained by mating different warheads to one of several fuzes. Common fuze functions may be described in one of the following ways:

- An impact, instantaneous, or point detonating (PD) fuze is designed to function on impact or a preset time after impact (impact delay). Detonation upon impact is selected for targets such as supply dumps when the main destructive energy desired is blast. For a building, a delayed detonation might be selected so a bomb can penetrate several floors before exploding. Figure 1.23 shows the FM 139/B electronic fuze, which can be

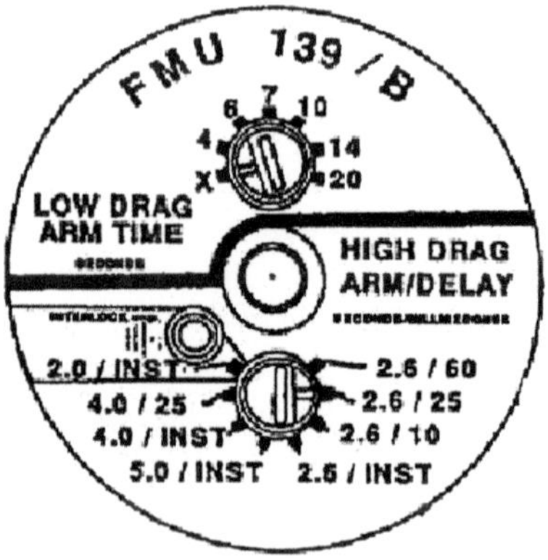

Fig. 1.23 Fuze setting options.

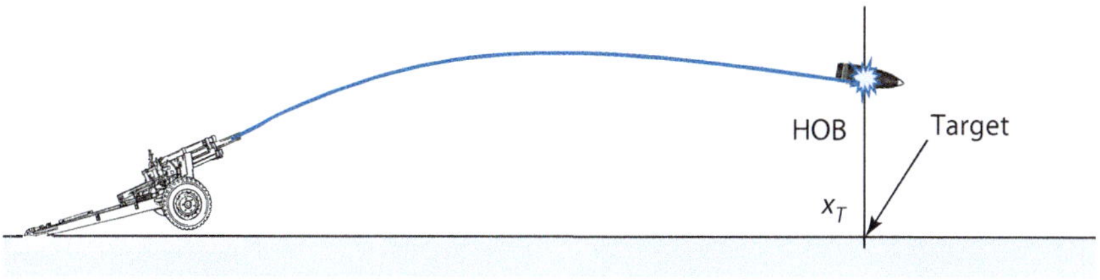

Fig. 1.24 Airburst with VT fuze.

fitted to low and high drag bombs and may be set for instantaneous or delay detonation with variable arming times after release.

- A proximity (PX) fuze is designed to detonate the warhead at some time after firing. A type of PX fuze is variable time (VT), in which a fixed time after firing is set and the fuze will initiate at that time irrespective of its location. This mechanism is a way to achieve an airburst over a target if accurate trajectory information is known. The fixed time after firing may be achieved using mechanical or electrical times.

  Other types of PX fuzes are sensor based, where closeness to the target will initiate detonation. A height of burst (HOB) fuze contains an altimeter (barometric or radar) that is designed to detonate the warhead a specific height above the ground (Fig. 1.24). Another sensor-triggered fuze contains a miniature Doppler radar set that senses distance from an object. When the explosion occurs above the ground it is called an *air burst,* and most of the destructive effect is caused by the bomb casing fragments. Proximity-fuzed bombs are used against targets such as troops in trenches, radars, vehicles, and in-flight aircraft. Most air-to-air missiles have proximity and impact fuzes.
- A hydrostatic fuze is employed in depth bombs used for underwater demolition work. The MK-36/40 Destructor is a special fuze with a sensor that can be mated to a bomb. It senses the presence of metallic objects such as trucks or ships, making it in effect a mine. These weapons can be used against either land or water targets.
- The Hard Target Smart Fuze (HTSF) is a microcontroller-based fuze designed to be physically and electrically compatible with several guided and unguided weapons. The HTSF was designed for current and future penetrator weapons to define the fuze function point as either a desired distance within a desired void or a depth of burial beneath a hard layer. It operates in one of three modes: hard-layer detection, void detection, and path-length integration. It also has an adjustable backup time delay that is set in 1-μs increments with a maximum delay of 250 μs. The HTSF uses a void sensing technique to count layers within a structure to initiate fuze function, a depth of burial mode that causes the fuze to function a preset distance after it senses a hard layer, and an integral time delay backup. The

layer counting method is shown in Fig. 1.25 for a weapon that, for example, has to detonate in a particular floor of a building.

Fuzes may be nose mounted; however, warheads penetrating hard materials will usually be tail mounted to avoid damage during the penetration process.

1.5.3 Delivery Accuracy

Estimating the accuracy with which a weapon is delivered to the target can play a crucial role in determining the damage sustained by the target. For example, if a nuclear warhead is directed against a civilian population of a city, accuracies of the order of several thousand feet are all that is required. A shaped charge warhead has to actually hit the target; therefore, high accuracy is required of such a weapon.

In some cases a human operator forms part of the delivery accuracy calculation; for example, if a tank gunner has to put an aiming cue (crosshairs) on the target, then the accuracy of the gunner in performing this task will form part of the overall weapon system accuracy. As such, human capabilities, particularly under combat conditions, need to be quantified.

In general, accuracies may be specified in terms of the ground plane for weapons that hit the ground (bombs, artillery shells) or in a plane normal to the trajectory of the projectile for weapons that impact the target in that direction (bullets, penetrators). Again, it is not possible to predict what a particular bomb or shell accuracy will be for a particular mission; all that can be done is to specify the statistics of the weapon miss distance. For example, from extensive testing we might be able to say "50% of the weapons tested

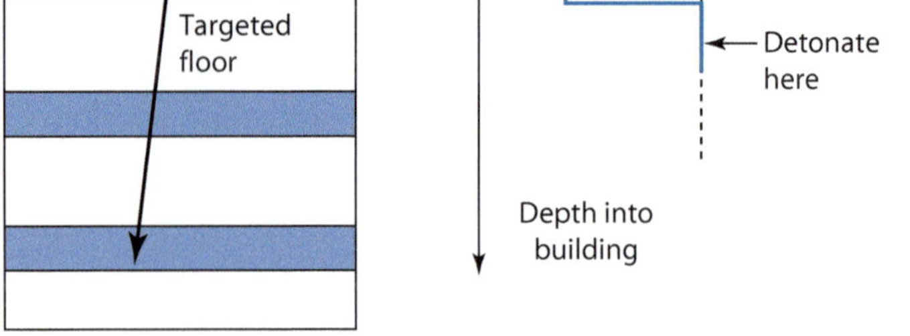

Fig. 1.25 Void counting fuze.

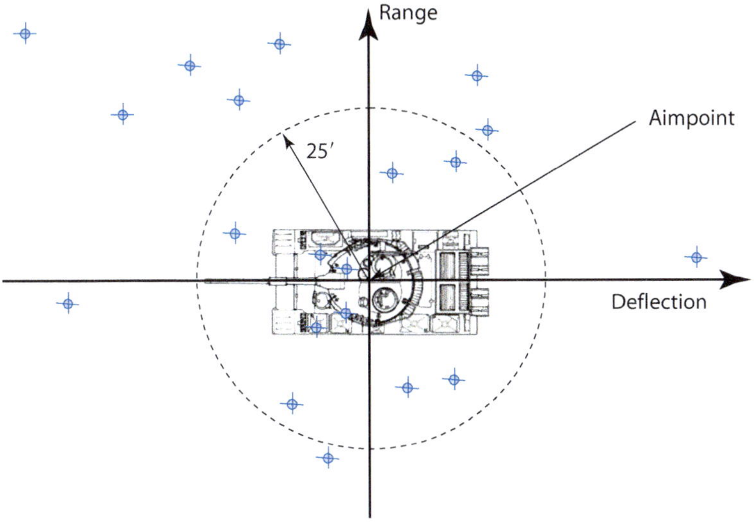

Fig. 1.26 Distribution of impact points.

fell within a circle of radius 25 ft of the aimpoint."* For the ground plane this is illustrated in Fig. 1.26.

In a particular mission, that weapon may fall within 1 ft of the target or 100 ft of the target. Clearly an understanding of statistical methods will be needed here. If we assume that the miss distance follows a known distribution (from testing), it will then be possible to estimate the probability of it landing within 1 ft or 100 ft of the target.

Ideally, delivery accuracy data would come from extensive testing for all possible applications of the weapon; however, this is prohibitively expensive for all but the most common weapons. Usually predicted accuracies are derived and supported by whatever test data may be available, although even this approach may not be feasible for very expensive weapon systems where actual combat data are the most plentiful source.

1.5.4 Arrival/Terminal Conditions

By arrival or terminal conditions, we mean the position and kinetics of the warhead/weapon at the time of detonation. If we consider a simple bomb as an example, the impact velocity is critical. Too fast an impact velocity and the bomb casing may break up before the explosive detonates; too slow a velocity will not produce the required penetration of the target. If the impact angle is too shallow, the round may ricochet off the target.

*This measure is known as the circular error probable (CEP).

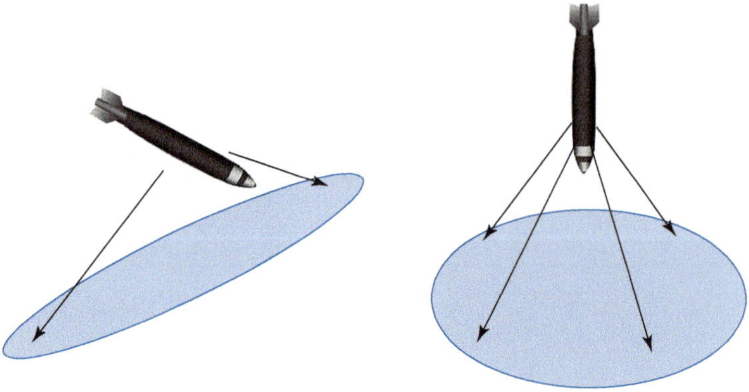

Fig. 1.27 Fragment spray pattern depends on impact angle.

Also, for weapons that produce fragments, these fragments are projected mainly from the side of the bursting case; therefore, the fragment spray pattern, or area of effects, will be short and very wide for low impact angles and almost circular for vertical impacts, as shown in Fig. 1.27.

The arrival conditions are usually determined from a trajectory program that for unguided weapons is the result of a solution of the differential equations of motion of a rigid body in a resistive material. For guided or powered weapons (missiles), a trajectory program based on the proprietary guidance laws of the weapon may be required. Trajectory programs increase in complexity when the weapon moves at transonic and supersonic speeds.

Some air-launched weapons dropped from low altitude are retarded by parachutes or deployable fins in order to provide high impact angles for penetration purposes, as indicated in Fig. 1.28. Although this mechanism

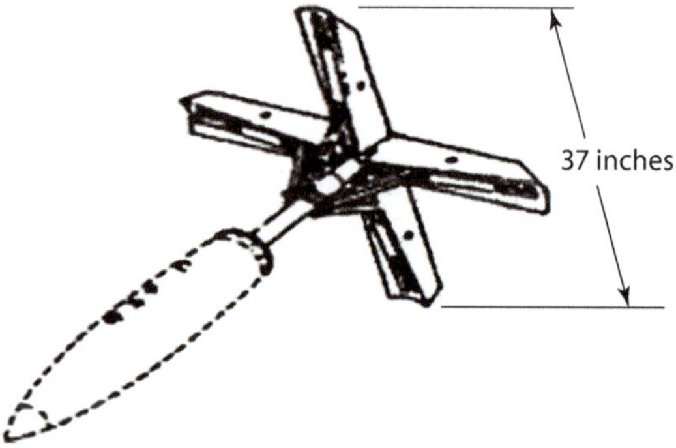

Fig. 1.28 Fin-retarded bomb.

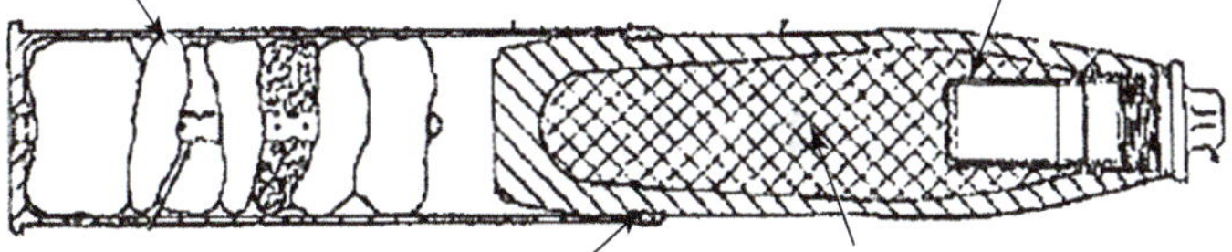

Fig. 1.29 Unitary/HE munition.

provides a steeper impact angle, these weapons' relatively large surface area may degrade their accuracy.

For unguided delivery of bombs from aircraft, the time of flight is relatively short; therefore, the bomb is almost directly below the releasing aircraft when it detonates. The retardation also allows the releasing aircraft to exit the region where the weapon detonates, thereby preventing damage to the aircraft from its own weapon (safe escape). These retarded weapons are typically used against concrete aircraft runways to create craters that prevent aircraft takeoff and landing.

1.5.5 Warhead Type

We have seen that a warhead may produce one or more damage mechanisms depending on its design and the type of target being attacked. In addition, different designs of warhead are available to the weaponeer to affect the intended damage. For example, Fig. 1.29 shows a single (unitary) high explosive (HE) cannon round with a single charge of explosive. Figure 1.30, however, shows a different cannon round that dispenses submunitions or grenades rather than detonating a single, large explosive charge. These weapons are known as cluster bombs when dropped from aircraft or

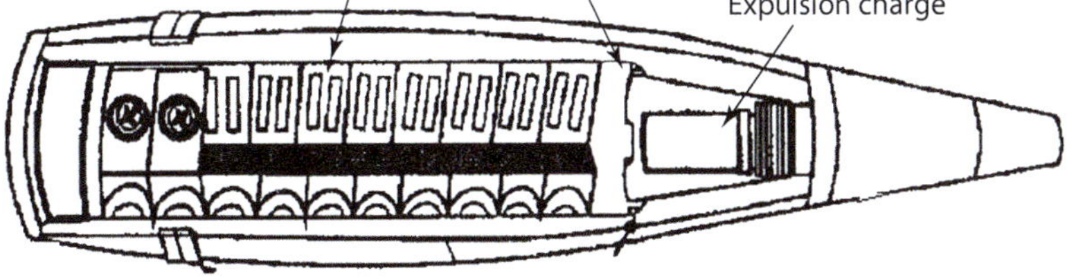

Fig. 1.30 Cluster/IC munition.

improved conventional munitions (ICMs) when referencing artillery- or tank-fired rounds.

Both warhead types—single charge and individual submunition—may damage target elements by blast, fragment, shaped charge, or a mixture of each mechanism.

1.6 Effectiveness Calculation Methodology

Now that we have all the components needed to calculate weapon effectiveness, what is needed is a method to produce the result. In terms of Fig. 1.6, we have filled in the various data needed by the program, so what does the program do to produce a measure of effectiveness when we click the "compute" button?

Much of this book attempts to answer that question, but for now a simple approach to calculating effectiveness will be given. Consider the scenario shown in Fig. 1.31 in which someone is throwing a Frisbee (with a hole in it) and trying to hit the center of a cross (target) drawn on the ground.

This Frisbee is actually a ring with a circular opening in the center, and we will call this inner opening our "weapon lethal area." We will show in Chapter 6 that this lethal area is a function of the weapon, target, and damage criteria, so a particular size opening might be annotated as follows:

- *Weapon:* AGM65D (Maverick missile)
- *Target:* T72 (Main battle tank)
- *Kill level:* K-Kill (unrepairable)

If the Frisbee is thrown such that the center of the cross is inside the opening, we define the target as killed; otherwise, it survives. The thrower positions themselves about 25 ft from the target, aims at the target, and throws the Frisbee. The result of whether the target is killed is recorded. This process is repeated 100 (N) times. If the number of kills recorded is N_K, then the

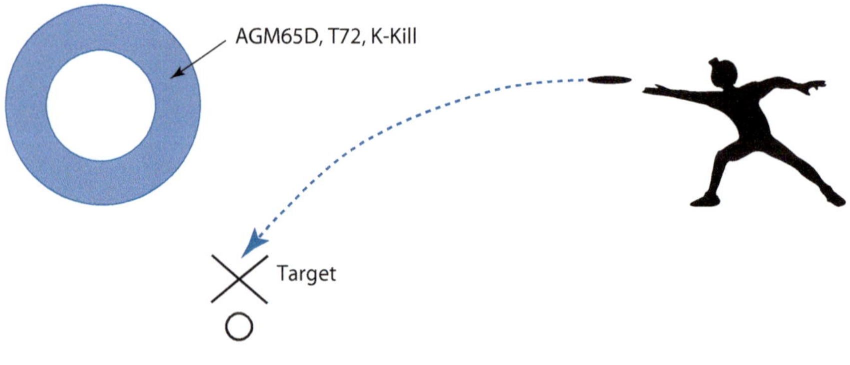

Fig. 1.31 Simplified effectiveness methodology.

probability of kill is calculated from the following equation:

$$P_K = \frac{N_K}{N} \tag{1.1}$$

The value of P_K is called the expected result of the test; in other words, it is the likelihood of killing the target if the mission were executed a large number of trials. It will not tell us if one particular trial will succeed or fail, only the expectation of success. This result may be interpreted in the following two ways:

1. It is the proportion of a large number of trials that will result in success, or
2. It is the probability a single trial will result in success.

What we have performed here is called a Monte Carlo simulation (i.e., a simulation that mimics the actual physical attack and then repeats it many times to obtain data from which conclusions may be drawn). The more trials we execute, the more accurate the result, but the longer the test takes to complete—so there is a tradeoff between accuracy of the result and the time needed to compute it.

The data resulting from these trials are usually presented using statistical methods, such as the average number of successes. We may model changes in the accuracy of the weapon system by moving the thrower closer to (improving accuracy) or further away from (degrading accuracy) the target. Similarly we can model increasing the weapon lethality/target vulnerability by using a Frisbee with a larger hole in it to represent a weapon with a larger and more effective warhead.

Clearly the development of effectiveness methods will be more complex than the one just described; however, this intuitive approach does provide a simple model that behaves in a predictable manner when the weapon characteristics and delivery accuracy measures are changed.

1.7 Implementing a Methodology Into a Model

The process just described has been described as a "model," and this term will be used extensively in this book. Therefore, a few comments on the nature of models seem appropriate. For our purposes, a model is an implementation of a methodology (i.e., a practical way to obtain a result, such as a P_K based on a Monte Carlo test). Models take several forms; however, Whitehouse [2] provides a convenient way to classify them based on their accuracy in reflecting the physical nature of the problem under study.

- A simple model is the least accurate in producing results and may be characterized as one that utilizes a graph or a limited number of equations. These models produce results rapidly but may be constrained to certain ranges of application; for example, the trajectory of an unguided bomb dropped from an aircraft in level flight can be predicted by a simple model by ignoring air resistance.† The resulting model is a differential equation that may be integrated directly, and the result is the one that most students have met in high school physics classes. It provides simple algebraic expressions for time of flight, distance traveled horizontally, impact angle, and so on that may be evaluated on a calculator. For example, if the time of flight t is of interest, an expression for this quantity becomes

$$t = \sqrt{\frac{2h_0}{g}} \tag{1.2}$$

where g is acceleration due to gravity and h_0 is the release altitude. The graph in Fig. 1.32 allows the time of flight to be obtained from the release altitude.

This graph is an example of a simple model that provides approximate answers very quickly.

- An engineering-level model is more complex than a simple model and therefore provides better accuracy in the result. It may still be based on

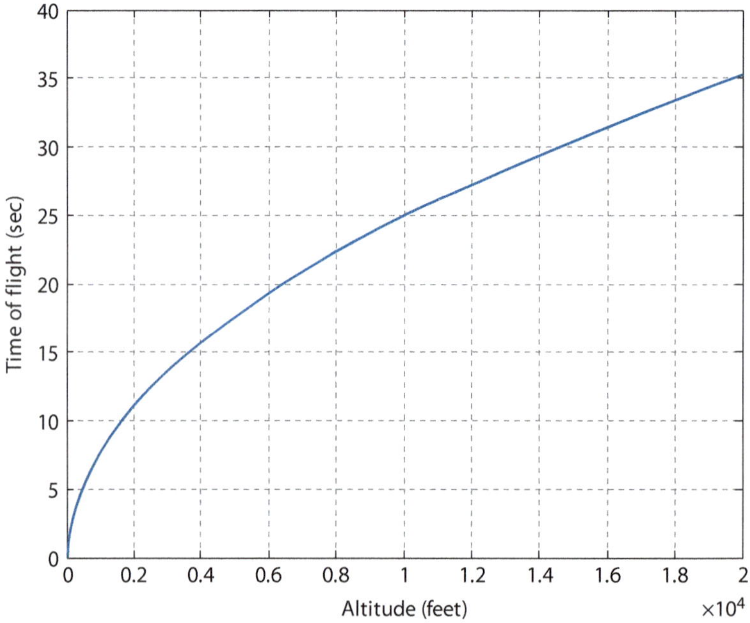

Fig. 1.32 Time of flight vs altitude.

†Detailed discussions of trajectory are dealt with in Chapter 3, but the results are used here without proof.

empirical data or simplified analysis and will in general take longer to compute a result than a simple model would. Continuing the trajectory example, an engineering model would introduce the concept of aerodynamic drag such that perhaps the drag force is proportional to the velocity of the bomb. This results in a differential equation of motion that can still be solved but not by direct integration. This time the expression for the time of flight is

$$h_0 + \left[\frac{g}{c_0{}^2}\right][1 - \exp(-c_0 t)] - \frac{gt}{c_0} = 0 \tag{1.3}$$

Here c_0 is the linear drag coefficient. This is more difficult to solve than the simple model and requires an iterative approach, although the solution may be implemented in a spreadsheet.

- The model that produces the best achievable accuracy is termed a first principle model because it will be based on best practice physics and mathematics that represent the actual process influencing the problem under study. For the trajectory problem, the drag force is now a function of the square of the velocity, and no expression can be derived for the time of flight. A solution is obtained by numerically integrating the nonlinear differential equation of motion of the projectile. Clearly this model requires more time to develop and execute.

Whitehouse also refers to *tempo-based requirements.* This term expresses how much time is available to the model user in order to achieve a result. It may be that for a particular problem, all three model types are available, as is the case in calculating the time of flight of a bomb dropped by an aircraft. However, if the result has to be achieved in a short time, then perhaps the first principles model cannot be used, and the engineering model is used instead. If the requirement to produce the result is even shorter, then perhaps only a simple model may be used.

In the weaponeering field, the methods we use, describe, and develop are all intended to support the warfighter, and on many occasions (such as the existence of time-sensitive targets), results are needed very quickly in order to plan and execute an attack. The general rule, therefore, is to use the highest fidelity model available consistent with the tempo-based requirement of the mission.

References

[1] Eidinger, J. M., and Ostrom, D., "Earthquake Loss Estimation Methods, Technical Manual, Electric Power Utilities," National Institute for Building Sciences, Rept. 23.04, June 1994.

[2] Whitehouse, S., "Change Assessment Theory (CAT)—A Conceptual Framework for Guiding the Development of Computer-Based Methods to Assess the Effects of Natural or Manmade Environments on Operational Systems," ARA Report, Albuquerque, NM, 31 July 2011.

Chapter 2 Introduction to Statistical Methods

2.1 Why are Statistics Needed?

Consider the situation shown in Fig. 2.1 where a self-propelled 155-mm artillery cannon is repeatedly firing single, unguided rounds at a target. Experience tells us that each round fired does not land in exactly the same place on the ground, but follows a slightly different trajectory contained within the curved conical envelope shown in the figure. This results in different impact points for each shell, as indicated by the plan view also shown in the figure. The impact point of the round relative to the target (accuracy) is a function of the variability of many parameters, such as:

- The properties of the atmosphere, and small changes in it (temperature, density, wind), that the projectile flies through
- The estimated position of the cannon and target
- The amount of wear of the gun tube
- The precise shape and surface roughness of the projectile
- The amount of propellant in the charge
- The manner in which the propellant burns when ignited

Considering the last two of these items, the amount of propellant may vary from round to round by 1 or 2 g, and the manner in which it burns may be very similar but not identical; however, both these effects will vary the pressure–time history at the back of the projectile as it is pushed down the gun tube.

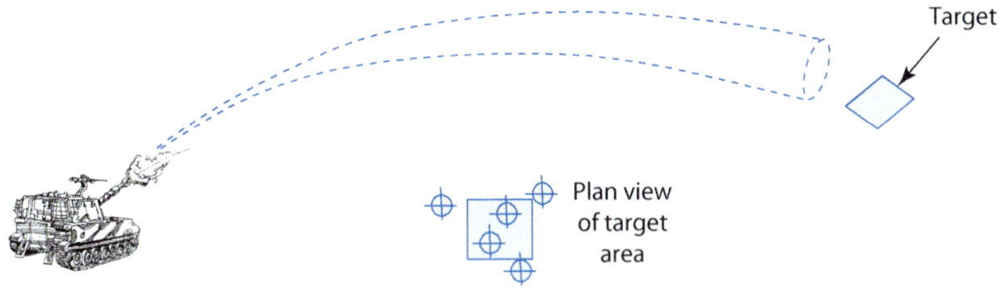

Fig. 2.1 Artillery firing at a target.

Because of the variability of these parameters, successive rounds will land in different locations on the ground. Even if we could model how these features would affect a particular projectile trajectory, it is not practically feasible to measure any of them prior to firing a particular round.

We have to accept that it is not possible to model all of the complex mechanisms that influence the trajectory of typical projectiles, which in turn lead to unique impact points for what are assumed to be identical projectiles. Instead we have to observe that the compound effect of these, and other unknown mechanisms, is a distribution of random impact points on the ground; the characterization of this distribution allows us to estimate the effect of a particular round on the damage sustained by the target.

It is the branch of mathematics known as statistics that deals with the characterization of random variables, such as the impact points just described, and allows us to determine the probability that a shell will land sufficiently close to the target in order to inflict the required amount of damage.

In the discussion of statistics that follows, the intent is not to provide the reader with a basic course in statistics, but to describe statistical techniques that will be used in the remainder of this book. Most of our discussion focuses on distributions of random variables associated with various aspects of weapon system performance.

2.2 Population and Sample

Consider any variable x that may be considered random in nature, such as the height of males over the age of 18. Clearly there is a large population of such males, in the order of several billion. We might take a sample of, say, 100 males at our place of work in order to estimate the average height of males in this age group. Our sample mean might be 71 in., which leads to the obvious question: Is this the mean of the population? A large part of estimation theory is directed at estimating statistics of the population from those of the sample.

For now we will be interested in two statistics of the population, namely the mean μ and variance σ^2. We are usually only able to compute the sample mean, although it is called the average and denoted by $\bar{x}$, and the sample variance S_x^2.

2.3 Univariate Distribution

Consider the situation shown in Fig. 2.2, where an aircraft is dropping unguided bombs on a road. In this scenario, the aircraft flies in the direction shown and drops a single bomb. It then circles around and drops another bomb, and so on.

A convention commonly used is to call the direction the aircraft is heading the range direction (or along-track), and the direction perpendicular

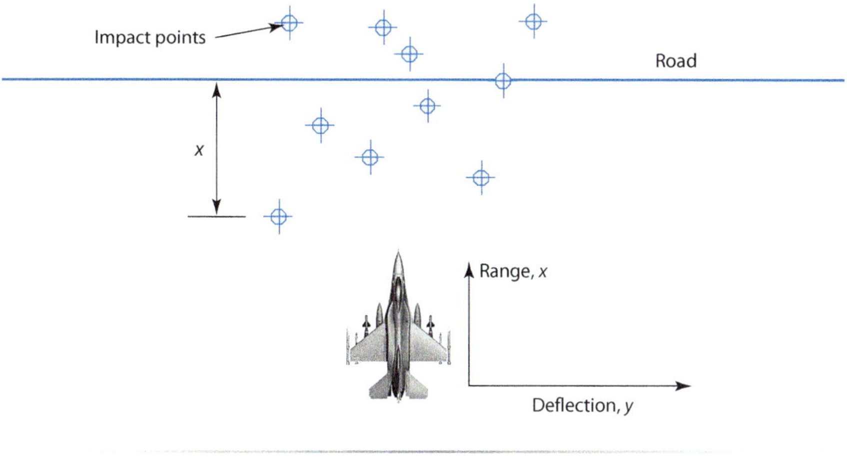

Fig. 2.2 Aircraft bombing a road.

to this the deflection (cross-track) direction. The following assumptions are made regarding the scenario shown in Fig. 2.2:

- The attack direction is perpendicular to the road.
- The road has negligible width.
- The only variable of interest is the miss distance of individual weapons in the range direction, x.

Suppose the aircraft, or others like it, make 100 bombing runs, and the miss distance x $(x_1, x_2, \ldots, x_{100})$ is recorded for each impact. Two variables may be calculated that describe the collective nature of the miss distances:

1. The mean miss distance, defined by

$$\overline{x} = \frac{1}{n}\sum_{i=1}^{n} x_i = \frac{1}{100}(x_1 + x_2 + \cdots + x_{100}) \tag{2.1}$$

2. The variance, defined by

$$\begin{aligned} S_x^2 &= \frac{1}{n-1}\sum_{i=1}^{n}(x_i - \overline{x})^2 \\ &= \frac{1}{99}\left[(x_1 - \overline{x})^2 + (x_2 - \overline{x})^2 + \cdots + (x_{100} - \overline{x})^2\right] \end{aligned} \tag{2.2}$$

The square root of the variance, S_x, is known as the standard deviation. It should be noted that some texts on statistics define the variance as follows:

$$\widehat{S}_x^2 = \frac{1}{n}\sum_{i=1}^{n}(x_i - \overline{x})^2 \tag{2.3}$$

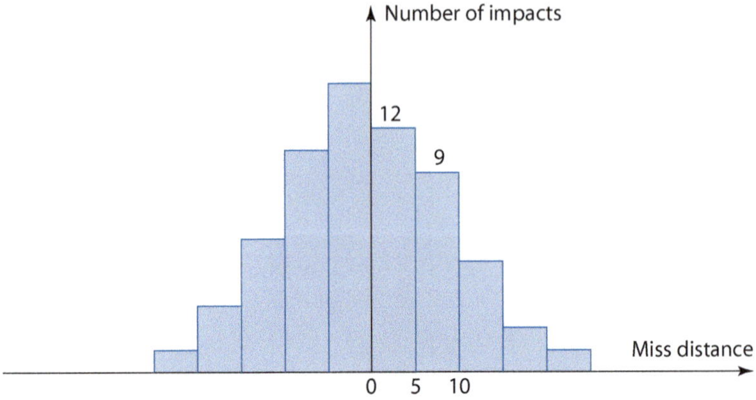

Fig. 2.3 Distribution of miss distances.

An unbiased estimation of a true statistic is one in which the mean value of the statistic is equal to the population statistic. The mean value given by Eq. (2.1) and the variance given by Eq. (2.2) are, in fact, unbiased estimates, whereas the variance given by Eq. (2.3) is not. In this book, therefore, the variance will be defined by Eq. (2.2).

Another useful quantity frequently used is the median. Like the mean, the median is a measure of centrality of a data set, and is obtained by ordering the data from smallest to largest. The median is the value of the single data point that has half the sample size smaller than it and half the sample size greater than it; for example, if a data set of 47 numbers is ordered from the smallest to the largest, the median is the value of the 24th element, because there are 23 values smaller and 23 values larger than this element. If the data set is even, the median is the mean of the middle two values; for example, if a data set of 48 numbers is ordered from the smallest to the largest, the median is the mean of the 24th and 25th elements.

A graphical means to display the so-called distribution of miss distances could be developed by grouping individual miss distances into a number of "bins" of arbitrary size. For example, suppose the bin size was chosen to be 5 ft; we could then group the number of impacts into the 0- to 5-ft bin, and another group of impacts into the 5- to 10-ft bin, and so on, remembering that negative miss distances have to be grouped into separate bins. This might produce a chart, or histogram, similar to the one shown in Fig. 2.3. In this figure, the number 12 indicates that 12 impacts fell in the range 0–5 ft, and 9 impacts fell in the range 5–10 ft. Clearly the sum of all these numbers is the total number of bombs dropped, 100 in this case.

Based on these data, it is reasonable to expect that for any bombing run, the probability that a single bomb will fall into the region 0–5 ft is 12 in 100 (or 12% or 0.12), and the probability a single bomb will fall into the region 5 – 10 ft is 0.09. It may also be observed that the probability a single bomb will fall in the range 0 – 10 ft is 0.12 + 0.09 or 0.21. Mathematically this may be written as

$$P(0 < x < 10) = 0.21 \tag{2.4}$$

What is more difficult to determine is the probability a single bomb will fall into the range, say 7 – 18 ft, because we would have to reinspect the original data set to answer that question. To solve this problem, the distribution shown in Fig. 2.3 is represented by a continuous function that will enable such probabilities to be calculated.

2.4 Univariate Normal Distribution

Many physical processes exhibit the bell-shaped histogram shown in Fig. 2.3. Such distributions are often represented by a continuous normal probability density function (PDF) of $f(x)$, given by

$$f(x) = \frac{1}{\sigma\sqrt{2\pi}} \exp\left[-(x-\mu)^2/2\sigma^2\right] \tag{2.5}$$

Here, the mean and variance would be calculated from the data set in the manner given in Eqs. (2.1) and (2.2). A graph of $f(x)$ as a function of x is shown in Fig. 2.4.

The shape of the PDF is similar to the histogram in Fig. 2.3, though not derived directly from it. The mean μ is the value of x for which the curve has a maximum, and the *spread* of the curve indicates its variance. A large variance produces a flatter curve, whereas a small variance produces more of a spike. A significant property of the distribution given by Eq. (2.5) is that the probability that an impact falls between any two values may be obtained by

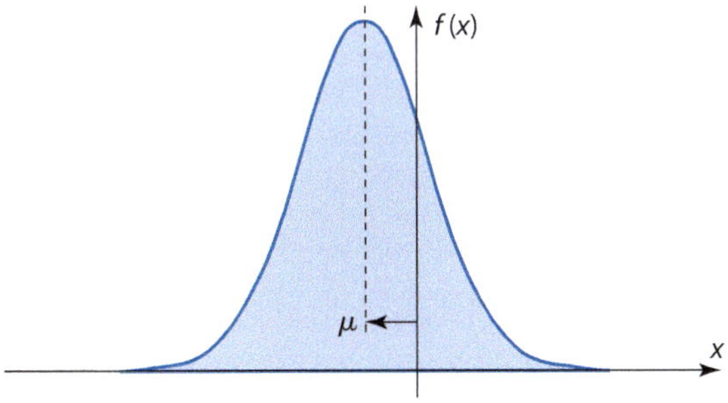

Fig. 2.4 Normal probability density function.

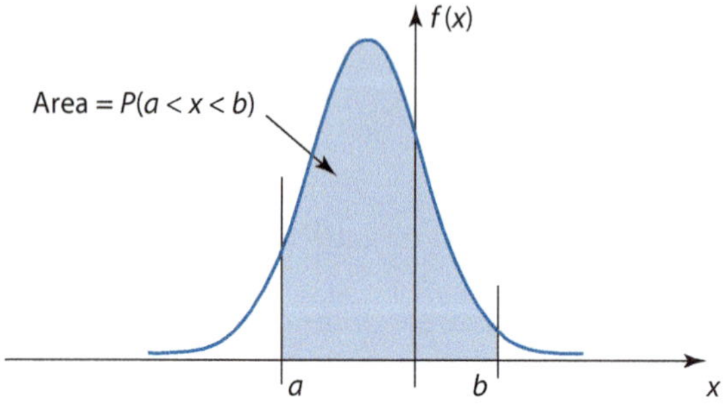

Fig. 2.5 Calculating probabilities from the PDF.

integrating the PDF between those values.

$$P(a < x < b) = \int_{x=a}^{x=b} f(x)\mathrm{d}x \tag{2.6}$$

This is shown in Fig. 2.5, where the integral is interpreted graphically as the area beneath the curve between the two values of interest.

It may be seen that

$$\int_{x=-\infty}^{x=+\infty} f(x)\mathrm{d}x = 1 \tag{2.7}$$

and that

$$\int_{x=-\infty}^{x=\mu} f(x)\mathrm{d}x = \int_{x=\mu}^{x=\infty} f(x)\mathrm{d}x = 0.5 \tag{2.8}$$

Unfortunately, Eq. (2.6) cannot be integrated directly; however, it may be computed using tabulated values of the cumulative distribution function (CDF), if available, defined by

$$\begin{aligned} F(X) &= P(x \le X) \\ &= \int_{x=-\infty}^{x=X} f(x)\mathrm{d}x = \int_{x=-\infty}^{x=X} \frac{1}{\sigma\sqrt{2\pi}} \exp\left[-(x-\mu)^2/2\sigma^2\right]\mathrm{d}x \end{aligned} \tag{2.9}$$

The cumulative density function for a particular value X is the integral of the PDF from minus infinity up to X, and is shown graphically for a typical distribution in Fig. 2.6. By varying the upper limit of integration X, the CDF may be plotted as shown in Fig. 2.7.

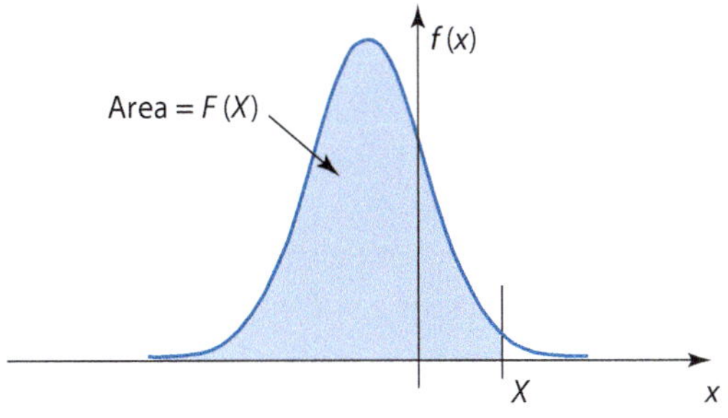

Fig. 2.6 Relationship between the PDF and CDF.

The CDF increases from zero to unity as X increases, passing through 0.5 at the mean value. If such tabulated values of the integral are available, it is seen from Fig. 2.5 that the desired area under the PDF may be written as

$$P(a < x < b) = \int_{x=a}^{x=b} f(x)\mathrm{d}x = F(b) - F(a) \tag{2.10}$$

Tabulated values for any mean and variance are not usually available because there is an infinite variety of mean and standard deviations that could be used in the integration. The question remains, therefore, of how to evaluate Eq. (2.10).

This is accomplished by introducing the concept of the standardized normal distribution function by mapping the physical domain variable x (e.g., bomb miss distance) using the following transformation:

$$z = \frac{x - \mu}{\sigma} \tag{2.11}$$

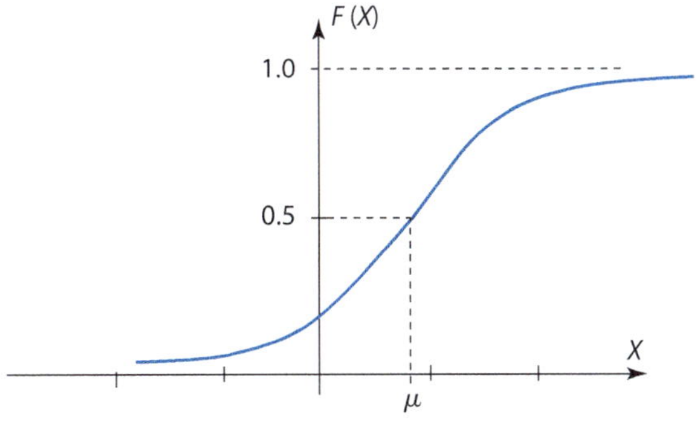

Fig. 2.7 Cumulative density function (CDF).

From this transformation, it is seen that if x is normal, so is z. In fact, the mean of z is zero, and the standard deviation is unity. The remarkable property of normal distributions is that whatever the physical domain mean and standard deviation, they all collapse to the same standard normal distribution having zero mean and unity standard deviation.

From Eq. (2.11) it may be appreciated that by subtracting the mean value μ from each value of x, the mean of z will indeed be zero. The fact that σ is unity, however, is a little less obvious, so it will now be proven. We have by definition that

$$\begin{aligned}\sigma_z^2 &= \frac{1}{n-1}\sum_{i=1}^{n}(z_i-\mu_z)^2 = \frac{1}{n-1}\sum_{i=1}^{n}(z_i)^2 \\ &= \frac{1}{n-1}\sum_{i=1}^{n}\left(\frac{x_i-\mu_x}{\sigma_x}\right)^2 \\ &= \frac{1}{\sigma_x^2(n-1)}\sum_{i=1}^{n}(x_i-\mu_x)^2 \qquad (2.12)\end{aligned}$$

But it is known that

$$\sigma_x^2 = \frac{1}{n-1}\sum_{i=1}^{n}(x_i-\mu_x)^2 \qquad (2.13)$$

so we could write that

$$\sum_{i=1}^{n}(x_i-\mu_x)^2 = \sigma_x^2(n-1) \qquad (2.14)$$

Substituting Eq. (2.14) into Eq. (2.12) leads to the required result

$$\sigma_z^2 = \frac{1}{\sigma_x^2(n-1)}\sum_{i=1}^{n}(x_i-\mu_x)^2 = \frac{1}{\sigma_x^2(n-1)}\sigma_x^2(n-1) = 1 \qquad (2.15)$$

The standard normal PDF for z now becomes

$$f(z) = \frac{1}{\sqrt{2\pi}}e^{-z^2/2} \quad (-\infty < z < \infty) \qquad (2.16)$$

Note that once more

$$\int_{z=-\infty}^{z=\infty} \frac{1}{\sqrt{2\pi}}e^{-z^2/2} = 1 \qquad (2.17)$$

and that

$$\int_{z=-\infty}^{z=\mu} f(z)\mathrm{d}z = \int_{z=\mu}^{z=\infty} f(z)\mathrm{d}z = 0.5 \tag{2.18}$$

The standard normal CDF is

$$F(Z) = P(z \leq Z) = \int_{z=-\infty}^{z=Z} \frac{1}{\sqrt{2\pi}} e^{-z^2/2}\mathrm{d}z \tag{2.19}$$

Note also that

$$p(Z_1 \leq z \leq Z_2) = \int_{z=Z_1}^{z=Z_2} f(z)\,\mathrm{d}z = F(Z_2) - F(Z_1) \tag{2.20}$$

Figure 2.8 shows the standardized PDF and CDF.

As before, the integral for the CDF cannot be evaluated in closed form; however, for symmetrical integral limits the Polya-Williams approximation may be used without incurring large errors.

$$\frac{1}{\sqrt{2\pi}} \int_{-x}^{+x} \exp\left(-\frac{t^2}{2}\right)\mathrm{d}t \approx \left[1 - \exp\left(\frac{2x^2}{\pi}\right)\right]^{1/2} \tag{2.21}$$

Figure 2.9 shows the error between the tabulated integrals of the standard PDF and the previous approximation as a function of the standardized variable Z.

For nonsymmetrical limits, the integral in Eq. (2.19) has been evaluated numerically and tabulated for incremental values of z, as shown

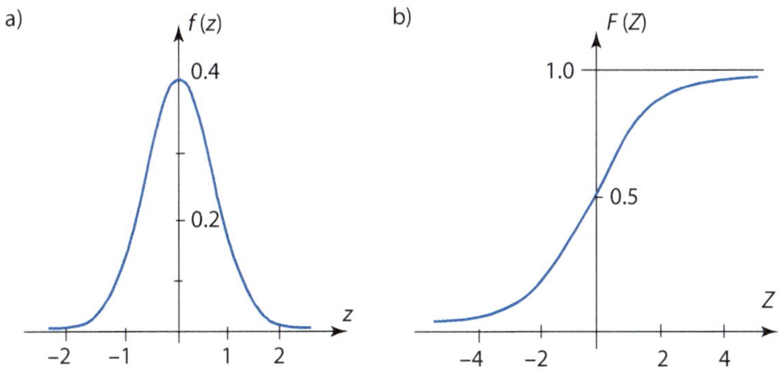

Fig. 2.8 (a) Standardized PDF and (b) standardized CDF.

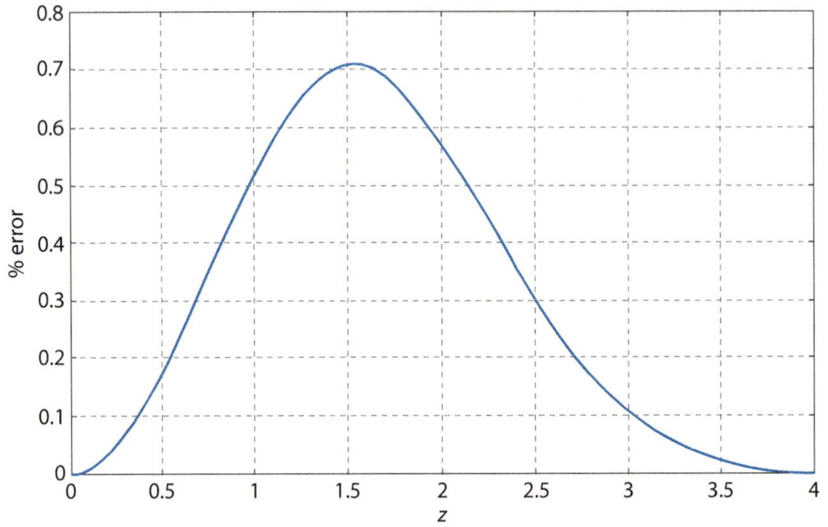

Fig. 2.9 Errors in the Polya-Williams approximation to the CDF.

in Table A.1 in Appendix A. Using this table, Eq. (2.6) may be evaluated indirectly through the variable Z. This process may be written as follows:

$$P(X_1 < x < X_2) = P(Z_1 < z < Z_2) = F(Z_2) - F(Z_1) \tag{2.22}$$

where

$$Z_1 = \frac{X_1 - \mu_x}{\sigma_x}, \quad Z_2 = \frac{X_2 - \mu_x}{\sigma_x} \tag{2.23}$$

This will be demonstrated in an example, but first it is worth noting that the integration in Eq. (2.19) may begin at either zero or minus infinity, resulting in a different table for each case, as shown in Fig. 2.10. Table A.1

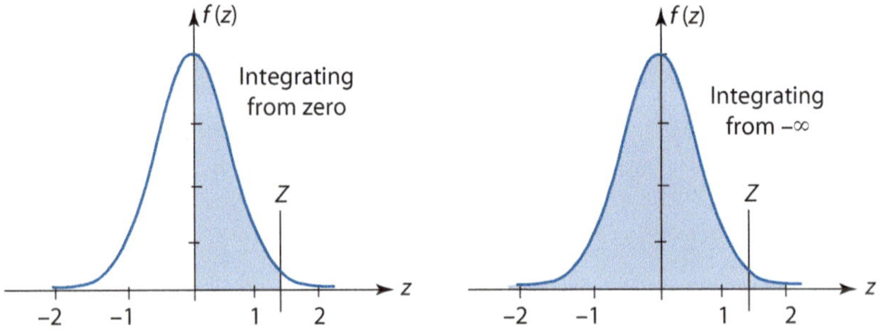

Fig. 2.10 Integration choices for the standard normal CDF.

corresponds to the right-hand case in the figure. Clearly, the tabulated values obtained by integration from minus infinity will be greater than those obtained by integrating from zero by one half.

Example 2.1

Suppose x is normally distributed with $\mu = 1.2$ and $\sigma = 0.7$. Find the probability that a sample of x lies between 1.0 and 2.0.

First, we find the two values of Z corresponding to the bounds of x. This gives

$$Z_1 = \frac{X_1 - \mu}{\sigma} = \frac{1 - 1.2}{0.7} = -0.286, \qquad Z_2 = \frac{2 - 1.2}{0.7} = \frac{X_2 - \mu}{\sigma} = 1.143 \tag{2.24}$$

The required area under the standard normal distribution is shown in Fig. 2.11.

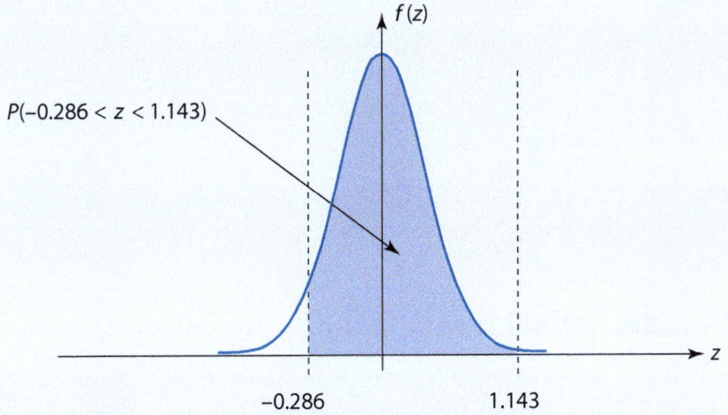

Fig. 2.11 Area under the standardized normal distribution.

Table A.1 does not have negative values of z, but we see that from the symmetry of the PDF

$$P(z < -0.286) = 1 - P(z < +0.286) = 1 - F(0.286) = 0.3859$$

Also from Table A.1, $F(1.143) = 0.8729$, hence

$$P(1.0 < x < 2.0) = P(-0.286 < z < 1.143) = F(1.143) - F(-0.286) = 0.487$$

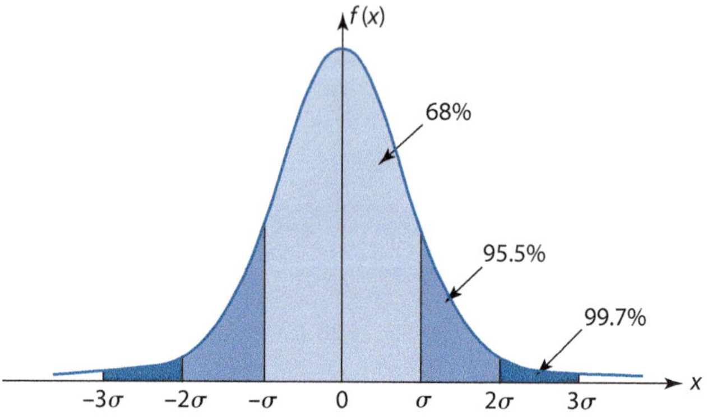

Fig. 2.12 Proportion of area under the normal distribution.

The following should be noted:

- The PDF is an approximation of the data represented by the histogram. In the bombing example given in Fig. 2.2, there were no impacts outside $\pm$ 25 ft, but the PDF predicts a finite probability this will happen.
- Because of this approximation, the PDF will not predict *exactly* 12 impacts in the range 0–5 ft, as indicated in the histogram, but it should be close when the data are approximately normal.
- From Table A.1, it may be seen that about 68% of the impacts should lie in the range $\pm\sigma$, 95.5% will lie in the range $\pm 2\sigma$, and 99.75% in the range $\pm 3\sigma$, as shown in Fig. 2.12.

These proportions and their related values of $n\sigma$ will be used in later chapters in the book.

2.5 Bivariate Normal Distribution

Suppose that we revisit the bombing problem in Fig. 2.2, but now we are interested in hitting a particular point on the road, perhaps a short over-bridge. In this case, not only the miss distance in the range direction is of importance, but also the miss distance y perpendicular to it, in the deflection direction, as shown in Fig. 2.13.

Now that the aircraft is aiming at a specific point, the term *desired point of impact* (*DPI*) is introduced to define this location. Sometimes this is referred to as the *desired mean point of impact* (*DMPI*), although this is strictly for the case where multiple weapons are released in a single pass.

Just as the univariate distribution in the range direction could be approximated by a normal distribution, so it could be assumed that the miss distance in the deflection direction is also normal, although there is no reason to expect that the statistics (μ and σ) are the same in both directions. In fact, it is usually assumed that the probability of a particular miss in the range

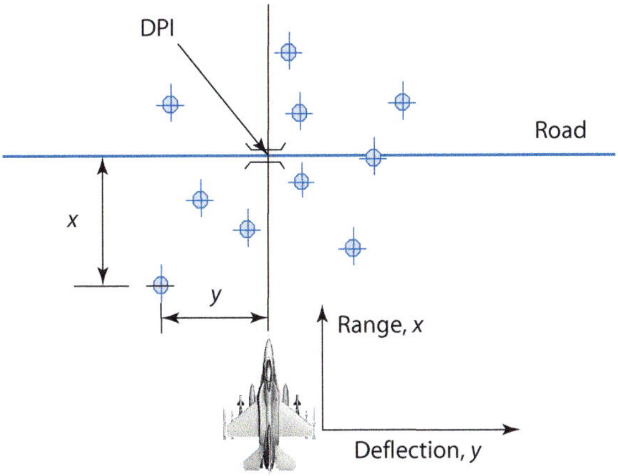

Fig. 2.13 Range and deflection miss distances.

direction is different, and independent of the miss in the deflection direction. Figure 2.14 shows such a bivariate distribution.

Given that the distributions in the x and y directions are different, we can write a bivariate PDF as

$$f(x, y) = \frac{1}{2\pi\sigma_x\sigma_y\sqrt{1-\rho^2}} \times \exp - \left[\frac{1}{(1-\rho^2)} \left\{ \frac{(x-\mu_x)^2}{2\sigma_x^2} + \frac{(y-\mu_y)^2}{2\sigma_y^2} - 2\rho \frac{(x-\mu_x)}{2\sigma_x} \frac{(y-\mu_y)}{2\sigma_y} \right\} \right] \tag{2.25}$$

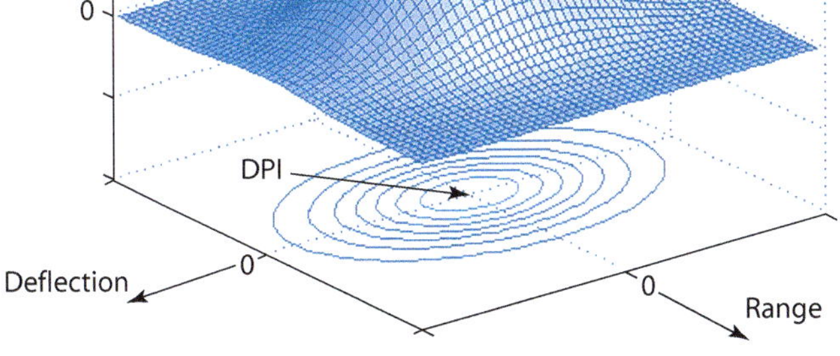

Fig. 2.14 Bivariate normal distribution.

In this equation it is assumed that x and y are not independent, and ρ is their correlation coefficient. If x and y are assumed independent, as is usual in most weaponeering methods, then Eq. (2.25) reduces to the following:

$$f(x,y) = \frac{1}{2\pi\sigma_x\sigma_y}\exp - \left[\frac{(x-\mu_x)^2}{2\sigma_x^2} + \frac{(y-\mu_y)^2}{2\sigma_y^2}\right] \tag{2.26}$$

This is known as the elliptical form of the distribution because contours of constant $f(x,y)$ are ellipses, as indicated in the base of Fig. 2.14. The bivariate CDF is obtained as before by integrating the PDF from minus infinity to some finite values of x and y.

$$F(X,Y) = \int_{x=-\infty}^{x=X}\int_{y=-\infty}^{y=Y} \frac{1}{2\pi\sigma_x\sigma_y}\exp - \left[\frac{(x-\mu_x)^2}{2\sigma_x^2} + \frac{(y-\mu_y)^2}{2\sigma_y^2}\right] dx\, dy \tag{2.27}$$

Note also that

$$\int_{-\infty}^{\infty}\int_{-\infty}^{\infty} f(x,y)\, dx\, dy = 1 \tag{2.28}$$

We can calculate the probability that x lies between two values and y lies between two other values from the CDF as follows:

$$P[(X_1 < x < X_2), (Y_1 < y < Y_2)]$$

$$= \int_{x=X_1}^{x=X_2}\int_{y=Y_1}^{y=Y_2} \frac{1}{2\pi\sigma_x\sigma_y}\exp - \left[\frac{(x-\mu_x)^2}{2\sigma_x^2} + \frac{(y-\mu_y)^2}{2\sigma_y^2}\right] dx\, dy \tag{2.29}$$

Equation (2.29) is rarely evaluated in the form shown here. In fact, because it is using the assumption of independence of samples in the range and deflection directions, we can treat the bivariate distribution as two separate, orthogonal univariate distributions. This allows us to compute the probability that a particular impact occurs inside a rectangle bordered by X_1, X_2 in the range direction and Y_1, Y_2 in the deflection direction. Before doing this, however, we will review some basic concepts in probability.

2.6 Probability

If E is some event that may or may not occur, the probability of the event occurring is written $P(E)$ and is represented by a number such that $0 < P(E) < 1$. $P(E) = 0$ means the event certainly does not occur, whereas

$P(E) = 1$ means that it certainly does. The complementary probability states that the probability of the event not happening is

$$P(\text{not}\,E) = 1 - P(E) \tag{2.30}$$

Suppose a single experiment has several possible outcomes $(1, \ldots, n)$, where the probability of each occurring is $p_1, \ldots, p_n$. Clearly,

$$\sum_{i=1}^{n} p_i = 1 \tag{2.31}$$

We now present some useful equations for probabilities that will be used later in the book. In doing so, the occurrence of two events, A and B, are used to illustrate the point; however, each theorem may be extended to more than two events.

Theorem 1

Suppose we perform an experiment in which the events A and B can occur with probabilities of occurrence $P(A)$ and $P(B)$, respectively. If A and B are independent events, then the probability of both A *and* B occurring is

$$P(A\,\text{and}\,B) = P(A) \times P(B) \tag{2.32}$$

Example:

If two coins are tossed, the probability of getting two tails is $1/2 \times 1/2 = 1/4$.

Theorem 2

If A and B are mutually exclusive (i.e., cannot both occur at the same time), then the probability of either A *or* B occurring is

$$P(A\,\text{or}\,B) = P(A) + P(B) \tag{2.33}$$

Example:

The chances of rolling a 4 or a 5 on a single toss of a die is $1/6 + 1/6 = 1/3$.

Theorem 3

If A and B are not mutually exclusive (i.e., both can occur at the same time), then the probability of either A *or* B occurring is

$$P(A\,\text{or}\,B) = P(A) + P(B) - P(A\,\text{and}\,B) \tag{2.34}$$

Example:

What is the probability a single card drawn from a deck of 52 cards is either a heart or a face card? $P\{\text{heart}\} = 13/52$, $P\{\text{face card}\} = 12/52$, and $P\{\text{heart and face}\} = 3/52$; hence, $P\{\text{heart and face}\} = 13/52 + 12/52 - 3/52 = 22/52 = 11/26$.

These theorems may be applied to the bombing example illustrated in Fig. 2.13 as shown in Example 2.2.

Example 2.2

Suppose x and y are normally distributed with $\mu_x = +3$ ft, $\sigma_x = 25$ ft, $\mu_y = -10$ ft, and $\sigma_y = 100$ ft. Find the probability a bomb will fall inside the rectangle shown in Fig. 2.15.

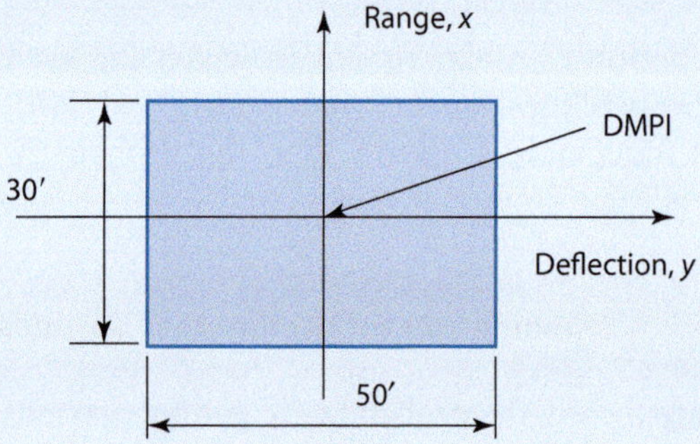

Fig. 2.15 Desired area for impact.

In the range direction we can write the following:

$$P(-15 < x < +15) = P(-0.72 < z_x < +0.48) = F(0.48) - (1 - F(0.72)) = 0.4486$$

Similarly, in the deflection direction

$$P(-25 < y < +25) = P(-0.15 < z_y < +0.35) = F(0.35) - (1 - F(0.15)) = 0.1964$$

Because in order to drop the bomb inside the rectangle, both events have to occur, and because they are assumed independent, theorem 1 applies.

$$P(-15 < x < 15 \text{ AND } -25 < y < 25) = 0.4486 \times 0.1964 = 0.0881$$

2.7 Circular Normal and Rayleigh Distributions

A special case of the bivariate distribution where both means are zero and the variances are equal is called the *circular normal distribution.* From Eq. (2.27) this PDF is given by

$$f(x, y) = \frac{1}{2\pi\sigma^2} \exp - \left[\frac{x^2}{2\sigma^2} + \frac{y^2}{2\sigma^2}\right] = \frac{1}{2\pi\sigma^2} \exp - \left[\frac{x^2 + y^2}{2\sigma^2}\right] \tag{2.35}$$

Unlike the previous example, we recognize that it may be less important to calculate the probability of a bomb falling within a rectangle than to find the probability it will fall within a given distance from the aimpoint (i.e., within a circle centered on the DPI). We define the radial miss distance r as

$$r^2 = x^2 + y^2 \tag{2.36}$$

Clearly for each impact point there will be a pair of range and deflection coordinates, and there will be a single radial miss distance computed from these coordinates. Even though x and y may be normally distributed, r is not. In fact, r will have what is defined as a *Rayleigh distribution,* which is only defined for zero mean and equal variances of x and y.

The Rayleigh distribution may be derived from the two univariate normal distributions in the range and deflection directions. Suppose the common variance for these normal distributions is known; the problem can then be posed in the form of finding the probability that an impact falls within a circle of radius R. Consider the geometry shown in Fig. 2.16.

We need to find

$$P(r \leq R) = P[(-R < x \leq R), (-R < y \leq R)] \tag{2.37}$$

subject to the constraint on x and y that

$$x^2 + y^2 = r^2 \tag{2.38}$$

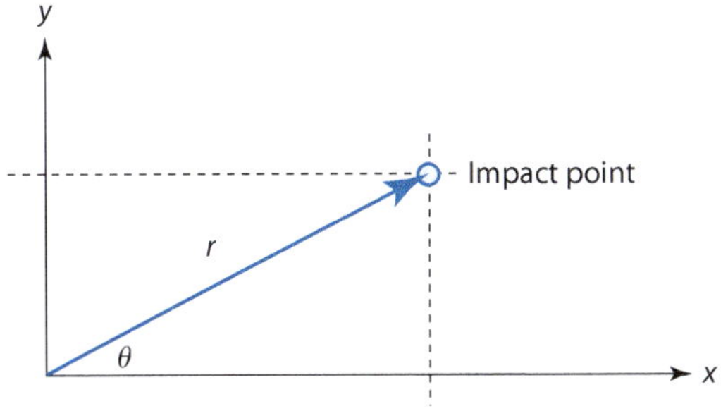

Fig. 2.16 Transformation of Cartesian to polar coordinates.

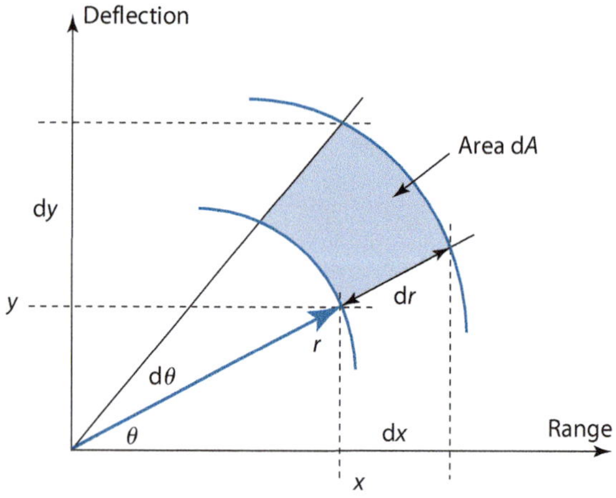

Fig. 2.17 Calculating elemental areas.

This may be evaluated from the bivariate PDF as

$$P(-R < x \leq R, -R < y \leq R) = \int_{-R}^{R}\int_{-R}^{R} f(x, y)\, \mathrm{d}x\, \mathrm{d}y \tag{2.39}$$

where $f(x,y)$ is given by Eq. (2.35) and the condition in Eq. (2.38) applies.

At this point we transform to polar coordinates, but first the relationship between elemental areas in both coordinate systems has to be derived using Fig. 2.17.

The elemental area used in the integration may be expressed as

$$\mathrm{d}A = \mathrm{d}x\, \mathrm{d}y = r\, \mathrm{d}\theta\, \mathrm{d}r \tag{2.40}$$

Substituting Eqs. (2.35), (2.38), and (2.40) into Eq. (2.39) gives*

$$P(x \leq R,\ y \leq R) = P(r \leq R) = \int_{r=0}^{r=R}\int_{\theta=0}^{\theta=2\pi} \frac{1}{2\pi\sigma^2}\exp\left[\frac{-r^2}{2\sigma^2}\right] r\, \mathrm{d}r\, \mathrm{d}\theta \tag{2.41}$$

Integrating with respect to θ first

$$P(r \leq R) = \int_{r=0}^{r=R} \frac{r}{2\pi\sigma^2}\exp\left[\frac{-r^2}{2\sigma^2}\right] [\theta]_0^{2\pi}\, \mathrm{d}r = \int_{r=0}^{r=R} \frac{r}{\sigma^2}\exp\left[\frac{-r^2}{2\sigma^2}\right] \mathrm{d}r \tag{2.42}$$

---

* The change in integral limits may be explained by considering the range of x and y in Eq. (2.39) with the constraint Eq. (2.38) as being the same as letting $0 < r < R$ and $0 < \theta < 2\pi$.

Now integrating with respect to the radius

$$
\begin{aligned}
P(r \leq R) &= \left[1 - \exp\left[\frac{-r^2}{2\sigma^2}\right]\right]_0^R \\
&= 1 - \exp\left[\frac{-R^2}{2\sigma^2}\right]
\end{aligned}
\tag{2.43}
$$

By definition this is actually the CDF; therefore, the expression inside the integral in Eq. (2.42) is the PDF, because integrating it gives the CDF. Hence we may write

$$
f(r) = \frac{r}{\sigma^2}\exp\left[\frac{-r^2}{2\sigma^2}\right] \tag{2.44}
$$

$$
F(R) = 1 - \exp\left[\frac{-R^2}{2\sigma^2}\right] \tag{2.45}
$$

Notice that the CDF is given analytically, unlike the normal distribution that has to be tabulated. Once again we have

$$
\int_{r=0}^{r=\infty} f(r)\mathrm{d}r = 1 \tag{2.46}
$$

This distribution, unlike the normal distribution, is asymmetrical; sample distributions are shown in Fig. 2.18. Some students find this distribution confusing because it appears that there is a larger probability a single round will land in the radial interval 0.5–1.0 than in the interval 0.0–0.5; however, the impacts are distributed circular normal about the origin on the ground plane, so it seems more would land closer to the origin than further away. It is left to the student (and instructor) to resolve this apparent paradox.

The form of the Rayleigh cumulative density function, Eq. (2.45), is shown in Fig. 2.19.

The probability that r lies between two values is obtained by direct integration of the PDF and can be calculated from Eq. (2.44) as

$$
P(r_1 < r < r_2) = F(r_2) - F(r_1) = \exp\left[\frac{-r_1^2}{2\sigma^2}\right] - \exp\left[\frac{-r_2^2}{2\sigma^2}\right] \tag{2.47}
$$

For $r_1 = 0$, Eq. (2.47) reduces to Eq. (2.45). As for any other distribution, the mean, median, and radial variance can be calculated in the usual manner from the radial miss distance data.

It is important to remember that in the preceding analysis, σ is the common standard deviation of the x and y data set, and not the standard deviation of the radial miss distance data set. Consider the following experiment where we generate 200 values of x and y from a normal distribution

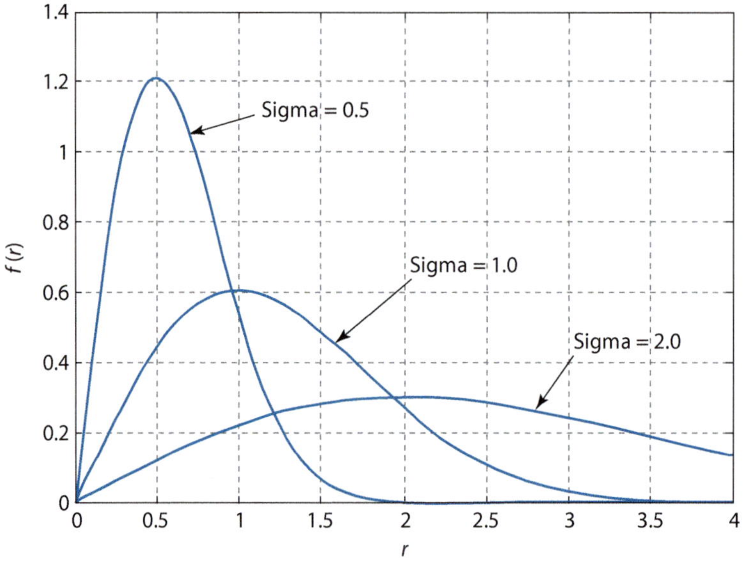

Fig. 2.18 Rayleigh PDF for some sample σ values

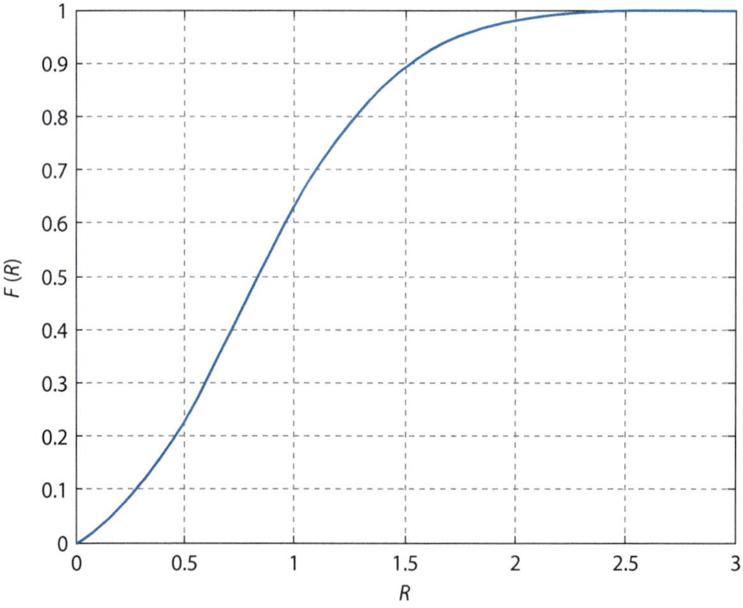

Fig. 2.19 Rayleigh CDF.

Table 2.1 Impact Point Samples

Sample #	x	y	r
1	0.5046	0.5500	0.7464
2	0.9641	−0.2416	0.9939
3	−0.1694	−2.1183	2.1250
4	−0.9390	−1.1046	1.4497
5	0.1359	0.1353	0.1918
6	0.7986	1.3874	1.6008
↓	↓	↓	↓
197	1.0354	−0.8141	1.3171
198	−1.3667	−0.3065	1.4006
199	−0.4985	0.4129	0.6473
200	−0.0752	1.2524	1.2546
Mean	*−0.0748*	*−0.0227*	*1.2842*
Stand. dev.	*1.0511*	*0.9918*	*0.6612*

with zero mean and standard deviation of unity. We root sum squared (RSS) x and y to get a circularly distributed value of r, and we calculate the statistics of x, y, and r as shown in Table 2.1.

The table shows the computed values of mean and standard deviation for x and y are sufficiently close to the known values of 0 and 1, but these are different to the mean and sigma of r. The value of sigma to be used in the previously derived formulae should be that of x and y (i.e., 1.0), not the sigma value of r, which is 0.6612.

The statistics of (x, y) are related to those of r, and it can be shown that the mean, mode, and variance of the radial miss distance data set are given by the following:

$$\mu_r = \sigma\sqrt{\frac{\pi}{2}} = 1.2533\sigma \tag{2.48}$$

$$\mathrm{r}_{\mathrm{median}} = \sigma\sqrt{\log_e(4)} = 1.1774\sigma \tag{2.49}$$

$$\sigma_r^2 = \left[\frac{4-\pi}{2}\right]\sigma^2 = 0.4292\sigma^2 \tag{2.50}$$

where σ is the common standard deviation of x and y, and the subscript r refers to the radial data set.

2.8 Uniform Distribution

Many random processes exhibit a uniform probability density function, or may be approximated by one. In this distribution, there is an equal

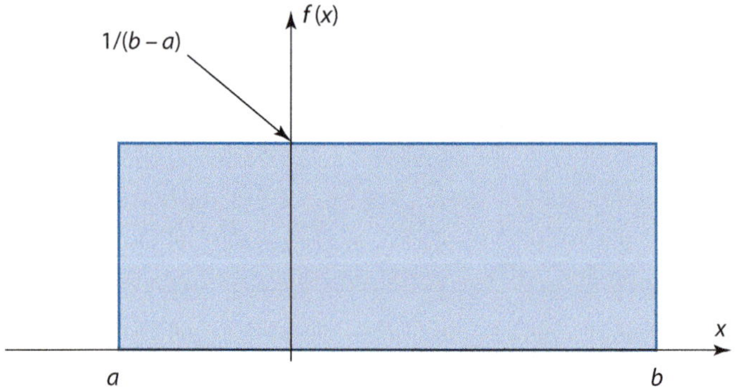

Fig. 2.20 PDF for uniform distribution.

probability a sample may lie between two values a fixed distance apart within the bounds of the distribution. For example, a cluster munition that opens up and releases mines attempts to distribute them uniformly along a road so that the probability of encountering one is equal anywhere between the beginning and end of the minefield. This can be represented by a continuous univariate uniform distribution, as shown graphically in Fig. 2.20.

In the previous section we discussed the radial miss distance as having a Rayleigh distribution that was derived from converting each impact point from an (x, y) Cartesian pair to an equivalent (r, θ) polar pair. It may be observed that for a circular normal distribution of impact points on the ground plane, as shown in Fig. 2.16, the angular position θ for each point is expected to be uniformly distributed in the region 0 to 2π. The reader is invited to consider if this would be true if the distribution is not circular normal.

Mathematically, the continuous univariate uniform PDF is written as follows:

$$f(x) = \frac{1}{(b - a)} \qquad \text{for } a < x < b \tag{2.51}$$

The CDF is obtained by directly integrating the PDF in the following manner:

$$F(X) = \int_{x=-\infty}^{x=X} f(x)\,\mathrm{d}x = \int_{a}^{X} \frac{1}{b-a}\,\mathrm{d}x = \left[\frac{x}{b-a}\right]_a^X = \frac{X-a}{b-a} \tag{2.52}$$

This is shown graphically in Fig. 2.21.

Another example of a uniform distribution is rolling a fair die. There is an equal probability that any number between 1 and 6 will result, and the

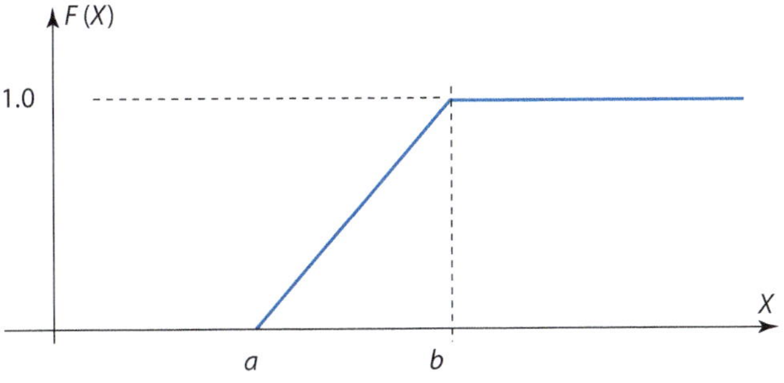

Fig. 2.21 CDF for uniform distribution.

probability that a particular value will be produced is 1/6. This is a discrete uniform distribution, and its PDF is shown in Fig. 2.22.

Suppose we plan a minefield to cover a rectangular area on the ground. It makes sense to distribute the mines in a uniform manner within the rectangular area rather than having a higher mine density in the middle of the field, making the probability of encountering a mine independent of the path through the minefield. In other words, the mines should be distributed uniformly rather than normally about the center of the field.

As another example, suppose a cluster munition releases 625 bomblets over an impact area on the ground measuring 25 ft by 25 ft. There will be an equal probability that one mine will fall into each 1 ft by 1 ft element of a grid placed on the impact area if the distribution is indeed uniform. This leads to the continuous bivariate uniform distribution with the PDF shown graphically in Fig. 2.23.

It is left to the reader to derive the bivariate PDF and CDF for this case.

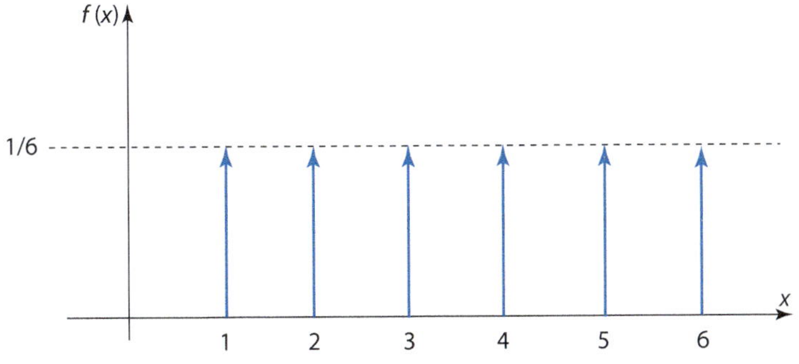

Fig. 2.22 Discrete uniform distribution.

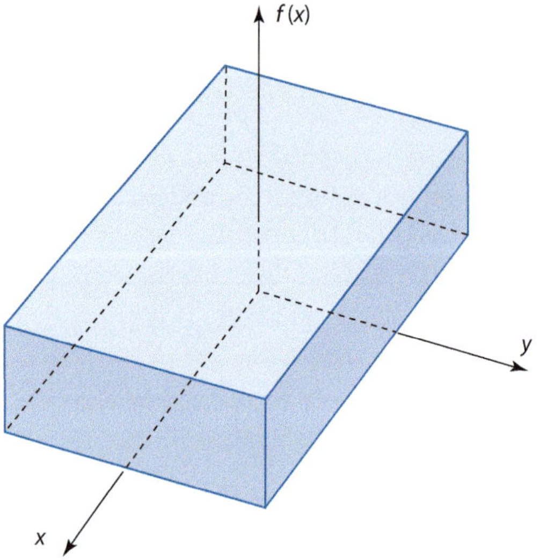

Fig. 2.23 Continuous bivariate uniform distribution.

2.9 Binomial Distribution

For many classes of problems, the outcome of a single trial may be either success or failure, that is, one of two possible states. An example would be attacking an armored vehicle with a shaped charge warhead; either the weapon hits the target and kills it or it misses and the target is undamaged. An experiment like this with successive trials is often called a *Bernoulli trial.* Another example might be the road bombing problem discussed earlier, except that instead of measuring the result of a particular impact in terms of a radial miss distance, we simply report whether the bridge was hit (success) or missed (failure).

An important restriction on Bernoulli trials is that they are independent. Consider the example of the rifleman shooting at a target, as shown in Fig. 2.24. If the shooter fires one shot, puts down the rifle, inspects the

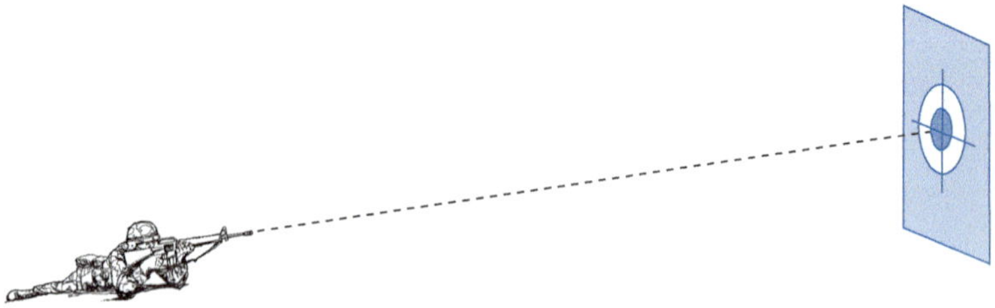

Fig. 2.24 Independent and dependent fire.

target, picks up the rifle, re-aims, fires again, and then repeats the process for a total of five shots, each shot is considered to be independent of the others. If, however, the shooter aims the rifle and fires a rapid burst of five shots, these shots are considered dependent. For example, in dependent fire, if the rifleman's elbow slips just as he is about to shoot, his aimpoint moves a considerable distance from the target and all five shots are displaced by the same amount. A Bernoulli trial requires that all events are independent.

A binomial distribution quantifies the statistics of a large number of Bernoulli trials in which the following parameters are defined:

$p =$ Probability that a single trial will result in success; it is usually obtained experimentally.
$n =$ Total number of independent trials.
$x =$ Required number of successes.

Given p, the probability of achieving exactly x successes in n trials is given by

$$\text{Prob}(x \text{ of } n) = f(x|n, p) = \frac{n!}{(n-x)!x!} p^x (1-p)^{n-x} \tag{2.53}$$

This is known as the binomial probability density function (PDF), because the coefficients in the fraction are successive terms of the familiar binomial series expansion. The distribution is written as $f(x|n, p)$ to indicate the dependence of x on n and p, and is a discrete, rather than a continuous, distribution because x may take only the integer values 0, 1, 2 ... and so on up to n.

As an example of the use of a binomial distribution, suppose that after considerable testing, it is found that the probability a single round fired by the rifleman in Fig. 2.24 hits the target p is 0.3. Equation (2.53) would be used to calculate the probability, for example, that exactly 3 (x) shots out of 8 (n) would hit the target. This would result in

$$P(3|8, 0.3) = \frac{8!}{(8-3)!3!} 0.3^3 0.7^5 = \frac{8 \times 7 \times 6}{3 \times 2} \times 0.027 \times 0.168 = 0.254 \tag{2.54}$$

Also of interest would be to predict the probability that at least three shots would hit the target. It can be seen that this would be given by the probability of three hits, or the probability of four hits, and so forth. Each outcome is exclusive, so we use Eq. (2.33) to write

$$\begin{aligned} P(\text{at least 3 of 8}) &= P(3|8, 0.3) + P(4|8, 0.3) + \cdots + P(8|8, 0.3) \\ &= \sum_{x=3}^{x=8} f(x|n, p) \end{aligned} \tag{2.55}$$

Because we know that something must happen in eight shots, that is,

$$\sum_{x=0}^{x=8} f(x|n, p) = 1 \tag{2.56}$$

we can write

$$\sum_{x=3}^{x=8} f(x|n, p) = 1 - \sum_{x=0}^{x=2} f(x|n, p) \tag{2.57}$$

The CDF is defined[†] as

$$F(x|n, p) = \sum_{k=0}^{k=x} f(k|n, p) \tag{2.58}$$

It can be interpreted as the probability of at most x successes in n trials. Equation (2.57) could therefore be written in the form

$$P(\text{at least 3 of 8}) = 1 - F(2|8, 0.3) \tag{2.59}$$

Equations (2.53) and (2.58) may be somewhat tedious to evaluate, particularly for large n, so precomputed tabular data are used, at least for the CDF. Table A.2 in Appendix A shows the values of the CDF for a limited range of n, k, and p.

Evaluating Eq. (2.59) with Table A.2, the probability of at least three shots out of eight will hit the target becomes

$$P(\text{at least 3 of 8}) = 1 - F(2|8, 0.3) = 1 - 0.5518 = 0.4482 \tag{2.60}$$

The CDF in Table A.2 can also be used to compute the PDF. Consider the previous example of determining the probability that exactly three out of eight shots hit the target, as shown in Eq. (2.54). We could also write

$$\begin{aligned} P(\text{exactly 3 of 8}) &= P(\text{at most 3 of 8}) - P(\text{at most 2 of 8}) \\ &= F(3|8, 0.3) - F(2|8, 0.3) \\ &= 0.8059 - 0.5518 \\ &= 0.2541 \end{aligned}$$

This is the same result obtained in Eq. (2.54). In general, therefore, we can write

$$f(x|n, p) = F(x|n, p) - F(x - 1|n, p) \tag{2.61}$$

Note the similarity of Eq. (2.61) to its continuous counterpart Eq. (2.10). Another common instance of using the binomial distribution is to solve a problem posed in the following form: If the probability of hitting a target

[†]Notice how a summation of the PDF is used for a discrete distribution, compared to an integral for a continuous distribution.

with one shot is p, what is the probability the target will be hit at least once by n shots?

First, the probability the target does not get hit can be written as:

$$\begin{aligned} P(\text{target not hit}) &= P(\text{zero hits in } n \text{ trials}) \\ &= f(0|n,\ p) \end{aligned} \tag{2.62}$$

Substituting $x = 0$ into the PDF, Eq. (2.53) results in

$$P(0|n, p) = \frac{n!}{n! \times 1}(1 - p)^n = (1 - p)^n \tag{2.63}$$

We can also write

$$P(\text{target hit}) = 1 - P(\text{target not hit}) \tag{2.64}$$

Substituting results in

$$P(\text{target hit}) = 1 - (1 - p)^n \tag{2.65}$$

Equation (2.65) is sometimes referred to as either the powering up or survivor rule. As an example, suppose the probability of hitting a target with a single shot is 0.3. We may ask: What is the probability the target is hit at least once when one, two, three, or more shots are fired? Using Eq. (2.65), Table 2.2 can be generated for n shots fired at the target.

Note how the probability of hitting the target approaches unity for increasing numbers of shots and also how the probabilities can be obtained from Table A.2 as

$$P(\text{target hit}) = 1 - F(0|n, p) \tag{2.66}$$

Because the powering up formula is derived from the binomial distribution, it must be kept in mind that all trials are considered independent of each other. In summary, given n and p, the binomial distribution can be used to calculate the probability of exactly x, at least x, and at least one success.

Table 2.2 Example of Powering Up

n	P(Target Hit)
1	0.300
2	0.510
3	0.657
4	0.760
5	0.832
7	0.918
10	0.972

2.10 Mott and Weibull Distributions

We have grouped these distributions together because they have both been used to describe the distribution of fragment mass produced by a naturally fragmenting warhead (i.e., one without a scored case). After collecting and weighing the resultant fragments, it is usual to describe test data in terms of a distribution function, as shown in Fig. 2.25.

In this figure, m is the fragment weight and $N(m)$ is the proportion of fragments having a weight greater than m. You can see that the CDF being the proportion of fragments having a weight less than m is given by

$$F(m) = 1 - N(m) \tag{2.67}$$

Historically, the Mott distribution has been fitted to the distribution shown in the figure as follows:

$$N(m) = \exp\left(-\frac{m}{\mu}\right)^{0.5} \tag{2.68}$$

where N_0 is the total number of fragments, and μ is the mean fragment mass. In the so-called generalized Mott distribution, the exponent is given the symbol λ, which for some cases is closer to $1/3$ than $1/2$.

The Weibull distribution has a continuous probability density function for a random variable x that takes the following form:

$$f(x|\lambda, k) = \frac{k}{\lambda}\left(\frac{x}{\lambda}\right)^{k-1} \exp\left[-\left(\frac{x}{\lambda}\right)^k\right] \tag{2.69}$$

In these equations, λ is a scale parameter, and k is a shape parameter. The corresponding CDF is

$$F(x|\lambda, k) = 1 - \exp\left[-\left(\frac{x}{\lambda}\right)^k\right] \tag{2.70}$$

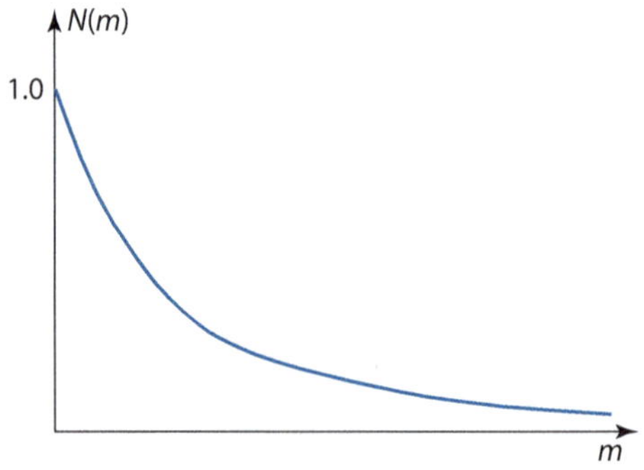

Fig. 2.25 Distribution of fragment weights.

Comparing Eqs. (2.67), (2.68), and (2.70), we see that the Weibull and Mott distributions are related, so we can write the following:

$$N(m) = 1 - F(x|\lambda, k) = \exp\left[-\left(\frac{x}{\lambda}\right)^k\right] \tag{2.71}$$

Note that with the appropriate selection of coefficients, Eqs. (2.68) and (2.71) fit the distribution of weights shown in Fig. 2.25. The manner in which these distributions would be used will be illustrated using Weibull distribution, where the mean and variance are known to be

$$\mu = \lambda\Gamma\left(1 + \frac{1}{k}\right) \tag{2.72}$$

$$\sigma^2 = \lambda^2\Gamma\left(1 + \frac{2}{k}\right) - \mu^2 \tag{2.73}$$

where Γ is the gamma function. Given a collection of fragments, the mean and variance can be calculated and Eqs. (2.72) and (2.73) solved for λ and k. These parameters completely specify the PDF and CDF. Reference [1] has a comparison among the Mott, Weibull, and other related distributions and test data. Note that when $k = 2$, the Weibull distribution is the same as the Rayleigh distribution.

2.11 Poisson Distribution

For some problems involving a binomial distribution, the value of n is large and the value of p is small. In this case, some mathematical simplifications to the PDF and CDF are possible and result in a Poisson distribution. If the product $n \cdot p = \alpha$, for the PDF from Eq. (2.53) we can write

$$f(x) = \frac{\alpha^x e^{-\alpha}}{x!} \tag{2.74}$$

and for the CDF

$$F(x) = \sum_{k=0}^{x} \frac{\alpha^x e^{-\alpha}}{x!} \tag{2.75}$$

Poisson distributions are usually used to determine the probability that zero, one, or more events occur in a finite time by putting $\alpha = \mu t$ in Eq. (2.74)

$$\text{Prob}\{\text{exactly } x \text{ events occur in time } t\} = e^{-\mu t}\frac{(\mu t)^x}{x!} \tag{2.76}$$

Here, μ is the mean rate at which the event occurs, and t is the time interval under consideration.

Example 2.3

Suppose that, on average, aircraft approach a carrier for landing at the rate of 1/min, and that the deck can handle a maximum of two aircraft at a time. What is the probability that during a 1-min interval an aircraft has to be turned away from landing due to a congested deck?

In this problem $\mu = 1/\text{min}$, and the time interval under consideration is 1 min. We can write

$$\text{Prob}\{\text{zero aircaft arrive in 1 min}\} = e^{-1}\frac{(1)^0}{0!} = 0.368$$

and that

$$\text{Prob}\{\text{one aircaft arrives in 1 min}\} = e^{-1}\frac{(1)^1}{1!} = 0.368$$

So we conclude that

$$\text{Prob}\{\text{at least two aircaft arrive in 1 min}\} = 1 - (0.368 + 0.368) = 0.264$$

Observe that the CDF, like the binomial distribution, can be interpreted as the probability of at most x events occurring in a time interval, and could have been used in the example given.

2.12 Testing Data for a Particular Distribution

When analyzing experimental data, it is usual to assume that the variable is random with a particular distribution (e.g., normal, Rayleigh, uniform, etc.). How can we measure the confidence of this assumption? In the field of statistics, this subject is known as *hypothesis testing*. The first test to apply is to simply observe the distribution in a histogram to give an idea of what analytical distribution may be present. For example, the histograms shown in Fig. 2.26 suggest perhaps uniform and normal distributions, respectively.

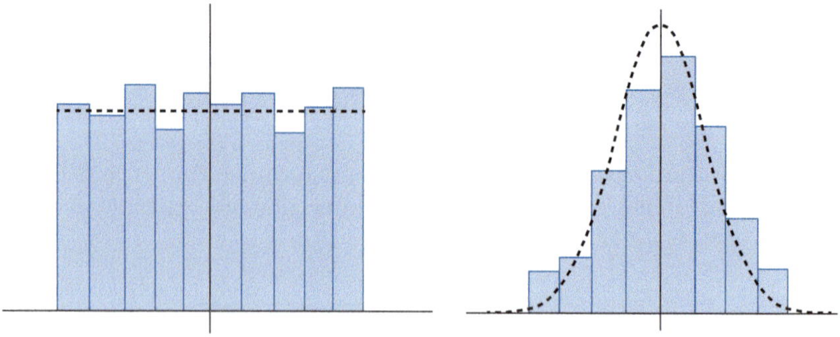

Fig. 2.26 Sample data sets grouped into histograms.

Table 2.3 Chi-Squared Test on Example Data Set

Range Error			
Bin Range	**Observed**	**Predicted**	**Chi Sq. Error**
Less than −9.71	5	2.91	1.496
−9.71 to −5.73	5	6.12	0.204
−5.73 to −3.16	5	6.59	0.383
−3.16 to −1.30	5	5.76	0.101
−1.30 to −0.04	5	4.13	0.185
−0.04 to 0.92	5	3.14	1.105
0.92 to 2.74	5	5.64	0.074
2.74 to 4.40	5	4.50	0.057
4.40 to 7.20	5	5.65	0.075
More than 7.20	5	5.56	0.057
Sum	50	50	3.74
		Chi-square test	0.9279

It is useful to obtain quantitative measures of how close the data are to the assumed distribution when results are presented. One example of such a process is the chi-squared test, which will be described briefly here. By way of an example, 50 data points were generated from a normal random number generator and the test applied using the following steps, the results being compiled in Table 2.3. Obviously we know the data are normal because we generated them, but ignore that for now.

1. Formulate the hypothesis, in this case that the 50 data points fit a normal distribution.
2. For the data set, calculate the statistics needed to define the suspected distribution; for example, if the set is thought to be normal, the mean and variance would be calculated.
3. The data points are ordered from the smallest to the largest.
4. The data are grouped into bins. Although the size of the bins is arbitrary, a rule of thumb is to ensure there are at least five data points in each bin. Note that in order to get five samples in each bin, the bin widths are not constant.
5. Using the calculated statistics for the assumed distribution, predict the number of samples that should occur in each bin using Table A.1. For example, in the bin range −9.71 to −5.73, the probability a single sample lies in this range would be

$$P(-9.71 < x < -5.73) = F(-5.73) - F(-9.71) \tag{2.77}$$

Therefore, in a sample of 50, we would expect to find 50 times the probability of finding a single sample, that is, the result of Eq. (2.77).

6. For each bin, calculate the chi square error statistic χ^2 from

$$\chi^2 = \frac{(\#\text{predicted} - \#\text{observed})^2}{\#\text{predicted}} \tag{2.78}$$

7. Sum the error for all bins. Intuitively, if the hypothesis is correct, the sum of the chi square errors will be "small," but what is small?
8. Compute from statistical tables (Table A.3 in Appendix A) the chi-squared test parameter from the sum of errors and the degrees of freedom (number of bins less one).
9. The closer the chi-squared test parameter is to unity, the closer the data set is to the assumed distribution.

The calculated values of mean and standard deviation are not shown in the table. Table 2.3 is an Excel spreadsheet that computes the chi-squared parameter automatically (0.9279), but it can also be derived from standard statistical tables, such as the one shown in Table A.3.

To obtain data from Table A.3, the number of degrees of freedom (DOF) has to be determined. In this case it is one less than the number of bins, or 9. Then the sum of the chi-squared parameters has to be found—3.74 from Table 2.3. Then from Table A.3, along the row corresponding to DOF = 9, the probability for the chi-square error parameter (3.74) has to be estimated. It appears to lie between the probabilities of 0.90 and 0.95, that is, the exact value of 0.9279 determined by Excel in Table 2.3. The result is interpreted as follows: There is a 92.79% probability the distribution is normal, that is, the hypothesis cannot be rejected.

2.13 Functions of Random Variables

Sometimes variables are derived from other variables that are random in nature, and it is of interest to find the statistics that describe the derived variable. For example, suppose x and y are random variables with mean μ_x and μ_y, and variances σ_x and σ_y. We now form a third variable z such that

$$z = x + y \tag{2.79}$$

The variable z will also be a random variable such that if x and y are uncorrelated (i.e., independent), then the mean and variance of z are given by

$$\mu_z = \mu_x + \mu_y \tag{2.80}$$

$$\sigma_z = \sqrt{\sigma_x^2 + \sigma_y^2} \tag{2.81}$$

This latter equation is sometimes referred to as a root sum squared (RSS) process for independent random variables. Apart from this particular case, it is difficult to predict the statistics of derived variables.

2.14 Mathematical Expectation

Suppose we have a discrete random variable x having the possible values $x_1, x_2, \ldots, x_n$. The expected value $E(x)$ is defined by

$$E(x) = x_1 P(x = x_1) + x_2 P(x = x_2) \cdots + x_n P(x = x_n)$$
$$= \sum_{i=1}^{i=n} x_i P(x = x_i) \quad (2.82)$$

The expected value can be thought of as the mean value the function $E(x)$ would take for a large number of independent random selections of x. Note that although x is a discrete variable, $E(x)$ may not be.

Consider the example of rolling a die where the probability of x taking a particular value (1 through 6) is equal at $1/6$. The expected value of the number shown by the die will be

$$E(x) = 1 \times \frac{1}{6} + 2 \times \frac{1}{6} + 3 \times \frac{1}{6} + 4 \times \frac{1}{6} + 5 \times \frac{1}{6} + 6 \times \frac{1}{6} = 3.5 \quad (2.83)$$

This would be the result of rolling the die a large number of times, recording each result, and finding the mean.‡

If we can write that $P(x = x_i)$, the value of the discrete PDF at $x = x_i$ in the following form:

$$P(x = x_i) = f(x_i) \quad (2.84)$$

then

$$E(x) = \sum_{i=1}^{i=n} x_i f(x_i) = \sum x f(x) \quad (2.85)$$

A special case of Eq. (2.85) occurs when the probability that x takes a particular value is equal for all trials

$$E(x) = f(x) \sum x = \frac{x_1 + x_2 + \cdots + x_n}{n} \quad (2.86)$$

and is called the arithmetic mean, which we dealt with earlier. In the case of rolling the die, the probability a particular number comes up is equal to $1/6$, so Eq. (2.86) could have been used to obtain the mean value of 3.5.

If x is now a continuous function, we can use Eq. (2.85) to write the expected value as

$$E(x) = \int_{-\infty}^{+\infty} x f(x) \mathrm{d}x \quad (2.87)$$

‡Notice how the expected value is not discrete and represents a number we would never see on the die.

Suppose now we consider a function of a random variable $y = h(x)$, where the random variable x has the probability density function $f(x)$. If x is a discrete random variable, the expected value of y from Eq. (2.82) becomes

$$\begin{aligned} E(y) = E(h(x)) &= h(x_1)f(x_1) + h(x_2)f(x_2) + \cdots + h(x_n)f(x_n) \\ &= \sum h(x)f(x) \end{aligned} \tag{2.88}$$

Similarly we can write from Eq. (2.87) for the continuous random variable x

$$E(y) = E(h(x)) = \int_{-\infty}^{+\infty} h(x)f(x)\mathrm{d}x \tag{2.89}$$

This is the expected value of a function of a continuous random variable. It may be thought of as the result of the following:

1. Drawing a random value of x knowing its distribution
2. Calculating the function $y = h(x)$ for that value of x
3. Storing the value of y
4. Repeating for a large number of samples of x
5. Calculating the mean of y

It can be seen that this process is equivalent to a Monte Carlo experiment, and in many cases computing the expected value is a way to avoid the long time needed to determine statistics using iterative techniques. Of course, computing the expected value involves performing the integration in Eq. (2.89), and depending on the functions $f(x)$ and $h(x)$ this is usually difficult at best, and in many cases impossible. However, considerable use of the expected value theorem will be made in later sections of the book, so this section should be understood thoroughly. This may be aided by the following example.

Example 2.4 Expected Value Theory

A ship passes along the center of a 1000-ft-wide channel. Suppose an aircraft has flown across the channel and dropped a mine such that the mine location may be anywhere between the banks with an equal (uniform) probability. If the ship is 100 ft wide, and the mine will detonate and damage the ship if it comes within 50 ft of it, as shown in Fig. 2.27, what is the probability the ship will be damaged?

Suppose the location of the mine from the left bank is x. We look at this problem by considering a "damage function," which is the probability the ship is damaged by the mine, and plotting this damage function as a function of x, as shown in Fig. 2.28.

(Continued)

Example 2.4 Expected Value Theory *(Continued)*

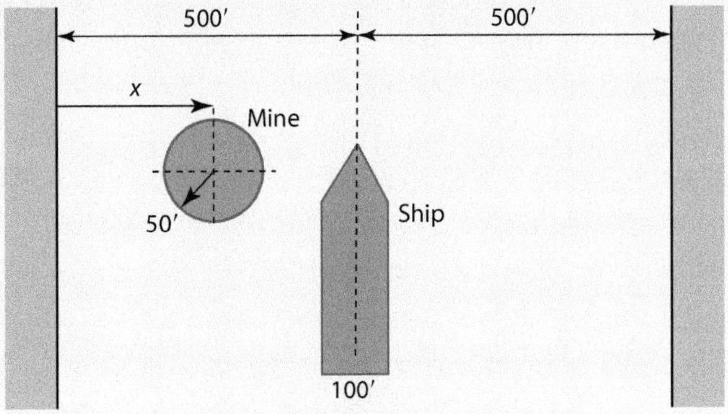

Fig. 2.27 Ship and mine geometry.

From this we can observe that if $400 < x < 600$, $P_D = 1$; otherwise, $P_D = 0$. Given that x is a random variable with a uniform distribution, we want to find the expected value of P_D for a large number of samples of x. This is evaluated from Eq. (2.89) as

$$E(P_D) = \int_{-\infty}^{+\infty} P_D(x)f(x)\mathrm{d}x = \int_{-\infty}^{+\infty} P_D(x)\frac{1}{1000}\mathrm{d}x = \int_{400}^{600} 1 \times \frac{1}{1000}\mathrm{d}x$$
$$= \left[\frac{x}{1000}\right]_{400}^{600} = 0.2 \tag{2.90}$$

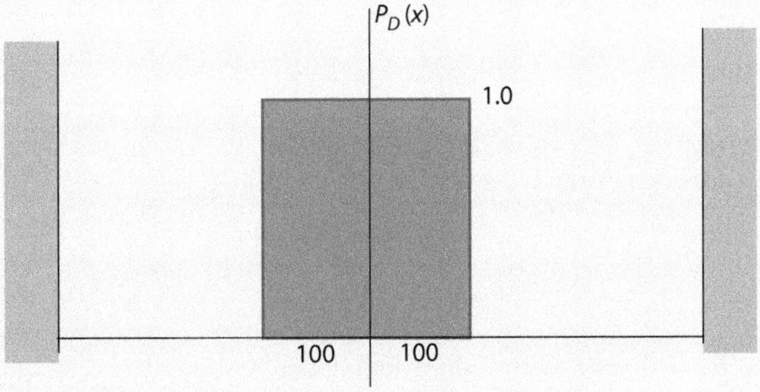

Fig. 2.28 Damage probability across the channel.

(Continued)

Example 2.4 Expected Value Theory *(Continued)*

An alternate way to solve this problem would be through simulation of the actual events that take place when the aircraft drops the mine and the ship traverses the channel. The following code segment is a MATLAB program to do this in which a random number is drawn from a uniform distribution for the mine location, which is then checked against Fig. 2.28 to determine whether that location damages the ship. Repeating this process for a large number of trials enables us to determine the proportion of mine locations that damage the ship, that is, the P_D.

```
%----------------------------------------------------
% Mine Example
n=100000;                              % number of iterations
Pk=0;                                  % initialize counter
i=1;                                   % first iteration
    while i<n,                        % conditional loop
         x=1000*rand;                  % get mine position
         if (x>400)                   % test left boundary
            if(x<600)                 % test right boundary
            Pk=Pk+1;                   % ship damaged
            end
         end
    i=i+1;                             % increment trial number
    end                                % done with iterations
Pd=Pk/n                                % calculate average
%----------------------------------------------------
```

Running the program for the 100,000 trials indicated produces the result $P_D = 0.2009$. This shows that the expected value theorem produces approximately the same result as the simulation.

It is important to note that if the Monte Carlo simulation is run repetitively without changing anything, different results would be obtained (e.g., 0.1987, 0.2015, 0.2012, 0.1996, etc.). The variability of these results depends on the number of Monte Carlo trials selected—more trials produce less variability but take more time to execute. This will be discussed in more detail in the next section. This variability is centered on the result produced by the expected value theorem, 0.2000.

Suppose now that the mine is dropped at a location specified not by a uniform probability distribution but with a normal distribution where $\mu = 500$ ft and $\sigma = 50$ ft. How does the calculation change? For the simulation, the only change will be to the calculation of the mine position. Because "randn" produces random samples of the standard variable z, and we know that

$$z = \frac{x - \mu}{\sigma} \tag{2.91}$$

(Continued)

Example 2.4 Expected Value Theory *(Continued)*

The mine position now becomes

```
x=(50* randn)+500;                    % get mine position
```

Making this modification and running the program gives $P_D = 0.9547$. The expected value calculation becomes

$$E(P_D) = \int_{-\infty}^{+\infty} P_D(x)f(x)\mathrm{d}x = \int_{400}^{600} 1 \times \frac{1}{\sqrt{2\pi}\sigma_x} \exp\left(\frac{(x-\mu_x)^2}{2\sigma_x}\right)\mathrm{d}x \tag{2.92}$$

This represents an integration of the normal PDF between 400 and 600. Using the standard normal variable z we find

$$z_1 = \frac{400-500}{50} = -2.0, \quad z_2 = \frac{600-500}{50} = 2.0 \tag{2.93}$$

Hence from Table A.1, Eq. (2.92) becomes

$$E(P_D) = F(2.0) - F(-2.0) = 0.9544 \tag{2.94}$$

The two results are very close; however, the "correct" answer is the one obtained from the expected value theorem.

2.15 Monte Carlo Simulations

In the example of the mine damaging the ship, a repetitive model of the scenario was executed using a Monte Carlo simulation. This involves conducting a large number of trials that simulate the randomness of the process of interest and then calculating the required statistics from the aggregate of results. For that example we used a random number generator to produce a set of n uniformly distributed random numbers from 0 ft to 1000 ft to simulate the mine position and determine the proportion that damage the ship. One obvious question not addressed, of course, is: How many repetitions are needed?

In general, it is not possible to predict this number because every case is different; however, some means to test whether a sufficient number of iterations have been performed can be obtained from studying the following case. Consider the previous example dealing with the probability a random sample drawn from a univariate normal distribution with a mean of 1.2 and standard deviation of 0.7 will lie in the range from 1 to 2. The result derived from the CDF earlier was 0.487.

We now solve this problem using a Monte Carlo simulation in MATLAB. We choose a fixed number of iterations, draw a random number from the specified distribution, and determine the proportion that lie between the

Table 2.4 Results of Monte Carlo Simulation

No. of Iterations	Run 1	Run 2	Run 3
100	0.360	0.520	0.450
1000	0.478	0.473	0.486
100,000	0.489	0.485	0.486

required bounds. This simulation is run three times for that number of iterations. Then the number of iterations is increased and the process repeated. The results are shown in Table 2.4.

We can see that for 100 iterations the simulation produces results that vary quite a bit but are in the region of the known answer. As the number of iterations increases there is less spread in successive runs of the simulation, until for 100,000 iterations the result is "stable" to two decimal places. Two observations may be drawn:

1. The number of iterations needed increases with the need for higher precision in the answer.
2. The Monte Carlo method may take considerable time to execute if high precision is needed.

The results of the closed form expected value solution may be considered the result obtained for an infinite number of iterations of the corresponding Monte Carlo simulation and is the preferred solution, as long as there are analytical methods available to calculate the integral in Eq. (2.89).

2.16 Generating Random Numbers from the Cumulative Distribution

As mentioned previously, a Monte Carlo simulation can be used to simulate a particular random process by drawing successive random numbers from the distribution assumed to govern the process. In the prior case, normally distributed random numbers were used. For some random processes the underlying distribution may not be clearly defined, and a means to simulate such a process in a Monte Carlo simulation using standard distributions becomes inaccurate.

In such cases, typical of experimentally generated data (test data), sequences of random numbers may be generated if the cumulative distribution function (CDF) has been obtained. Consider the CDF shown in Fig. 2.29.

To obtain random numbers for this distribution, we perform the following steps:

1. Generate uniformly distributed random numbers n in the range 0.0 to 1.0.
2. Mark each random number on the y-axis of Fig. 2.29.

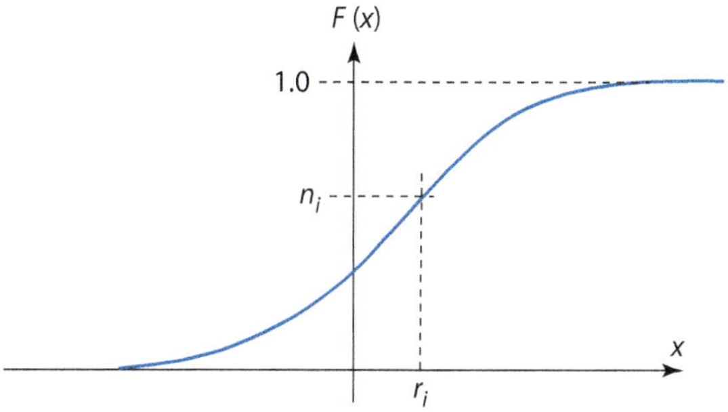

Fig. 2.29 Experimentally derived CDF.

3. Read across until it intersects the CDF curve.
4. Read down to the x-axis to obtain the required random number r of the distribution.

This process can be automated by curve fitting the CDF to a user-selected polynomial.

2.17 Confidence Levels, Limits, and Intervals

Recall our discussion of a univariate, normal distribution in Section 2.4 regarding the area under the PDF corresponding to $\pm\sigma$, $\pm 2\sigma$, and so forth. From tables of the cumulative distribution function for a normally distributed random variable, we have seen that about 68% of the samples will lie in the range $\pm\sigma$, 95.5% will lie in the range $\pm 2\sigma$, and 99.7% in the range $\pm 3\sigma$. This allows us to illustrate the use of confidence intervals and levels in the following manner:

- There is a 68% chance that a random sample of the variable x will lie within $\pm 1\sigma$ of the mean.
- There is a 95.5% chance that a random sample of the variable x will lie within $\pm 2\sigma$ of the mean.
- There is a 99.75% chance that a random sample of the variable x will lie within $\pm 3\sigma$ of the mean.

The percentage value is the confidence level bounded by the upper and lower confidence limits. This interval can be expressed in the form of an upper U and lower L limit, where

$$U = \mu + z_C\sigma \tag{2.95}$$

and

$$L = \mu - z_C\sigma \tag{2.96}$$

Table 2.5 Values of z_C for Different Confidence Levels

Confidence Level %	99.75	99	98	96	95.5	95	90	80	68	50
z_C	3.00	2.58	2.33	2.05	2.00	1.96	1.645	1.28	1.00	0.6745

Values of confidence coefficients z_C for different confidence levels are given in Table 2.5. From this table we might say, for example, that our 95% confidence level has an interval of 1.96σ with an upper limit and lower limit given by Eqs. (2.95) and (2.96) about whatever the mean of the distribution happens to be.

2.17.1 Distribution of the Mean

Suppose we have a normally distributed random variable x where the (unknown) population mean is μ_x and the variance is σ_x^2. A sample size n is drawn from the population, and the sample mean $\overline{x}$ is calculated. If we do this a number of times (for the same n) we will obtain a different mean value each time, and hence will get a distribution of $\overline{x}$. What can we say about the distribution of means? Using the central limit theorem, it is possible to show that

- The expected[§] mean of a set of sample means $\overline{x}$ is μ_x the population mean, and the sample mean is normally distributed for a large ($n > 30$) sample size.
- The expected variance of the set of means is also normally distributed for large n and may be expressed in terms of the variance of x by

$$\sigma_{\overline{x}}^2 = \frac{\sigma_x^2}{n} \tag{2.97}$$

Because the means are normally distributed, we are now able to define a confidence limit and level for the mean $\overline{x}$. If the confidence level is selected at, say, 95%, then the confidence interval of the mean is

$$(L, U)_{0.95} = \mu_{\overline{x}} \pm 1.96\,\sigma_{\overline{x}} \tag{2.98}$$

This is stated as: We are 95% confident that a sample of the mean $\overline{x}$ will be within (L, U) of the true mean, or stated another way, we are 95% confident that the true mean is within (L, U) of a sample of the mean $\overline{x}$.

However, using Eq. (2.97) we write

$$(L, U)_{0.95} = \mu_{\overline{x}} \pm 1.96\,\sigma_{\overline{x}} = \mu_{\overline{x}} \pm 1.96\,\frac{\sigma_x}{\sqrt{n}} \tag{2.99}$$

[§] The usual interpretation of the expected value is the mean; therefore, the expected value of the mean reads "the mean of the means."

Because the population mean and variance are unknown, the unbiased estimators $\overline{x}$ and S_x are used, giving

$$(L, U)_{0.95} = \overline{x} \pm 1.96 \frac{S_x}{\sqrt{n}} \tag{2.100}$$

The more general expression is

$$(L, U) = \overline{x} \pm z_c \frac{S_x}{\sqrt{n}} \tag{2.101}$$

As an example, suppose the mine problem discussed earlier is reconsidered, and it is run once each for 100, 1000, and 100,000 iterations. Each run will generate 100, 1000, and 100,000 samples of P_D for which we calculate the sample mean and standard deviation. We are interested in a 95% confidence level for the mean P_D calculated from the set of P_D values generated. The results are summarized in Table 2.6, where the half-interval is calculated from Eq. (2.101).

The case for 1000 iterations may be explained as: We are 95% confident that the true P_D (mean) is within 0.172 ± 0.019 (i.e., in the range 0.153 to 0.191). Clearly, as we take more iterations the half-interval decreases, the sample mean approaches the population mean, and the sample variance approaches the population variance. Note that for a fixed number of iterations we will, in general, get different sample means and confidence limits for the same confidence level.

It may be concluded that for any Monte Carlo simulation of the type described previously, a confidence interval for a specified confidence limit can be calculated, but that these values apply only to the simulation completed and will be different if the simulation is run again for the same inputs.

2.17.2 Confidence Intervals for Small Samples of the Mean

If the number of samples of the mean is small, typically less than about 25, then the confidence coefficients given in Table 2.5 cannot be used. This is a common problem for some weapons testing procedures where the cost of each weapon is high, so few trials can be performed. Instead of using the normal distribution, the so-called t-distribution has to be used. This distribution has a similar shape to the normal distribution, but is a function of an additional parameter called the degrees of freedom ν, which equals the

Table 2.6 Single Run of Mine Example

n	$P_D = \overline{x}$	S_x	Half-Interval
100	0.153	0.785	0.154
1000	0.172	0.302	0.019
100,000	0.177	0.308	0.002

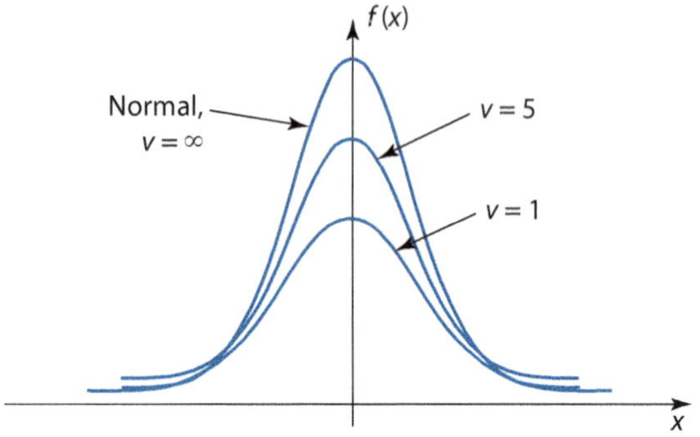

Fig. 2.30 The normal and t-distributions.

sample size minus 1. The probability density function for different degrees of freedom is shown in Fig. 2.30.

The t-distribution tends to resemble the normal distribution as the number of degrees of freedom increases. In fact, the number of degrees of freedom is determined as the number of samples minus one, so this is why for small sample sizes the t-distribution has to be used instead of the normal distribution. Just as the integral of the normal probability density function (i.e., the CDF) is tabulated, so is the integral of the t-distribution.

The confidence level and limits work the same way as for the normal distribution. Figure 2.31 shows a t-distribution for a specific number of degrees of freedom together with a confidence limit (L, U) for a sample of the random variable x.

For a normal distribution, the range (L, U) would be given by Eq. (2.101), but for a t-distribution it is given by

$$(L, U) = \overline{x} \pm t_c \frac{S_x}{\sqrt{n-1}} \tag{2.102}$$

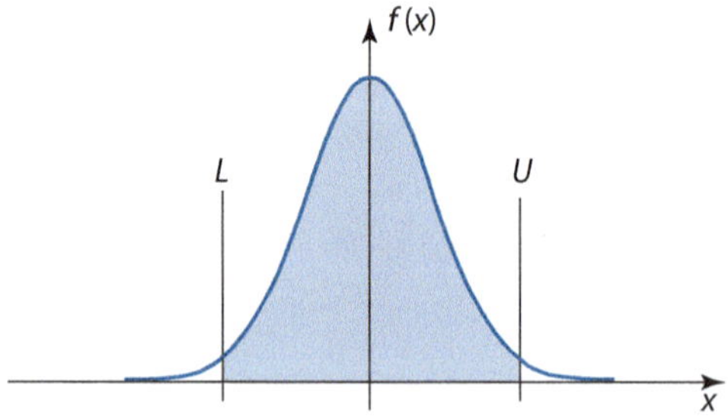

Fig. 2.31 Confidence limits for a t-distribution.

Table 2.7 Confidence Coefficients t_c for t-Distribution

ν	80%	90%	95%	99%
1	3.08	6.31	12.71	63.66
2	1.89	2.92	4.30	9.93
3	1.64	2.35	3.18	5.84
4	1.53	2.13	2.78	4.60
5	1.48	2.01	2.57	4.03
10	1.37	1.81	2.23	3.17
30	1.31	1.70	2.04	2.75
∞	1.28	1.65	1.96	2.58

Here the mean and variance are calculated from the small sample. As before, the confidence coefficient t_c is obtained from the area beneath the PDF, but this time it will be dependent on the number of degrees of freedom for which Fig. 2.31 is drawn. It is usual to tabulate the area under the PDF for a single "tail" (i.e., t_c is given only for the upper interval U). What this means is that if we require, for example, the 95% confidence interval, then the sum of the area under the PDF less than L and greater than U must be 5% of the total area; hence, the confidence coefficient needed in Eq. (2.102) is $t_{0.975}$. Table 2.7 shows values of the confidence coefficients for different confidence levels and degrees of freedom.

Usually we assume we are dealing with "large" samples, so the normal distribution is a reasonable assumption; however, this section provides the necessary tools to make the adjustments if the sample size is small.

2.18 Required Number of Monte Carlo Iterations

We saw in the previous section how it is possible to run a Monte Carlo simulation for a fixed number of iterations, analyze the data resulting from each iteration, and determine the confidence interval of the result. It is frustrating, however, to take, for example, an hour to produce 100 iterations that, when analyzed, give an unacceptable confidence bound, because the simulation would have to be run again for more than 100 iterations.

It would be much more useful if an estimate of the number of iterations required to satisfy a given confidence interval could be determined a priori. It was stated earlier that this cannot be done, but a running estimate can be generated if the simulation is executed in a "free-run" mode, whereby the user can stop the simulation when the desired confidence limits have been satisfied.

Consider an example where we have a closed form solution to the problem (i.e., the exact answer is known to be 0.176 in this case). Suppose we can also solve the same problem using a Monte Carlo approach; this is executed using the following steps:

1. Select a trial value of n, and run the simulation for that number of iterations.
2. For each iteration, record the result.
3. Calculate the mean of the n results.
4. Knowing the "correct" answer (0.176), we then calculate the percentage error of the mean.

The result of running the Monte Carlo simulation for n iterations is shown in Table 2.8.

It is also noted that the population standard deviation is not known, but may be accurately estimated by running the simulation a large number of times, say 10^7. The result of doing this is

$$\sigma_x \approx S_x = 0.3073 \tag{2.103}$$

The approach suggested by Table 2.8 is that if we can specify a maximum acceptable percentage error for the mean, we can inspect the table and determine the required number of iterations needed to achieve that error. By considering the confidence interval to represent twice this maximum error, we can write

$$\text{error}_{\text{max}} = \frac{z_c S_x}{\sqrt{n}} \tag{2.104}$$

The percentage error of the mean becomes

$$E = \frac{100 \times z_c S_x}{\overline{x}\sqrt{n}} \tag{2.105}$$

Transforming for n yields

$$n = \left[\frac{100\, z_c S_x}{E\overline{x}}\right]^2 \tag{2.106}$$

So, for the example used, suppose the required confidence level is 95%, $z_c = 0.196$, $\overline{x} = 0.176$, $S_x = 0.3073$, and $E = 5$, the required number of iterations becomes 4684.

In words this reads: If the simulation is run for 4684 iterations, we are 95% confident that the calculated result will not differ by more than 5% from the true result.

Table 2.8 Percentage Error as a Function of Iterations

n	$\overline{x}$	% Error
100	0.153	13.1
1000	0.172	2.3
100,000	0.177	0.1

Running the simulation 10 times for 4684 trials produced percentage errors of 4.89, 4.99, 4.92, 4.92, 5.1, 5.0, 4.95, 4.98, 5.00, and 4.94, so it appears to work.

The only problem in this approach is that the separately calculated population statistics μ_x and σ_x were used in Eq. (2.106), and these are not known. We now consider what happens if we use the sample statistics instead of the population statistics.

Consider an iterative Monte Carlo simulation that has no upper bound to the number of iterations to be performed; that is, it is in "free run" mode and would continue indefinitely unless interrupted. As the iterations proceed, the results accumulate into a sample of increasing size. As more iterations occur, the sample statistics approach the population statistics. Table 2.8 shows how the sample mean approaches the population mean; this also is true of the variance, although it is not shown in the table.

We therefore calculate the sample mean and variance continuously as the simulation proceeds, and use these values in Eq. (2.106) to determine how many iterations are needed to achieve a specified maximum percentage error with a specified confidence level. It has been observed that this number of iterations converges quickly, and even for sample sizes an order of magnitude lower than the number required, the calculation of that number is quite stable.

Consider again the previous example where the exact number of iterations was determined to be 4684. We can calculate this number as a function of the number of iterations actually performed. This estimation is shown in Fig. 2.32.

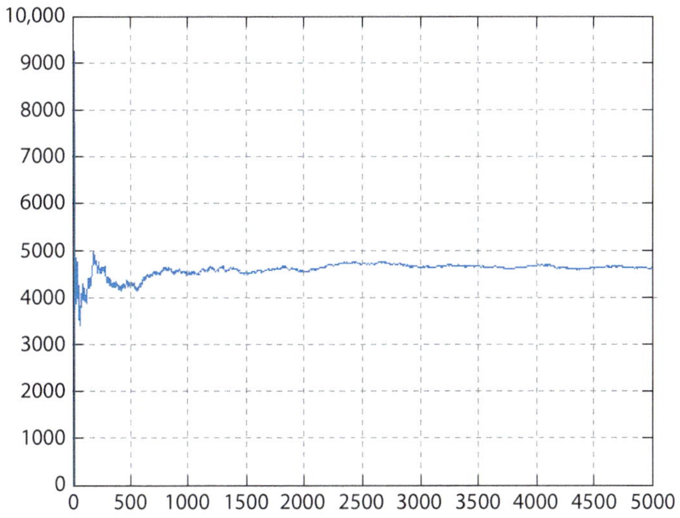

Fig. 2.32 Number of iterations required vs. iteration number.

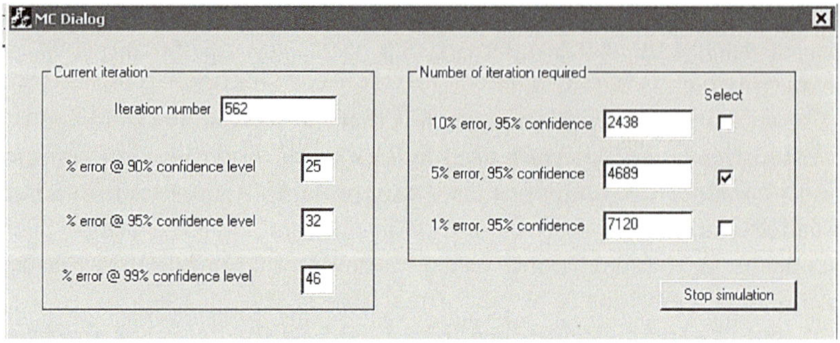

Fig. 2.33 Suggested dialog box to display confidence levels.

This shows how the number of iterations needed stabilizes within about 500–1000 iterations, and within about 100 iterations it is accurate to 20%. In practice, a dialog box such as that shown in Fig. 2.33 would be incorporated into the Monte Carlo simulation being used; the result from each iteration is recorded and analyzed as the simulation proceeds in free-run mode.

If this dialog box were displayed during the Monte Carlo simulation, it would enable the user to determine how accurate the simulation is so far and give an estimate of how many iterations are required to achieve a specific error with known confidence levels. The user may then stop the simulation if sufficient accuracy has been achieved, or let it run to a specific terminal error criterion.

2.19 Application of Statistics to Weaponeering

To summarize, weaponeering uses statistical principles to describe physically observed phenomena that cannot be modeled, hence predicted. Examples of variables described statistically include the following:

- The impact point of a particular weapon, leading to statistical descriptions of miss distance. This results in delivery accuracy data being obtained from tests.
- Distribution of submunitions from cluster bombs or artillery shells within a specified ground pattern.
- Errors associated with weapon fire control systems and instruments contained therein [e.g., cannon elevation angles, laser rangefinder measurements, FLIR pointing angle measurements, aircraft state variables (position, orientation), etc.].
- For artillery fire, the location of a target being attacked and the location of the gun(s) firing at them.
- The effect of terrain on the outcome of an engagement.

If we use statistics to calculate an effect, that result represents what we might reasonably expect to happen, not what will happen on a particular occasion. For example, suppose we fire a missile at a target, and we predict the probability of killing the target is 0.9. In an actual attack, the outcome is either kill or not kill; we cannot predict which will occur, just that a kill is more likely. It is important to state that we must not say that the target is 90% killed.

Reference

[1] Jaramaz, S., and Predrag, E., "Size Distribution of Fragments by Detonation of Fragmenting Warheads," *Proceedings of the 23rd International Symposium on Ballistics*, Tarragona, Spain, April 2007.

Chapter 3 Weapon Trajectory

3.1 Introduction

We have seen that the arrival conditions of a weapon at the target play a significant role in determining its effectiveness. In order to find these arrival conditions, it is necessary to determine how the weapon got to that point. These conditions can be summarized as follows:

- The impact angle of a weapon may determine its effectiveness. Striking a target at a shallow impact angle may cause the weapon to be susceptible to ricochet or broaching, resulting in a failure to penetrate the target.
- For a fragmentation warhead, the impact angle will also determine the fragment spray pattern around the target, again influencing its ability to inflict damage.
- The impact velocity of the weapon may affect its ability to penetrate a target, particularly if the target is underground or hardened in some manner. Too high an impact velocity may result in fracturing of the casing.
- The down-range distance traveled determines whether the target can be reached.
- When multiple munitions are released from an aircraft or fired from a cannon, the trajectory and intervals between individual warheads determines the area on the ground covered by all of the munitions. This area may or may not have overlapping regions producing different effects on areas of target elements.
- The trajectory geometry may affect the accuracy with which munitions (particularly unguided) can be placed on the target.

These variables will be available from a weapon trajectory; however, computing trajectory data may be a difficult task, particularly for guided weapons, depending upon the level of detail contained within the model. As usual with physical modeling problems, there is a tradeoff among the level of complexity of the model, the ease with which results can be computed, and the correspondence between those results and experimental or test data.

Reasonable answers can be obtained from fairly simple models, at least in order to illustrate weaponeering principles; therefore, simple models will be developed and used in this context. For more accurate work, highly detailed and validated models exist, and these are also described. Much of this chapter details the trajectory of unguided weapons, usually termed *ballistics*, although a discussion of guided weapon trajectory is also presented.

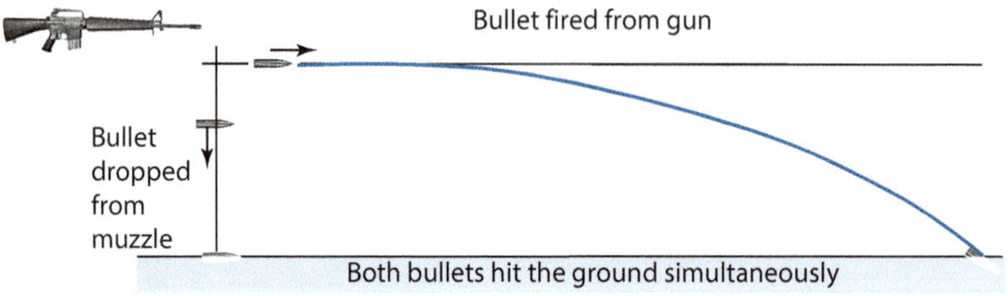

Fig. 3.1 Independent vertical and horizontal motions.

3.2 Initial Release or Firing Velocities

In the trajectory models that follow, we assume that the motions in the vertical and horizontal directions are independent. To illustrate this, consider the situation shown in Fig. 3.1.

Here we have gun firing a bullet horizontally; as it exits the muzzle, an identical bullet is released from rest and falls to the ground. Ignoring air resistance for both, they will impact the ground at the same time, illustrating the vertical motion is independent of the horizontal motion. To begin the analysis, assume that the initial velocities in these directions are known. For air-launched weapons, these initial velocities will be functions of the dive angle of the aircraft θ, the ejection velocity of the weapon v_e, and the forward velocity of the aircraft v_a at release. This situation is shown in Fig. 3.2.

It is assumed that the initial weapon axial velocity is the same as that of the aircraft releasing it. Positive ejection is necessary to project the bomb through the turbulent boundary layer that exists close to the aircraft, especially at high speeds. This layer causes unpredictable motion of the weapon and will lead to large delivery accuracy errors for unguided

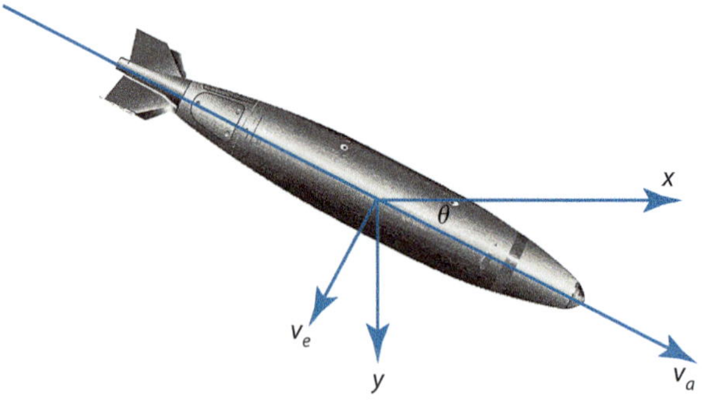

Fig. 3.2 Air-launched weapon velocities at release.

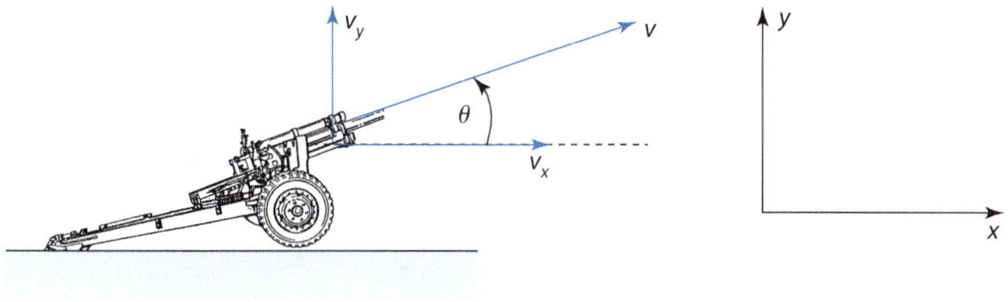

Fig. 3.3 Surface-launched weapon velocities at release.

munitions, or a break in the seeker lock on the target for guided weapons. Ejection is achieved by detonating small cartridges on the release rack, with typical ejection velocities in the region of 5–10 ft/s.

Surface-launched munitions are represented in Fig. 3.3 and do not have an ejection velocity normal to the muzzle velocity. Note also that for historical reasons the coordinate frames are usually different.

We will develop models based on the air-launched weapons because they have additional velocity at release, but the resulting analysis is easily adapted for the surface-launched case.

The initial horizontal velocity is obtained from the geometry of Fig. 3.2 as

$$v_{0h} = v_a \cos\theta - v_e \sin\theta \tag{3.1}$$

and the initial vertical velocity (down) is

$$v_{0v} = v_a \sin\theta + v_e \cos\theta \tag{3.2}$$

We will now look at three trajectory models, each with increasing complexity, that use these initial velocities and other release parameters to compute impact, or terminal conditions.

3.3 Zero-Drag, Point Mass Model

In this model, the projectile is represented by a point in space (zero dimensions) and experiences no drag as it moves through the air mass. The projectile is moving with initial horizontal velocity v_{0h} and initial vertical velocity v_{0v} released from altitude h_0 at time $t = 0$. Figure 3.4 shows this situation, together with the sign convention and a simple representation of the trajectory to impact.

We will calculate the following parameters:

- Range and horizontal velocity at any time t
- Altitude and vertical velocity at any time t

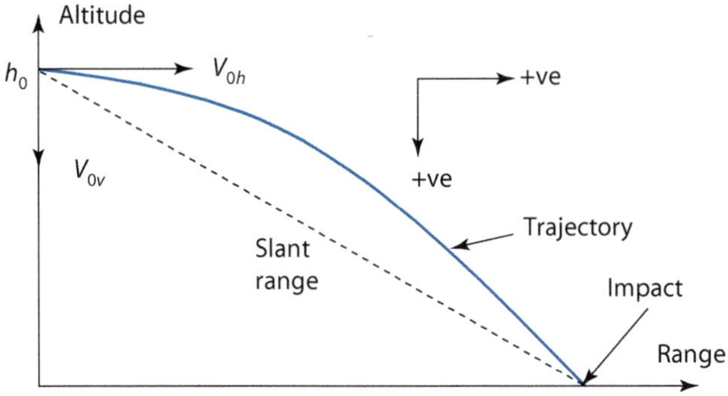

Fig. 3.4 Trajectory with sign convention—air launched.

From these parameters, the following outputs can be calculated:

- Time of flight to impact
- Range at impact
- Impact velocity
- Impact angle
- Slant range

By analyzing the free body motion of the projectile, equations for the first three parameters listed will be produced, allowing equations for the others to be derived. The analysis begins with a free-body diagram of the projectile, represented in Fig. 3.5 as a simple point mass. Note that here the vertical displacement y is measured positive downwards from the release point.

In the vertical direction, Newton's second law of motion gives

$$mg = m\ddot{y} \tag{3.3}$$

Canceling the mass and integrating with respect to time yields

$$\dot{y} = gt + a \tag{3.4}$$

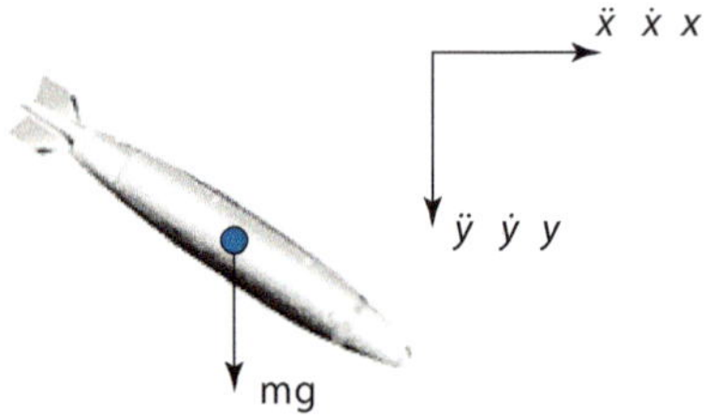

Fig. 3.5 Free-body diagram of zero-drag projectile.

The constant a is obtained by considering that at $t = 0$, the velocity is v_{0v} downward, giving $a = v_{0v}$. So,

$$\dot{y} = v_{0v} + gt \tag{3.5}$$

Integrating again gives

$$y = v_{0v}t + \frac{1}{2}gt^2 + b \tag{3.6}$$

The initial conditions again give that $y = 0$ when $t = 0$, which yields $b = 0$. Substituting gives the following result:

$$y = v_{0v}t + \frac{1}{2}gt^2 \tag{3.7}$$

At the general time t, the altitude of the projectile h is related to y and the release altitude h_0 by the equation

$$h = h_0 - y \tag{3.8}$$

Substituting into Eq. (3.7) gives

$$h = h_0 - v_{0v}t - \frac{1}{2}gt^2 \tag{3.9}$$

In the horizontal direction, no forces act, so the initial velocity is constant, that is,

$$\dot{x} = v_{0h} \tag{3.10}$$

Integrating with respect to time gives

$$x = v_{0h}t + a \tag{3.11}$$

The initial conditions reveal that at $t = 0$, $x = 0$, so $a = 0$, giving

$$x = v_{0h}t \tag{3.12}$$

Equations (3.5), (3.9), (3.11), and (3.12) give the vertical and horizontal velocities, range, and altitude at any time $t > 0$.

The derived parameters are found by first calculating the time of flight (TOF) by setting the altitude in Eq. (3.9) to zero. This gives the following quadratic equation in t:

$$\frac{1}{2}gt^2 + v_{0v}t - h_0 = 0 \tag{3.13}$$

the solution of which is given by

$$t_i = \text{TOF} = \frac{-v_{0v} + \sqrt{{v_{0v}}^2 + 2gh_0}}{g} \tag{3.14}$$

The horizontal and vertical velocities at impact (v_{ih}, v_{iv}) can be found from Eqs. (3.5) and (3.10), while the distance traveled downrange, sometimes called the bomb fall range R_B, is found from Eq. (3.12) with $t = t_i$.

$$R_B = v_{0h} \times \text{TOF} \tag{3.15}$$

The slant range is calculated as follows:

$$\text{SR} = \sqrt{{h_0}^2 + {R_B}^2} \tag{3.16}$$

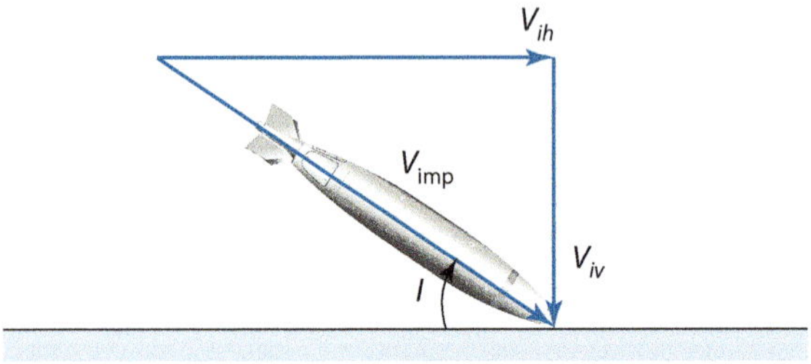

Fig. 3.6 Impact geometry.

Consider now the impact geometry shown in Fig. 3.6.

The projectile velocity at impact is given by

$$v_{\text{imp}} = \sqrt{{v_{ih}}^2 + {v_{iv}}^2} \tag{3.17}$$

The impact angle measured relative to the horizontal is obtained from

$$I = \tan^{-1}\left[\frac{v_{iv}}{v_{ih}}\right] \tag{3.18}$$

The advantage of this model is that the solution is in a simple closed form to compute impact conditions. For low drag munitions, it will give results within about 5% of higher fidelity models, but for high drag munitions, the errors increase considerably, as would be expected. For surface-launched projectiles, the trajectory is shown in Fig. 3.7.

The characteristics of this trajectory are as follows:

- It is symmetrical about a vertical line through the apex.
- The trajectory lies on a vertical plane containing the initial velocity.

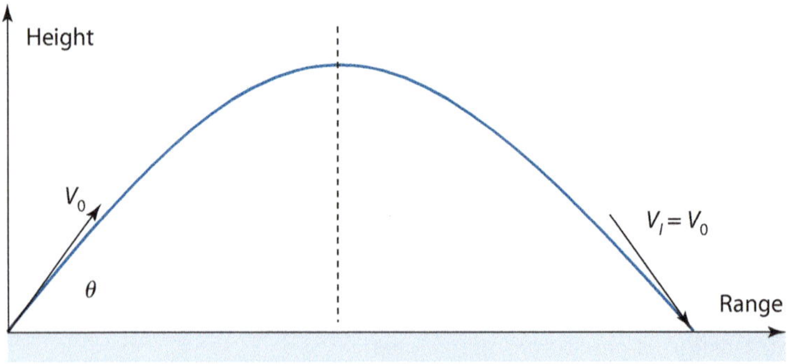

Fig. 3.7 Surface-launched trajectory.

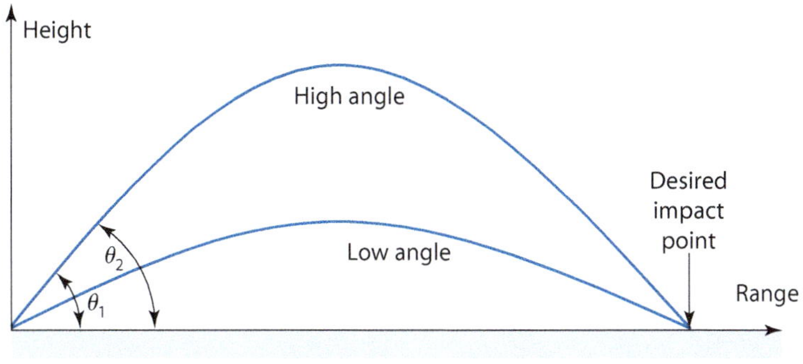

Fig. 3.8 Two elevation angle for the same range.

- For a given muzzle velocity, the range increases with elevation angle up to 45 deg, after which it decreases.
- The angle of firing equals the angle of impact (fall), and the firing and impact velocities also are the same.
- The trajectory is independent of the shape and weight of the projectile.

It should be noted that if a fixed range for the projectile is defined, two elevation angles can be calculated to give that range, one below 45 deg and one above. This is shown in Fig. 3.8.

These angles are referred to as the high- and low-angle firing solutions. It may be shown from the zero drag projectile equations that the angle required for a fixed range R is given by the following:

$$\theta = \frac{1}{2} \sin^{-1} \left[\frac{gR}{v_0{}^2} \right] \tag{3.19}$$

This leads to two solutions corresponding to angles θ_1 and θ_2 in Fig. 3.8. From a tactical perspective, low angle is preferred unless the projectile has to clear an obstruction.

3.4 The Characterization of Air Resistance

Clearly the air through which the weapon travels will resist its motion (i.e., exert drag on it), and this will affect the trajectory. Aerodynamic drag has many components, and the development of each that follows is based on the work of Farrah and Fleming [1]. In order for the projectile to move through the atmosphere, the air ahead of it has to be displaced; the energy needed to do this comes from the initial kinetic energy of the projectile. As shown in Fig. 3.9, the projectile produces a compression wave ahead of it.

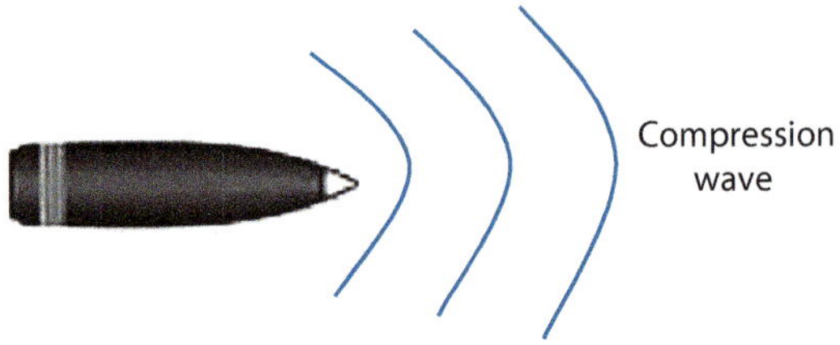

Fig. 3.9 Compression wave ahead of projectile.

This wave is essentially a sound wave propagated in the direction of travel of the projectile, and therefore moves at the acoustic speed in air. When the projectile is moving slower than the acoustic speed, the wave propagates ahead of it; this is the case for most air-delivered weapons. Artillery projectiles, however, are typically supersonic, at least in the early stages of motion; therefore, the projectile moves faster than the acoustic speed, resulting in two shockwaves being generated at the nose and rear of the projectile, as shown in Fig. 3.10.

The angle the shockwave makes with the axis of the projectile is given by

$$\sin\theta = \frac{1}{M} \tag{3.20}$$

where M is the Mach number of the projectile. The generation of the compression wave and the shock wave lead to the fore-body drag—the force needed to displace the air ahead of the projectile. This component of drag is considerable at high Mach numbers. Another form of drag is the base drag, which is at the rear of the projectile. This occurs because the air flowing around the projectile cannot immediately fill in the region behind the projectile; therefore, this region becomes one of low pressure, approaching a vacuum at high velocities. This low pressure exerts a retarding force on the projectile and hence slows it down. Some projectiles, including air-launched bombs, reduce the base drag by tapering the rear of the projectile, allowing the air to flow more easily into this region. This is known as *boat-tailing* and is illustrated in Fig. 3.11.

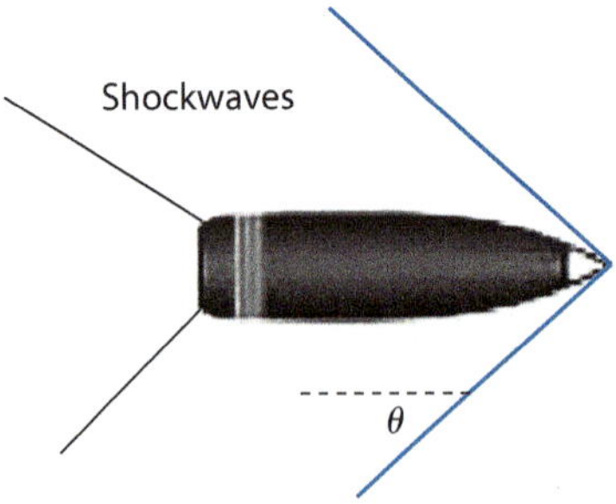

Fig. 3.10 Shockwaves generated by projectile.

Artillery projectiles may have this geometry too, but more effective ways to reduce the base drag are to burn some of the propellant located in the base of the projectile or to ignite a small flare. This is known as *base bleed* or *base burn*, and it produces gases that

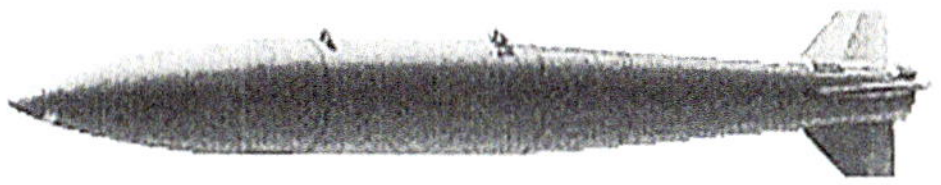

Fig. 3.11 Unguided bomb showing boat-tailing at the rear.

expand into the region to the rear of the projectile, relieving the low pressure and thereby reducing the base drag.

Another component of drag is *skin friction*, which is attributable to the boundary layer of the air as it passes along the sides of the projectile. For most projectiles this component is relatively small. A final component of drag is sometimes referred to as *excrescence drag* and is due to protuberances around the projectile; for example, artillery shells are spun as they travel along the tube by means of rifling grooves cut into the bore. A rotating band fixed to the outside of the shell engages in these grooves and imparts the necessary rotation, but after the shell leaves the tube, the rotating band will generate additional drag because it projects above the smooth shell casing.

Figure 3.12 shows how the drive band has been smoothed into the surface to reduce this component of drag. The unguided bomb shown in Fig. 3.11 retains the mounting lugs at the top, which will have the same effect. Figure 3.13 indicates the way the three most significant components of drag vary with projectile velocity.

The effect of drag on the trajectory of a ground-launched projectile is shown in Fig. 3.14 with the symmetrical zero drag case included for comparison.

Clearly the total drag experienced by a projectile is composed of these three components, and although each varies with projectile velocity, the

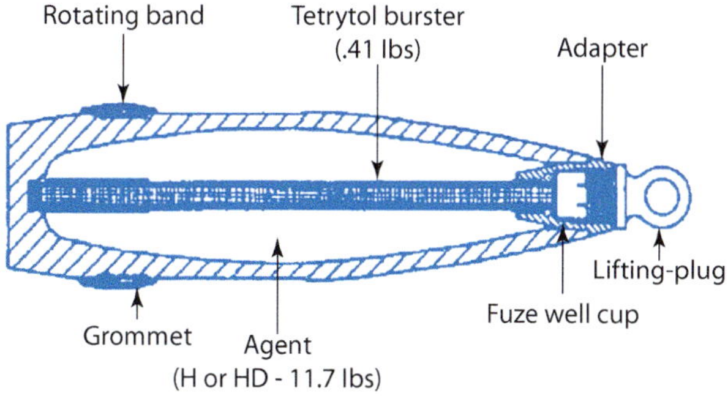

Fig. 3.12 Artillery shell showing the rotating or drive band.

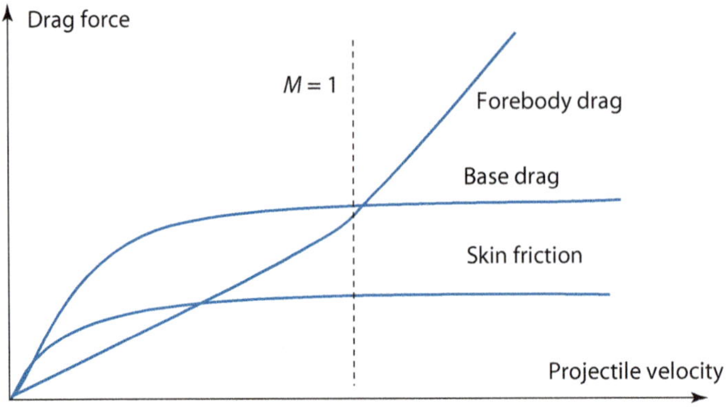

Fig. 3.13 Resistance as a function of velocity.

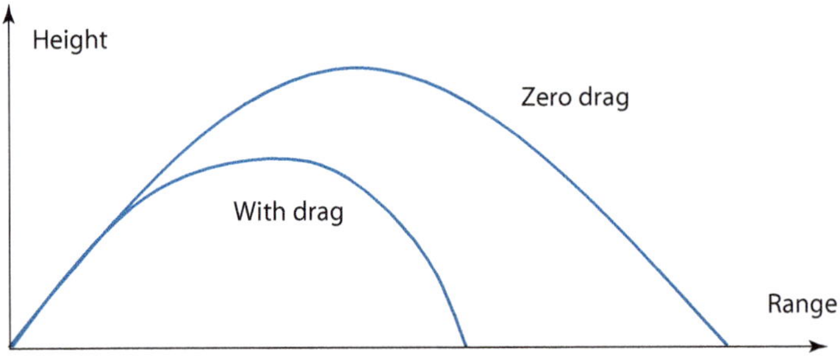

Fig. 3.14 Effect of drag on trajectory.

form of the variation can be quite complex. It is usual, therefore, to aggregate the resistance into a single term with an assumed variation with velocity. Some representations will be more accurate than others when compared to measured trajectories.

Because the drag force will appear somewhere in the differential equation of motion for the projectile, it will have to be integrated to determine the trajectory parameters of interest. As before, there will be a tradeoff between accurate representation of the drag force and the ability to obtain a solution of the resulting equations of motion.

3.5 Trajectory in a Standard Atmosphere

The resistance of the air to projectile movement depends on the air speed over the projectile, air density, and temperature. The velocity of the air over the projectile depends on two independent velocities:

1. The velocity of the projectile relative to the ground
2. The velocity of the air relative to the ground (wind)

As a starting point, we assume fixed values of air density and air temperature (sea level) with no wind. This is called the *standard atmosphere*, and is defined by the ICAO U.S. Standard Atmosphere 1962:

- Air temperature: 59°F
- Air density: 1225 g/m^2
- No wind

The most apparent difference between the trajectory in a vacuum and the trajectory in the standard atmosphere is a net reduction in the range achieved by the projectile. A trajectory in the standard atmosphere, compared to a trajectory in a vacuum, will be shorter and lower at any specific point along the trajectory for the following reasons:

- Horizontal velocity is not a constant value; it decreases with time.
- Vertical velocity is affected by both gravity and the effects of the atmosphere on the projectile.
- The highest altitude achieved, or apogee, in a vacuum is midway between the origin and the level point; in the standard atmosphere, it is actually nearer the target point.
- The angle of fall in a vacuum is equal to the angle of elevation; in the standard atmosphere, it is greater.

Actual atmospheric conditions for a particular firing will never equate to standard conditions. These deviations from standard conditions, if not corrected for when computing firing data, will cause the projectile to impact at a point other than the desired location. Corrections for nonstandard conditions are made to improve accuracy; details about these are deferred to Volume 2, Chapter 5. Some of the deviations from standard conditions affecting range are the following:

- Muzzle velocity
- Projectile weight
- Range direction wind
- Air temperature
- Air density
- Rotation of the earth

Some of the deviations from the standard conditions affecting deflection are the following:

- Drift
- Crosswind
- Rotation of the earth

As well as specifying atmospheric conditions at sea level, it is known that temperature and density vary with altitude. These variations can also be specified as a standard, and can be obtained from most texts on

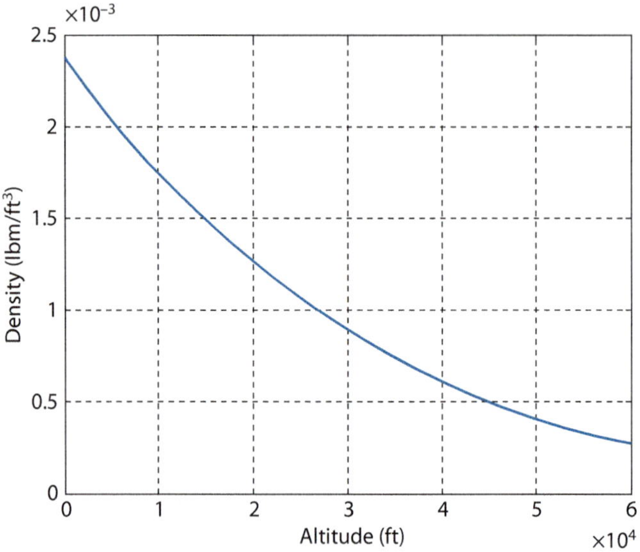

Fig. 3.15 Variation of density with altitude.

thermodynamics [2]. Figures 3.15 and 3.16 show how density and temperature, respectively, vary with altitude.

Now that the nature of air resistance has been discussed, we investigate how this effect can be incorporated into a simple trajectory model.

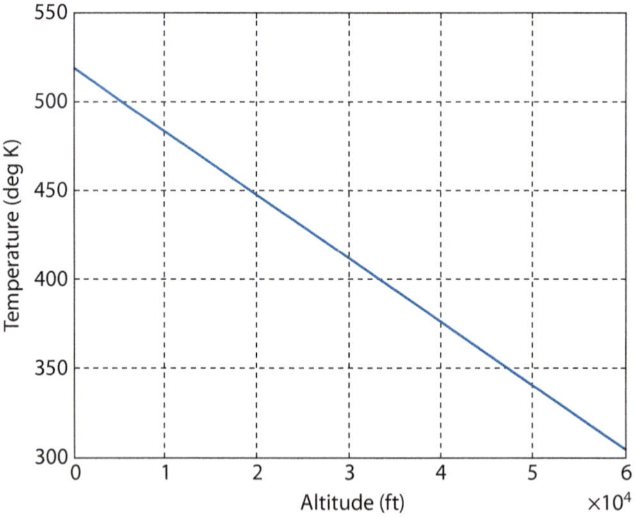

Fig. 3.16 Variation of temperature with altitude.

3.6 Linear-Drag, Point Mass Model

A more accurate model than the zero-drag model will result if the effects of air resistance are taken into account. The simplest way to do this is to assume that the drag is proportional to the velocity of the weapon. This has the effect of changing the free-body diagram shown in Fig. 3.5 to the one shown in Fig. 3.17.

In this diagram, C_d is the drag coefficient, the constant of proportionality of the drag force, which in the free-body diagram has been decomposed into its vertical and horizontal components. It is seen that this is a simplification of the effects of air resistance on the projectile, ignoring the effects of air density and temperature variation as the projectile changes altitude.

Again we will analyze the vertical motion independent of the horizontal motion.

3.6.1 Vertical Motion

Newton's second law applied to the free-body diagram of Fig. 3.17 gives

$$mg - C_d v_v = m\dot{v}_v \tag{3.21}$$

where v_v is the vertical velocity. Rearranging results in the differential equation

$$m\dot{v}_v + C_d v_v - mg = 0 \tag{3.22}$$

This first-order, linear differential equation will be solved using Laplace transforms, recalling that at $t = 0$, $v_v = v_{0v}$.

$$m[sV_v - v_{0v}] + C_d V_v - \frac{mg}{s} = 0 \tag{3.23}$$

In this equation, s is the Laplace operator, and the uppercase variables are the Laplace transform of the lowercase time domain variables, that is,

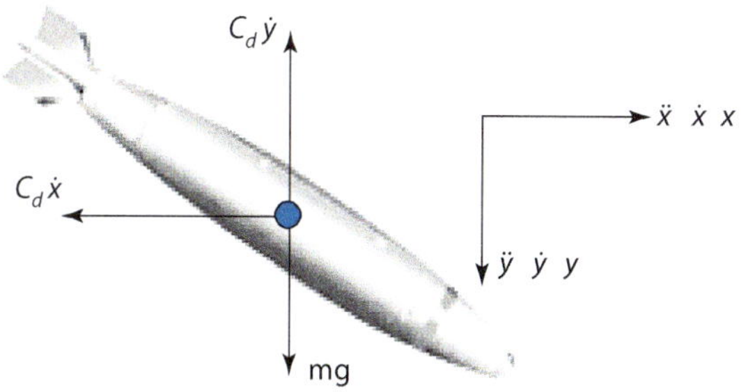

Fig. 3.17 Free-body diagram for linear drag model.

$V_v(s) = L[v_v(t)]$. Rearranging gives

$$V_v[ms + C_d] = mv_{0v} + \frac{mg}{s} \tag{3.24}$$

$$V_v[s + c_0] = v_{0v} + \frac{g}{s} \tag{3.25}$$

where $c_0 = C_d/m$. Manipulating Eq. (3.25) gives

$$V_v = \frac{v_{0v}}{s + c_0} + \frac{g}{s(s + c_0)} = \frac{v_{0v} - g/c_0}{s + c_0} + \frac{g/c_0}{s} \tag{3.26}$$

Taking the inverse Laplace transform results in

$$v_v = \left[v_{0v} - \frac{g}{c_0}\right] \exp(-c_0 t) + \frac{g}{c_0} \tag{3.27}$$

This expression is for the vertical velocity at any time t and can be compared to Eq. (3.5) for the zero-drag case. Equation (3.27) indicates that as time gets large, the vertical velocity approaches the value g/c_0, known as the terminal velocity.

Equation (3.27) can be integrated to obtain the vertical displacement from the release position

$$y = \int_0^t v_v dt = \left[v_{0v} - \frac{g}{c_0}\right]\left[-\frac{1}{c_0}\right] \exp(-c_0 t) + \frac{gt}{c_0} + \gamma \tag{3.28}$$

at $t = 0$, $y = 0$, giving

$$\gamma = \left[v_{0v} - \frac{g}{c_0}\right] \frac{1}{c_0} \tag{3.29}$$

Substituting for γ gives

$$y = \left[v_{0v} - \frac{g}{c_0}\right]\left[\frac{1}{c_0}\right][1 - \exp(-c_0 t)] + \frac{gt}{c_0} \tag{3.30}$$

As before, this can written in terms of the altitude using Eq. (3.8) as

$$h = h_0 - \left\{\frac{1}{c_0}\left[v_{0v} - \frac{g}{c_0}\right][1 - \exp(-c_0 t)] + \frac{gt}{c_0}\right\} \tag{3.31}$$

Equations (3.27) and (3.31) represent the vertical velocity and altitude at any time between release and impact.

3.6.2 Horizontal Motion

Newton's second law applied in the horizontal direction is

$$-C_d v_h = m\dot{v}_h \tag{3.32}$$

Here, v_h is the horizontal velocity. This differential equation can be written in the form

$$\dot{v}_h + c_0 v_h = 0 \tag{3.33}$$

Again, taking Laplace transforms gives

$$sV_h - v_{0h} + c_0 V_h = 0 \tag{3.34}$$

This results in the following:

$$V_h = \frac{v_{0h}}{s + c_0} \tag{3.35}$$

which, after taking inverse Laplace transforms, gives

$$v_h = v_{0h} \exp(-c_0 t) \tag{3.36}$$

integrating to get the horizontal displacement in the range direction

$$x = \int_0^t v_h \, \mathrm{d}t = -\frac{1}{c_0} v_{0h} \exp(-c_0 t) + \beta \tag{3.37}$$

Because at $t = 0$, $x = 0$, $\beta = v_{0h}/c_0$, substituting gives

$$x = \frac{v_{0h}}{c_0}[1 - \exp(-c_0 t)] \tag{3.38}$$

In the horizontal direction, the initial velocity tends exponentially to zero. Again, Eqs. (3.36) and (3.38) give the horizontal velocity and displacement between release and impact. Impact conditions can be calculated by setting $h = 0$ in Eq. (3.31), to give

$$h_0 - \frac{1}{c_0}\left[v_{0v} - \frac{g}{c_0}\right][1 - \exp(-c_0 t)] - \frac{gt}{c_0} = 0 \tag{3.39}$$

Unfortunately, Eq. (3.39) is not as easy to solve as Eq. (3.13) for the time of flight; however, if it is put in the form $t = f(t)$, it becomes

$$t = a[1 - \exp(-c_0 t)] + c_0 h_0 / g \tag{3.40}$$

where the constant a is given by

$$a = \frac{1}{c_0} - \frac{v_{0v}}{g} \tag{3.41}$$

Table 3.1 Spreadsheet Implementation of Linear-Drag Model

TRAJECTORY FOR LINEAR DRAG MODEL				
initial height	1000		**zero drag solution**	
aircraft speed (knots)	500	844		
Dive angle	0	0		
Ejection velocity (ft/sec)	7			
initial horiz vel. (ft/s)	844			
initial vert velocity (ft/s)	7		vert impact vel	253.87
drag constant	0.005		v_impact	881.35
gravity	32.2		time to impact	7.67
constant a	199.78			
x (time of flight)	7.72			
f(x)	7.72			
horiz vel @ impact (ft/s)	812.05			
vert vel @ impact (ft/s)	250.54			
impact velocity (ft/s)	849.82			
impact angle (rad/deg)	0.30	17.15		
ground range (ft)	6514.61			
slant range (ft)	6590.92			

Now it is possible to use iterative numerical methods to solve for the TOF. In fact, spreadsheets can be configured to do this iteration automatically. Table 3.1 shows such an implementation. The basic process used in the zero-drag case is repeated here, namely find the time of flight and from that calculate the vertical and horizontal velocities and displacement to determine the terminal conditions. For comparison, Table 3.1 also shows the corresponding computation for the time of flight corresponding to the zero-drag case.

Notice that the drag constant c_0 has the value 0.005. This has been set to an average value for low-drag munitions by comparing the time of flight obtained from the spreadsheet with that from a high-fidelity trajectory program (see next section) for a range of delivery altitudes, velocities, and dive angles. It gives trajectory data close enough to the high-fidelity model for the purposes of these notes. A value for of drag constant for high-drag munitions might be in the region $c_0 = 0.15$.

The linear-drag model is obviously an improvement over the zero-drag model in terms of accuracy of results. It does require an iterative technique to converge to a solution of Eq. (3.40), but it can be implemented in a spreadsheet without resorting to a higher level computational code. Compared to high-fidelity models, it will give results accurate to about 3% for low-drag munitions and 5–10% for high-drag. As explained before, it requires calibration with respect to a high-fidelity model to obtain a single value of c_0 in order to work correctly.

3.7 High-Fidelity Point Mass Model

We will now analyze trajectories in more detail, leading to a more accurate estimation of ballistic data. This model is known as the point mass model and now includes a drag force that depends on the square of the axial velocity. Although the example shown here is for a nonrotating, air-launched weapon, the same analysis is applicable to some nonrotating, surface-launched projectiles such as air- or surface-launched rockets. The free body diagram for this situation is shown in Fig. 3.18.

It is assumed here that the weapon axis and velocity vector are aligned (i.e., zero angle of attack). The drag force is computed as

$$F_d = \frac{1}{2} s C_d \rho V^2 \tag{3.42}$$

where

s = cross sectional area of the munition $= \pi D^2/4$
C_d = coefficient of drag
ρ = density of the air
V = munition velocity

The drag coefficient C_d is not constant, but a function of Mach number expressed as

$$C_d = f(M) = f\left(\frac{V}{v_M}\right) \tag{3.43}$$

Typical variations of C_d with Mach number are shown in Fig. 3.19 with the maximum occurring at about $M = 1.2$.

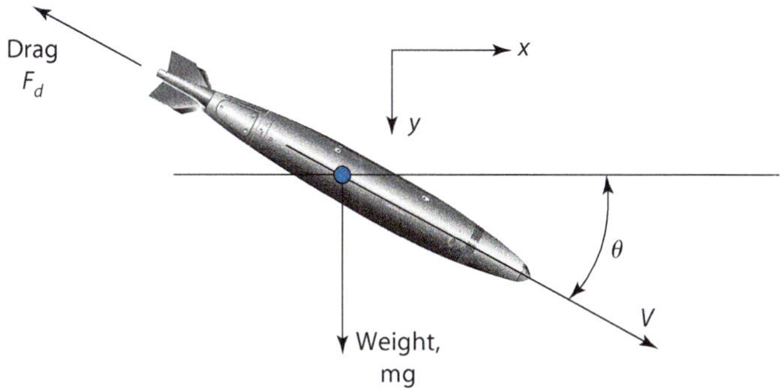

Fig. 3.18 Free body diagram of projectile.

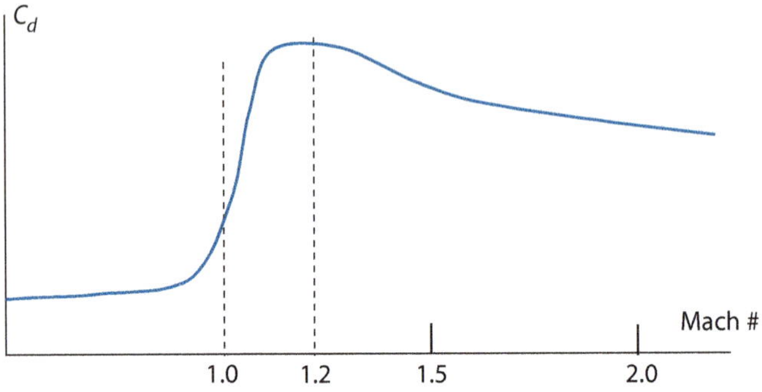

Fig. 3.19 Variation of drag coefficient with Mach number.

This variation is obtained by recognizing that the acoustic speed v_M is given by

$$v_M = \sqrt{kRT} \tag{3.44}$$

where

k = the ratio of specific heats c_P/c_V of air
R = the difference in specific heats $c_P - c_V$ of air
T = absolute temperature

Figure 3.16 show how the temperature of the atmosphere varies with altitude, so as the projectile moves along its trajectory passing through different altitudes, the corresponding temperature, and hence acoustic speed, can be determined.

Usually the C_d – Mach number relationship is expressed in tabular form for Mach numbers in the range 0.5 to about 10 for specific projectiles, and is obtained from wind tunnel testing. In Eq. (3.42) the air density is also altitude dependent (see Fig. 3.15), and again this variation along the trajectory has to be accounted for. Proceeding as before, Newton's second law, applied in the horizontal direction, yields the following:

$$-F_d \cos\theta = ma_x \tag{3.45}$$

In the vertical direction we get

$$mg - F_d \sin\theta = ma_y \tag{3.46}$$

Here, a_x and a_y are the horizontal and vertical accelerations, respectively. Substituting for the drag force and rearranging to obtain a_x and a_y gives

$$a_x = -\frac{1}{2}\frac{\pi D^2}{4} C_d \rho V^2 \cos\theta \Big/ m \tag{3.47}$$

and

$$a_y = g - \frac{1}{2}\frac{\pi D^2}{4} C_d \rho V^2 \sin\theta \Big/ m \tag{3.48}$$

The basic solution algorithm integrates Eqs. (3.47) and (3.48) numerically using Euler's method with a small time step. Suppose at time $t = 0$ we know the horizontal and vertical velocities of the munition and the release altitude; the air density and C_d then may be determined. The release angle of the munition is also known, because the horizontal and vertical velocities were derived from it. Equations (3.47) and (3.48) enable the accelerations at this instant to be calculated. The horizontal velocity at the end of the first time step $v(t + \mathrm{d}t)$ is computed from the horizontal acceleration as

$$a_x = \frac{\Delta v}{\Delta t} = \frac{v_x(t + \mathrm{d}t) - v_x(t)}{\Delta t} \tag{3.49}$$

Rearranging gives

$$v_x(t + \mathrm{d}t) = v_x(t) + a_x \mathrm{d}t \tag{3.50}$$

Similarly, in the vertical direction

$$v_y(t + \mathrm{d}t) = v_y(t) + a_y \mathrm{d}t \tag{3.51}$$

The new total velocity of the munition at the end of the time step becomes

$$V^2(t + \mathrm{d}t) = {v_x}^2(t + \mathrm{d}t) + {v_y}^2(t + \mathrm{d}t) \tag{3.52}$$

In the horizontal direction we may write the velocity as

$$v_x(t + \Delta t) = \frac{\Delta x}{\Delta t} = \frac{x(t + \mathrm{d}t) - x(t)}{\Delta t} \tag{3.53}$$

At the beginning of the first time step, $x(t) = 0$. Rearranging, the distance moved horizontally at the end of the first time step becomes

$$x(t + \mathrm{d}t) = x(t) + v_x(t + \mathrm{d}t)\mathrm{d}t \tag{3.54}$$

while the new altitude of the munition is

$$h(t + \mathrm{d}t) = h(t) - v_y(t + \mathrm{d}t)\mathrm{d}t \tag{3.55}$$

The new angle that the munition makes with the horizontal is given by

$$\theta(t + \mathrm{d}t) = \tan^{-1}\left(\frac{v_y(t + \mathrm{d}t)}{v_x(t + \mathrm{d}t)}\right) \tag{3.56}$$

Figure 3.20 shows how the trajectory variables change over this small increment of time $\mathrm{d}t$.

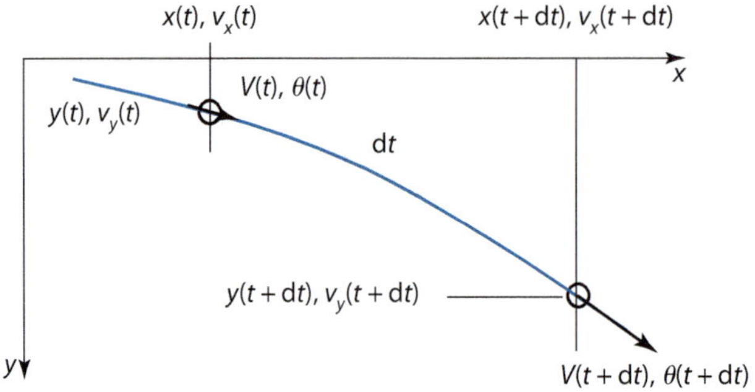

Fig. 3.20 Change of variables over time increment dt.

This completes the first cycle of the integration, and after making the following substitutions,

$$\begin{aligned} v_x(t) &= v_x(t + \mathrm{d}t) \\ v_y(t) &= v_y(t + \mathrm{d}t) \\ x(t) &= x(t + \mathrm{d}t) \\ h(t) &= h(t + \mathrm{d}t) \\ \theta(t) &= \theta(t + \mathrm{d}t) \end{aligned} \tag{3.57}$$

the cycle is repeated. The munition trajectory is computed through successive time steps until it impacts the ground, a condition obtained from Eq. (3.55) as

$$h(t + \mathrm{d}t) \leq 0 \tag{3.58}$$

When this condition is satisfied, the impact conditions are given by the current values of v_x, v_y, V, x, and θ. In order to implement this method, certain physical data need to be known, such as the diameter of the weapon, its mass, the density of air, and the drag coefficient C_d. As previously mentioned, the drag coefficient is not treated as a constant, but as a coefficient that varies with Mach number based on the munition velocity V through the air. In turn, the Mach number is dependent on air temperature, which varies with altitude. These $C_d - M$ data are available for most conventional munitions.

For simulation purposes, this methodology requires a procedural language, such as Basic, C, or Fortran, and may not be easily computed using simple spreadsheet functions. It is possible to use Excel, however, by defining macros written in Visual Basic to implement the method.

This model accurately predicts the two degrees of freedom (2-DOF) planar motion of the projectile; that is, all motion takes place in a vertical plane containing the target and either the ground-based shooter or the aircraft releasing the weapon. It will now be shown, however, that if the projectile is spinning, then an out-of-plane motion will result, requiring a more sophisticated model to allow for such motions.

3.8 General Stability of Projectiles

As a rotationally symmetrical projectile moves through the air mass, all of the aerodynamic forces can be resolved into a single force and moment acting at any prescribed point along the axis of the projectile. If this point is selected such that there is no resultant moment, then this point is known as the *center of pressure (CP)*. The stability of a projectile is determined by the position of the CP relative to the center of gravity (CG), as shown in Fig. 3.21.

If the CP is behind the CG, then the projectile is stable; otherwise, it is unstable and will tumble end over end along the trajectory. This has been recognized ever since the development of crude weapons: metal-tipped spears are stable because the mass is concentrated at the front, and arrows are stabilized by placing feathers (fins) at the rear. In the first case, the CM is moved ahead of the CP; in the second, the CP is moved behind the CM.

Projectiles having the tapered shape shown in Fig. 3.21 will always be unstable. Because it is desirable to keep the nose of the weapon pointed along the trajectory and have the axis tangential to the trajectory at all times, a stable motion is required. A modern projectile is usually stabilized in one of two ways.

- *Fin stabilized*: In this method, similar to the arrow principle just described, fixed or moveable fins are located at the rear of the projectile to provide an additional aerodynamic force aft of the center of gravity so that perturbations to the flight cause steering corrections to be made automatically. Such systems are usually used for bombs, as illustrated in Fig. 3.22 (left).
- *Spin stabilized*: This method is usually used for artillery projectiles and smaller caliber bullets. This is because the projectile has to sustain the violent explosion of the propellant, and therefore the rear of the shell is thick and heavy. This design feature moves the CM towards the rear and aft of the CP. The shell has to fit inside the bore, so any fin design must be

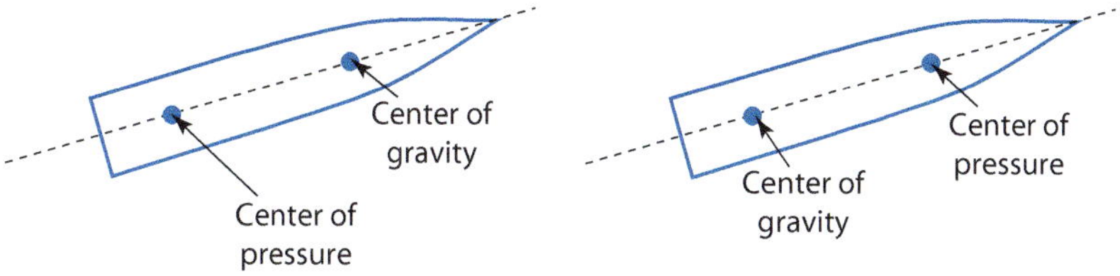

Fig. 3.21 Stable (left) and unstable (right) projectiles.

Fig. 3.22 Fin- (left) and spin- (right) stabilized munitions.

deployable once the shell exits the tube, and moveable fins are difficult to design for the harsh environment in the breach of a cannon when detonation takes place. For these reasons, projectiles are usually spin stabilized—see Fig. 3.22 (right).

A projectile that is rotating about its axis of symmetry resists disturbances due to the angular momentum about this axis, thereby stabilizing the projectile. Spin is imparted on the projectile as it moves along the firing tube by rifling grooves inside the barrel and a matching rotator ring on the shell. Typical spin velocities are on the order of 250 rpm. Most artillery pieces are rifled, but some mortars are smooth bore, and therefore their munitions have fins.

High-fidelity trajectory programs for surface-launched, fin-stabilized projectiles are basically the same as for air-launched weapons as long as a drag coefficient vs. Mach number relationship for the projectile is available;

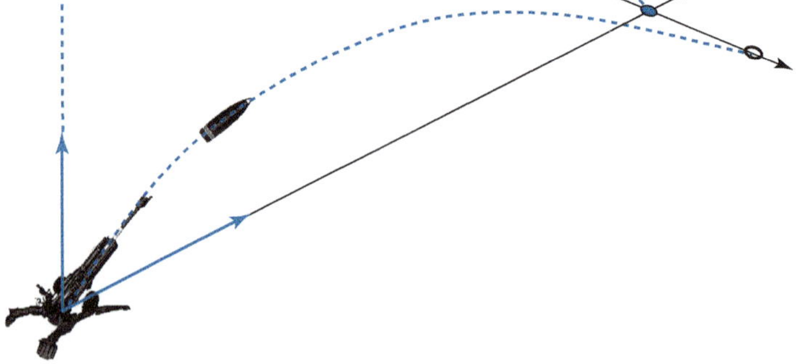

Fig. 3.23 Fin- and spin-stabilized trajectories.

however, substantial differences occur if the projectile is spinning. Spinning projectiles have trajectories that are more complex than fin-stabilized munitions; the reasons for this and their effect on the projectile motion are dealt with in Volume 2, Chapter 2.

In the absence of a crosswind, the trajectory of a fin-stabilized munition is contained in a vertical plane that includes the release platform (cannon or aircraft) and the target. A spinning projectile, however, generates several forces other than gravity that will move the round out of this vertical plane and cause the impact point to be to the right of the target, as shown in Fig. 3.23. For the latter it is important to have an accurate trajectory model so that this offset, or drift, can be calculated so that the gun may be traversed to the left to compensate. These issues are discussed fully in Volume 2, Chapter 2.

3.9 Effects of Wind on Ballistic Trajectory

It should be apparent that for an unguided munition, wind may have a significant effect on its trajectory. It is difficult to model wind effects accurately, because atmospheric conditions will vary greatly from location to location, so some simplifying assumptions are usually made.

In general, winds will be three dimensional, but for convenience they are separated into three Cartesian directions.

1. *Range wind:* In the range direction
2. *Crosswind:* In the deflection direction
3. *Vertical wind:* Usually ignored

We will begin our discussion with the effects of range wind. Aircraft avionic systems will have a wind model to enhance their trajectory algorithm in order to compute ballistics. For artillery projectiles, attempts are made to obtain meteorological data in the vicinity of the firing to obtain, among other things, a profile of wind speed and direction with altitude.

For unguided air-launched weapons, the aircraft is able to estimate the wind speed and direction from on-board instrumentation. Basically, the aircraft makes two measurements:

1. *True air speed (TAS)*: Obtained from a Pitot tube located at the front of the aircraft. This measures the velocity of the aircraft relative to the air mass through which it is traveling.
2. *Ground speed (v_X)*: Obtained from the inertial measurement unit (IMU) or GPS/INS. This measures the velocity of the aircraft relative to the ground.

Therefore, the wind speed in the range direction W_x is obtained from the vector difference of these two measurements

$$W_x = \text{TAS} - v_x \tag{3.59}$$

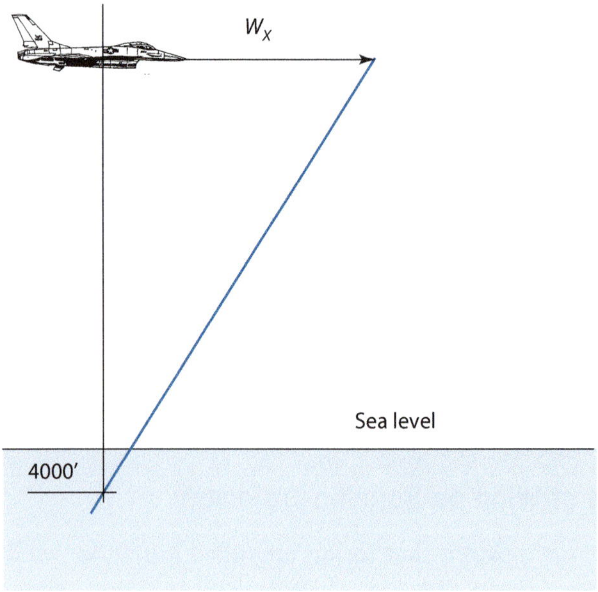

Fig. 3.24 Representative vertical wind profile.

Note that this gives the wind speed at the altitude of the aircraft, sometimes referred to as winds aloft, but gives no information on how this may vary between this point and the ground. Some avionic systems assume the wind speed to be constant between altitude and the ground, but other more sophisticated systems have exponentially decreasing wind speed. Other factors, such as whether the weapon will pass through the jet stream during its descent, may be of significance, though unpredictable. For air-launched weapons, a model that has been found reasonably accurate is a linear decrease of winds measured at altitude to zero at an altitude of −4000 ft, as shown in Fig. 3.24.

Whatever the wind model, the effect on the trajectory may be described as follows, at least for the high-fidelity model discussed previously. It is assumed that the wind is in the horizontal direction only, positive in the x-direction, and its effect is to change the speed of the air over the surface of the weapon, as indicated in Fig. 3.25. Here we assume positive wind speed in the x-direction.

Any axisymmetric body moving in air will align itself with the "relative wind," which is the vector of airflow over the body. We obtain this vector using the kinematic relative velocity equation. If we use the subscripts A = air, P = projectile, and G = ground, this equation becomes

$$v_{A/P} = v_{A/G} + v_{G/P} = v_{A/G} + \left(-v_{P/G}\right) = W_x + (-V) \tag{3.60}$$

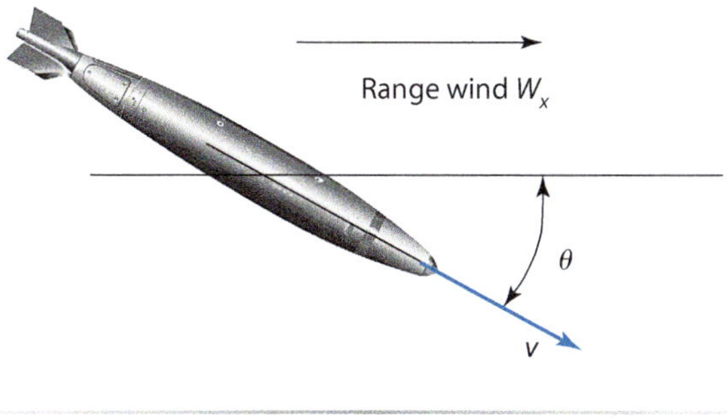

Fig. 3.25 Effect of range wind.

From Fig. 3.25, this vector equation is implemented graphically as shown in Fig. 3.26.

It is this relative wind velocity that is used in Eq. (3.43) to compute the drag coefficient. Equation (3.60) shows how wind modifies the air mass velocity from the calculated velocity, and although small in magnitude compared to the weapon velocity, when integrated over the time of flight may cause considerable deviations from a model not allowing for winds. A similar analysis may be done for surface-launched projectiles

We now look at crosswinds, and for an example we will look at the case of an unpowered, nonspinning ballistic missile fired at a target downrange, as shown in Fig. 3.27.

In this example, the missile is fired from a tube located at the origin with initial velocity v_0, and there is a positive (to the right) crosswind of speed W_z. Assuming that the missile immediately aligns with the relative wind, at some time t later it will be in the position shown. At this position, it has the velocities, accelerations, and applied forces shown. We can use the linear drag model to write the following equations of motion in the horizontal plane:

$$m\ddot{x} = -c_d\dot{x} \tag{3.61}$$

$$m\ddot{z} = -c_d(\dot{z} - W_z) \tag{3.62}$$

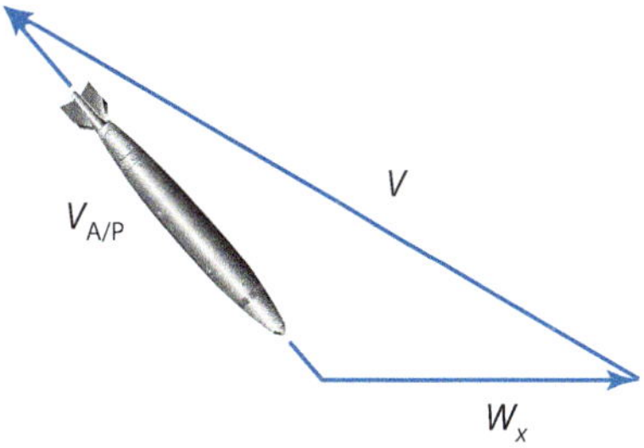

Fig. 3.26 Determining the relative wind.

Eliminating the drag coefficient results in the following:

$$\ddot{z} = \frac{\ddot{x}}{\dot{x}}(\dot{z} - W_z) \tag{3.63}$$

Expanding we get

$$\ddot{x}\dot{z} - \dot{x}\ddot{z} = \ddot{x}W_z \tag{3.64}$$

Dividing by $\dot{x}^2$

$$\frac{\ddot{x}\dot{z} - \dot{x}\ddot{z}}{\dot{x}^2} = \frac{\ddot{x}W_z}{\dot{x}^2} \tag{3.65}$$

Integrating this equation results in the following:

$$\left[\frac{\dot{z}}{\dot{x}}\right]_{t=0}^{t} = \left[\frac{W_z}{\dot{x}}\right]_{0}^{t} \tag{3.66}$$

Substituting the limits gives

$$\dot{z}(t) = W_z\left[1 - \frac{\dot{x}(t)}{\dot{x}_0}\right] \tag{3.67}$$

Integrating again gives

$$z(t) = W_z\left[t - \frac{x(t)}{\dot{x}_0}\right] \tag{3.68}$$

At impact this equation becomes

$$Z = W_z\left[\mathrm{TOF} - \frac{R}{v_0}\right] \tag{3.69}$$

In this equation, Z is the total lateral drift due to crosswind, TOF is the time of flight of the projectile, R is the down range distance, and v_0 is the muzzle velocity. The first term in Eq. (3.69) is the distance moved by the air while the projectile is in flight; the second term is how far to the right the projectile would have moved in still air if it was pointed at an angle to the right of the gun-target line given by $\tan^{-1}(W_z/v_0)$.

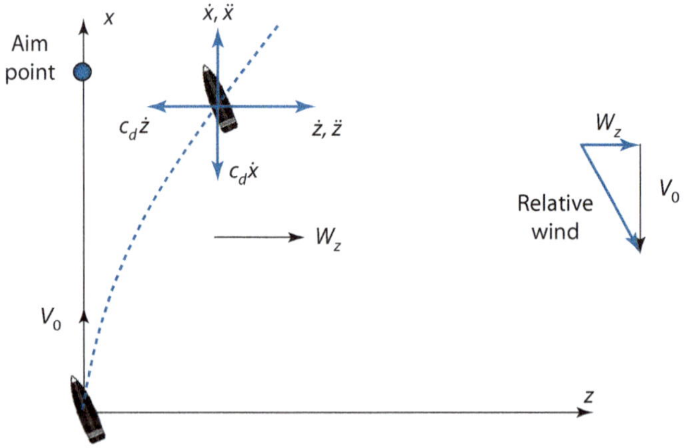

Fig. 3.27 Effect of crosswind.

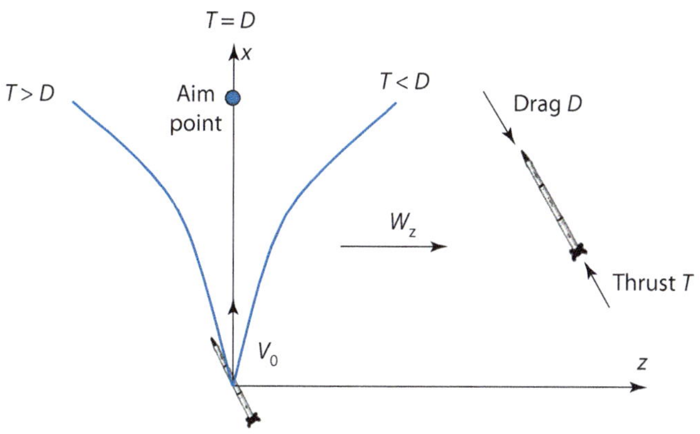

Fig. 3.28 Rocket trajectory.

Now consider the case of a projectile that may have thrust (i.e., a rocket). Possible trajectories are shown in Fig. 3.28.

From the figure, the following will be noted:

- The projectile will always align with the relative wind (i.e., point to the left) immediately after launch.
- For the unpowered trajectory ($T = 0$, hence $T < D$), wind blows the projectile to the right as shown previously for the shell, missing the aimpoint to the right.
- For a powered projectile, there will be a net component of force ($T - D$) acting to the left, and therefore the projectile moves in this direction and will miss the aimpoint to the left.
- When $T = D$, or if the projectile is unpowered and there is no wind, the projectile moves to the aimpoint from the launch point in a vertical plane containing both.

3.10 Using Trajectory Models in Fire Control Systems (FCSs)

A reasonably accurate trajectory model is used either directly or indirectly in fire control systems in order to establish the correct release conditions for unguided air-launched weapons or firing solution (pointing angles) for artillery projectiles.

For air-launched weapons, the high-fidelity trajectory model described in Section 3.7 is incorporated into the on-board FCS. The pilot selects the weapon to be dropped, and the correct drag coefficient/Mach number data are loaded. The FCS reads the aircraft state (altitude, speed, dive angle, estimated wind) to provide the initial conditions for the trajectory. In this case, the trajectory model is used directly in determining the release conditions. Details of how the trajectory model is used to compute the correct release point depends on the mode of delivery; this will be dealt with in detail in Volume 2, Chapter 4.

For artillery fire of unguided munitions, the trajectory model is used to compute tabular firing tables (TFTs), which are generated offline for all cannon/shell/charge combinations that are practical. These tables are then used to determine the tube quadrant elevation (QE) and azimuth pointing angles to bring fire onto the target. As for air-launched weapons, the resultant accuracy depends on the technique of fire implemented; again, a discussion is deferred until Volume 2, Chapter 5. However, it is useful at this point to provide an overview of the generation of the widely used firing tables.

3.11 Artillery Tabular Firing Tables

Tabular firing tables (TFTs) are based on test firings and computer simulations of a weapon and its ammunition, correlated to a set of conditions that are defined and accepted as standard. These standard conditions are points of departure for actual firings. Corrections are used to compensate for variables in the weather-weapon-ammunition combination that are known to exist at a given instant and location. The atmospheric standard accepted in U.S. firing tables reflects the mean annual condition in the North Temperate Zone—ICAO model, as defined previously.

TFTs are developed for weapons ranging from crew-served medium caliber guns to heavy artillery. The format for artillery firing tables is based on standardized agreements and, with small exceptions, are very similar between services and countries.

The principal elements measured in experimental firings include the following:

- Angle of elevation
- Angle of departure
- Muzzle velocity
- Achieved range
- Drift
- Concurrent atmospheric conditions

The main purpose of the TFT is to provide the data to bring effective fire on a target under any set of conditions. Data for firing tables are obtained from test firings of a weapon conducted at various quadrant elevations. Computed trajectories are based on the equations of motion and are compared with the data obtained in the firings. The computed trajectories are adjusted to the measured results, and data are tabulated. Firing table data define the performance of a projectile of known properties under standard conditions.

The tables contain considerable data regarding adjustments that have to be made to various inputs to the FCS that is used to compute a firing solution. For example, the effect the variability of temperature and density from standard has on drag forces is listed for a particular projectile/weapon

Table 3.2 Part of the Tabular Firing Table (TFT)

1	2	3	4	5	6	7	8	9
RANGE	ELEV	FS FOR GRAZE BURST FUZE M564	DFS PER 10 M DEC HOB	DR PER 1 MIL D ELEV	FORK	TIME OF FLIGHT	AZIMUTH CORRECTIONS	
							DRIFT (CORR TO L)	CW OF 1 KNOT
M	MIL			M	MIL	SEC	MIL	MIL
0	0.0			20	1	0.0	0.0	0.00
100	5.1			20	1	0.3	0.0	0.01
200	10.1			20	1	0.6	0.0	0.01
300	15.2			20	1	1.0	0.1	0.01
400	20.3			20	1	1.3	0.1	0.02
500	25.4			19	1	1.6	0.2	0.02
600	30.6	1.9	1.06	19	1	1.9	0.3	0.03
700	35.8	2.2	0.91	19	1	2.3	0.4	0.03
800	41.1	2.5	0.79	19	1	2.6	0.5	0.04
900	46.4	2.8	0.71	19	1	2.9	0.6	0.04
1000	51.7	3.2	0.63	19	1	3.2	0.7	0.04
1100	57.1	3.5	0.57	19	1	3.6	0.8	0.05
1200	62.5	3.8	0.53	18	1	3.9	0.8	0.05
1300	67.9	4.2	0.48	18	1	4.3	0.9	0.05
1400	73.4	4.5	0.45	18	1	4.6	1.0	0.06
1500	78.9	4.9	0.42	18	1	4.9	1.1	0.06

combination, allowing more accurate data to be input to the FCS. Of particular interest here are the data shown in Table 3.2.

This table shows, in the two right-most columns, deflection errors due to drift (rotational effects) and crosswind as a function of range. This allows the gun crew to correct for these effects and place fire more accurately on target. Note also how the time of flight is used for fuze settings. These numbers are derived from actual firings and are used to calibrate the trajectory model for the munition.

3.12 Trajectory for Guided Weapons and Munitions

The preceding sections address unguided munitions only, because these follow a ballistic trajectory. For guided munitions, a different approach is taken depending on what kind of weapon is being used. Theoretically, we may assume an air-launched weapon such as a guided missile follows a straight-line path from the launch aircraft to the target, as shown in Fig. 3.29.

This figure shows how the weapon attempts to follow the straight line corresponding to the line of sight (LOS) from launch to impact. The quality of the control system determines the fluctuation of the path from the LOS, which if considerable results in an unpredictable impact angle. In low-cost munitions such as legacy laser-guided bombs (LGBs) produced in large quantities, the control system is typically of the inexpensive "bang-bang" type where the control surfaces are hard over in one direction or the other as it attempts to align the weapon with the LOS. If the weapon is not powered, then it is easier to fall than it is to climb, and therefore it may spend more time below the LOS than above. For this reason, such weapons tend to fall short.

Bang-bang control systems produce considerable excursions from the assumed straight line path. Newer munitions, such as cruise missiles and second-generation LGBs, tend to have more accurate, proportional control systems, which reduce considerably the deviations from the LOS.

In practice, aircrew may release unpowered guided munitions, and LGBs in particular, as close to the ballistic point as possible (i.e., treating the weapon as unguided). In doing so, the target may not even be illuminated prior to release, which may then be considered a true unguided delivery. The reason for doing this is not only to avoid the target knowing it is being designated, but also to conserve the kinetic energy of the unpowered weapon. Designating the target a few seconds before impact allows the guidance system to make relatively minor corrections to the trajectory to bring it accurately onto the target. As far as trajectory is concerned, this delayed lasing mode is almost identical to an unguided delivery.

Some munitions have a trajectory shaping capability that enables the desired impact angle to be input into the control system before launch. Trajectories of this kind, which are common for weapons that need high impact angles to penetrate a hard target, are shown in Fig. 3.30. Such pop-up maneuvers are usually initiated by an on-board altimeter.

The deviation from the straight LOS trajectory occurs at a fixed altitude. This capability enables the munition to impact the target at a high incidence angle, thereby maximizing its penetration capability.

For guided munitions in general, the impact angle is computed as either the LOS angle θ shown in Fig. 3.29, the desired impact angle I as indicated in

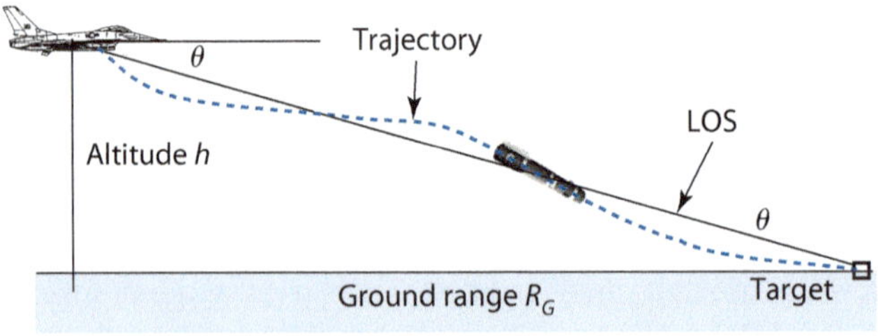

Fig. 3.29 Assumed trajectory of a missile to target.

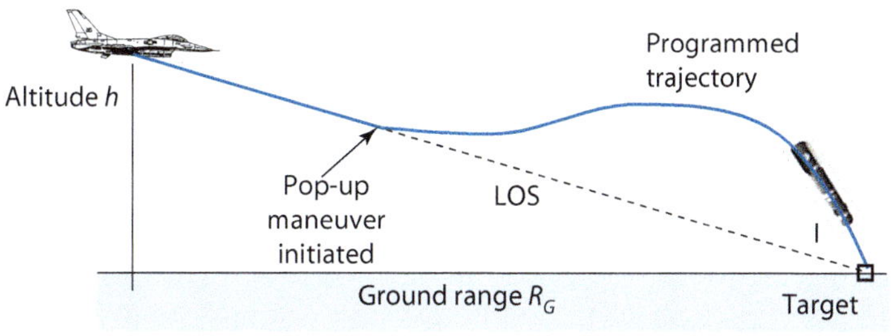

Fig. 3.30 Trajectory shaping for required impact angle.

Fig. 3.30, or the impact angle resulting from a quasi-ballistic trajectory if that delivery technique is utilized.

If a guided weapon does not have its impact conditions specified, doesn't follow an essentially ballistic trajectory, or doesn't move in a straight line from the firing point to the target, the impact conditions are usually difficult to predict. Munitions that fall into this category might be INS/GPS guided munitions or missiles attacking a moving target. Under these circumstances, the weaponeer must estimate the impact angle (and any other parameters that might be needed), making one of the assumptions just discussed. If precise impact conditions are required, a trajectory program for the specific weapon employed (fly-out model) may be needed for the planned attack. For some weapons this has been incorporated into fast-running weaponeering tools such as JWS.

3.13 Ballistic Partial/Unit Effects from Trajectory Programs

The ballistic partial, sometimes referred to as a unit effect, is a quantity that plays a role in the errors expected from air- and surface-launched unguided ordnance. The concept may be best explained by representing the trajectory program as a simple "black box" that accepts inputs, processes them, and produces outputs, as shown in Fig. 3.31.

Suppose we consider an air-launched unguided weapon having nominal release conditions such as 10,000-ft altitude, 500 kt initial speed, and

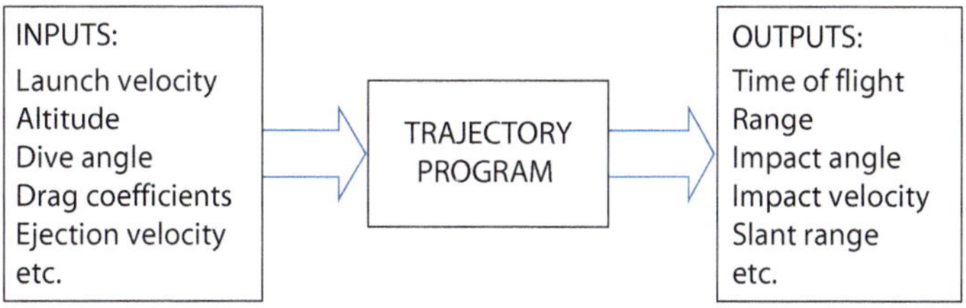

Fig. 3.31 Functional form of a trajectory program.

horizontal flight. These nominal inputs will produce a set of corresponding outputs. If we now perturb one of the inputs, say the altitude becomes 10,010 ft, all the outputs will change. If we select one particular output, say ground range, and measure the change in this parameter from the nominal value, represented by ΔR, we can form the ballistic partial from the following:

$$\frac{\partial R}{\partial h} \cong \frac{\Delta R}{\Delta h} \tag{3.70}$$

The left side of this equation is the definition of the ballistic partial, and it is written as a partial derivative because it is assumed that when we vary one of the input parameters (h in this case), all others (release velocity, dive angle) remain unchanged. The right side of the equation represents the manner in which the ballistic partial is calculated, simply the resulting change in an output parameter divided by the perturbation in the input parameter that caused the change, as shown in Fig. 3.32. We can see from Fig. 3.31 that many ballistic partials can be calculated, but only certain ones are needed in the study of weapon accuracy.

In general, the trajectory models used to compute ballistic partials are high-fidelity models, because these give the most accurate results. It is possible, however, to obtain analytical expressions for demonstrating a ballistic partial from simpler trajectory models. Consider, for example, computing the ballistic partial given in Eq. (3.70) using a zero-drag trajectory model discussed in Section 3.3. From Eq. (3.7) in the vertical direction, setting $v_0 = 0$ and $y = h$, we get

$$t = \sqrt{\frac{2h}{g}} \tag{3.71}$$

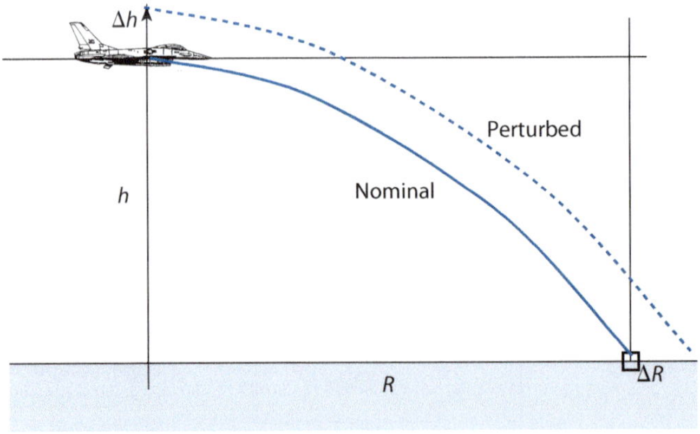

Fig. 3.32 Change in h produces a change in R.

The ground range in the horizontal direction is given by Eq. (3.12) as

$$R = v_{0h}\, t \tag{3.72}$$

Substituting Eq. (3.71) into Eq. (3.72) gives

$$R = v_{0h}\sqrt{\frac{2h}{g}} = \sqrt{\frac{2{v_{0h}}^2}{g}}\, h^{\frac{1}{2}} \tag{3.73}$$

which is in the form

$$R = f(h) \tag{3.74}$$

Differentiating with respect to h gives

$$\frac{\partial R}{\partial h} = \frac{1}{2}\sqrt{\frac{2{v_{0h}}^2}{g}}\,\frac{1}{\sqrt{h}} = \sqrt{\frac{{v_{0h}}^2}{2gh}} \tag{3.75}$$

However, because there is no drag on the bomb, it may be considered a conservative system, and we can write from conservation of energy that in the vertical direction

$$\frac{1}{2}mv_v^2 = -mgy = mgh \quad \Rightarrow \quad 2gh = v_v^2 \tag{3.76}$$

Recalling that the horizontal velocity remains unchanged, substituting Eq. (3.76) into Eq. (3.75) gives

$$\frac{\partial R}{\partial h} = \frac{v_{0h}}{v_v} \tag{3.77}$$

Finally, considering the moment of impact, this equation becomes

$$\frac{\partial R}{\partial h} = \frac{v_{0h}}{v_v} = \frac{v_{ih}}{v_{iv}} = \frac{1}{\tan I} \tag{3.78}$$

where, from Fig. 3.6, I is the impact angle. Equation (3.78) is often seen in the literature as a means to calculating this particular ballistic partial, but it has to be remembered that it is only correct for a zero-drag weapon.

The most common way to obtain ballistic partial data for unguided delivery accuracy estimations, a high-fidelity trajectory program has to be run twice to compute two values of a particular output ballistic parameter for a small change in a single input parameter, as illustrated by the following scenario corresponding to Fig. 3.32.

Consider a delivery at 500 kt, 6000-ft altitude with zero dive angle at release. In order to evaluate the partial derivative in Eq. (3.70) we will increment the release altitude from 6000 ft to 6100 ft and use a high-fidelity trajectory model to determine the resulting incremental increase in range. The results are summarized in Table 3.3 for a low-drag weapon.

Table 3.3 Ballistics for an Air-Launched, Low-Drag Weapon

Release Altitude (ft)	Impact Angle (deg)	Down Range (ft)	Time of Flight (s)
6000	37	15,556	19.5
6100	38	15,683	19.7

From these data, the ballistic partial is evaluated from Eq. (3.70) as follows:

$$\frac{\partial R}{\partial h} \cong \frac{\Delta R}{\Delta h} = \frac{15683 - 15556}{6100 - 6000} = 1.27000 \tag{3.79}$$

while Eq. (3.78) gives

$$\frac{\partial R}{\partial h} = \frac{1}{\tan I} = 1.27994 \tag{3.80}$$

In this case, the approximation is a good one, because low-drag munitions have trajectories close to zero-drag munitions. If, however, we consider the same situation for high-drag munitions, larger differences would appear, and accurate ballistic partials would have to be computed by running a trajectory program twice and computing the results as in Eq. (3.79).

To expedite this process for artillerymen, unit corrections may be precalculated for specific projectile and charge combinations and included in firing tables. A portion of one is shown in Table 3.4.

This table indicates, for the M107 projectile and M232A1 charge, the unit corrections for muzzle velocity, wind in the range direction, air temperature, and air density as a function of range. Two values are quoted, one for increasing the value of the parameter and one for decreasing; usually an average value is used.

3.14 Air-Launched Loft and Dive Toss Deliveries

A specific release tactic for air-launched weapons can be implemented for unguided or LGB deliveries that allows the aircraft to release the weapon at longer standoff ranges than normal. Originally developed for nuclear weapon delivery where large standoff distances are desired, the weapon is released while the aircraft executes a constant acceleration maneuver in the vertical plane. If the aircraft is in a dive when the pull-up maneuver is begun, it is known as a *dive toss*, whereas if the aircraft is in level flight it is called a *loft*. The analysis is much the same for both deliveries, so only the loft maneuver is presented here.

At some distance from the target, as shown in Fig. 3.33, the aircraft will climb to some altitude to designate the target. This is done so that the designation device (such as a targeting pod) will not have a too shallow look-

Table 3.4 Extract from Firing Table F

FT 155-AM-3 — TABLE F — CHARGE 4H M232A1

PROJ, HE, M107 — CORRECTION FACTORS

FUZE, PD, M739A1

1	10	11	12	13	14	15	16	17	18	19
RANGE	RANGE CORRECTIONS FOR									
	MUZZLE VELOCITY 1 M/S		RANGE WIND 1 KNOT		AIR TEMP 1 PCT		AIR DENSITY 1 PCT		PROJ WT OF 1 SQ (4 SQ STD)	
	DEC	INC	HEAD	TAIL	DEC	INC	DEC	INC	DEC	INC
M	M	M	M	M	M	M	M	M	M	M
14000	22.8	-22.5	15.4	-12.6	-0.8	-3.2	-67.0	63.5	21	-18
14100	22.9	-22.6	15.6	-12.8	-0.4	-3.5	-67.4	64.0	22	-18
14200	23.0	-22.7	15.9	-13.0	0.1	-3.9	-67.9	64.5	22	-18
14300	23.0	-22.8	16.2	-13.2	0.5	-4.3	-68.4	64.9	22	-19
14400	23.1	-22.8	16.4	-13.3	1.0	-4.6	-68.8	65.4	23	-19
14500	23.2	-22.9	16.7	-13.5	1.5	-5.0	-69.3	65.8	23	-19
14600	23.3	-23.0	17.0	-13.7	1.9	-5.4	-69.7	66.3	23	-20
14700	23.4	-23.1	17.2	-13.9	2.4	-5.8	-70.2	66.8	24	-20
14800	23.5	-23.2	17.5	-14.1	2.9	-6.2	-70.6	67.2	24	-20
14900	23.6	-23.2	17.8	-14.3	3.4	-6.6	-71.1	67.7	24	-21
15000	23.7	-23.3	18.1	-14.5	3.8	-7.0	-71.5	68.2	25	-21
15100	23.8	-23.4	18.4	-14.7	4.3	-7.4	-72.0	68.6	25	-21
15200	23.8	-23.5	18.7	-14.9	4.8	-7.8	-72.4	69.1	25	-22
15300	23.9	-23.5	19.0	-15.1	5.3	-8.2	-72.9	69.7	26	-22
15400	24.0	-23.6	19.3	-15.3	5.7	-8.6	-73.4	70.2	26	-22
15500	24.1	-23.7	19.6	-15.6	6.2	-9.0	-73.8	70.7	26	-23
15600	24.2	-23.8	19.9	-15.8	6.7	-9.4	-74.3	71.3	27	-23
15700	24.3	-23.9	20.2	-16.0	7.2	-9.8	-74.7	71.8	27	-23
15800	24.4	-23.9	20.5	-16.2	7.6	-10.2	-75.2	72.4	27	-23
15900	24.5	-24.0	20.8	-16.4	8.1	-10.5	-75.7	73.0	28	-24
16000	24.6	-24.1	21.1	-16.6	8.6	-10.9	-76.2	73.6	28	-24
16100	24.6	-24.2	21.5	-16.8	9.0	-11.3	-76.6	74.2	28	-24

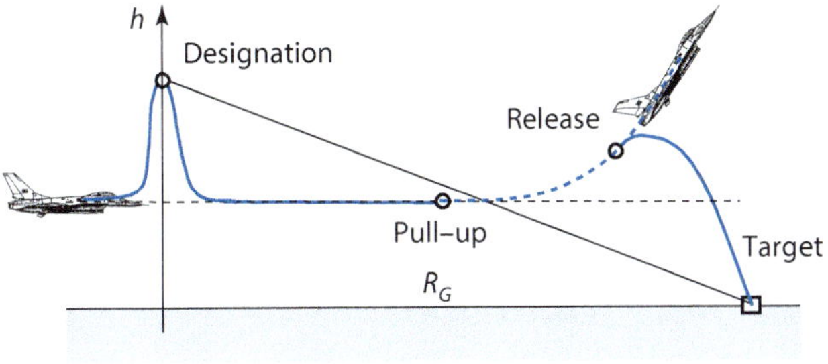

Fig. 3.33 Loft release.

down angle leading to inaccurate designation measurements. Immediately after this, the aircraft descends to a run-in altitude to approach the target. At the pull-up point, the aircraft executes a constant acceleration (g) climb, causing it to move along a circular path in the vertical plane. As the aircraft climb angle gets larger, the FCC computes the trajectory of the weapon, hence the ballistic range. As the aircraft moves around the circle, the range to target decreases and the ballistic range increases. When the two are equal (within certain limits) the weapon is released automatically.

Aircraft state variables that are measured by the IMU include altitude, R_G at designation, and speed, whereas the FCC computes ballistic range R_B and ground range to target R_G. Because the pull-up point is defined by an elapsed time after designation, $T_{D/PU}$, it is assumed that this is the start point for the range calculations.

Given a constant pull-up acceleration in g's ($=ng$), the radius of the circle the aircraft moves along is given by

$$R = \frac{v^2}{a} = \frac{v_A^2}{ng} \tag{3.81}$$

Consider the aircraft moving from the pull-up point A to B along this circular path, rotating through the angle θ, as shown in Fig. 3.34. We calculate the change in altitude and distance moved horizontally as follows:

$$dh = R(1 - \cos\theta) \tag{3.82}$$

$$dx = R\sin\theta \tag{3.83}$$

Suppose the weapon is released at B, and the ballistic range R_B is calculated from a trajectory program for the instantaneous release conditions at B,

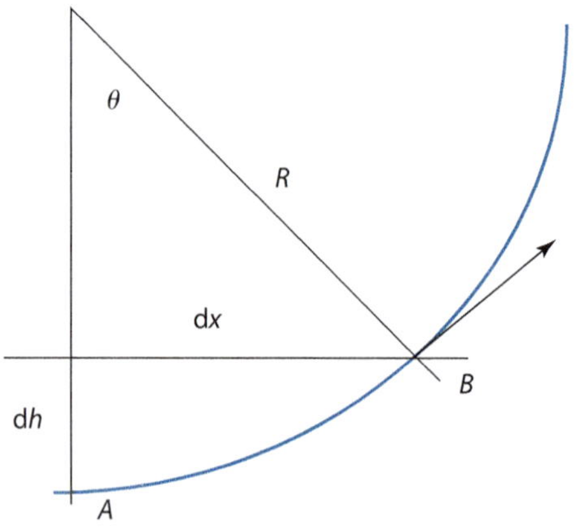

Fig. 3.34 Geometry from pull-up to release.

namely the altitude, release angle, and speed. The correct release point is reached when

$$R_G - v_A T_{D/PU} - dx = R_B \tag{3.84}$$

The method just described will provide the release point conditions, but is not computationally efficient. The method implemented is to iterate to find the loft angle at the ballistic point by computing the change in range and altitude from Eqs. (3.82), (3.83), and (3.84) to determine when the latter is satisfied. This usually takes less than 50 iterations.

Another approach requires knowledge of the time between designation and pull-up, but this is not as simple as it seems. It can be seen from Fig. 3.33 that if designation takes place at a considerable distance from the target and a pull-up maneuver is executed almost immediately, there is no certainty that the weapon can reach the target as it moves around a circle in the vertical plane. It follows, therefore, that there is a minimum value of $T_{D/PU}$ that must be exceeded. This time may be calculated by recognizing that there is a maximum loft ballistic range that occurs when the release angle is approximately 45 deg. The range from designation to pull-up in this case may be obtained from Eq. (3.84) as

$$R_{D/PU} = R_G - \hat{R}_B - R \sin 45 \tag{3.85}$$

This gives the minimum time between designation and pull-up as

$$T_{D/PU}(\text{min}) = \frac{R_G - \hat{R}_B - R \sin 45}{v_A} \tag{3.86}$$

Similarly, there is a maximum time between designation and pull-up, because if the aircraft continues too long in horizontal flight, it is possible for it to fly beyond the ballistic point. This maximum time is calculated by assuming the aircraft flies horizontally to the pull-up point, and the weapon is released immediately before it moves around the circular path. The ballistic range is calculated for a horizontal release at the pull-up altitude, and the maximum time between designation and release becomes

$$T_{D/PU}(\text{max}) = \frac{R_G - R_B}{v_A} \tag{3.87}$$

It should also be mentioned that there are two points on the circle shown in Fig. 3.34 where the weapon can be released and reach the target in a ballistic manner. This situation is analogous to the low/high-angle firing solution for artillery shells. The methods described here will give the first point on the path, leading to a shallower trajectory for the weapon.

3.15 Fuze Models

In order to calculate the effects of a warhead on a target, their relative position has to be known when detonation occurs. A brief overview of fuze types was given in Chapter 1, but more details regarding the influence trajectory has on fuzing will now be discussed, and is based on [3].

So far, only the path the projectile takes as it moves through the atmosphere towards the target has been considered, but the precision with which the fuze operates will affect this relative position and hence the effectiveness of the munition. In treating how fuze function affects the location of the warhead at detonation, we will frame the discussion in terms of a Monte Carlo approach; that is, we will observe where the warhead is located when the randomness of the fuze function is allowed to vary over the principle features of the fuze mechanism. In general, the three types of fuze fitted to a warhead that we will consider here are:

1. Impact, or point detonating (PD)
2. Proximity (PX)
3. Time delay

3.15.1 Point-Detonating (PD) Fuze

A PD, or contact or direct action, fuze is one that triggers the warhead when it comes into contact with something reasonably hard, like the ground, a building, or a vehicle. It may be located in the nose of the weapon, as in the case of a general purpose Mk 82 unguided bomb, or in the tail for a bomb designed to penetrate hard materials such as concrete where a nose fuze would be damaged. Different fuze locations produce different detonation waves in the explosive fill and can result in varying damage effects. From a Monte Carlo perspective, assuming instantaneous fuzing with contact, the fuze variability is simply the accuracy with which the weapon impacts the ground relative to the target, as shown in Fig. 3.35.

This figure shows impacts in the ground plane where the variability in impact point about the nominal trajectory is due to effects that cannot be modeled, as discussed earlier. Quantifying these effects will be dealt with in the delivery accuracy chapter (Chapter 4).

3.15.2 Proximity (PX) Fuze

This fuze type triggers when a particular threshold is exceeded, and is further divided into the following categories:

- Variable time (VT)
- Sensor trigger

The VT fuze is one where the threshold is a fixed time after firing where the timer may be mechanical (MT) or electronic. For the MT fuze, the time is set

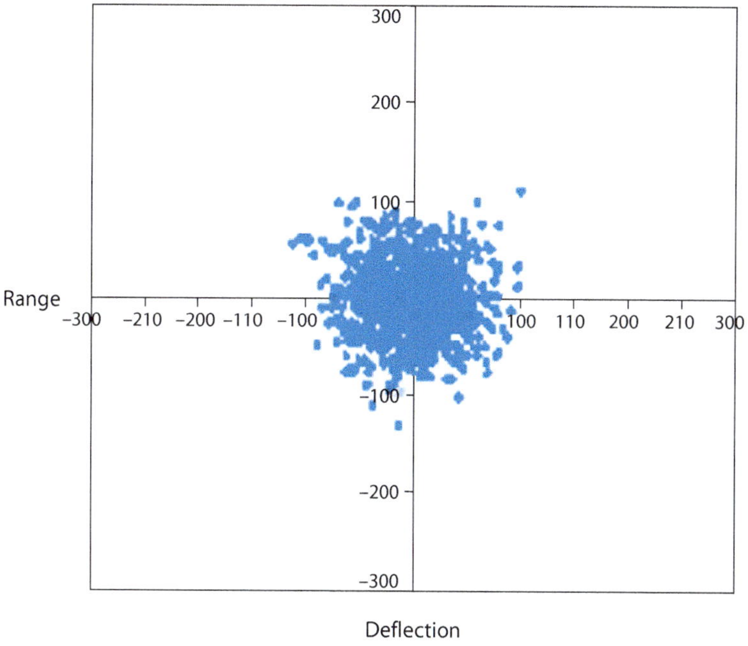

Fig. 3.35 Ground plane impact points relative to the target.

manually prior to firing the round, or an electronic timer may be set with software. Early VT fuzes were designed for air-bursting artillery shells in World War I, where it was observed that lethality was increased considerably compared to PD fuzed shells. In order to correctly set the fuze, some knowledge of the shell trajectory and desired height of burst (HOB) would need to be known, as illustrated in Fig. 3.36.

In sensor fuzes, the threshold needed to detonate the warhead comes from a more sophisticated device compared to a simple timer. In many cases the sensor is a small radar system that estimates the distance to the target (or ground) and initiates detonation when this distance is smaller

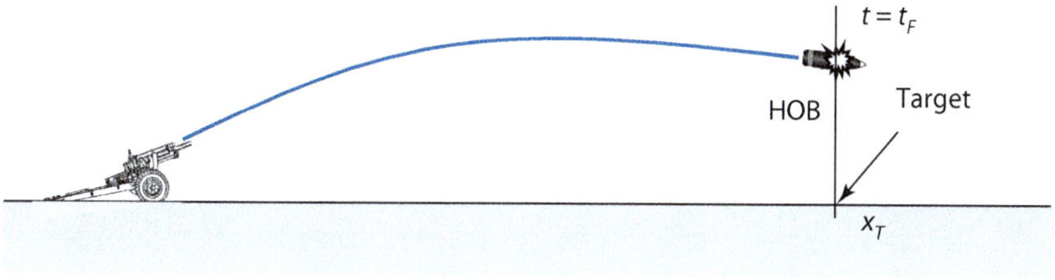

Fig. 3.36 Airburst with PX fuze.

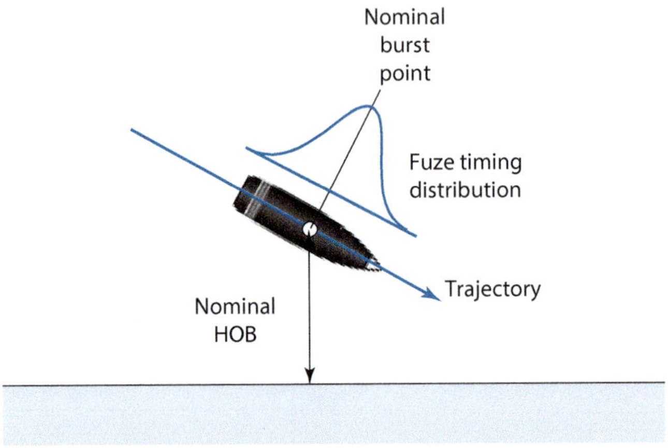

Fig. 3.37 VT fuze model.

than the set threshold value. These fuzes are typically used in antiaircraft weapons or the air-bursting scenario shown in Fig. 3.36, where there is less dependency on having accurate trajectory information.

Given the trajectory of the projectile as in Fig. 3.36, the time of flight to the point above the target can be calculated, and this would be the VT fuze setting. The variability is therefore in the accuracy of the timer in the fuze about its nominal or desired value. Assuming this variability is Gaussian in nature, it may be represented by Fig. 3.37.

Given the randomness of delivery shown in Fig. 3.35, and adding in the variability in fuze timing, the expected detonation points for a Monte Carlo simulation would be as shown in Fig. 3.38. In these graphs, the effect of fuze variability is shown in the horizontal and vertical planes with the nominal burst at coordinates (0, 0, 2), that is, 2 m above the ground.

Note that for long fuze time errors, an airburst is not achieved and the round impacts the ground plane.

For a sensor-triggered air-bursting round, again the objective is to achieve an airburst of the type shown in Fig. 3.36, but the employment of an altitude sensor may give better accuracy at detonation, as shown in Fig. 3.39.

Again, the variability of detonation point in the horizontal and vertical directions may be simulated, as shown in Fig. 3.40 for a nominal 10-m HOB.

For the case shown where the desired burst point is at (0, 0, 10), there is greater spread for the HOB fuze compared to the VT fuze, as would be expected for shallow approach angles of the warhead.

For an antiaircraft round, the model is more complex than the previous ones because we need to represent the probability the target is detected within the fuze sensing envelope. The assumption here is that the detection envelope may be represented by a cone by specifying a cone angle and

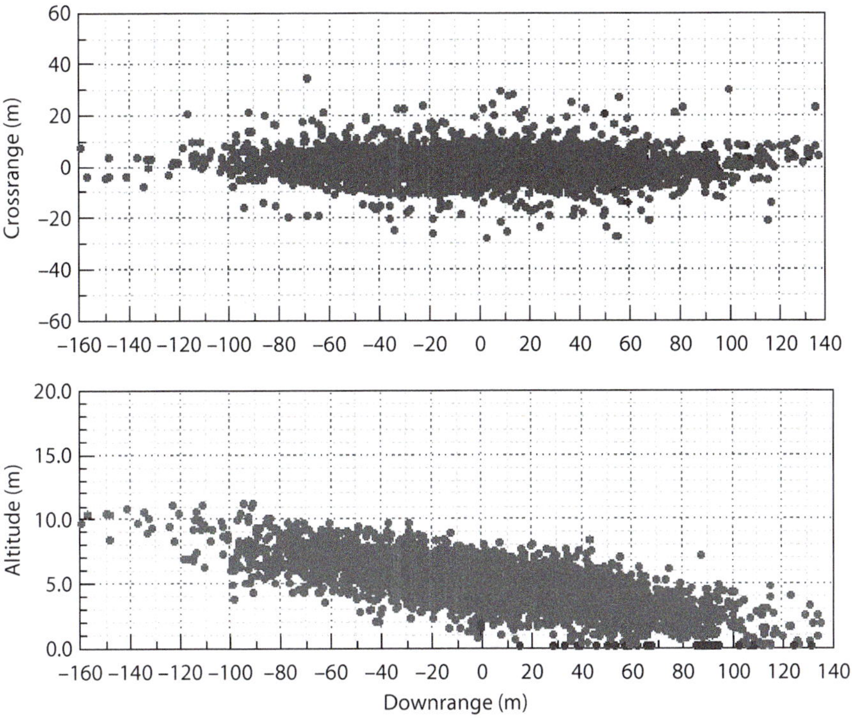

Fig. 3.38 Simulated detonation points for VT fuze model.

maximum radius, and the target is detected when it is inside the cone, as illustrated in Fig. 3.41. In this figure, the target may consist of a physical object, in which case its geometry is required to determine the first part of the target to trigger the fuze, or simply the target centroid.

Clearly the process of determining when the fuze will function is more complex than for other fuze types. Typically the velocity vector of the

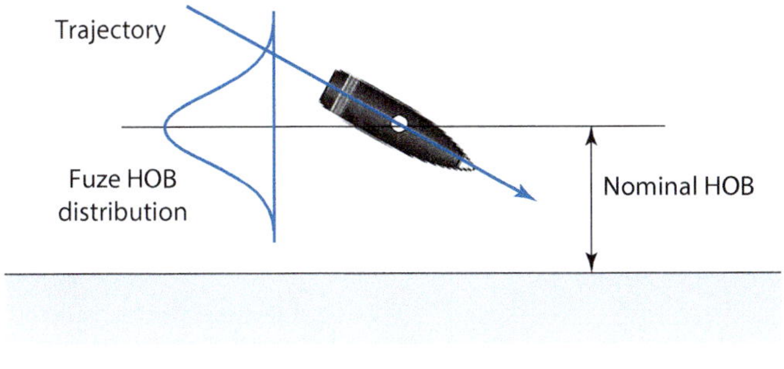

Fig. 3.39 HOB fuze model.

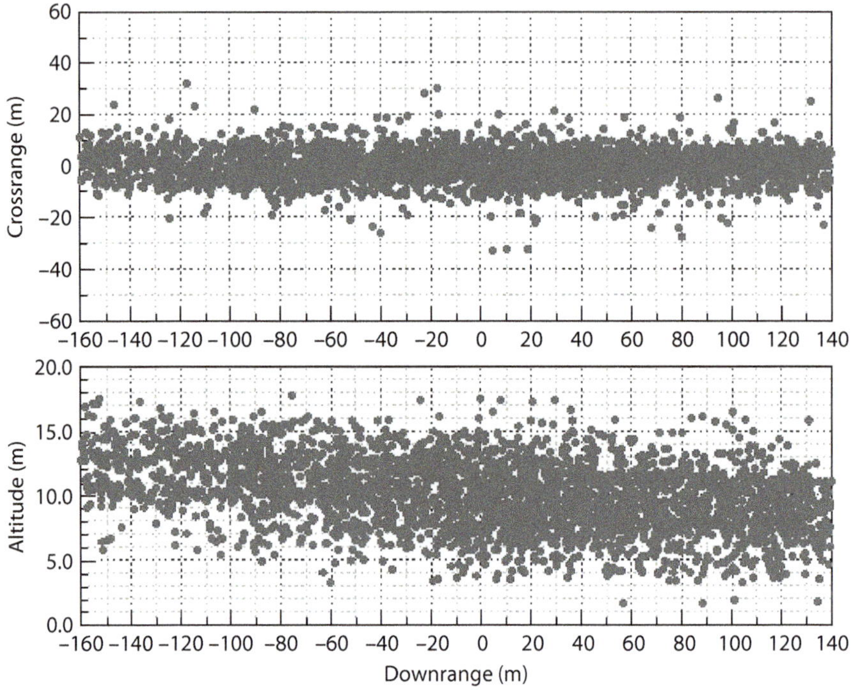

Fig. 3.40 Simulated detonation points for HOB fuze model.

weapon in its terminal state is determined from trajectory calculations. Then the weapon is considered fixed and its velocity imposed on the target. The fuze functions when the target intersects the conical volume of the envelope.

Figure 3.42 shows simulated detections where the solid area represents the detection volume and the individual dots represent target locations that do not fuze the warhead.

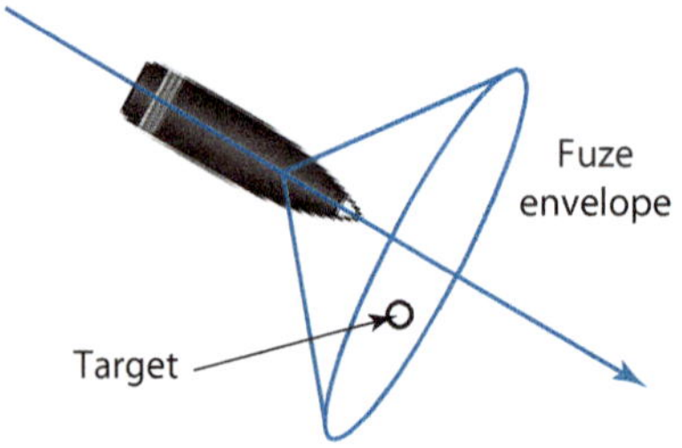

Fig. 3.41 Proximity fuze detection envelope.

3.15.3 Time Delay Fuze

A delay fuze is basically one with a fixed time delay following impact and the subsequent initiation of the PD fuze. Again, the delay mechanism may be set mechanically or electrically. These fuzes are used against targets where a small delay following impact can increase the effectiveness of the warhead against the target. For example:

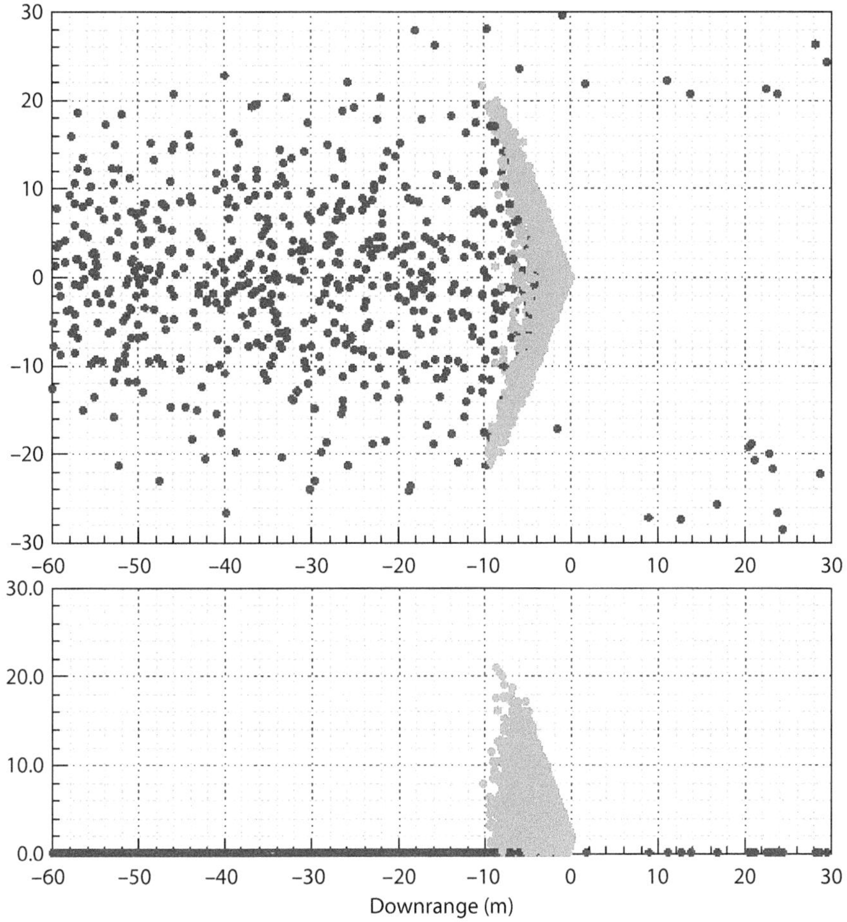

Fig 3.42 Simulated detonation points for PX fuze model.

- For a building target, delaying detonation until the warhead is inside the structure rather than detonating at impact on the roof may result in preventing the functions inside the building without its total demolition.
- For a bridge target, detonation just below the span initially lifts it up, but then the downward motion imparts a greater loading than an identical detonation on top of the span. This may result in the whole span collapsing rather than simply making a hole in the deck.
- For a ship target, greater damage will result from an internal detonation compared to a surface contact detonation and may initiate additional damage mechanisms such as fire or secondary explosions from a magazine or fuel tanks.

You can see from the previous discussions that the type of fuze used to detonate a warhead has to be specified along with the variability in fuze mechanism it uses to operate. As an example, a warhead traveling at 1000 m/s with a variable time fuze having a 50-ms delay error in a shallow approach angle will overshoot the target by 50 m. This will have a considerable effect on the damage sustained by the target.

References

[1] Farrah, C. L., and Fleming, D. W. (eds.), *Military Ballistics—A Basic Manual*, Brassey's Battlefield Weapons Systems & Technology, Vol. X, Oxford Press, New York, 1983.

[2] Hill, P., and Peterson, C. (eds.), *Mechanics and Thermodynamics of Propulsion*, 2nd ed, Addison Wesley, New York, App. 3, Table 2.

[3] Rowles, S., *Joint Gun Effectiveness Model with Air-Bursting Methodology—Users Manual*, Systems Concepts and Analysis Branch, Naval Surface Warfare Center, Dahlgren, VA, April 2006.

Chapter 4 Delivery Accuracy: Overview

4.1 Introduction

Delivery accuracy may be defined as a quantitative measure of the capability of a weapon system to place ordnance on its intended target. It is measured using a variety of statistics to be described in the following sections, and it is assumed that readers have a working knowledge of the statistical methods described in Chapter 2.

Basically we are interested in the dispersion of rounds directed at a target. If a test or simulation produces impact points on the ground like those shown in Fig. 4.1, we could express the accuracy of such a weapon system in terms of the dispersion in x and y (i.e., as σ_x and σ_y). The smaller the sigma, the more accurate the weapon system.

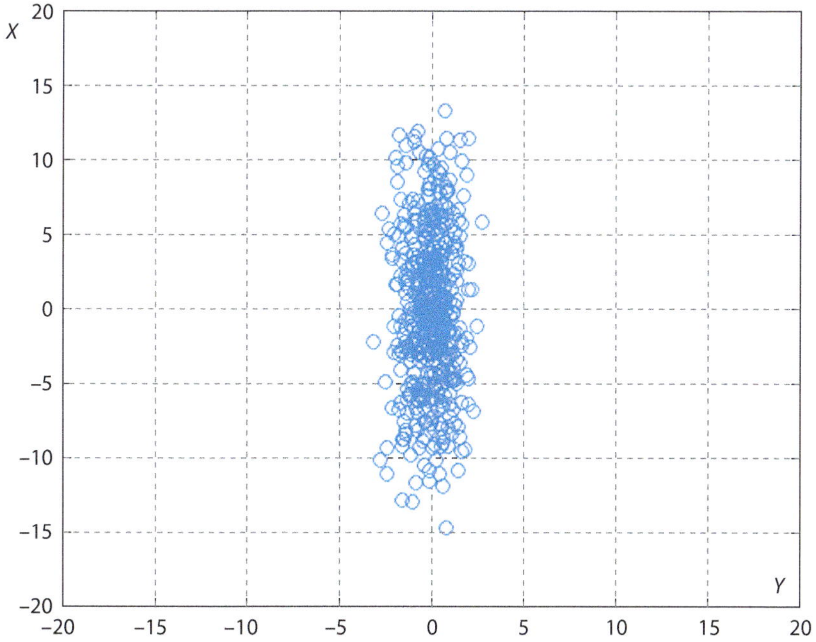

Fig. 4.1 Impact point distributions on the ground plane.

4.2 Error Types

The case outlined in the previous section is perhaps the simplest measure of delivery accuracy and corresponds to a large number of independent firings of a single weapon. Unfortunately, things get more complicated when multiple rounds are fired at the target, such as bullets fired from a machine gun or a stick of unguided bombs dropped from an aircraft. In general, weapons or munitions will have more than one accuracy measure, and many weapon systems can be described by three classes of error.

1. *A fixed bias error:* This first type of error is not random, but represents a systemic error of the weapon system that should not be present. For example, for the reasons given in the previous chapter, one might say that a laser-guided bomb (LGB) with a bang-bang control system will always fall short of the target by 25 ft. As long as the bias is known, then the aimpoint can be adjusted accordingly, that is, aim 25 ft past the target. Few weapons exhibit an uncorrected bias, although the JMEM Weaponeering System (JWS) database can accept inputs for those that do have it.
2. *Occasion-to-occasion and round-to-round errors:* Consider the case in Fig. 4.2 where a soldier is firing a machine gun at a target on a wall. He aims at the center of the target, fires a burst of six rounds, inspects the impact points, re-aims, fires another six-round burst, and so on. The impact points for the first three bursts are shown in Fig. 4.3.

For each burst, the mean point of impact (MPI) can be determined. Its distance from the origin (MPI error) represents the ability of the shooter to correctly aim the rifle at the center of the target. As expected, not all shots in the burst disappear into the same hole, but six individual impact points on the wall will be observed. The dispersion of these six points

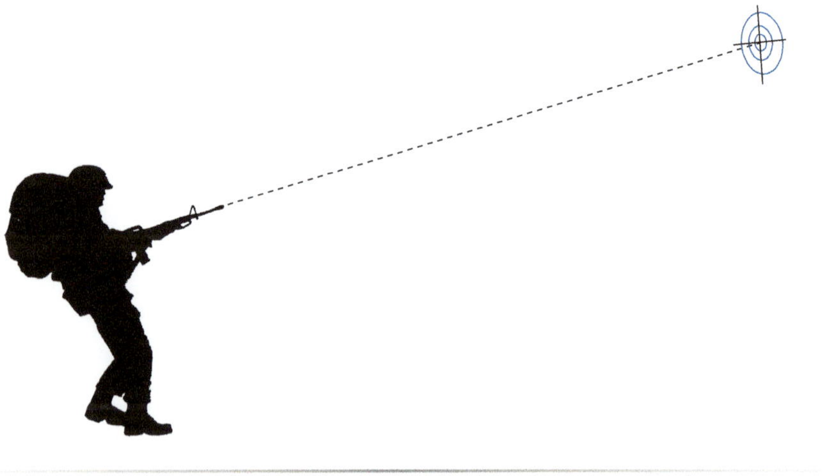

Fig. 4.2 Automatic weapon firing at a target.

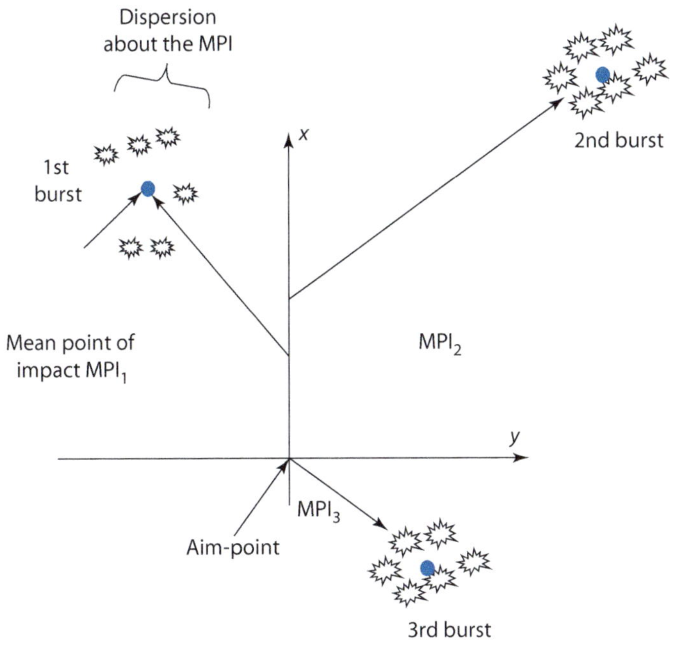

Fig. 4.3 Pattern of shots on the vertical plane.

about the MPI is called the precision error and represents the relative accuracy of each round with respect to the other rounds fired.

We call each burst an "occasion" and it is characterized by an aiming or re-aiming of the rifle, where it will be noted that the time between occasions (bursts) is long relative to the time during which the bullets are fired during the burst. Therefore, the MPI error may also be called the occasion-to-occasion error. It will also be seen that the MPI error characterizes the weapon delivery system (rifle), whereas the precision error characterizes the bullets themselves.

In terms of accuracy measures, we now have a dispersion in the x and y directions as before, but corresponding to both the MPI and precision errors. The accuracy statistics characterizing the scenario are therefore $\sigma_{x-\text{MPI}}$, $\sigma_{y-\text{MPI}}$ and $\sigma_{x-\text{prec}}$, $\sigma_{y-\text{prec}}$.

The concept of two error types (MPI and precision) is found in other weapon systems. As an example, consider firing multiple rounds from an artillery cannon in the space of, say, 1 min. This period of 1 min corresponds to an "occasion" and defines a single independent aiming of the cannon that is then undisturbed while the rounds are fired rapidly. Therefore, an artillery cannon aimed at a target and firing three rounds within 1 min would be considered one occasion.

Towing the cannon to a different position an hour later, re-aiming at the same target, and firing three more rounds would be considered a different occasion. Moving the cannon again, re-aiming the gun at the target, and firing three more rounds would be the third occasion. During the occasion, some ballistic properties are assumed to remain constant. For example, the atmosphere through which the three projectiles fly is assumed to remain constant; therefore, it has the same effect on all rounds fired during that occasion. An hour later conditions have changed; therefore, it is a different occasion. The duration of an occasion is somewhat subjective; however, it is of long duration compared to the time taken to fire the rounds for a single occasion. The impact points corresponding to the three occasions just described, and the three rounds fired per occasion, are shown in Fig. 4.4, where the impacts are shown in the ground plane.

For air-launched weapons, both classes of error also will be present. For example, an aircraft drops a stick of four unguided bombs on a target where each bomb is designed to be at the corner of a square with 200-ft sides. The center of the stick, which should ideally be located on the target, will be offset due to inaccurate aiming, as shown in Fig. 4.5. Notice how the whole stick pattern is displaced by the MPI error, called here the aiming error. Also shown is the effect of the precision errors perturbing the impact locations of the individual bombs from their ideal impact points at the corner of the square to some individually randomly located impact points.

A further example would be when multiple GPS-guided weapons are directed to a set of coordinates that are thought to point to the target of interest. Depending on how such coordinates are obtained, there is usually an error associated with their specification known as the target location error (TLE). The resultant impacts might look as shown in Fig. 4.6.

In this case, the occasion-to-occasion error is the TLE, whereas the round-to-round (precision) error is attributable to differences in the

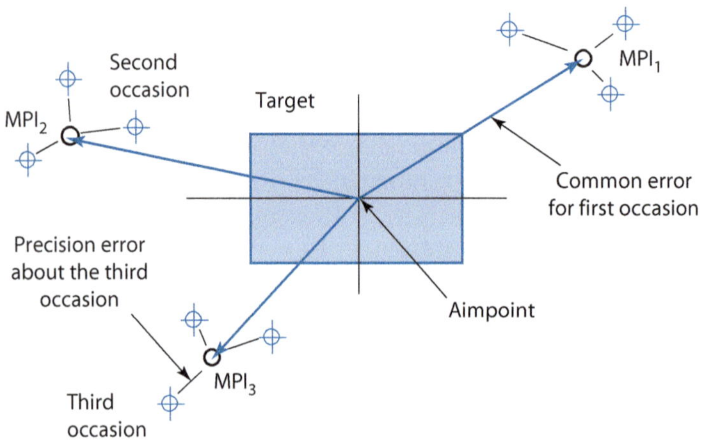

Fig. 4.4 MPI and precision errors.

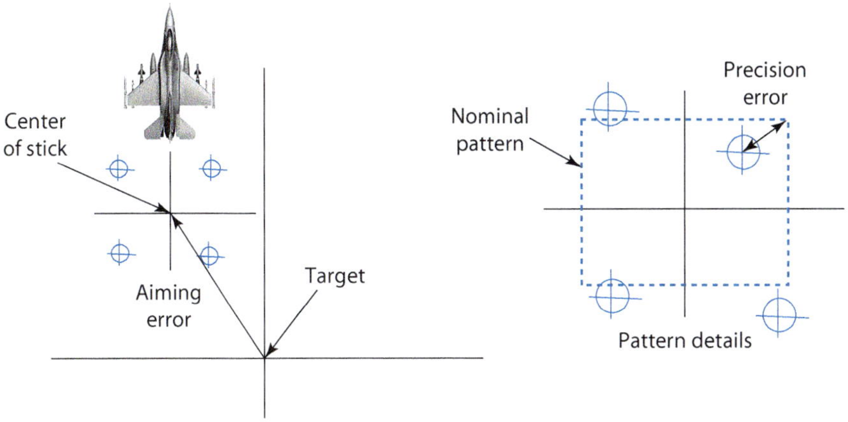

Fig. 4.5 Aiming and precision errors.

guidance and control system performance of each weapon, resulting in different flight paths to the specified common coordinates.

The reader may have noticed some differences in terminology in describing the same type of error; that is, MPI, aiming error, and TLE all describe the same thing. For historical and service-specific reasons, different communities use different expressions to describe the same types of error; efforts for different operational communities to agree on a common terminology have not succeeded. Among analysts, however, MPI error types are usually referred to as dependent errors, whereas precision-type errors are called independent errors. Table 4.1 lists terminology among the different operational communities.

Some care has to be exercised in using the word *dependent* because this is not accurate in the strict statistical sense. Within a burst, or

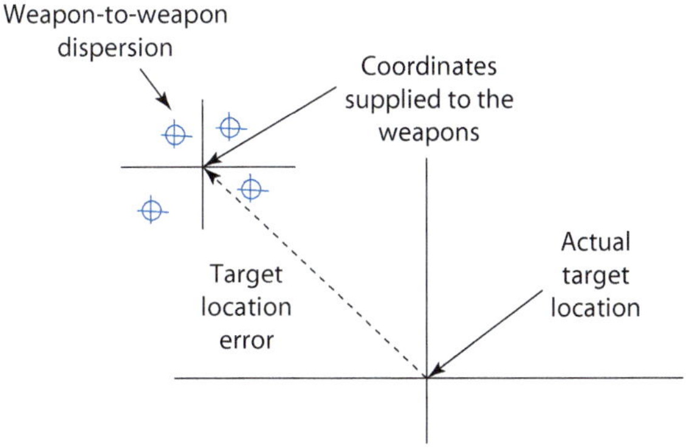

Fig. 4.6 GPS-guided weapons with TLE.

Table 4.1 Alternative Descriptions of Error Type

Community	Dependent Error	Independent Error
Artillery	Mean point of impact (MPI)	Precision
Air-to-surface	Aiming	Ballistic dispersion
Direct fire	Burst-to-burst	Within burst
GPS	Target location error (TLE)	Guidance and control
Surface-to-surface	Occasion-to-occasion	Round-to-round

occasion, one round does not depend on where a previous round landed, but they do share a common error, such as the same target coordinates or manually sighted aim point. A better description is to say the rounds are correlated; however, due to common usage, the term *dependent* will be used.

While on the topic of terminology, it is worth noting that units used by communities differ, too. Air-to-surface operators use imperial units, whereas surface-to-surface operators use the metric system, and some Navy systems use yards for range distances.

It is important to note that both dependent and independent errors will not be constant for a given weapon system but have to be further qualified before meaningful accuracy measures can be used. Consider, for example, an M240 machine gun firing the same round but with different mounts, as shown in Fig. 4.7.

Of the two cases shown, the vehicle-mounted weapon will be more stable, and therefore the dispersion of bullets on the target should be smaller. Note that the aiming system that governs the MPI error is also different for the two cases shown. Therefore, both MPI and precision errors will be different for these systems even though it is the same gun firing the same round.

In each case, the different precision error will contain the same fundamental ballistic dispersion of the round, which is dependent only on

Fig. 4.7 Different mountings of an M240 machine gun.

Fig. 4.8 Mann barrel used to determine ballistic dispersion.

the type of projectile being fired. It is worth asking: What is the actual dispersion of a .50-caliber bullet, and how is it measured? It is determined in a laboratory setting by using a Mann barrel, an example of which is shown Fig. 4.8.

This is a piece of test equipment comprising a heavy, thick-walled barrel of the correct caliber for the munition under study, and a breach mechanism to load and fire the round. The complete firing mechanism is fixed to a heavy plate that, in turn, is attached to a large, rigid mount, usually made of concrete. Rounds are fired at targets located at fixed ranges, and the ballistic dispersion is determined from impact data. This would be the smallest ballistic dispersion of the round, independent of the gun used to fire it or how the gun was mounted.

It is important to categorize a specific error type as either dependent or independent because they are treated differently in the effectiveness calculations that require accuracy data. This will be dealt with in greater detail in Chapter 10. Only in the case of single weapon delivery or a single shot can dependent and independent errors be combined directly in order to obtain the correct probability of damage; otherwise, incorrect damage estimations will result.

In general, independent errors can be measured using the same parameters as dependent errors and combined in a way that depends on the scenario being considered. Both errors can be considered bivariate random variables in the ground or normal plane, an example of which for a single shot is shown in Fig. 4.9.

The aimpoint is usually the target when it is small, or the center of a larger target. When single munitions are used against a target, the miss distances

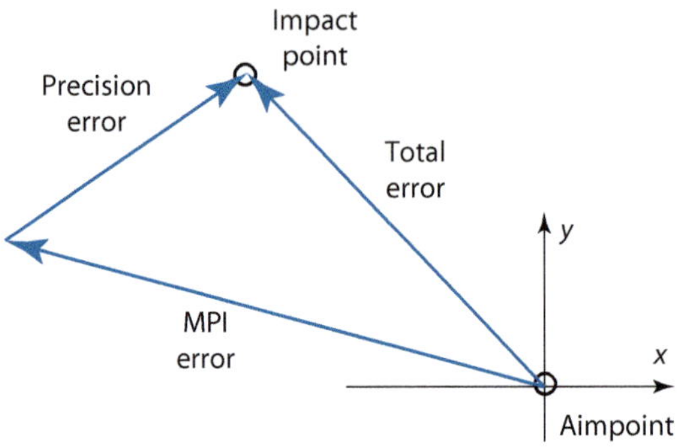

Fig. 4.9 Single-shot aiming, precision, and total errors.

actually measured are due to both MPI and precision errors and would be added as follows:

$$x_{DEP} + x_{IND} = x_{MISS} \tag{4.1}$$

$$y_{DEP} + y_{IND} = y_{MISS} \tag{4.2}$$

Because the MPI and precision errors are assumed to be independent random variables, the standard deviation of the measured miss distances would be obtained from the root sum square (RSS) process, as discussed in Eq. (2.81).

$$\sigma_{DEP}{}^2 + \sigma_{IND}{}^2 = \sigma_{MISS}{}^2 \tag{4.3}$$

In calculating effectiveness, Eq. (4.3) may be used only for the delivery of a single munition and not for multiple munitions such as air-delivered sticks, volleys fired from artillery cannons, or multiple rounds from a machine gun. It will be shown later that although the two error types have to be kept separate for the purpose of calculating effectiveness, they may be combined using Eq. (4.3) even for multiple weapondelivery if only accuracy (not effectiveness) is of interest.

4.3 Unguided and Guided Weapon Accuracy

The accuracy of unguided munitions will depend on the accuracy of locating the delivery system relative to the target, that is, the ability of a pilot to place the aircraft releasing an unguided bomb at the correct ballistic point, or determining the location of an artillery cannon firing an unguided shell—see Fig. 4.10. These factors are a large component of the dependent accuracy, or MPI.

Guided weapons or munitions are more accurate than unguided types because they use a guidance or seeker system that provides modifications to the trajectory from the simple ballistic form in order to reduce the miss distance. Guidance systems may use laser light reflected from the target; a visual, thermal, or radar image of the target itself; or guidance to a point in space where the target is or is thought to be located. As such, the resulting accuracy depends only on the performance of the guidance system and the ability of the munition to respond to commands that it issues. For example, the accuracy of a GPS-guided bomb will be the same whether it is launched from a B-2 stealth bomber or pushed out the back of a transport aircraft, assuming the weapon has the same set of target coordinates in each case and is close enough to the target to reach it.

The accuracy with which guided weapons are delivered is a function of several factors including the weapon guidance system and estimation of the target location, where the latter is usually considered independent of the aircraft delivering the weapon. Guided weapon accuracies are not subject to precision error, although there is an equivalent quantity called the guidance and control error.

Unlike an unguided weapon, which has to be released at the exact ballistic point in order to hit the target, a guided weapon has a volume in space such that if the weapon is released anywhere within this volume it will reach the target. This volume is referred to as the launch acceptable region (LAR), or basket, and is represented in Fig. 4.11 for a typical guided weapon. For surface-launched weapons, the equivalent of a LAR is a region on the ground plane in which the cannon may be located in order to hit a specified target.

The existence of a LAR rather than a unique ballistic point places less stress and precision requirements on the aircrew. In practice, however, aircrew may attempt to release guided weapons (LGB in particular) at the ballistic point if possible, so as to reduce the demands placed on the weapon control system, which in turn tends to reduce miss distances.

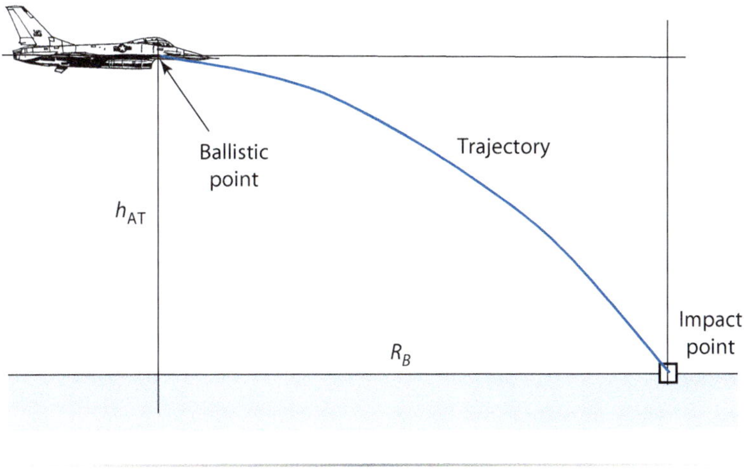

Fig. 4.10 Bomb release at ballistic point.

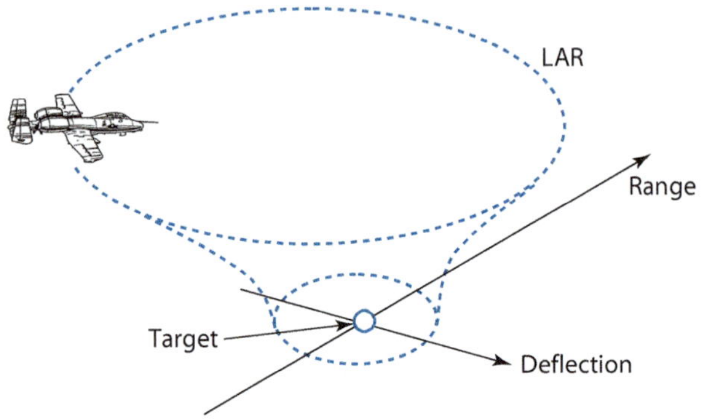

Fig. 4.11 Launch acceptable region for a guided weapon.

It is possible to define several LARs for a guided weapon, of which the two most common are:

1. *In-range LAR:* The weapon can reach the target from any point inside the defined region.
2. *In-zone LAR:* The weapon can reach the target with specific minimum impact conditions (impact velocity and angle) from anywhere inside the defined region.

The in-zone LAR will be smaller than the in-range LAR.

For specific weapons, some aircraft avionic systems using current system state measurements can display an instantaneous LAR to the pilot in the cockpit, which together with the current aircraft position symbol tells the pilot if they can launch.

Sometimes a LAR may be used to generate a footprint. The footprint of a weapon is a region on the ground that a weapon can reach relative to the current aircraft location. The footprint is usually the reflection of a horizontal section of the LAR centered on the target, as shown in Fig. 4.12. The

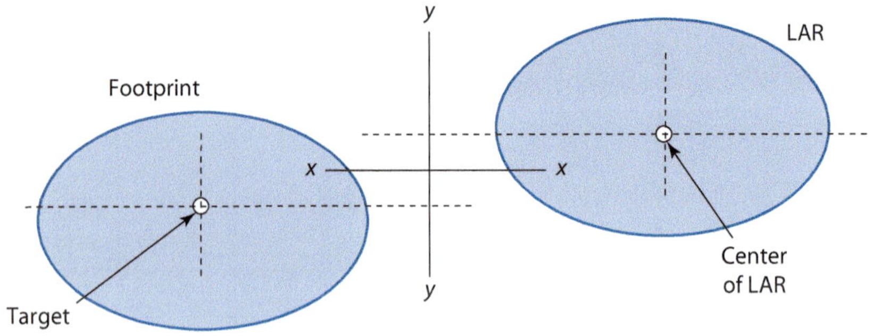

Fig. 4.12 Footprint generated from LAR.

lines $x-x$ and $y-y$ are midpoint between the center of the LAR (at a fixed altitude) and the target. The LAR is reflected about these lines to obtain the footprint. The LAR can be thought of as "at a given altitude what is the horizontal area where the weapon can reach the target," whereas the footprint can be thought of as "from this point in space where I am now, where can the weapon reach."

LARs are generated using a fly-out simulation model of the weapon and take into account all subsystem performance characteristics (i.e., target signature, seeker, guidance system, actuators, and missile body dynamics). Precomputed LARs or real-time generation using simplified models are usually available in the cockpit, so the aircrew know that the aircraft is within the target-centered LAR, and therefore the weapon can reach it.

4.4 How Accuracy Data are Obtained

The principal mission of delivery accuracy in the weaponeering process is to provide statistical parameters describing the expected error distribution of munitions launched from given conditions against a defined target. In an ideal world one might think these statistical parameters would be obtained from field trials of the weapon concerned; however, readers will realize this is impractical from both a testing and fiscal perspective.

Consider the accuracy of a single unguided bomb dropped from an aircraft. There are so many combinations of release parameters such as altitude, aircraft velocity, dive angle, ejection velocity, mode of release, and so forth that all possible combinations cannot be tested. Instead, the approach used is to develop predictive models that take into account all these effects and then validate the model by a relatively small number of trials. Because the development of these models and the validation tests may be quite lengthy, delivery accuracy data usually follow a series of steps that refine the accuracy estimates over time.

1. Specification (spec) data are found in the contractor's original specification of weapon performance. When a company bids on supplying a weapon, its performance has to meet the requirements of the customer, and one of these requirements will be accuracy. Usually this value is a combined value for single weapon delivery only, and as we have stated, multiple weapon deliveries require both dependent and independent accuracy values.
2. Developmental/operational testing is a process in which an initial batch of weapons is produced for the customer and tested either by the developer (DT) or by the military operational services that will use the weapon (OT). Clearly this accuracy must meet or exceed the spec value.
3. Predictive model development is a later activity because DT/OT data will be available for a preliminary validation. These models cover a wide range of use of the weapon for perhaps different ranges, release conditions, and

so forth that are not covered by the spec value. The spec value may be as simple as "the weapon will have a CEP* of 25 ft." Using the weapon under different circumstances may result in an accuracy of half, or maybe twice that of the spec value, something the warfighter needs to know. Usually these models apply to a class of weapons (e.g., unguided artillery rounds or GPS-guided bombs). This means the methodology should apply to all weapons of that type, and a new weapon is characterized only by the data the model needs to compute its specific accuracy.

4. It is tempting to use battle damage assessment (BDA) data to validate accuracy metrics; however, this is generally unreliable. A bomb crater 50 ft from the target tells us nothing about where the laser designator was pointed or what coordinates were fed into a GPS-guided munition, and we probably only have a sample of one or two impacts.

The methodologies used to describe unguided and guided weapon accuracy will be dealt with separately based on these observations, but first some general descriptions of delivery accuracy measures will be discussed.

4.5 General Measures of Dispersion

Suppose we are presented with a set of impact points on the ground plane, as shown in Fig. 4.13. The data shown in the figure may be interpreted

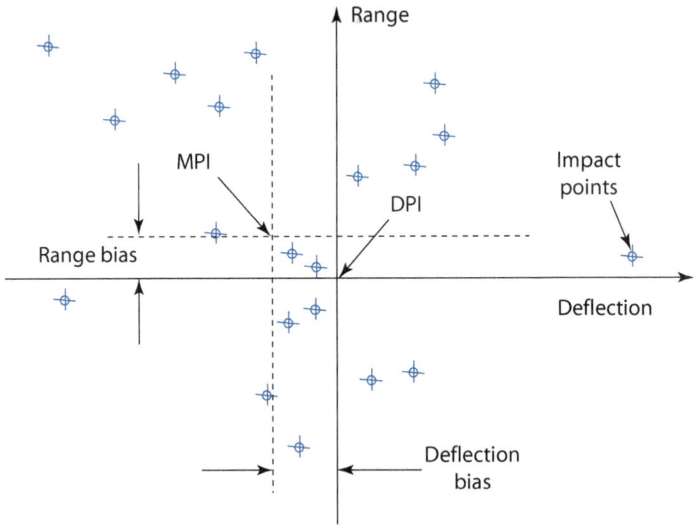

Fig. 4.13 Distribution of impact points around the DPI.

*See the next section for a definition of CEP.

in different ways; for example, the points may represent impacts of singly released bombs from multiple aircraft, in which case the figure shows independently released munitions where the miss distance for each point contains both aiming and ballistic errors. However, the same figure might represent successive impacts of artillery shells fired from the same cannon within a short amount of time, and therefore the bias would represent the single MPI error and the dispersion around it showing the effects of precision error. Just given holes in the ground, we cannot say without further information how they got there and how the miss distances should be interpreted.

We will restrict the discussion to where we know each round is independently fired at the target, that is, the weapon is re-aimed between individual rounds as would be obtained, for example, by one aircraft dropping one bomb followed by another aircraft with another bomb and so on. It is also assumed that all impacts occur in a relatively short time, that is, on the order of minutes rather than hours.

Clearly it is impossible to predict what happens to a particular bomb; all that may be done is to describe the behavior, in a statistical sense, of the sample of impact points. The desired point of impact (DPI) is the location of the aim point on the ground. We will see later that in studying the effects of collateral damage, for example, the DPI might be some distance from the "target" in order to assess the effect the weapon has if it lands close to this offset aimpoint.

Two axes in the ground plane are defined as before as the range and deflection directions. The range direction is defined as the projection of the velocity vector of the weapon at impact/detonation, onto the ground plane. The deflection direction is perpendicular to it, to the right. The resulting z-axis points upwards. A usual approximation of the range direction for air-launched unguided munitions is the velocity vector of the aircraft when the weapon is released; for artillery and direct fire weapons it is the direction of the velocity vector of the munition when it is fired. This implies the trajectory is contained in a vertical plane containing both the aircraft/shooter and target.

One of the simplest statistical tasks that may be performed is to find the mean of the range and deflection coordinates for each impact point. The resulting point is the mean point of impact (MPI). The MPI may be offset from the DPI by the range and deflection fixed bias values, as shown in Fig. 4.13.

To accompany the mean values just outlined, it would be useful to have some measure of the dispersion or distribution of the impact points. Although the variance or standard deviation, discussed in Chapter 2, might appear a suitable choice, weaponeers prefer to use the range error probable (REP) and the deflection error probable (DEP). The REP is the distance from the Desired Mean Point of Impact (DMPI) to one of a pair of lines perpendicular to the range direction, equidistant from the DPI such that 50% of the impact points lie between them. This is shown in Fig. 4.14. Similarly, the

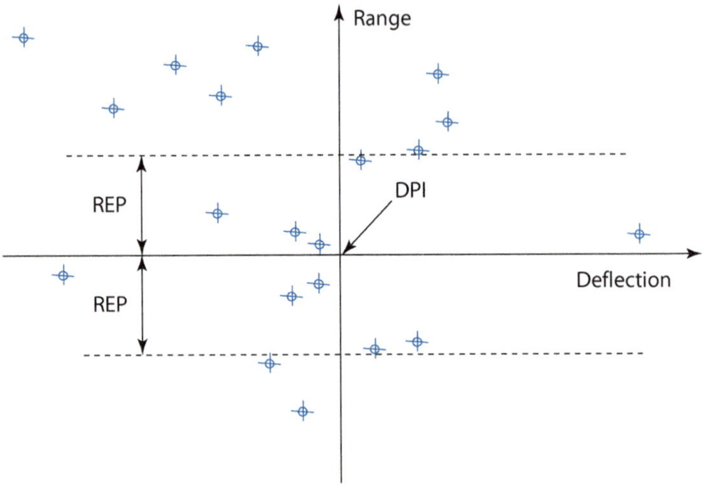

Fig. 4.14 Definition of REP.

DEP is the distance from the DMPI to one of a pair of lines parallel to the range direction, equidistant from the DMPI such that 50% of the impact points lie between them, as shown in Fig. 4.15.

Although REP and DEP give an idea of dispersion in the range and deflection directions, another commonly used measure is the circular error probable (CEP). The CEP is the radius of a circle, centered on the DPI such that 50% of the impact points lie within it, as shown in Fig. 4.16.

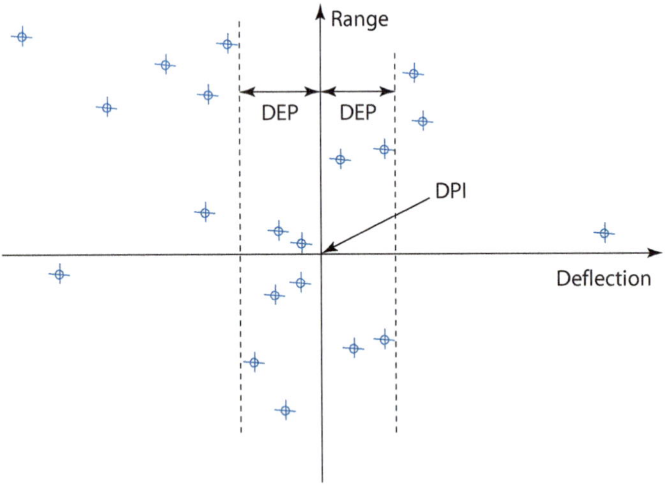

Fig. 4.15 Definition of DEP.

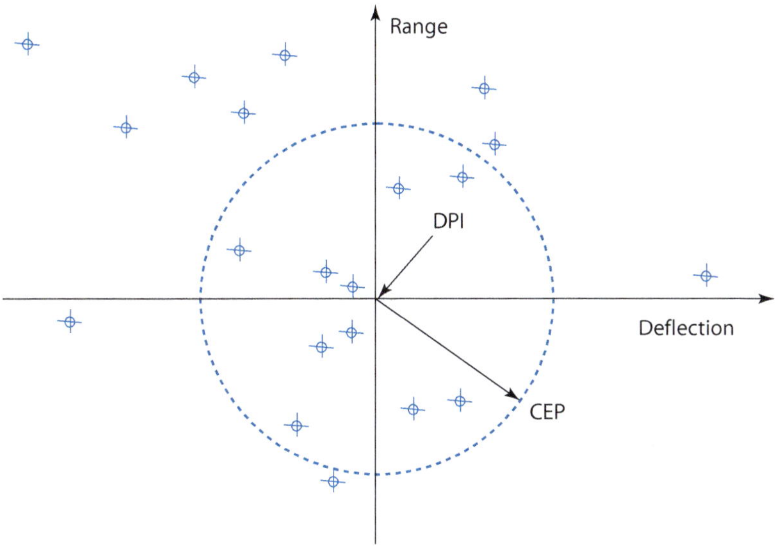

Fig. 4.16 Definition of circular error probable (CEP).

As for the variance, the error probable can be interpreted two ways:

1. The distance inside which 50% of the weapons may be expected to fall
2. The distance inside which a single weapon has a 50% chance of falling

Note that these metrics do not assume any specific distribution associated with the miss distances and may be determined graphically as indicated in the preceding figures from impact points plotted on graph paper.

An issue that regularly arises in the analysis of field data is whether any significant fixed bias can be ignored, and some judgment is required here. If the bias is considered to be a function of the test (e.g., wind effects), then the bias is usually removed so as to not penalize the weapon accuracy by a feature associated with the test. If, however, biases occur each time the test is performed over a long period of time and has the same value and direction, then it may be a systemic characteristic of the weapon, and a fixed bias should be reported as part of the delivery accuracy data. The effect on the resulting CEP of removing a significant bias is shown in Fig. 4.17.

If the test data were from impacts that are not independent, such as artillery impacts taken over several occasions, the removal of the bias would essentially remove the MPI error for each occasion, leaving only the precision error to be measured. For this reason, most testing addresses the independent error only. Ideally the impact points should be analyzed in a manner that allows the MPI error and precision error to be determined. Suppose the test results are as shown in Fig. 4.18.

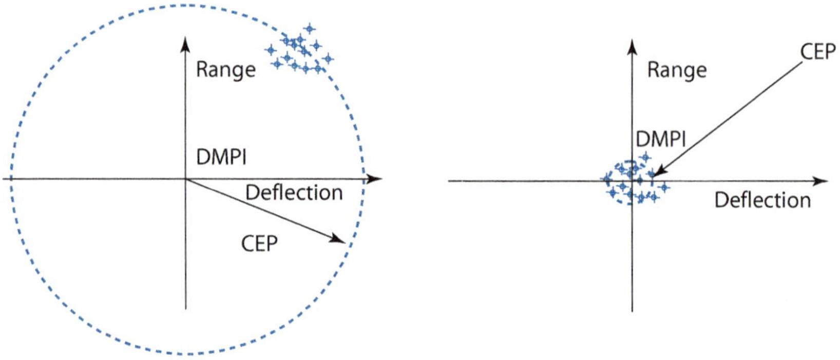

Fig. 4.17 Effect of removing the bias on a CEP calculation.

The CEP can be calculated graphically from the impact points as shown in the figure, but usually REP and DEP of the MPI error would be calculated from the sample of MPI errors, and another REP and DEP for the precision error would be calculated from the impacts about each MPI. These results would be reported separately for MPI and precision errors, allowing the effectiveness calculations to proceed correctly.

REP and DEP are more useful than CEP, because it will be shown how CEP can be derived from REP and DEP. The opposite is not true, however, at least not without making a simplifying assumption.

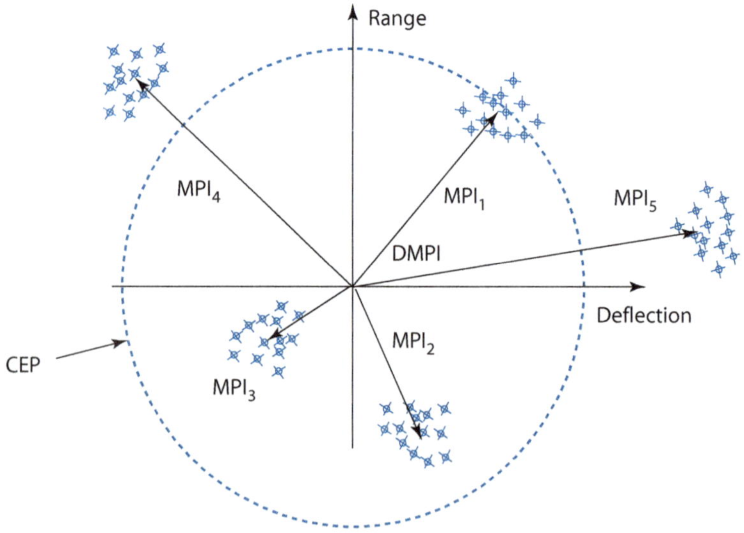

Fig. 4.18 Example test results.

4.6 Vertical Errors

So far, only miss distances in the ground plane have been considered, but some targets may be represented by a vertical target plane, in which case the distribution of impacts vertically has to be considered. This is shown in Fig, 4.19. In this case the height error probable (HEP) is defined as the distance between two horizontal lines equidistant from the DPI enclosing 50% of the impacts.

As stated earlier, all the measures of accuracy discussed so far can be estimated from experimental data without any assumptions being made about the statistical nature of the data. If some assumptions are made, however, the computational analysis becomes simpler. In the next section we will discuss the types of assumptions that may be made, and how the statistics of accuracy may be calculated.

4.7 Commonly Used Equations Based on Normal Data

The usual assumption made in the analysis of accuracy data, supported by analysis of test data, is that the distributions of impact points in range, deflection, and height are Gaussian, or normal, and there is no cross-correlation (i.e., they are independent in these directions). It is also assumed that either the fixed bias values are zero or the data have any bias removed. Given these assumptions, the analysis described in the previous chapter on statistics (Chapter 2) may be applied. For example, from tabulated values of the normal PDF (Table A.1, Appendix A), it can be seen that because $\pm\sigma$ of a zero mean distribution contains about 68% ($\pm$34%) of the

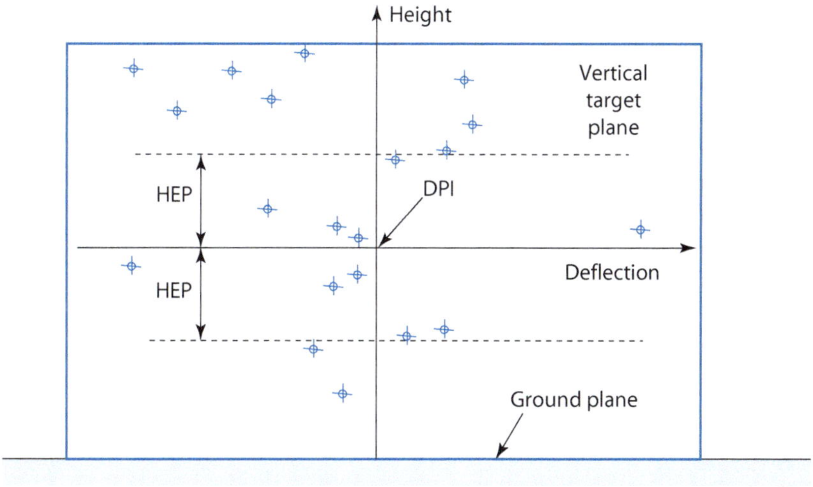

Fig. 4.19 Example test results on vertical target.

samples ($Z = 1$), REP, DEP, and HEP containing 50% of the samples will correspond to a fraction of the variance. In fact 50% ($\pm 25\%$) will correspond to $F(Z) = 0.75$, or $Z = 0.6745$. Hence we can write

$$\text{REP} = Z \times \sigma_x = 0.6745\sigma_x \tag{4.4}$$

$$\text{DEP} = Z \times \sigma_y = 0.6745\sigma_y \tag{4.5}$$

$$\text{HEP} = Z \times \sigma_z = 0.6745\sigma_z \tag{4.6}$$

In the ground plane, if $\sigma_x = \sigma_y = \sigma$ (circular normal), and the mean values of miss distance are both zero, the CEP can be obtained from the Rayleigh distribution described earlier in Eq. (2.47) and reproduced here:

$$P(r_1 < r < r_2) = \exp\left[\frac{-r_1^2}{2\sigma^2}\right] - \exp\left[\frac{-r_2^2}{2\sigma^2}\right] \tag{4.7}$$

By setting $r_1 = 0$, $r_2 = \text{CEP}$, and $P(r < \text{CEP}) = 0.5$, we get the following:

$$0.5 = 1 - \exp\left[-\frac{\text{CEP}^2}{2\sigma^2}\right] \tag{4.8}$$

This reduces to[†]

$$\text{CEP} = 1.1774\sigma \tag{4.9}$$

Recalling the discussion of the Rayleigh distribution earlier, it is reassuring to note that the CEP is simply the median radial miss distance in Eq. (2.49). We can find the relationship among REP, DEP, and CEP by writing

$$\text{CEP} = 1.1774\sigma = 1.1774\frac{\text{REP}}{0.6745} = 1.7456\,\text{REP} \tag{4.10}$$

Because REP and DEP are equal in a circular normal distribution, we also obtain

$$\text{CEP} = 1.7456\,\text{DEP} \tag{4.11}$$

Alternatively we may write

$$\text{REP} = \text{DEP} = 0.573 \times \text{CEP} \tag{4.12}$$

For some applications it is necessary to find the radius of a circle that contains 90% of the impacts, sometimes referred to as the CEP_{90} or CE_{90}. Equation (4.8) becomes

$$0.9 = 1 - \exp\left[-\frac{\text{CEP}_{90}^2}{2\sigma^2}\right] \tag{4.13}$$

[†]Notice how for the univariate distribution the error probable measures are smaller than sigma, whereas for the bivariate distribution it is larger.

leading to

$$\mathrm{CEP}_{90} = 2.146\sigma \tag{4.14}$$

Clearly this may be extended to finding the radius of a circle containing any proportion N of impacts by generalizing Eq. (4.8) to the following form:

$$N = 1 - \exp\left[-\frac{\mathrm{CEP}_N^2}{2\sigma^2}\right] \tag{4.15}$$

leading to the following:

$$\mathrm{CEP}_N = \left[\sqrt{-2 \times \log_e(1 - N)}\right]\sigma \tag{4.16}$$

From Eqs. (4.9) and (4.14) we get the useful scaling

$$\mathrm{CEP}_{90} = 1.8227 \times \mathrm{CE}_{50} = 1.8227 \times \mathrm{CEP} \tag{4.17}$$

Another useful form of Eq. (4.7) is when we express the radius as a proportion u of the common sigma.

$$p_u = 1 - \exp\left(\frac{-(u\sigma)^2}{2\sigma^2}\right) = 1 - \exp(-u^2/2) \tag{4.18}$$

For example, putting $u = 1.1774$ from Eq. (4.9) gives $p_u = 0.5$. In the vertical direction, the 90th percentile is a linear error, as opposed to the circular error in Eq. (4.13). It is obtained in a similar manner to Eqs. (4.4) through (4.6) by recognizing $F(Z) = 0.95$, and hence $Z = 1.6449$. Therefore, we can write for the vertical direction

$$\mathrm{HEP}_{90} = 1.6449 \times \sigma_Z = \frac{1.6449}{0.6745} \times \mathrm{HEP}_{50} = 2.483 \times \mathrm{HEP} \tag{4.19}$$

In all the equations developed here, it is assumed that the range, deflection, and height miss distances are normally distributed with zero means and equal variances. For actual field data, it is usual to apply a chi-squared test in order to check this assumption is within acceptable confidence limits before analysis proceeds.

We now revisit the Rayleigh distribution discussed in Chapter 2, Section 2.7. Consider 100 impact points in the ground plane, and suppose the means in range and deflection are zero and the standard deviations are equal and have the value of 25 ft. This is the σ that should be used in the preceding equations, as indicated by Eq. (4.9), for example.

As pointed out before, for each impact point, a radial miss distance r can be calculated, and there will be 100 of those data points. The radial miss distance has a Rayleigh distribution, and common statistics of this 100-sample data set can be calculated; for example, we can calculate a mean, median, and variance of the radial miss distance. It may be recalled from Eqs. (2.48) through (2.50) that the statistics of the radial miss distance data set may be related to the standard deviation of the linear distribution σ.

$$\mu_r = \sigma\sqrt{\frac{\pi}{2}} = 1.2533\sigma \tag{4.20}$$

$$r_{\text{median}} = \sigma\sqrt{\log_e(4)} = 1.1774\sigma \tag{4.21}$$

$$\sigma_r^2 = \left[\frac{4-\pi}{2}\right]\sigma^2 = 0.4292\sigma^2 \tag{4.22}$$

Comparing Eq. (4.21) with Eq. (4.9) it can be seen that the CEP is the median radial miss distance. Further, setting $r_1 = 0$ and $r_2 = \sigma$ in Eq. (2.47) yields $P(0 < R < \sigma) = 0.393$. This states that 39.3% of the radial impact points are contained in a circle of radius σ_r. Substituting Eq. (4.22) into Eq. (2.47) gives $P(0 < R < \sigma_R) = 0.193$, showing that only 19.3% of the radial miss distances are contained inside a circle of radius σ_R. One might expect to find 67% of the impacts contained within σ_r, but remember this proportion applies to a normal distribution, and the radial miss distance is Rayleigh distributed.

A question regarding semantics: What is the difference between CE_{50} and the CEP? Usually, the *P* notation means that the number is derived from data by assuming a specific probability density function, such as Gaussian or Rayleigh. We have seen, however, that the CEP can be calculated directly from the data set without using any assumed distribution, as shown in Fig. 4.16. Strictly speaking, therefore, the CEP is derived from a data set using an assumed, specific distribution, whereas a CE_{50} is derived from raw data. The same also applies to the 90% metric.

4.8 Noncircular Distributions

In many instances, the distribution of miss distances may be Gaussian in range and deflection, but is not circular, so the assumption that the radial miss distance is Rayleigh distributed is weak. For these cases we will consider the ratio of the smallest to largest standard deviations, which will be less than unity. Mathematically we define the smallest and largest standard deviations as

$$\sigma_s = \min(\sigma_x, \sigma_y) \tag{4.23}$$

$$\sigma_l = \max(\sigma_x, \sigma_y) \tag{4.24}$$

Typically, noncircular data are assumed to be sufficiently circular as long as the ratio of standard deviations is not less than about 0.5 (σ_s divided by σ_l). If this condition is met we can use the analysis applicable to truly circular distributions, but the question is, what value of σ should be used? There are some commonly used approximations to a single σ that are used, for example, in Eq. (4.9).

$$\mathrm{CEP} = 1.1774\frac{\sigma_x + \sigma_y}{2} = 0.5887(\sigma_x + \sigma_y) \tag{4.25}$$

$$\mathrm{CEP} = 1.1774\sqrt{\frac{\sigma_x^2 + \sigma_y^2}{2}} = 0.8326\sqrt{\sigma_x^2 + \sigma_y^2} \tag{4.26}$$

$$\mathrm{CEP} = 1.1774\sqrt{\sigma_x \sigma_y} \tag{4.27}$$

These equations correspond to calculating an arithmetic mean, root mean square, and geometric mean standard deviation, respectively. By comparing to an exact integration of the bivariate normal distribution, Freeman [1] found that Eq. (4.25) gave smaller errors than the other two as the distribution became less circular.

When the ratio of standard deviations is smaller than 0.5, however, other more precise methods have to be used based on the numerical integration of the bivariate normal distribution given in Eq. (2.26). A useful empirical approximation to this integral for calculating CEP is given by Pitman's equation

$$\mathrm{CEP} = 0.562 \times \mathrm{MAX}(\sigma_x, \sigma_y) + 0.617 \times \mathrm{MIN}(\sigma_x, \sigma_y) \tag{4.28}$$

Using this equation instead of direct integration of the bivariate distribution will result in an error of less than 1%. Figure 4.20 shows the difference between Pittman's equation and direct integration of the bivariate distribution for calculating CEP when the miss distance is not circular, as a

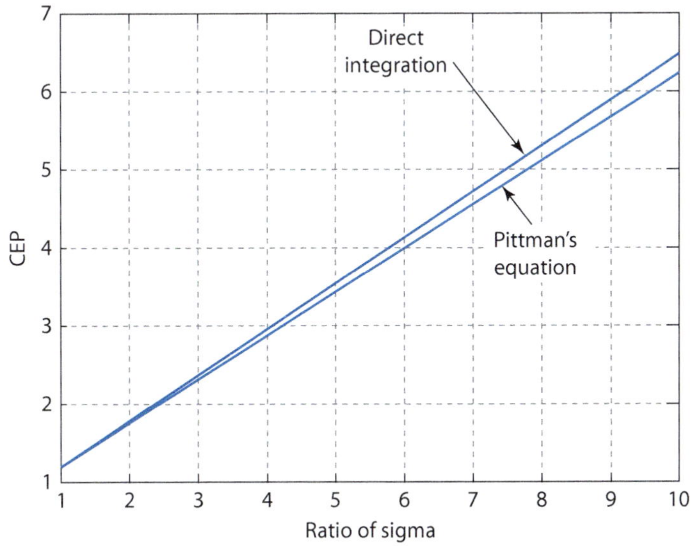

Fig. 4.20 Difference between Pittman's equation and Eq. (4.9).

function of the ratio of largest to smallest standard deviations. Note that although Pitman's equation allows for different variances, the means must still be zero or close to it.

4.9 Accuracy Measures Including a Bias

As mentioned previously, some weapons (e.g., LGBs) have a fixed bias associated with their impact points; for example, data such as those shown in Fig. 4.13 might have resulted from single weapon impacts over multiple occasions with appreciable time between each. Because the bias is not zero, it cannot be attributed to occasion-specific inaccuracies such as wind speed, because these would be expected to average out over all occasions with appreciable time between them. It is concluded, therefore, that the observed bias is systemic and will occur for every weapon used.

For such weapons we have shown that the fixed bias forms part of the delivery accuracy in addition to the calculated disturbance; however, for some applications it may still be necessary to estimate the CEP relative to the target, which includes the effect of the bias, rather than relative to the MPI. Equation (4.9) will underestimate the CEP because it is based only on the standard deviation, which is independent of any bias present. This problem has been addressed by Grubbs [2], and the CEP is calculated in the following manner:

1. From the raw data, the variances are calculated in the usual way.

$$\sigma_x^2 = \frac{1}{n-1}\sum_{i=1}^{n}(x_i - \bar{x})^2 \qquad \sigma_y^2 = \frac{1}{n-1}\sum_{i=1}^{n}(y_i - \bar{y})^2 \tag{4.29}$$

2. If (a, b) is the aimpoint, usually (0, 0), then we calculate the following quantities:

$$\sigma = \sqrt{\sigma_x^2 + \sigma_y^2} \tag{4.30}$$

$$m = 1 + \frac{1}{\sigma^2}\left[(a-\bar{x})^2 + (b-\bar{y})^2\right] \tag{4.31}$$

$$v = 2\left\{\frac{\sigma_x^4 + \sigma_y^4}{\sigma^4} + 2\left[\frac{\sigma_x^2(\bar{x}-a)^2 + \sigma_y^2(\bar{y}-b)^2}{\sigma^4}\right]\right\} \tag{4.32}$$

3. The CEP is given by

$$\text{CEP} \approx \sigma \times \sqrt{m} \times \left[1 - \frac{v}{9m^2}\right]^{3/2} \tag{4.33}$$

Note that this method also allows for different range/deflection variances; however, if we now assume circular normal with zero means, Eq. (4.33) reduces to

$$\text{CEP} = \sigma_x \sqrt{2}\left[1 - \frac{1}{9}\right]^{/2} = 1.1849\,\sigma_x \tag{4.34}$$

which is close to the exact value of 1.1774 given in Eq. (4.9). Note this process is not necessary for JWS because a bias may be included separately in its database.

4.10 Accuracy in the Normal Plane

Accuracy in the ground plane is measured in feet or meters, but the metrics used when impacts occur in the normal plane are angular units such as the milliradian (mil). Consider the soldier shooting at a target on the vertical wall shown in Fig. 4.21.

The normal plane is defined as a plane perpendicular to the velocity vector of the projectile at impact, so if we assume a straight line trajectory of the bullet, the vertical plane is the wall. The angular measure θ is related to the linear distance at the target using Fig. 4.22.

If θ is the angular error, and assuming the arc length is approximately the same as the chord length, the linear error in the normal plane is given by the following equation:

$$x \approx c = R\theta \tag{4.35}$$

One advantage of angular measures for the error is that it scales linearly with distance, as shown in Fig. 4.23.

The main difficulty in using Eq. (4.35) for weapons that do not have a straight line trajectory is what to use for the range R. Some care is needed to identify the normal plane if the munition follows a curved trajectory before impact, as shown in Fig. 4.24.

Here the normal plane is defined normal to the line of sight between the aircraft and the target at the moment of weapon release (i.e., normal to the

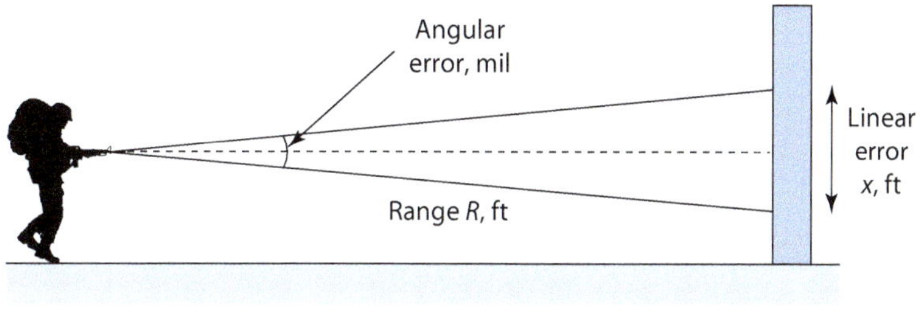

Fig. 4.21 Normal plane angular and linear errors.

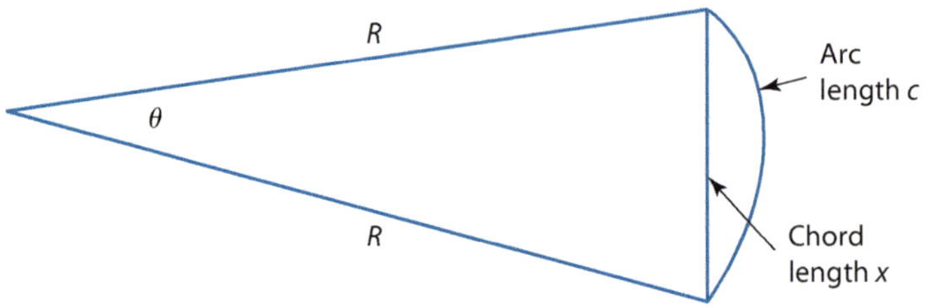

Fig. 4.22 Converting angular measures to linear measures.

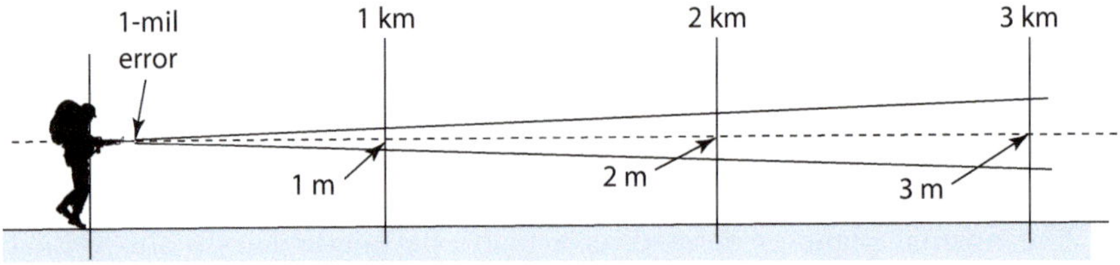

Fig. 4.23 Linear error scales with range.

slant range), where the angle H is known as the Harp angle. Note the definition of the impact plane, normal to the impact velocity. If variations in terrain are considered, the concept of the "ground plane" becomes unclear. In these cases, the target plane is defined as a horizontal plane that passes through the target.

In applying Eq. (4.35) to a curvilinear trajectory, the value of R should be the path length, and the value of θ set equal to the impact angle; however, for all but a straight line or dragless, parabolic path, the trajectory length is difficult to calculate. Therefore, for air-to-surface applications, x is set to the slant range SR and θ is the harp angle H.

The surface-to-surface case is shown in Fig. 4.25. Here the normal plane is perpendicular to the velocity vector of the projectile at impact (i.e., the

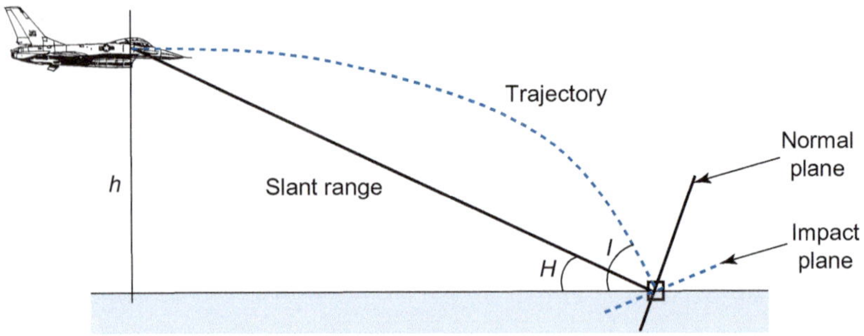

Fig. 4.24 Normal and impact plane for air-launched weapons.

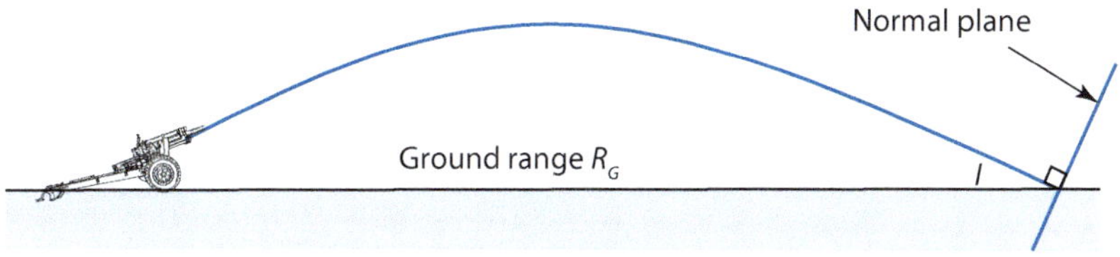

Fig. 4.25 Normal plane for surface-fired munitions.

impact plane). In this case, the value for R is set equal to the ground range R_G, and θ is equal to the impact angle I.

The choice of defining accuracy in the ground or normal plane is also dependent on the munition. Usually, projectile-type weapons (bullets, rockets, powered missiles) are specified in the normal plane whereas weapons that fall on the ground (bombs, mortar and artillery shells) are usually specified in the ground plane. Hence, the normal plane is usually associated with direct fire systems whereas the ground plane is associated with indirect fire. Note that Eq. (4.35) also provides a simple way to convert between angular and linear normal plane accuracy metrics.

4.11 Definitions of the Angular Mil

Some care has to be exercised when discussing the quantity termed the *mil* introduced in the previous section. Its usual connotation is one thousandth of a radian, which means there are 2000π or about 6283 milliradians in a circle. There are other definitions, however, that are in use:

1. 6400 mils in a circle—NATO countries
2. 6000 mils in a circle—former Soviet Union and Finland
3. 6300 mils in a circle—Sweden

The rational for these alternative definitions is that a circle is difficult to divide into many equal parts when the number of parts is 6283, whereas for the NATO definition there are 1600 mils in a right angle, 800 mils in 45 deg, and so on. The last two definitions are rarely seen, but the first one is in common use in the U.S. Army artillery community.

4.12 Converting Between Ground and Normal Plane Accuracy Values

Delivery accuracy data can be specified in the ground plane or the normal plane, so it may be necessary to convert from one to the other. Consider first the air-launched case shown in Fig. 4.24. Instead of representing individual impact points, we will mark the calculated REP and DEP values on the ground plane. Note that for clarity, only one of the two lines for REP and

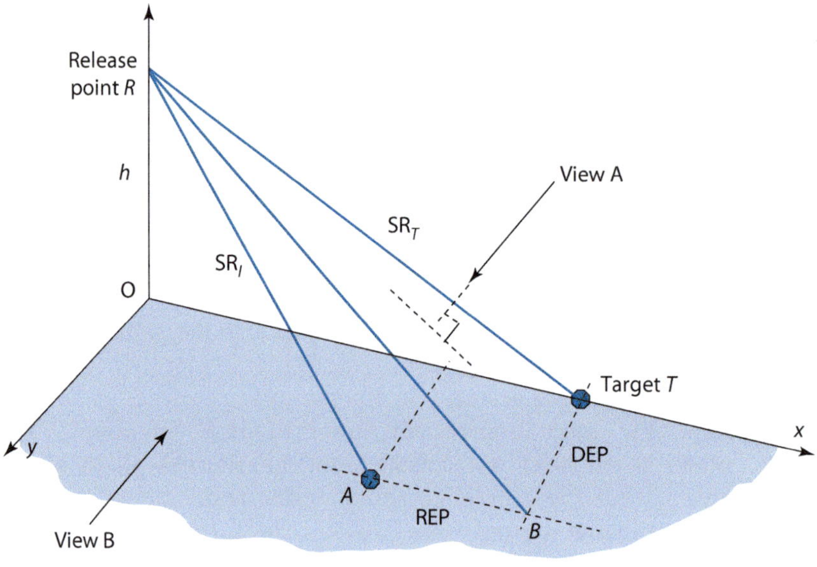

Fig. 4.26 Converting from ground to normal plane.

DEP has been drawn to the right of the x-axis, rather than the total spread shown in Fig. 4.21.

The relationship between REP and DEP in the normal and ground planes can be analyzed by considering range and deflection separately, beginning with the deflection direction by looking in the direction perpendicular to triangle *RBT* indicated by view A in Fig. 4.26 (see Fig. 4.27).

DEP in mils is defined as the angle subtended by DEP in feet in the ground plane over the slant range SR. In most cases SR >> DEP (ft), so we get from the geometry of Fig. 4.27

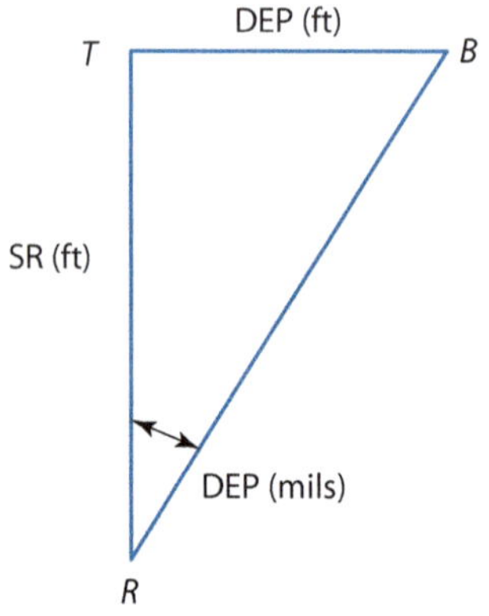

Fig. 4.27 DEP in the ground and normal planes (view *A* in Fig. 4.26).

$$\text{DEP (ft)} = \frac{\text{SR} \times \text{DEP (mils)}}{1000} \tag{4.36}$$

Now considering the range direction, the simplified geometry shown in Fig. 4.28 is obtained from view B of Fig. 4.26.

Once again, REP in mils is defined as the angle subtended by REP in feet in the ground plane over the slant range SR. We can determine how the distance x in the normal plane translates into the ground plane from the geometry of Fig. 4.28, again

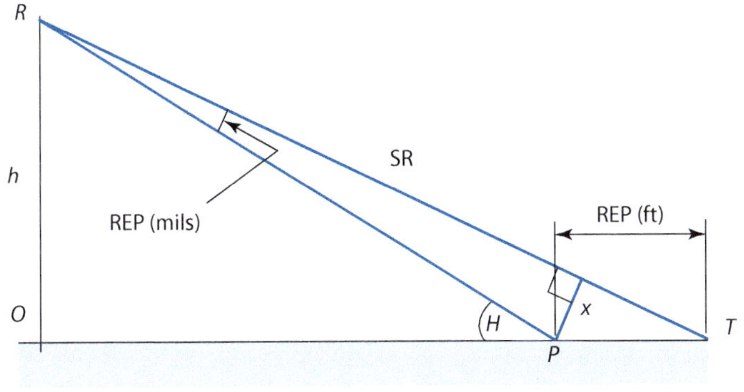

Fig. 4.28 REP in the ground and normal plane (view *B* in Fig. 4.26).

assuming SR >> REP (ft).

$$x = \text{SR} \times \text{REP (mils)}/1000 \tag{4.37}$$

But the distance TP = REP (ft), and for most operational situations, angle RTO $\cong H$, the harp angle, so we get

$$\sin H \approx \sin(\text{RTO}) = \frac{x}{\text{TP}} \tag{4.38}$$

leading to

$$\sin H = \frac{\text{SR} \times \text{REP (mils)}}{1000 \times \text{REP (ft)}} \tag{4.39}$$

Because it is also known that

$$\sin H = \frac{h}{\text{SR}} \tag{4.40}$$

we finally get

$$\text{REP (ft)} = \text{REP (mils)} \frac{\text{SR}^2}{1000h} \tag{4.41}$$

Equations (4.36) and (4.41) are used for the conversion between the ground plane and normal plane for air-launched weapons. If the miss distances are circular normal in the normal plane, we get from Eq. (4.12)

$$\text{REP (mils)} = \text{DEP (mils)} = 0.573\text{CEP (mils)} \tag{4.42}$$

Equations (4.36) and (4.41) give

$$\text{DEP (ft)} = \frac{0.573 \times \text{SR} \times \text{CEP (mils)}}{1000} \tag{4.43}$$

$$\text{REP (ft)} = \frac{0.573 \times \text{SR}^2 \times \text{CEP (mils)}}{1000h} \tag{4.44}$$

For surface-launched weapons, the distance subtending DEP in mils should be the path length of the round; however, the ground range R is used instead, hence Eq. (4.36) for the deflection direction becomes

$$\text{DEP (ft)} = \frac{R \times \text{DEP (mils)}}{1000} \tag{4.45}$$

In the range direction, the value of REP in mils uses the same approximation.

$$\text{REP (ft)} = \frac{R \times \text{REP (mils)}}{1000} \tag{4.46}$$

These equations allow conversion of ground plane accuracies to the normal plane and vice versa. Care has to be used in this conversion process, however, because the assumption of normality in one plane does not necessarily imply normality in the converted plane.

4.13 Treatment of Precision Errors in the Normal Plane

Usually precision errors are defined by σ_b, the standard deviation (in mils) in the normal plane, and are assumed Gaussian. This standard deviation can be converted to a range and deflection error (in feet) in the ground plane to be combined with other ground plane errors. The same ranges are used to convert the ballistic errors from the normal plane to the ground plane, with one exception, as indicated in Fig. 4.24.

For air-to-surface attacks, the precision error conversion uses the impact angle I rather than the harp angle H that was used in Fig. 4.28, although the slant range is used as an approximation to the path length traveled by the weapon as before.

If σ_b is a true standard deviation in mils ($\pm 34\%$) and not an error probable measure ($\pm 25\%$), then from a figure similar to Fig. 4.27 and Eq. (4.36) we can write in the deflection direction as follows:

$$x_{bd}\text{ (ft)} = \frac{\text{DEP}_b \times \text{SR}}{1000} = \frac{0.6745 \times \text{SR} \times \sigma_b}{1000} \tag{4.47}$$

In the range direction we use Eq. (4.39), except we use the weapon impact angle rather than the harp angle.

$$x_{br}\text{ (ft)} = \frac{\text{REP}_b \times \text{SR}}{1000 \sin(I)} = \frac{0.6745 \times \text{SR} \times \sigma_b}{1000 \sin(I)} \tag{4.48}$$

The impact angle is used for both air-to-surface and surface-to-surface cases. For the purpose of calculating just accuracy (not effectiveness), these error probable values in feet may be combined with the error probable

MPI errors in feet to give an equivalent REP and DEP, which includes precision errors as follows:

$$\text{REP}' = \sqrt{\text{REP}^2 + x_{br}^2} \tag{4.49}$$

$$\text{DEP}' = \sqrt{\text{DEP}^2 + x_{bd}^2} \tag{4.50}$$

Having defined the accuracy metrics that may be used for any kind of weapon system, we now look at how these are applied to guided and unguided weapons for both air-to-surface and surface-to-surface applications.

4.14 Nonnormal Distributions

In the analysis so far we have assumed that miss distances can be represented by normal, or Gaussian, distributions. When testing guided weapons, it became apparent that the data from the test range did not always fit a normal distribution[‡] and that further characterization was required. It should be apparent that any attempt to predict the effectiveness of a weapon will depend on its accuracy distribution, so mathematical ways to characterize nonnormal distributions are necessary.

It can be argued that it should be possible to find a distribution (Weibull, skewed normal, etc.) that fits the test data adequately; however, the original JMEM effectiveness methods were based on normal accuracy data and are therefore unable to accept anything else. Rather than develop effectiveness methodologies to accept different distributions, a formulation was made that was nonnormal in nature, but allowed the original effectiveness methods based on assumed normal distributions to be used.

Guided weapons test data have resulted in range and deflection miss distance distributions shown by the histogram in Fig. 4.29 and, as confirmed by hypothesis testing, is clearly nonnormal. This distribution has the following features:

- A number of gross errors
- A portion of impacts that appear to have a normal distribution
- A number of direct hits on the target in excess of the previous item

Guided weapons may exhibit gross errors characterized by impacts a considerable distance from the target. Such errors are usually due to some catastrophic failure of the guidance system, such as loss of seeker lock at firing for missiles, disruption of the laser guidance beam for LGBs, or failure of one of the control surfaces. Gross errors are sometimes referred to as an in-flight reliability issue.

[‡]Only for test data do predictive models always produce normal distributions.

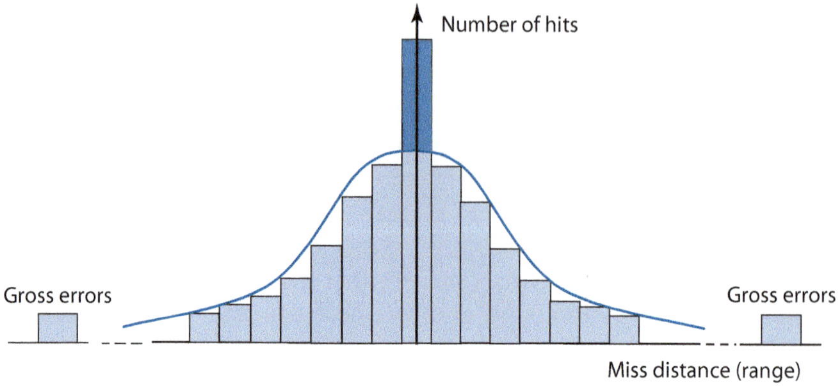

Fig. 4.29 Non-Gaussian distribution of guided weapons.

In addition, there is a disproportionate number of direct hits (i.e., weapons that impact within a histogram bin that is relatively small, centered on the aimpoint). This is a good thing; after all, this is why guided weapons are expensive—they hit the target more often.

Considering the data represented in Figure 4.29, we begin our analysis by first removing the gross errors. A gross error is defined by calculating the standard deviation σ of the whole data set and defining impacts outside $\pm 4\sigma$ to be gross errors. Figure 2.12 indicates that the probability any impacts outside these boundaries are part of the Gaussian distribution is negligible. These are therefore not deleted, just removed from the distribution.

Next we attempt to determine how non-Gaussian the remaining data are by considering the remaining data to be Gaussian plus an excess of direct hits. We could treat direct hits as having a miss distance of zero in both range and deflection, and calculate the mean and standard deviation for all the data. Performing a chi-squared test would then reveal how close our miss distances were to an assumed Gaussian distribution, and would result in a relatively poor result (i.e., not close to Gaussian). This is because of the large number of hits in the bin containing the target.

If we now remove a few of the direct hit data points and repeat the test, the hypothesis test results would probably improve because the peak in the histogram is reduced and the remaining data become more normally distributed. The question is: How many direct hits should be removed to obtain the best fit to a normal distribution for the remaining data?

As we remove more of the "excess" direct hits, the closeness to a Gaussian distribution would improve up to a point corresponding to removing the shaded part of the peak, but removing more direct hits will then degrade the chi-square test results because what was a peak now becomes a trough. This is illustrated by Fig. 4.30, which shows how the distribution varies with increasing numbers of direct hits removed.

By removing a particular number of direct hits N_{HIT}, the remaining data get as close to Gaussian as they can. These remaining data may now have a revised mean and standard deviation computed. Any data points outside the new $\pm 4\sigma$ limit will be defined as additional gross errors (see Fig. 4.29), removed from the data set, and a new standard deviation computed.

In theory, this becomes an iterative process because each time more gross errors are removed, the remaining data set requires recalculation of the standard deviation; however, after a few iterations, no more gross errors have to be removed. From this final standard deviation, REP and DEP are calculated in the usual manner to describe the subset of the data that is as close to normal as we can get it.

We now consider three quantities expressing the proportion of impacts falling into three groups: gross errors; direct hits; and those remaining in the normal distribution, termed near misses. If N_{GE} is the number of gross errors removed from the original sample, N_{HIT} the number of direct hits that had to be removed from the center bin to form the "best" normal distribution for the remaining data, and N_{NM} is the number of impact points remaining, the following ratios are defined as:

$$P_{GE} = \frac{N_{GE}}{N} \tag{4.51}$$

$$P_{HIT} = \frac{N_{HIT}}{N} \tag{4.52}$$

$$P_{NM} = \frac{N_{NM}}{N} \tag{4.53}$$

Because

$$N_{NM} + N_{HIT} + N_{GE} = N \tag{4.54}$$

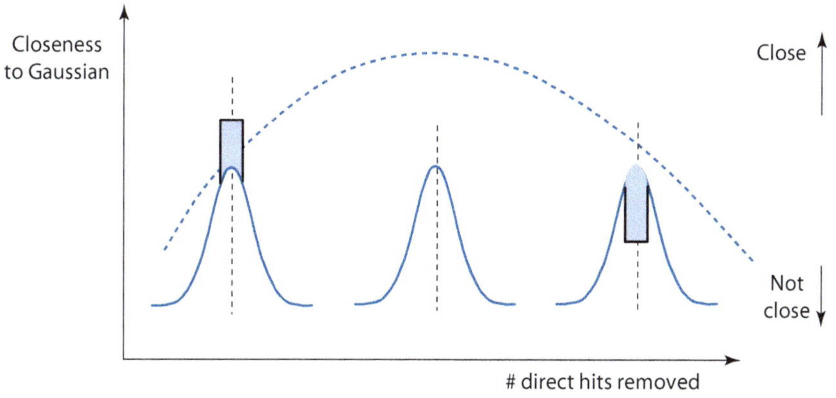

Fig. 4.30 Effect on distribution on removing direct hits.

then

$$P_{NM} + P_{HIT} + P_{GE} = 1 \tag{4.55}$$

For example, suppose we have 100 impacts, of which 5 are gross errors. In order to make the resulting histogram (after the gross errors have been removed) as close as possible to normal, we have to remove 25 direct hits on the target. We calculate the following:

$$\begin{aligned} P_{GE} &= \frac{5}{100} = 0.05 \\ P_{HIT} &= \frac{25}{100} = 0.25 \\ P_{NM} &= \frac{100 - 3 = 25 - 5}{100} = 0.70 \end{aligned} \tag{4.56}$$

We now calculate REP, DEP, or perhaps the CEP of the 70 data points making up the normally distributed part of the data set. The resulting data would be reported in terms of four parameters (REP, DEP, P_{HIT}, and P_{NM}) or in terms of three parameters (CEP, P_{HIT}, and P_{NM}). A typical specification might be

$$P_{HIT} = 0.25, \quad P_{NM} = 0.70, \quad \text{CEP} = 25\,\text{ft} \tag{4.57}$$

How these data are used to calculate effectiveness will be dealt with later. First we will consider the following specifications for a particular weapon:

$$\text{CEP} = 15\,\text{ft}, \quad P_{HIT} = 0.05, \quad P_{NM} = 0.95$$

This would indicate that the weapon is better than 95% Gaussian, and the CEP value can be used as it would for an unguided weapon. If, however, we have

$$\text{CEP} = 15\,\text{ft}, \quad P_{HIT} = 0.95, \quad P_{NM} = 0.05$$

it would indicate that the weapon has a greater than 95% probability of hitting the target and to a first approximation will not miss. The value of CEP is therefore of little consequence, because it applies to less than 5% of the weapon data. It is a common mistake to regard the quoted CEP as the weapon CEP; it is not. It is the CEP of the normal portion of the accuracy distribution only. The overall weapon CEP will be calculated in a moment.

Although Fig. 4.30 illustrates the principle of finding N_{HIT}, it is obviously a tedious process to implement, and a more efficient approach is used. If we assume a finite bin width for the histogram shown in Fig. 4.29, we can predict the number of excess hits in the center bin. For example, if each bin is $0.5 \times \sigma$ wide, the probability a sample lies in that bin can be obtained from Eq. (4.18) by setting $u = 0.5$, resulting in $p_u = 0.1175$. For an exact Gaussian distribution with the gross errors removed, the center bin would be expected to contain $0.1175(N - N_{GE})$ impact points. We therefore obtain the following expression for the excess direct hits:

$$N_{HIT} = (N - N_{GE}) - 0.1175 \times (N - N_{GE}) = 0.8825(N - N_{GE}) \quad (4.58)$$

Of course, removing these excess hits will affect the value of σ, so an iterative process is used as follows.

1. Calculate the σ for the whole data set $(N - N_{GE})$.
2. Check to see if there are any excess direct hits by using Eq. (4.58).
3. If there are, remove the smallest miss distance and repeat steps 1–3 until the center bin contains 11.75% of the remaining data set.
4. The total number of impacts removed is N_{HIT}.

Using the P_{HIT}/P_{NM} approach, and for reasons of simplicity, it is possible to represent all weapons using a common set of data using Eq. (4.57), but including for completeness any fixed bias.

- Fixed bias X
- Fixed bias Y
- P_{HIT} and P_{NM}
- (REP and DEP) or CEP

Note the following:

- If the weapon miss distances are normal in range and direction and there are no gross errors (typically unguided weapons), then $P_{HIT} = 0$, $P_{NM} = 1$, and REP and DEP are preferred to CEP. If only CEP is available, we assume circular normal so that REP = DEP.
- If the weapon miss distances are not normal or the radial miss distances are not Rayleigh distributed (typically guided weapons), use P_{HIT} and P_{NM} values provided, and again REP and DEP are preferred to CEP.

4.15 Processing Test Data Using P_{HIT}/P_{NM} Methodology

The analysis of the previous section works well for a univariate distribution, but in general, impact data are provided in range and deflection directions, usually in the ground plane. The method has to be modified for analysis of bivariate data, because a single impact point could, for example, be considered a direct hit in range, but a gross error in deflection.

The modified method uses the radial miss distance as a metric for both gross errors and direct hits, although once completed, range and deflection measures may be calculated separately. The process is as follows:

1. Convert ground plane impact location relative to the desired mean point of impact (MPI) to range and deflection miss distances in the normal plane. Also compute the radial miss distance by taking the square root of the sum of the squares of the miss distances in each direction.

2. Analysis will be centered on MPI.
3. Determine standard deviations in both range and deflection.
4. Rank order the radial miss distance data in increasing order. Iteratively discard impact points with radial miss distances in the normal plane outside 4σ from the MPI, where

$$\sigma = \frac{\sigma_r + \sigma_d}{2} \tag{4.59}$$

5. Recompute the standard deviations after each point is removed. Continue removing points outside the 4σ interval until none remains. The number of discarded impact points N_{GE} is retained for calculating gross error. The remaining impact points make up the sample $N_{\mathrm{HIT}} + N_{\mathrm{NM}}$.
6. Calculate the MPI in range and deflection. This will give the fixed bias, if present or significant.
7. If necessary, remove impact points near the aimpoint. The removed impact points are considered direct hits N_{HIT} and will be used to calculate the P_{HIT} value. The process for removing impact points near the aimpoint is as follows: Set $N_{\mathrm{HIT}} = 0$ and N_{NM} equal to the total number impact points less N_{GE}.
   a) Let σ_x and σ_y be the standard deviations in range and deflection (normal plane), respectively, and then

$$\sigma = \frac{\sigma_r + \sigma_d}{2} \tag{4.60}$$

   b) For a bivariate normal (Rayleigh) distribution, the expected number of hit points that are within $1/2$ standard deviation from the DMPI is 0.1175 times the number of points remaining. Compare this to the actual number of impact points that are less than $\sigma/2$ from the MPI to the expected number.
8. If the actual number of impact points exceeds the expected number of impact points in the band, iteratively remove the impact point with the smallest miss distance, update N_{HIT} and N_{NM}, and recalculate standard deviation and Rayleigh distribution statistics.
9. Continue until the actual number of points with miss distances less than $\sigma/2$ from the MPI is less than or equal to the expected number.
10. The statistics of the Gaussian portion of the data may be calculated in the usual way to provide σ_x, σ_y, REP, DEP, or CEP. These would be reported along with P_{HIT}, P_{NM}, and any fixed bias values.

One further point should be noted regarding accuracy specifications for guided weapons: the definition of the range (and hence deflection) direction. In Fig. 4.13, for example, it might be assumed that if the impacts are due to artillery fire, the shooter is towards the bottom of the figure firing up, or in the case of the air-launched weapon, the delivery aircraft is flying in the

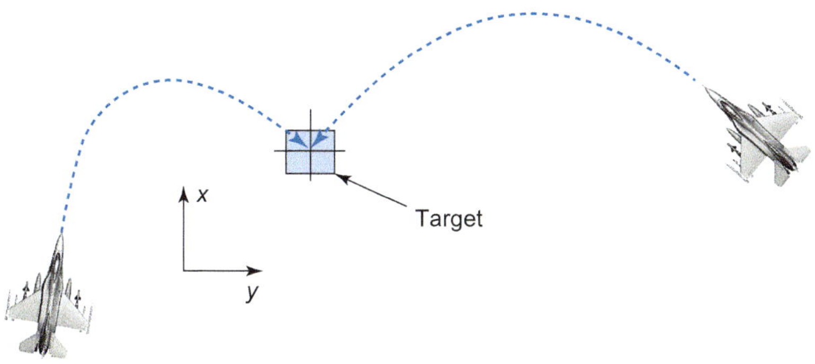

Fig. 4.31 Two releases of a guided weapon.

indicated range direction when it drops the weapon. Although this is a good assumption for unguided weapons, guided weapons may impact the target from a variety of directions. Consider the situation shown in Fig. 4.31 where two separate releases of a guided weapon are shown, as is the north/east coordinate frame (x, y). The question is: Where is the range direction?

The answer is that the range direction is in the direction of the horizontal component of the weapon velocity vector at impact. It is not in the direction of a ground fixed coordinate frame (x, y), nor is it in the direction the aircraft is heading when it releases the weapon. It will, in fact, be different for each weapon released. For unguided weapons, the trajectory lies approximately in a vertical plane containing the aircraft velocity vector and the target; hence the range direction is the aircraft heading at release.

However, guided weapons may move considerable distances outside this vertical plane; hence the range direction and direction of the aircraft at release may differ considerably. This has implications for guided weapon testing because any data collected for range and deflection miss distances must be obtained by taking into account the direction the weapon was heading at impact. It also has implications for collateral damage estimates.

In concluding this section on nonnormal accuracy distributions, it should be stated that the method outlined, commonly called the $P_{\mathrm{HIT}}/P_{\mathrm{NM}}$ *methodology*, represents the legacy JMEM approach developed several years ago. Since then, different approaches have been developed along with automated methods to process accuracy test data that avoid the repetitive, manual approach just described. More information about the methods is found in Volume 2, Chapter 3.

4.16 CEP for Nonnormal Miss Distance Distributions

Given that for some guided weapons, the specification of accuracy requires P_{HIT}, P_{NM}, REP, DEP, or CEP, some care has to be exercised in the interpretation of CEP. It is commonly understood that the CEP is the radius of a circle that will contain 50% of the impacts, or alternatively is the distance to the target within which a single weapon has a 50% chance of landing. This is no longer true of the quoted CEP for nonnormal distributions.

Consider the previous case of 100 samples with the accuracy defined by Eq. (4.57), and ask the question: How many weapons land inside a circle of radius equal to the CEP, 25 ft? Because the CEP is based on the portion of impacts that are normally distributed (70), then 35 of these impacts land inside the CEP, but so will the 25 direct hits that had to be removed. Therefore, a total of 60 weapons will land inside the CEP, not the 50 expected.

Assuming the radial miss distances are Rayleigh distributed, the probability that an impact r lies within a circle of radius R is given by the following:

$$P(r < R) = 1 - \exp\left(\frac{-0.693R^2}{\text{CEP}^2}\right) \tag{4.61}$$

where the CEP is given in terms of the (x, y) miss distance standard deviation by

$$\text{CEP} = 1.1774\sigma \tag{4.62}$$

Given that the Rayleigh distribution addresses only the normally distributed portion of the weapon impacts, the total number of weapons impacting inside a circle of radius R is given by N_R, where

$$N_R = N_{\text{HIT}} + N_{\text{NM}}\left[1 - \exp\left(\frac{-0.693R^2}{\text{CEP}^2}\right)\right] \tag{4.63}$$

Assume for now that $N_{\text{GE}} = 0$, then $N_{\text{HIT}} = N - N_{\text{NM}}$, and we may write

$$\begin{aligned} N_R &= N - N_{\text{NM}} + N_{\text{NM}}\left[1 - \exp\left(\frac{-0.693R^2}{\text{CEP}^2}\right)\right] \\ &= N - N_{\text{NM}}\exp\left(\frac{-0.693R^2}{\text{CEP}^2}\right) \end{aligned} \tag{4.64}$$

Dividing by the total number of impacts gives the proportion of impacts contained within a circle of radius R as follows:

$$P_R = 1 - P_{\text{NM}}\exp(-0.693\,a^2) \tag{4.65}$$

where the parameter a is the ratio R/CEP. The proportion of impacts (probability of a single weapon) landing inside the scaled radius R as a function of P_{NM} is shown in Fig. 4.32, assuming P_{GE} is zero.

Equation (4.65) is commonly used in two ways.

1. The overall weapon CEP can be found by setting $P_R = 0.5$ and calculating R; for example, if $P_{\text{GE}} = 0$ and $P_{\text{NM}} = 0.6$, CEP $= 20$ ft, and then from the

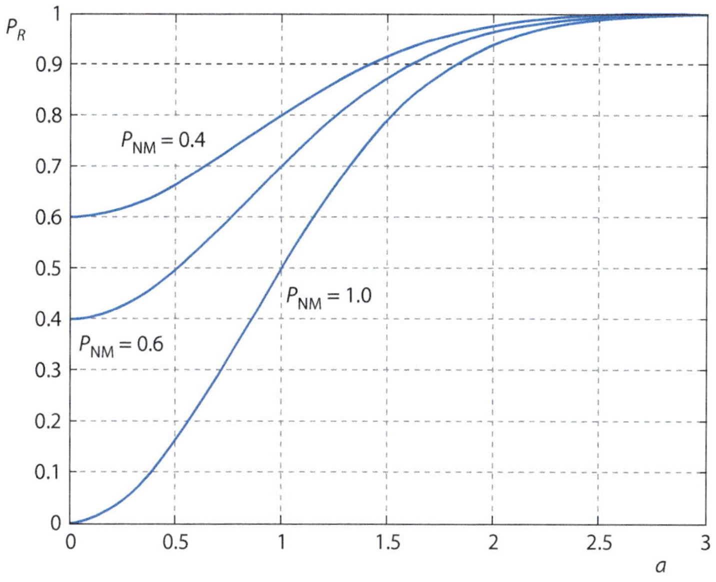

Fig. 4.32 Proportion of impacts inside a circle of given radius.

chart in Fig. 4.32, we can see that 50% of the impacts land inside a circle of radius where $a = 0.5$ (i.e., a radius of 10 ft). Hence, the CEP of the normal part of the distribution is 20 ft, but the overall weapon CEP is 10 ft due to the additional direct impacts.

2. For collateral damage studies, it is often useful to calculate a CE_{90} (i.e., the radius of a circle containing 90% of the impacts). For the same case as in item 1, using $P_R = 0.9$, the chart gives $a \approx 1.6$, so $R = 32$ ft. Therefore, there is a 10% chance a single bomb will land more than this distance from the target.

Note that the simple scaling between CEP and CEP_{90} defined by Eq. (4.17) does not apply for non-Gaussian weapons unless $P_{NM} = 1$. In Eq. (4.65), by putting $P_R = 0.9$ and realizing that R then becomes CEP_{90}, the equation may be put in the form

$$\frac{CE_{90}}{CE_{50}} = \sqrt{1.443 \log_e \frac{P_{NM}}{0.1}} \tag{4.66}$$

Figure 4.33 shows the scale factor between CE_{90} and CE_{50} as a function of P_{NM}, approaching the value given in Eq. (4.17) when the distribution becomes normal. These results will be modified somewhat when P_{GE} is not zero.

Interpreting data that include a nonzero P_{HIT}, such as those found in the JWS database, requires understanding of the P_{HIT}/P_{NM} methodology; otherwise, incorrect interpretation of the CEP will result. As indicated previously,

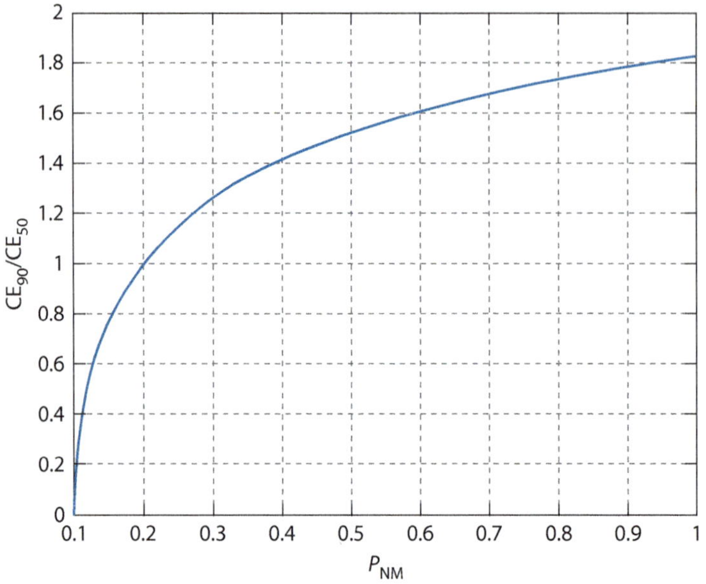

Fig. 4.33 Scaling between CE_{90} and CE_{50} for nonnormal distributions.

weapons that have a true Gaussian miss distance distribution may be represented in the same format by setting $P_{NM} = 1$.

4.17 Generating Random Miss Distances for P_{HIT}/P_{NM} Distributions

If an effectiveness methodology uses a Monte Carlo approach, and one of the randomized variables is the impact point, samples for the impact point are simply the result of a Gaussian random number generator. The question arises, however, as to how we might generate such data if the miss distance distribution is not normal, but characterized by the P_{HIT}/P_{NM} methodology. Suppose we have P_{HIT}, P_{NM}, P_{GE}, μ (usually zero), and σ. One way to generate random numbers for a non-Gaussian distribution would be as follows:

1. Convert the REP, DEP, or CEP into an equivalent σ.
2. Draw a normally distributed random number z with zero mean and unity variance.
3. Draw a random number z' from a uniform distribution in the range $0 < z' < 1$.
4. By representing Eq. (4.55) as shown in Fig. 4.34, the miss distance depends on z' and is computed as follows:

```
If z' < P_GE, then x = μ + 4σ + 1              ------ Gross error
   Else if z' > P_NM + P_GE, then x = 0        ------ Direct hit
      Else x = μ + σz                          ------ Gaussian miss
```

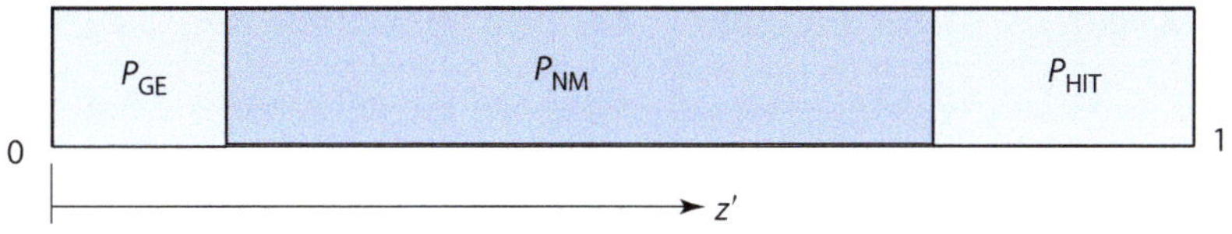

Fig. 4.34 Linear representation of P_{HIT}, P_{NM}, and P_{GE}.

4.18 Overview of Weapon System Accuracies

Most weapon systems currently in use may be categorized as follows:

- Air launched—unguided
- Air launched—guided
- Surface launched—unguided
- Surface launched—guided

Analysis of unguided weapons becomes somewhat complicated, so they are deferred to later chapters; however, guided weapons can be handled using methods described in this chapter. For completeness and uniformity, however, the general approach to representing weapons in each category will be discussed.

4.19 Accuracy of Air-Launched, Unguided Weapons

The delivery accuracy of air-launched, unguided weapons depends mainly on the aircraft releasing them and may be considered a form of direct fire, because the pilot attempts to place an aiming cue on the target before releasing the weapon. Clearly the results of field data might generate REP, DEP, and CEP values, but these relate only to the release or firing conditions of the test, and do not provide estimates for different parameters.

As discussed earlier, the general approach for air-launched unguided weapons is to develop a predictive methodology for generating REP, DEP, and CEP values, and then to use what limited test data are available to verify the method for specific release conditions. If good agreement is found, the predictive method may be used with confidence for a wide range of release conditions. The basis of this predictive method is deferred until Section 4.23.

It should be noted that this predictive model determines the aiming error only, that is, the dependent error that would be, for example, the common miss distance of a stick of unguided bombs. The independent error—that is, the unique miss distance of each bomb from its intended place in the stick—is obtained from test data or SME recommendations.

4.20 Accuracy of Air-Launched, Guided Weapons

The accuracy with which guided weapons are delivered is a function of several factors including the weapon guidance system and estimation of the target location, and is usually considered independent of the aircraft delivering the weapon. Guided weapon accuracies are not subject to the same mechanism that produces a precision error, although there is an equivalent quantity called the guidance and control error.

The accuracy of multiple guided weapon releases is not usually addressed in the manufacturer's specification; any quoted accuracy value is assumed to be only the independent error, that is, the variability observed between multiple independent firings of a single round. For some guided weapon systems, however, it is possible to identify both a dependent and independent component.

- *Laser-guided weapons:* In this case, any error between the DMPI and the laser spot would affect all weapons released by the same amount, as shown in Fig. 4.35. In this case, the dependent error would be the laser designation error, and the independent error would be the G&C weapon error.
- *GPS-guided weapon:* A similar scenario occurs when inaccurate coordinates are supplied to several GPS-guided weapons, which then fly precisely to the wrong coordinates. In this case, the dependent error would be the target location error (TLE), and the independent component is again the weapon G&C error.

In both cases, the dependent error cannot be attributed to the weapon itself, although operational users are more interested in the total accuracy to be expected of the weapon.

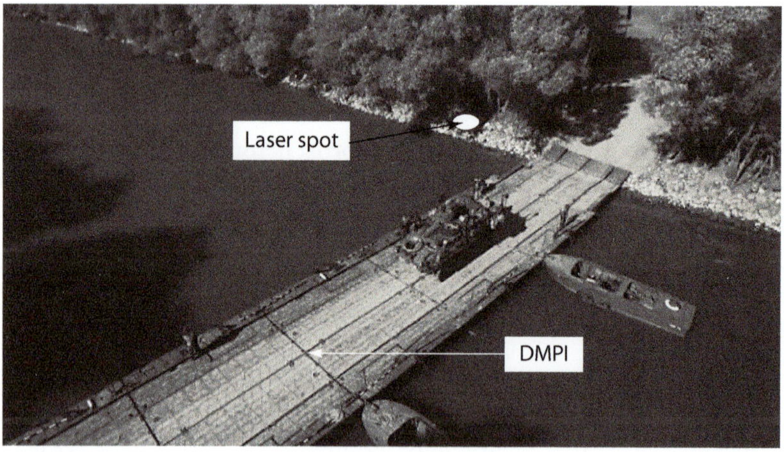

Fig. 4.35 Laser designation of target.

4.21 Accuracy of Surface-Launched, Unguided Weapons

For these weapon systems, a distinction has to be made between direct and indirect fire systems when discussing accuracy.

- *Direct fire* occurs when the shooter places an aiming cue on the target before firing. Examples include a tank main gun and smaller caliber burst fire weapons from both aircraft and ground platforms. Because the effectiveness calculations for such systems usually take place in the normal plane, REP and DEP for dependent and independent errors in this plane are usually required.

  In the case of gun-type systems, if they are relatively small caliber and short range, they are usually inexpensive to physically test to obtain accuracy data. It is not too difficult to devise trials to determine MPI and precision errors; however, the previously discussed issue must be kept in mind—"system" accuracy will include the mount and rigidity of the firing platform. This adds to the complexity of the test and proliferation of data for what initially seems like a simple system. In most cases, therefore, we get the accuracies $\sigma_{x-\mathrm{MPI}}$, $\sigma_{y-\mathrm{MPI}}$, $\sigma_{x-\mathrm{PREC}}$, and $\sigma_{y-\mathrm{PREC}}$ from the test range.
- *Indirect fire* occurs when the target is not visible to the firing platform, but both shooter and target coordinates are available. Examples are ground artillery cannons, mortars, and unguided rockets. For the case of unguided artillery projectiles, an accurate ballistics model is used to produce firing orders (i.e., direction and elevation of the gun barrel). Predictive models are used to estimate the accuracy of these systems and will be dealt with in Volume 2, Chapter 5. Usually accuracies are specified in the ground plane, and again, dependent and independent error values of $\sigma_{x-\mathrm{MPI}}$, $\sigma_{y-\mathrm{MPI}}$, $\sigma_{x-\mathrm{PREC}}$, and $\sigma_{y-\mathrm{PREC}}$ are predicted from this model.

It will be shown in Volume 2, Chapter 5 that because these predictive methods are somewhat complex and could not produce results in a few seconds—a warfighter requirement—the program was run offline and the results presented in tabular form. An example is shown in Table 4.2.

Clearly we are faced with the generation of large quantities of tabular data covering all possible combinations of munition, weapon, range, and charge, and a fast-running predictive capability is required.

4.22 Accuracy of Surface-Launched, Guided Weapons

Surface-launched, guided weapons are fewer in number than their air-launched counterparts, and because many of them are GPS-guided, the discussion on their accuracy will be deferred until a later chapter. Again surface-to-surface guided weapons are divided into direct and indirect fire types.

Table 4.2 Sample Delivery Accuracy Data

				Probable Error (50%)					
				Precision Error (m)		MPI Error (m)			
Impact Conditions						Observer Adjusted		Nonadjusted	
Range	Charge	Angle	Velocity	Range	Defl.	Range	Defl.	Range	Defl.
3000	4 (M67)	13	246	13.5	1.3	22.3	7.4	29.7	10.1
5000	5 (M67)	19	263	12.1	2.7	21.6	7.4	44.5	12.8
7000	6 (M67)	26	269	11.5	1.3	21.6	6.7	70.1	20.2
9000	7 (M67)	31	278	15.5	2.0	22.9	7.4	80.3	33.7
11,000	7 (M67)	44	283	18.2	3.4	25.0	8.1	108.6	51.9
13,500	8 (M200)	49	300	35.0	8.0	40.0	12.0	131.0	73.0

Fig. 4.36 TOW guided missile.

Accuracy of guided, direct fire weapons is specified in terms of the MPI (aiming error) in the manner described earlier in this section. Examples of direct fire guided weapons include:

- Tube-launched, optically tracked, wire-guided (TOW) missile (Fig. 4.36) in which the shooter maintains the DPI on the target with a tracker while the weapon flies out from the launch point.
- Javelin (Fig. 4.37) is an anti-tank weapon with which the shooter places the DPI on the target, and this is passed to the missile before launch. Once fired, the missile uses an infrared tracker to guide to the target without the further assistance of the shooter.

Fig. 4.37 Javelin guided missile.

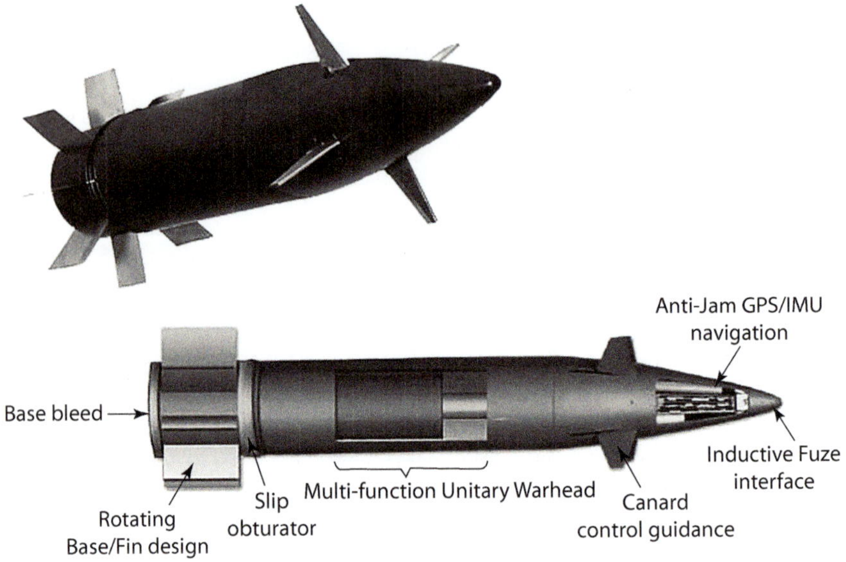

Fig. 4.38 Excalibur GPS/INS 155-mm artillery shell.

Indirect fire guided weapon accuracy is dealt with in a similar way where the direct fire aiming error is replaced by the indirect fire target location error. Examples of such weapons are:

- The M982 Excalibur 155-mm artillery round (Fig. 4.38) is a GPS/INS high explosive round with a bomblet-type warhead used against armored vehicles, hard targets, or personnel.
- The M270 guided multiple launch rocket system (GMLRS) uses a GPS/INS guidance system.
- The M39 Army tactical missile system (ATACMS) is a GPS/INS guided rocket capable of being launched from the multiple launch rocket system (MLRS) platform.

The method of describing the accuracy of surface-launched guided weapons is the same as for air-launched, where the spec value may be regarded as the independent error, and the dependent error is a function of the guidance system (e.g., laser or GPS).

4.23 General Approach to Accuracy Predictions: Error Budgets

This technique is at the heart of most methods developed to predict accuracy and is used extensively for guided and unguided weapons for both air and surface deliveries. The basic assumption is that the total error of the impact points relative to the DPI is due to contributions from several independent error sources. This being the case, and expressing the individual error variability in terms of its variance, the total error variance is the combination (RSS) of the variances of these sources.

Consider a generalized variable ε that affects the impact point x of a weapon. If there is an error in the assumed value of ε, what is the corresponding error (miss distance) in x? We may write

$$\frac{\partial x}{\partial \varepsilon} = \frac{\Delta x}{\Delta \varepsilon} \tag{4.67}$$

which may be written as

$$\Delta x = \frac{\partial x}{\partial \varepsilon} \Delta \varepsilon \tag{4.68}$$

In this equation, the partial derivative is the sensitivity of impact point x to a variability in the parameter ε, a concept identified previously in Section 3.13 as a ballistic partial or unit effect. If we can calculate the unit effect and we know the error in ε, the corresponding error in x can be calculated. It will now be shown how this is applied to a particular case and generalized to all identical weapon systems rather than a single example.

Example 4.1

Consider the accuracy of a cannon firing at a downrange target, as shown in Fig. 4.39. If we know the range to the target and we know the muzzle velocity of the projectile (for a particular gun, charge, and projectile selection) we can calculate the elevation of the gun tube needed to hit the target precisely from a trajectory program. Our simple ballistics program tells us that in order to land a shell on target we need to know the muzzle velocity and elevation angle of the gun tube.

However, there will be variations in each of the parameters associated with the mission for a particular gun and projectile. For example, if the gun we use has considerable wear to the bore, the muzzle velocity will be less than that for

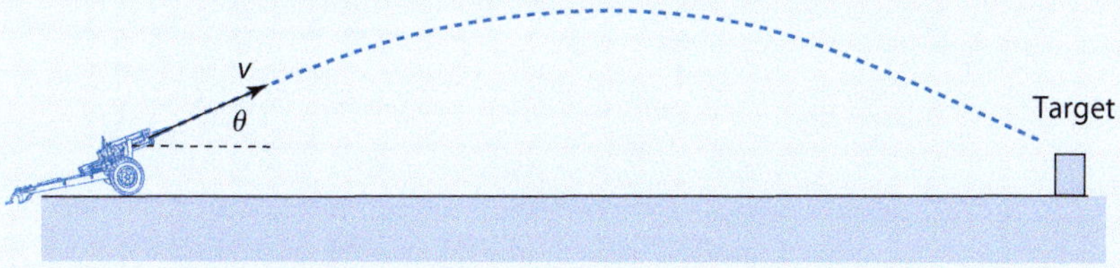

Fig. 4.39 Nominal trajectory for artillery shot.

(Continued)

Example 4.1 *(Continued)*

a new tube. If the charge we use has a few grams of propellant missing, the muzzle velocity will again be less than expected. Also, each gun elevation mechanism will have an accuracy associated with it, so that if the mission calls for an elevation of 30 deg, perhaps the gun is actually elevated 29 deg. The question is, how does an error in muzzle velocity and/or an error in gun elevation angle affect accuracy?

Consider the muzzle velocity first. Suppose that to hit the target at a range of 10 km, and for a particular charge such that the resultant muzzle velocity is 1000 m/s, a quadrant elevation (QE) of 600 mils is required. This is called the nominal firing solution. If, however, our particular gun produces a muzzle velocity of 980 m/s, the shell will fall short, but by how much? We can apply Eq. (4.68) to this case to give

$$\Delta x_V = \frac{\partial x}{\partial V}\Delta V \tag{4.69}$$

The left hand side of this equation is the difference in where the shell lands relative to the target due to a difference in muzzle velocity V, and ΔV is the actual difference in V for this particular gun (i.e., 2 m/s). How can we calculate the unit effect? Starting with the nominal trajectory, we keep the QE at 500 mils but perturb the muzzle velocity by 1 m/s and run the trajectory program again. Suppose the range corresponding to $V = 1001$ m/s is 10,050 m, then the unit effect becomes

$$\frac{\partial x}{\partial V} = \frac{\Delta x}{\Delta V} = \frac{10{,}050 - 10{,}000}{501 - 500} = 50 \tag{4.70}$$

In words, the sensitivity of impact point (range error) is 50 m for each m/s error in V. For our particular cannon, therefore, having a muzzle velocity error of -2 m/s, our shells will always fall 100 m short of the target. Unfortunately, it may not be possible to know the muzzle velocity error for each gun because it will be different for each. Assuming this error is a random variable, it might be possible to find the statistics of this variability through testing. The mean should be the nominal value of 500 m/s and the sigma σ_V might be, say, 5 m/s. We can generalize Eq. (4.69) from a single gun to the population of guns by writing

$$\sigma_{XV} = \frac{\partial x}{\partial V}\sigma_V \tag{4.71}$$

In this equation, σ_{XV} is the standard deviation of miss distance in the x direction due to variability in muzzle velocity. Given the data presented, we would get $\sigma_{XV} = 50 \times 5 = 250$ m as the standard deviation of miss distance in range, from which REP could be calculated.

(Continued)

Example 4.1 *(Continued)*

Now assume the muzzle velocity is fixed but there is an error in elevating the gun barrel to the required QE. Following the same arguments, we will get for a particular cannon

$$\Delta x_{\text{QE}} = \frac{\partial x}{\partial \text{QE}} \Delta \text{QE} \tag{4.72}$$

The ballistic partial would be calculated as before by subtracting the nominal range from that obtained by perturbing the QE to perhaps 601 mils and finding the resulting range is 10,025 m. The unit effect is

$$\frac{\partial x}{\partial \text{QE}} = \frac{\Delta x}{\Delta \text{QE}} = \frac{10{,}025 - 10{,}000}{601 - 600} = 25 \tag{4.73}$$

that is, 25 m in range for each mil in QE. Moving from considering a single cannon to the population of cannons, we can write

$$\sigma_{X\text{QE}} = \frac{\partial x}{\partial \text{QE}} \sigma_{\text{QE}} \tag{4.74}$$

As before, the σ_{QE} would be obtained from testing. Because the two error sources can be considered independent of each other, the total error due to errors in both muzzle velocity and QE can be obtained by the RSS of the individual errors.

$$\sigma_X^2 = \left[\frac{\partial x}{\partial V} \sigma_V\right]^2 + \left[\frac{\partial x}{\partial \theta} \sigma_\theta\right]^2 \tag{4.75}$$

Here we are simply root sum squaring the standard deviations of the independent error sources. This concept may be extended to include as many contributors to the overall miss distance as can be modeled. This approach is called an *error budget* because the aggregated miss distance is the root sum squared (RSS) of the miss distances produced by each independent error source. It should be noted that the ballistic partial is not fixed and will be different for each firing condition V and QE.

4.24 Treatment of Vertical Errors

Vertical errors can be addressed in two ways, depending on whether the target being attacked is horizontal or vertical. Horizontal targets are usually considered to be materiel, personnel, or infrastructure targets (buildings, bunkers, bridges, etc.), which can be considered essentially planar and located on the ground plane. Some targets, however, require the weapon to detonate while in almost horizontal flight. Such targets include tunnel openings, caves, or windows on the side of a building. Munitions that dispense bomblets sometimes function while in horizontal flight also.

Attacking vertical targets is usually accomplished with guided weapons because it is difficult to control the trajectory of unguided weapons near the impact point. For example, a coordinate seeking weapon (GPS) will attempt to fly to a point defined by a fixed height above a defined point on the ground. The uncertainty with which the weapon control system can achieve this height is one contributor to the total vertical error. Another contributor might be the uncertainty of knowing the elevation of the target being attacked.

Some weapons, such as cruise missiles, are designed to fly horizontally a fixed distance above the Earth's surface, and simply run into the target while in horizontal flight. This reduces the complexity of the guidance system from three dimensions to two. Another example would be a sea-skimming antiship guided missile.

For horizontal targets, the objective is to determine the accuracy of munitions in terms of REP and DEP in the ground plane. The vertical component of error is treated by resolving into an equivalent horizontal plane error through the trajectory of the weapon, as shown in Fig. 4.40.

In this figure, the aggregated height error can be represented by the HEP and has to be known for the weapon in question. The aggregated corresponding equivalent range error can be represented by REP′. Although a trajectory program should be used to translate the vertical error into a horizontal error, it is usually assumed that the distances involved are small, so the terminal trajectory is approximately a straight line. This allows the quantities to be related through the impact angle as follows:

$$\text{REP}' = \frac{\text{HEP}}{\tan I} \tag{4.76}$$

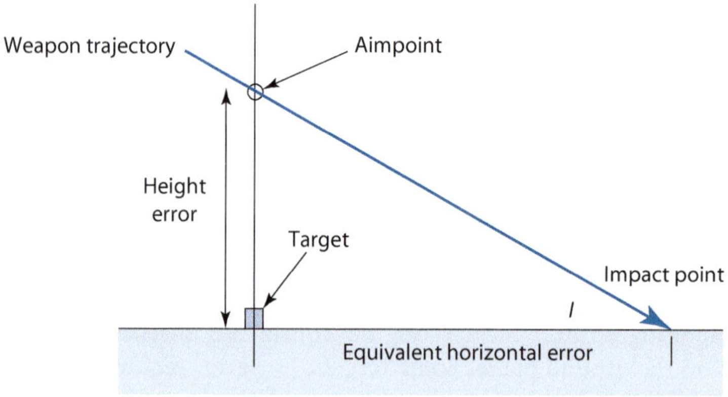

Fig. 4.40 Resolving of elevation error to horizontal error.

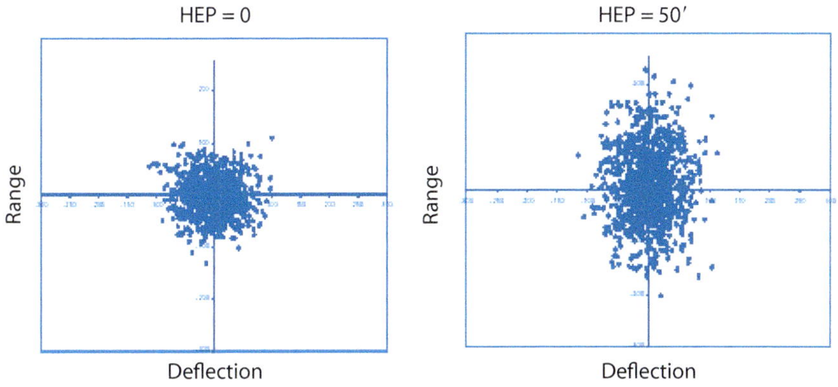

Fig. 4.41 Effect of height error on ground plane impacts.

The HEP produces a resolved error in the ground plane in range direction—only REP′, which may then be RSS with the existing REP value.

$$(\text{REP}_{\text{TOTAL}})^2 = (\text{REP}')^2 + (\text{REP})^2 \tag{4.77}$$

Figure 4.41 shows the effect of 1000 impacts of a weapon, first with zero elevation error and then with a 50-ft elevation error (sigma) on the impact points on the ground. Clearly these height errors are magnified for small impact angles and are mitigated for high impact angles. It is usual to plan for guided weapons to impact at high angles to reduce this effect as much as possible.

Note that the errors in the deflection direction are unaffected by impact angle; therefore, we may write

$$\text{DEP}_{\text{TOTAL}} = \text{DEP} \tag{4.78}$$

If required, a CEP may be calculated by converting $\text{REP}_{\text{TOTAL}}$ and $\text{DEP}_{\text{TOTAL}}$ to standard deviations and then using Pitman's equation.

$$\text{CEP} = 0.562 \times \text{MAX}(\sigma_x, \sigma_y) + 0.617 \times \text{MIN}(\sigma_x, \sigma_y) \tag{4.79}$$

For weapons that are directed against vertical targets, or that dispense submunitions while in horizontal flight, the vertical and horizontal plane errors have to be kept separate. Consider a weapon that attempts to fly into a tunnel, as shown in Fig. 4.42.

The magnitude of the height and deflection errors (HEP, DEP) together with the opening dimensions will determine the probability the weapon enters the tunnel. Just as the height error was resolved into the ground plane and aggregated with REP to get the total ground plane errors, the ground plane range error has to be resolved into the vertical plane and

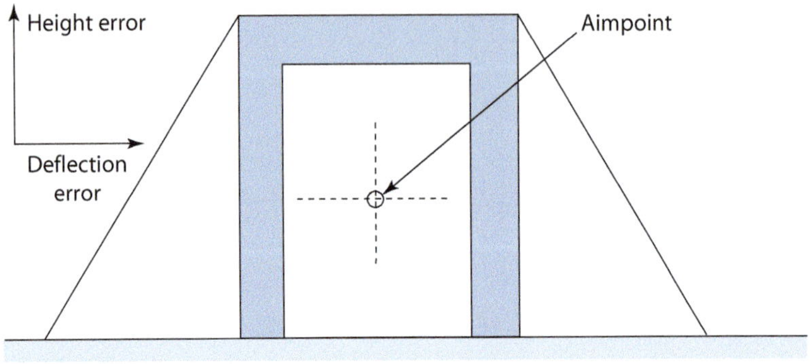

Fig. 4.42 Vertical target.

aggregated with HEP to get the total vertical plane errors. Rotating the range error using Fig. 4.40 gives

$$\text{HEP}' = \text{REP} \times \tan I \tag{4.80}$$

This is then combined (RSS) with the height error.

$$(\text{HEP}_{\text{TOTAL}})^2 = (\text{HEP}')^2 + (\text{HEP})^2 \tag{4.81}$$

Again the deflection errors are unchanged. Just as the remedy for eliminating the equivalent range error for a given elevation error was to impact the target vertically, elimination of the equivalent height error due to a given range error is achieved by impacting the target horizontally.

It will be seen in later chapters that for dispenser munitions, the height error (HEP) will produce some variability in the submunition pattern dimensions, whereas the horizontal plane errors (REP, DEP) affect the accuracy with which the pattern center is placed over the target.

4.25 Spherical Errors: SEP

For some scenarios, such as describing the miss distance of an air-to-air (AA) missile against an airborne target, the intuitive concept of a three-dimensional miss distance has to be considered. It follows that such a miss distance will be random in nature with a corresponding three-dimensional statistical distribution. This concept follows the extension of one-dimensional linear miss distance measures such as REP and DEP through two-dimensional radial miss distance metrics such as CEP to the three-dimensional miss distance characterized by the spherical error probable (SEP).

Following this concept of progression through multidimensional space, we can write for one dimension that the proportion of impacts in the zero mean interval L is given by the integral of the zero mean, univariate normal PDF over this interval.

$$P(L) = \int_{-L/2}^{+L/2} \frac{1}{\sigma_x\sqrt{2\pi}} \exp - \left[\frac{x^2}{2\sigma_x^2}\right] dx \tag{4.82}$$

For two dimensions, the proportion of impacts contained inside a circle of radius R is given by the double integral of the zero mean, uncorrelated, bivariate PDF transposed to polar coordinates.

$$P(R) = \int_{r=0}^{r=R} \int_{\theta=0}^{\theta=2\pi} \frac{1}{2\pi\sigma_x\sigma_y} \exp - \left[\frac{(r\cos\theta)^2}{2\sigma_x^2} + \frac{(r\sin\theta)^2}{2\sigma_y^2}\right] r\, dr\, d\theta \tag{4.83}$$

As was shown earlier, if certain restrictions apply, this integral may be calculated analytically and results in the Rayleigh distribution; otherwise, it has to be calculated numerically, although the set of numerical solutions may be approximated by the empirical Pittman equation. If $P(R)$ is set to 0.5, then the corresponding R is known as the CEP.

Extending this concept to three dimensions, we can write that the proportion of points within a sphere of radius R is given by the integral of the zero mean, uncorrelated, trivariate normal distribution expressed in polar coordinates

$$P(R) = \frac{1}{(2\pi)^{3/2}\sigma_x^2\sigma_y^2\sigma_z^2} \int_0^R \int_0^{\pi} \int_0^{2\pi} f(r, \phi, \theta)\, r^2 \sin\theta\, d\theta\, d\phi\, dr \tag{4.84}$$

where

$$f(r, \phi, \theta) = \exp - \left[\frac{r^2 \sin^2\phi \cos^2\theta}{2\sigma_x^2} + \frac{r^2 \sin^2\phi \sin^2\theta}{2\sigma_y^2} + \frac{r^2 \cos^2\phi}{2\sigma_z^2}\right] \tag{4.85}$$

Again, this integral has to be calculated numerically; however, Del Marco [3] has a series approximation to its solution that is shown to be accurate to within a fraction of a percent. If $P(R)$ is set to 0.5, then the corresponding value of R is defined to be the SEP.

A literature search indicates there is a simple empirical equation for the solution to the integral shown, which can be considered analogous to Pitman's equation for the two-dimensional case. This equation takes the following form:

$$\text{SEP} = 0.513\left[\sigma_x + \sigma_y + \sigma_z\right] \tag{4.86}$$

In checking his series solution against numerical integration, Del Marco has a table that we will use to compare his result, which is the same as a

Table 4.3 Comparison of SEP Values

σ_X	σ_X	σ_X	*R* (calc)	*P*(*R*)	*R* (approx)
1	3	5	4.8	0.503	4.6
1	5	8	7.7	0.503	7.2
1	6	6	7.0	0.486	6.6

numerical integration, against Eq. (4.86). Unfortunately, [3] only provides three cases close to $P(0.5)$; these are shown in Table 4.3.

In this table, the *R* (calc) value is Del Marco's series solution, and *R* (approx) is Eq. (4.86). It can be seen that the approximation to the SEP is close to the correct value, but not as close as Pitman's equation is for the two-dimensional case.

Recalling the problem of extracting the two-dimensional distribution from a single CEP value, we see that the same challenge occurs here trying to extract the three-dimensional distribution given by Eq. (4.86) should only a single value of SEP is provided. One can assume the distribution is spherically normal, resulting in a common standard deviation obtained as follows:

$$\sigma = 0.65 \times \text{SEP} \tag{4.87}$$

Reconstructing the CEP we get

$$\text{CEP} = 1.1774 \times \sigma = 0.765 \times \text{SEP} \tag{4.88}$$

Due to spherical normality, this CEP will be the same in the vertical plane, horizontal plane, and normal plane for any weapon impact angle.

References

[1] Freeman, B., "Computation of Target Location Error and Weapon Circular Error Probable," MSEE Thesis, George Mason University, Aug. 2007.

[2] Grubbs, F., "Statistical Measures of Accuracy for Riflemen and Missile Engineers," privately published, Havre de Grace, MD, 1964.

[3] Del Marco, S., "A Serial Representation of the Spherical Error Probability Integral," *IEEE Transactions in Aerospace and Electronic Systems*, Vol. 29, No. 4, Oct. 1993, pp. 1349–1355.

Chapter 5 Accuracy of GPS/INS Guided Munitions

5.1 Overview of the Global Positioning System (GPS)

The GPS is a system of 24 satellites that orbit the Earth at an altitude of approximately 12,000 miles, repeating the same path over the ground twice a day, and is operated by the U.S. Department of Defense (DOD). At each location on the Earth's surface there is a minimum of 5 and maximum of 10 satellites in view at any one time. Each satellite transmits a microwave signal that contains information including the satellite location and precise time the signal was sent. This signal is received by a GPS receiver, and the time of arrival is determined by the clock in the receiver. Because the microwave transmission travels at the speed of light c, the distance or range from the satellite to the receiver is given by

$$L = c \times \Delta t \tag{5.1}$$

Here Δt is the transmission time as determined by the difference in the two clock times, including Einstein's relativistic effects due to the motion of the satellite and receiver. The satellite orbits are precisely controlled so that at any instant the position of each satellite is known relative to a set of coordinate axes that are Earth centered (geocentric), Earth fixed (ECEF). In this frame, which rotates with the Earth, the x-axis points from the center of the Earth through the point of intersection of the equator and the prime meridian (0 deg longitude). The y-axis is at a right angle to the x-axis in the equatorial plane, and the z-axis makes up the right-handed orthogonal frame (see Fig. 5.1).

The satellite signal also contains other information, such as its identification number, its "health," and where it is in x-y-z coordinates relative to the ECEF frame. The GPS receiver takes these signals from the satellites and calculates the range to each one. This can then be used to trilaterate* the position of the receiver with respect to the satellites, hence with respect to the ECEF coordinate axes, as indicated in Fig. 5.2.

If the ranges are calculated accurately, then only three satellites are needed for the trilateration calculation, but we will see later that this case

*Trilateration means using three distances to determine a position, as opposed to triangulation, which uses three angles.

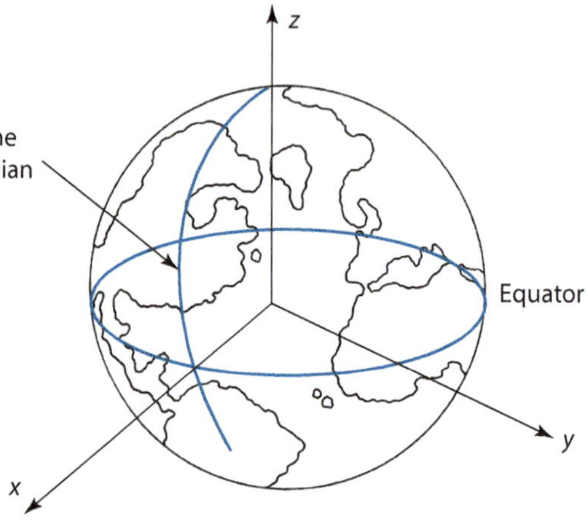

Fig. 5.1 The ECEF frame.

is never possible. It is instructive, however, to consider the error-free case first to understand how the trilateration process is implemented.

5.2 Determining Position with no Errors Present

Figure 5.3 shows the ECEF frame, three satellites, a user receiver, and ranges calculated by the receiver to the three satellites. The analysis that is presented follows that of Tsui [1].

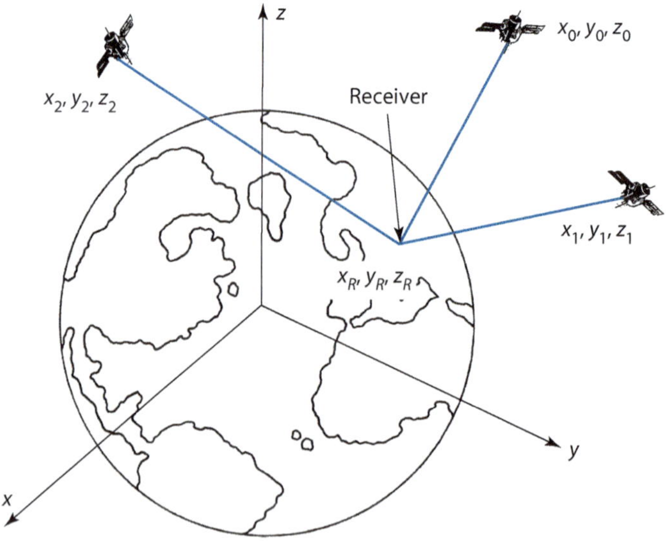

Fig. 5.2 Locating the receiver.

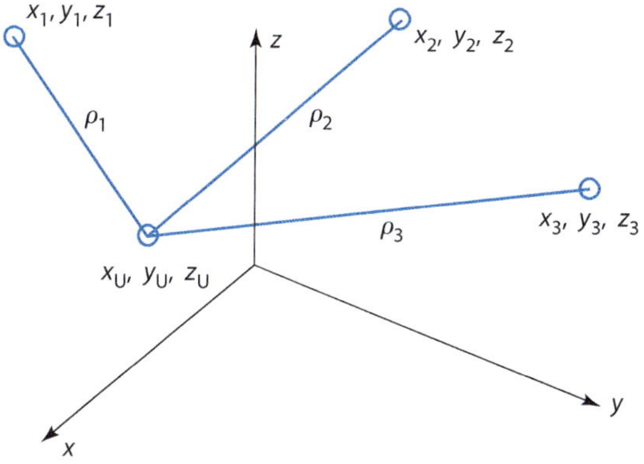

Fig. 5.3 Ranging from three satellites to a user receiver.

In the absence of any ranging errors, we can write the following relationships for the location of the receiver (required), the location of the satellites (known), and the ranges (calculated):

$$\rho_1 = \sqrt{(x_1 - x_U)^2 + (y_1 - y_U)^2 + (z_1 - z_U)^2} \tag{5.2}$$

$$\rho_2 = \sqrt{(x_2 - x_U)^2 + (y_2 - y_U)^2 + (z_2 - z_U)^2} \tag{5.3}$$

$$\rho_3 = \sqrt{(x_3 - x_U)^2 + (y_3 - y_U)^2 + (z_3 - z_U)^2} \tag{5.4}$$

Collectively these equations can be written as

$$\rho_i = \sqrt{(x_i - x_U)^2 + (y_i - y_U)^2 + (z_i - z_U)^2} \tag{5.5}$$

where $i = 1, 2, 3$. If the positions of the satellites and ranges are known, these three equations can be solved for the three unknowns x_U, y_U, and z_U. This solution is difficult, however, because of the nonlinear nature of the equations, so a different approach is used. This method will be discussed shortly, but first we explain why four rather than three satellites are needed.

5.3 Pseudorange and Clock Errors

Considerable effort is expended to ensure that the atomic clocks in each of the satellites are synchronized, but it is impossible to do the same for the clocks in the user's receiver. From Eq. (5.1) it may be seen that very accurate measurement of Δt is needed; in fact, a 1-ns error in this quantity will result in about 1 ft of error in the range calculation. It is inevitable, therefore, that although the receiver clock will have adequate precision, it will not be synchronized with the satellite clocks. Therefore, the constant error between the receiver and satellite clocks has to be considered an unknown because it cannot be determined from any data the satellite transmits.

We assign this error the symbol b_{UT} as the bias in the user receiver clock, and it will result in an error in the range calculation. Equation (5.5) has to be rewritten as

$$\rho_i = \sqrt{(x_i - x_U)^2 + (y_i - y_U)^2 + (z_i - z_U)^2} + b_U \tag{5.6}$$

Here ρ_i is termed the pseudorange, and b_U is the error in receiver clock bias measured in units of distance, hence

$$b_U = c \times b_{UT} \tag{5.7}$$

Because b_U is unknown, we are going to need four not three equations to solve for the position of the receiver and the clock bias; hence, pseudoranges from four satellites are needed. We can still use Eq. (5.6), but we must write $i = 1, \ldots, 4$.

5.4 User Position from four Satellite Pseudoranges

We will now solve Eq. (5.6) not directly, but by assuming that we have an initial, known position of the receiver and calculating a change from this position to determine the current position. To do this it is necessary to find an expression relating the change in the pseudorange $\delta\rho$ corresponding to a small change in the position of the receiver (δx, δy, δz). This is essentially linearizing Eq. (5.6) about the assumed, known position and may be calculated from the following:

$$\delta\rho_i = \frac{\partial\rho_i}{\partial x_U}\delta x_U + \frac{\partial\rho_i}{\partial y_U}\delta y_U + \frac{\partial\rho_i}{\partial z_U}\delta z_U \tag{5.8}$$

An example of the first partial derivative obtained by differentiating Eq. (5.6) would be

$$\begin{aligned}\frac{\partial\rho_i}{\partial x_U} &= \frac{1}{2}\left[(x_i - x_U)^2 + (y_i - y_U)^2 + (z_i - z_U)^2\right]^{-\frac{1}{2}} 2(x_i - x_U)(-1) + \frac{\partial b_U}{\partial x_U} \\ &= \frac{-(x_i - x_U)}{\sqrt{(x_i - x_U)^2 + (y_i - y_U)^2 + (z_i - z_U)^2}} + \frac{\partial b_U}{\partial x_U}\end{aligned} \tag{5.9}$$

Evaluating the other two derivatives and substituting into Eq. (5.8) results in the following:

$$\begin{aligned}\delta\rho_i &= -\frac{(x_i - x_U)\delta x_U + (y_i - y_U)\delta y_U + (z_i - z_U)\delta z_U}{\sqrt{(x_i - x_U)^2 + (y_i - y_U)^2 + (z_i - z_U)^2}} + \delta b_U \\ &= -\frac{(x_i - x_U)\delta x_U + (y_i - y_U)\delta y_U + (z_i - z_U)\delta z_U}{\rho_i - b_U} + \delta b_U\end{aligned} \tag{5.10}$$

In this equation, the unknowns are δx_U, δy_U, δz_U, and δb_U if x_U, y_U, and z_U are assumed known as well as x_i, y_i, z_i, and $\delta\rho_i$. Equation (5.10) may be written in the following form:

$$\begin{bmatrix}\delta\rho_1 \\ \delta\rho_2 \\ \delta\rho_3 \\ \delta\rho_4\end{bmatrix} = \begin{bmatrix}\alpha_{11} & \alpha_{12} & \alpha_{13} & 1 \\ \alpha_{21} & \alpha_{22} & \alpha_{23} & 1 \\ \alpha_{31} & \alpha_{32} & \alpha_{33} & 1 \\ \alpha_{41} & \alpha_{42} & \alpha_{43} & 1\end{bmatrix}\begin{bmatrix}\delta x_U \\ \delta y_U \\ \delta z_U \\ \delta b_U\end{bmatrix} \tag{5.11}$$

where

$$\alpha_{i1} = -\frac{x_i - x_U}{\rho_i - b_U}, \quad \alpha_{i2} = -\frac{y_i - y_U}{\rho_i - b_U}, \quad \alpha_{i3} = -\frac{z_i - z_U}{\rho_i - b_U} \tag{5.12}$$

In shorthand form, Eq. (5.11) becomes

$$\delta\rho = A \times \delta P \tag{5.13}$$

This leads to the solution for the four-dimensional change in δP as

$$\delta P = A^{-1}\delta\rho \tag{5.14}$$

Problems may occur if the A matrix is ill-conditioned, meaning that its inverse is large. This happens when the satellites ranging to the receiver are almost aligned; however, the GPS constellation is designed such that this effect is minimized. Now consider an absolute error quantity defined by the following:

$$\delta v = \sqrt{\delta x_U^2 + \delta y_U^2 + \delta z_U^2} \tag{5.15}$$

The receiver position is determined from these equations in an iterative manner as follows:

1. Set an initial state for position and clock bias such as x_U, y_U, z_U, b_U all zero (i.e., the center of the Earth, or the last known position, and zero clock bias).
2. Given the four satellite positions' x_i, y_i, z_i, use Eq. (5.6) to determine the estimated pseudoranges.

3. From the measured pseudoranges, determine the vector $\delta\rho$.
4. Calculate the A matrix from the calculated ρ_i and Eq. (5.12).
5. Use Eq. (5.14) to find δx_U, δy_U, and δz_U.
6. From Eq. (5.15), find the total error δv.
7. Compare δv to an arbitrary threshold.
   a) If it is larger, then add δx_U, δy_U, and δz_U to the initial values of x_U, y_U, z_U, and b_U, and return to step 2. Repeat until the error threshold is met.
   b) If it is smaller, then the current values of x_U, y_U, z_U, and b_U are the receiver position and clock bias.

5.5 User Position from more than Four Satellites

An alternative, improved approach is to use pseudoranges from more than four satellites, because the resulting receiver position will be determined more accurately. Some GPS weapons describe this approach as an "all in view" mode. The equations for the pseudoranges will be the same as in Eq. (5.6), but now $i = 1, \ldots, n$, where n is the total number of satellites used. Equation (5.11) then becomes

$$\begin{bmatrix} \delta\rho_1 \\ \delta\rho_2 \\ \vdots \\ \delta\rho_n \end{bmatrix} = \begin{bmatrix} \alpha_{11} & \alpha_{12} & \alpha_{13} & 1 \\ \alpha_{21} & \alpha_{22} & \alpha_{23} & 1 \\ \vdots & \vdots & \vdots & \vdots \\ \alpha_{n1} & \alpha_{n2} & \alpha_{n3} & 1 \end{bmatrix} \begin{bmatrix} \delta x_U \\ \delta y_U \\ \delta z_U \\ \delta b_U \end{bmatrix} \tag{5.16}$$

In shorthand form we still get Eq. (5.13)

$$\delta\rho = A \times \delta P \tag{5.17}$$

but $\delta\rho$ is $n \times 1$, A is $n \times 4$, and δP is still 4×1. Because A is no longer square, we have to solve Eq. (5.17) using the pseudoinverse of A.

$$\delta P = \left(A^T A\right)^{-1} A^T \times \delta\rho \tag{5.18}$$

The comments regarding the ill conditioning of the A matrix in the previous section also apply here to the pseudoinverse. The iterative solution for the change in position vector proceeds as before.

5.6 User Position in Geodetic and Local Coordinates

By itself, (x, y, z) data relative to the ECEF frame as derived previously are of little use, so the receiver converts them to the more convenient latitude, longitude, and height. This geodetic frame may be considered a

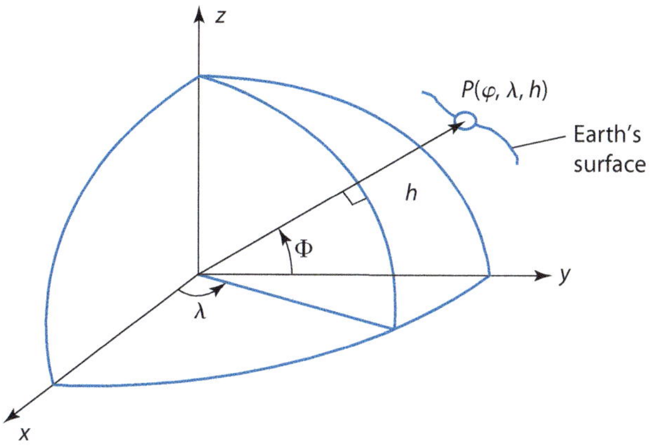

Fig. 5.4 Latitude, longitude, and height coordinate frames.

three-dimensional polar coordinate frame, as opposed to the Cartesian ECEF frame (see Fig. 5.4). This geodetic frame is also referred to as a global reference frame.

In order to do this transformation, some assumptions have to be made regarding the shape of the Earth. The Earth is not a sphere; it is slightly "flattened" at the poles, so a convenient mathematical representation of the Earth is an ellipsoid. There are many geodetic ellipsoidal representations of the Earth going back to the early 1800s, but the one used by GPS is the World Geodetic System—1984, abbreviated to WGS-84. This ellipsoid may be considered to be a smooth version of the Earth's surface and is defined by the equatorial and polar radii defined by a and b, respectively, which for WGS-84 are $a = 6{,}378{,}137.0$ m and $b = 6{,}356{,}752.3$ m. From these data, two other properties of the ellipsoid can be calculated, the flattening factor f and eccentricity e.

$$f = \frac{a - b}{a} \tag{5.19}$$

$$e^2 = 2f - f^2 \tag{5.20}$$

This allows the calculation of latitude φ, longitude λ, and height above ellipsoid (HAE) or h to be calculated from the ECEF (x, y, z) coordinates as follows. First we calculate

$$p = \sqrt{x^2 + y^2}, \quad \theta = \tan^{-1}\left(\frac{z \times a}{p \times b}\right), \quad N(\varphi) = \frac{a}{\sqrt{1 - e^2 \sin^2\varphi}} \tag{5.21}$$

Then the latitude φ, longitude λ, and height h are derived as follows:

$$\varphi = \tan^{-1}\left[\frac{z + e^2 b \sin^3\theta}{p - e^2 a \cos^3\theta}\right] \tag{5.22}$$

$$\lambda = \tan^{-1}\left(\frac{y}{x}\right) \tag{5.23}$$

$$h = \frac{p}{\cos\varphi} - N(\varphi) \tag{5.24}$$

Note that height is height above ellipsoid (HAE), so in order to specify the height of a specific point on the Earth's surface we would need to know the height of that point relative to the WGS-84 ellipsoid. In fact, some points on the Earth's surface are below the ellipsoid whereas some are above, which is why the ellipsoid may also be thought of as a smoothed average representation of the surface of the Earth.

It is worth considering the vertical reference further when attempting to specify a location on or above the Earth's surface. The WGS-84 ellipsoid is the datum used by GPS, and all height data that the GPS calculates are HAE, which may be positive or negative. HAE is measured along a normal to the ellipsoid passing through the point of interest. Many systems such as aircraft reference mean sea level (MSL) as the vertical datum. Actual sea level at a particular location will vary with time due to tides and also due to variations in gravity over the surface of the Earth.† The WGS-84 ellipsoid does not coincide with MSL. One of the functions of the National Geospatial-Intelligence Agency (NGA, formerly NIMA) is to provide data on HAE for all points over the surface of the Earth. These reference heights are illustrated in Fig. 5.5.

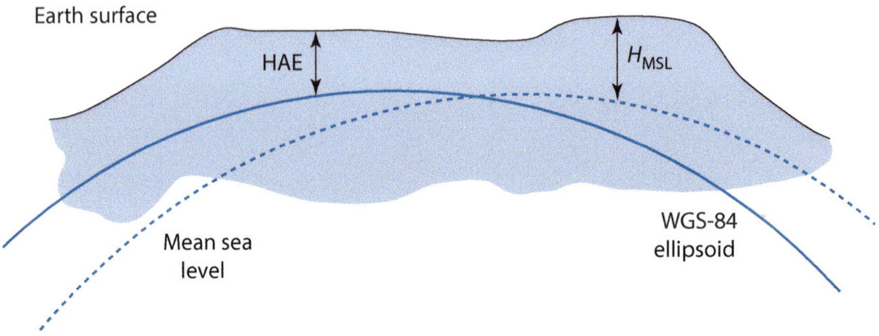

Fig. 5.5 Height relative to different data.

†Even in the absence of waves, the surface of the ocean is not locally flat.

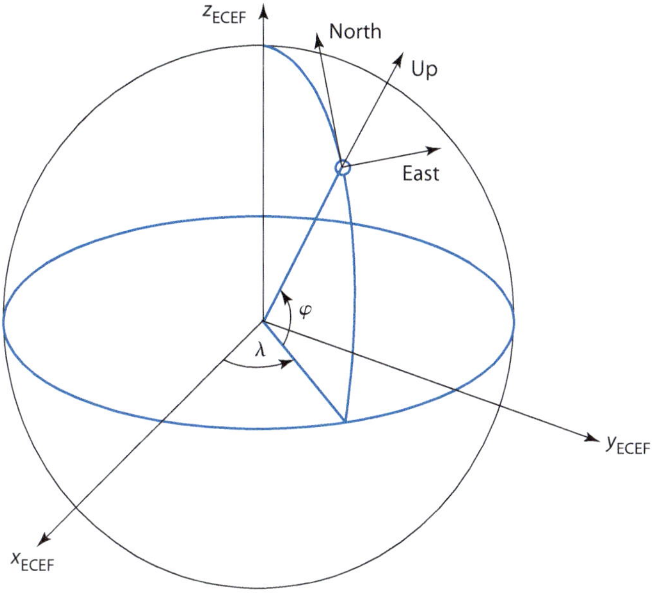

Fig. 5.6 ECEF, global, and local coordinate frames.

Another useful frame of reference needed to analyze GPS/INS guided weapons is termed the *local frame*. This reference frame origin is located at some reference point of interest, such as a target on the surface of the Earth with axes pointing east, north, and up (ENU). This frame is needed when discussing weapon accuracy, because a statement such as, "the weapon missed the target by 50 ft" is meaningless in anything other than the local frame. Sometimes an airborne local frame may be specified as north, east, and down (NED). The three most common frames used are ECEF (x, y, z), geodetic (lat, long, HAE), and local (east, north, up); these are shown in Fig. 5.6.

The transformation from ECEF to geodetic coordinates is defined in Eqs. (5.19) through (5.24). To transform from ECEF coordinates to the local coordinates we need a local reference point, such as the location of a target. If the target is located at a point x_T, y_T, z_T and a weapon is at x_p, y_p, z_p, both measured in the ECEF frame, then the vector pointing from the target to the weapon in the ENU frame is given by the following transformation:

$$\begin{bmatrix} x \\ y \\ z \end{bmatrix} = \begin{bmatrix} -\sin\lambda & \cos\lambda & 0 \\ -\sin\phi\cos\lambda & -\sin\phi\sin\lambda & \cos\phi \\ \cos\phi\cos\lambda & \cos\phi\sin\lambda & \sin\phi \end{bmatrix} \begin{bmatrix} x_P - x_T \\ y_P - y_T \\ z_P - z_T \end{bmatrix} \tag{5.25}$$

If the origin of the ENU frame is located at the target, then all points in that frame are referenced relative to the target, and Eq. (5.25) reduces to

$$\begin{bmatrix} x \\ y \\ z \end{bmatrix} = \begin{bmatrix} -\sin\lambda & \cos\lambda & 0 \\ -\sin\phi\cos\lambda & -\sin\phi\sin\lambda & \cos\phi \\ \cos\phi\cos\lambda & \cos\phi\sin\lambda & \sin\phi \end{bmatrix} \begin{bmatrix} x_P \\ y_P \\ z_P \end{bmatrix} \tag{5.26}$$

Considerable care must be taken in specifying target and weapon coordinates because their reference frames must also be stated and, if necessary, converted to the WGS-84 axes.

5.7 Accuracy of GPS-Measured Locations

There are a number of other error sources apart from receiver clock bias that directly affect the accuracy of the pseudorange calculation from the satellite to the receiver, including the following:

- *Ionosphere and troposphere delays:* The satellite signal slows as it passes through the atmosphere, so the GPS system uses a built-in model that calculates an average amount of delay to partially correct for this type of error.
- *Signal multipath:* This occurs when the GPS signal is reflected off objects such as tall buildings or terrain surfaces before it reaches the receiver. This increases the travel time of the signal, thereby causing ranging errors.
- *Orbital errors:* Also known as ephemeris errors, these are inaccuracies in the satellite's reported location.
- *Number of satellites visible:* The more satellites a GPS receiver can "see," the better the accuracy. Buildings, terrain, electronic interference, or sometimes even dense foliage can block signal reception, causing position errors or possibly no position reading at all. GPS units typically will not work indoors, underwater, or underground.

Typical ranging errors are of the order of 15 m, which given the altitude of the satellite corresponds to an error of about 0.00001%. In practice, all errors are random variables, so the 15-m error mentioned may be thought of as the standard deviation of ranging errors. This quantity is usually referred to as the user equivalent range error (UERE).

Each of the satellites used in the trilateration calculation will have a similar error, and the way they combine depends on the geometry of the position of the satellites relative to each other.

Figure 5.7 shows two satellites ranging to a receiver, each of which has a radial ranging accuracy dr associated with it. In the case on the left, the two satellites are close together, resulting in a large region of uncertainty in which the receiver may be located. The figure on the right shows that when the satellites are more spread out, the region of uncertainty diminishes, resulting in less uncertainty regarding the location of the receiver. This effect of satellite

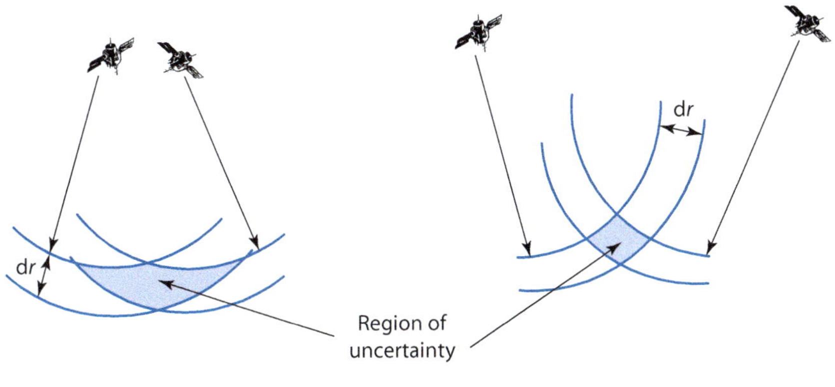

Fig. 5.7 Effect of satellite geometry on trilateration accuracy.

geometry leads to the definition of a dilution of precision (DOP) in calculating the accuracy of a receiver location on the Earth's surface from the individual satellite ranging errors.

The values of DOP are functions only of the satellite/receiver geometry. The following analysis follows that of Kaplan [2] and begins by reconsidering the expression for the change in user position.

$$\delta P = \left(A^T A\right)^{-1} A^T \times \delta\rho \tag{5.27}$$

The elements of $\delta\rho$ are random variables that may be considered Gaussian in nature with zero mean; therefore, so will the elements of δP. We are interested in forming an expression that relates variances of the elements of $\delta\rho$ to the variances of the elements of δP. It is shown in [2] that the covariance matrix of δP is related to the covariance matrix of $\delta\rho$ by the following equation:

$$\mathrm{cov}(\delta P) = \left[A^T A\right]^{-1} \sigma_\rho^2 \tag{5.28}$$

This can be written in the form

$$\begin{bmatrix} \sigma_x^2 & & & \\ & \sigma_y^2 & & \\ & & \sigma_z^2 & \\ & & & \sigma_{bu}^2 \end{bmatrix} = \begin{bmatrix} x_{\mathrm{DOP}}^2 & & & \\ & y_{\mathrm{DOP}}^2 & & \\ & & z_{\mathrm{DOP}}^2 & \\ & & & t_{\mathrm{DOP}}^2 \end{bmatrix} \sigma_\rho^2 \tag{5.29}$$

where the off-diagonal terms have been omitted for clarity. The DOP terms are therefore obtained directly from the diagonal of the $[A^T A]^{-1}$ matrix and are seen to be multipliers of the single, scalar pseudorange variance σ_ρ. This leads to the following expressions for the variances in the four-dimensional receiver position:

$$\sigma_x = x_{\text{DOP}}\sigma_\rho \tag{5.30}$$

$$\sigma_y = y_{\text{DOP}}\sigma_\rho \tag{5.31}$$

$$\sigma_z = z_{\text{DOP}}\sigma_\rho \tag{5.32}$$

$$\sigma_t = t_{\text{DOP}}\sigma_\rho \tag{5.33}$$

Because the DOP values are dependent on the elements of the A matrix, they must be geometry dependent, as indicated in Eq. (5.12). Some more commonly used dilution of precision parameters are h_{DOP} and v_{DOP} for horizontal and vertical DOP values, respectively. These are derived as follows:

$$h_{\text{DOP}} = \frac{\sigma_h}{\sigma_\rho} = \frac{\sqrt{\sigma_x^2 + \sigma_y^2}}{\sigma_\rho} \tag{5.34}$$

$$v_{\text{DOP}} = z_{\text{DOP}} = \frac{\sigma_z}{\sigma_\rho} \tag{5.35}$$

The positional dilution of precision is defined as the following:

$$p_{\text{DOP}} = \sqrt{h_{\text{DOP}}^2 + v_{\text{DOP}}^2} \tag{5.36}$$

The overall geometric DOP is given by G_{DOP}, where

$$\begin{aligned} G_{\text{DOP}} &= \frac{\sqrt{\sigma_x^2 + \sigma_y^2 + \sigma_z^2 + \sigma_t^2}}{\sigma_\rho} \\ &= \sqrt{t_{\text{DOP}}^2 + p_{\text{DOP}}^2} \end{aligned} \tag{5.37}$$

Note also that the velocity of the receiver may be determined from the Doppler shift of the transmitted signal carrier frequency, which is particularly useful for aircraft avionic systems and for weapons attacking a moving target.

Table 5.1 Average Over Space and Time DOP Values

DOP Type	Average Value
x_{DOP} (East)	0.6
y_{DOP} (North)	0.7
v_{DOP} (up)	1.8
t_{DOP}	1.1
h_{DOP}	1.0
P_{DOP}	2.0
G_{DOP}	2.3

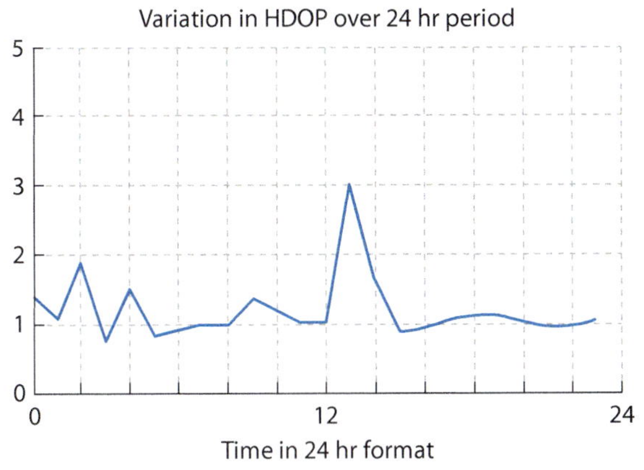

Fig. 5.8 Variation of HDOP with time.

Clearly the DOP is constellation specific, so approximate location and date/time are required for its calculation. For some attack situations, such as time-sensitive targets (TSTs) or targets of opportunity, this information may not be available in a timely manner. A single-value "average" value of DOP may be made [3] by averaging DOP values over the surface of the Earth and at varying times. The results are shown in Table 5.1, which indicates that DOP values less than unity are achievable. This implies that the accuracies in the ENU frame are better than the ranging accuracy to a single satellite.

At a fixed location on the Earth's surface, the DOP values will vary with time due to the changing constellation around that point. A typical variation in HDOP over a fixed point on the Earth's surface is shown in Fig. 5.8.

For particularly critical strike missions that use GPS/INS guided weapons, and where the variation of DOP is in excess of 2:1, the time of attack may be influenced by this variation, that is, attacking when the weapon accuracy is maximized (minimum DOP).

5.8 GPS Error Budgets

We have seen how the accuracy in GPS coordinates are obtained by multiplying the pseudorange error by the appropriate DOP value. Now we look more closely at what the total ranging error is and those component errors that contribute to it. Some of these have been mentioned in the previous section, but the list of principle errors will be defined more completely. They are as follows:

- *Satellite clock:* Although care is taken to ensure that all of the satellite clocks are accurate and synchronized, small differences are inevitable.

Table 5.2 Typical Commercial GPS Accuracy Error Budget

Error Source	GPS Ranging Error (m)	Error Type
Satellite clock	1.0	Bias
Orbit ephemeris	1.0	Bias
Troposphere	1.0	Bias
Ionosphere	10.0	Bias
Receiver noise	1.0	Random
Multipath	0.5	Random
Total RMS error	10.2	

- *Ephemeris:* Again, care is taken to provide an accurate orbit to each satellite; however, some drift is present between intermittent orbital corrections.
- *Troposphere and ionosphere:* These are the two regions of the upper atmosphere that have a dominant effect of delaying satellite transmissions. Receivers have models for each, but local variations will introduce errors.
- *Receiver noise:* Random electrical noise in the receiver circuitry may cause errors in the calculation of pseudorange.
- *Multipath:* Reflections of the satellite/receiver transmissions from surfaces near the user.

GPS error components may be classified using the same terminology as weapon errors in the sense that some are occasion-to-occasion (MPI) type and some are random for a given occasion (precision). Here we will refer to them using terms common for describing GPS errors as bias and random, respectively. We can also use the same error budget approach to accumulation of independent error sources into a total error using the error budget approach described in earlier chapters. A typical commercial GPS error budget is shown in Table 5.2.

This table shows that the error in the satellite ionosphere model dominates the other error sources. The total ranging error would be multiplied by a DOP value to determine the local plane errors, and is known as the navigation or NAV error. For example, if the value of H_{DOP} was 2.0, then the GPS NAV error would be 20.4 m circular error in the ground plane, from which a CEP could be calculated.

5.9 Differential GPS

Differential GPS (DGPS) is used to increase the accuracy of a GPS receiver by correcting some of the error quantities listed in the previous section. To implement DGPS, a receiver station, sometimes referred to as a beacon, is located at an accurately mensurated point on the ground, and the error between where GPS says the location is and the known, correct position is

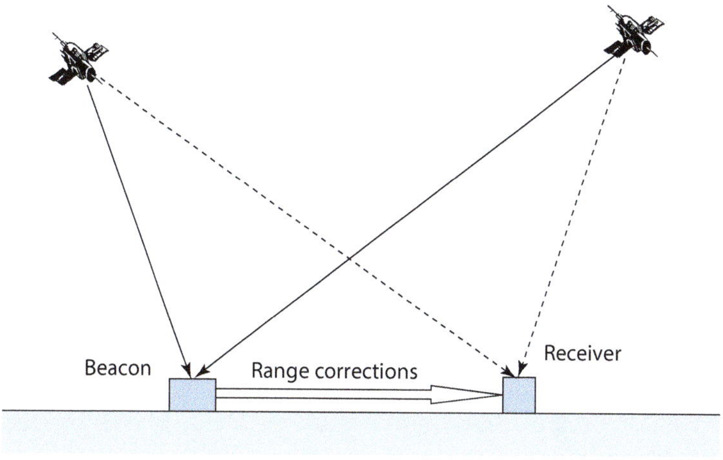

Fig. 5.9 Differential GPS geometry.

determined. Alternatively, the beacon may be a high quality GPS receiver that ranges to a large number of satellites over a period of several hours and therefore determines a mean position through self-survey.

Although it would be possible to send this position error directly to all other receivers in the vicinity of the beacon, a different approach is used. The beacon station computes corrections to the range from every satellite that is visible to it and transmits this information to the other receivers (see Fig. 5.9). This allows all visible satellites to be corrected rather than just those used to determine the beacon position, and extends the distance over which such corrections will have an effect. Receivers have to have the capability to receive such signals in order to take advantage of DGPS.

DGPS results in better accuracy by eliminating many, though not all, of the error sources associated with the satellite and receiver. Table 5.3 gives an indication of the accuracy that may be achieved using commercial

Table 5.3 Typical Commercial DGPS Accuracy Improvement

Error Source	GPS Ranging Error (m)	DGPS Ranging Error (m)	Error Type
Satellite clock	1.0	0.0	Bias
Orbit ephemeris	1.0	0.0	Bias
Troposphere	1.0	0.0	Bias
Ionosphere	10.0	0.0	Bias
Receiver noise	1.0	1.0	Random
Multipath	0.5	0.5	Random
Total RMS error	10.2	1.5	

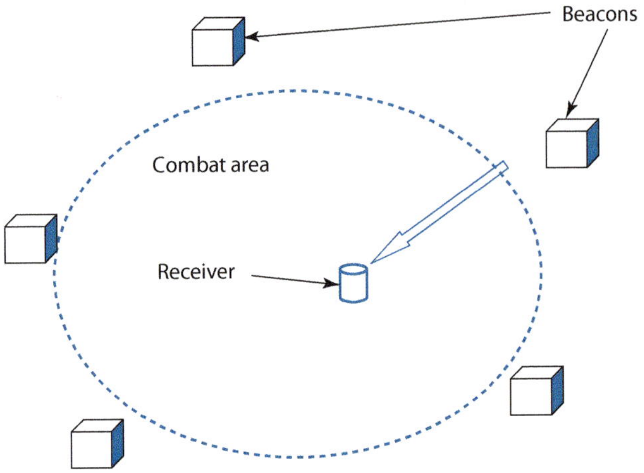

Fig. 5.10 Wide area DGPS system.

receivers. It shows that DGPS eliminates the bias errors shown previously in Table 5.2.

Military systems both for aircraft and weapon navigation and for geolocation use higher quality receivers that typically use two clock frequencies broadcast by satellites. Such dual frequency units all but eliminate the ionospheric component of error, the largest one present.

In addition, differential GPS is implemented differently than for commercial systems because it is not always feasible to locate beacon stations in combat zones. A method known as wide area differential GPS (WADGPS) uses a set of beacons located around the area of interest where the separation distances may be on the order of several hundred miles, as shown in Fig. 5.10.

A representative error budget for military systems is shown in Table 5.4.

Clearly, the accuracy of GPS systems such as that shown in Table 5.4 is very good; however, this NAV error is but one component that contributes

Table 5.4 Military GPS Accuracy

Error Source	GPS Ranging Error (m)	DGPS Ranging Error (m)	Error Type
Satellite clock	0.4	0.0	Bias
Orbit ephemeris	0.4	0.0	Bias
Troposphere	0.5	0.0	Bias
Ionosphere	0.5	0.0	Bias
Receiver noise	0.4	0.4	Random
Multipath	0.5	0.5	Random
Total RMS error	1.11	0.64	

to the overall accuracy of a GPS guided weapon or an aircraft navigating using GPS.

5.10 GPS/INS Guided Munitions

Although some weapons are described as GPS guided, they are actually GPS/INS guided, which means that the weapon is steered to the target by a conventional inertial guidance system. The term *GPS aided* is more accurate. The shortcoming of INS guided systems is their drift over time, so the GPS receiver constantly updates the INS with accurate position and velocity information, thereby eliminating the drift problem. Functionally, the guidance system is as shown in Fig. 5.11.

Typically the update rate is several time per second, but obviously in the presence of GPS jamming the accuracy will degrade to that of the INS system alone, drifting over the time from when jamming began.

GPS/INS weapons will maneuver to a specified target with a total error that is determined largely by three independent components:

1. The inherent accuracy of the on-board weapon GPS system, termed the navigation error (NAV)
2. The ability of the guidance and control system of the weapon to maneuver the weapon to the target coordinates (G&C)
3. The error with which the coordinates of the target are known, termed the *target location error* (*TLE*).

Estimates of all three error sources in both horizontal and vertical directions are needed in order to predict the total accuracy of GPS guided munitions. Because the three error sources are independent, we may write the notional equation for the total error as follows:

$$\text{Error}_{\text{TOTAL}}^2 = (\text{NAV})^2 + (\text{G\&C})^2 + (\text{TLE})^2 \tag{5.38}$$

In this equation, the error components must be in consistent measures such as standard deviations (sigma), CE(50)/CEP or CE(90).

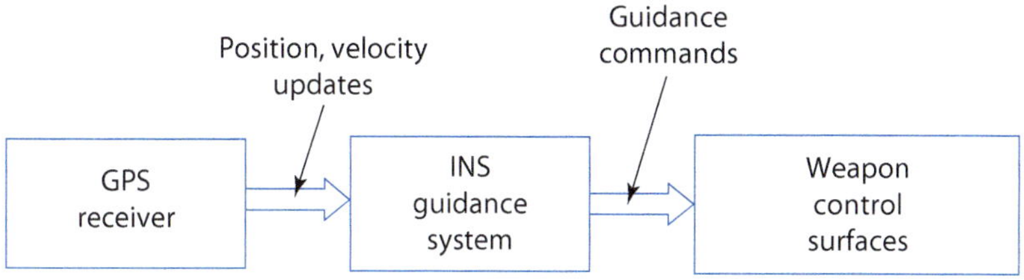

Fig. 5.11 GPS/INS guidance system.

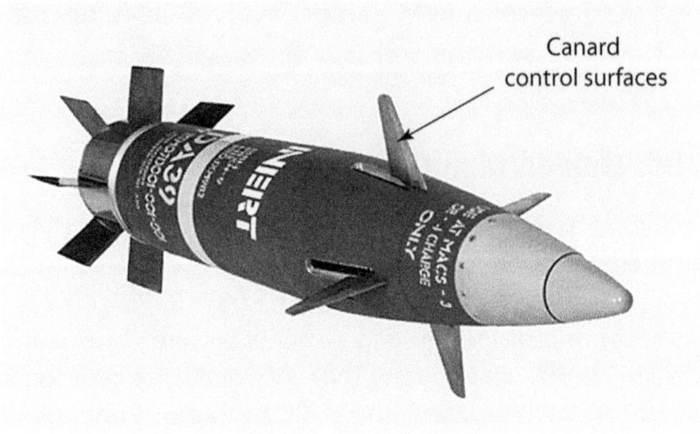

Fig. 5.12 Canard control surfaces on XM982 Excalibur round.

The first of these error sources, NAV, is dependent on the location and number of satellites from which the position of the weapon GPS derives its location; this has been discussed and characterized in the previous sections of this chapter.

The second error source is the G&C error. Typically the weapon guidance system uses GPS to determines the difference between its position and that of the target and sends this information to the control system that modifies the aerodynamic surfaces in order to adjust the trajectory. Some munitions have small control surfaces at the front of the projectile known as canards, such as the 155-mm Excalibur artillery round shown in Fig. 5.12. Other

Fig. 5.13 Control surfaces on the GBU-43/B (MOAB).

Table 5.5 TLE Category (CAT) Definitions

Category (CAT)	Range (m)
1	0–6
2	7–15
3	16–30
4	31–91

weapons, such as the GBU-43 shown in Fig. 5.13, have tail-mounted control surfaces.

The ability of the control surfaces to change the trajectory of the weapon will depend on their surface area and the inertial properties of the weapon. The inability of the weapon to maneuver in response to such commands, particularly in the terminal phase where relative position changes quickly, may result in missing the aimpoint. These errors are difficult to quantify or measure, so the weapon manufacturer usually supplies them. They are based on aggregating many high fidelity engagement simulations.

The third error source, TLE, is dependent on the source of target coordinates. This may vary considerably from several hundred feet if targets are identified from a map to a few feet if the target has been previously mensurated. Obviously in combat situations where targets are located in enemy territory, accurate target GPS coordinates may be difficult to obtain; several interesting schemes have been developed to overcome the TLE.

Operationally, TLE accuracy is partitioned into several accuracy ranges or categories as a way of indicating the quality of the TLE source. These categories (CAT) are defined in Table 5.5.

This terminology often relates to the rules of engagement where, for example, a target may not be engaged with a coordinate-seeking weapon unless the TLE is no larger than a specified CAT level.

5.11 Near-Precision GPS Guided Munitions

It should be mentioned at this point that there are different ways to implement GPS/INS guidance on a weapon. The one discussed so far is guidance to a point in space based on the error between the target coordinates and estimated, instantaneous weapon position. Another, simpler implementation is a path-following control system approach in which the ideal trajectory for an unguided weapon is calculated from shooter to target and corrections are made to the projectile flight when deviations from this ideal path are detected. This is shown in Fig. 5.14.

When viewed in the normal plane from the rear at a particular instant, the position of the weapon relative to the ideal trajectory is as shown in Fig. 5.15.

In this instant, the missile would be commanded to reduce the error between its current location and the ideal trajectory, which would be two

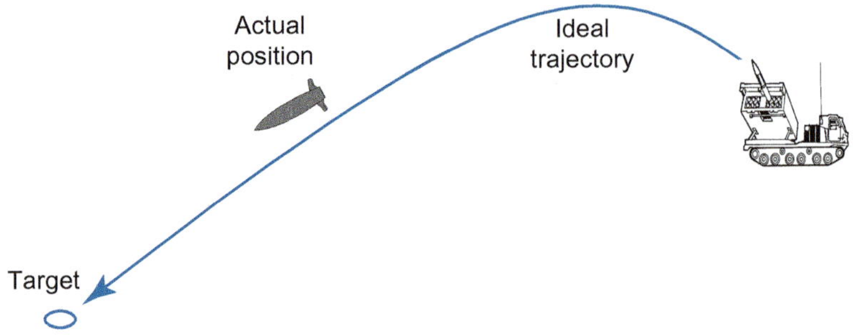

Fig. 5.14 Trajectory-following control.

steering commands (left and down) rather than three commands needed to get it to the target (Δx, Δy, Δz). This simpler implementation is typically used for weapons such as smaller artillery shells or mortar rounds where the trajectory is more predictable.

This method is sometimes referred to as "near precision" because it gives precision accuracies on the order of 50 m, which is better than the 150-m to 200-m accuracy of unguided munitions but an order of magnitude worse that a true coordinate-seeking weapon. It is, however, less expensive and is easier to install by simply replacing the fuze unit in the nose of the projectile, as shown in Fig. 5.16. This shows the rear-mounted GPS receiver and

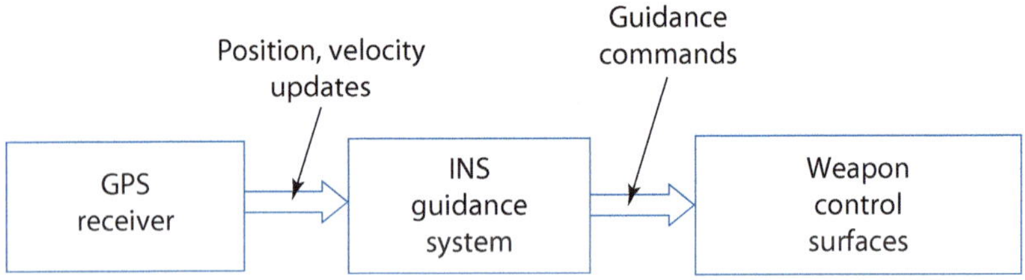

Fig. 5.15 Path-following position error.

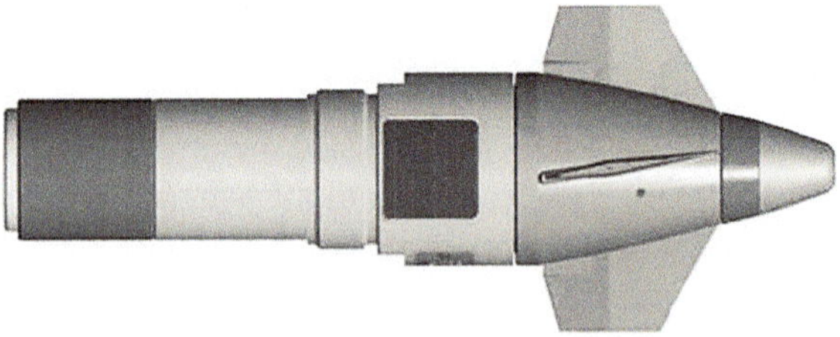

Fig. 5.16 Course-correcting fuze system.

forward control surfaces in one unit; for obvious reasons this is sometimes referred to as a course-correcting fuze. The resultant accuracy will only be as good as the ideal trajectory that the projectile is trying to follow.

5.12 Effect of Vertical Errors on Total Ground Plane Accuracy

This was discussed in general terms in Chapter 4, Section 4.24, and the only modification necessary to expand this approach to include GPS guided weapons is to recognize that each error contribution (NAV, G&C, TLE) will have its own horizontal and vertical error component.

First, the three horizontal, linear components are RSS together to obtain the overall horizontal error, which is assumed circular normal.

$$\sigma_H^2 = \sum_{i=1}^{3} \sigma_{Hi}^2 \tag{5.39}$$

Similarly, the vertical error components are RSS together, assuming they are linear normal in this direction.

$$\sigma_V^2 = \sum_{i=1}^{3} \sigma_{Vi}^2 \tag{5.40}$$

Then the total vertical error is resolved into the ground plane through the impact angle, as shown in Fig. 4.40, and RSS with the horizontal error to obtain the total range error.

$$\sigma_{R-\text{TOTAL}}^2 = \sigma_H^2 + \left(\frac{\sigma_V}{\tan I}\right)^2 \tag{5.41}$$

Note that if the horizontal error is given as a circular error, such as a CEP, the range and deflection linear errors have to be extracted from the following equation:

$$\sigma_H = \frac{\text{CEP}}{1.1774} \tag{5.42}$$

As before, the deflection error is unaffected by the impact angle.

$$\sigma_{D-\text{TOTAL}} = \sigma_H \tag{5.43}$$

From this REP, DEP and HEP are obtained from the following:

$$\text{REP} = 0.6745 \times \sigma_{R-\text{TOTAL}} \tag{5.44}$$

$$\text{DEP} = 0.6745 \times \sigma_{D-\text{TOTAL}} \tag{5.45}$$

$$\text{HEP} = 0.6745 \times \sigma_V \tag{5.46}$$

If required, the equivalent horizontal plane CEP may be obtained using Pitman's equation.

$$\text{CEP} = 0.562 \times \text{MAX}(\sigma_{R-\text{TOTAL}}, \sigma_{D-\text{TOTAL}}) + 0.617 \times \text{MIN}(\sigma_{R-\text{TOTAL}}, \sigma_{D-\text{TOTAL}}) \tag{5.47}$$

5.13 Targeting Modes

GPS guided munitions can be employed in different ways in order to reduce errors inherent in the system. The simplest way to use such munitions is shown in Fig. 5.17, and is termed *bombing on coordinates* (*BOC*) if the attack is from the air or *predicted fire* if it is from the ground. These two modes are analogous to their unguided weapon counterparts.

Figure 5.17 shows a target being attacked by either a surface-launched or air-launched GPS/INS guided weapon. In such an attack, all three error sources listed in the previous section would be present. One improvement to such a scenario would be to utilize differential GPS if it is available in order to eliminate the bias components of the NAV error—see Table 5.3.

A different approach is to designate the target from either an aircraft or an observer on the ground. A targeting pod is shown in Fig. 5.18; it basically uses an aircrew-operated optical system and laser rangefinder to determine the position of the target relative to the designator.

Because the designator knows its GPS coordinates, the coordinates of the target can be computed as an offset, as shown in Fig. 5.19. The calculated GPS coordinates are then loaded into the munition, and it flies to the target.

The accuracy associated with target-designated releases may be different depending on whether the designation is relative or absolute.

- In a *relative ranging* designation, the aircraft GPS position is compared to the weapon GPS position, and any error is compensated for in calculating the target coordinates. This has the effect of eliminating the bias

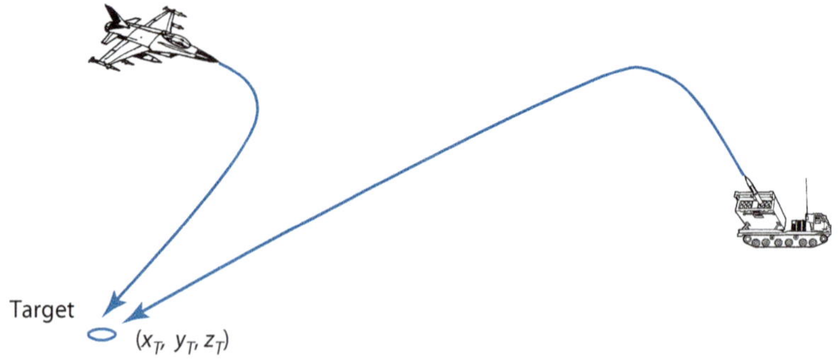

Fig. 5.17 Bombing on coordinates/predicted fire mode.

Fig. 5.18 Airborne targeting pod.

component of the NAV error, improving this accuracy estimate considerably. Not all systems are capable of supporting this weapon/aircraft interaction, but if it is, it is the preferred implementation. A typical relative designation is when the aircraft designating the target also releases the weapon.

- If the designation uses *absolute ranging*, no compensation of error between the weapon GPS and aircraft GPS occurs. Each obtains coordinates usually from a different set of satellites, perhaps at different times (i.e., at different occasions). In these circumstances, the bias components of the NAV error are additive rather than canceling each other out. An example of an

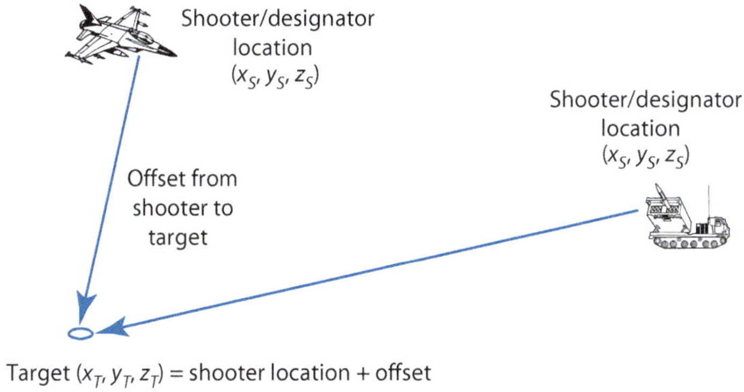

Fig. 5.19 Designating the target.

absolute ranging designation is when target coordinates are obtained by one airborne platform and the attack occurs several hours later from another platform.

In designating the target, there will be an associated error, as indicated in Fig. 5.20 for an aircraft-mounted device that measures pointing angle and slant range. This error then becomes the TLE and is designation geometry dependent.

Consider the situation shown in Fig. 5.21. The Cartesian coordinate frame is attached to the aircraft, and two pointing angles in azimuth θ, elevation φ, and the slant range SR are measured from a pod such as the one shown in Fig. 5.18. If each measurement is subject to error, we can determine the corresponding errors in target coordinates x_T, y_T, and z_T.

From this geometry, we find that the target position is given by

$$x = \mathrm{SR}\cos\phi\cos\theta \tag{5.48}$$

$$y = \mathrm{SR}\cos\phi\sin\theta \tag{5.49}$$

$$z = \mathrm{SR}\sin\phi \tag{5.50}$$

Differentiating each of these equations with respect to SR, ϕ, and θ leads to the following relationship between the errors:

$$\begin{bmatrix} \Delta x \\ \Delta y \\ \Delta z \end{bmatrix} = \begin{bmatrix} \cos\phi\cos\theta & -\mathrm{SR}\sin\phi\cos\theta & -\mathrm{SR}\cos\phi\sin\theta \\ \cos\phi\sin\theta & -\mathrm{SR}\sin\phi\sin\theta & \mathrm{SR}\cos\phi\cos\theta \\ \sin\phi & \mathrm{SR}\cos\phi & 0 \end{bmatrix} \begin{bmatrix} \Delta\mathrm{SR} \\ \Delta\phi \\ \Delta\theta \end{bmatrix} \tag{5.51}$$

The pod errors (ΔSR, $\Delta\varphi$, and $\Delta\theta$) would come from test data or contractor specifications, allowing the geometry-dependent designation errors to be

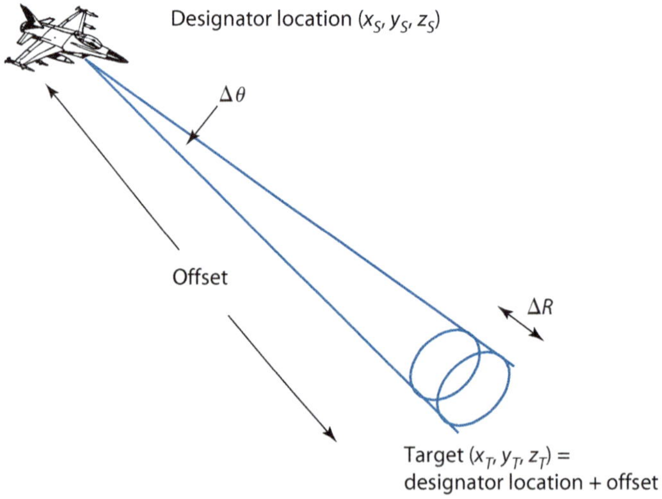

Fig. 5.20 TLE errors through designation.

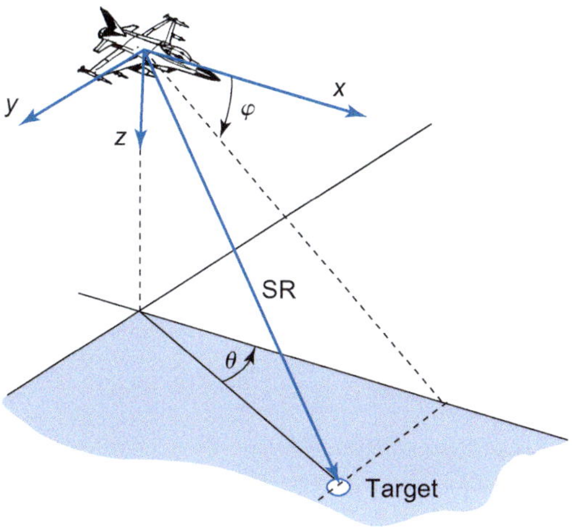

Fig. 5.21 Designation geometry.

calculated from Eq. (5.51). Again, interpreting the error terms as standard deviations for the fleet rather than a specific instrument, the error budget becomes

$$\mathrm{TLE}_X^2 = (\cos\phi\cos\theta)^2\sigma_{\mathrm{SR}}^2 + (\mathrm{SR}\sin\phi\cos\theta)^2\sigma_\phi^2 + (\mathrm{SR}\cos\phi\sin\theta)^2\sigma_\theta^2 \tag{5.52}$$

$$\mathrm{TLE}_Y^2 = (\cos\phi\sin\theta)^2\sigma_{\mathrm{SR}}^2 + (\mathrm{SR}\sin\phi\sin\theta)^2\sigma_\phi^2 + (\mathrm{SR}\cos\phi\cos\theta)^2\sigma_\theta^2 \tag{5.53}$$

$$\mathrm{TLE}_Z^2 = (\sin\phi)^2\sigma_{\mathrm{SR}}^2 + (\mathrm{SR}\cos\phi)^2\sigma_\phi^2 \tag{5.54}$$

These calculations are usually performed in the pod itself. If the pod has its own GPS/INS system, the absolute target coordinates are provided directly to the weapon without intermediate communication with the aircraft avionic system. Clearly there are many combinations of aircraft/weapon and targeting pod, allowing a multitude of information paths between them. Each will have its own accuracy, perhaps in different coordinate systems, which is challenging to the weaponeer to quantify.

An alternative to using a pod to designate the target is to use conventional or high-resolution radar. An example of a synthetic aperture radar (SAR) image is shown in Fig. 5.22.

Points can be selected in the image for targeting purposes, and their location relative to the aircraft determined from the display scale. Again, this offset in range and deflection or lat/long may be added to the aircraft's

Fig. 5.22 High altitude SAR image.

known GPS position to determine the target's GPS coordinates in the horizontal plane.

Obviously what is missing is target elevation information; this would have to be estimated by other means, such as subtracting the height above target as measured by a radar altimeter from the GPS height above ellipsoid of the aircraft when it is directly above the target. Alternatively, the image may be registered with a digital terrain elevation database (DTED), which provides lat/long/HAE coordinates for any selected point in the image.

Whatever sensor is used to designate the target, errors are introduced into the overall accuracy of target coordinates; however, with advanced sensors these are usually smaller than map-based coordinates. For time-sensitive targets, localized designation may be the only means of using coordinate-seeking weapons for an attack.

The description of relative ranging just given, although theoretically correct, has some practical difficulties. For example, when comparing the aircraft GPS position with the weapon GPS position to obtain the relative error,

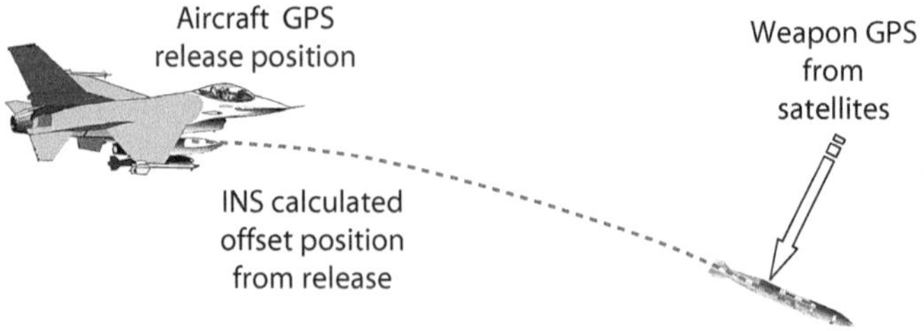

Fig. 5.23 Practical relative ranging implementation.

Table 5.6 Applicable Error Components by Mode

Mode	Designator GPS NAV		Receiver GPS NAV		G&C	TLE	Designation Error
	Bias	Rand	Bias	Rand			
BOC/predicted fire			X	X	X	X	
BOC w/DGPS				X	X	X	
Relative ranging		X		X	X		X
Absolute ranging	X	X	X	X	X		X

it is usually the case that the weapon is mounted beneath the aircraft wing. Because the weapon GPS antenna is on top of the weapon, the GPS constellation is shielded, and its position cannot be determined. In this case, the following process, shown schematically in Fig. 5.23, is usually employed.

1. The aircraft position at release is transferred to the weapon through transfer alignment (TXA).
2. After release, the weapon INS calculates its position, using the TXA data as a starting point, up to the time when the weapon GPS locks on to the visible satellites, usually about 30 s after release.
3. The difference between the INS calculated position and the measured GPS position corresponds to the GPS bias and may then be eliminated.

Clearly there will be some weapon INS drift error between release and acquiring the satellites for navigation, but this is assumed small enough to neglect. In summary, therefore, Table 5.6 shows the error components present for all the targeting modes discussed.

5.14 GPS Jamming

It is relatively easy to jam signals received by a GPS device because they are at very low power levels. These signals propagate from satellites with limited power storage and recharging capabilities, resulting in transmitted signals approximately in the 35-W power level. Because the satellite orbit is around 20,000 km and some loss occurs due to transmission through the ionosphere and troposphere, signal strength on the Earth surface is very small, about the same as that received from a 100-W lightbulb on the surface of the moon.

Although this signal is below the level of naturally occurring background radiation, digital signal processing techniques allow the GPS signal to be extracted by the receiver from this white noise background. Even so, it is not difficult for an external white noise transmitter to emit a signal that even at relatively low milliwatt power overcomes the GPS signal the receiver is trying to detect. This is the basis of the simplest form of GPS jamming.

Before discussing jamming, it is necessary to understand the nature of the signal transmitted from a satellite to a GPS receiver. Each GPS satellite transmits data on two carrier frequencies, L1 (1575.42 Mhz) and L2 (1227.60 MHz). The atomic clocks aboard the satellite produce the fundamental L-band frequency, 10.23 Mhz, and then the L1 and L2 carrier frequencies are generated by multiplying the fundamental frequency by 154 and 120, respectively. Two pseudorandom noise (PRN) codes, along with satellite ephemerides (broadcast ephemerides), ionospheric modeling coefficients, status information, system time, and satellite clock corrections, are superimposed onto the carrier frequencies, L1 and L2. As discussed earlier, the measured travel times of the signals from the satellites to the receivers are used to compute the pseudoranges.

The course-acquisition (C/A) code, sometimes called the standard positioning service (SPS), is a pseudorandom noise code that puts modulated GPS data onto the L1 carrier. The precision (P) code, sometimes called the precise positioning service (PPS), is modulated onto the L1 and L2 carriers, allowing for the removal of the first-order effects of the ionosphere, the largest source of pseudorange errors. The P code is referred to as the Y code if encrypted. Most commercial GPS receivers use the C/A code, whereas military systems use the more accurate P(Y) code.

In discussing the capability of jammers to disable a GPS receiver, there are several parameters to consider.

- The jammer-to-signal strength J/S expressing the signal power generated by the jammer relative to the power of the satellite signal at the receiver.
- The maximum power output of the jammer—the more powerful the jammer, the more effective it is.
- The distance from the receiver over which the jammer has an effect. This is a maximum at the receiver and decays exponentially with separation distance.
- The code the receiver is utilizing—civilian C/A code or military P(Y). The latter is more resistant to jamming.
- What function of the receiver is disabled—initial acquisition of the satellites or the ability to receive signals once acquisition has occurred.

5.15 Receiver–Jammer Interaction

Figure 5.24 shows the relationship among the parameters listed in the previous section [4].

The horizontal dotted lines correspond to representative J/S values to inhibit receiver functions that utilize different codes. For example, the upper line shows about a 55-dB jammer-to-signal ratio is needed to cause the loss of a P(Y) receiver to track GPS satellites once acquisition has occurred. This must now be read in conjunction with the diagonal lines to determine for what range a jammer of known input will be effective for

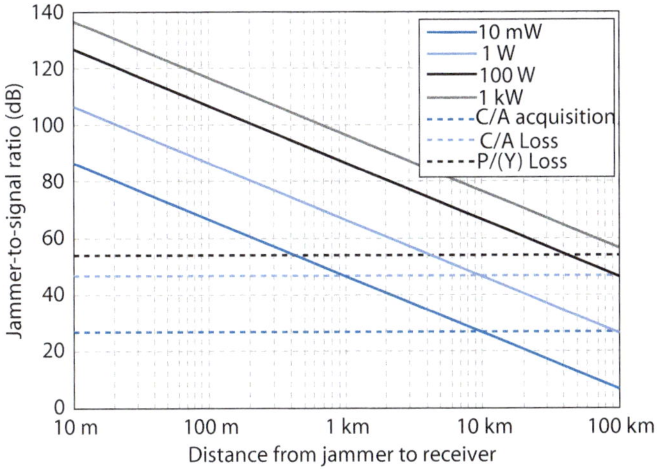

Fig. 5.24 Receiver/jammer performance.

that jamming capability. The graph shows that a 1-W jammer will be effective over a distance of about 4.5 km. Given this conclusion, it is instructive to look at commercially available jammers.

5.16 GPS Jammer Performance

In most countries the public use of jamming devices is illegal, but this has not stopped such devices from being available from the usual retail outlets at a low price. The devices can be either handheld or fit into a vehicle and are about the size of a cell phone (see Fig. 5.25). The in-car unit on the left jams

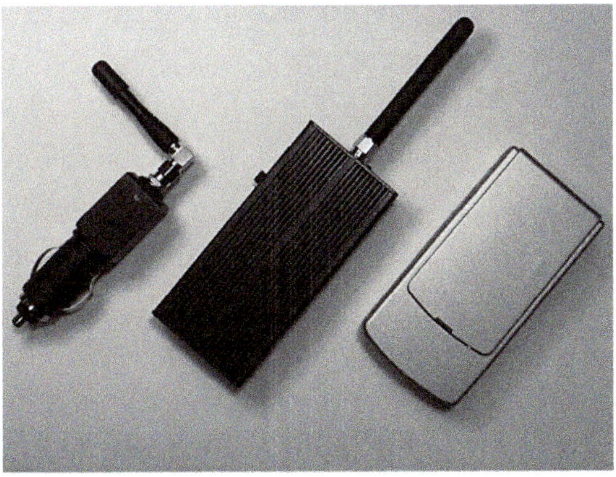

Fig. 5.25 Commercial GPS jamming devices.

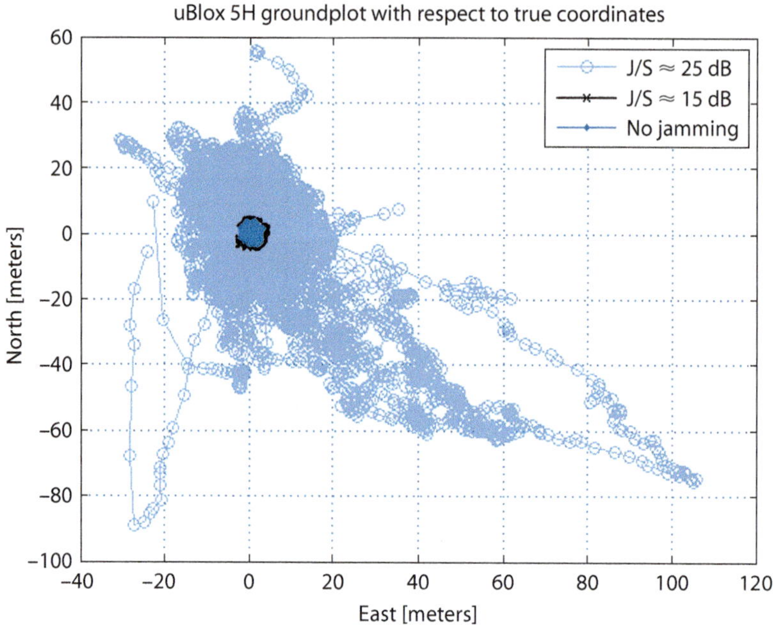

Fig. 5.26 Degradation of positional accuracy with jamming.

the L1 band (C/A) receiver and has a power of 125 mW, which from Fig. 5.24 gives it an effective radius in excess of 1 km. The other jammers shown have a power output of 500 mW. Although P(Y) receivers are more resistant to jamming, it should be obvious that GPS jamming is easily achieved by relatively high-power, simple, compact, and inexpensive devices.

The effect increased noise level has on the positional accuracy of a receiver can be interpreted using Table 5.4 by observing the contribution to the error budget by noise in the receiver. Increasing the power output of the jammer would degrade the positional accuracy of the receiver, as indicated in Fig. 5.26. This map shows the range of effects of jammers of various power levels.

5.17 Jamming Mitigation

There are several improvements to the initial GPS system that attempt to make receivers more resistant to jamming signals, particularly for military systems.

- *M-Code:* A new encrypted signal is being developed for the U.S. military called the M-Code. It is also transmitted on the L1 and L2 carrier, providing better anti-jam capability by using higher power levels than the P(Y) code. The challenge is to ensure this higher power signal does not interfere with existing signals on the same carrier frequency.

- *Directional antenna:* As discussed earlier, the minimum DOP (accuracy) is achieved by reading as many spatially diverse satellites as possible; therefore, GPS antennas observe a full hemisphere of the sky. Unfortunately this allows jamming signals from any direction to influence the receiver. A directional antenna receives signals from the known directions of the broadcasting satellites while generating null-reception areas everywhere else, thereby blocking any jamming signals. Electrically steerable directional antennas, also known as controlled-reception pattern antennas (CRPAs) or "smart" antennas, also are effective.
- *INS navigation:* As we have learned, GPS guided weapons are really INS guided with the GPS updating the INS to periodically eliminate gyroscopic drift. It will be shown in the next section how such drift increases the CEP over time, but the amount of increase depends on the quality (hence cost) of the INS. This is shown in Fig. 5.27.[‡] As a point of reference, the cost of these INS systems is shown in Table 5.7.

  Because the high-cost INS systems are those found in military aircraft, of which there are relatively few, the INS systems used in high production weapons with limited flight time will have to consider a tradeoff between accuracy and cost.
- *Integration:* Mission planning tools currently exist that allow the user to lay out a map of the engagement scenario so that mobile GPS receivers can be moved through an environment of user-defined GPS jammers. As the

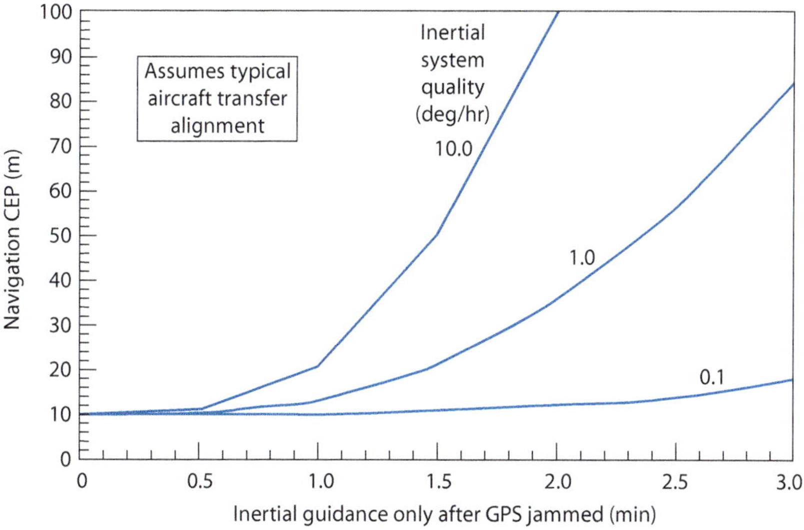

Fig. 5.27 Effect of INS drift on weapon CEP.

[‡]Based on work by Gilmore and Delaney of Lincoln Laboratory, Lexington, Massachusetts.

Table 5.7 Cost of INS vs. Performance

Drift rate (deg/hr)	Cost ($)
10	1000
1	2000–5000
0.1	20,000–50,000
0.01	100,000–200,000

receiver moves along a specified path, degradation of receiver performance can be determined. The GPS Interference and Navigation Tool (GIANT) is such a system (Fig. 5.28), simulating many receivers vs many jammers. It produces maps and charts of DOP and CEP as well as J/S values along the route.

- *Spoofing:* Spoofing is different from jamming and is more challenging. In this technique, fake satellite signals are sent to a GPS receiver simulating genuine signals expected from the satellites, causing the receiver to estimate a location different from where it actually is. It is then possible to make the receiver navigate the weapon to a different location than the intended target. Spoofing requires sophisticated electronics to track the weapon trajectory and simulate satellite transmission signals, which can be difficult to achieve for the encoded P(Y) code.

Fig. 5.28 GIANT mission planning tool.

5.18 INS-Only Accuracy

Weapons GPS systems can monitor the signal-to-noise ratio or the J/S ratio. If jamming is present, and if it exceeds a prescribed level, it will ignore subsequent GPS location data until they drop to an acceptable level. If GPS is switched out or denied to the weapon guidance system, then it must navigate using its internal inertial navigation system (INS). In general, the INS system is less accurate than when GPS is present due to inherent drift; therefore, additional errors are introduced. In order to estimate the magnitude of such errors, a model of the weapon-specific INS is required.

The sources of INS errors are sensor uncertainties, processor limitations, and TXA effects. As described previously, the transfer alignment is the process by which the weapon INS is aligned with the aircraft INS so as to share a common reference frame. The INS errors are modeled by nine linear differential equations, three each for position (linear and angular) and velocity. These equations have a periodic solution with a period of 84 min known as the Schuler period. The TXA errors provide the initial conditions for the solution of these equations.

Because the flight time for most GPS/INS weapons is considerably shorter than the Schuler period, the INS error equations can be simplified to yield the following solution for position error:

$$s(t) = s_0 + v_0 t + \frac{1}{2}(-g\varepsilon_0 + a_b)t^2 - \frac{1}{6}g\omega_b t^3 \tag{5.55}$$

where

$t =$ time of flight
$s_0, v_0, \varepsilon_0 =$ initial position, velocity, and tilt errors
$a_b, \omega_b =$ accelerometer and gyro biases
$g =$ acceleration due to gravity

As a first approximation, the dominant errors are initial velocity and accelerometer bias, leading to the following simplified quadratic form:

$$s(t) = s_0 + v_0 t + \frac{1}{2}a_b t^2 \tag{5.56}$$

This may be written in the following form, where a_0 and a_1 characterize the INS drift:

$$\text{CEP} = \text{CEP}_{\text{GPS}} + a_0 t + a_1 t^2 \tag{5.57}$$

It should also be recognized that GPS/INS guided weapons begin their movement towards the target guided by INS only. This is because the receiving antenna on the weapon may not be exposed to the satellites until the munition is fired from a cannon or released from an aircraft, where it is

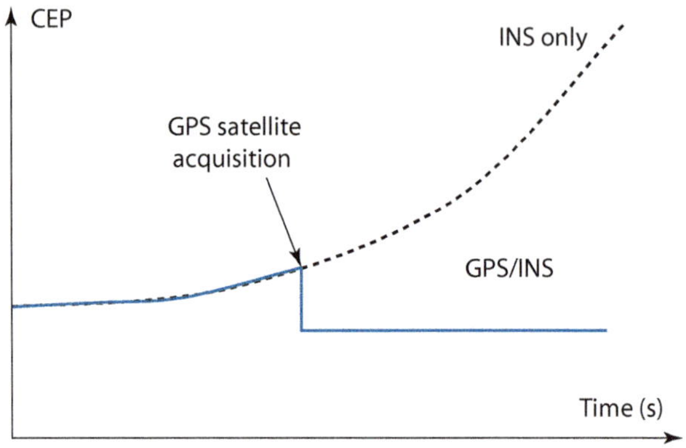

Fig. 5.29 Acquiring GPS after launch.

shielded by the wing or fuselage. Figure 5.29 shows how the equivalent ground plane CEP of a GPS/INS weapon varies with time from release.

In this figure, the improvement in accuracy can be seen once GPS satellites have been acquired and positional information is used to update the INS system. If satellite acquisition does not occur, the weapon is guided by INS control only where the quadratic degradation of accuracy is governed by Eq. (5.56).

This quadratic function can also be used to predict the effect of GPS jamming on accuracy by determining the errors incurred over that portion of the weapon's time of flight subject to jamming, as indicated in Fig. 5.30.

Clearly, the longer the flight time before impact that GPS is not available, the worse the weapon accuracy will be. For error budget purposes, the INS

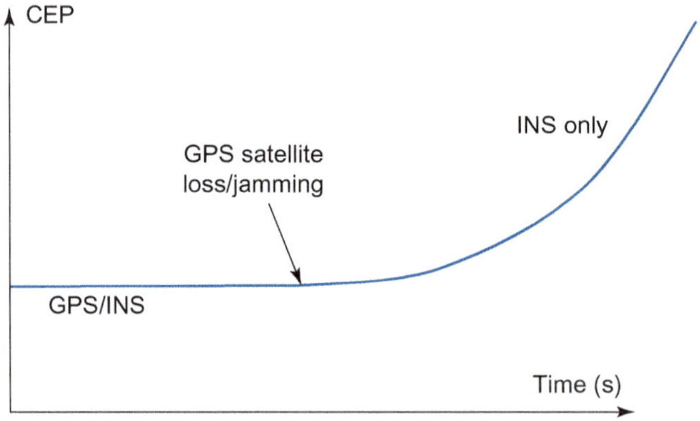

Fig. 5.30 INS drift after GPS jamming.

drift may then be used in the form that follows:

$$\sigma_{\text{DRIFT}-X} = \sigma_{\text{DRIFT}-Y} = s_0 + v_0 t + a_b t^2 \tag{5.58}$$

The coefficients s_0, v_0, and a_b have to be obtained from contractor or program offices for the weapon of interest. As usual, some care needs to be exercised in interpreting weapon data because some weapon systems show a growth of CEP with time whereas others predict a growth of a standard deviation, as shown in Eq. (5.58).

5.19 Low-Level Wind (LLW) Effects

Low-level wind effects are primarily an issue for cluster munitions or ICM that release submunitions over the target area. Air-dropped or cannon-fired munitions may have GPS/INS guidance, but this will only allow the dispenser to accurately navigate to the dispense point. Once the submunitions are released, they are subject to wind effects between the release point and the ground, particularly because they may be attached to parachutes or streamers in order to avoid tumbling in flight.

As a result, although the dispense point may be accurate to a few feet, the pattern as a whole may be blown hundreds of feet from its intended location; therefore, this effect has to be considered in the overall accuracy of the total weapon delivery system. Clearly, the higher the dispense point, the longer the submunitions are subjected to the effects of wind and the larger the resulting error. The most general way to account for LLW effects is by relating the airburst/dispense accuracy to the ground plane accuracy by the following equations:

$$\sigma_{\text{XG}}^2 = \sigma_{\text{XB}}^2 + \left(\frac{\partial x}{\partial(\text{LLW})}\right)^2 \sigma_{\text{LLW}}^2 \tag{5.59}$$

$$\sigma_{\text{YG}}^2 = \sigma_{\text{YB}}^2 + \left(\frac{\partial y}{\partial(\text{LLW})}\right)^2 \sigma_{\text{LLW}}^2 \tag{5.60}$$

$$\sigma_{\text{ZG}}^2 = \sigma_{\text{ZB}}^2 + \left(\frac{\partial z}{\partial(\text{LLW})}\right)^2 \sigma_{\text{LLW}}^2 \tag{5.61}$$

where σ_{XG}, σ_{YG}, and σ_{ZG} are the ground place accuracies; σ_{XB}, σ_{YB}, and σ_{ZB} are the air burst accuracies; and σ_{LLW} is the expected error between the measured wind state and when the weapon is deployed.

The partials could be obtained from a submunition trajectory model by perturbing the wind velocity in the usual manner, although these models would require C_D vs Mach number data for the submunition. An alternative is to supply previously calculated or measured values for the partials relating to a specific submunition as a function of altitude, as indicated in Table 5.8.

Table 5.8 Submunition Sensitivity to Low-Level Wind

Altitude (m)	d/d(LLW) (m/knot)
1200	40
1100	30
1000	20
900	15
800	5

Table 5.9 Effect of Staleness on Wind Error

Staleness (hr)	σ_W (knots)
0	0.8
1	4.0
2	4.9
4	7.2
unknown	11.0

However the partial is derived, it has to be multiplied by a standard deviation σ_{LLW}, representing the error between the assumed wind condition between the dispense point and the ground, and the actual wind condition. In the artillery community it is usual to obtain meteorological (MET) conditions near the location of the gun firing the round; however, in most cases it is not possible to accurately predict or measure the wind state at the target location.

Under these circumstances, we approximate σ_{LLW} to the quantity σ_W representing the standard deviation of the error in wind state between actual conditions and assumed conditions at the target. The value of this standard deviation increases as the time between measured conditions and actual firing increases; this effect is known as *staleness*. An acceptable estimate of the effect of staleness on wind error may be seen from Table 5.9.

Hence, if the staleness between the wind measurement and weapon dispense can be estimated, a value of σ_{LLW} can be obtained from Table 5.9, and together with the partials in Table 5.8 can be used in Eqs. (5.59) through (5.61) to get the pattern errors.

5.20 Nonplanar Trajectories

So far, it might have been assumed that the trajectory of the weapon lies in a vertical plane that contains both the releasing aircraft and the target. Although this is generally true for unguided weapons, it is not so for guided weapons and for GPS/INS guided weapons in particular. This gives rise to some ambiguity regarding the definition of the range and deflection directions. This issue was outlined in Chapter 4, Section 4.15 and is expanded further here.

The range direction is defined to be the direction of the horizontal component of the weapon velocity at impact. For example, if a weapon is going to impact long (past the target), then it is in this direction that the miss distance will occur. GPS/INS guided weapons may be released at considerable distances from the target and will navigate in three-dimensional space to the impact point. In some cases, a guided bomb released from an aircraft may reverse direction in order to guide to the target.

It follows, therefore, that the range direction may not be obvious at the moment of release or firing, or when the attack is planned. This has implications in most GPS modes but particularly in designation modes. Reconsider the point-and-range designation system shown in Fig. 5.20 where the geometry is now specified in terms of a slant range SR, depression angle φ, and slew angle θ. Equation (5.51) shows the relationship between the pod error budget and the resulting geometry-dependent designation error—for example, the error in the range direction x as a function of the pointing angle θ is as follows:

$$\Delta x = -\mathrm{SR} \sin\theta \cos\phi \, \Delta\theta \tag{5.62}$$

Hence if θ is a random variable, the standard deviation of range error is

$$\sigma_X = -\mathrm{SR} \sin\theta \cos\phi \, \sigma_\theta \tag{5.63}$$

If each parameter is a random variable whose standard deviation is known, then each will contribute an error in the x, y, and z directions of the aircraft frame. The aircraft may hand off the calculated GPS coordinates of the target to the weapon, and these will contain errors due to the designation errors, but the transformation from aircraft frame to GPS frame (lat/long) depends on the aircraft heading at designation ε, as shown in Fig. 5.31.

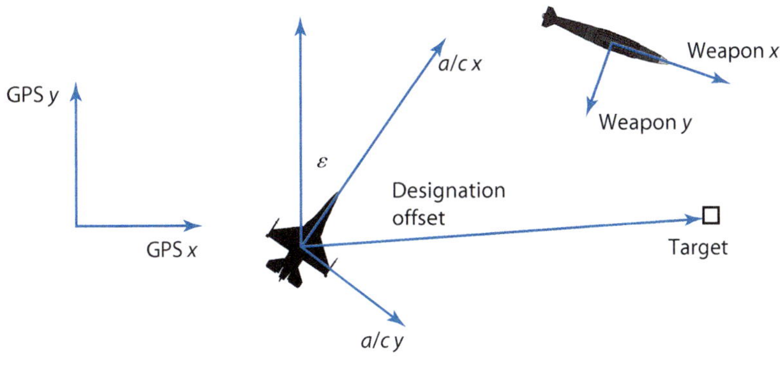

Fig. 5.31 Aircraft, weapon, and GPS frames.

Assuming the aircraft is flying level, the corresponding errors in GPS coordinates (lat/long) would be as follows:

$$\sigma_{X-\mathrm{GPS}} = \sigma_{X-A/C} \cos \varepsilon - \sigma_{Y-A/C} \sin \varepsilon \tag{5.64}$$

$$\sigma_{Y-\mathrm{GPS}} = \sigma_{X-A/C} \sin \varepsilon + \sigma_{Y-A/C} \cos \varepsilon \tag{5.65}$$

More advanced pods have an internal IMU that measures the aircraft orientation in the lat/long coordinate frame and performs the calculations in Eqs. (5.64) and (5.65) before passing the target coordinates to the weapon, now referenced to lat/long. Legacy systems may not do this.

Finally, the error measures needed to compute effectiveness are measured in the weapon frame at impact, also shown in Fig. 5.31. Hence, the aircraft heading and weapon heading are needed to translate the designation errors into the weapon frame. Also, the weapon heading at impact must be known in order to transform other error sources such as NAV, G&C, and TLE into the weapon frame. Fortunately, these latter errors are usually considered circular normal; therefore, they are the same in any direction and so the weapon heading is not needed, but the designation errors are not circular normal. The approach is to convert the bivariate distribution of designation error into an equivalent circular error by first calculating the CEP from Pittman's equation.

$$\mathrm{CEP} = 0.562 \times \mathrm{MAX}(\sigma_X, \sigma_Y) + 0.617 \times \mathrm{MIN}(\sigma_X, \sigma_Y) \tag{5.66}$$

Then a common REP and DEP is calculated that gives the same CEP.

$$\mathrm{REP} = \mathrm{DEP} = 0.573 \times \mathrm{CEP} \tag{5.67}$$

This allows the designation error to be treated the same as the other error sources.

As mentioned before, advanced designation pods now supply target coordinates in terms of latitude/longitude and height above ellipsoid, so the issue of transforming aircraft coordinates to GPS coordinates is eliminated. However, knowing the weapon range and deflection directions at impact is still advisable, particularly if collateral damage is a concern.

5.21 Computational Tools to Generate GPS Accuracies

A simple spreadsheet that illustrates the combination of the NAV, G&C, and TLE errors is shown in Table 5.10. In this spreadsheet, the weapon impact angle is input into the shaded cells together with the navigation (NAV), guidance and control (G&C), and target location errors (TLEs). These errors are specified as circular ground plane and linear vertical measures, where the TLE is provided at the more common 90% level. The calculations proceed as follows:

Table 5.10 Simple GPS Weapon Accuracy Calculator

GPS Weapon accuracy calculator 1				
Impact angle (deg)	65	tangent (impact)	2.14	
Errors	Ground plane (ft)		Vertical (ft)	
NAV (with DOP)	15	Circular 50%	20	Linear 50%
G&C	5	Circular 50%	5	Linear 50%
TLE	25	Circular 90%	35	Linear 90%
Convert TLE to 50%	13.72	Circular 50%	14.10	Linear 50%
Convert CEP to REP, DEP	REP	DEP		
NAV	8.59	8.59		
G&C	2.86	2.86		
TLE	14.32	14.32		
Projection of height error				
NAV vertical	9.33			
G&C vertical	2.33			
TLE vertical	6.57			
RSS total	20.54	16.95		
CEP	32.62			

1. The 90% TLE values are converted to 50% values using Eq. (4.17) for the circular measure and Eq. (4.19) for the vertical, linear measure.
2. The 50% circular measures are converted to REP and DEP using Eq. (4.12).
3. The three vertical error components are projected into the ground plane using the tangent of the impact angle and Eq. (4.76) to produce equivalent errors in the range direction only.
4. The component REP and DEP errors are RSS together and a CEP calculated using Pittman's equation.

Although REP and DEP are preferred in any effectiveness calculations, as mentioned in the previous section, it may not be obvious what these directions are relative to the target, so using the CEP is an acceptable alternative in this case.

The next spreadsheet, shown in Table 5.11, is a more detailed version of the previous one in that it provides the error budget for the NAV component and user-supplied values for certain DOP quantities. It implements the methodologies discussed in this chapter to determine GPS weapon accuracy and allocates each error into a bias or random component, which may be combined for accuracy purposes but kept separate in order to calculate effectiveness correctly.

The user again has to supply data in the shaded cells, although representative values discussed earlier in the chapter have been used. The results are valid for bombing on coordinates/predicted fire mode with no GPS jamming.

Table 5.11 Detailed GPS weapon accuracy calculator

GPS Weapon Accuracy Calculator 2 - (inputs are green cells) Distances in meters						
	Bias	Random				
Satellite clock	0.400					
Ephemeris	0.400			HDOP	1	
Troposphere	0.500			VDOP	1.8	
Ionosphere	0.500			GDOP	2.06	
Noise		0.400		Impact θ	70	
Multipath		0.500				
Receiver	0.954					
Space and clock (S&C)	0.566					
RSS	0.906	0.640				
UERE (pseudo range sigma)	1.109					
	Horizontal			Vertical		
	Bias	Random	Total H	Bias	Random	Total V
Nav error H (sigma)	0.640	0.453	0.784	1.630	1.153	1.996
G&C H (sigma)		1.000	1.000		1.000	1.000
TLE H (sigma)	3.000		3.000	5.000		5.000
RSS (sigma)	3.068	1.098	3.258	5.259	1.526	5.476
	Range			Deflection		
Summary-horizontal plane	Bias	Random	Total R	Bias	Random	Total D
REP/DEP H plane	3.616	1.230	3.819	3.068	1.098	3.258
Single weapon CEP H plane	4.157					
	Vertical			Deflection		
Summary-vertical plane	Bias	Random	Total V	Bias	Random	Total D
HEP/DEP V plane	9.934	1.526	10.051	3.068	1.098	3.258
Single weapon CEP V plane	7.659					

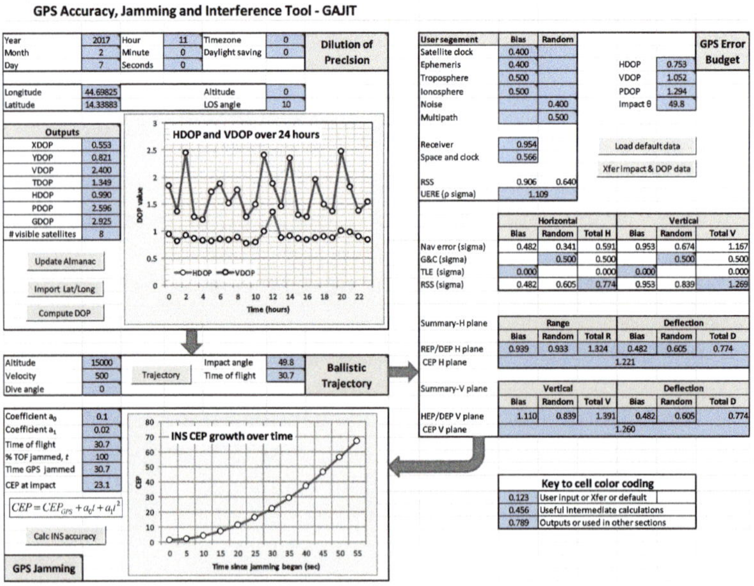

Fig. 5.32 Integrated GPS weapon accuracy calculator.

Finally, many of the features discussed can be combined into a single computational tool such as the one shown in Fig. 5.32.

- Starting in the top left, the user can specify the time, date, and location of the planned attack and the point value as well as the variation in DOP is generated. This includes being able to obtain the latest almanac data and the ability to import target locations from resources such as Google Earth or Google Maps.
- Moving down, the user inputs trajectory information. This may be a simple ballistics calculation using user-supplied inputs such as altitude, velocity, and dive angle, or specific weapon terminal conditions obtained offline.
- The next step on the right is the GPS accuracy calculator shown in Table 5.11, where the user supplies the GPS receiver or uses the default. In addition, the DOP and impact data can be transferred from the previous sections. This produces the various accuracy metrics REP, DEP, and CEP in both the horizontal and vertical directions.
- The final section allows accuracy in a GPS-denied environment to be calculated and plotted as a function of time. The user has to supply the INS drift characteristics in Eq. (5.58) and the duration of the INS-only navigation. This is expressed as a percentage of the weapon flight time from the previous section.

References

[1] Tsui, J. B., *Fundamentals of Global Positioning System Receivers*, Wiley Interscience, New York, 2000.
[2] Kaplan, E. D. (ed.), *Understanding GPS Principles and Applications*, Artech House, Boston, MA, 1996.
[3] Yuen, M. F., "Dilution of Precision (DOP) Calculations for Mission Planning Purposes," MS Thesis, Naval Postgraduate School, March 2009.
[4] Jones, M., "The Civilian Battlefield: Protecting GNSS Receivers from Interference and Jamming," *Inside GNSS*, March/April, pp. 40–49, 2011.

Chapter 6 Introduction to Vulnerability and Effectiveness

6.1 Introduction

This chapter investigates the destructive effects that weapons have on targets, and involves both the lethality of the weapon and the vulnerability of the target. In Volume 1 of the book, weapon effectiveness is discussed in terms of an effectiveness index (EI). In Volume 2 of the book, more detailed effects metrics will be presented as higher fidelity methods are described in detail. Some of the concepts in this topic can be somewhat complex when seen for the first time, so a two-step approach will be taken.

1. In this chapter, the terminology and basic approach will be explained, and by making some simplifying assumptions, the calculation of lethality will be demonstrated.
2. In the next chapter, some of the simplifying assumptions will be discarded and reference will be made to specific tools used by the Joint Munitions Effectiveness Manual (JMEM) community to calculate effectiveness.

6.2 Effectiveness Indices

The vulnerability of a particular target to a weapon will be of interest to a number of analysts, each with their own perspective on the requirements for data and measures of damage. From the weaponeering point of view, the principal reason for studying vulnerability is to determine a numerical quantity that measures the effectiveness of a particular weapon against a specific target. One such measure is known as an effectiveness index (EI), which may also be referred to as a lethal area or a damage function. An EI is defined only for a given degree of damage by a specific weapon to a particular target.

Examples of EIs include:

- Mean area of effectiveness due to fragments (MAE_F, ft^2)
- Mean area of effectiveness due to blast (MAE_B, ft^2)
- Vulnerable area (A_V, ft^2)

Table 6.1 Examples of Kill Definitions

Target Type	Kill Definitions
Land vehicles	K: catastrophic kill (not repairable) M0: mobility kill (cannot move, immediately) M40: mobility kill (cannot move within 40 minutes) F: firepower kill (cannot fire)
Parked aircraft	PTO: Repairs requiring at least 5 min PTO_4: Repairs requiring at least 4 h PTO_{24}: Repairs requiring at least 24 h
Personnel (standing)	Defense (prevent) within 30 s Assault (prevent) within 30 s Assault (prevent) within 5 min Supply (prevent) within 12 h

- Bridge effectiveness index (BEI, ft)
- Mean area of effectiveness for a building (MAE_{BLDG}, ft^2)
- Effective miss distance (EMD, ft)
- Disruptive diameter, for example, a crater (D_C, ft)

The specific nature of damage inflicted on the target is usually referred to as a damage definition and varies for different types of targets. Sometimes the word *kill* is used instead of *damage,* and the words *level* or *criterion* are used instead of *definition.* Therefore, phrases such as *damage definition, kill level,* or *damage criterion* are synonymous and used interchangeably. Examples of kill definitions for some target types are shown in Table 6.1.

Appendix B lists most of the JTCG/ME target types vulnerable to conventional weapons and the kill definitions associated with them. Also shown are recommendations for the types of weapons considered to be effective against a specific target type.

To illustrate the variability of EI with target, weapon, and kill criteria, consider the effect of both an Mk-82 500-lb bomb and an Mk-84 2000-lb bomb with a fuze that detonates the weapon on impact. Representative MAE_F values (in ft^2) are shown in Table 6.2 for various target types and kill definitions.

Readers should understand the trends associated with the EI value as the weapon, target, and kill definition vary.

Table 6.2 Example EI Values

Target ►	Tank	Tank	Troops
Kill Definition ►	M-kill	K-kill	5-min assault
Weapon ▼			
Mk-82	600	450	3000
Mk-84	1200	550	12,000

6.3 Requirements for the Computation of Effectiveness Indices

Calculating an EI requires considerable time, knowledge, expertise, and data. In general, the following elements will be needed:

- Definition of the kill levels required
- Collection of the physical, geometrical, and functional descriptions of the target and the components that are contained within it
- Knowledge of the critical components of the target, that is, which components need to be killed in order to achieve the kill level(s) required
- The warhead characteristics (blast, frag) of the weapon to be used against the target
- An understanding of what damage level is required to kill each of the critical components
- A methodology to compute the appropriate EI for the given combination of weapon, target, and kill level

To determine the effect of a particular weapon on a target of interest, two individual studies are usually done for each kill definition of interest:

1. *Vulnerability assessment*: The target vulnerability to a wide range of user-selected warhead characteristics is made. This is followed by:
2. *Effectiveness assessment*: The computation of an EI for a specific weapon and detonation scenario is made.

The rationale for this approach is that if the vulnerability analysis is comprehensive enough (i.e., it considers ranges of fragment mass, fragment velocity, blast radii, etc. that are sufficiently wide), it will cover all current and possible warhead characteristics for future weapons. This means it will only have to be performed once. As a new weapon or warhead is developed, the vulnerability of all existing targets to that warhead may be extracted as a subset of the vulnerability study based on the warhead characteristics. This concept is shown in Fig. 6.1.

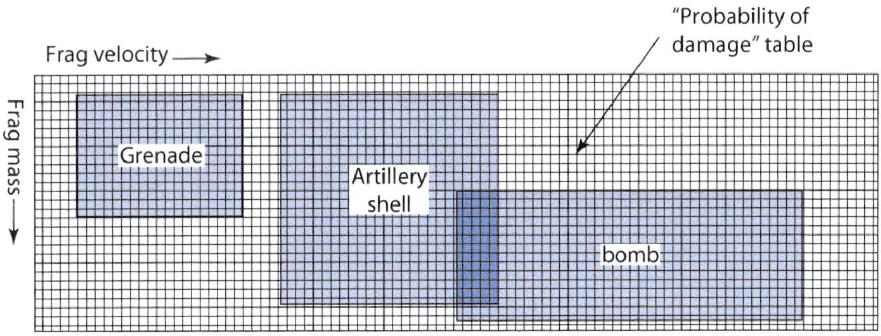

Fig. 6.1 Relationship between vulnerability data and specific weapon characteristics.

In order to study the vulnerability of specific targets to conventional weapons, we will consider the effects of the primary damage mechanisms a weapon may produce, including the following:

- Fragments
- Blast
- Shaped charge and explosively formed projectile (EFP)

The nature of these damage mechanisms was described in detail in Chapter 1; however, a particular warhead may produce one or more of these effects, so the combined effect of each damage mechanism on the target will have to be determined. In terms of modeling complexity, fragments present the most challenges, so they will be dealt with first.

6.4 Collecting and Characterizing Warhead Fragment Data

In order to determine the effectiveness of fragments on materiel and personnel targets, the characteristics of such fragments have to be measured for each warhead/bomb type. This characterization is usually achieved through testing, and the process has matured into a standardized approach. The process is usually referred to as *arena testing* because the static warhead is surrounded by various devices that determine the number, weight, velocity, and region of origin of fragments produced when it detonates. In planning the test, or in the preliminary stages of weapon design, some estimate of the various fragment characteristics needs to be made in order to provide a starting point for the collection of data. Theoretical analysis of a bursting cylindrical shell, often referred to as the Gurney equations, yields the following results.

6.4.1 Initial Velocity of Fragments for a Cylindrical Shell

The theoretical velocity of fragments immediately following detonation is given by

$$V_0{}^2 = \frac{2(C/M)A^2}{2 + (C/M)} \tag{6.1}$$

where

V_0 = initial fragment velocity (ft/s)
A = a constant depending on the explosive (~8400)
C/M = ratio of the weight of explosive to the weight of the metal casing calculated from the following:

$$\frac{C}{M} = \frac{P_C}{P_M[(D_0{}^2/D_i{}^2) - 1]} \tag{6.2}$$

D_0 = outside diameter of shell (in.)
D_1 = inside diameter of shell (in.)
P_C = density of explosive (lb/in.^{3})
P_M = density of metal parts (lb/in.^{3})

6.4.2 Initial Velocity of Fragments for a Spherical Shell

Although not applicable to conventional weapons, the following is provided for completeness:

$$V_0{}^2 = \frac{5(C/M)A^2}{5 + 3(C/M)} \tag{6.3}$$

6.4.3 Average Mass of Fragments

Fragment masses are usually measured in grains where 7000 grains = 1 lbm. Although there will be a distribution of fragment masses, the average mass is given by

$$m_0 = \frac{C D_0{}^2}{V_0{}^2} \tag{6.4}$$

where

m_0 = average mass of fragments (grains)
C = a constant = 60.10^6

For a standard shell, this does not allow for the large fragments that come from the nose or fuze base.

6.4.4 Number of Fragments Having a Specific Mass

This is usually known as the two-dimensional Mott distribution law—see [1] and Chapter 2, Section 2.10.

$$N = \frac{M \exp\left[-(2m/m_0)^{0.5}\right]}{m} \tag{6.5}$$

where

N = number of fragments
M = weight of metal case (grains)
m = weight of smallest fragment considered (grains)

6.4.5 Spatial Distribution of Fragments

Fragments are not evenly distributed around the warhead because the main spray from the side has the greatest number of fragments whereas the nose and base have a smaller number of large or very large fragments emanating from the fuze and base plate. The main spray is not centered about the 90-deg line but is deflected by the explosive detonation wave, about 5 deg backwards for a nose fuze and by the same amount forward by a tail or base fuze.

The fragment characteristics are defined within zones that are conical regions demarcated by boundaries radiating out from the warhead centroid, as shown in two dimensions in Fig. 6.2. A typical set of zones for a weapon is shown in Fig. 6.3 for 18 zonal increments of 10 deg.

In three dimensions for an air-bursting munition, the conical zones will spray fragments onto the ground plane in the manner shown in Fig. 6.4. It is assumed that the fragment spray pattern is radially symmetrical about the weapon axis; therefore, the test arena is configured in the manner shown in Fig. 6.5, where one side is designed to collect the fragments themselves while the other side determines their velocity.

The fragment recovery area consists of fiberboard panels in which the fragments become embedded. The fragments are then removed, counted, and weighed. The velocity panels are coated with a piezoelectric material that produces an electrical signal when impacted by a fragment. The time taken for a fragment to travel from the warhead to the panel may therefore be calculated.

This process is also recorded by high speed cameras behind the flash panels, which for a given frame rate allow the fragment velocity to be determined. Of course, the panels have to be sufficiently far from the detonation so that they are not damaged by the blast wave.

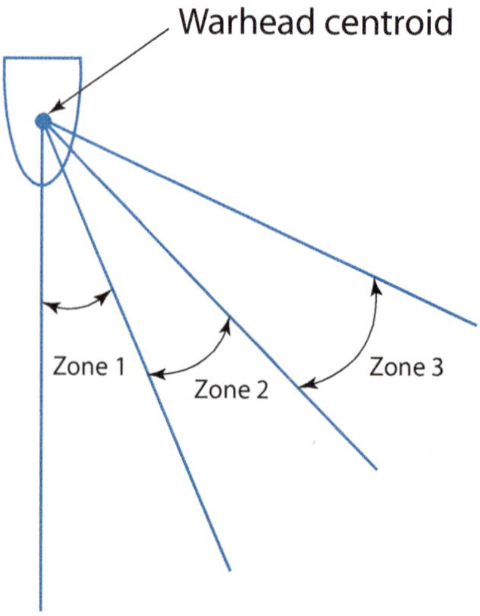

Fig. 6.2 Weapon static fragmentation zones.

The collection panels have the fragmentation zones projected onto them to determine which fragments belong in which zones, as indicated in Fig. 6.6.

6.4.6 Data Analysis

Although the photographic velocity V_P is defined as the distance traveled by the fragment divided by the time it travels, it has to be recognized that the actual fragment velocity varies over this short distance. What is needed to characterize the fragments is the measured [as opposed to Eq. (6.1)] initial velocity V_0, that is, the velocity with which the fragment leaves the shell. This is

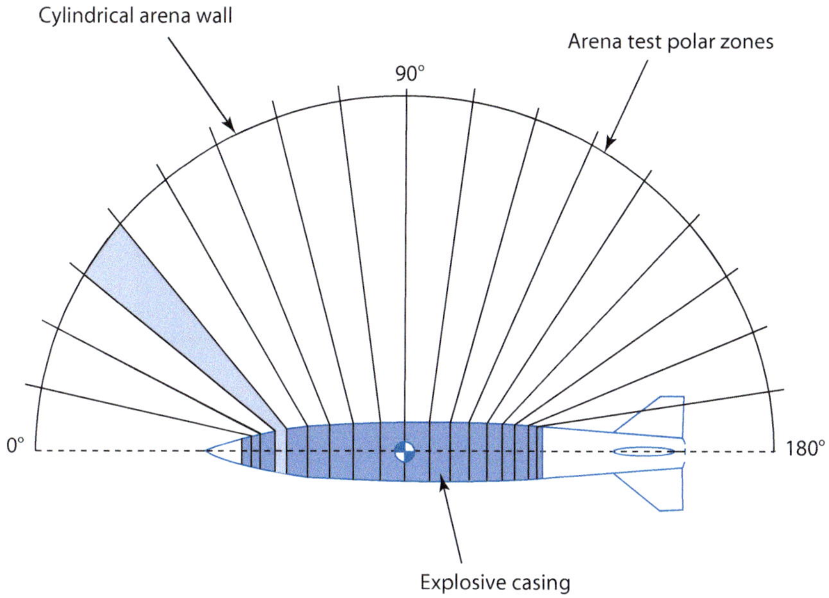

Fig. 6.3 Complete set of fragmentation zones.

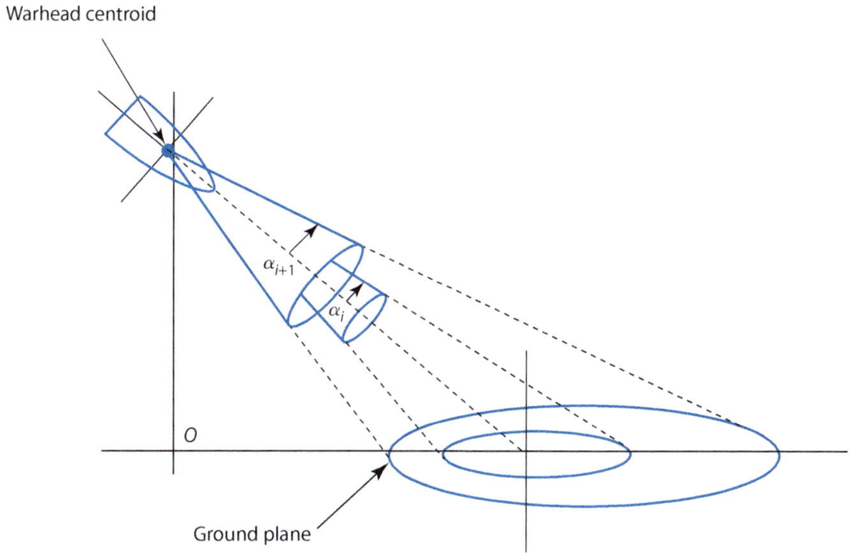

Fig. 6.4 Fragment spray pattern on the ground.

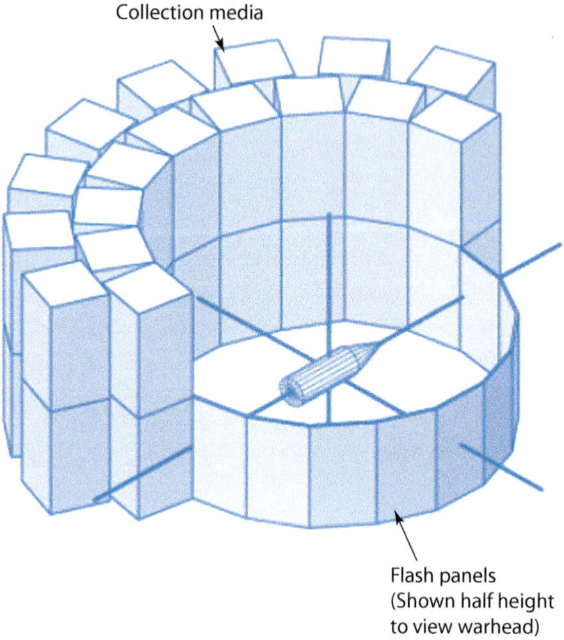

Fig. 6.5 Typical test arena configuration.

computed as follows:

$$V_0 = V_P \frac{\exp\left[\dfrac{a\,r}{m^{1/3}}\right] - 1}{\dfrac{a\,r}{m^{1/3}}} \tag{6.6}$$

where

$$a = 12K_d\rho(1/K)^{2/3}\ (\text{grains}^{1/3}/\text{ft})$$
$$K = m/A_M{}^{3/2}\ (\text{grains}/\text{in.}^3)$$
$$r = \text{distance traveled (ft)}$$

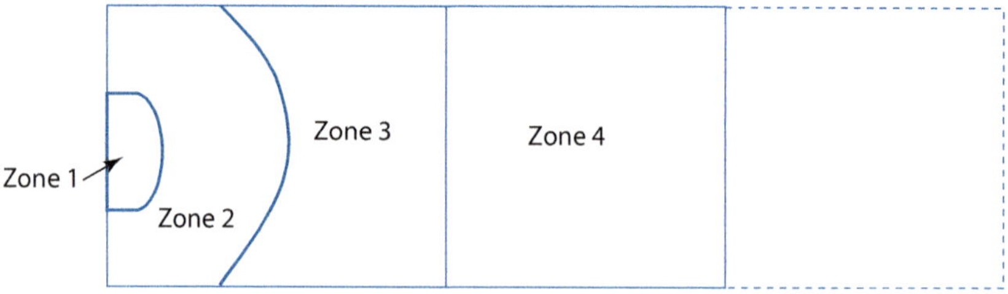

Fig. 6.6 Fragment zones projected onto the recovery panels.

K_d = drag coefficient (approximately 0.66)
ρ = air density (grains/in.^{3})
A_M = average presented area of the fragment (in.^{2})
m = weight of the fragment (grains)

This calculation is tedious because a different drag coefficient is needed for each fragment, so instead a "representative fragment" is selected and the calculation performed just for that value. The resulting velocity is then assumed to be constant for all fragments recovered in that zone.

The mass and number of fragments are more easily handled than velocity, however, and these data are also grouped, usually aggregated for more than one test, into each zone. Then, within each zone the weights are grouped into bins, and the average weight in each bin is determined. The average number of fragments is again averaged over several tests; with this, the average weight is then calculated.

The data for each zone have to be scaled to account for the fact that only a proportion of fragments are captured. This scaling is done by calculating the zone multiplication factor Z_F. Although this factor is calculated using the three-dimensional geometry of the fragment zone and fragment capture panel, a two-dimensional illustration of the concept will be presented.

Consider Fig. 6.7, which shows a view along the horizontally mounted munition axis at detonation. A vertical capture panel of height h is located at a distance r from the munition axis.

Assuming that the N fragments are projected radially in a uniform manner, the proportion of fragments impacting the panel will be given by

$$p = \frac{h}{2\pi r} \tag{6.7}$$

Therefore, the zone multiplication factor is given by the following:

$$Z_F = \frac{1}{p} = \frac{2\pi r}{h} \tag{6.8}$$

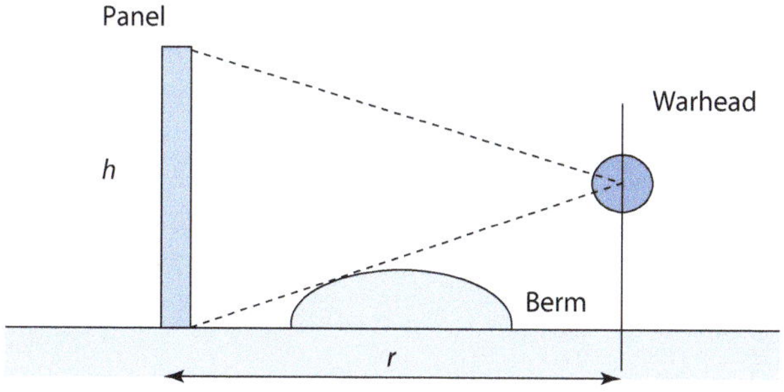

Fig. 6.7 Partial fragment capture.

Table 6.3 Fragment Characteristics for One Zone

α's			Velocities			# Wts
0.000	20.000	10.000	5500	4770	5010	5

Frag weight →	0.100	4.336	16.303	33.862	104.337
# of frags →	2.927	52.692	40.128	34.273	16.709

An 8-ft panel located 25 ft from the warhead would have a $Z_F = 19.6$. Therefore, if the total number of fragments captured by the panel in that zone is N_C, then the number of fragments produced by the warhead is

$$N = N_C \times Z_F \tag{6.9}$$

The total number of fragments captured within each zone is scaled by the zone-dependent Z_F to estimate the total number of fragments produced by the warhead; however, because not all fragments are recovered from the test, the total number of fragments shown in Eq. (6.5) is scaled from the number collected in the test as follows:

$$K_{Z\text{–DATA}} = \frac{W}{W_T} K_{\text{TEST}} \tag{6.10}$$

where W is the total mass of the shell casing, and W_T is the total weight of fragments recovered in the test from all zones. This final scaling shown in Eq. (6.10) is applied to each fragment zone.

The specification of fragment mass, velocity, and number of fragments is used to define the zonal or Z-data file for each zone. The Z-data file is the end result of the test and is required for the effectiveness calculation. The Z-data file has the format shown in Table 6.3.

The first three numbers give the angular boundaries; for this example they start at 0 deg (weapon axis-nose) and extend to 20 deg with a midpoint of 10 deg. The fragment velocities at these angles are 5500, 4700, and 5010 ft/s, respectively. The midpoint velocity of 5500 ft/s corresponds to that measured from the test as described previously, and the outer velocities are interpolated from the next zone. Their use will be explained in Chapter 7.

The number 5 indicates there are five fragment weight groups in this zone. The first group has an average weight of 0.1 grains, and there are, on average, 2.927 fragments at this weight; the next group comprises 52.692 fragments weighing an average of 4.336 grains, and so on. Clearly, fractional values of the average number of fragments indicates data from more than one test; in fact, the JTCG/ME produces a testing document recommending the number of weapons to be tested, their orientation, the analysis procedure, and so forth.

6.5 Compartment vs Component Vulnerability

Early vulnerability models divided whole targets into compartments so that, for example, a vehicle would have a crew compartment, an ammunition compartment (magazine), an engine compartment, and so forth, as shown in Fig. 6.8.

In such a representation, the probability of killing the engine equated to the probability of a fragment penetrating the compartment, essentially an empirical value. As computer capability increased, it became possible to model the vulnerability to a higher fidelity by representing the actual components within the compartment. The engine, for example, could be represented by its constituent parts such as the oil pump, main block, fuel injectors, and so on. This allowed more accurate assessment of kill levels associated with the compartment, such as mobility and fire. For the remainder of this book, only component-level modeling will be addressed.

6.6 Vulnerability Assessment for Fragmentation Warheads

An interesting observation can be made here regarding the differences between the terms *survivability* and *vulnerability*. If we are interested in studying the survivability of the M1A1 tank to a warhead from country *X*, we have full knowledge of all the components in the target (tank) but only an estimate of the warhead characteristics. On the other hand, if we want to study the effects of a U.S. Army standard 155-mm shell against a missile launcher from country *Y*, we know exactly the characteristics of the warhead and probably very little about the components inside the missile launcher.

Obtaining detailed information on targets is a common though difficult problem for weaponeers because targets of interest are rarely available for inspection and analysis; yet, as we will see, such detail is needed if meaningful results regarding the vulnerability of the target are to be obtained. As

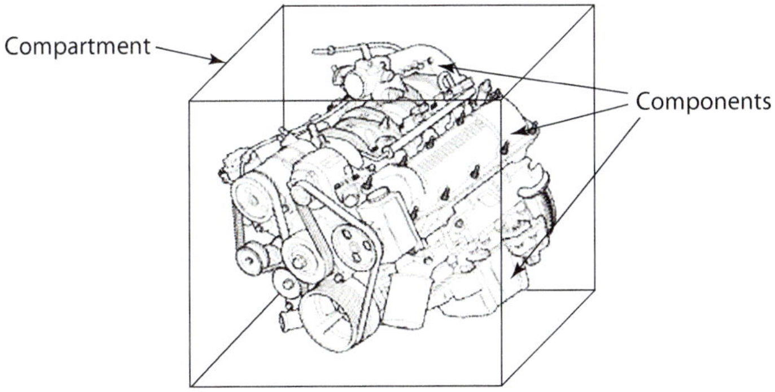

Fig. 6.8 Engine compartment with internal components.

weaponeers, we have to rely on whatever data are available regarding potential enemy targets, and these data are very dependent on intelligence information. In the absence of hard data, surrogation methods may have to be used by equating a target about which little is known to a similar target about which more information is known.

The general approach to vulnerability assessment is to assess whether the kill level has been achieved by determining how many and which components that make up the target have been individually killed. The definition of exactly what comprises a component is somewhat vague but depends upon the level of accuracy needed in the analysis. For example, if we consider the target to be a vehicle, one component of the target is the propulsion system; however, this can be divided into smaller components such as the engine, gearbox, oil cooling system, electrical controls, and so on. As we have seen in the previous section, the engine also can be further divided into its subcomponents.

One important parameter produced by a vulnerability analysis is $P_{K/H}$. This is the probability that the object is killed (to a specified level) if it is hit. The word *object* is used, because $P_{K/H}$ could refer to an individual component or to the whole target. Another term in common usage is *survivability*. For example, the probability that the target survives a hit $P_{S/H}$ is the complement of the probability the target is killed if it is hit $P_{K/H}$. We write this as

$$P_{S/H} = 1 - P_{K/H} \tag{6.11}$$

At this point we will introduce some notational conventions used in this chapter; these are the same as those used by Ball [2], allowing us to differentiate between complete targets and the components contained within them. In this convention, uppercase symbols refer to a whole target, whereas lowercase ones refer to a particular component. Some sample terms to be used later are listed in Table 6.4.

Table 6.4 Definition of Variables Used in Vulnerability Analysis

Description	*i*th Component	Target
Probability of killing the *i*th component given a hit on the *i*th component, or probability of killing the target given a hit on the target	$P_{ki/hi}$	$P_{K/H}$
Probability of survival of the *i*th component/target given a hit on the target	$P_{si/H}$	$P_{S/H}$
Probability of hitting the *i*th component given a hit on the target	$P_{hi/H}$	
Probability of killing the *i*th component given a hit on the target	$P_{ki/H}$	
Vulnerable area of the *i*th component/target	A_{vi}	A_V
Presented area of the *i*th component/target	A_{pi}	A_P

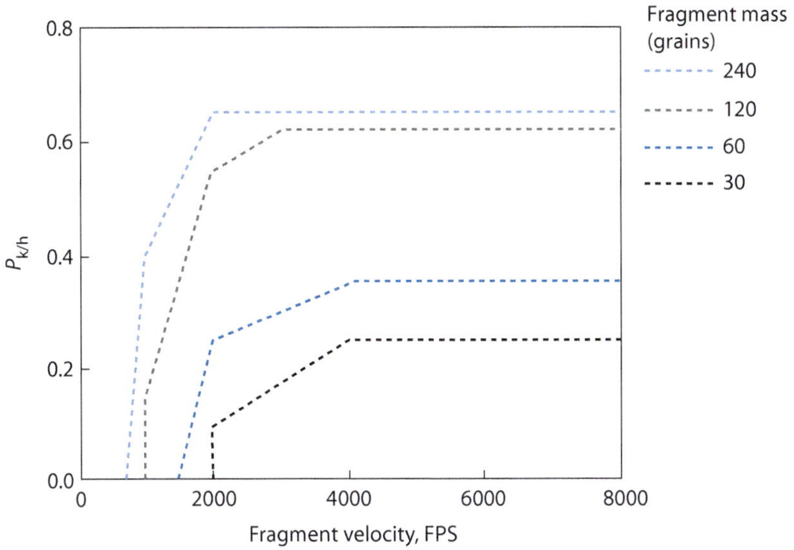

Fig. 6.9 Fragility graph for a single component.

The most basic data used in a vulnerability assessment are $P_{ki/hi}$, and this value may be determined experimentally by firing single fragments of different weights at different velocities at the component from different directions. An alternative approach to testing is to run analytical models that predict penetration of a component and its subsequent loss of function. However it is determined, typical data that might result from this process are shown in Fig. 6.9; this is usually known as a fragility curve. For a component, there will be one fragility curve for each kill level.

Another common form for the fragility curve is shown in Fig. 6.10 for a given fragment mass.

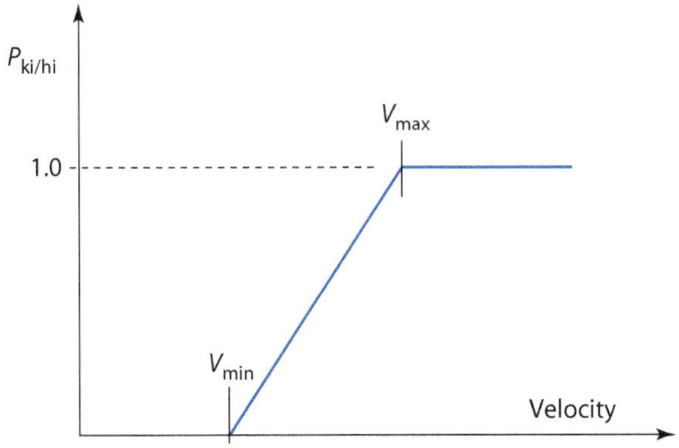

Fig. 6.10 Simplified fragility graph.

These forms are generally used when the minimum energy needed to cause any damage and the maximum energy needed to totally fail the component are estimated, and a linear relationship in between is assumed.

6.7 Presented and Vulnerable Areas

The presented area is simply the target silhouette area when viewed from any given aspect; the vulnerable area is the fraction of the presented area that will result in a defined kill level when impacted by a fragment characterized by its material, shape, weight, and velocity.

The vulnerable area is a quantitative measure of vulnerability that may be applied to a single component or a complete target. When applied to a whole target, it is an effectiveness index (EI), specifically A_V or sometimes VA_N. The vulnerable area is defined as the product of the presented area in a plane normal to the velocity vector of the fragment, and the probability of kill given a hit. An example of the presented area would be the area inside the boundary of the engine shown in Fig. 6.11.

We may write for a target

$$A_V = A_P \times P_{K/H} \tag{6.12}$$

or for the ith component

$$A_{vi} = A_{pi} \times P_{ki/hi} \tag{6.13}$$

The relationship between vulnerable areas and presented area for a target and component is shown in Fig. 6.12.

Note that in this figure, the target and component presented areas correspond to a physical object (i.e., the target and the component); however, the vulnerable area has no assumed location or shape at this point.

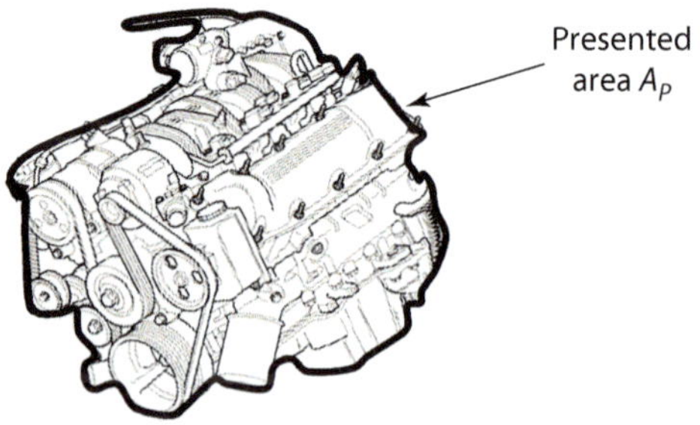

Fig. 6.11 Presented area of a component.

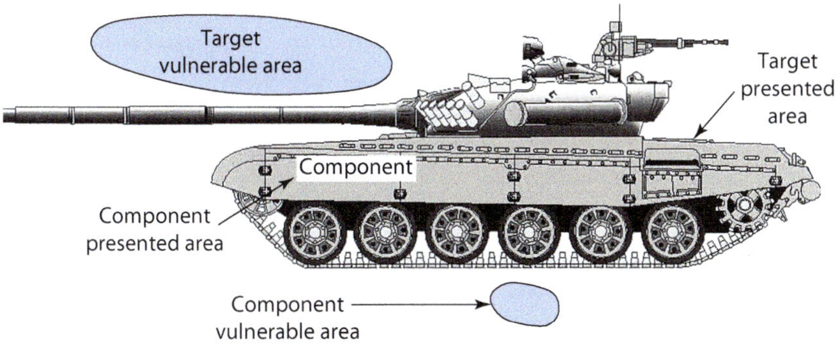

Fig. 6.12 Presented and vulnerable areas.

6.8 Critical and Noncritical Components

Components of a target may be defined as either critical or noncritical. A critical component is one that, if killed, causes a target kill, whereas a noncritical component does not lead directly to a target kill. It is important to include noncritical components in a vulnerability analysis, however, because they may shield critical components to such an extent that the critical component does not absorb sufficient damage to kill it, thereby preventing a target kill. A good example of a noncritical component is armor plate.

6.9 Redundant and Nonredundant Critical Components

A component is said to be nonredundant if killing that component alone results in a target kill. Redundant components require that several be killed before a target kill results. Redundant components are usually referenced as a group, or system, such that a group kill results in a target kill; however, it may not be necessary for all redundant components to be killed in order to produce a group kill. For example, an aircraft may have four engines and is able to fly on two or more. The engines comprise the propulsion group, and are individually redundant, because killing one or two does not result in a group kill. Killing three engines does, however, result in a group kill, and therefore a target kill. These components are linked by a kill tree, sometimes referred to as a disablement diagram or a failure analysis and logic tree (FALT), to be discussed shortly.

6.10 Critical Components, Kill Definitions and Failure Analysis Logic Trees (FALTs)

Although it may be appreciated that a given target will contain many critical components, it is important to understand that different critical components are associated with different kill definitions. For example,

Fig. 6.13 Critical components for a mobility kill.

Fig. 6.13 shows some critical components associated with a mobility kill, such as the driver's compartment, engine, and track drive system. On the other hand, Fig. 6.14 shows a different set of components corresponding to a firepower kill. In this case these components might be the main gun, magazine, engine (for power), and the gunner and loader in the main crew compartment.

In order to determine whether a particular target kill has been achieved, we will be investigating which critical components associated with that kill definition have been killed. It is necessary, therefore, to have a formal association of kill definitions and those critical components needed to achieve it. This association may be described by several terms such as a failure analysis logic tree (FALT), failure modes and effects analysis (FMEA) diagram, damage modes and effects analysis (DMEA), or a disablement diagram.

Whatever the terminology, this association is represented by a logic flow diagram in which the critical components are connected by paths typically defined by the AND and OR operations. For example, if we consider the mobility kill shown in Fig. 6.13 and assume the critical components are

- One driver
- Two fuel tanks
- One engine
- Two track systems (track, drive sprocket, idler wheels, etc.)

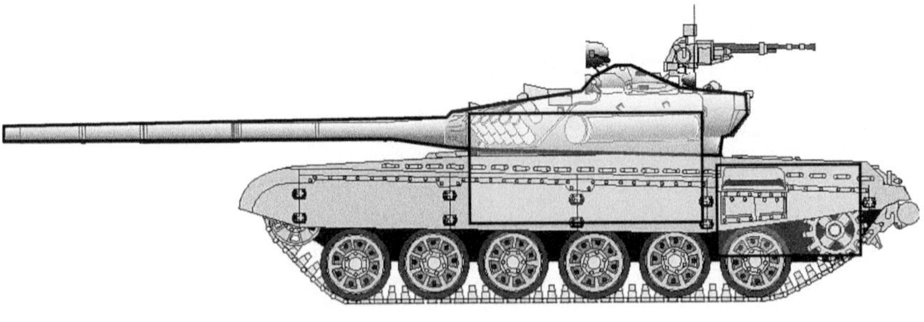

Fig. 6.14 Critical components for a firepower kill.

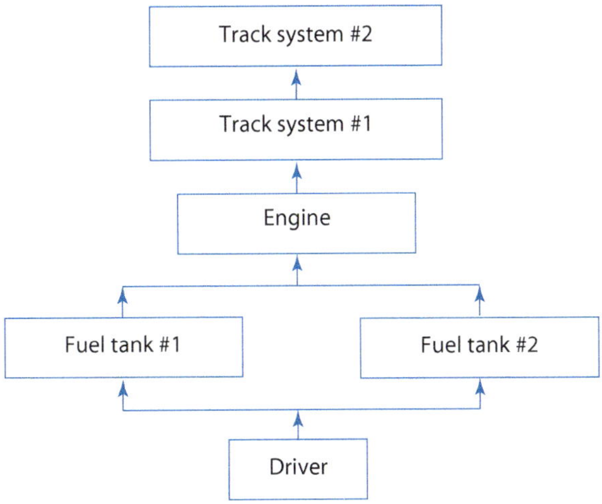

Fig. 6.15 FALT for a mobility kill.

A mobility kill is achieved if

- The driver is killed, OR
- Both fuel tanks are killed, OR
- The engine is killed, OR
- Either track system is killed.

The FALT tree would take the form shown in Fig. 6.15.

Traditionally* the flow starts from the bottom and moves upwards; if the flow is unbroken, the target survives. It can be seen that the OR operation translates to elements in series whereas the AND operation becomes parallel paths.

6.11 Target Vulnerability to Single Fragments

We will now analyze the vulnerability of a target to the penetration of a single fragment, and hence calculate $P_{K/H}$. In doing so, it is understood that the following conditions apply:

- The appropriate kill level has been defined.
- The direction the fragment approaches the target is given.
- The mass and velocity of the fragment is known.

Recall that this is the first step in our analysis—the vulnerability study, where we select an appropriate fragment mass and velocity. The probability a target survives a single-fragment hit on the target is: (the probability the first critical

*JTCG/ME documentation has flow from top to bottom.

component survives) AND (the probability the second critical component survives) AND . . . , and so on. If the target comprises N nonredundant critical components only, this condition may be written mathematically as

$$P_{S/H} = P_{s1/H} \times P_{s2/H} \times \cdots \times P_{sN/H} = \prod_{i=1}^{N} P_{si/H} \tag{6.14}$$

where $P_{si/H}$ is the probability the ith critical component survives a hit on the target. We can write

$$P_{S/H} = \prod_{i=1}^{N} P_{si/H} = \prod_{i=1}^{N} (1 - P_{ki/H}) \tag{6.15}$$

The probability the target is killed, given the target is hit, is

$$P_{K/H} = 1 - P_{S/H} = 1 - \prod_{i=1}^{N} (1 - P_{ki/H}) \tag{6.16}$$

The probability that the ith component is killed, given that the target is hit, can be expressed in the form

$$P_{ki/H} = P_{hi/H} \times P_{ki/hi} \tag{6.17}$$

where $P_{hi/H}$ is the probability the ith component is hit, given the target is hit, and $P_{ki/hi}$ is the probability the ith component is killed if the ith component is hit.

As mentioned before, the value of $P_{ki/hi}$ may be determined from the fragility curves obtained experimentally by firing different fragment weights at different velocities at the component. If the probability of a fragment hitting the ith component is uniformly distributed over the presented area of the target, we can write

$$P_{hi/H} = A_{pi}/A_P \tag{6.18}$$

Because it is also known from the definition of the vulnerable area in Eq. (6.13) that

$$P_{ki/h} = A_{vi}/A_{pi} \tag{6.19}$$

then substituting Eqs. (6.18) and (6.19) into Eq. (6.17) gives

$$P_{ki/H} = \frac{A_{pi} \times P_{ki/hi}}{A_P} = \frac{A_{vi}}{A_P} \tag{6.20}$$

Substituting into Eq. (6.16) results in

$$P_{K/H} = 1 - \prod_{i=1}^{N} \left(1 - \frac{A_{pi} \times P_{ki/hi}}{A_P}\right) \tag{6.21}$$

Another useful form of Eq. (6.21) is

$$P_{K/H} = 1 - \prod_{i=1}^{N} (1 - P_{hi/H} \times P_{ki/hi}) \tag{6.22}$$

These equations will now be applied to four cases that deal with the possible combinations of overlapping and nonoverlapping redundant and nonredundant components. Remember that in the analysis that follows, we are considering the effect of a single fragment of known mass and velocity impacting the target from a defined direction. The analysis of multiple fragments will be dealt with later.

6.12 Case A: Target Composed of Nonredundant Critical Components with no Overlap

Consider the target shown in Fig. 6.16 comprising an armored vehicle viewed from the left side with three nonredundant critical components: the magazine, the engine, and a fuel tank.

The objective of the analysis is to determine the probability the target can be killed by a single fragment approaching it from the left side, the aspect shown in Fig. 6.16. The data needed to compute this value consist of the presented area A_{pi} of each of the vulnerable components, the corresponding values of $P_{ki/hi}$ from their fragility curves, and the presented area of the target A_P. We could use Eq. (6.21) to compute $P_{K/H}$, but a simpler computational form will be derived in a moment. For illustrative purposes, the individual component vulnerable areas will be computed from Eq. (6.13).

Before doing this, however, there are some modifications to Eq. (6.21), which enforces the specification that we have nonoverlapping, nonredundant

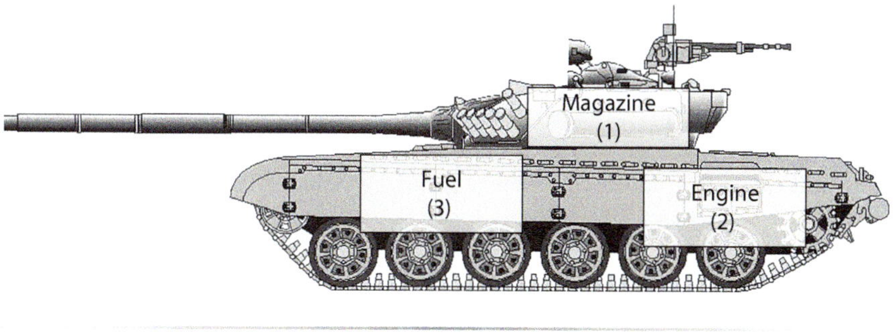

Fig. 6.16 Target with three nonoverlapping, nonredundant components.

components. For this system, $N = 3$, so expanding Eq. (6.16), we get

$$P_{K/H} = (P_{k1/H} + P_{k2/H} + P_{k3/H}) - (P_{k1/H}P_{k2/H} + P_{k1/H}P_{k3/H} + P_{k2/H}P_{k3/H}) + P_{k1/H}P_{k2/H}P_{k3/H} \tag{6.23}$$

Because there is only one fragment and the critical components do not overlap, all products of probabilities must be zero, so

$$P_{K/H} = (P_{k1/H} + P_{k2/H} + P_{k3/H}) = \sum_{i=1}^{3} P_{ki/H} \tag{6.24}$$

Suppose we are interested in a K-kill and have identified the three critical components that will independently produce this level of damage. It is possible to define a kill tree for this target in terms of those critical components, as shown in Fig. 6.17.

The serial nature of the kill tree is perhaps counterintuitive, but it is the accepted way to represent the logic. It indicates that the target is killed if the magazine is killed OR the fuel system is killed OR the engine is killed. Combining events with individual probabilities of occurrence using an OR statement is achieved by adding their probabilities, as indicated by Eq. (6.24).

There are other expressions that may be derived from Eq. (6.24); for example, substituting for $P_{k/Hi}$ from Eq. (6.20),

$$P_{K/H} = \sum_{i=1}^{3} \frac{A_{vi}}{A_P} = \frac{1}{A_P} \sum_{i=1}^{3} A_{vi} \tag{6.25}$$

This allows us to compile a table listing for each critical component in the presented area, the probability of a kill given a hit, vulnerable area, and that component's contribution to the probability of killing the complete target. Table 6.5 shows both input data and computed values for a specific fragment.

In this table, the presented areas and $P_{ki/hi}$ entries represent known data; the rest is computed. The column for A_{vi} is computed from Eq. (6.13) and the $P_{ki/H}$ from Eq. (6.20). The final $P_{K/H}$ can be computed from Eq. (6.24) or Eq. (6.25). The end result of this process is the computation of vulnerable area for the specified fragment mass, velocity, and direction (i.e., 68 ft^2), although the value of $P_{K/H}$ is also used.

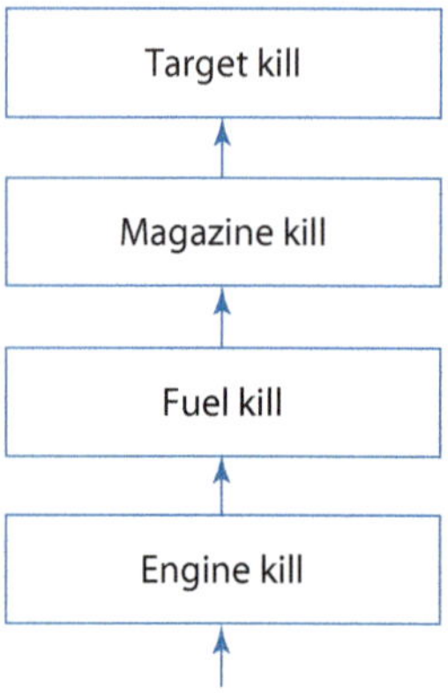

Fig. 6.17 Kill tree for case A.

Table 6.5 Computations for Case A

Component	A_{pi} (ft^2)	$X\ P_{ki/hi}$	$= A_{vi}$ (ft^2)	$P_{ki/H}$
1. Magazine	20.0	1.0	20.0	0.0667
2. Engine	50.0	0.6	30.0	0.1000
3. Fuel	60.0	0.3	18.0	0.0600
$A_P = 300.0$ ft^2			$\Sigma A_{Vi} = 68.0$	$P_{K/H} = \Sigma P_{ki/H} = 0.2267$

6.13 Case B: Target Composed of Nonredundant Critical Components that Overlap

When critical components overlap in a particular view, as shown in Fig. 6.18, a simplifying assumption, will be made, namely that when a fragment passes through the first component it does not lose mass or velocity. This implies it does not matter which overlapping component is struck first.

Because we are still dealing with nonredundant critical components, the kill tree is the same as in Fig. 6.17; however, Eq. (6.24) cannot be used, because the overlap region makes some of the products of probabilities in Eq. (6.23) nonzero. For example, a single fragment could kill both the engine and fuel tank, or either one.

The process for dealing with this case is to treat the overlap region as a separate nonredundant component (denoted as #4 in the following list) with its own presented area and $P_{ki/hi}$. The area is the common area of overlap between the fuel tank and engine for this aspect, which we will say is 20 ft^2, but the question is: What is its $P_{ki/hi}$? Consider a fragment passing through the overlap region. It will cause one of the following four events to occur:

1. The engine is killed, but the fuel tank survives.
2. The fuel tank is killed, but the engine survives.
3. Both the engine and fuel tank are killed.
4. Both the engine and fuel tank survive.

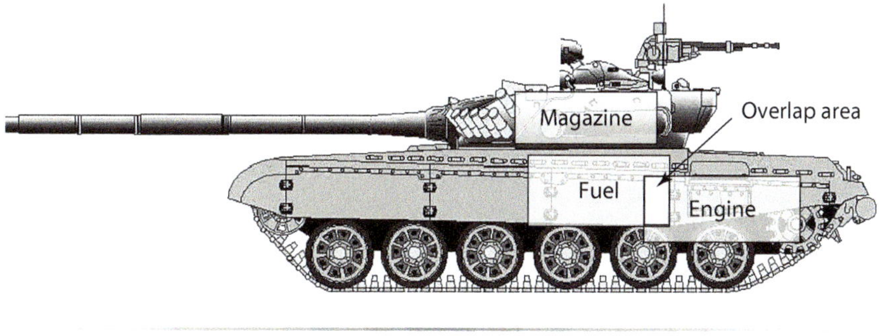

Fig. 6.18 Target with two overlapping nonredundant components.

Table 6.6 Computations for Case B

Component	A_{pi} (ft^2)	$X\ P_{ki/hi}$	$= A_{vi}$ (ft^2)	$P_{ki/H}$
1. Magazine	20.0	1.0	20.0	0.0667
2. Engine	30.0	0.6	18.0	0.0600
3. Fuel	40.0	0.3	12.0	0.0400
4. Overlap	20.0	0.72	14.4	0.0480
$A_P = 300.0$ ft^2			$\Sigma A_{Vi} = 64.4$	$P_{K/H} = \Sigma P_{ki/H} = 0.2147$

The fourth possible outcome, the probability that the engine and fuel system both survive, is the only one not leading to a target kill. It may be observed that the probability that one of the first three events occurs is equal to 1 – {the probability that both engine and fuel tank survive a single fragment hit in the overlap area}. Hence we can write

$$\text{Probability\{engine survives a hit\}} = 1 - P_{k2/h2}$$

$$\text{Probability\{fuel tank survives a hit\}} = 1 - P_{k3/h3}$$

$$\text{Probability\{engine AND fuel tank survive a hit on the overlap\}} = (1 - P_{k2/h2})(1 - P_{k3/h3})$$

$$\text{Prob\{engine OR fuel OR both are killed by a hit on the overlap\}} = 1 - (1 - P_{k2/h2})(1 - P_{k3/h3}) \tag{6.26}$$

So from Eq, (6.26) and the data in Table 6.5 we can calculate

$$P_{k4/h4} = 1 - 0.7 \times 0.4 = 0.72 \tag{6.27}$$

This is the $P_{ki/hi}$ used for the common overlap area. Now we can produce a revised table (Table 6.6) similar to Table 6.5 for this case. Note how the presented areas of the remaining fuel and engine have been reduced by the overlap area.

We see from Table 6.6 that the total $P_{K/H}$ has gone down, as has the vulnerable area of the target.

6.14 Case C: Target Composed of Some Redundant Critical Components with no Overlap

The target now consists of two fuel tanks, as shown in Fig. 6.19, such that the target can move with only one fuel tank not killed, but in the view under consideration, they do not overlap.

It will be assumed that the second fuel tank has the same presented area as the first, but to distinguish it we will further assume that $P_{ki/hi} = 0.4$. The presence of a redundant component changes the kill tree (FALT), as shown in

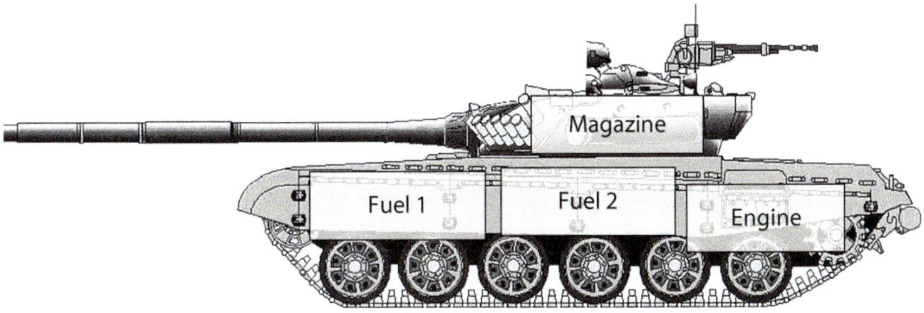

Fig. 6.19 Target with two nonoverlapping redundant components.

Fig. 6.20. The parallel paths through the two fuel tanks indicate that both tanks have to be killed in order to kill the target. Note that this kill tree is also dependent on the kill criteria considered.

The two fuel tanks may be thought of as comprising the fuel supply group. The probability the target survives a single hit is given by a modified form of Eq. (6.14)

$$P_{S/H} = P_{s1/H} \times P_{s2/H} \times P_{sF/H} \tag{6.28}$$

where $P_{s/HF}$ is the probability that the fuel supply group survives a single hit. However, because there is no overlap, the fuel supply group must survive a single hit, because for the view shown, a single fragment cannot kill both fuel tanks. Hence, $P_{s/HF} = 1$, and Eq. (6.28) is used to write the following:

$$\begin{aligned} P_{K/H} &= 1 - P_{S/H} = 1 - P_{s1/H} \times P_{s2/H} \\ &= 1 - (1 - P_{k1/H})(1 - P_{k2/H}) \end{aligned} \tag{6.29}$$

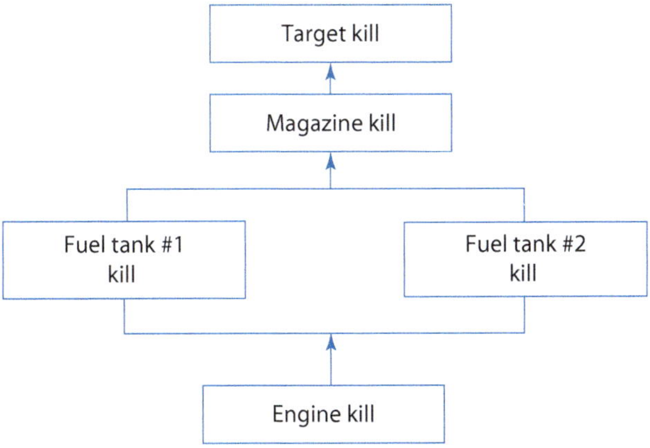

Fig. 6.20 Kill tree for case C.

Table 6.7 Computations for Case C

Component	A_{pi} (ft^2)	X $P_{ki/hi}$	= A_{vi} (ft^2)	$P_{ki/H}$
1. Magazine	20.0	1.0	20.0	0.0667
2. Engine	50.0	0.6	30.0	0.1000
3. Fuel tank 1	60.0	0.3	18.0	0.0600
4. Fuel tank 2	60.0	0.4	24.0	0.0800
$A_P = 300.0$ ft^2			$\Sigma A_{Vi} = 50.0$	$P_{K/H} = \Sigma P_{ki/H} = 0.1667$

This means that for a single fragment, only the magazine and engine contribute to the target kill, and not the fuel tanks. This enables us to compile Table 6.7 for this case.

Note that in this table, the $P_{ki/H}$ and A_{vi} values corresponding to the redundant components (shaded cells) are not included in the summation for the whole target $P_{K/H}$ and vulnerable area, just the magazine and engine. Clearly if more than one fragment hits the target, the possibility of killing both fuel tanks would have to be considered.

6.15 Case D: Target Composed of Some Redundant Critical Components that Overlap

The vehicle shown in Figure 6.21 represents this situation. The nonoverlapping parts of the fuel tanks will be treated as redundant components that contribute nothing to the target vulnerability or vulnerable area as in the previous case. The overlap region will, however, contribute to both.

As before, the overlap region will be treated as a separate component (#5) where its presented area is measured from the view considered. Assume again this overlap area is 20 ft^2. The same question remains: What is the appropriate $P_{ki/hi}$ for this "component"?

In order for the overlap area to contribute to the target vulnerability, both fuel tanks must be killed by a single fragment, because they form the critical

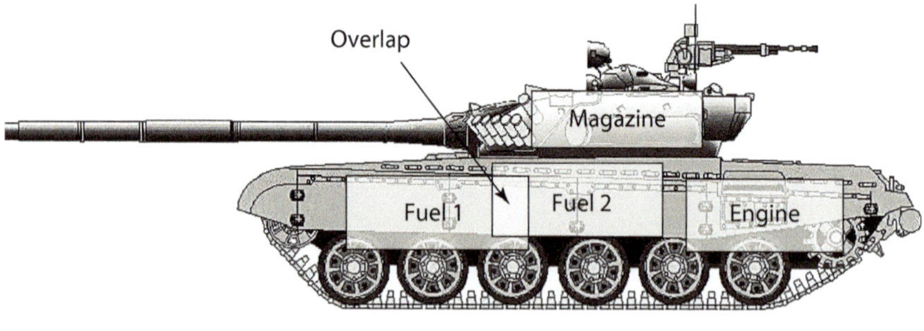

Fig. 6.21 Target with two overlapping redundant components.

Table 6.8 Computations for Case D

Component	A_{pi} (ft^2)	X $P_{ki/hi}$	= A_{vi} (ft^2)	$P_{ki/H}$
1. Magazine	20.0	1.0	20.0	0.0667
2. Engine	50.0	0.6	30.0	0.1000
3. Fuel tank 1	40.0	0.3	12.0	0.0400
4. Fuel tank 2	40.0	0.4	16.0	0.0533
5. Overlap	20.0	0.12	2.4	0.0080
$A_P = 300.0$ ft^2			$\Sigma A_{Vi} = 52.4$	$P_{K/H} = \Sigma P_{ki/H} = 0.1747$

components of the fuel supply group. The probability this will occur is simply the product of the individual $P_{ki/hi}$ values (probability fuel tank 1 AND fuel tank 2 killed). Hence if component #5 is considered to be the overlap region, we can write directly

$$P_{k5/h5} = P_{k3/h3} \times P_{k4/h4} = 0.4 \times 0.3 = 0.12 \tag{6.30}$$

Using 20 ft^2 for the overlap area, Table 6.8 may be compiled as before.

In cases such as this, a contribution to the target vulnerability will occur only if all redundant components can be killed by a single fragment.

It is important to remember that in the preceding analysis of the four cases, the $P_{K/H}$ computed is the single impact probability of kill, that is, a vulnerability measure assuming the target is impacted by a single fragment. These four cases will now form the basis for investigating what happens when the target is struck not by one, but by multiple fragments.

6.16 Multiple-Hit Vulnerability

In studying the multiple-hit vulnerability of a target, the example that will be considered is the nonoverlapping redundant case given in Section 6.12. Recall Eq. (6.28)

$$P_{S/H} = P_{s1/H} \times P_{s2/H} \times P_{sF/H} \tag{6.28}$$

If the target is hit by multiple fragments, we cannot write $P_{s/HF} = 1$ because both fuel tanks in the fuel supply group could be killed by different fragments. Instead of Eq. (6.28), we have to write

$$P_{S/H} = (1 - P_{k1/H})(1 - P_{k2/H})(1 - P_{kF1/H}P_{kF2/H}) \tag{6.31}$$

where the last term is interpreted as the probability that either tank 1 (F1) or tank 2 (F2) or both survive given a hit on the target. If we extend this to n hits, we can write

$$P^{(n)}_{S/H} = \left(1 - P^{(n)}_{k1/H}\right)\left(1 - P^{(n)}_{k2/H}\right)\left(1 - P^{(n)}_{kF1/H}P^{(n)}_{kF2/H}\right) \tag{6.32}$$

Consider just the first critical component. In this equation, $P^{(n)}_{k1/H}$ is the probability that component 1 will be killed after the target has been hit n times. We can write this in terms of $P_{k1/H}$, the probability the component is killed after one hit on the target,z by powering up the single probability of kill given a hit on the target.

For one hit

$$P_{s1/H} = 1 - P_{k1/H} \tag{6.33}$$

For n hits

$$P^{(n)}_{s1/H} = \prod_{j=1}^{n} (1 - P_{k1/Hj}) \tag{6.34}$$

where $P_{k1/Hj}$ is the probability component 1 is killed by the jth of n hits on the target. Because $P_{k1/Hj}$ is independent of j for identical fragments, we may write

$$P^{(n)}_{s1/H} = (1 - P_{k1/H})^n \tag{6.35}$$

leading to

$$P^{(n)}_{k1/H} = 1 - (1 - P_{k1/H})^n \tag{6.36}$$

Similarly, we obtain for the other terms of Eq. (6.32)

$$P^{(n)}_{k2/H} = 1 - (1 - P_{k2/H})^n \tag{6.37}$$

and

$$P^{(n)}_{kF1/H} P^{(n)}_{kF2/H} = [1 - (1 - P_{kF1/H})^n][1 - (1 - P_{kF2/H})^n] \tag{6.38}$$

Finally, we obtain from Eq. (6.32)

$$P^{(n)}_{K/H} = 1 - P^{(n)}_{S/H} = 1 - (1 - P_{k1/H})^n (1 - P_{k2/H})^n \\ \times \left\{1 - [1 - (1 - P_{kF1/H})^n][1 - (1 - P_{kF2/H})^n]\right\} \tag{6.39}$$

Note that this gives similar, though not identical, results as the simple powering up formula, also shown in the table.

$$P^{(n)}_{K/H} = 1 - P^{(n)}_{S/H} = 1 - (1 - P_{k1/H})^n \tag{6.40}$$

Using the data for $P_{ki/H}$ from Table 6.7 derived earlier, we can compile Table 6.9, which shows how the $P_{K/H}$ increases from the single fragment case when the number of identical fragment impacts increases for both Eqs. (6.39) and (6.40).

Table 6.9 $P_{K/H}$ for Multiple Hits

Multiple Hit $P_{K/H}$							
$P_{K1/H} = 0.1667$		$P_{K2/H} = 0.1000$		$P_{KF1/H} = 0.0600$		$P_{KF2/H} = 0.0800$	
$n \rightarrow$	2	3	4	5	10	20	30
Eq. (6.39)	0.307	0.430	0.533	0.620	0.871	0.987	0.999
Eq. (6.40)	0.306	0.422	0.518	0.598	0.839	0.974	0.996

In most cases, the simpler powering up formula of Eq. (6.40) is used without significant error.

6.17 Effectiveness Assessment for a Specific Weapon

So far, the analysis has focused on a vulnerability assessment of the target, that is, calculating the vulnerability to a hypothetical single fragment, or multiple fragments of the same mass and velocity, as specified by the analyst. Now attention is turned to the second stage of the analysis defined in Section 6.3, the effectiveness study.

Knowing the $P_{K/H}$ as a function of the number of hits, it is possible to calculate the probability of kill for a particular fragmentation warhead. To do this, it is necessary to calculate the expected number and mass of fragments impacting the target, and clearly the characteristics of the warhead and its proximity to the target at detonation have to be taken into account. In order to emphasize the nature of the process, a simplified treatment of weapon/target interaction will be given here, and a more detailed analysis provided in the next chapter.

The fragment pattern projected from a particular warhead is specified in terms of the Z-data file discussed previously, another sample of which is reproduced in Table 6.10.

Now consider the relationship between the weapon at detonation and the target, shown in a plan view in Fig. 6.22.

We make the following simplifying assumptions regarding the scenario shown in Fig. 6.22:

- The fragmentation zone corresponds approximately to that shown in Table 6.10, except we further assume that fragments smaller than 5 grains do not penetrate the skin of the target and may be ignored.

Table 6.10 Fragment Characteristics

α's			Velocities			# Wts
0.000	20.000	10.000	5500	4770	5010	5

Frag weight →	0.100	4.336	16.303	33.862	104.337
# of frags →	2.927	52.692	40.128	34.273	16.709

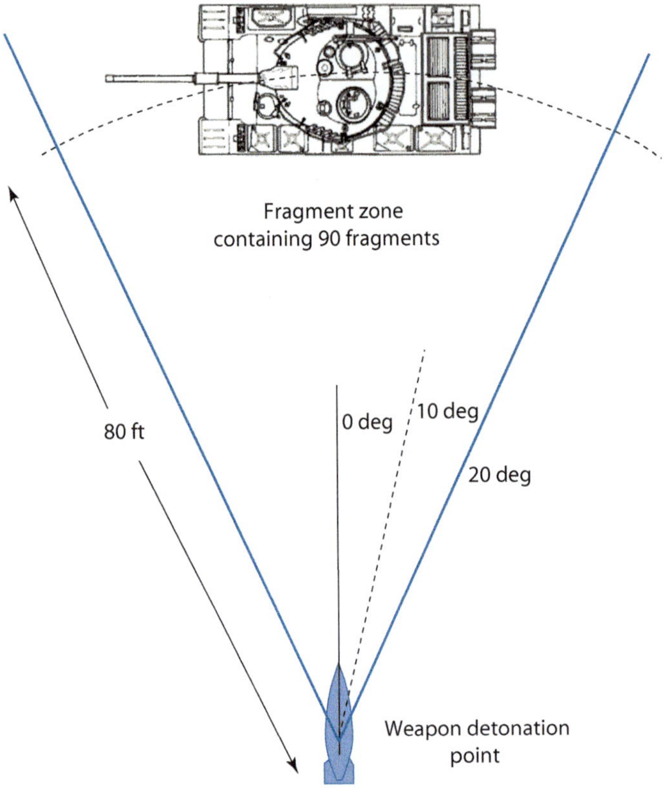

Fig. 6.22 Engagement geometry.

- From the Z-data file, we assume a total of $K = 90$ fragments (40 + 34 + 16) are contained in the fragment zone that includes the target.
- The target corresponds to that in Section 6.12, case A, the nonredundant, nonoverlapping model having $A_V = 68\ \text{ft}^2$.
- Although Table 6.10 shows fragments of different weights within the zone, we assume that all fragments have the same mass and velocity corresponding to the values used to derive Table 6.5.
- The weapon detonates at a distance $r = 80$ ft from the target and is pointed towards the target at detonation.
- The velocity of a fragment is not reduced as it travels from the detonation point to the target.

In calculating the number of hits on the target, the three-dimensionality of the fragment spray pattern has to be considered by using some simple elements of solid geometry. Consider a cone formed by a solid angle Ω and a spherical cap on it, as shown in Fig. 6.23.

The solid angle Ω measured in steradians (sr) in terms of the conical half-angle φ in radians is defined by the following:

$$\Omega = 2\pi(1 - \cos\varphi) \tag{6.41}$$

The area of the spherical cap subtending the solid angle Ω at the apex is

$$A_S = \Omega r^2 \tag{6.42}$$

If we have a hollow cone, such as fragment zone 2, where the angular boundaries are defined by φ_1 and φ_2, Eqs. (6.41) and (6.42) lead to

$$A_S = 2\pi r^2(\cos\varphi_1 - \cos\varphi_2) \tag{6.43}$$

where φ_1 and φ_2 are the angles defining the zone measured from the warhead axis. It is assumed that all fragments in the zone lie on the spherical cap.

For the case considered in Fig. 6.22, the 90 fragments are contained within a 40-deg cone. This cone corresponds to a solid angle [Eq. (6.41)] of $\Omega = 0.379$ steradians. At a range of 80 ft from the detonation point, this zone will project a spherical area A_Z where, from Eq. (6.42), $A_S = 2425$ ft^2. Recognize though that not all of the 90 fragments contained on the spherical cap will impact the target. Assuming the fragments are uniformly distributed over the spherical cap area, the number hitting the target is seen to be

$$n = \frac{A_P}{A_S} \times K = \frac{300}{2425} \times 90 \approx 11 \tag{6.44}$$

Here we make a somewhat sweeping assumption that will be relaxed later: Because there are several fragment weight groups represented in the uniform distribution of the fragment zone, it is assumed that there is a single, "average" fragment weight represented by the 90 fragments in the zone. It is further assumed that this single fragment weight is the same weight used to generate Table 6.5.

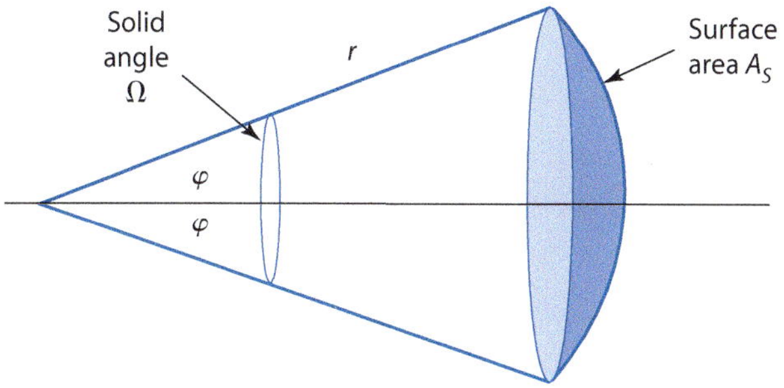

Fig. 6.23 Solid angle of a cone.

The $P_{K/H}$ could be computed from a table similar to Table 6.9 derived for the case being considered,† resulting in the value being close to, but greater than, 0.871; however, a simplified analytical approach will be given here. We can write that the probability the target survives a warhead detonation $P_{S/D}$ resulting in n fragments is given by

$$P_{S/D} = P^{(1)}_{S/H} \times P^{(2)}_{S/H} \times P^{(3)}_{S/H} \times \cdots \times P^{(n)}_{S/H} \tag{6.45}$$

Here $P^{(j)}_{S/H}$ is the probability the target survives the jth hit. Alternatively, in terms of the probability of kill given a detonation

$$\begin{aligned} P_{K/D} &= 1 - P_{S/D} = 1 - \left(1 - P^{(1)}_{K/H}\right)\left(1 - P^{(2)}_{K/H}\right) \cdots \left(1 - P^{(n)}_{K/H}\right) \\ &= 1 - \prod_{j=1}^{n} \left(1 - P^{(j)}_{K/H}\right) \end{aligned} \tag{6.46}$$

Now we use the Morse-Kimball approximation, which shows that for small x,

$$\prod_{j=1}^{n} (1 - x_j) \approx \exp\left(-\sum_{j=1}^{n} x_j\right) \tag{6.47}$$

Applying this approximation to Eq. (6.46), we get

$$P_{K/D} = 1 - \exp\left(-\sum_{j=1}^{n} P^{(j)}_{K/H}\right) \tag{6.48}$$

Because all fragments striking the target are assumed to be identical, $P_{K/H}$ is constant, allowing Eq. (6.48) to be written in a computationally simple form

$$P_{K/D} = 1 - \exp\left(-nP_{K/H}\right) = 1 - \exp\left(-n\frac{A_V}{A_P}\right) \tag{6.49}$$

Using Eq. (6.44), it follows that

$$\begin{aligned} P_{K/D} &= 1 - \exp\left(-K\frac{A_P}{A_S} \times \frac{A_V}{A_P}\right) = 1 - \exp\left(-\frac{KA_V}{r^2\Omega}\right) \\ &= 1 - \exp\left(-\rho A_V\right) \end{aligned} \tag{6.50}$$

In Eq. (6.50), ρ is the range specific fragment density over the spherical surface. It has units of fragments/ft^2 and is calculated from Eq. (6.50) as

$$\rho = \frac{K}{r^2\,\Omega} = \frac{90}{80^2 \times 0.379} = 0.0371 \tag{6.51}$$

---

†The equation used instead of Eq. (6.39) would be

$$P^{(n)}_{K/H} = 1 - \left(1 - P_{K1/H}\right)^n \left(1 - P_{K2/H}\right)^n \left(1 - P_{K3/H}\right)^n$$

Using the vulnerable area of the target as 68 ft^2 from Table 6.5, the probability of damage given a detonation becomes

$$P_{K/D} = 1 - \exp(-0.0371 \times 68) = 0.920 \tag{6.52}$$

If we wished to use the number of fragments striking the target to compute $P_{K/D}$, Eq. (6.49) would have given the same result.

6.18 Centroid of Vulnerability and Different Fragment Weights

In the preceding analysis, there should be a couple of issues troubling the reader if we reconsider the relationship between the target and warhead shown in Fig. 6.22.

- From the Z-data file, it is seen that even if we assume a single fragment mass for the 90 fragments, the fragment velocity depends on the angular offset into the zone α. Because the target spans the zone, the fragment velocity will vary; however, the vulnerable area of 68 ft^2 assumes a single velocity. What velocity should be used?
- If the target is moved closer to the detonation point, it will eventually spill over into the next fragment zone, and we have to consider how we will account for part of the target being in zone 1 while part is in zone 2. It becomes mathematically intractable to divide the target into zones as we have done for the warhead.

Clearly a different approach of looking at the target/weapon interaction is needed to resolve these issues, and in doing so the nature of the target will be changed.

We now change the physical representation of the target from a three-dimensional object of fixed dimensions to a single point, known as the centroid or center of vulnerability (COV), together with the associated vulnerable area given in Table 6.5. This transformation is shown in Fig. 6.24.

In doing this transformation, the target loses all geometric and functional form developed in the vulnerability study; however, the $P_{K/D}$ may still be calculated using Eq. (6.52).

$$P_{K/D} = 1 - \exp(-\rho_F A_V) \tag{6.53}$$

This transformation resolves the two issues previously listed: as a point, all fragments will have a common velocity defined by a single offset into the zone, and as a point, it cannot span two zones. In addition, it might have been noticed that the common fragment velocity depends only on the angular offset into the zone, which would vary for a target of finite dimensions, but not for a point.

We are now in a position to revisit the rather drastic assumption that all fragments hitting the target have the same mass. The Z-data file shown in Table 6.10 indicates three out of the five total fragment weight groups are

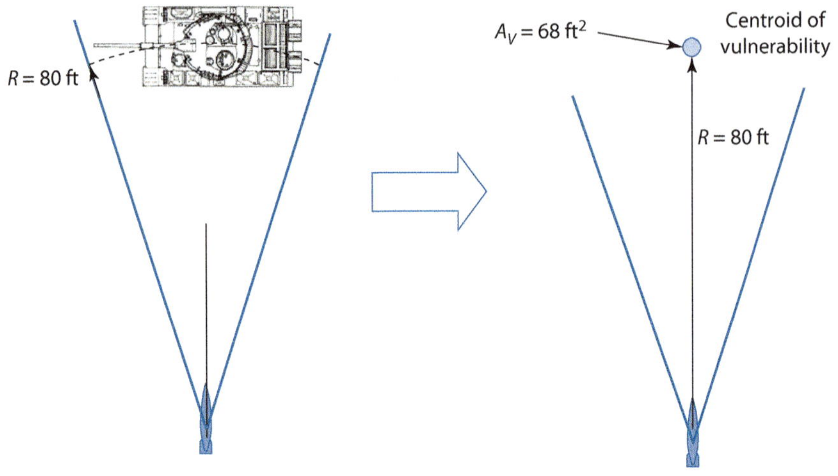

Fig. 6.24 Transforming target to the centroid of vulnerability.

used (>5 grains), leading to three different fragment densities.

$$\rho_F(j) = \frac{K(j)}{r^2\,\Omega} \tag{6.54}$$

The densities are different because there are different numbers of fragments belonging to each group, and this leads to a different $P_{K/D}$ for the weight group given by Eq. (6.53). Generalizing Eq. (6.50) by applying it to each weight group in the Z-data file, rather than a single aggregated fragment weight, we can write

$$P_{K/D}(j) = 1 - \exp(-\rho_F(j)A_V(j)) \tag{6.55}$$

Observe that not only the fragment density is dependent on the weight group due to varying numbers of fragments, but also the vulnerable area of the target due to the varying weights. For example, the 68 ft^2 assumed vulnerable area for all common fragment weights would now have three vulnerable areas corresponding to the average fragment weight shown in the Z-data file. These would have to be evaluated by repeating the process given in Section 6.12 for the fragment weights in the Z-data file. The total $P_{K/D}$ would be given by using the survivor rule for all fragment weight groups

$$P_{K/D} = 1 - \prod_{j=1}^{N}\left(1 - P_{K/D}(j)\right) \tag{6.56}$$

where there are a total of N fragment weight groups. Note that in this approach, no use of the presented area of the target is made. Equation

(6.56) would also allow us to relax the condition on minimum fragment weights and include all those shown in the Z-data file.

6.19 Damage Matrix and Lethal Area

The value of $P_{K/D}$ just computed is only valid at the detonation standoff of 80 ft. At a larger range, the $P_{K/D}$ will decrease because fewer fragments impact the target (i.e., the fragment density will decrease). For example, at 120-ft range we get $P_{K/D} = 0.455$. It is possible to plot $P_{K/D}$ against detonation range for this case and obtain the graph shown in Fig. 6.25.

Suppose we consider the warhead fixed and the location of the target to vary along the weapon axis. This figure shows how the probability of killing the target varies with distance from the point of detonation to the centroid of vulnerability in the direction of the weapon axis.

The question now arises: What if the target, again represented by the center of vulnerability, is located relative to the weapon at some point off the weapon axis?

This situation is shown in Fig. 6.26, which is a plan view of the ground plane and the weapon/target engagement scenario. The weapon is fixed at the origin and the target placed relative to it. Consider covering the ground plane with a grid of cells of selected dimensions. We place the target at coordinates (x, y) at the center of an arbitrary cell having a fixed size dx × dy ft.

For this location, the value of $P_{K/D}$ is calculated in the same way as before except that now the target may be located in a different fragment zone. If the cell size is small, it may be assumed that the calculated value of $P_{K/D}$ is constant within it; however, in an adjacent cell, the value of $P_{K/D}$ would probably be different. Representing the target by the centroid of vulnerability ensures that it will be located within a single fragmentation zone. By placing the target at the center of all cells in the ground plane and calculating the $P_{K/D}$ in each cell, we generate the so-called damage matrix (Fig. 6.27).

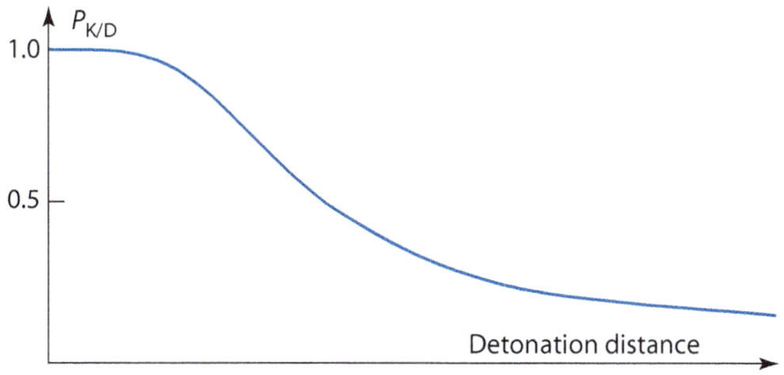

Fig. 6.25 $P_{K/D}$ plotted against detonation distance.

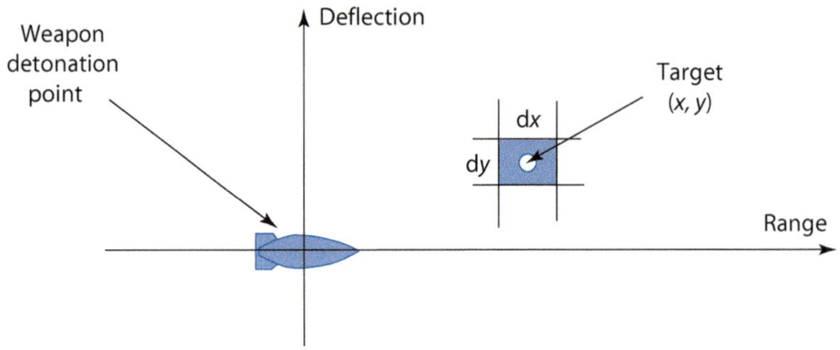

Fig. 6.26 Calculation of $P_{K/D}$ at a point (x, y).

Note how, in practice, the value of $P_{K/D}$ becomes zero at considerable distance from the detonation point. This is because when air resistance on the fragments is accounted for, they slow to such a small velocity they fail to penetrate the target. The lethal area matrix will therefore be bounded.

The lethal area is calculated by multiplying each cell area by the value of $P_{K/D}$ inside it and summing over all cells with a nonzero $P_{K/D}$. This area may be written mathematically as

$$A_L = \text{MAE}_F = \sum_{x=x_{\min}}^{x=x_{\max}} \sum_{y=y_{\min}}^{y=y_{\max}} P_{K/D} \, \Delta y \, \Delta x \tag{6.57}$$

This represents the final result from our vulnerability study and produces a single lethal area number from Eq. (6.57) with units of square feet. These are how the lethal area numbers shown in Table 6.2 are derived, and how the lethal area is equated to the mean area of effectiveness MAE_F, one of the EI values discussed earlier.

Observe that the effectiveness has been represented in two ways.

0	0	0	0	0	0	0	0	0	0	0	0	0	0
0	0	0	0	0	0.001	0	0	0.001	0	0	0	0	0
0	0	0	0	0.0	0.002	0.003	0.003	0.002	0.001	0	0	0	0
0	0.001	0.002	0.003	0.004	0.007	0.014	0.014	0.007	0.004	0.003	0.002	0.001	0
0	0	0	0.002	0.008	0.04	.227	.227	0.04	0.008	0.002	0	0	0
0	0	0	0	0.001	0.002	0.049	0.049	0.002	0.001	0	0	0	0
0	0	0	0	0	0	0.002	0.002	0	0	0	0	0	0
0	0	0	0	0	0	0	0	0	0	0	0	0	0

Fig. 6.27 Damage matrix.

1. The lethal area matrix shows the distribution of effectiveness of the weapon against the target for specific placement of the latter.
2. The EI is a single number representation of the lethal area lacking any information about the shape of the effectiveness distribution.

It is also important to remember the following points:

- The vulnerable area, $P_{K/H}$, and lethal area are dependent upon the fragment weight and velocity. Recall Fig. 6.9 to realize that changing the weight and velocity will change $P_{ki/hi}$, thus changing all the computations from Section 6.12 onwards.
- All computations are dependent on the aspect of the target. The cases considered assume fragments approach the target from the left-hand side. If this direction is changed to, say, from above the target, the $P_{ki/hi}$ will also change because different noncritical components may shield the critical components to a greater or lesser degree, depending upon the physical layout of the target. In addition, components that may overlap in one view may not overlap in another view (e.g., the two fuel tanks).
- All the measures discussed are dependent on the kill level of interest. Again recall that the fragility curves shown in Fig. 6.9 are defined for a specific kill level. If the kill level becomes more severe, for example, the $P_{k/h}$ values will change and probably decrease.

Clearly, some way has to be found to take these factors into account when determining the lethal area values required by the weaponeering process. Unfortunately this requires employing more detailed techniques than those described in this chapter, and numerical answers may only be obtained by using computer programs to implement these more complex methodologies. That being said, we will see in the next chapter that these methodologies have their basis in the somewhat simple methods described here.

6.20 Vulnerability Assessment for Blast Warheads

In this section we consider a target that is vulnerable to blast. This may be due to one of the following:

- A warhead that produces only blast effects
- A general-purpose warhead that produces blast effects in addition to other damage mechanisms (e.g., fragments)

In the latter case, each damage mechanism will be considered separately and then combined, as explained in the next chapter. For now, we consider damage due to blast without the contribution of any other mechanism.

Superficially, blast appears simpler to model than fragments because the blast wave propagates spherically outward from the point of detonation with decreasing amplitude. It is reasonable to argue, therefore, that the amount of

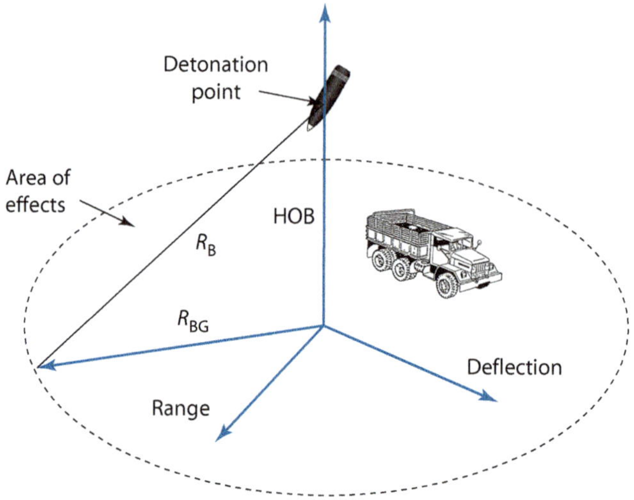

Fig. 6.28 Effects area for blast warhead.

damage sustained by a target is a function only of its distance from the detonation point, as shown in Fig. 6.28.

For a given combination of warhead, target, and kill definition, we define a distance R_B within which the kill is achieved and outside of which the target survives (i.e., a spherical cookie cutter). This distance is simply the slant range SR between the warhead centroid and the target center of vulnerability.

For an air burst weapon, the equivalent ground plane area would be a circle of radius R_{BG} in which

$$R_{BG}{}^2 = R_B{}^2 - \text{HOB}^2 \tag{6.58}$$

where HOB is weapon height of burst. Clearly for HOB > R_B, there will be no ground plane effect due to blast. If there is a ground plane effect, then we can define another effectiveness index—the mean area of effectiveness due to blast (the MAE_B)—where

$$\text{MAE}_B = \pi(R_{BG})^2 \tag{6.59}$$

The next chapter will present details of how R_B is calculated.

References

[1] Grady, D., "The Statistical Fragmentation Theory of N. F. Mott," *Proceedings of the Conference of the American Physical Society Topical Group on Shock Compression of Condensed Matter*, Proceedings of the Conference of the American Physical Society Topical Group on Shock Compression of Condensed Matter. AIP Conference Proceedings, Volume 706, 2004, 455–460.

[2] Ball, R., *The Fundamentals of Aircraft Combat Survivability Analysis and Design*, 2nd ed., AIAA Education Series, AIAA, Reston, VA, 2003.

Chapter 7 Vulnerability Assessment: Implementation

7.1 Introduction

The previous chapter provided an overview of the concepts and methodology to determine the effects of fragments and blast warheads on single targets comprising critical and noncritical components. This was divided into a vulnerability study that produced vulnerable area data to user-specified fragments followed by an effectiveness assessment where the effect of a specific warhead was introduced. Because of the complexity in implementing these two phases, this chapter will focus on only the first of these, and the next chapter will deal with effectiveness calculations.

In the examples used previously, the target was considered to comprise three critical components only, and simplifying assumptions were made regarding the penetration of the target by fragments. Some of these assumptions were:

- The effects of noncritical components were not included.
- No velocity or mass reduction as the fragment penetrated the target was allowed for.
- The order of overlapping components was not considered.
- The fragments penetrated the target from one direction only.

Clearly these restrictions will have to be addressed when real targets are considered, as will the target/weapon geometry.

The material presented in the next two chapters will transition the simple, generalized methods introduced in the previous chapter into those dealing with more realistic target and weapon systems, and in doing so will frame the analysis in terms of those tools used by the U.S. DoD to conduct these analyses.

7.2 Complex Targets Sensitive to Fragments

As in the previous chapter, the vulnerability assessment for more complex targets consists of generating the vulnerable area of the target to single fragments of specified weight traveling at a specified speed and direction when impacting the target.

General methods were developed in order to compute the lethal area of a weapon for a given target; however, more advanced methods are required to deal with more complex targets. For example, subtle changes in target design may provide substantially more shielding of vulnerable components, leading to a reduction in the vulnerable area and mean area of effectiveness (MAE). Similarly, the introduction of redundant components or subsystems may also bring about a reduction in target vulnerability.

The example shown in Fig. 7.1 clearly illustrates a target that is more complex and may be made up of thousands of individual components.

For a target such as this, it is not surprising that the specification of the target, groups, subsystems, individual components, their fragility curves, and kill tree relationships takes considerable time. The level of detail in specifying these data is always traded off against the accuracy of the resulting vulnerability assessment. Higher fidelity models produce more accurate predictions but take more resources to develop.

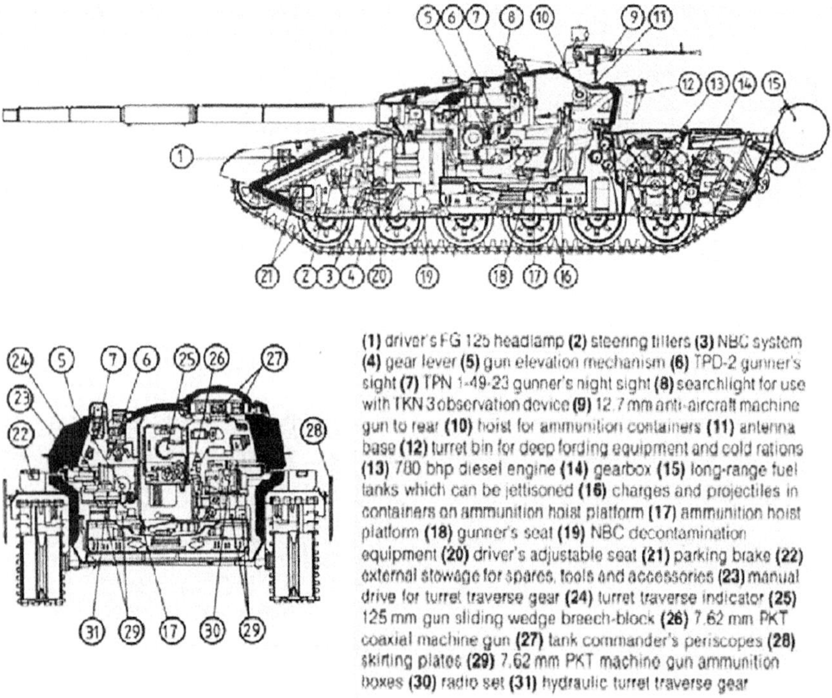

Fig. 7.1 Example of complex target.

In addition, the availability of detailed data on newly fielded "red" systems has to be considered. It may be that only photographs of the target might be available, providing no information on the internal design and layout. In such cases, the analyst may have to resort to surrogation methods from what is perceived to be a close match to a known target for which data are available.

7.3 Overview of Tools Used for Vulnerability Analysis

Although the process has become more complex, the same basic process described in the last chapter will be repeated; that is, a user-defined threat (fragment, projectile, shaped charge) is imposed on a target, resulting in a vulnerable area for a given direction of approach. By initially considering the threat to be fragments, as in the previous chapter, the following requirements now have to be considered:

- Because the target is more complex, a formal description of the target geometric model (TGM) has to be developed in order to include all critical and noncritical components, the materials that make up the components, and the spatial relationship between them within the target itself.
- An equally complex failure analysis logic tree (FALT) has to be developed in order to identify which critical components lead to different kill definitions associated with the target under study.
- Fragment impact directions have to be specified so that the vulnerability from fragments coming from any direction may be determined.
- The loss of fragment velocity and mass as it penetrates through the target and the components it encounters has to be calculated.

These requirements have existed for over 50 years and have been met by a set of tools assisted by increasingly capable computational support. Although many of these tools may be operated in a stand-alone manner, their combined use to achieve overall vulnerable area calculations has resulted in several packages including the Computation of Vulnerable Area Tool (COVART) and the Advanced Joint Effectiveness Model (AJEM). We begin by describing in detail COVART; AJEM is dealt with in Section 7.12.

COVART has been developed with service-wide support to provide a consistent assessment of a specific target's vulnerability to different threats. It contains several modules that address the requirements just listed, and is shown in functional form in Fig. 7.2. Although we will describe COVART in terms of its constituent modules, a more complete description may be found in [1].

In the next few sections we will address each of the four previously listed requirements needed to undertake complex target vulnerability assessment in terms of the modules in COVART that perform the equivalent task.

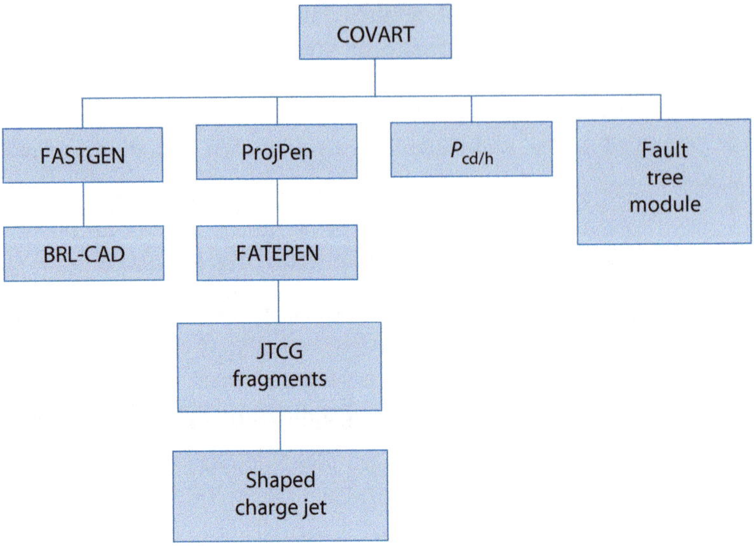

Fig. 7.2 COVART structure.

Again, we will restrict the discussion to those details needed to understand the module function; if more details are needed, the reader is directed to [1] and [2].

7.4 Target Geometric Model (TGM)

The formal representation of a target in terms of its constituent components is in the form of a computer-aided design (CAD) model incorporated into a programming environment known as BRL-CAD [3], as indicated in Fig. 7.2. This system is a powerful open source programming structure running on a variety of platforms. Modeling may be achieved using several commonly accepted modeling techniques, such as

- Constructive solid geometry (CSG) objects based on a library of primitive shapes
- Explicit boundary representation (BREP)
- Non-uniform rational B-spline (NURBS) surfaces
- N-manifold polygonal mesh geometry (NMG)
- Triangle mesh (BoT) geometry

By way of example of the first of these [4], a set of primitive shapes, such as the ones shown in Fig. 7.3, can be used with suitable Boolean operations to form more complex shapes, such as those shown in Fig. 7.4.

Using this process, targets, subsystems, and individual components may be constructed from engineering drawings or translated from other

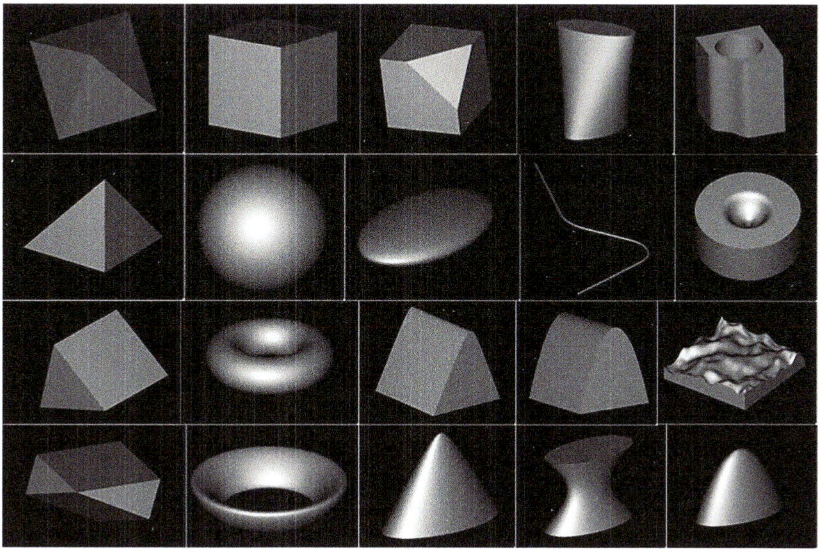

Fig. 7.3 Examples of BRL-CAD primitive shapes.

commercial CAD formats such as CATIA, Solidworks, and Pro/Engineer, as shown in Fig. 7.5. This allows both surface models (Fig. 7.6) and interior component models (Fig. 7.7) to be generated.

The important consequence of modeling targets in this manner is that when a fragment approach direction is defined, and the target model

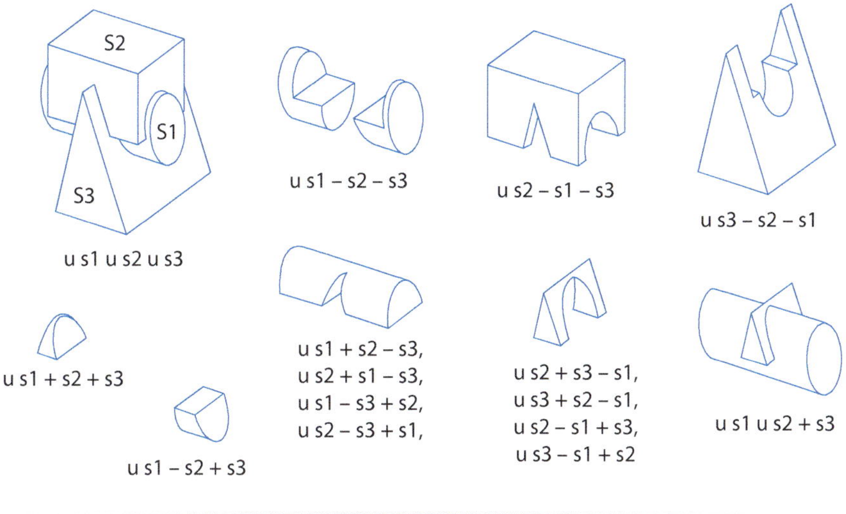

Fig. 7.4 CSG complex shape generation from primitives.

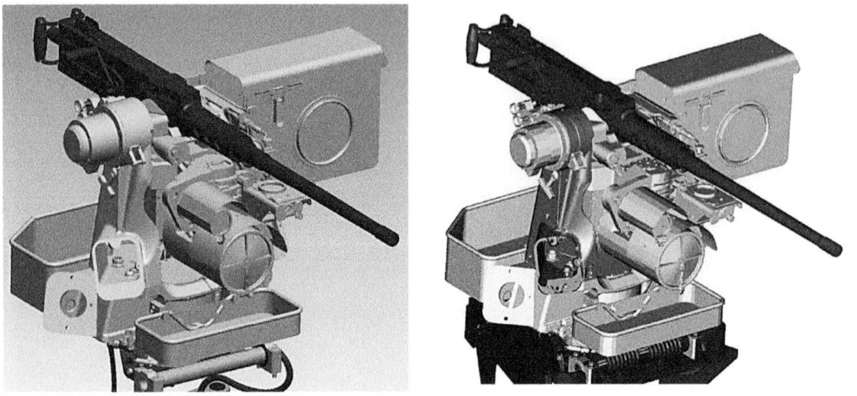

Fig. 7.5 Commercial (left) CAD file converted to BRL-CAD (right).

is rotated so that the direction is normal to the viewing plane (Fig. 7.7), all the internal components are also rotated. This means that if a line of sight for fragment penetration is visualized normal to the viewing plane, all components along that line of sight may be extracted from the CAD model. This line of sight is usually called a *shotline* and will be dealt with shortly.

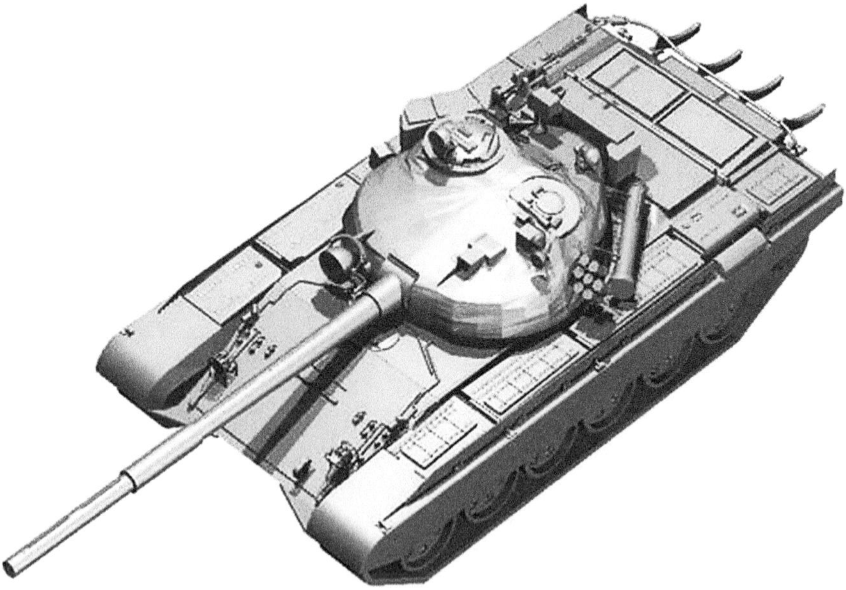

Fig. 7.6 BRL-CAD target surface model.

Fig. 7.7 BRL-CAD target interior model.

7.5 Failure Analysis Logic Tree (FALT) Diagrams

Just as the TGM increases in complexity when real targets are considered, so do the FALT diagrams associated with the appropriate failure modes. This is represented as the fault tree module in Fig. 7.2. Because of the complexity of a FALT diagram for a complete system, it is usually structured in a hierarchical manner where the top level corresponds to the major systems leading to a kill, the next level addresses for each system those subsystems that would kill the parent system, and so on.

In order to illustrate this approach and to allow unlimited distribution of this material, individual critical components will not be given their real names but defined by a four-digit number such as 5213 or 0204. Consider the mobility kill definition for the target shown in Fig. 7.7. This level of damage may be achieved by either fragments, blast, penetrators, or projectiles. For fragment damage, there is a hierarchy of systems, subsystems, and components leading to a target kill. Using the convention discussed in Chapter 6, Section 6.10, this may be written in the logic tree form shown in Fig. 7.8.

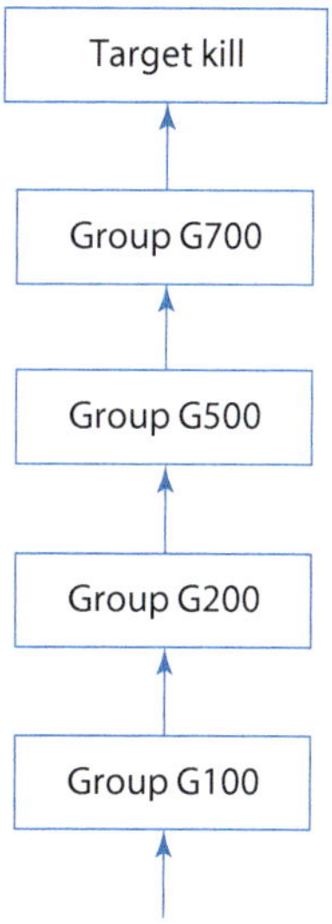

Fig. 7.8 Top-level FALT diagram for fragment kill (mobility).

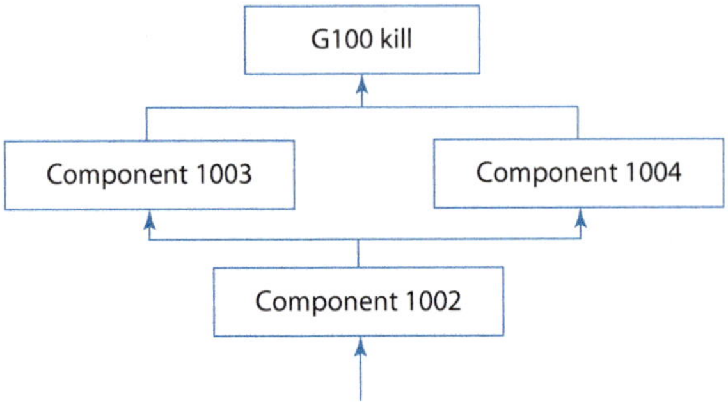

Fig. 7.9 FALT logic for group G100.

These groups might correspond to engine, tracks, driver, and so on. This logic tree produces a mobility kill in the following manner:

Target kill if [G100 killed OR G200 killed OR G500 killed OR G700 killed]

In each subgroup—G100, G200, G500, and G700—there are 3, 24, 138, and 5 critical components, respectively. A corresponding logic diagram for group G100 would be as shown in Fig. 7.9.

Note that in this example there are two redundant components, 1003 and 1004, which correspond with the following statement:

G100 kill if [Component 1002 killed OR
(Component 1003 AND Component 1004) killed]

For blast there are five critical components, any of which would produce a mobility kill (Fig. 7.10).

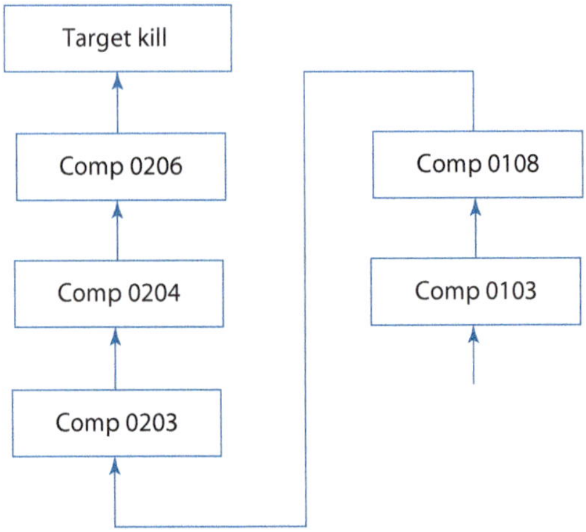

Fig. 7.10 FALT diagram for blast kill.

This corresponds to the logic statement

Target kill if [0103 killed OR 0108 killed OR ... OR 0206 killed]

The complexity of the target failure modes and associated FALT diagrams will depend on the knowledge of the internal components of the target and how each contributes to a different kill definition.

7.6 Definition of Fragment Impact Angles

The simple example dealt with in the previous chapter assumed all fragments impacted the target horizontally from the left side, and now this has to be generalized. To do this we allow impacts from discrete angles in both azimuth and elevation, as shown in Fig. 7.11.

It can be seen that a number of increments in azimuth and elevation are chosen where each combination will produce different values of A_V, $P_{K/H}$,

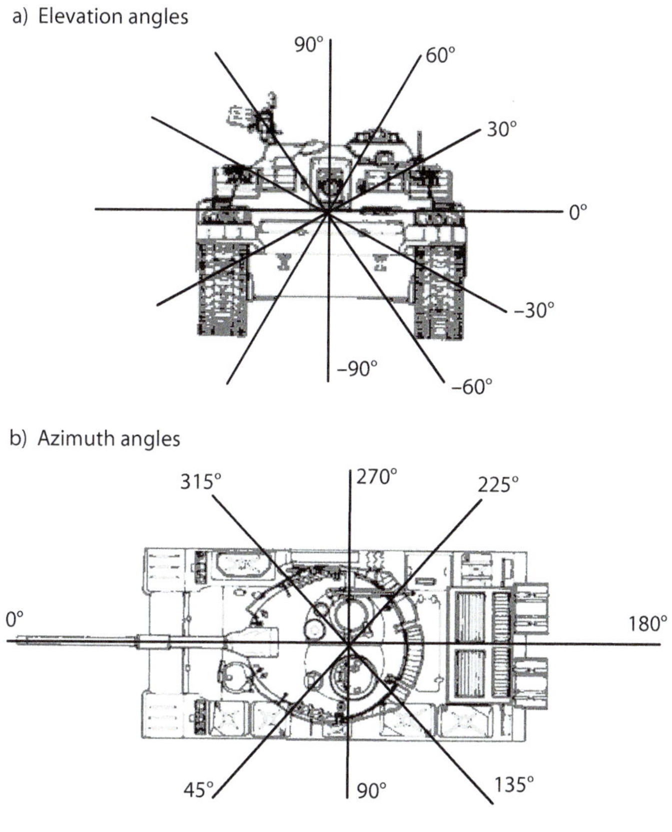

Fig. 7.11 Directions from which fragments approach target.

and lethal area. It should also be noted that a target may appear externally to be symmetrical about a longitudinal axis, but internally the layout of components is not symmetrical. For this reason, such symmetry should not be assumed, and all azimuth angles should be considered. Negative elevation angles will also be defined because these relate to elevated targets such as a radar dish on a tower, or to evaluate the effect of mines on the bottom of targets passing over them.

7.7 Shotline Analysis

The last requirement needed to analyze complex targets is usually termed shotline analysis and deals with how a user-defined fragment traveling in a straight line through a target interacts with the various critical and noncritical components it encounters. For a given fragment approach direction (azimuth and elevation), the BRL-CAD model of the target is rotated to the resulting aspect and overlaid with a grid of cells of uniform dimensions with each cell containing one shotline, as shown in Fig. 7.12.

Although the target shown in Fig. 7.12 shows the external surface rotated to the required aspect, all the internal components will be rotated, too. A shotline is defined as the line of sight perpendicular to the normal plane passing through the target.

Grid-line spacing is user defined, but obviously more closely spaced grids will require more computational time. For many targets, a grid of 4 in. by 4 in. has been found adequate, although grids as small as 1 in. by 1 in. have been used. Once the grid has defined the shotlines, it is assumed that the weapon detonates some distance from the target so that the individual fragment paths and therefore the shotlines themselves are parallel.

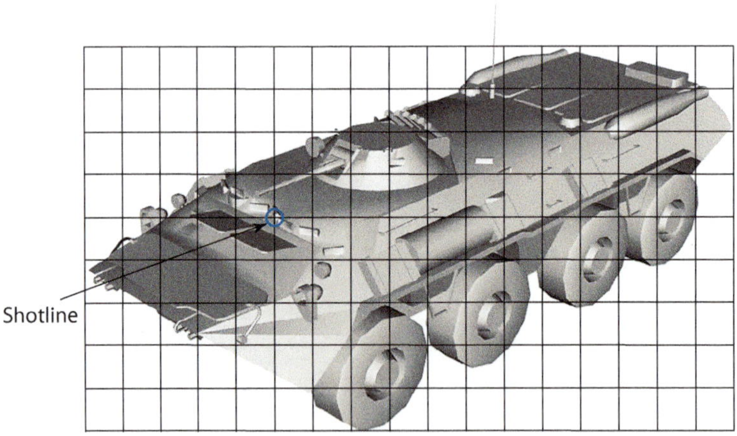

Fig. 7.12 Shotline grid overlaying target.

The module in COVART that generates shotlines through the target as represented by the TGM is the Fast Shotline Generator (FASTGEN), as shown in the overall COVART architecture in Fig. 7.2.

The shotline may be located randomly within the cell or fixed at the center of each grid element, or cell, and from the transformed TGM is able to establish which components lie along each shotline. This defines not only which components are impacted by the fragment, but also in which order. Shotline analysis then has three calculations to make:

1. How does the fragment mass and velocity change as it propagates down the shotline and encounters a component?
2. What is the probability of killing that component?
3. What is the resulting probability of killing the target?

Note that in the vulnerability study we do not consider the reduction in fragment velocity as it moves from the point of detonation to the target, so only the velocity with which the fragment first impacts the target, the striking velocity, is of interest.

The first of these questions is probably the most involved one to answer, because it raises two further issues: How do we represent the penetrability of components located along a shotline, and how does that representation affect the mass and velocity of a fragment as it passes through it?

7.8 Component Representation Along a Shotline

Using FASTGEN to determine, for particular azimuth and elevation directions, those components a fragment will encounter, it is now necessary to calculate the damage done to each component due to a specified fragment mass and striking velocity, which is defined as the fragment velocity when it encounters the first component. Each component is represented by two different mechanisms, a fragility curve and an equivalent plate.

The fragility curve has been discussed in Chapter 6, Section 6.6; it expresses the probability of killing the component when it is impacted by a fragment of given mass and velocity. A fragility curve is required for all critical components of the target, as indicated in the $P_{cd/h}(=P_{ki/hi})$ module in COVART (Fig. 7.2).

The equivalent plate is an attribute associated with both critical and noncritical components and is used to determine the loss of fragment mass and velocity as it passes through. Most components are made up of collections of smaller physical parts, so trying to assess the interaction of a fragment with these individual parts becomes intractable, at least in terms of how much mass is removed and how the fragment slows down; therefore, these effects are embodied into a single plate.

For example, a vehicle battery might be considered to be a critical component for an M-kill, and therefore a fragility curve would have to be

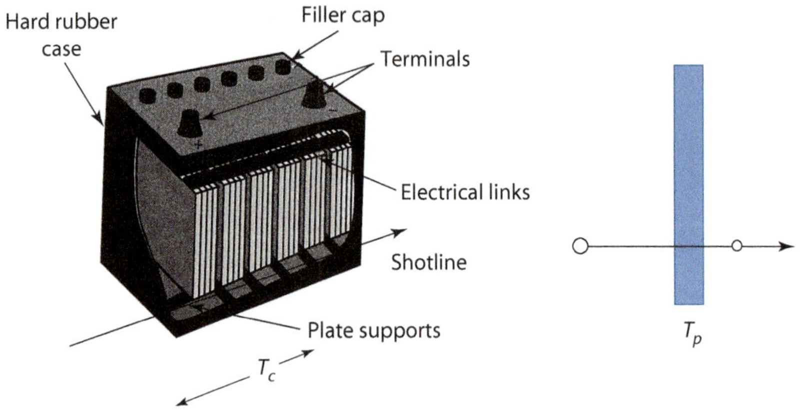

Fig. 7.13 Component represented by a single plate.

determined. In addition, if this battery was on a shotline of interest and a fragment passes through it, that fragment would experience a reduction in mass and velocity as it does so. Consider a fragment passing along a horizontal shotline through the battery, as shown in Fig. 7.13. It will encounter the case, a series of lead plates separated by the electrolyte, and again the case as it exits.

Instead of considering the individual parts, the battery as a whole is represented for penetration purposes by a single plate. In defining the plate, the analyst may select the dominant material, such as lead, and a thickness factor to relate the physical path length through the component T_C to the equivalent plate thickness T_P. If the width of the battery (i.e., the shotline length) is 12 in. and the thickness factor is 10% (T_P/T_C), then the battery is represented by a lead plate of thickness 1.2 in.

Some components such as the protective armor on the side of a tank may be accurately represented by a single plate, but others such as the battery require subject matter expert (SME) assessment. Assuming all critical and noncritical components are represented by equivalent plates, we can now determine the loss of mass and velocity of a penetrating fragment through a series of plates using penetration equations.

7.9 Penetration of Components Along a Shotline

Once FASTGEN has identified the components along a particular shotline, the Fast Air Target Encounter Penetration Program (FATEPEN) tracks a user-supplied fragment (mass and velocity) as it moves down it. In this process, the fragment will lose mass and velocity as it perforates each component until it reaches a component it cannot penetrate, or it can penetrate but not exit. If none of these conditions occurs, the fragment will exit the target at the end of the shotline.

Within the overall COVART structure (Fig. 7.2), FATEPEN deals with fragments whereas ProjPen and Shaped Charged Jet do the same for different penetrating threats. Focusing on fragments, FATEPEN uses a combination of fundamental principles of mechanics and experimental data. It allows for fracture of a component into a debris cloud, which it then tracks to determine its loading on subsequent plates.

In order to calculate mass and velocity degradation, the JMEM uses two methodologies for this process: THOR equations [5] and an internal FATEPEN model [6]. Results were obtained from extensive testing of regular-shaped fragments fired at plate arrays, as shown in Fig. 7.14. Based on results from such tests, the loss of speed and mass of a fragment may be predicted.

Although the THOR model worked well for relatively slow-moving fragments, once a material-dependent critical impact velocity is exceeded, the fragments tend to shatter rather than remain intact, and the damage mechanism changes. Sample results for residual mass and velocity for a particular combination of fragment and plate, together with corresponding test data, are shown in Figs. 7.15 and 7.16 [5, 6].

The FATEPEN model replaces the THOR model for faster fragments and predicts the new plate damage observed in these impacts. FATEPEN is applicable over the full range of impacts of interest to the JTCG/ME and provides a uniform predictive methodology over the complete range of impact speeds.

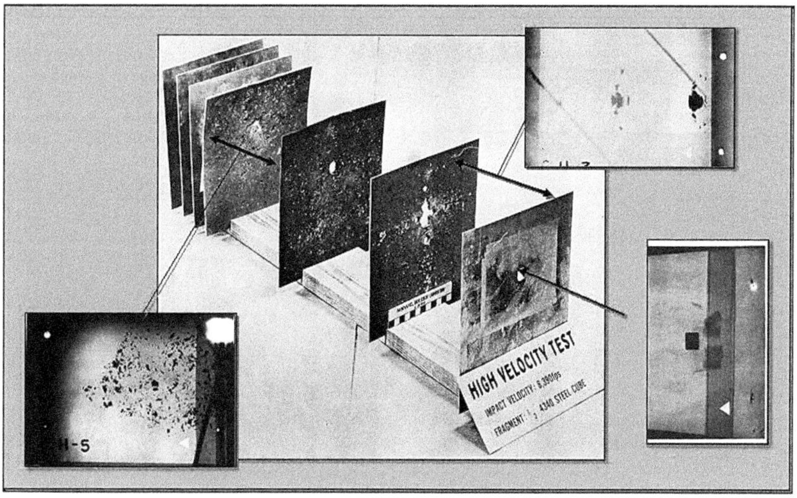

Fig. 7.14 Experimental fragment penetration test.

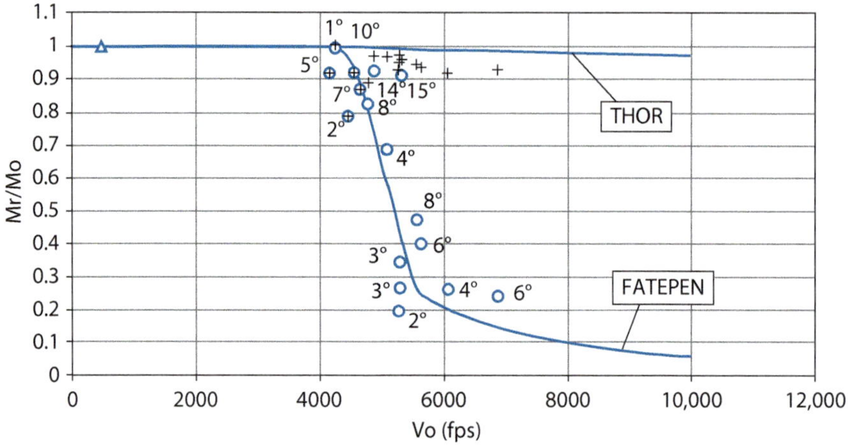

Fig. 7.15 Normalized residual fragment mass vs velocity.

FATEPEN is described as "a physics based model anchored with test data." An example of this is to note that the model predicts a debris cloud of small particles moving from the rear surface of a plate to the front of the next plate where the shape of the cloud is elliptical. The actual size of the ellipse is obtained from high speed imagery of a test.

This resulted in a common set of penetration equations described in the FATEPEN User and Analyst Manuals that may be used for a number of vulnerability models including COVART and AJEM/Modular Unix-based Vulnerability Estimation Suite (MUVES S-2).

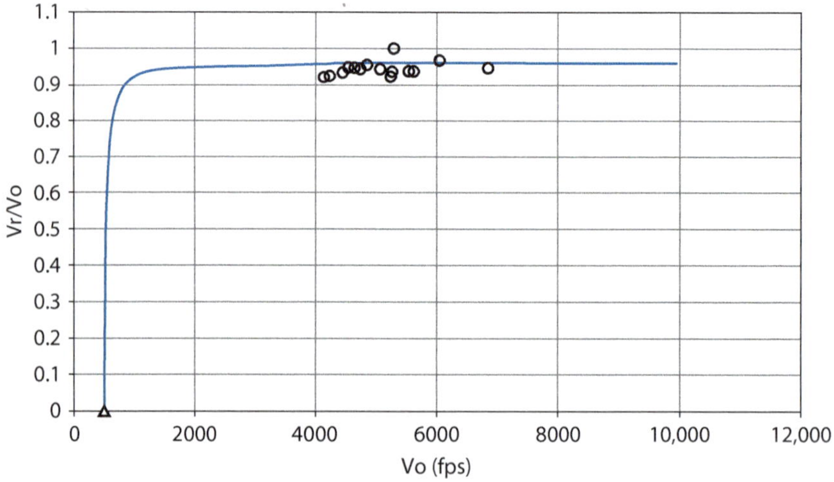

Fig. 7.16 Normalized residual fragment velocity.

Table 7.1 Fragment Presented Area by Shape

Natural	Compact	$A_P = \frac{d^2\sqrt{2}}{2}\left[1+\frac{\pi}{4}\right]$
	Noncompact	$A_P = \frac{d^2\sqrt{2}}{2}\left[\frac{L}{d}+\frac{\pi}{4}\right]$
Controlled	Diamond	$A_P = \frac{wd}{2}\left[\frac{\sqrt{2}}{2}\right]$
	Sphere	$A_P = \frac{\pi d^2}{4}$
	Cube	$A_P = \frac{3d^2}{2}$
	Parallelepiped	$A_P = \left[\frac{L}{d}+0.5\right]d^2$

FATEPEN supports fragments of different shapes: sphere, cube, cylinder, and parallelepiped, where each shape may have different sizes. The sizes are defined by up to three dimensions, depending on the shape: diameter d, length L, and thickness/width w. Table 7.1 shows how the presented areas of each shape are calculated. Cylinders can also have a pointed or hemispherical shaped nose. For each shape, the user can also input orientation with respect to the shotline, which is critical for both the penetration process and the breakup of the impactor.

Examples of fragment geometry are given in Fig. 7.17. Presented area may also be obtained from a fragment mass as follows:

$$A_P = k \times m^{2/3} \tag{7.1}$$

Here, a characteristic length is derived from the cube of the fragment mass, squared to obtain a characteristic area, and then multiplied by an empirical constant that depends on the shape and material.

Considering components to be represented by equivalent plates, Fig. 7.18 shows what can happen to a fragment as it propagates down a shotline, damaging components along the way.

As the fragment impacts the first plate, the penetration equations calculate whether the plate is penetrated, the nature of the debris cloud generated, and what changes to the fragment mass, shape, and velocity occur.

At low velocities, the fragment remains intact and pushes a plug of plate material ahead of it, but at higher velocities the initial fragment will shatter into several smaller pieces. The spall material produced by a fragment penetrating a plate is sometime referred to as behind armor debris (BAD).

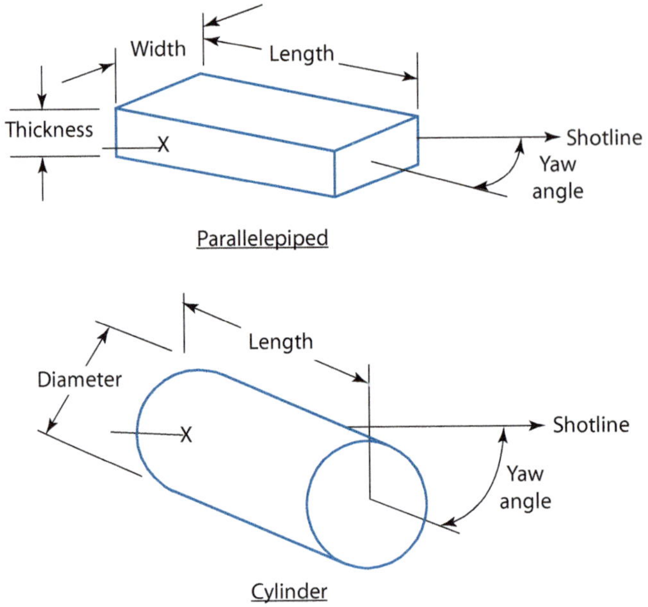

Fig. 7.17 Fragment shapes and dimensions.

Impacting the next component will be the spall cloud and the modified fragment pieces. At impact on the second plate, the effect of the spall fragments and primary fragment pieces is considered. The spall may damage the second plate, but after that its effect is ignored, so only one spall cloud at a time has to be considered.

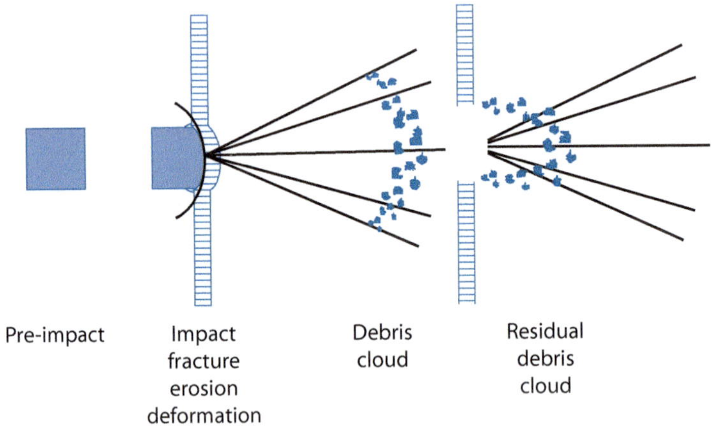

Fig. 7.18 Fragment damage process and subsequent propagation.

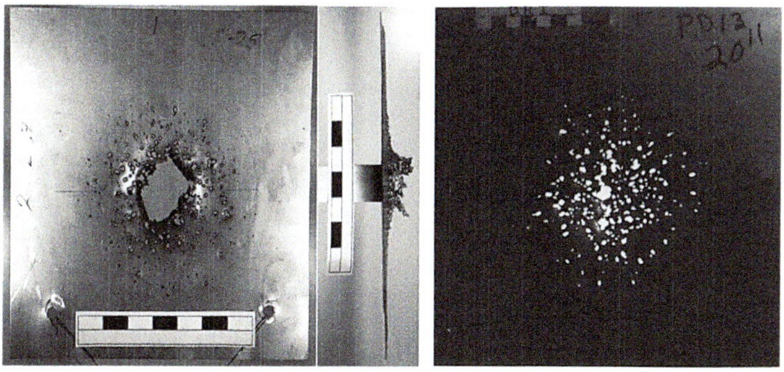

Fig. 7.19 Plate damage due to primary and secondary fragments.

Figure 7.19 shows the effect on plate penetration of a primary fragment and debris cloud. Note that the fragment shape is considered only up to impact on the first component; after that, it is assumed to take the shape of a compact fragment.

7.10 Calculating the Shotline $P_{K/sh}$

It is assumed that the fragments passing along the shotlines are parallel to each other and no ricochet occurs. For example, a horizontal shotline passing through the target shown in Fig. 7.20 might encounter the components indicated. Symbolically, these components may be represented by C_1, C_2, . . . , C_n, as shown in Fig. 7.21.

The fragment mass and velocity when it impacts the first component is known, and from the TGM we know the physical composition of the first component, or plate equivalent. We can then use the penetration equations described earlier to determine the exit velocity of the fragment as well as any loss of material (mass) due to its passage through component C_1. The process is repeated until the fragment fails to enter/exit a component, or it exits the target.

Because fragility curves are a required input for COVART for all critical components, knowing the fragment weight and velocity as it impacts each component, we may interpolate the fragility curve to determine a $P_{ki/hi}$ value for each critical component on the shotline. If the components on the shotline are nonredundant, critical components, the cumulative $P_{k/sh}$ is computed by considering Chapter 6, Section 6.13 (i.e., the overlapping case). A generalized form of Eq. (6.26) is simply the survivor formula and may be written for all n components on the shotline as

$$P_{K/sh} = 1 - \prod_{i=1}^{n} (1 - P_{ki/hi}) \tag{7.1}$$

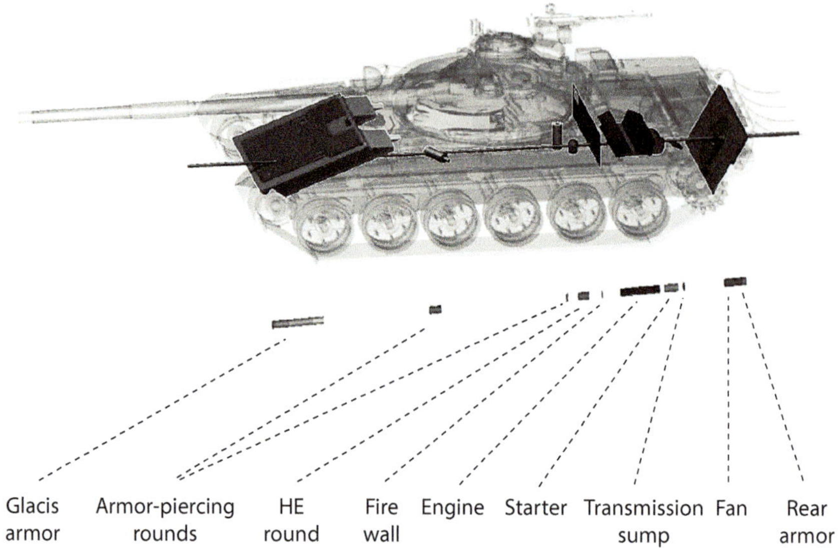

Fig. 7.20 Components encountered along a shotline.

In this equation, $P_{k/sh}$ is the shotline $P_{k/h}$ and may be considered the $P_{K/H}$ if the fragment passes along that shotline. Should a component on the shotline be noncritical, the reduction of mass and velocity still has to be calculated, but such a component will not contribute to the shotline P_K. If a redundant component is encountered, one of two possibilities may occur.

1. The component is part of a group where the other redundant components lie on the same shotline. In this case, the probability that the group is killed by the fragment is given by an extension of Eq. (6.30)

$$P_{K/G} = \prod_{j=1}^{m} P_{kj/hj} \tag{7.2}$$

This simply states that the probability of killing the group is the probability of killing component 1 in the group AND killing component 2 AND killing component 3, and so on up to m, which is the number of

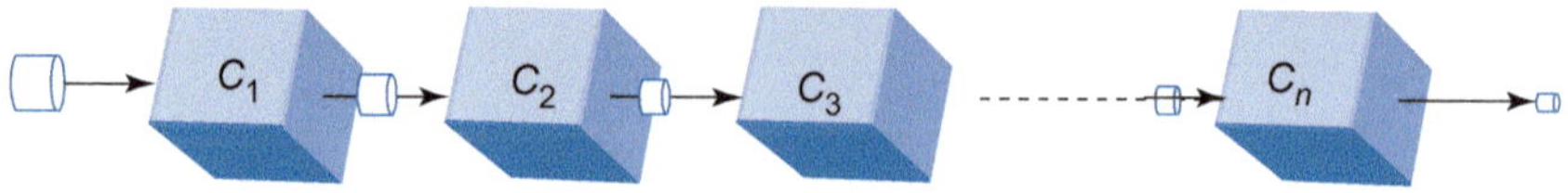

Fig. 7.21 Shotline passing through multiple components.

redundant components needed to kill the group. All the redundant components in the group are then replaced by a single nonredundant component with $P_{ki/hi} = P_{k/G}$.

2. If the redundant component is part of a group where at least one other redundant component is not on the shotline, the redundant component on the shotline in question does not contribute to the shotline $P_{k/sh}$. This is a significant limitation of the methodology—see Chapter 9.

Once the shotline $P_{k/sh}$ has been accumulated, the shotline vulnerable area of the cell is calculated from the following equation:

$$A_{\text{v-cell}} = P_{k/sh} \times A_{\text{p-cell}} \tag{7.3}$$

Here, $A_{\text{p-cell}}$ is the area of the target contained within the cell. This will usually be the same as the cell area, but there will be some cells on the edge of the target where the presented area of target is less than that of the cell (see Fig. 7.12).

The vulnerable area of the target is then found from summing the cell A_vS over all of the p cells that contain the target

$$A_V = \sum_{i=1}^{p} A_{\text{v-cell}} \tag{7.4}$$

Finally, the target P_K is found from

$$P_K = \frac{A_V}{A_P} \tag{7.5}$$

As before, A_P is the presented area of the target for this aspect. Note the difference between $P_{K/H}$ and P_K. Whenever we compute shotline probability of kill this way, we shall use the notation P_K because it is assumed that the target is hit by a single fragment at the cell considered.

7.11 Computation of Vulnerable Area Tool (COVART)

Now that the shotline analysis is complete, we can describe the methodology to compute vulnerable areas of a complex target for a range of user-specified aspect angles, fragment weights, and fragment velocities. This process may be defined by the following steps:

1. The model selects from ranges of values input by the user an azimuth and elevation angle to define the directions from which fragments approach the target. Typical initial values might be 90 deg and 0 deg, respectively, corresponding to the left-hand view of the target analyzed in the previous chapter.
2. The model selects from user inputs a fragment weight and velocity.

3. The target vulnerable area A_V is computed using the techniques described in Section 7.10.
4. The next fragment weight is selected and the process repeated.
5. When all fragment weights have been processed, the fragment velocity is incremented and the process repeated, again for all fragment weights.
6. When all fragment velocities have been processed, the azimuth angle is incremented by 45 deg (see Fig. 7.11), and the whole process is repeated. At this point we have a table of vulnerable areas for user-specified ranges of fragment weights and masses, for every combination of azimuth and elevation angles.
7. The vulnerable areas corresponding to a particular fragment weight and velocity, and a fixed elevation angle are averaged over the eight azimuth angles, where only the nonzero vulnerable areas are averaged.
8. For each vulnerable area, a corresponding probability of exposure (PE) is computed from

$$\mathrm{PE} = \frac{\text{Number of azimuth angles where } A_V \neq 0}{\text{Number of azimuth angles for which } A_V \text{ was computed } (=8)} \tag{7.6}$$

   For example, consider Table 7.2, which shows the vulnerable area as a function of azimuth angle.

   The average vulnerable area that we calculate is $A_V = 42.4/6 = 7.1$ (not $42.4/8$), and the PE is $6/8 = 0.75$. A question often asked is why not take a simple arithmetic mean and not use PE? This issue will be addressed shortly.
9. The elevation angle is incremented by 30 deg, and steps 2–8 are repeated.

Table 7.3 shows a sample output from this process for a particular elevation angle, noting that we now have two tables, one for vulnerable areas and a corresponding one for probabilities of exposure.

Table 7.2 Sample A_V for One Elevation Angle

Azimuth (deg)	Vulnerable Area (ft^2)
0	0.0
45	5.2
90	8.0
135	7.5
180	0.0
225	7.2
270	8.5
315	6.0
Sum of A_V	42.4

Table 7.3 Sample Output for One Elevation Angle

Vulnerable Areas						**Weight (grains)**
Velocity ►	**500**	**1000**	**3000**	**5000**	**7000**	**▼**
	0.00	0.00	0.00	0.00	0.00	5
	0.00	0.00	0.00	2.85	4.65	120
	0.00	0.00	3.66	7.38	7.99	500
	0.00	0.57	8.96	10.89	11.47	5000
Probabilities of Exposure						**Weight (grains)**
Velocity ►	**500**	**1000**	**3000**	**5000**	**7000**	**▼**
	0.00	0.00	0.00	0.00	0.00	5
	0.00	0.00	0.00	1.00	1.00	120
	0.00	0.00	1.00	1.00	1.00	500
	0.00	1.00	1.00	1.00	1.00	5000

In this table the numbers 500 1000, . . . , 7000 refer to the fragment velocities in ft/s, and the tabular values of vulnerable areas are in ft^2. For this case the trend is as expected—vulnerable areas increases with higher fragment weights and velocities, and below certain values of weight and velocity, the target is not vulnerable at all. As an example, the value of 7.99 ft^2 would correspond to the average of the eight azimuth angles for this combination of weight and velocity. The corresponding PE = 1 states that a nonzero vulnerable area was computed for all eight azimuth angles.

A typical table of vulnerable areas will include all eight azimuth angles for elevation angles of −60, −30, 0, 30, and 60 deg. Together with the azimuth angles of ±90 deg, this ensures that the target is completely defined as to its vulnerability for all fragment approach angles. Table 7.4 shows complete results for user-selected elevation angles. Sometimes data in this form are referred to as an RCAS file.

It is important to note that the data shown in the table are for a specific target kill definition. As described in the previous chapter, the target is now represented by the centroid of vulnerability, together with the vulnerable area and probability of exposure data given in Table 7.4. A more complete table is shown in Table 7.5.

Let us now return to the question of why we average the nonzero A_V and introduce PE rather than taking a straight arithmetic average of all eight values. In most cases, very little difference will occur; however, for some cases there is a substantial difference. Consider the case in Fig. 7.22 where there is only one nonzero A_V.

Recall that the P_K is given by the following equation:

$$P_{K-FRAG} = PE[1 - \exp(-\rho A_V)] \tag{7.7}$$

Table 7.4 Output Data from COVART—The RCAS File

	Vulnerable Areas					Weight (grains) ▼
Elevation Angle ▼	500	1000	3000	5000	7000	
−90 deg	0.0000	0.0000	0.0000	0.0000	0.0000	5
	0.0000	0.0000	0.0000	2.8500	4.6500	120
	0.0000	0.0000	3.6600	7.3800	7.9900	500
	0.0000	0.5700	8.9600	10.890	11.470	5000
0 deg	0.0000	0.0000	0.0000	0.0000	0.0000	5
	0.0000	0.0000	0.0000	0.0263	0.0400	120
	0.0000	0.0000	0.0526	0.0667	0.4800	500
	0.0000	0.0833	1.1034	3.6800	5.0600	5000
30 deg	0.0000	0.0000	0.0000	0.0000	0.0000	5
	0.0000	0.0833	0.0267	0.1494	0.2874	120
	0.0000	0.0833	0.1494	0.8900	1.7800	500
	0.0400	0.0345	2.9600	6.6400	8.5200	5000
60 deg	0.0000	0.0000	0.0833	0.0400	0.0400	5
	0.0000	0.0400	0.0667	1.1100	2.0800	120
	0.0000	0.0800	1.5400	4.1600	5.5100	500
	0.0800	0.0690	7.7400	12.310	14.680	5000
90 deg	0.0000	0.0000	0.0100	0.0200	0.0300	5
	0.0000	0.0300	0.0700	2.2000	3.0600	120
	0.0300	0.0600	2.2810	4.8500	6.5600	500
	0.0800	0.0900	8.6300	10.850	12.580	5000

	Probabilities of Exposure					Weight (grains) ▼
Elevation Angle ▼	500	1000	3000	5000	7000	
−90 deg	0.0000	0.0000	0.0000	0.0000	0.0000	5
	0.0000	0.0000	0.0000	1.0000	1.0000	120
	0.0000	0.0000	1.0000	1.0000	1.0000	500
	0.0000	1.0000	1.0000	1.0000	1.0000	5000
0 deg	0.0000	0.0000	0.0000	0.0000	0.0000	5
	0.000	0.0000	0.0000	0.3750	0.5000	120
	0.0000	0.0000	0.3875	0.7500	1.0000	500
	0.0000	0.1250	0.8750	1.0000	1.0000	5000
30 deg	0.0000	0.0000	0.0000	0.0000	0.0000	5
	0.0000	0.1250	0.7500	0.8750	0.8750	120
	0.0000	0.1250	0.8750	1.0000	1.0000	500
	0.2500	0.8750	1.0000	1.0000	1.0000	5000
60 deg	0.0000	0.0000	0.1250	0.2500	0.2500	5
	0.0000	0.2500	0.7500	1.0000	1.0000	120
	0.0000	0.2500	1.0000	1.0000	1.0000	500
	0.2500	0.8750	1.0000	1.0000	1.0000	5000
90 deg	0.0000	0.0000	1.0000	1.0000	1.0000	5
	0.0000	1.0000	1.0000	1.0000	1.0000	120
	1.0000	1.0000	1.0000	1.0000	1.0000	500
	1.0000	1.0000	1.0000	1.0000	1.0000	5000

Table 7.5 Composite A_V and PE Table

90-Degree Elevation									M-Kill(C-Category)
Fragment Mass (Grains)	Fragment Impact Velocity for $A_V = 0$	Fragment Impact Velocity (FPS)							
		500		1000		2000		3000	
		$\bar{A}_V$	PE	$\bar{A}_V$	PE	$\bar{A}_V$	PE	$\bar{A}_V$	PE
2	7000	0.00	0.00	0.00	0.00	0.00	0.00	0.00	0.00
5	7000	0.00	0.00	0.00	0.00	0.00	0.00	0.00	0.00
10	2000	0.00	0.00	0.00	0.00	0.00	0.00	5.56	1.00
15	1734	0.00	0.00	0.00	0.00	1.58	1.00	5.81	1.00
30	1203	0.00	0.00	0.00	0.00	8.01	1.00	11.04	1.00
60	1203	0.00	0.00	0.00	0.00	11.11	1.00	13.94	1.00
120	859	0.00	0.00	1.44	1.00	13.37	1.00	15.21	1.00
240	641	0.00	0.00	8.82	1.00	15.14	1.00	17.49	1.00
500	641	0.00	0.00	11.65	1.00	18.23	1.00	20.83	1.00
1000	391	0.20	1.00	13.96	1.00	22.03	1.00	30.30	1.00
2000	391	0.72	1.00	17.31	1.00	27.19	1.00	35.87	1.00
5000	391	0.74	1.00	19.50	1.00	28.95	1.00	37.04	1.00
10,000	391	1.52	1.00	20.52	1.00	30.35	1.00	37.59	1.00

(Continued)

Table 7.5 Composite A_V and PE Table *(Continued)*

90-Degree Elevation									M-Kill(C-Category)
Fragment Mass (Grains)	Fragment Impact Velocity for $A_V = 0$	Fragment Impact Velocity (FPS)							
		4000		5000		6000		7000	
		$\bar{A}_V$	PE	$\bar{A}_V$	PE	$\bar{A}_V$	PE	$\bar{A}_V$	PE
2	7000	0.00	0.00	0.00	0.00	0.00	0.00	0.00	0.00
5	7000	0.00	0.00	0.00	0.00	0.00	0.00	0.00	0.00
10	2000	5.56	1.00	5.56	1.00	5.56	1.00	5.56	1.00
15	1734	7.45	1.00	5.58	1.00	5.59	1.00	5.56	1.00
30	1203	13.22	1.00	13.54	1.00	13.57	1.00	13.19	1.00
60	1203	14.73	1.00	15.00	1.00	15.39	1.00	14.80	1.00
120	859	16.66	1.00	17.44	1.00	17.34	1.00	17.15	1.00
240	641	21.51	1.00	23.55	1.00	24.00	1.00	23.82	1.00
500	641	24.08	1.00	27.98	1.00	28.81	1.00	28.76	1.00
1000	391	34.59	1.00	38.24	1.00	38.83	1.00	37.76	1.00
2000	391	40.33	1.00	43.54	1.00	43.74	1.00	44.19	1.00
5000	391	41.23	1.00	44.49	1.00	45.07	1.00	45.29	1.00
10,000	391	41.70	1.00	44.88	1.00	45.58	1.00	46.60	1.00

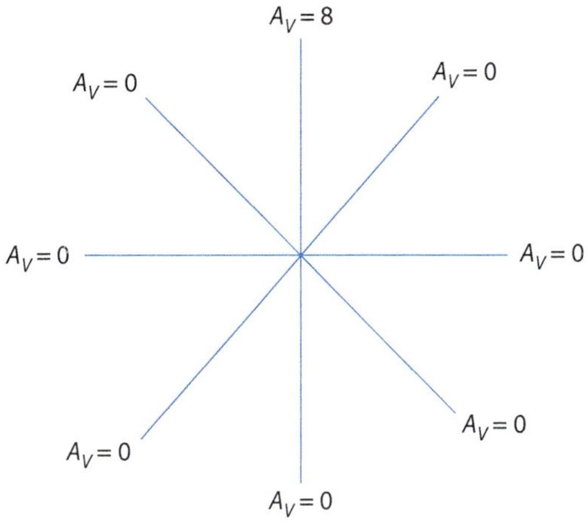

Fig. 7.22 Distribution of A_V with azimuth.

Suppose the fragment density is 1 frag/ft^2, then we get the following computation:

$$P_{K-FRAG} = \frac{1}{8}[1 - \exp(-1 \times 8)] \approx 0.125 \quad (7.8)$$

If, however, we ignore PE and just average all eight azimuth A_V values, we get

$$P_{K-FRAG} = [1 - \exp(-1 \times 1)] = 0.632 \quad (7.9)$$

Because the target is invulnerable from seven directions, and the maximum P_K can be unitary, the first value in Eq. (7.8) is intuitively closer to what might be expected in averaging the vulnerable area around the azimuth. Hence the approach using PE should be used in preference to a simple arithmetic average.

It is important to note that the process of calculating a shotline P_K does not directly involve the FALT diagram. A simple value is calculated for each shotline and then aggregated over the whole target for those aspect angles. The FALT tree, however, is used to define which are the critical components along the shotline corresponding to the selected damage criterion.

This final form of vulnerability data is suitable for input to the effectiveness assessment, which will examine the target's interaction with a specific weapon.

7.12 Advanced Joint Effectiveness Model (AJEM)

AJEM is a more comprehensive program than COVART, but for JMEM purposes it is used for basically the same end, namely to produce vulnerability

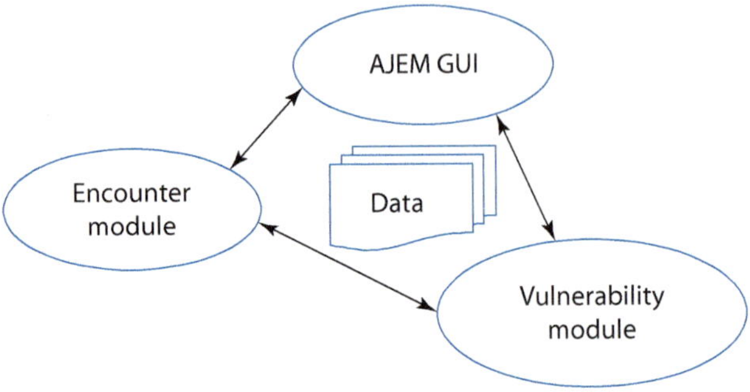

Fig. 7.23 AJEM functionality.

data. AJEM may be described as an end-to-end model in that it first produces vulnerability data and then uses them in an end-game scenario to compute effectiveness. The model focuses on airborne targets, although ground mobile targets may be modeled within its framework. JMEM use a different effectiveness model than the one in AJEM, called JMAE, which will be described in Chapter 8. The basic functionality of AJEM, as far as the vulnerability part, is shown in Fig. 7.23.

Figure 7.24 shows a typical end-game scenario analyzed by AJEM resulting in the estimation of several P_K values for the airborne target. Because AJEM was originally developed for air targets, that scenario is depicted in the figure.

For our purposes in looking at the vulnerability part of AJEM, it is noted that this function is performed by another set of program modules collectively known as the Modular Unix-based Vulnerability Estimation Suite (MUVES S-2). With the exception of buildings, bunkers, and ships, JWS uses AJEM to assess the vulnerability of all targets.

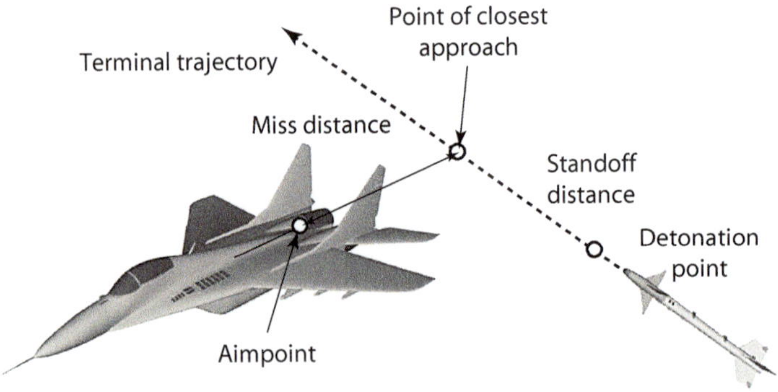

Fig. 7.24 AJEM end-game or encounter scenario.

7.13 Modular Unix-Based Vulnerability Suite (MUVES S-2)

MUVES S-2 dates to the 1980s and was developed by the Army Research Laboratory (ARL) with the intent to provide a common methodology to estimate vulnerability/survivability of blue and red targets. It will be seen that MUVES S-2 uses a methodology very similar to COVART regarding target, weapon, and encounter data. MUVES S-2 uses a common target description via the BRL-CAD format, a standard shotline module, JTCG/ME penetration equations, and a suite of weapon damage mechanisms, or threats, allowing the user to combine them for specific weapon effectiveness evaluations. Examples of threats used in MUVES S-2 include fragments, blast, penetrators, shaped charge, explosively formed projectiles (EFPs), and armor-piercing incendiary (API) rounds.

In order to analyze the probable outcomes of an encounter between a target and a threat (i.e., weapon), MUVES S-2 uses a series of data files to describe the target, threat, and initial conditions of the interaction. For kinetic energy weapons, this means defining the impact direction, threat speed, and so forth. For bursting munitions, the burst point information is also provided as input. One can manually enter burst point locations or use the Endgame Module to predict burst point locations using various weapon trajectories and fusing algorithms.

A threat input file (or series of files, depending on the threat) contains a description of the initial condition of the threat in terms of its physical characteristics such as mass, velocity, length, diameter, and so on. Several target input files are used to describe the geometry, criticality, and functionality of the combat system (i.e., target) being analyzed. The geometry is stored in BRL-CAD format. The BRL-CAD file is accompanied by a text file that assigns physical characteristics such as material type, density, hardness, and so forth to each geometric region in the target description. Geometric regions are grouped into components, which are themselves grouped into categories. A component's category (e.g., component, fluid, etc.) determines how the component is analyzed. Combinations of critical components compose target system definitions (i.e., fault trees), which are used to measure the resulting target response to a threat.

Each threat is initially assigned a threat path, which contains an origin, direction, and threat packet describing the current threat conditions. The threat path is passed to the ray-tracer to compute intersections with the target geometry. The shotline grid shown in Fig. 7.12 is therefore a characteristic of both COVART and MUVES S-2. If the threat path interacts with a target component, the component's category determines an appropriate interaction module (IM) to be used. An IM is responsible for calculating the results of the interaction, which could be a degraded (or defeated) threat, a redirected threat path, damage to the component, or the production of secondary threats (i.e., spall, broken threat pieces, etc.). Unless defeated, a degraded or redirected threat continues to interact with subsequent target geometry until it exits the target. If secondary threats are created, they are assigned initial conditions and a threat path and passed to the ray-tracer. This recursive process continues until the threat and all subsequent threats it has created have been exhausted.

An IM records damage to a component in a damage packet for later evaluation after all threat interactions have been completed. Damage packets are expressed in terms of physical parameters, such as the hole diameter of the impacting threat, energy imparted, hit location, and so on. When damage evaluation occurs, the damage packets are sorted by component, and an evaluation module (EM) is called. An EM is responsible for combining multiple damage packets to produce a user-defined metric of the functionality of the component. Possible metrics (as defined in the states file) include probability of kill P_K, hit/not-hit, and binary kill states (killed/not killed). This process can be represented by a damage vector (list of damaged components), which can be used with a system definition (fault tree) to produce a degraded capability vector.

7.14 Vulnerability Assessment for Bullets, Penetrating Rod, and Shaped Charge Warheads

One might think these are three different weapons with different damage mechanisms; however, they share sufficient similarities that their effect on a target have more in common than their differences [7].

Bullets may consist of a solid projectile (ball roaund) or one that has a case that breaks up and ignites, such as an API. A simple bullet of small to medium caliber is shown in Fig. 7.25.

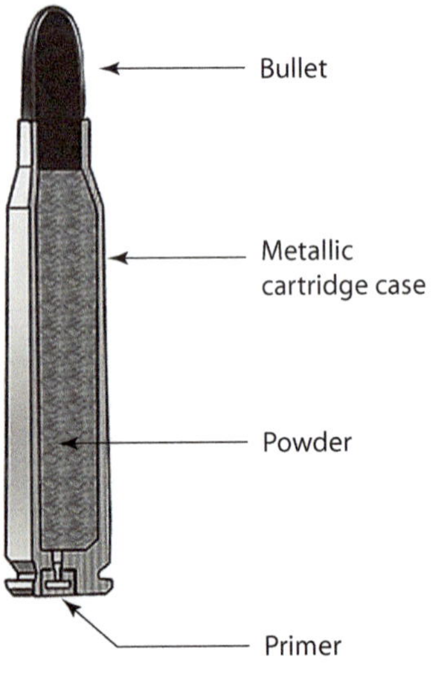

Fig. 7.25 Simple bullet construction.

The cartridge or casing contains a propellant that is initiated by a primer charge, usually by a mechanically generated impact from a firing pin. This propels the bullet along a barrel, which imparts a stabilizing spin though the spiral grooves machined on the inside bore. The bullet itself consists of a core (usually lead) surrounded by a casing, or jacket, commonly made of copper.

Figure 7.26 shows two 120-mm rounds, both of which may be fired from the M1A1 tank main gun. The upper one is a discarding sabot round that projects a solid depleted uranium (DU) flechette; the bottom, high explosive anti-tank (HEAT) round has a shaped charge (SC) warhead.

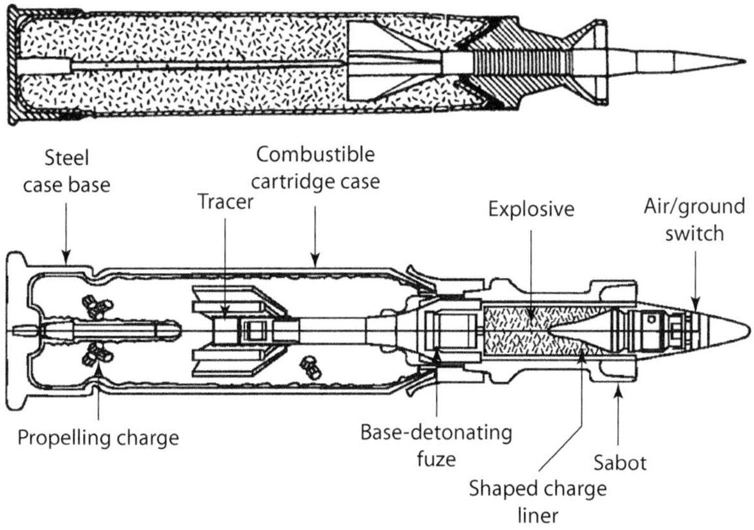

Fig. 7.26 Penetrator (above) and HEAT (below) 120-mm rounds.

Notice that both of these rounds are subcaliber sabot types; therefore, neither is rifled. In each case, stabilization of flight is achieved using fins.

The former is sometimes referred to as a kinetic energy (KE) round and the latter a chemical energy (CE) round. In both cases, the penetrating object may be considered a metal cylinder that impacts the target at high speed, typically greater than 1 mm/μs (= 1 km/s).

If we consider smaller caliber rounds used against personnel, the resistance of soft tissue to a ball round is negligible unless part of the skeletal system is encountered. In most cases, therefore, the projectile may slow down, but unlike a materiel target, loss of mass is not usually considered. Vulnerability of personnel to both fragments and single projectiles is dealt with in Volume 2, Chapter 13.

Turning to the effect of projectiles on materiel targets, we can represent the projectile by a cylinder impacting an armored plate approximated by a flat surface, as shown in Fig. 7.27.

Assuming for now that the cylinder behaves as an incompressible fluid, we can use Bernoulli's equation to relate the pressures, velocities, and elevations before and after impact as follows:

$$p_1 + \frac{1}{2}\rho_1 v_1^2 + gz_1 = p_2 + \frac{1}{2}\rho_2 v_2^2 + gz_2 \tag{7.10}$$

For a steel rod impacting a thick armor plate and ignoring changes in elevation, this equation may be used to estimate the stagnation pressure:

$$p_2 = \frac{1}{2}\rho v_1^2 = 0.5 \times 7850 \times (1000)^2 = 3.925 \times 10^9\ \mathrm{N/m^2} \tag{7.11}$$

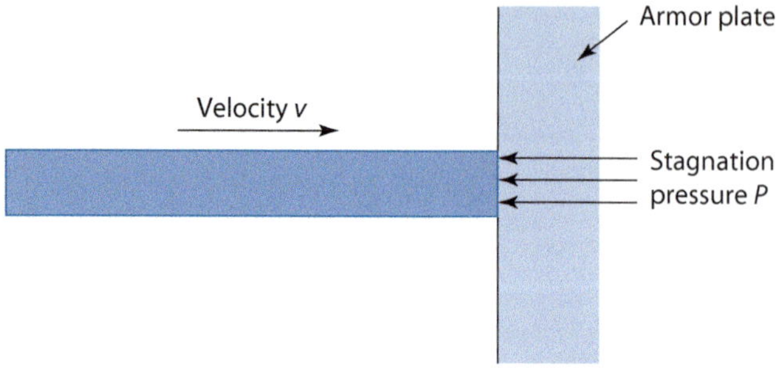

Fig. 7.27 Cylinder brought to rest.

This is much larger than the yield strength of both the penetrator and the target armor, so the assumption of the rod and armor plate behaving like a fluid is justified. If a shaped charge jet is made of copper, the density is slightly larger than that of steel.

It is also important to note that the static strength of the armor is irrelevant here because it is adequately represented as a fluid; therefore, the ability of the rod to penetrate will depend more on the inertia of the target rather than its strength (i.e., its ability to be displaced by the penetrator).

Some differences will be noted regarding rods and jets; a typical rod will arrive at around 2 mm/μs, and obviously the whole rod will have the same velocity. A shaped charge jet, however, has a frontal speed of 8 mm/μs and a trailing speed of 1 mm/μs; that is, the jet is being stretched and has a velocity gradient along it. Eventually, in the absence of a target, the jet will break up into smaller pieces.

We now consider the jet penetrating into a monolithic material where the jet velocity is v and the velocity of the penetration front is u, as shown in Fig. 7.28.

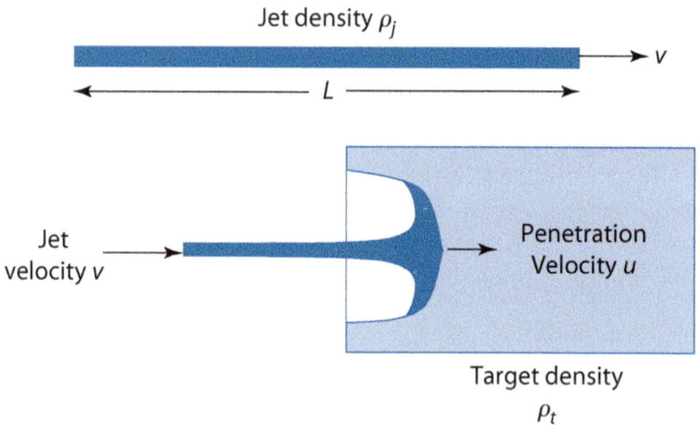

Fig. 7.28 Shaped charge and target geometry.

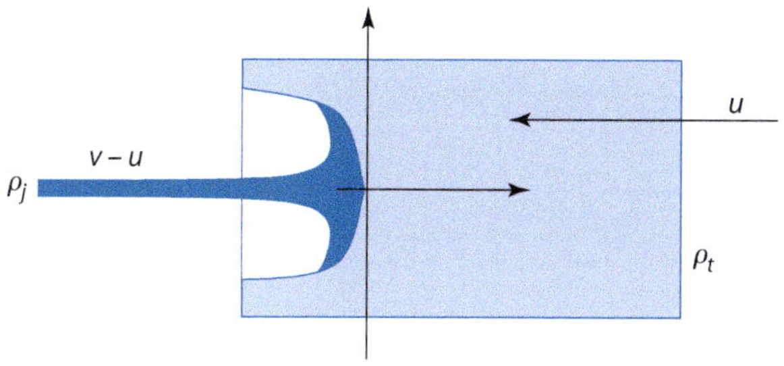

Fig. 7.29 Stationary interface coordinate frame.

If we now superimpose a velocity of u to the left on the whole system, the penetration front becomes stationary, the target is moving to the left at velocity u, and the jet is moving to the right at velocity $(v - u)$. This is shown in Fig. 7.29.

Because the pressures at the interface must be equal, and the system shown in the figure is in steady state, Bernoulli's Eq. (7.10) may be applied to the jet for the left-hand side and the target on the right-hand side, leading to the following:

$$\frac{1}{2}\rho_j(v - u)^2 = \frac{1}{2}\rho_t u^2 \tag{7.12}$$

Note, however, from Fig. 7.28 that the jet (or rod) is of finite length; therefore, the maximum penetration P will occur when the rear of the jet reaches the origin of the fixed coordinate frame. This time is given by

$$t = \frac{L}{(v - u)} \tag{7.13}$$

During this interval, the target will have moved to the left a distance equal to the penetration depth P, hence for the target

$$P = u \times t = u\frac{L}{v - u} \tag{7.14}$$

Substituting Eq. (7.12) into Eq. (7.14) gives the density law

$$P = L\sqrt{\frac{\rho_J}{\rho_t}} \tag{7.15}$$

Table 7.6 shows the relative densities of possible penetrator materials and a steel target plate for use in Eq. (7.15).

Table 7.6 Shaped Charge and Target Material Densities

Material	Specific Gravity
Steel	7.87
Copper	8.96
Aluminum	2.70
Depleted uranium	19.10
Lead	11.35

From Eq. (7.14) we obtain the penetration velocity.

$$u = \frac{v}{1 + \sqrt{\frac{\rho_t}{\rho_j}}} \tag{7.16}$$

The stagnation pressure at the jet/target interface is obtained from the following equation:

$$p_s + \frac{1}{2}\rho_t u^2 = \frac{\frac{1}{2}\rho_j v^2}{1 + 2\sqrt{\frac{\rho_j}{\rho_t}} + \frac{\rho_j}{\rho_t}} \tag{7.17}$$

This is a somewhat simplistic approach but provides an estimate of the penetration expected in a single plate. A more detailed description of the development of penetration models may be found in [8].

Now that the basic mechanics of penetration have been described, the question arises as to how this might be incorporated into a weaponeering tool. Early methods used Eq. (7.15) together with a compartment model to determine whether a crew compartment (K-kill) or engine compartment (M-kill) in an armored vehicle was penetrated [9]. The damage mechanism is not simply penetration into the compartment, but that there is sufficient residual jet to cause detonation of fuel and/or munitions inside. A rule of thumb was that about 2.5 in. of residual jet was sufficient to accomplish this.

When the JMEM moved from a compartment to component level of vulnerability analysis, a different approach was needed. This revised process begins with Fig. 7.30, showing typical penetration depth as a function of standoff distance for a particular penetrator/target combination.

In this figure, the built-in standoff is the distance from the target the warhead is designed to detonate, and this is achieved by mounting a standoff sensor ahead of the warhead by a distance that maximizes penetration, as shown in Fig. 7.31.

The higher fidelity method we now describe is known as the theory of residual penetration and is attributable to Fireman and Pugh [9].

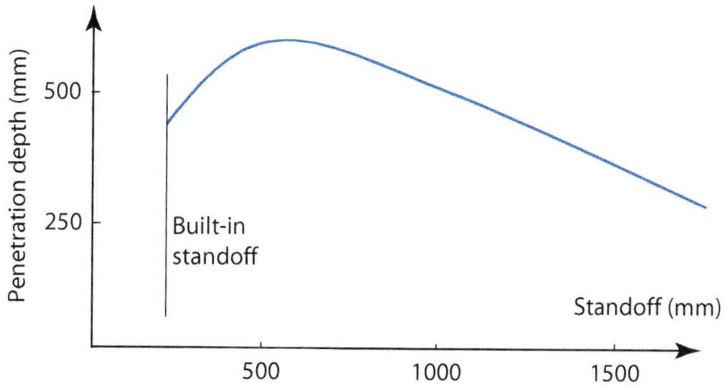

Fig. 7.30 Nominal shaped charge penetration depth vs standoff.

One may see that the simple expression in Eq. (7.15) suggests that for most practical penetrator and target materials (i.e., metals), the density ratio is close to unity; therefore, the penetration depth is approximately equal to the jet length. Figure 7.30, however, shows that the penetration depth depends on the standoff distance. The initial increase in penetration depth is caused by the stretching of the jet and its effect in terms of Eq. (7.15). The resultant decrease in penetration is as a result of radial dispersion, particulation, and subsequent breakup of the jet.

Figure 7.32 may be used to understand an important characteristic of armor in protecting the hull of an armored vehicle.

In this case, the armor does not just provide a strong plate that resists the shaped charge jet, thereby preventing it from interacting with the hull; rather,

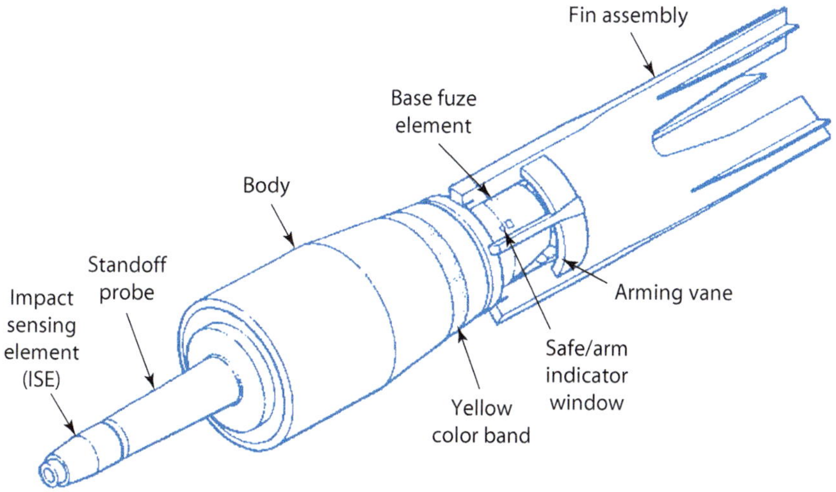

Fig. 7.31 Shaped charge warhead with standoff trigger.

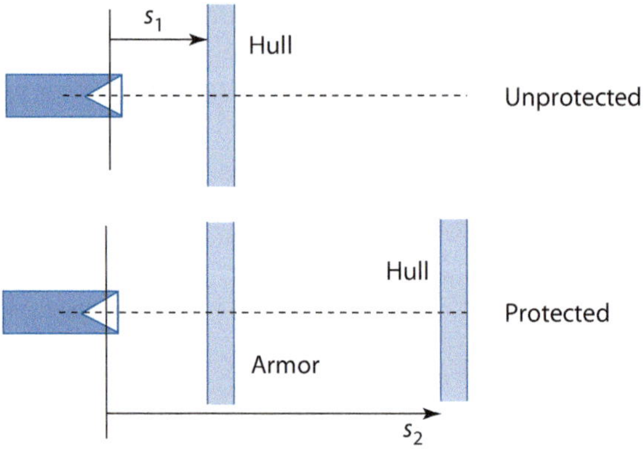

Fig. 7.32 Effect of armor in protecting hull.

it detonates the SC warhead a sufficiently large distance from the hull (s_2) that the penetration depth is small due to the large standoff when the jet finally reaches it. This may be achieved with relatively low cost protective objects such as sandbags or a chain mesh strung around the outside of the vehicle.

The higher fidelity approach we will now describe is based on implementing the Fireman-Pugh technique using the FATEPEN model described in Section 7.10, where we consider a shotline through a sequence of components. Each is considered to be represented by a semi-infinite plate of different thicknesses with a total standoff from the warhead as shown in Fig. 7.33. Note the similarity between this figure and Fig. 7.21 used for fragment penetration.

In its most simplistic form, the residual jet theory states that if the initial jet is, say, 12 in. long and the first plate it encounters is 1 in. thick, there will be a residual jet of 11 in. emerging from the rear of the first plate. Successive plates will be penetrated until the original 12-in. jet has been dissipated. This is also similar to fragment penetration where its mass and velocity are degraded as it passes through components.

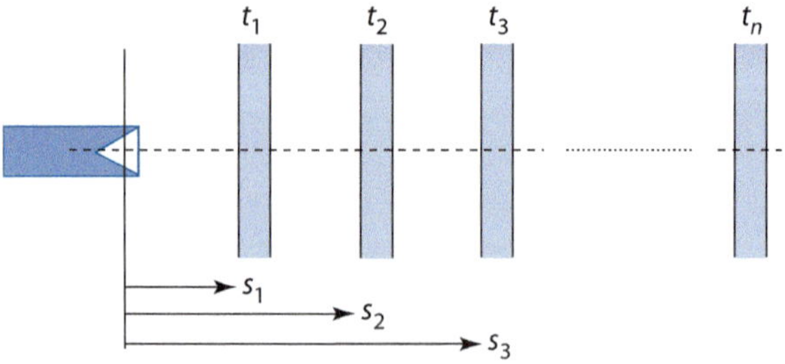

Fig. 7.33 Shotline components penetration by SC jet.

For spaced plates, however, we cannot simply add the plate thicknesses to determine when the jet has been used up, as might be implied by the previous paragraph. This is because:

- The increase in standoff for successive plates has not been accounted for.
- Any dispersion and particulation of the jet in penetrating the preceding plates has also not been considered.

For example, penetration of the third plate in Fig. 7.33 without the presence of the first two plates would require more than the subtraction of t_3 from the initial jet length because the larger standoff s_3 and Fig. 7.30 suggest a different penetration process, and hence different penetration depth. It is also clear that passing through the first two plates degrades the jet, and therefore not all of the initial jet length would be available to penetrate the third plate.

Held [10] suggests that the residual penetration P_{RES} is proportional to the reduced penetration capacity obtained from the following equation:

$$P_{RES} = \frac{P_0 - t}{P_0} \times \eta \times P_S \tag{7.18}$$

where

P_0 = the penetration capacity at the built-in standoff
t = plate thickness
η = a constant degradation factor (0.85–0.9)
P_S = penetration capacity at standoff distance s

To see the application of this equation, consider Fig. 7.30 combined with the spaced plate case shown in Fig. 7.33. The result is shown in Fig. 7.34.

Assuming the warhead fuzes correctly, the first plate has the built-in standoff, so Eq. (7.18) would give the following for the maximum, unscaled penetration capacity at the front of the first plate:

$$P_{RES1} = \frac{P_0 - t_1}{P_0} \times \eta \times P_0 = (P_0 - t_1) \tag{7.19}$$

Notice here we set $\eta = 1$ because there are no preceding plates. As long as $P_0 > t_1$, the first plate (component) is penetrated, and reference to its fragility curve will provide its $P_{ki/hi}$. For the second plate we have

$$P_{RES2} = \frac{P_0 - t_2}{P_0} \times \eta \times P_2 \tag{7.20}$$

For the third plate

$$P_{RES3} = \frac{P_0 - t_3}{P_0} \times \eta \times P_3 \tag{7.21}$$

and so on. This incremental process is represented graphically in Fig. 7.35.

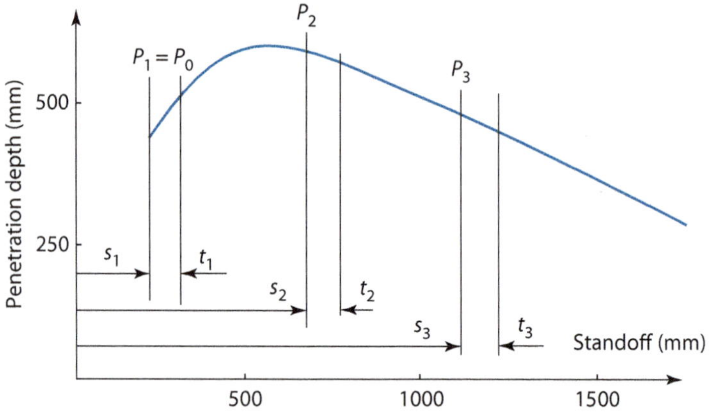

Fig. 7.34 Plate example.

As each plate is encountered, the residual penetration capacity is calculated using Eqs. (7.19) through (7.21). The jet is considered to propagate along the shotline in a manner similar to a single fragment, being degraded by each component until it either reaches a component it cannot penetrate or exits the rear of the target. The shotline P_K may be calculated in the same way as for a fragment and represented either as a cell-by-cell file or aggregated into a whole target vulnerable area, as previously described.

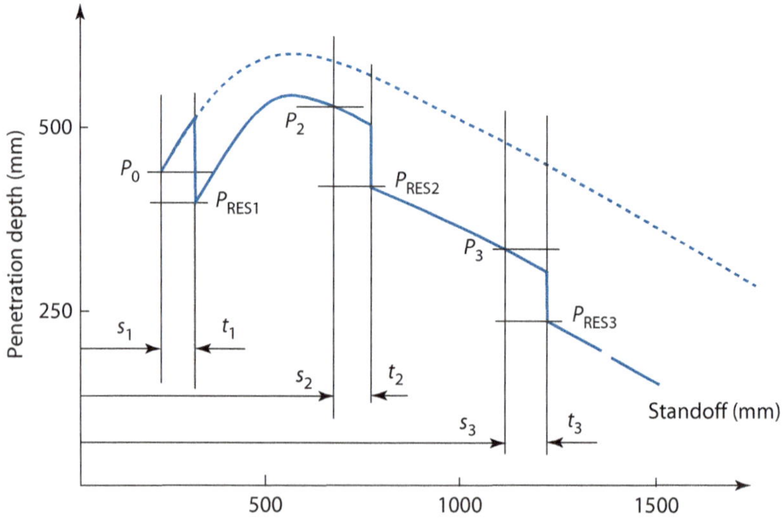

Fig. 7.35 Residual jet propagating through components.

7.15 Vulnerability Data for Projectiles

From an analytical perspective, the same process used to generate vulnerability data due to fragments is used for single penetrators (i.e., COVART and AJEM). The TGM and FALT diagrams are essentially unchanged; however, the fragility curves will reflect the greater energy imparted to the critical components due to the larger mass of the impacting object. Recall Fig. 7.2 showing the COVART modules ProjPen and Shaped Charge Jet to accommodate these threats. Note that FATEPEN is used for both fragments and monolithic projectiles whereas ProjPen is used for nonmonolithic, or segmented projectiles.

In terms of specific data to be used in an effectiveness study, tables similar to the RCAS file (Table 7.4) are generated by COVART and AJEM. An example of nominal data for small caliber ball rounds is shown in Table 7.7.

Here the vulnerable area in square feet depends on the impacting angle of the bullet and the initial impacting velocity. Once more, the vulnerable areas are averaged around the azimuth.

Corresponding vulnerable area data for a shaped charge warhead are shown in Table 7.8. The weapon is characterized by a nominal diameter, and the vulnerable area of a specific target is shown for two kill definitions, each being a function of weapon elevation angle relative to the target.

Table 7.7 Vulnerable Area Data for Small Caliber Bullets

	Impact Velocity	Impact Angle (deg)			
Caliber	(ft/s)	0	15	45	90
0.50 in.	1000	0.00	0.03	0.14	0.22
	2000	0.40	0.54	4.34	9.72
	3000	2.84	2.88	9.02	13.07
	4000	5.90	5.07	11.93	13.54
	5000	7.78	7.10	13.12	14.24
25 mm	1000	0.58	1.22	5.77	9.99
	2000	7.69	6.67	14.60	13.69
	3000	11.23	11.14	17.36	15.51
	4000	13.17	14.21	19.06	16.08
	5000	14.51	16.11	20.29	16.64
30 mm	1000	0.36	0.53	5.70	9.46
	2000	7.80	7.17	14.17	14.40
	3000	12.42	12.31	17.49	15.40
	4000	14.30	15.36	19.05	16.41
	5000	15.61	16.98	20.47	16.84

NOTIONAL DATA

Table 7.8 Vulnerable Area Data for Shaped Charge

Kill Definition	Diameter (in.)	Elevation Angle (deg)			
		0	30	60	90
K-kill	1.00	1.70	1.37	1.72	4.10
	2.00	3.52	2.52	3.48	6.31
	3.00	4.70	4.18	5.51	6.94
	4.00	5.19	5.46	6.58	7.89
	5.00	5.62	6.37	7.74	8.20
	6.00	5.62	6.58	7.79	8.49
M-kill	1.00	6.70	6.17	5.99	11.01
	2.00	11.15	11.71	12.25	18.24
	3.00	17.98	21.00	23.95	31.77
	4.00	26.89	36.00	38.86	46.24
	5.00	36.08	50.65	55.71	60.40
	6.00	38.45	54.40	60.35	61.97
NOTIONAL DATA					

In both these cases the trends are as expected—as the projectile and shaped charge get bigger, so does the vulnerable area. The increasing projectile impact velocity also increases the target's vulnerable area, but for the shaped charge warhead, the target vulnerability is independent of the impacting velocity. It might be argued, however, that in the latter case, the breakup of the weapon casing and its subsequent impact on the target could cause additional damage to that caused by the shaped charge jet by itself.

References

[1] "Vulnerability Toolkit," DSIAC Bulletin, 10-2017.

[2] Staley, T., "COVART 6: Modularization of Vulnerability Models," *Aircraft Survivability*, Fall 2009, 28–31.

[3] Hornung, S., "The Impact of High Accuracy Target Geometry in Modeling and Simulation to Support Live-Fire Test and Evaluation," U.S. Army Ballistics and NBC Division, ARL, SLAD, 16 March 2011.

[4] Butler, L. A., Edwards, E. W., and Kregel, D. L., "BRL-CAD Tutorial Series: Volume III—Principles of Effective Modeling," ARL Report SR-119, Sept. 2003.

[5] "The Resistance of Various Metallic Materials to Perforation by Steel Fragments; Empirical Relations for Fragment Residual Velocity and Residual Weight," Project THOR, Ballistics Analysis Laboratory, Johns Hopkins University, Technical Rept. 51, April 1963.

[6] Yatteau, J., Zernow, R., Recht, R., Dickinson, D., and Wasmund, T., "FATEPEN, A Model to Predict Terminal Ballistic Penetration and Damage to Military Targets," Naval Surface Warfare Center, Dahlgren Div., Dahlgren, VA, 1999.

[7] Dehn, J., "A Unified Theory of Penetration," Ballistics Research Lab., Technical Rept. BRL-TR-2770, Aberdeen, MD, Dec. 1986.
[8] Walthers, W. P., and Zukas, J. A., *Fundamentals of Shaped Charges*, Wiley, New York, 1989.
[9] Pugh, E.M., Fireman, E. L., "Fundamentals of Jet Penetration", Carnegie Institute of Technology, Monthly Report OT9-4 (OSRD-4357), November 1944.
[10] Held, M. "Explosive Reactive Armour Systems", Propellants, Explosives, Pyrotechnics, Vol. 24, pp. 70–75, 1999.

Chapter 8 Effectiveness Assessment: Implementation

8.1 Effectiveness Assessment: Computing the Damage Matrix

This section reviews the basic process of the effectiveness assessment resulting in the computation of a damage or lethal area matrix. This process is embedded in the General Full Spray Materiel (GFSM) program. As in Chapter 6, the vulnerability data of the target are combined with a warhead of known characteristics in a defined geometric relationship to the target. The process is best understood by considering the following sequential steps:

1. Consider the specific weapon–target interaction geometry shown in Fig. 8.1. This scenario represents an air-bursting weapon that propagates a blast wave and fragments from the detonation point. The point O directly below the warhead centroid is defined as ground zero, the origin of the ground plane coordinate frame. Assuming the velocity vector of the weapon is coincident with the weapon axis, projecting this axis to the ground plane results in point R. The line OR defines the x, or range axis; the y, or deflection axis, is perpendicular to it with z pointing up.
2. The target, represented by the center of vulnerability, will be located on the ground plane at coordinates (x, y) relative to ground zero of the weapon detonation point. You can see in Fig. 8.1 that the target is located at a particular deflection angle α relative to the weapon, and the weapon is located at a particular elevation angle θ relative to the target. This has two implications:
   (a) The deflection angle determines the fragment zone in which the target is located.
   (b) The elevation angle determines in which section of the RCAS file the weapon is located, thereby identifying the vulnerable area and probability of exposure tables.
3. Suppose the deflection angle happens to be 45 deg. The static fragment zone containing this angle can be extracted from the Z-data file, as shown in Table 8.1. This indicates that all fragments, although of different masses, are traveling with the same velocity of 5000 ft/s. If the deflection angle were 40 deg, the velocity would be interpolated from the 30-deg and 45-deg values.

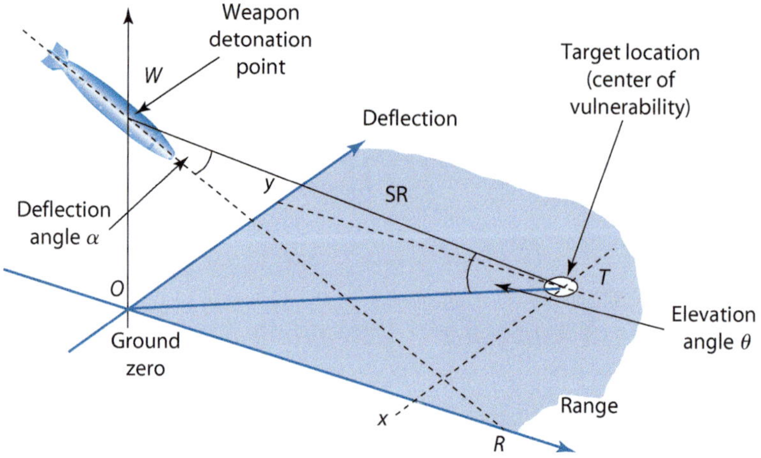

Fig. 8.1 Weapon/target interaction geometry.

4. Suppose the elevation angle is 30 deg. The vulnerable area/PE data are taken directly from the RCAS file (Table 7.4) for this angle and reproduced in Table 8.2. If the elevation angle did not correspond to one of those in the table, such as 50 deg, the data would have to be linearly interpolated from Table 7.4 between two adjacent elevation angles, 30 deg and 60 deg in this case.
5. Because all fragments are traveling with the same velocity (5000 ft/s), that column from Table 8.2 may be extracted to obtain values of A_V and PE for this velocity (Table 8.3). If the velocity didn't coincide exactly with one listed in the table, it would again have to be interpolated.

   At this point we have found the vulnerable area for the scenario elevation angle corresponding to the common velocity of all fragments for the appropriate fragmentation zone. The remaining discrepancy is that Table 8.3 does not contain A_V and PE values for the fragment weights of the weapon under consideration, and interpolation for the weight is necessary.
6. For the ith weight group of the weapon fragments, we interpolate Table 8.3 to obtain a vulnerable area A_{Vi} and probability of exposure PE_i. For example, the fragment weight of 50.5 grains would be interpolated

Table 8.1 Static Fragment Zone for 45-Deg Elevation

α's			Velocities			# wts
30.000	60.000	45.000	4500	6050	5000	4

frag weight →	0.440	8.500	50.500	200.61
# of frags →	88.96	17.79	17.790	35.590

Table 8.2 Vulnerable Area/PE for an Elevation Angle of 30 Deg

Vulnerable Areas						Weight (grains)
Velocity ►	500	1000	3000	5000	7000	▼
	0.0000	0.0000	0.0000	0.0000	0.0000	5
	0.0000	0.0833	0.0267	0.1494	0.2874	120
	0.0000	0.0833	0.1494	0.8900	1.7800	500
	0.0400	0.0345	2.9600	6.6400	8.5200	5000
Probabilities of Exposure						**Weight (grains)**
Velocity ►	**500**	**1000**	**3000**	**5000**	**7000**	**▼**
	0.0000	0.0000	0.0000	0.0000	0.0000	5
	0.0000	0.1250	0.7500	0.8750	0.8750	120
	0.0000	0.1250	0.8750	1.0000	1.0000	500
	0.2500	0.8750	1.0000	1.0000	1.0000	5000

between the 5- and 120-grain fragments in Table 8.3 to get the A_V and PE values.

Based on Eq. (6.50), we can write for the ith fragment weight

$$P_{\text{K-FRAG}(i)} = \text{PE}_i \left[1 - \exp\left(\frac{-K_i \times A_{Vi}}{\text{SR}^2 \times \Omega} \right) \right] \tag{8.1}$$

PE_i = Probability of exposure for the ith weight
K_i = # of fragments of weight i
A_{Vi} = Vulnerable area for the ith fragment weight
SR = Slant range from the weapon to the target
Ω = Fragment zone solid angle = $2\pi(\cos\varphi_1 - \cos\varphi_2)$

The probability of exposure PE_i is included to allow for some aspect ratios having a zero-valued vulnerable area. For example, if $PE = 0.5$,

Table 8.3 Vulnerable Area/PE Data for 5000 ft/s

A_V	Weight
0.0000	5
0.1494	120
0.8900	500
6.6400	5000
PE	
0.0000	5
0.8750	120
1.0000	500
1.0000	5000

only half the azimuth aspects have a nonzero vulnerable area, hence the P_K has to be scaled by 0.5 to reflect this.

7. The total probability of surviving all N fragment weight groups becomes

$$P_{\text{S-FRAG}} = \prod_{i=1}^{N} (1 - P_{\text{K-FRAG}(i)}) \tag{8.2}$$

leading to the probability of killing the target by one or more fragment weight groups [Eq. (6.56)] as

$$P_{\text{K-FRAG}} = 1 - \prod_{i=1}^{N} (1 - P_{\text{K-FRAG}(i)}) \tag{8.3}$$

8. We now add in the effects of blast. Although this was dealt with in a simplistic form in Chapter 6, Section 6.20, it is modified slightly into the form shown in Fig. 8.2 where there is a linear decrease in P_K rather than an abrupt cutoff. Because this damage mechanism is symmetrically spherical, its probability of kill is represented by two blast radii, RB_1 and RB_2.

   Given the slant range from Fig. 8.1, the value of $P_{\text{K-BLAST}}$ may be read directly from the graph.

9. The probability that the target survives both blast and fragments is given by

$$P_S = [1 - P_{\text{K-BLAST}}] \times [1 - P_{\text{K-FRAG}}] \tag{8.4}$$

giving the total P_K due to blast and fragments as

$$P_K = 1 - \{[1 - P_{\text{K-BLAST}}][1 - P_{\text{K-FRAG}}]\} \tag{8.5}$$

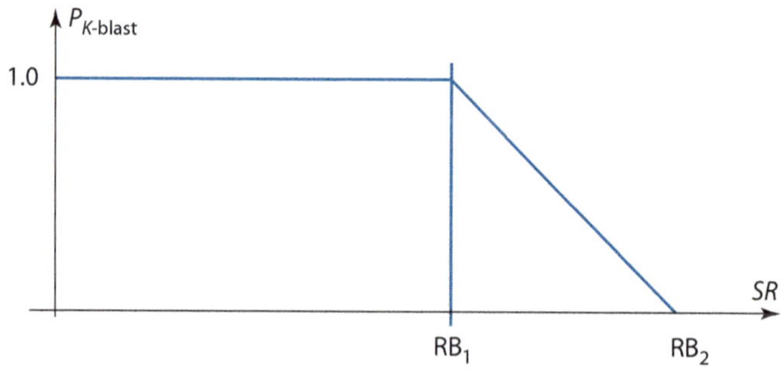

Fig. 8.2 Blast radius characteristics.

10. This process is repeated for all target locations on a grid centered beneath the weapon detonation point, as long as the resulting P_K is nonzero. Observe that as the target moves into different cells relative to the weapon, it will move into different fragmentation zones and different elevation angles. Typical values of P_K at these grid points are shown in Table 8.4.

    Only part of the matrix is shown here; it extends further in the deflection direction, getting smaller in size until it disappears at about 600 ft. In addition, the matrix is symmetrical in the deflection direction, so only half is shown here. The coordinates of the grid points are noninteger values because for the legacy GFSM, the grid size is fixed at 40×40 cells, so the damage matrix has to be divided into this number of cells. The matrix form of P_K shown may be referred to as a P_K matrix, damage matrix, or lethal area matrix.
11. Given the damage matrix, the lethal area specified as MAE_F in this case would be obtained from Eq. (8.6) by multiplying the P_K for each cell by the area of the cell and summing over the matrix for which there is a nonzero P_K.

$$A_L = \text{MAE}_F = \sum_{i=1}^{n} P_K(i) \times A_{\text{cell}} \tag{8.6}$$

    For example, for the partial damage matrix shown in Table 8.4, applying Eq. (8.6) gives $\text{MAE}_F = 1135$ ft^2; however, this is only half of the matrix, so the total EI is $\text{MAE}_F = 2270$ ft^2.

    Note that the damage matrix will be a function of the weapon impact angle due to varying fragment spray patterns, as indicated in Fig. 8.1, and that although the effectiveness index is written as MAE_F, it contains the effect of blast, too. This is because in most fragmenting weapons it is impossible to separate out the damage due to blast from that due to fragments; however, some weapons have only blast effects and little or no damage contributions from fragments.

This section has provided an overview of how to compute a damage or P_K matrix and lethal area that are principal components in the weaponeering process. These data are used for a variety of purposes, of which the JMEM toolset is but one. A later Section 8.15 shows how these data are used to compute simplified damage functions.

The GFSM program computes lethal areas for a variety of applications, including weaponeering. In many cases, the overview just presented provides sufficient understanding of the process without overcomplicating the issue, and the reader may now proceed to Section 8.14. For those requiring an in-depth explanation of GFSM, Sections 8.2 through 8.13 should be studied.

Table 8.4 Damage Matrix Computations

Deflection ► Range ▼	37.9	75.8	113.7	151.6	189.5	227.4	265.3	303.2	341.1	379.0
−114.4	0	0	0	0	0	0	0	0	0	0
−100.1	0	0	0	0	0	0	0	0	0	0
−85.8	0	0	0	0	0	0	0	0	0.0001	0.0001
−71.5	0.0001	0	0	0	0	0	0.0001	0.0002	0.0001	0.0001
−57.2	0.0011	0	0	0	0	0.0003	0.0004	0.0002	0.0001	0.0001
−42.9	0.0028	0	0	0	0.0009	0.0008	0.0004	0.0002	0.0001	0.0001
−28.6	0.0064	0.0001	0.0006	0.0029	0.0017	0.0009	0.0005	0.0002	0.0001	0.0001
−14.3	.1402	0.0059	0.0099	0.0042	0.0019	0.0009	0.0005	0.0002	0.0001	0.0001
0	.5571	0.0459	0.0127	0.0045	0.0019	0.0009	0.0005	0.0002	0.0001	0.0001
14.3	.6794	0.0891	0.0156	0.0045	0.0019	0.0009	0.0005	0.0002	0.0001	0.0001
28.6	.1741	0.0927	0.0325	0.0116	0.0041	0.0012	0.0005	0.0002	0.0001	0.0001
42.9	0.0060	0.0186	0.0258	0.0128	0.0063	0.0034	0.0016	0.0006	0.0002	0.0001
57.2	0.0007	0.0050	0.0105	0.0118	0.0061	0.0032	0.0017	0.0010	0.0006	0.0003
71.5	0	0.0024	0.0015	0.0072	0.0056	0.0031	0.0017	0.0010	0.0006	0.0004
85.8	0	0.0010	0.0012	0.0011	0.0045	0.0028	0.0017	0.0009	0.0005	0.0003
100.1	0	0.0003	0.0009	0.0005	0.0012	0.0025	0.0015	0.0009	0.0005	0.0003
114.4	0	0	0.0006	0.0004	0.0002	0.0011	0.0014	0.0009	0.0005	0.0003
128.7	0	0	0.0003	0.0003	0.0002	0.0001	0.0009	0.0007	0.0004	0.0003
143.0	0	0	0.0001	0.0003	0.0001	0.0001	0.0001	0.0006	0.0004	0.0003
157.3	0	0	0	0.0002	0.0001	0.0001	0	0.0002	0.0004	0.0002
171.6	0	0	0	0.0001	0.0001	0.0001	0	0	0.0002	0.0002
185.9	0	0	0	0.0001	0.0001	0.0001	0	0	0	0.0001
200.2	0	0	0	0	0.0001	0	0	0	0	0
214.5	0	0	0	0	0.0001	0	0	0	0	0
228.8	0	0	0	0	0	0	0	0	0	0
243.1	0	0	0	0	0	0	0	0	0	0

8.2 Detailed Description of the General Full Spray Materiel Model (GFSM)

This model defines the interaction of a specific weapon with the target. The coordinate system is defined as a polar set, shown in the plan view of Fig. 8.3, where the position of the target with respect to the weapon is given by (r, γ), where the angle is measured with respect to the deflection direction.

The simplified equation giving the lethal area for a small incremental target area may be written as

$$\Delta A_L = P_K \Delta A_P \tag{8.7}$$

We wish to calculate the lethal area for all possible placements of the target with respect to the weapon. For the position shown in Fig. 8.3, the variation in elemental area for small variations in r and γ is illustrated in Fig. 8.4. For this specific location of the target, we can write Eq. (8.7) as

$$\Delta A_L = P_K(r, \gamma) r \delta r \delta \gamma \tag{8.8}$$

For all possible locations of the target, we integrate the lethal area to obtain

$$A_L = \int_{r=0}^{\infty} \int_{\gamma=0}^{2\pi} P_K(r, \gamma) r dr d\gamma \tag{8.9}$$

From the symmetry of Fig. 8.3, this simplifies to

$$A_L = 2 \int_{r=0}^{\infty} \int_{\gamma=-\pi/2}^{\pi/2} P_K(r, \gamma) r dr d\gamma \tag{8.10}$$

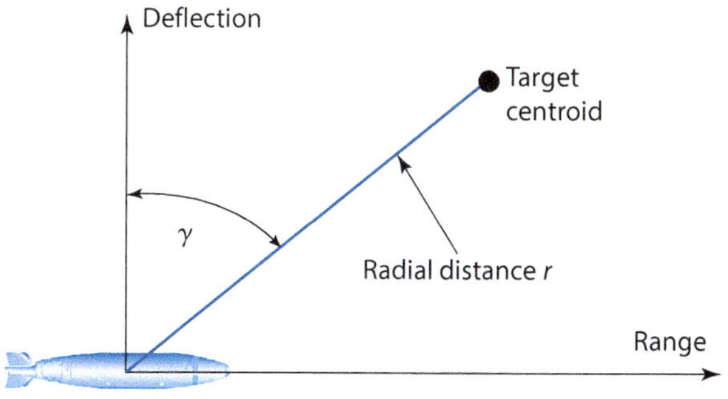

Fig. 8.3 Coordinate system for weapon/target interaction.

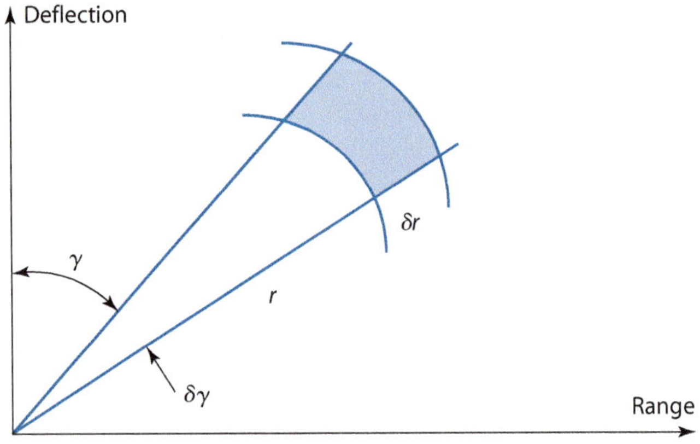

Fig. 8.4 Elemental area for variations in r and γ.

It will be seen in a moment that the computation uses A_V, which is obtained from COVART by averaging the vulnerable areas around the azimuth direction. No loss of accuracy results, therefore, in averaging the P_K around the azimuth direction to also produce $\overline{P}_K$, where using the mean value theorem we get

$$\overline{P}_K(r) = \frac{1}{\pi} \int_{-\pi/2}^{\pi/2} P_K(r, \gamma) d\gamma \tag{8.11}$$

Substituting Eq. (8.11) into Eq. (8.10) produces

$$A_L = 2\pi \int_{r=0}^{\infty} \overline{P}_K(r) r dr = 2\pi \int_{r=0}^{R} \overline{P}_K(r) r dr \tag{8.12}$$

In Eq. (8.12), R is the maximum effective range of the weapon computed by GFSM. This equation is the basic equation from which lethal areas are computed. Before examining methods to evaluate this integral, other input parameters to GFSM will be discussed.

8.3 Orientation of the Weapon in GFSM Scenario

Figure 8.3 shows the general relationship between the target and weapon, but GFSM allows the user to specify the altitude above the target, the angle of inclination to the range direction (weapon impact angle), and the weapon velocity at detonation. Table 8.5 shows an example of such data where PBH = 2.1 (ft) is the height of the burst point above the ground, OMEGA = 30 (deg) is the inclination, and VELM = 900 (ft/s) is the velocity

Table 8.5 Weapon Input Parameters: Sample Case 1

NTP	1.0
	3.0
PBH	2.1
OMEGA	30
VELM	900.0
Z-DATA	2.0
RCAS	1
CDRAG	1.0
INTRVC	1
MATRIX	40 40 1 21
BLT	20.7 20.8 2.1
COTM	4.99

of the weapon at detonation. The number 3.0 below the NTP line is the height of the warhead centroid above the ground.

8.4 GFSM Target Description

This is simply the output from COVART rearranged for input to GFSM, shown in Table 7.4. GFSM accepts this format for the target directly, in the form of an RCAS file.

8.5 GFSM Warhead Description

The weapon characteristics are described in terms of static fragment zones. Due to the symmetry of the weapon, only zones in the 0- to 180-deg range have to be specified; these follow the same Z-data format shown previously. Table 8.6 shows a complete Z-data file. It is assumed that for a particular angle within the static fragment zone, all fragments have the same velocity, which may be obtained by interpolating between the two angles enclosing the boundary.

8.6 GFSM Fragment Drag Data

Because the target will be located at various distances from the weapon, the fragments will slow down in the air between them. Data supplied by the user enable the program to calculate this reduction in velocity so that accurate impact conditions may be determined. These data are contained in the CDRAG file. The drag curve, C_D vs Mach number, is input for the type of fragment produced by the weapon, such as random, spherical, or cubic. Typical general purpose (GP) bombs produce fragments of random shape.

Table 8.6 Complete Z-Data File

Angles (deg)			Velocities (ft/s)			# Groups	Zone
0	20	10	5500	4770	5010	5	1
0.100	4.336	16.303	33.862	104.337			
2.927	52.692	35.128	29.273	11.709			
20	40	30	4770	6450	5545	2	2
21.991	50.232						
24.100	24.100						
40	55	47.5	6450	4295	4510	4	3
0.440	8.500	50.50	200.61				
88.963	17.793	17.793	35.585				
155	180	167.5	6170	5865	5160	7	9
0.519	3.772	16.967	35.709	102.63	169.196	357.34	
82.845	372.81	62.134	93.201	10.356	20.711	20.711	

In addition, the INTRVC1 file contains weights and corresponding velocities for which no target damage may be expected, known as the cutoff velocity. This allows some reduction in computation by ignoring light and/or slow-moving fragments that produce no $P_{K/H}$ on the target.

8.7 GFSM Fragmentation Zones

The first calculation GFSM does is to add vectorially the weapon velocity at detonation to the static fragmentation zone velocity data. This is the reason the velocities of the boundaries in each fragment zone are needed; the weapon velocity changes not only the velocities in the zone but the also the angles at the boundaries. Figure 8.5 shows this vector addition of the two velocities. In this diagram, V is the weapon velocity along its axis, and V_1 is the velocity of the fragments at a zone boundary. This has the effect of changing the boundaries of the fragment zones in terms of both the fragment velocities and the angles defining the zone boundaries.

These new zones are called *vectored fragmentation zones*, and in most cases the effect is to collapse the static fragmentation zones towards the weapon axis, as indicated in Table 8.7. However, depending on the fragment velocities at the boundaries, the zone may reverse itself, as indicated by the two zones shown in Fig. 8.6.

In such a case, it can be seen that the second zone now partially overlaps the first zone, and therefore there is a region in which fragments from two zones are present. This leads to the definition of a set of nonoverlapping fragment zones known as dynamic fragment zones, as indicated in Fig. 8.7.

It will be seen that the combination of fragments from static zones 1 and 2 in dynamic zone 2 will have to be allowed for. If a zone does not reverse itself,

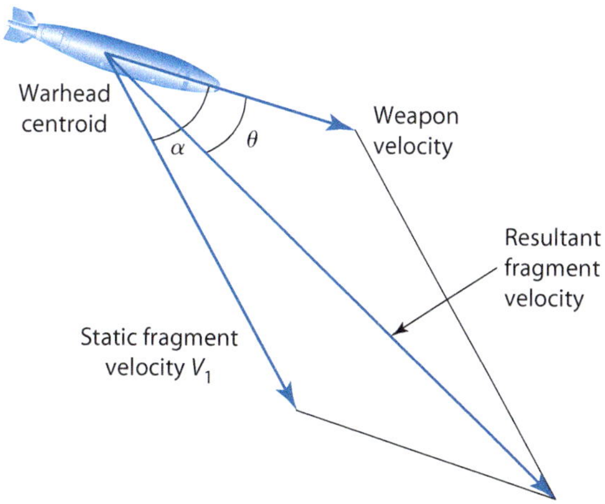

Fig. 8.5 Vector addition of weapon and fragment velocities.

Table 8.7 New Vectored Fragment Zones

Static Zone Angle (deg)	Static Velocity (ft/s)	Dynamic Zone Angle (deg)	Dynamic Velocity (ft/s)
0	5500	0	5974
20	4770	16.86	6375
40	6450	35.37	8765
▼	▼	▼	▼
180	5865	180	5391

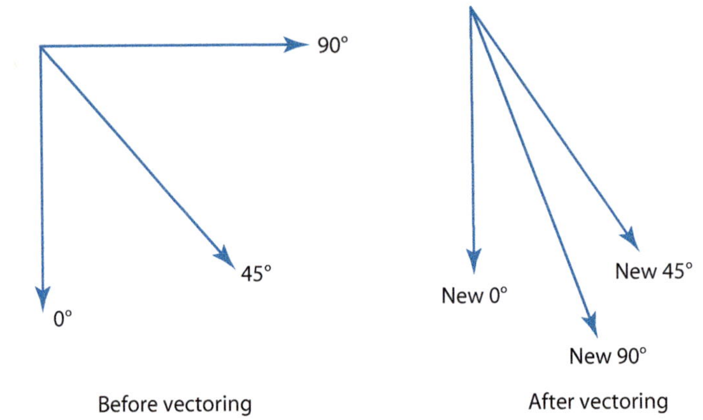

Fig. 8.6 Overlapping vectored fragmentation zones.

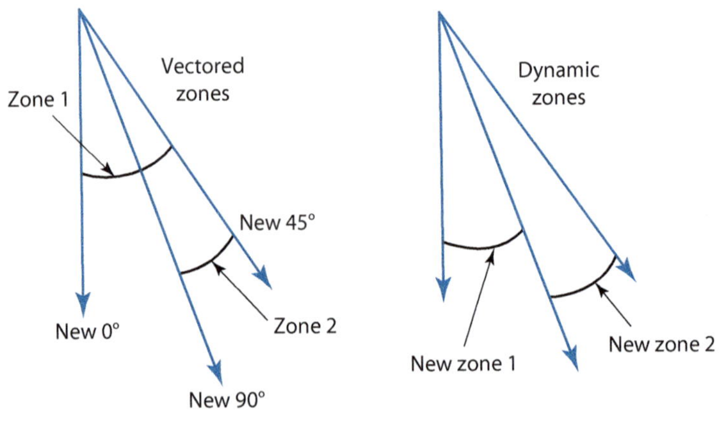

Fig. 8.7 Dynamic zones from overlapping vectored zones.

then the dynamic fragment zones are obtained from the vectored fragment zones, and no overlapping occurs.

8.8 Fragment Velocity/Maximum Effective Range

The user may specify a cutoff velocity for each fragment weight; inspection of the RCAS file for zero vulnerability will indicate suitable values. This cutoff velocity may be used to reduce the number of computations needed to calculate P_K values and is also used to determine the maximum effective range of the weapon.

GFSM takes into account the drag on each fragment in a specific weight group so that the actual striking velocity on the centroid of vulnerability may be calculated. The governing equation of motion is

$$C_D \rho \bar{A} V^2 = -m \frac{\mathrm{d}V}{\mathrm{d}t} = -m \frac{\mathrm{d}V}{\mathrm{d}s} \frac{\mathrm{d}s}{\mathrm{d}t} = -mV \frac{\mathrm{d}V}{\mathrm{d}s} \tag{8.13}$$

where

$\bar{A}$ = mean presented area of the fragment
C_D = drag coefficient
ρ = air density
m = fragment mass
V = fragment velocity

As for bombs and projectiles, a C_D–Mach number table is available for particular shaped fragments such as spherical, cuboid, and so forth. Equation (8.13) is written in the form

$$\mathrm{d}s = \frac{-m\,\mathrm{d}V}{C_D \rho \bar{A} V} \tag{8.14}$$

Integrating this equation yields

$$\int_0^{S_i} \mathrm{d}s = \int_{V_0}^{V_S} \frac{-m}{C_D \rho \bar{A} V} \mathrm{d}V \tag{8.15}$$

In this equation, S_i has to be calculated by numerical integration of the righthand side from the initial fragment velocity up to the cutoff velocity, because C_D will be a function of V. For simplicity, only the largest value of m for each fragment zone is considered; the maximum effective slant range is obtained from the maximum value of S_i from all fragmentation zones.

$$\mathrm{SR}_{\mathrm{MAX}} = \max(S_i) \tag{8.16}$$

The maximum effective range in the ground plane is then calculated from this slant range from

$$R_{MAX} = \sqrt{SR_{MAX}^2 - (h - t)^2} \tag{8.17}$$

where h and t are the warhead and target centroid heights, respectively. For a particular weapon/target geometry, the striking velocity can also be determined from Eqs. (8.13) and (8.14) by integrating ds up to the target and determining the resultant velocity at the target location. It is this velocity that is input into the RCAS file to obtain A_V and PE data.

8.9 GFSM Computation of P_K

The first stage is to evaluate Eq. (8.11) in the discrete form.

$$\overline{P}_K(r) = \frac{1}{\pi} \int_{-\pi/2}^{\pi/2} P_K(r, \gamma) d\gamma = \frac{1}{\pi} \sum_{j=1}^{m} P_K(r, \gamma_j) \Delta\gamma_j \tag{8.18}$$

What this equation states is that we want to find the mean P_K, for a fixed r, when γ varies from -90 to $+90$ deg. This is computed by summing the product of a constant P_K and an angle $\Delta\gamma$ over which the P_K is constant. To understand this further, consider the weapon spraying fragments on a horizontal plane containing the target center of vulnerability, as shown in Fig. 8.8. The dynamic fragment zones will project into nonconcentric ellipses in the horizontal plane, as shown in Fig. 8.9.

Figure 8.9 shows a plan view of these elliptical zones together with a circle of radius r, which intersects the elliptical contours, defining the dynamic fragmentation zone boundaries. We will compute an average P_K as the target

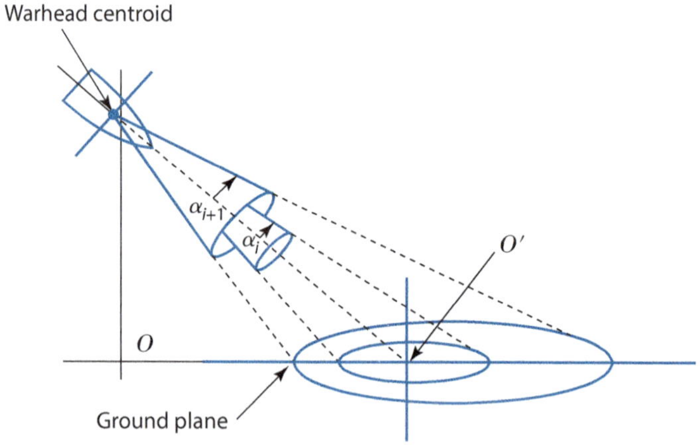

Fig. 8.8 Fragment spray on a horizontal plane.

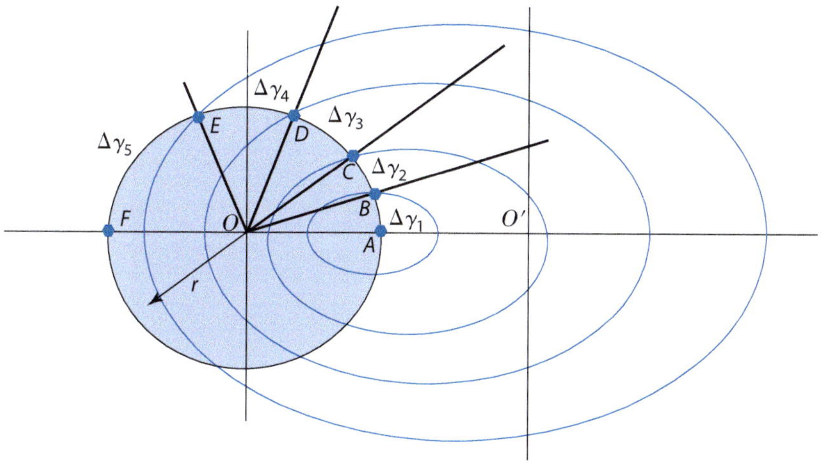

Fig. 8.9 Dynamic fragment zones intersected by a circle.

moves around the circle shown in Fig. 8.9, crossing the dynamic fragment zones as it does so. The angle at which the fragments strike the target is shown in Fig. 8.10 as α; as the target moves around the circle, this angle remains constant.

This angle is the same elevation angle described in the COVART model for which vulnerable areas are provided, as shown in Table 7.4, so we can conclude that

- For a constant fragment weight, and
- For a constant fragment velocity, and
- At a constant radius r, and
- Within a single fragmentation zone

the vulnerable area, and hence the P_K, is constant.

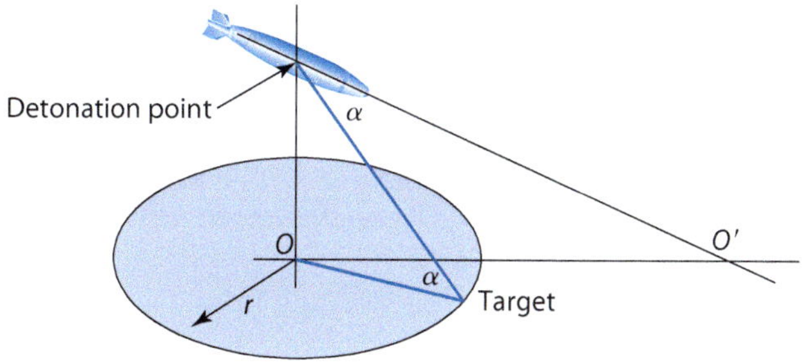

Fig. 8.10 Fragment striking angle.

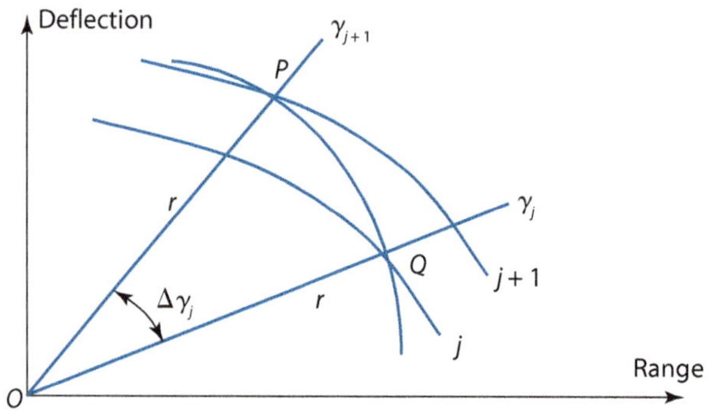

Fig. 8.11 Plan view of dynamic fragmentation zones.

Figure 8.11 shows the jth and $j+1$ zone boundaries at points Q and P, respectively. Because the target is represented as a single point (centroid of vulnerability), P_K will be constant within a dynamic fragment zone (i.e., all along PQ). The problem, therefore, is to determine the angular increment $\Delta\gamma_j$ over which the P_K value remains constant. The GFSM User and Analyst Manual has analysis to compute these angles. In Eq. (8.18) there will be m individual $\Delta\gamma_j$'s, and they will add up to 180 deg. Attention is now directed to the calculation of the $P_K(r, \gamma_j)$ for each of the $\Delta\gamma_j$ regions.

8.10 GFSM Calculation of $P_K(r, \gamma_j)$

GFSM computes these values from the previously derived Eq. (8.5)

$$P_K(r, \gamma_j) = 1 - \left\{[1 - P_{K-\mathrm{BLAST}}(r, \gamma_j)][1 - P_{K-\mathrm{FRAG}}(r, \gamma_j)]\right\} \tag{8.19}$$

where $P_{K-\mathrm{BLAST}}(r, \gamma_j)$ is the probability of kill due to blast alone at ground range r, constant over $\Delta\gamma_j$, and $P_{K-\mathrm{FRAG}}(r, \gamma_j)$ is the probability of kill due to fragments alone at ground range r, constant over $\Delta\gamma_j$.

The righthand side is one minus the probability of survival from both blast and fragmentation, and GFSM includes both of these effects simultaneously. Blast is modeled by a simple graph of $P_{K/H-\mathrm{BLAST}}$ plotted against radius from the target, as shown in Fig. 8.2. The user has to supply values of RB_1 and RB_2. Table 8.5 shows input values of BLT parameters corresponding to $RB_1 = 20.7$ ft and $RB_2 = 20.8$ ft. In GFSM, RB_2 defaults to $RB_1 + 0.1$ unless otherwise specified.

For the blast contribution, the slant range from the weapon detonation point to the target is computed using

$$\mathrm{SR} = \sqrt{r^2 + (h - t)^2} \tag{8.20}$$

where r is the radius of the circle on which the target lies, h is the height above the ground where the weapon detonates, and t is the height of the target above the ground. The slant range computed from Eq. (8.20) is used in conjunction with Fig. 8.2 to determine $P_{\mathrm{K-BLAST}}$.

For the fragmentation part, the $P_{\mathrm{K-FRAG}}$ value for a particular fragment weight group i and dynamic fragment zone j is computed using Eq. (8.21)

$$P_{\mathrm{K-FRAG}(i,j)}(r,\gamma) = \mathrm{PE}_i\left[1 - \exp\left(\frac{-K_{ij} \times A_{Vi}}{\mathrm{SR}^2 \times \Omega_j}\right)\right] \tag{8.21}$$

PE_i = Probability of exposure for the ith weight
K_{ij} = # of fragments of weight i, jth dynamic fragment zone obtained from the Z-data file
A_{Vi} = Target vulnerable area for the ith fragment weight
SR = Slant range from the weapon to the target
Ω_j = # of steradians subtended by the jth dynamic fragment zone at a distance r

Because in the general case the jth dynamic fragment zone may have more than one vectored static fragment zones in it (see Fig. 8.7), the total probability of kill in this zone due to all N fragment weight groups in this zone is

$$P_{\mathrm{K-FRAG}}(r,\gamma_j) = 1 - \prod_{k=1}^{L}\prod_{i=1}^{N}\left[1 - P_{\mathrm{K-FRAG}(i,j)}(r,\gamma_j)\right] \tag{8.22}$$

where L is the number of vectored static zones contributing to the jth dynamic fragment zone.

8.11 Review of Computational Procedure

The various computational stages may be summarized as follows:

1. Consider the target located with respect to the coordinate frame shown in Fig. 8.3, with r set at some value and $\gamma = 90$ deg (i.e., on the positive range axis).
2. Given the terminal conditions of the weapon, calculate the new dynamic fragmentation zones.
3. Determine the elevation angle of the target with respect to the weapon (Fig. 8.10) so that the weapon dynamic fragment zone containing the target may be determined. This determines the initial fragment velocity.
4. Take the smallest fragment weight above the cutoff value, and for the given elevation, find the A_{Vi} from the RCAS file.
5. Compute fragment density from the slant range to the target, the solid angle subtended by the dynamic fragment zone, and the number of fragments of that weight.

Table 8.8 Average P_K at Different Ranges from Target

Range ►	0.5	1.0	20.7	21.5	28	40	60	80	100	102.3
P_K►	1.0	1.0	1.0	0.15	0.10	0.05	0.02	0.01	0.01	0
MAE_F	2270.67									

6. Calculate $P_{K-FRAG(i,j)}$ from Eq. (8.22).
7. Repeat steps 2–6 for all fragment weights to find $P_{K-FRAG}(r, \gamma)$ from Eq. (8.22).
8. For this value of r, find $P_{K-BLAST}$ from Fig. 8.2 and substitute into Eq. (8.19) with the value from step 6 to obtain $P_K(r, \gamma)$.
9. Find $\Delta\gamma$ for which this radius spans one dynamic fragment zone, as shown in Fig. 8.9.
10. Multiply the results of steps 7 and 8 as in Eq. (8.18), repeating steps 3–8 and summing to obtain $P_K(r)$, at which point the target will be located on the negative range axis.

This completes the computation for a single range value, which may now be incremented and the whole process repeated. Table 8.8 shows the results of this computation and tabulates the mean P_K at ranges from 0.5 ft from the target out to a maximum range of 851 ft.

Once the average P_K has been computed at various range values, the lethal area is calculated from Eq. (8.12) using numerical methods to evaluate the integral. This produces the matrix lethal area shown at the bottom of Table 8.8, resulting in 2270.67 ft^2 for this example.

Examination of the data in Table 8.8 shows that the P_K is unity out to a range of 20.7 ft. Considering Eq. (8.19) and Fig. 8.2, this should be expected, because it may be observed that in this region blast is the dominant damage mechanism. Blast disappears after a range of 20.8 ft, however, so beyond this range, fragments are the damage mechanism. The P_K quickly falls to zero at a range of 102.3 ft, where the fragments no longer penetrate the target.

8.12 Worked Problem for GFSM Methodology

Because the computational procedure may seem complex, a simplified problem will be evaluated to demonstrate the stages involved. The following assumptions are made:

- The $\Delta\gamma_j$'s in Fig. 8.9 are given.
- Table 7.4 applies to our target.
- The weapon detonates 20 ft above the ground.

We will look in detail at the first $\Delta\gamma$ (i.e., $\Delta\gamma_1$). Assume that the value of r we are interested in is $r = 34.6$ ft. Assume further that the first two weight groups will be considered in the first dynamic fragmentation zone (Table 8.9).

Table 8.9 Fragment Weight and Velocity Values

Group #	# of Fragments	Fragment Weight	Impact Velocity
1	50	500	5000
2	60	5000	5000

Equation (8.18) is written in the form

$$\overline{P_K}(r) = \frac{1}{\pi}\{P_K(r, \gamma_1)\Delta\gamma_1 + P_K(r, \gamma_2)\Delta\gamma_2 + \cdots\} \tag{8.23}$$

We will calculate $P_K(r, \Delta\gamma_1)$. From Fig. 8.9, $\Delta\gamma_1$ is measured to be 18 deg, or 0.314 rad. What is the corresponding $P_K(r, \Delta\gamma_1)$? We use Eq. (8.19) recognizing that at this radius the effects of blast are negligible.

$$P_K(r, \gamma_1) = 1 - (1 - P_{\text{K-FRAG}}(r, \gamma_1)) = P_{\text{K-FRAG}}(r, \gamma_1) \tag{8.24}$$

From Eq. (8.22), for two weight groups we get

$$P_{\text{K-FRAG}}(r, \gamma_1) = 1 - \left\{\left(1 - P_{\text{K-FRAG}(1,1)}\right)\left(1 - P_{\text{K-FRAG}(2,1)}\right)\right\} \tag{8.25}$$

From Eq. (8.21), we can write for the first weight group

$$P_{\text{K-FRAG}(1,1)}(r, \gamma_1) = \text{PE}_1\left[1 - \exp\left(\frac{-N_{1,1} \times A_{v1}}{\text{SR}^2 \times \Omega_1}\right)\right] \tag{8.26}$$

The slant range may be computed from Fig. 8.12.

This results in $\text{SR} = 40$ ft. Also $N_{1,1} = 50$. Suppose we assume that $\Omega_1 = 0.628$ steradians, corresponding to dynamic fragmentation zone 1. We can use Table 7.4 with elevation angle = 30 deg, weight = 500 grains, and $v = 7000$ ft/s to obtain $A_{v1} = 0.89$ and $\text{PE}_1 = 1.0$. Hence

$$P_{\text{K-FRAG}(1,1)}(r, \gamma_1) = \left[1 - \exp\left(\frac{-50 \times 0.89}{1600 \times 0.628}\right)\right] = 0.043 \tag{8.27}$$

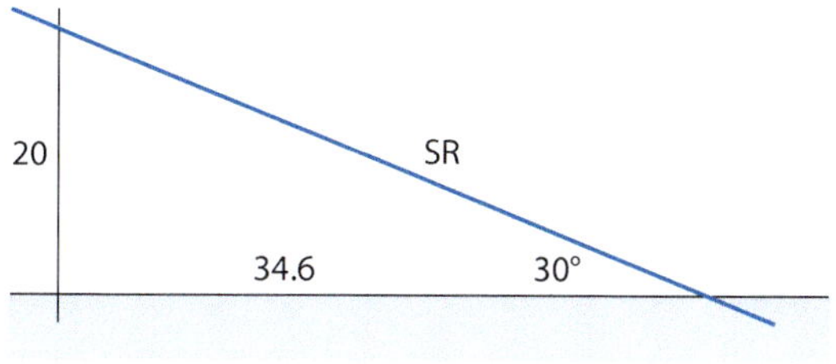

Fig. 8.12 Computation of slant range.

Again, for the second fragment weight group

$$P_{K-FRAG(2,1)}(r,\gamma_2) = PE_2\left[1 - \exp\left(\frac{-N_{2,1}\times A_{v2}}{SR^2\times\Omega_2}\right)\right] \tag{8.28}$$

Suppose $\Omega_2 = 0.628$ again, from Table 7.4, with elevation angle = 30 deg, weight = 5000 grains, and $v = 5000$ ft/s, to obtain $A_{v2} = 6.64$ and $PE_2 = 1.0$. Hence

$$P_{K-FRAG(2,1)}(r,\gamma_2) = \left[1 - \exp\left(\frac{-60\times 6.64}{1600\times 0.628}\right)\right] = 0.298 \tag{8.29}$$

Substituting back into Eq. (8.25) we get

$$P_{K-FRAG}(r,\gamma_1) = 1 - 0.957\times 0.702 = 0.328 \tag{8.30}$$

So we now compute using Eq. (8.24)

$$P_K(r,\gamma_1)\Delta\gamma_1 = 0.328\times 0.314 = 0.103 \tag{8.31}$$

This corresponds to the first term in Eq. (8.23). The other terms are computed in the same way, but using different dynamic fragmentation zones. Suppose we perform these computations, and the other terms evaluate to 0.13, 0.37, 0.02, and 0.24. Note there are five $\Delta\gamma_j$'s from Fig. 8.9. We may then write

$$\overline{P_K}(r) = \overline{P_K}(34.6) = \frac{1}{\pi}\sum P_K(r,\gamma_j)\Delta\gamma_j = 0.275 \tag{8.32}$$

Now we increment r by Δr and repeat the whole process. The results are shown in Fig. 8.13.

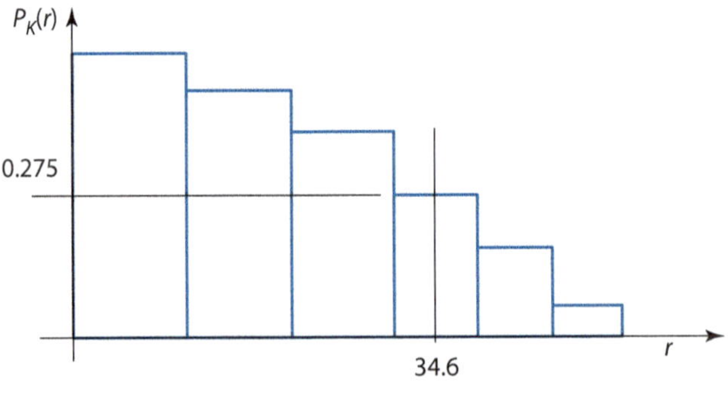

Fig. 8.13 Discrete values of $P_K(r)$.

Hence, from Eq. (8.12), the lethal area would be obtained from

$$A_L = 2\pi \int_{r=0}^{\infty} \overline{P}_K(r) r dr = 2\pi \sum_{i=1}^{i=N} \overline{P}_K(r_i) r_i \Delta r_i \tag{8.33}$$

8.13 Computation of the P_K Matrix

The previous analysis may be thought of as a polar calculation, that is, fixing the range and moving the target in a circle around the weapon calculating average P_K's and then incrementing the range. For some targets, it is useful to generate additional data that divide the horizontal plane containing the target into a grid and compute a P_K for each grid element. This value is the probability that if the target is located in that cell it will be damaged to the level of the selected damage criteria. Figure 8.14 shows the relationship between the polar and rectangular geometry.

The computation of the P_K for the cell from the radial value is somewhat complex, although the GFSM User and Analyst Manual describes it in detail. The process is based on one or more arcs passing through each cell, and taking the weighted average of the P_K multiplied by the arc length within the cell boundaries. Because the weapon fragment characteristics are symmetrical about its longitudinal axis, only the positive deflection cell P_K values are given. An example of the results of these computations is given in Table 8.10.

Only part of the matrix is shown here; it extends further in the deflection direction, getting smaller in size until it disappears at about 600 ft. The total lethal area may be found from summing the product of the P_K and the cell areas, and will produce the same value as for the polar integration,

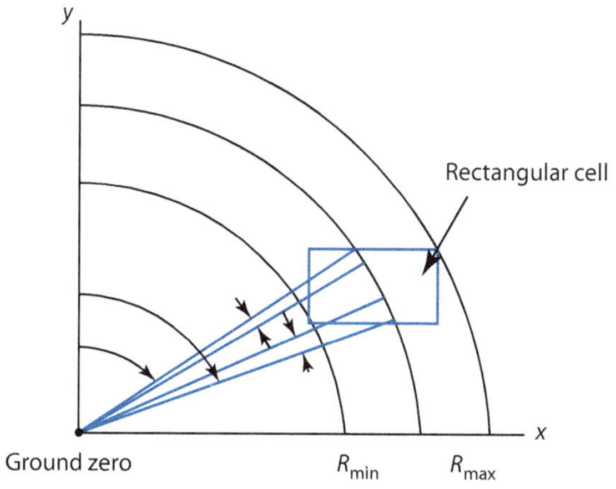

Fig. 8.14 Definition of rectangular cell.

Table 8.10 Damage Matrix Computations

	Deflection (ft)									
Range▼	**37.9**	**75.8**	**113.7**	**151.6**	**189.5**	**227.4**	**265.3**	**303.2**	**341.1**	**379.0**
−114.4	0	0	0	0	0	0	0	0	0	0
−100.1	0	0	0	0	0	0	0	0	0	0
−85.8	0	0	0	0	0	0	0	0	0.0001	0.0001
−71.5	0.0001	0	0	0	0	0	0.0001	0.0002	0.0001	0.0001
−57.2	0.0011	0	0	0	0	0.0003	0.0004	0.0002	0.0001	0.0001
−42.9	0.0028	0	0	0	0.0009	0.0008	0.0004	0.0002	0.0001	0.0001
−28.6	0.0064	0.0001	0.0006	0.0029	0.0017	0.0009	0.0005	0.0002	0.0001	0.0001
−14.3	0.1402	0.0059	0.0099	0.0042	0.0019	0.0009	0.0005	0.0002	0.0001	0.0001
0	0.5571	0.0459	0.0127	0.0045	0.0019	0.0009	0.0005	0.0002	0.0001	0.0001
14.3	0.6794	0.0891	0.0156	0.0045	0.0019	0.0009	0.0005	0.0002	0.0001	0.0001
28.6	0.1741	0.0927	0.0325	0.0116	0.0041	0.0012	0.0005	0.0002	0.0001	0.0001
42.9	0.0060	0.0186	0.0258	0.0128	0.0063	0.0034	0.0016	0.0006	0.0002	0.0001
57.2	0.0007	0.0050	0.0105	0.0118	0.0061	0.0032	0.0017	0.0010	0.0006	0.0003
71.5	0	0.0024	0.0015	0.0072	0.0056	0.0031	0.0017	0.0010	0.0006	0.0004
85.8	0	0.0010	0.0012	0.0011	0.0045	0.0028	0.0017	0.0009	0.0005	0.0003
100.1	0	0.0003	0.0009	0.0005	0.0012	0.0025	0.0015	0.0009	0.0005	0.0003
114.4	0	0	0.0006	0.0004	0.0002	0.0011	0.0014	0.0009	0.0005	0.0003
128.7	0	0	0.0003	0.0003	0.0002	0.0001	0.0009	0.0007	0.0004	0.0003
143.0	0	0	0.0001	0.0003	0.0001	0.0001	0.0001	0.0006	0.0004	0.0003
157.3	0	0	0	0.0002	0.0001	0.0001	0	0.0002	0.0004	0.0002
171.6	0	0	0	0.0001	0.0001	0.0001	0	0	0.0002	0.0002
185.9	0	0	0	0.0001	0.0001	0.0001	0	0	0	0.0001
200.2	0	0	0	0	0.0001	0	0	0	0	0
214.5	0	0	0	0	0.0001	0	0	0	0	0
228.8	0	0	0	0	0	0	0	0	0	0

2270 ft^2. The matrix form of P_K shown in Table 8.10 is often referred to as either a damage matrix, a P_K matrix, or a lethal area matrix, which may be represented by a damage function.

It may be observed that the maximum P_K is 0.6794, and yet we might expect 1.0 due to blast at the detonation point. A possible explanation is that if the scenario involved an air-bursting weapon, a target at ground zero might be further away than RB_2. Also, due to the relatively large grid size, a target located at the center of the cell closest to ground zero might again be further away than RB_2. In both cases there is no blast effect

It may also be noticed that the cell dimensions (37.9 ft by 14.3 ft) are an unusual choice; in fact, these are not chosen but are determined by the total number of cells GFSM can process, which is 40 by 40. Recall the program dates back to the 1970s, when computer memory and processing speed were both very inferior to contemporary levels, so the total cells processed were limited to these fixed values and the cell size adjusted so the total area of effects would be covered by the same number of cells.

Given the area covered by the lethal area of matrix in Table 8.10, we can surmise it corresponds to a relatively large weapon extending several hundred feet in both range and deflection. If a small warhead weapon were processed into the same 40 by 40 matrix, the dimensions of each cell would be correspondingly smaller. When JMAE replaced GFSM, there were no restrictions on cell size or numbers.

We can represent the lethal area matrix as a surface above the horizontal plane where the height is proportional to the P_K in each cell. Such a surface is shown in Fig. 8.15.

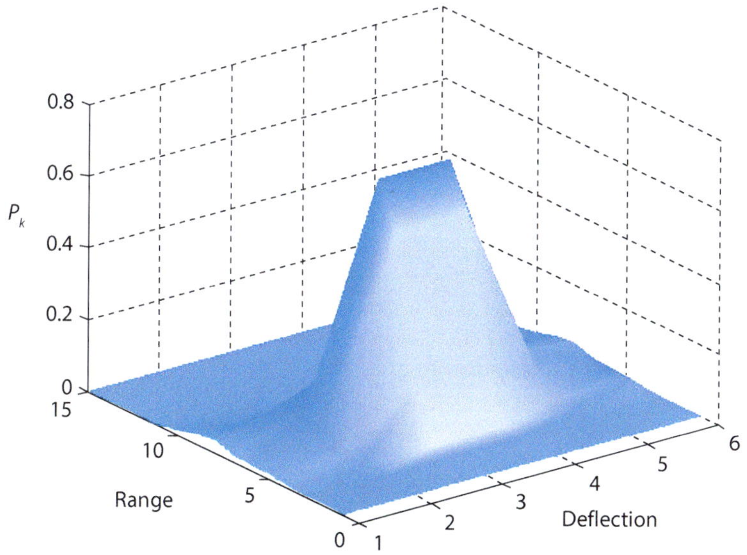

Fig. 8.15 Lethal area matrix as a three-dimensional surface.

8.14 Joint Mean Area of Effects (JMAE) Model

JMAE is a program similar to GFSM in that it can produce a lethal area matrix, or surface, from which an MAE may be calculated; however, it has more capability than GFSM and somewhat higher fidelity in how the target is represented. The description of JMAE will be made by emphasizing its differences from GFSM, so it is assumed that the student is familiar with that program, as described in the previous sections of this chapter.

8.14.1 Target/Weapon Spatial Relationship

In GFSM, the weapon remains fixed at the burst height above ground zero while the P_K is calculated for different placements of the target relative to this location—see Fig. 8.1. In JMAE, the situation is reversed, and the target is fixed and the weapon is placed relative to the target location, as shown in Fig. 8.16.

Also note the following from Fig. 8.16:

- The coordinate frame is defined relative to the target.
- The following impact conditions for the weapon have to be specified:
    - Attack heading
    - Impact angle
    - Impact velocity
- Weapon placement is conducted following a circular pattern within the horizontal plane containing the burst height.

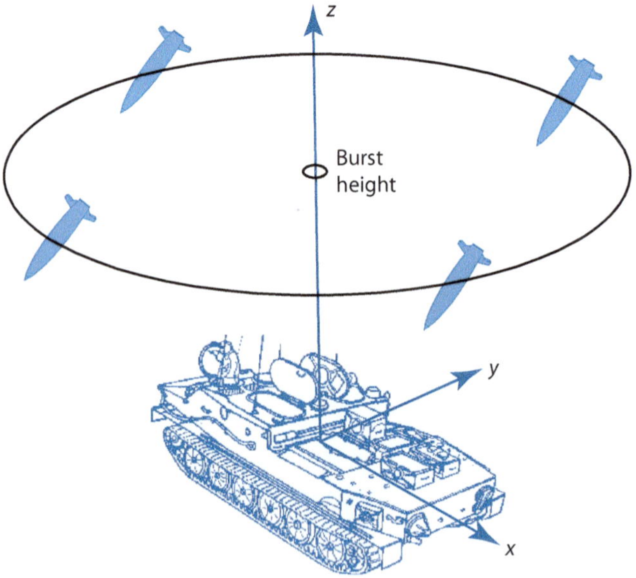

Fig. 8.16 Weapon/target relationship.

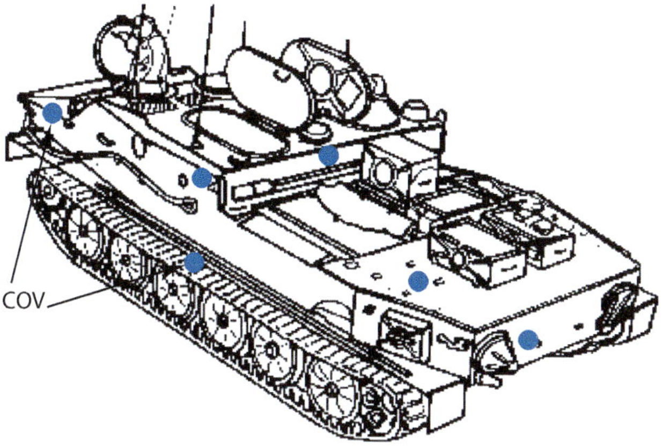

Fig. 8.17 Target with centers of vulnerability (COVs)

8.14.2 Target Representation

GFSM represents a unitary target with a single center of vulnerability (COV), whereas JMAE is able to use several COV if those data are available (Fig. 8.17). JMAE also includes a simple surface modeling capability that is used to represent the physical surfaces of the target. These surfaces affect the weapon burst distribution and are the primary means to introduce direct hit effects.

Each COV typically represents one or more components and is specified in terms of its coordinates relative to the target reference frame. Each component in the target has its vulnerability specified by up to three independent damage mechanisms:

1. Fragment vulnerable area A_V tables
2. User-defined blast envelope
3. Direct hit P_K

8.14.2.1 Fragments

The A_V data may be averaged around the azimuth, as is the case for GFSM with the RCAS files, or specified by azimuth. In the case of the former, probability of exposure data are also included, and the fragmentation model uses the Z-data files as before. The vectored fragment zones are calculated using the weapon impact velocity as for GFSM; however, the dynamic fragment zones are not. Each COV can be subject to multiple vectored zone fragments without consolidating them, allowing the component P_D to be calculated for each zone's fragments and then aggregated using the survivor rule.

8.14.2.2 Blast

Blast may be modeled as in GFSM with the cookie cutter parameters RB_1 and RB_2, except that a blast envelope can be defined for each blast-sensitive component as well as for the target as a whole. In addition, more complex blast envelopes may be constructed if data are available to generate the resulting surface. Two blast envelopes are provided, as shown in Fig. 8.18.

The double cylinder is typically used to describe system-level blast vulnerability, and as such does not represent vulnerability associated with a single point on the target. Therefore, the component COV associated with this envelope will be located anywhere along the center line of the envelope. The larger lower cylinder is meant to allow for Mach stem (ground reflection) effects. Lastly, a sphere, centered z_2 off the ground with radius r_3, is dimensioned to bound the vertical space of the upper cylinder. For the sphere, the COV is located at the center of the sphere. Figure 8.18 shows the RB_1 envelopes, and there is a linear reduction beyond this to RB_2.

8.14.2.3 Direct Hit

The direct mechanism accommodates situations where the weapon impact on the target imparts damage that is not adequately captured by the fragment or blast damage mechanisms. This mechanism is most commonly used when a hit-to-kill weapon such as a penetrator or shaped charge is being modeled. JMAE computes direct hit effects by assigning direct hit P_K values (user specified) to the simple elements representing the target surfaces. Examples of a target surface model are shown in Fig. 8.19.

JWS uses a very simple representation of a surface model, namely the one shown on the left of Fig. 8.19, and is referred to as bounding, or shoe box. More detailed surface models may be used in JMAE, as shown on the right

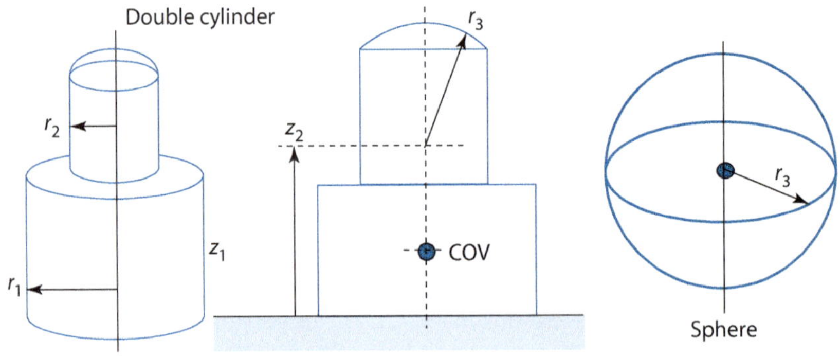

Fig. 8.18 JMAE blast envelopes.

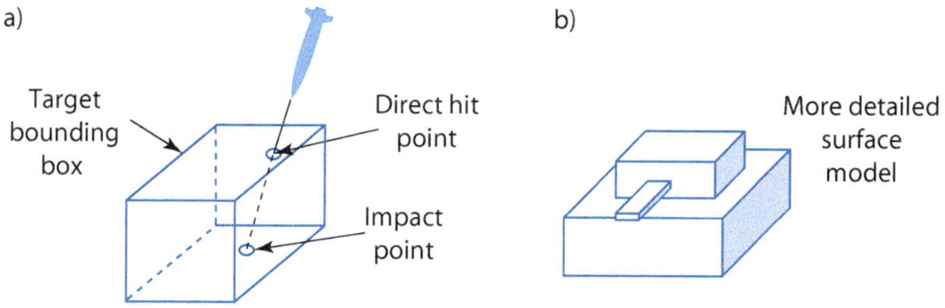

Fig. 8.19 JMAE target surface representation.

of the figure. As each weapon placement is considered according to the height of burst specified, the projection of the weapon backward along its trajectory from the burst point is evaluated to determine any intercept by surface elements representing the target. The first surface element so encountered will identify the corresponding $P_{K/H}$ and establish the direct hit effect for that weapon "sample." Typically, the probability of kill given a hit will be an average over the whole target that is assigned to the several target surface elements. This $P_{K/H}$ is frequently obtained from the following relationship:

$$P_{K/H} = \frac{A_{\mathrm{VN}}}{A_{\mathrm{PN}}} \tag{8.34}$$

For more detailed models, some elements such as a gun tube may be associated with particular kill definitions, such as a firepower kill.

The direct hit P_K can also be obtained by using the Advanced Diverging Ray Module (ADRAM) found in both COVART and JMAE. For warheads that intersect the target surface, individual fragments (as specified by the Z-data file) are randomly "flown out" from the impact point on the target surface, diverging shotlines propagated through the target, and a P_K calculated. The process is repeated several times, and the average P_K is assigned the direct hit P_K for that surface.

8.14.3 Damage Criteria

The damage criteria tree is a combinational circuit that defines a particular level of target damage (kill) in terms of the target components. Each kill of interest will have a corresponding tree defined in the kill definition file. It is typical to build up to an overall kill by initially defining subkills for smaller logical subsystems of the target (i.e., crew mobility, crew firepower, communication equipment, firepower only, etc.). At an appropriate level these subkills can be combined to form the system-level kills that are ultimately of interest. There will be a user learning curve associated with complicated definitions that can sometimes be subtle.

8.14.4 Constructing the Lethal Area Matrix (LAM)

The lethal area matrix, and hence the mean area of effectiveness (MAE), is derived from a large number of points for which the discrete P_K is calculated. JMAE refers to this process as sampling the P_K surface. The sampling is done using polar coordinates by establishing a series of concentric rings with a number of sampling points uniformly distributed around each ring. This was indicated in Fig. 8.16 but is shown in more detail in Fig. 8.20.

These sampling circles lie in the horizontal plane associated with the height of burst and centered on the target frame of reference.

The size/location of the circles is determined by JMAE. First, radii in increments of 1000 ft are considered and the P_K determined for all points on the circle. If the mean P_K falls below a value specified by the user, no larger circles are considered, and the last of these radii become the upper bound of the search process. This final radius is termed the integration distance (ID). Close to the target, circles of 0 ft, 0.5 ft, and 1.0 ft are drawn. In between 1 ft and the ID, the radii increase logarithmically such that the total number equals a user-specified limit.

Once the sampling radii have been determined and the sample weapon locations placed around the horizontal circle with that radius, the P_K values are determined in the following manner.

For each weapon detonation point, the P_K for each vulnerable component is determined for each of the applicable damage mechanisms. Given the warhead terminal conditions, the dynamic fragment zone in which the component falls may be determined and the P_{K-FRAG} computed similarly to GFSM. In addition, any blast envelopes that capture the burst

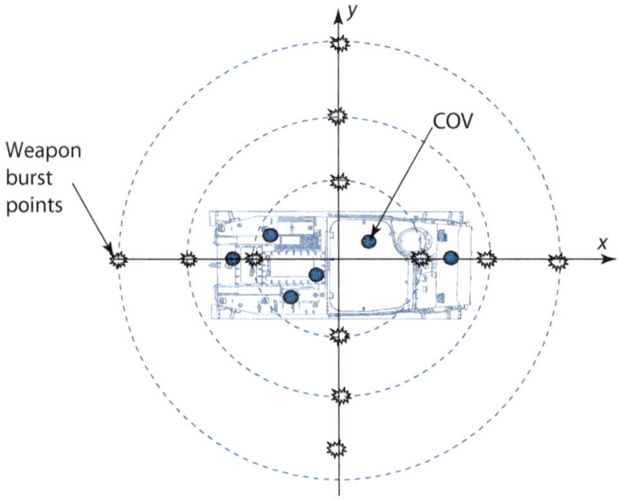

Fig. 8.20 Sample points for the P_K surface.

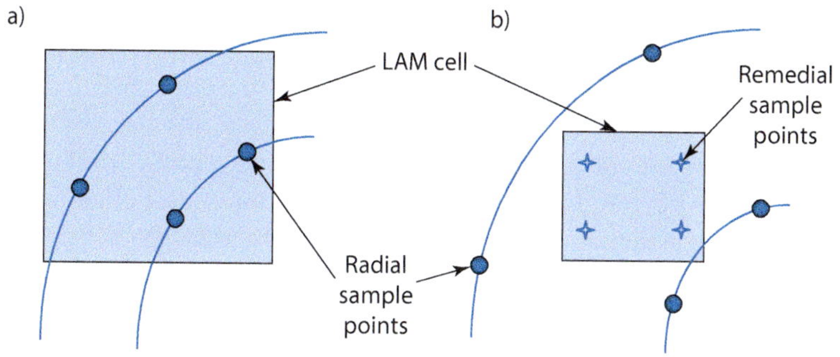

Fig. 8.21 Grid P_K from radial P_K.

may be checked for $P_{K-BLAST}$ contribution, and the probability of direct hit included for any surface encounter. The individual component P_K values may be combined through the fault tree and overall system-level kill criteria determined. This is the P_K associated with that point.

Once the sampling is complete, the generation of the lethal area matrix proceeds in a manner that is the same as for GFSM, and has been describe in Section 8.13. In review, the user specifies a grid on the ground plane that forms the basis of the matrix where the grid may take the following forms:

- Full specification
- Fixed number of cells
- Fixed cell size
- User-defined gridlines
- Exponential or logarithmic grid line spacing

If the selected cell encloses a number of P_K values from the radial samples, the aggregated cell P_K is calculated as for GFSM (Fig. 8.21a). If, however, a cell falls inside a set of samples, points within the cell are used to calculate a set of P_K values from which an aggregated P_K is derived. This process is termed *remedial sampling* and is shown in Fig. 8.21b.

From this LAM the MAE may be determined in the usual manner with the following equation:

$$A_L = \text{MAE} = \sum_{i=1}^{n} P_K(i) \times A_{\text{cell}} \tag{8.35}$$

8.14.5 The Weapon-Centered Lethal Area Matrix

The MAE derived from Eq. (8.35) is independent of an association with either the target or the weapon. The lethal area matrix described in the previous section will be different from that for GFSM. This is because the GFSM matrix is weapon centric whereas that of JMAE is target centric. Although

JMAE produces a target-centric lethal area matrix by the process just described, it then transforms it to a weapon-centric lethal area matrix as for GFSM. This is done by first considering the target/weapon geometry shown in Fig. 8.22.

In this figure we consider a single weapon sample point located on the sampling circle at the location shown in the figure together with the weapon heading at detonation. The P_K is determined in the usual manner, and after converting from a polar coordinate system to a Cartesian coordinate system in order to produce a lethal area matrix, that P_K is associated with the coordinates (x_T, y_T). The question is: What coordinates in the weapon frame are that P_K associated with? The origin of the target frame relative to the weapon frame is given by the following transformation:

$$T = \mathrm{rot}(z, \ -(90+\psi)) = \begin{bmatrix} \cos\theta & -\sin\theta & 0 \\ \sin\theta & \cos\theta & 0 \\ 0 & 0 & 1 \end{bmatrix} \tag{8.36}$$

where $\theta = -(90+\psi)$. For example, if the weapon is located at (–10, 10) in target frame coordinates and the heading is 45 deg, the target in the weapon frame is at the following coordinates:

$$\begin{bmatrix} x_T \\ y_T \\ z_T \end{bmatrix} = \begin{bmatrix} \cos(-135) & -\sin(-135) & 0 \\ \sin(-135) & \cos(-135) & 0 \\ 0 & 0 & 1 \end{bmatrix} \begin{bmatrix} 10 \\ 10 \\ z_W \end{bmatrix} = \begin{bmatrix} -14.14 \\ 0 \\ z_W \end{bmatrix} \tag{8.37}$$

Graphically, this transformation is shown in Fig. 8.23.

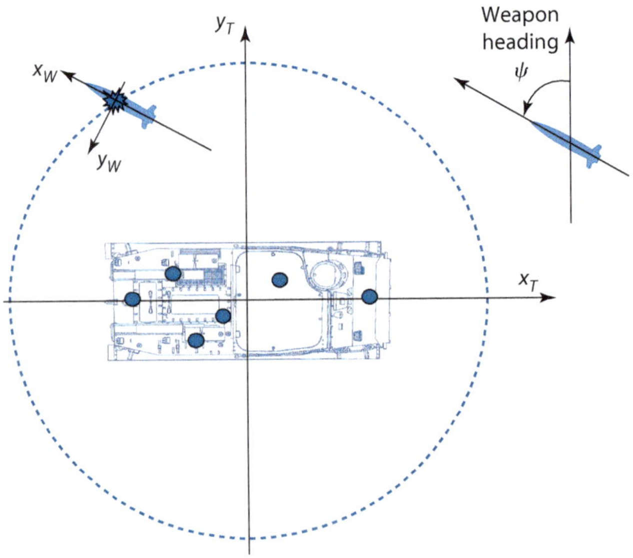

Fig. 8.22 Rotating and translating the weapon impact point.

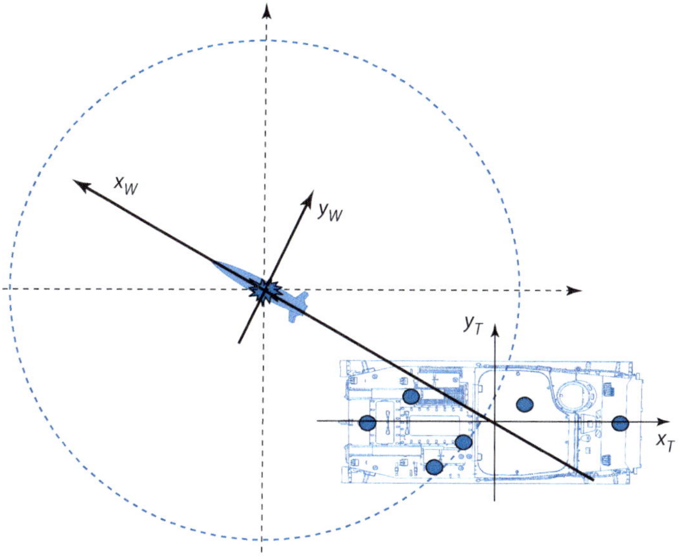

Fig. 8.23 Weapon-centric frame.

8.14.6 Program Output

JMAE produces a similar output to GFSM, that is, a lethal area matrix and a derived MAE. In the case of JMAE, however, the LAM may be weapon or target centric, and the expectation is that the resulting matrix and area of effects is more accurate because the target vulnerability is modeled to a higher fidelity than GFSM. A sample lethal area matrix is shown in Fig. 8.24.

Note how the cells are now user defined, and in this example logarithmically distributed about the impact point.

8.14.7 JMAE in a Monte Carlo Environment

JMAE may be used in any effectiveness methodology that employs a Monte Carlo approach in a manner that doesn't require the generation of a lethal area matrix. Because the program is weapon centric, and in a Monte Carlo approach to effectiveness the weapon impact point is obtained by randomizing the delivery errors, a P_K unique to that iteration's specific impact point may be calculated. This is illustrated in Fig. 8.25.

Given the P_K value for each impact point, the average or expected value can be obtained after all iterations of the simulation have been completed, without the need to compute or utilize a lethal area matrix.

In summary, the overall JMAE process can be represented as in Fig. 8.26.

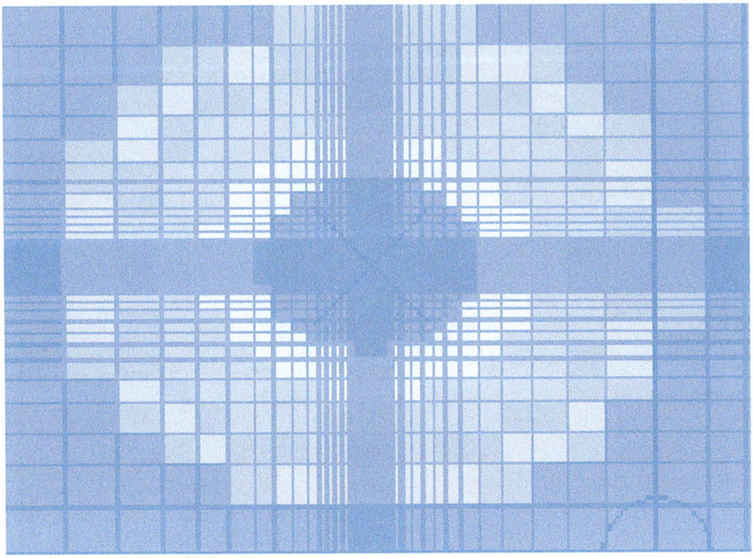

Fig. 8.24 JMAE-generated LAM cells.

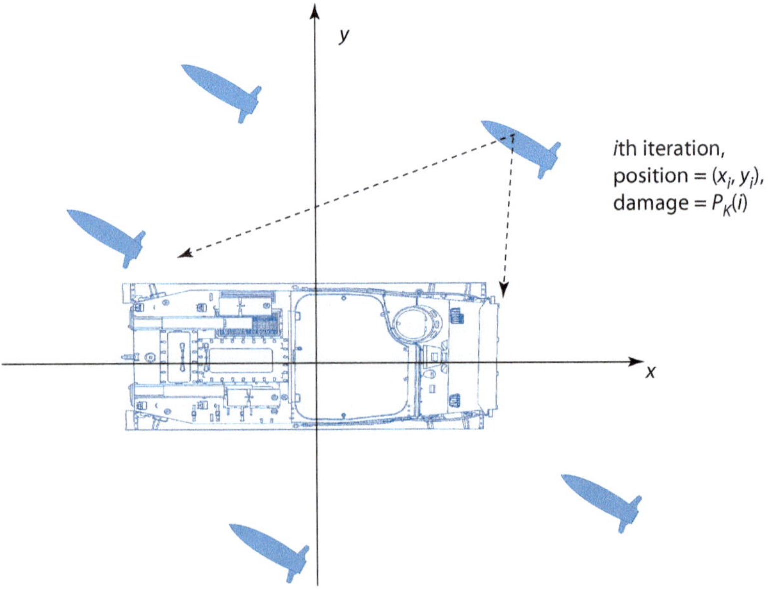

Fig. 8.25 Weapon/target geometry for specific impact point.

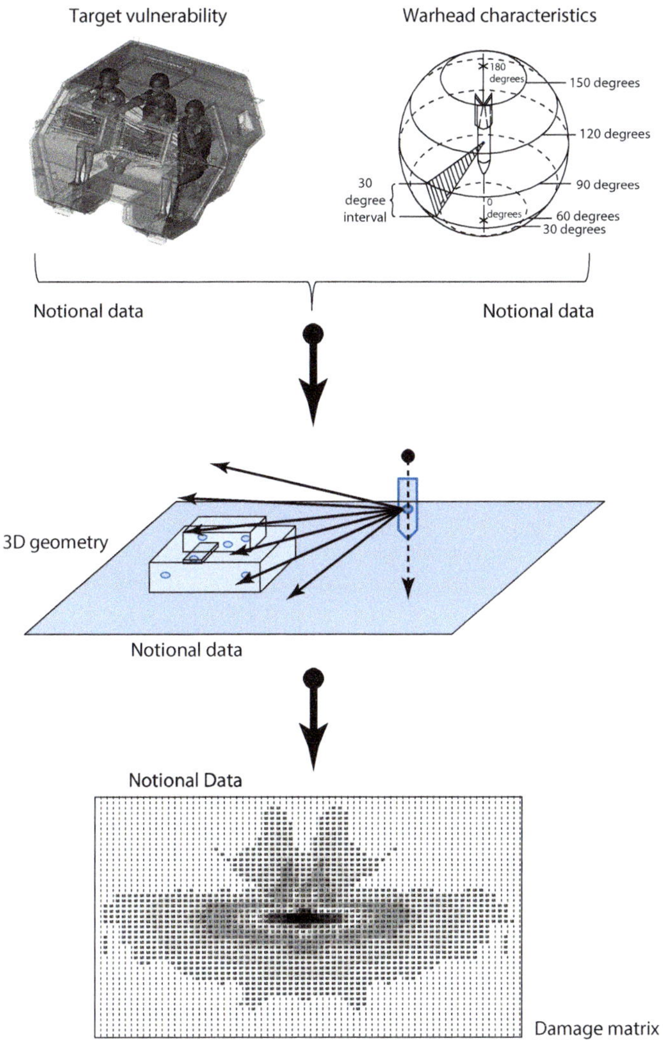

Fig. 8.26 Overall JMAE process.

8.15 Simplification of Damage Matrix

Although the damage matrix shown in Table 8.4 represents the P_K values in cells around the weapon, it is too cumbersome to use in simple weaponeering methodologies, and approximations have to be used to describe it. From Table 8.4 we could plot contours of constant values of P_K around the weapon detonation point. Such a plot is shown in Fig. 8.27 and represents the damage function for blast- and fragment-sensitive

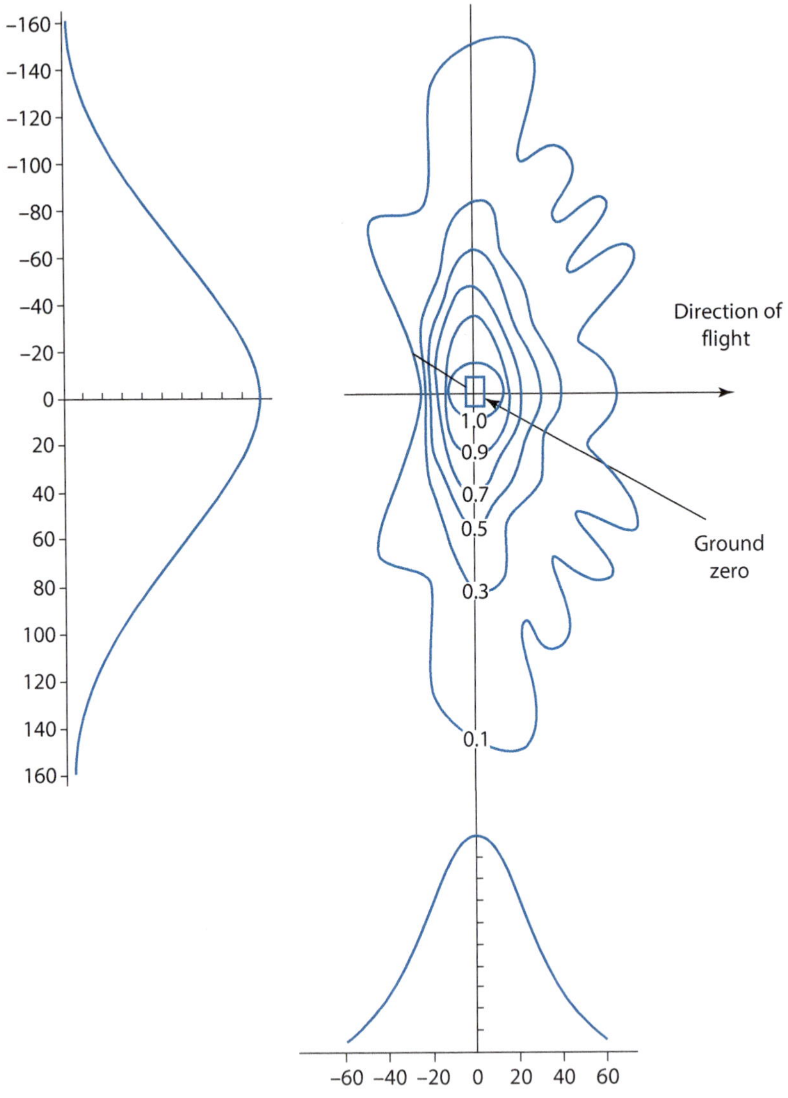

Fig. 8.27 P_K contours with gaussian approximations.

targets. Sometimes this representation is termed the "bug splat." With some imagination, these contours appear approximately elliptical.

Also shown in the figure are Gaussian-like approximations to the actual contour profiles in the range and deflection directions. Note how the parameters describing the spread of these distributions are different. This observation leads to an analytical expression for an approximation to the tabulated damage function (P_K's) and the bug splat contour, which is referred to as the

Carlton damage function and takes the form

$$P(x, y) = \exp - \left[\frac{x^2}{\mathrm{WR}_r^2} + \frac{y^2}{\mathrm{WR}_d^2} \right] \tag{8.38}$$

where WR_r and WR_d are defined as the weapon radius in the range and deflection directions, respectively. Note that the Carlton damage function is similar in shape to the bivariate Gaussian, or normal distribution, except that it has value of unity at the point (0, 0), which the Gaussian distribution does not.

Another difference is that for the Carlton damage function, it is the value of the function itself that is of interest, whereas for the bivariate distribution it is the area beneath the function that is important. Figure 8.28 shows a three-dimensional view of the Carlton damage function and the elliptical nature of horizontal sections (constant P_K contours) through it. Consider how Fig. 8.28 is a smoothed version of Fig. 8.15.

We do not know the values of WR_r and WR_d because they depend on the rate at which P_K decreases, but note that in general they will be different. The lethal area is obtained by summing the products of cell area and cell P_K values, but if we have an analytical value for the damage function we may also write the following:

$$A_L = \mathrm{MAE}_F = \sum_{i=1}^{n} P_K(i) \times A_{\mathrm{cell}} = \int_A P_K \mathrm{d}A \tag{8.39}$$

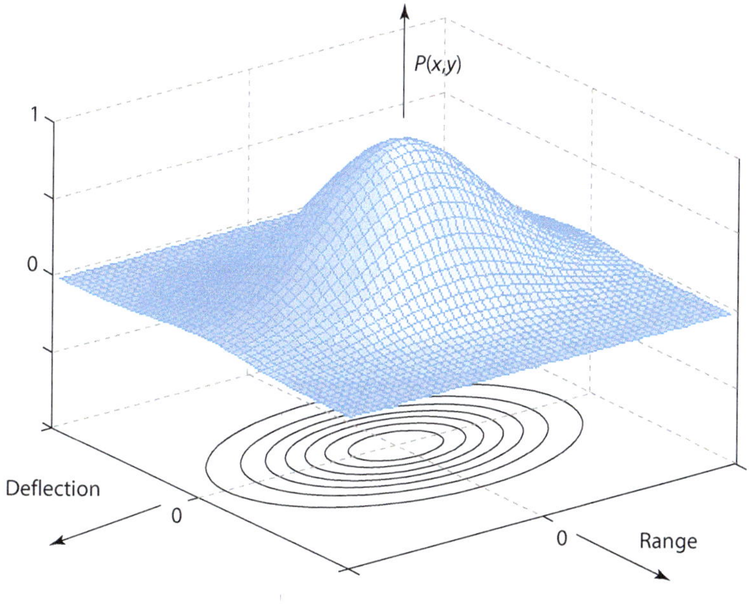

Fig. 8.28 Carlton damage function.

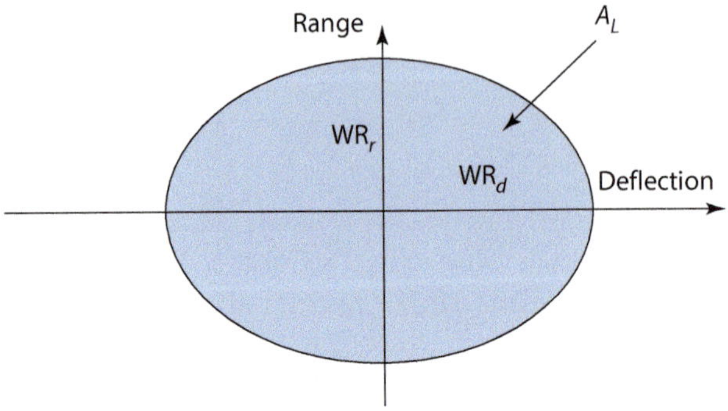

Fig. 8.29 Elliptical representation of the lethal area.

So if the P_K's are to be correctly represented by the Carlton damage function, we would expect that

$$A_L = \mathrm{MAE}_F = \int_{-\infty}^{\infty}\int_{-\infty}^{\infty} \exp - \left[\frac{x^2}{\mathrm{WR}_r^2} + \frac{y^2}{\mathrm{WR}_d^2}\right] \mathrm{d}x\,\mathrm{d}y \tag{8.40}$$

It may be shown that the integral evaluates to

$$A_L = \mathrm{MAE}_F = \pi \times \mathrm{WR}_r \times \mathrm{WR}_d \tag{8.41}$$

which is the equation for the area of an ellipse with radii WR_R and WR_D. The lethal area may therefore be equated to the area of the ellipse shown in Fig. 8.29.

This damage function is different from those discussed previously in that the lethality is concentrated inside a specified boundary and is zero outside the boundary. This type of damage function is known as a "cookie cutter" for obvious reasons. Hence the damage matrix may be represented either by the Carlton damage function extending over the whole of the ground plane or by a cookie cutter elliptical damage function where the P_K inside is unity and zero outside. We can define the ratio of the weapon radii as

$$a = \frac{\mathrm{WR}_r}{\mathrm{WR}_d} \tag{8.42}$$

and, in general, this aspect ratio is found to be a function of the weapon impact angle. It is found to be given by the empirical equation

$$a = \mathrm{MAX}[1 - 0.8\ \cos(I),\ 0.3] \tag{8.43}$$

where I is the impact angle in degrees. This enables the weapon radii to be computed from the MAE_F rather than the tabular P_K data.

$$\mathrm{MAE}_F = \pi \times \mathrm{WR}_R \times \frac{\mathrm{WR}_R}{a} \tag{8.44}$$

$$\mathrm{WR}_r = \sqrt{\mathrm{MAE}_F \times a/\pi} \tag{8.45}$$

$$\mathrm{WR}_d = \mathrm{WR}_r/a \tag{8.46}$$

Sometimes the lethal area is described in terms of the effective target length L'_{ET} and effective target width W'_{ET} rather than the weapon radii, where $L'_{\mathrm{ET}} = 2\mathrm{WR}_r$ and $W'_{\mathrm{ET}} = 2\mathrm{WR}_d$. They may be computed from Eqs. (8.45) and (8.46) as

$$L'_{\mathrm{ET}} = 2 \times \mathrm{WR}_r = 1.128\sqrt{\mathrm{MAE}_F \times a} \tag{8.47}$$

$$W'_{\mathrm{ET}} = 2 \times \mathrm{WR}_d = \frac{L'_{\mathrm{ET}}}{a} \tag{8.48}$$

We can also write the effective target area A_{ET} as

$$A_{\mathrm{ET}} = \mathrm{MAE}_F \tag{8.49}$$

The Carlton damage function may be expressed in terms of the parameters L'_{ET} and W'_{ET}, occasionally seen in JMEM documentation as

$$P(x, y) = \exp - \left[\frac{4x^2}{{L'_{\mathrm{ET}}}^2} + \frac{4y^2}{{W'_{\mathrm{ET}}}^2} \right] \tag{8.50}$$

Unfortunately for some applications, even the Carlton damage function or the elliptical cookie cutter damage function shown in Fig. 8.29 is not sufficiently simple to allow a mathematically tractable solution, so further modifications are necessary. To do this, a rectangular approximation is made to the elliptical lethal area, as shown in Fig. 8.30.

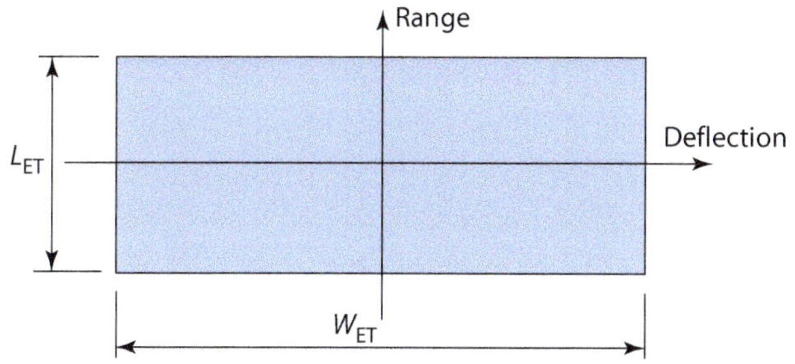

Fig. 8.30 Rectangular approximation to the lethal area.

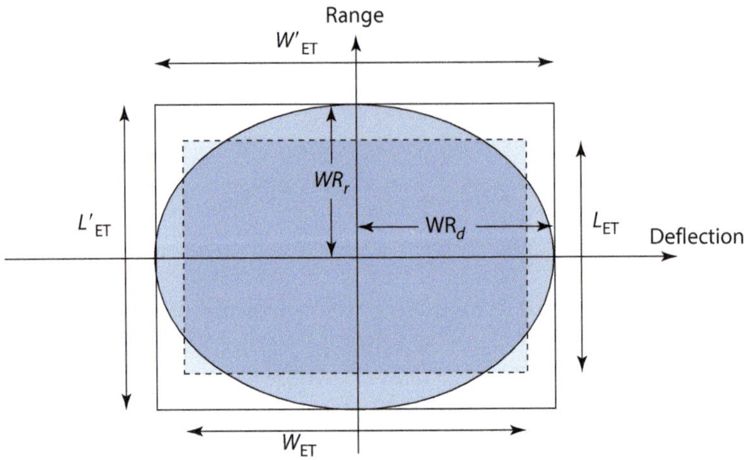

Fig. 8.31 Definition of effective target length and widths.

In this case, the effective target length and width* are given by

$$L_{\mathrm{ET}} = \sqrt{\mathrm{MAE}_F \times a}, \quad W_{\mathrm{ET}} = L_{\mathrm{ET}}/a \tag{8.51}$$

This damage function is another cookie cutter function because the value of P_K inside the function is unity and zero outside. It is also bounded by L_{ET} and W_{ET}, whereas the Carlton damage function extends to infinity in every direction and varies in value over the ground plane. Cases where the Carlton damage function and its rectangular approximation may be applied will be dealt with in later Chapters 9–11. Note that the aspect ratio parameter a applies to the rectangular cookie cutter, too, allowing Eq. (8.42) to be extended to the following variables:

$$a = \frac{\mathrm{WR}_r}{\mathrm{WR}_d} = \frac{L'_{\mathrm{ET}}}{W'_{\mathrm{ET}}} = \frac{L_{\mathrm{ET}}}{W_{\mathrm{ET}}} \tag{8.52}$$

The relationship among L'_{ET}, L_{ET}, W'_{ET}, W_{ET}, and the elliptical damage function shown in Fig. 8.29 are shown in Fig. 8.31.

Although confusing, it should be remembered that L'_{ET} and W'_{ET} are associated with the Gaussian or Carlton damage function whereas L_{ET} and W_{ET} are the effective target length and width, respectively, and define an equivalent rectangular damage function. Note the following:

$$\mathrm{MAE}_F = L_{\mathrm{ET}} \times W_{\mathrm{ET}} \neq L'_{\mathrm{ET}} \times W'_{\mathrm{ET}} \tag{8.53}$$

*Note that $A_{\mathrm{ET}} \neq L'_{\mathrm{ET}} W'_{\mathrm{ET}}$, but that $A_{\mathrm{ET}} = L_{\mathrm{ET}} W_{\mathrm{ET}}$.

8.16 Conserving Lethality for Different Damage Functions

Suppose we have a lethal area matrix such as the one shown in Table 8.10 for a weapon impact angle of, say, 65 deg and for a specified damage criteria. We can calculate the lethal area using Eq. (8.39) by multiplying each cell P_K by the area of the cell and summing over the ground plane for all cells with a nonzero P_K. The result is $\text{MAE}_F = 2270\ \text{ft}^2$.

We can determine the equivalent Carlton damage function by calculating the weapon radii using Eqs. (8.43) through (8.46). These equations yield

$$a = \text{MAX}[1 - 0.8\ \cos(65),\ 0.3] = 0.662 \tag{8.54}$$

$$\text{WR}_r = \sqrt{2270 \times 0.662/\pi} = 21.87' \tag{8.55}$$

$$\text{WR}_d = 21.87/0.662 = 33.04' \tag{8.56}$$

The resulting Carlton damage function becomes

$$P(x, y) = \exp - \left[\frac{x^2}{21.87^2} + \frac{y^2}{33.04^2}\right] \tag{8.57}$$

Lethality is conserved by noting that the integral of the Carlton damage function shown in Eq. (8.57) over the ground plane would yield $\text{MAE}_F = 2270\ \text{ft}^2$.

Representing the damage function as an elliptical cookie cutter as shown in Fig. 8.29 would result in an ellipse of major radius 33.04 ft and minor radius of 21.87 ft. Because the probability of damage inside the cookie cutter is unity and zero outside, the lethal area is simply the area of the ellipse, or

$$\text{MAE}_F = \pi \times 21.87 \times 3.04 = 2270 \tag{8.58}$$

Again this indicates preservation of lethality. Finally, if the damage function is represented by a rectanglular cookie cutter as shown in Fig. 8.30, the dimensions may be calculated from Eq. (8.51)

$$L_{\text{ET}} = \sqrt{2270 \times 0.662} = 38.76\ \text{ft}, \quad W_{ET} = 38.76/0.662 = 58.56\ \text{ft} \tag{8.59}$$

The lethal area is $L_{\text{ET}} \times W_{\text{ET}} = 2270\ \text{ft}^2$ as expected. This analysis shows how modifying the damage function from the most detailed (lethal area matrix) to the simplest (rectangular cookie cutter) preserves the lethality of the weapon/target/damage criteia.

8.17 Lethal Area Calculation for Targets Sensitive to Blast

So far, only fragment-sensitive targets have been discussed, although it is important to remember that the P_K matrices shown in Table 8.4 include the effect of blast as well as fragments. In some cases, however, the target may be vulnerable only to blast, and damage functions for this mechanism have to be represented in mathematical form for calculating effects.

This is accomplished by defining the mean effective area for blast MAE_B and then representing that number by a damage function. The usual representation is an area on the ground plane within which the P_K is unity and outside of which P_K is zero. This is, therefore, a cookie cutter EI similar to the one described in the previous section. This EI may be defined in terms of the weapon blast radii shown in Fig. 8.2 and takes the form of a truncated cone, as shown in Fig. 8.32.

The $P_{\mathrm{K-BLAST}}$ could be determined for each cell and the MAE_B determined as for the fragmentation case. Any cell closer to ground zero than RB_1 would have a $P_K = 1$ whereas any cell further away than RB_2 would have a $P_K = 0$. Cells between RB_1 and RB_2 would have the P_K determined from Fig. 8.33. The blast equivalent to the Carlton damage function for MAE_F would be the truncated cone centered over ground zero shown in Fig. 8.32.

The symmetry of the blast envelope, however, lends itself to a simpler calculation for the MAE_B from the general expression for the lethal area in Eq. (8.39)

$$\mathrm{MAE}_B = \int P_K \mathrm{d}A \tag{8.60}$$

The variation in $P_{\mathrm{K-BLAST}}$ with radius is shown in Figure 8.33 as a function of radius, where it is assumed in the linear region that the P_K is constant within a small annular ring of width dr.

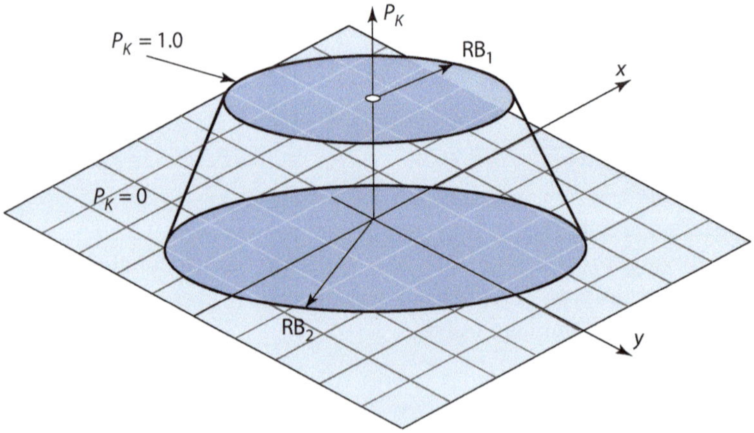

Fig. 8.32 Blast envelope placed over ground plane cells.

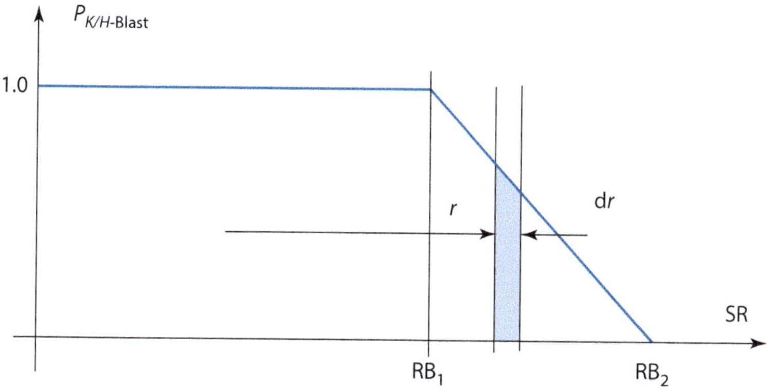

Fig. 8.33 Variation in $P_{K/H}$ for blast.

Equation (8.60) may be integrated directly if we put it in the polar form

$$\mathrm{MAE}_B = \int_{r=0}^{r=\infty} P_K(r) \times 2\pi r\,\mathrm{d}r \tag{8.61}$$

This integral may be evaluated in two regions, $P_K < \mathrm{RB}_1$ and $\mathrm{RB}_1 < P_K < \mathrm{RB}_2$.

$$\begin{aligned}\mathrm{MAE}_B &= 2\pi\left[\int_{r=0}^{r=\mathrm{RB}_1} r\mathrm{d}r + \int_{r=\mathrm{RB}_1}^{r=\mathrm{RB}_2}\left(\frac{\mathrm{RB}_2 - r}{\mathrm{RB}_2 - \mathrm{RB}_1}\right) r\mathrm{d}r\right] \\ &= 2\pi\left[\frac{\mathrm{RB}_1^2}{2} + \frac{1}{\mathrm{RB}_2 - \mathrm{RB}_1}\left[\frac{\mathrm{RB}_2 r^2}{2} - \frac{r^3}{3}\right]_{\mathrm{RB}_1}^{\mathrm{RB}_2}\right]\end{aligned} \tag{8.62}$$

After some manipulation, this reduces to

$$\mathrm{MAE}_B = \frac{\pi}{3}\left[\mathrm{RB}_1^2 + \mathrm{RB}_1\mathrm{RB}_2 + \mathrm{RB}_2^2\right] \tag{8.63}$$

For weaponeering purposes, it is more convenient if the damage function is represented in the ground plane, this time as a square defined again by L_{ET} and W_{ET}, where

$$L_{\mathrm{ET}} = W_{\mathrm{ET}} = \sqrt{\mathrm{MAE}_B} \tag{8.64}$$

Graphically, this cookie cutter damage function may be represented in the ground plane as shown in Fig. 8.34.

Note the similarity between Eqs. (8.47) and (8.64). The difference is the absence of the aspect ratio parameter a, because blast propagates radially

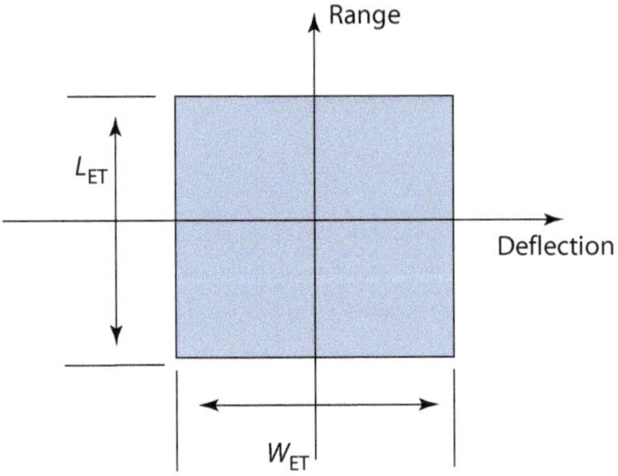

Fig. 8.34 Blast cookie cutter damage function.

from the point of detonation and will therefore be independent of the weapon impact angle.

In summary, the highest fidelity representation of the interaction between a weapon and a target has been derived in the form of the lethal area matrix. Although this may be conveniently represented by a single number, the MAE_F, the distribution of lethality around the weapon is lost when this is done. For some fast running methods to calculate effectiveness, the lethal area matrix cannot be used directly and has to be simplified into a form that is mathematically tractable. Examples of this simplification include the Carlton and cookie cutter–type damage functions.

8.18 Combining Lethal Area Matrices: The AVMAT Program

Recall Fig. 8.1 where an air-bursting weapon was considered and the geometry of the munition at detonation relative to the target was used to derive the lethal area matrix. Given the discussion in Chapter 3, Section 3.15 on fuze variability and the resultant height of burst (HOB), it may be asked what effect such uncertainty has on the resulting lethal area matrix. Clearly if the airburst was 10 ft higher we would get a different matrix than we would if the height of burst were 10 ft lower than that indicated in the figure. How should this variability be handled?

In practice we may generate lethal area matrices for a range of burst heights above the target, say every 20 ft. This set of matrices can then be used in two ways.

1. For Monte Carlo models such as those illustrated in Chapter 3, Section 3.15, where the delivery and fuzing errors are randomized, the detonation height will determine which lethal area matrix is to be used for that

iteration. In the case of the height not coinciding with one of the set, interpolation between the bracketing matrices may be used.
2. For analytical models, the lethal area matrices are combined or averaged into a single matrix using a program called AVMAT to average the matrices.

Consider a series of lethal area matrices calculated at different burst heights, as shown in Fig. 8.35, where each matrix lies directly on top of the previous one.

There are several options for producing a single lethal area matrix by aggregating each cell of the HOB matrices in different ways:

- An *arithmetic average* of each cell. If there are n LAMs stacked on top of each other, then the P_K in cell (i, j) of the averaged single matrix becomes

$$P_K(i,j) = \frac{1}{n}\sum_{q=1}^{n} P_K(i,j,q) \tag{8.65}$$

- By *combining* cells using the survivor rule, the effect of LAMs from different warheads may be aggregated in the following manner:

$$P_K(i,j) = 1 - \prod_{q=1}^{n} (1 - P_K(i,j,q)) \tag{8.66}$$

- A *weighted average* may be calculated by assuming, for example, a Gaussian distribution for fuze HOB about the nominal burst height, as shown in Fig. 8.36.

In this figure there is a nominal height of burst for an altitude fuzed warhead and an associated normal distribution of the error in fuze functioning height. Each lethal area matrix at a height different from the nominal has

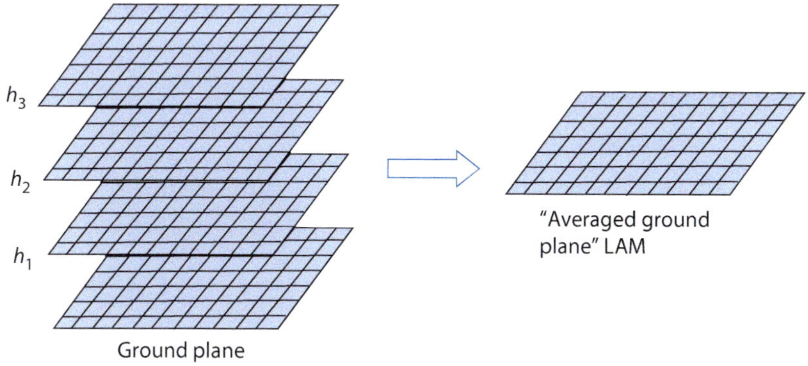

Fig. 8.35 Aggregating a set of LAMs using AVMAT.

to be weighted by the probability the fuze detonates at that height, or more specifically it detonates between the bracketing LAMs, and this is obtained from the PDF of the normal distribution. Consider the LAM at a height h_i relative to the nominal HOB h_0. The probability the fuze functions between altitude h_{i-1} and h_{i+1} is given by

$$p(h_i) = p\{h_{i+1} > h > h_{i-1}\} = \int_{h_{i-1}}^{h_{i+1}} f(h)\mathrm{d}h$$

$$= \int_{h_{i-1}}^{h_{i+1}} \frac{1}{\sqrt{2\pi}\sigma_h} \exp\left[\frac{-(h-h_0)^2}{2\sigma_h^2}\right] \mathrm{d}h \tag{8.67}$$

where the value σ_h is obtained from the population of fuze timers through testing. The P_K in the cell (i, j) of the aggregated, single LAM is then weighted by this probability, and is given by

$$P_K(i, j) = \frac{1}{n}\sum_{q=1}^{n} P_K(i, j, q) \times p(h_q) \tag{8.68}$$

In this manner, a single LAM is derived that may be considered to rest on the ground plane, from which it may be used directly in any non–Monte Carlo effectiveness methodology requiring a LAM.

Note that in the description of the AVMAT program we have assumed for simplification purposes exact alignment of the various matrices that have been combined; however, the program allows for this requirement to be relaxed. In such cases, a fractional coverage calculation is performed to allow for the misalignment. For example, Eq. (8.65) for a simple average would become

$$P_K(i, j) = \frac{1}{n}\sum_{q=1}^{n} P_K(i, j, q) \times F_C(i_1, j_1, i_2, j_2) \tag{8.69}$$

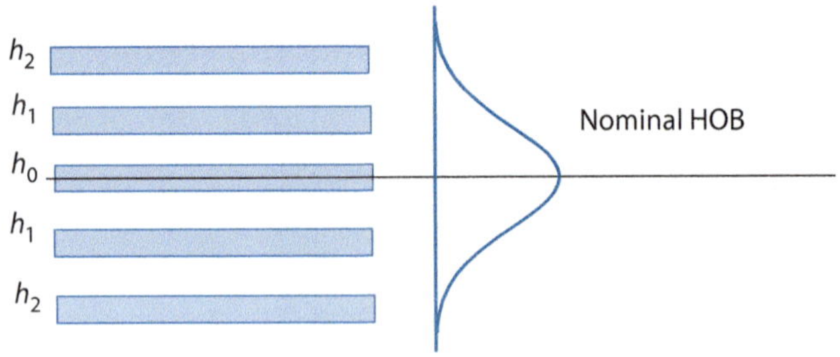

Fig. 8.36 Weighted average of lethal area matrices.

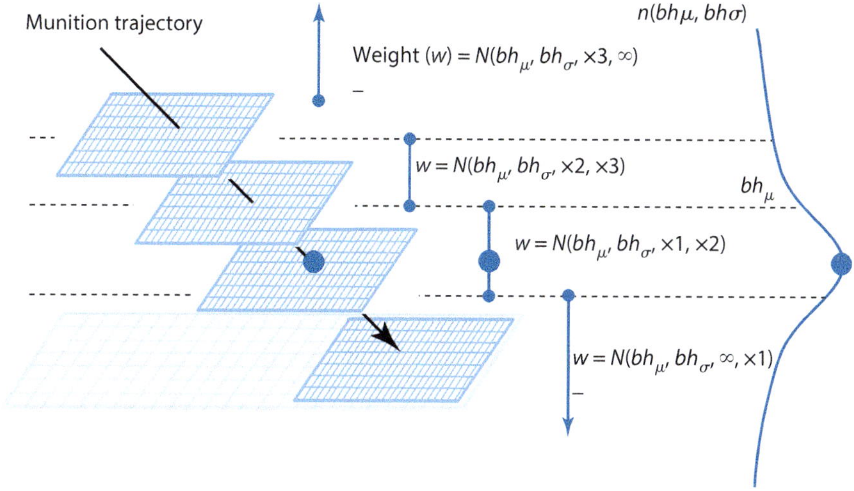

Fig. 8.37 Range-shifted, weighted lethal area matrices.

In this equation, $F_C(i_1, j_1, i_2, j_2)$ is the fractional coverage of the (i, j) cell of the qth LAM on the corresponding (i, j) cell on the resultant, combined matrix. The program also allows for a LAM with logarithmically spaced cells such as that generated by JMAE, user-specified weighting functions for Eq. (8.68), and range shifting to allow for a weapon's terminal trajectory, as shown in Fig. 8.37.

The lethal area, MAE_F, may be computed from the averaged lethal area matrix in the usual manner.

Chapter 9 Calculating Damage Probabilities: Monte Carlo Methods

9.1 Introduction

The preceding chapters have now developed the basic tools needed to calculate the effectiveness of a weapon directed towards a target. To begin with, the simplest possible case will be considered, that is, a single or unitary weapon directed towards a single target. We are interested in the value of the probability of damage caused by a single weapon, defined as PD_1, using:

- Measures of delivery accuracy
- Weapon-specific warhead effects such as fragments, blast, and shaped charge
- Target vulnerability to these effects

This assumes knowledge of range error probable (REP), deflection error probable (DEP), or circular error probable (CEP), and a damage function represented by either a lethal area matrix, a Carlton damage function, or a cookie cutter of specified size and shape. The exact method of determining PD_1 will vary somewhat depending on the specific data supplied, but the basic methodology will be the same.

Because we are dealing with single weapon releases, the two principle error sources—mean point of impact (MPI) and precision—may be combined into a single term, as shown in Eq. (4.3). Therefore, as far as delivery accuracy is concerned, we have single values of σ_x and σ_y or their error-probable equivalents. The most general expression for effectiveness may then be written as

$$\text{Prob}\{\text{damage}\} = \text{Prob}\{\text{hit}\} \times \text{Prob}\{\text{damage if hit}\} \tag{9.1}$$

In the righthand side of Eq. (9.1) the first term addresses the accuracy with which the weapon may be delivered, and the second term is a measure of the effectiveness of the weapon against the target if it is hit. Both of these have been considered in detail in earlier chapters.

In this equation the word *hit* has to be interpreted with caution because we know a target can be damaged given a sufficiently close detonation without it actually being hit. Perhaps a better word would be *detonation.*

9.2 Basic Approach to Effectiveness Calculation

The simplest way of explaining how the PD_1 would be calculated is to consider a Monte Carlo simulation where the damage function is a rectangular cookie cutter, and single error parameters REP and DEP are known. The target is a point located at the origin of a coordinate frame on the ground plane, and a weapon is directed at the target, as shown in Fig. 9.1. If we were to program the Monte Carlo simulation, it would consist of the following steps:

1. Initialize a counter, in this case the P_K counter, to zero.
2. Using REP and DEP and a normal random number generator, calculate the weapon impact point on the ground plane (see Fig. 9.1).
3. Draw the lethal area represented by the rectangular cookie cutter centered on the impact point.
4. Determine if the target, represented as a point, is enclosed within the cookie cutter.
5. If it is, then the target is killed, so increment the P_K counter by one,
6. Repeat from step 2.
7. After a sufficiently large number of steps, terminate the simulation.
8. Calculate the PD_1 from the following equation:

$$PD_1 = \frac{P_K \text{ counter}}{\text{number of iterations}} \tag{9.2}$$

A sample iteration of this simulation where the target is not damaged is shown in Fig. 9.1.

So, if we perform 100,000 iterations, and 47,396 of them resulted in the point target being enclosed by the lethal area, Eq. (9.2) would give the following:

$$PD_1 = \frac{47{,}396}{100{,}000} = 0.474 \tag{9.3}$$

Note that this method would work for a blast/fragment warhead MAE_F damage function represented by a rectangular cookie cutter, or a blast warhead MAE_B represented by a square cookie cutter.

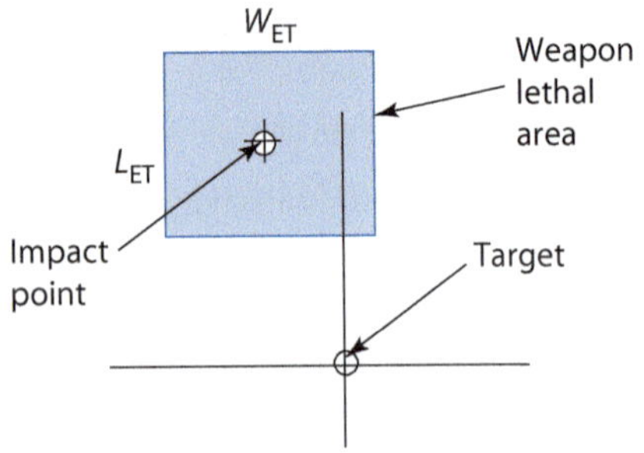

Fig. 9.1 One iteration of a Monte Carlo simulation.

9.3 Interpretation of the Weapon Lethal Area

The relationship between the target and weapon lethal area may be viewed in two ways and has often led to some confusion. For example, in the previous chapter the rectangular cookie cutter damage function was derived from the effectiveness of the weapon, and yet it was given the dimensions L_{ET} by W_{ET}—the effective target length and width. So does the rectangle represent the weapon or the target? The answer is both, because the lethal area represents both the damage effects of the warhead and the target's vulnerability to those effects.

Calculating effectiveness can take advantage of this duality. Consider the situation shown in Fig. 9.1 (reproduced in Fig. 9.2a), where the point target is fixed at the origin and the weapon lethal area is centered on the randomly derived impact point. Figure 9.2b shows an alternative view of the interaction where the target is represented by the lethal area rectangle fixed at the origin, and the weapon, represented as a point, impacts at some randomly determined point relative to it.

This again begs the question: Which representation is correct? Is the target a point and the weapon lethal area a rectangle, or is the target a rectangle and the weapon represented by a point? As before, the answer is that both are correct interpretations of the interaction, and in calculating the effectiveness it is seen that the probability of placing the lethal area rectangle on a fixed point target is the same as the probability of hitting a fixed rectangle with a single impact point.

Historically weaponeers have used both representations interchangeably when discussing effectiveness, and care must be exercised when determining which representation is being analyzed.

Using this alternative interpretation makes it easier to solve the example given in the previous section using a Monte Carlo simulation by fixing the weapon lethal area at the origin of the coordinate frame and determining the probability the impact point lies within it.

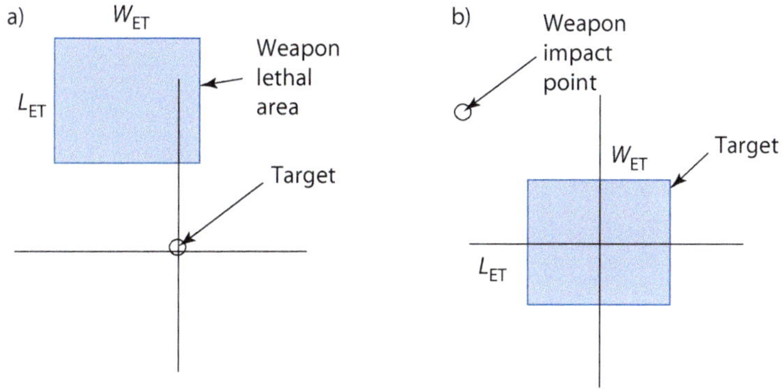

Fig. 9.2 Alternate representation of weapon and target.

Consider the example where the target has a vulnerable area that is a 10 ft × 10 ft rectangle. We are dealing with single weapon deliveries, so the combined dependent and independent accuracies in range and deflection are equal and represented by a zero mean normal distribution with a standard deviation of 5.385 ft. The following MATLAB code implements this Monte Carlo process:

```
%-------- Probability of Damage for One Weapon -----------------
clear all;
LET=10.0;           % target size - range
WET=10.0;           % target size - deflection
sigx=5.385;         % miss distance sigma, range
sigy=5.385;         % miss distance sigma, deflection
m=500000;           % number of iterations
count=0;            % initialize counter
%------------- Monte Carlo loop ------------------------------
   for j=1:m;                                  % single loop
      x_impact=sigx*randn;                     % impact point range
      y_impact=sigy*randn;                     % impact point deflection
      if ((abs(x_impact)<=LET/2) & (abs(y_impact)<=WET/2));
         count=count+1; % impact point inside rectangle
      end
   end
   PD1=count/(m); % final prob. damage
t=['Single Weapon Monte Carlo=',num2str(PD1)];
disp(t)
```

The function *randn* generates a normally distributed random variable with zero mean and unity standard deviation (i.e., the standard normal variable z). The physical impact point is obtained from the following mapping:

$$x = \mu + \sigma z \tag{9.4}$$

Running this program for the specified number of iteration produces the following result:

L_{ET} = 10 ft
W_{ET} = 10 ft
σ_X = 5.385 ft
σ_Y = 5.385 ft

⇒ PD_1 = 0.4194

recognizing that when the EI is MAE_F, the aspect ratio of the lethal area depends on the weapon impact angle. We use this together with the lethal area to determine the dimensions of the cookie cutter. The appropriate equations from Eq. (8.43) are as follows:

$$a = \max[(1 - 0.8 \times \cos I), 0.3] \tag{9.5}$$

$$L_{ET} = \sqrt{MAE_F \times a} \tag{9.6}$$

$$W_{ET} = \frac{L_{ET}}{a} \tag{9.7}$$

We also express accuracy in the more commonly used error probable rather than sigma values. Using these equations to determine L_{ET} and W_{ET} and modifying the Monte Carlo code shown previously to include the MAE_F and an impact angle produces the following result:

$MAE_F = 2270$ ft^2
$I = 65$ deg
REP = 20 ft
DEP = 10 ft

→

$PD_1 = 0.4655$

Note the result here (0.4655) is not comparable with the previous program result (0.4194) because in one case L_{ET} and W_{ET} are defined whereas in the other case they are calculated from the MAE_F and the impact angle.

9.4 Using other Damage Functions

So far the damage function used has been the simple rectangular cookie cutter, so it prompts the question: How would other damage functions be used to calculate effectiveness? In answering this, the same Monte Carlo approach is considered again using the representation shown in Fig. 9.2b where the damage function is fixed at the origin, and we attempt to hit it with an accuracy defined by REP and DEP. Also note that we will use the weapon impact angle as an input.

9.5 Carlton Damage Function

Assume the damage function is given as

$$P(x, y) = \exp - \left[\frac{x^2}{WR_r^2} + \frac{y^2}{WR_d^2} \right] \tag{9.8}$$

where the weapon radii WR_r and WR_d are known. The ground plane impact point relative to the damage function (x', y') are obtained from random draws for REP and DEP, and substituted for x and y, respectively, in Eq. (9.8) to calculate $P(x', y')$. This is repeated for as many Monte Carlo iterations as necessary to achieve a stable solution, and the average $P(x', y')$ calculated. A single iteration is shown graphically in Fig. 9.3.

Recall that here we are utilizing two similar three-dimensional Gaussian-shaped functions—one for the Carlton damage function and the other for the bivariate accuracy distribution. Given, for example, the same data used for the rectangular cookie cutter (i.e., $MAE_F = 2270$ ft^2) and a 65-deg impact

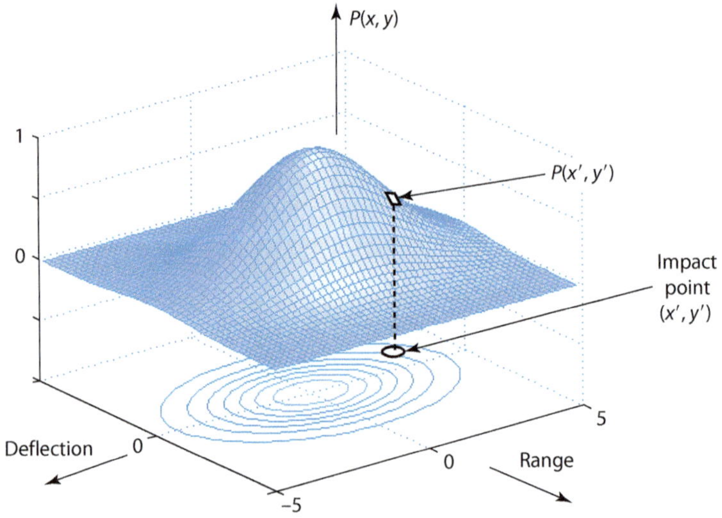

Fig. 9.3 Monte carlo iteration for carlton damage function.

angle, we can calculate the following:

$$a = \max[(1-0.8 \times \cos I), 0.3] = 0.6618 \tag{9.9}$$

$$L_{\mathrm{ET}}^{\mathrm{ft}} = 2 \times \mathrm{WR}_r = 1.128\sqrt{\mathrm{MAE}_F \times a} = 43.72\,\mathrm{ft} \tag{9.10}$$

$$W_{\mathrm{ET}}^{\mathrm{ft}} = 2 \times \mathrm{WR}_d = \frac{L_{\mathrm{ET}}^{\mathrm{ft}}}{a} = 66.06\,\mathrm{ft} \tag{9.11}$$

The Carlton damage function may be written as

$$P(x, y) = \exp - \left[\frac{4x^2}{43.72^2} + \frac{4y^2}{66.06^2}\right] \tag{9.12}$$

A random sample of x' and y' would be substituted into this equation to get the $P(x, y)$. Writing a MATLAB code to do this and setting REP = 20 ft and DEP = 10 ft produces the following result:

$\mathrm{MAE}_F = 2270\ \mathrm{ft}^2$ $I = 65$ deg REP = 20 ft DEP = 10 ft	⇒	$\mathrm{PD}_1 = 0.3915$

9.6 Lethal area Matrix

A similar approach may be used for the lethal area matrix damage function, which again would be fixed on the ground plane at the origin and a random weapon impact point on it determined by REP and DEP, as shown in Fig. 9.4. This would yield an impact point within the bounds of a single cell, resulting in a single sample of P_K for that Monte Carlo iteration.

Calculating the average of the P_K values over all iterations would yield the PD_1, noting that an impact point outside the lethal area matrix would have a cell $P_K = 0$. Assuming the lethal area matrix is given by Table 8.4 in Chapter 8, writing a MATLAB code that accepts the lethal area matrix as an array input, generating random impact points, extracting the corresponding cell P_K, and averaging over a sufficient number of iterations yields the following result:

$MAE_F = 2270\ ft^2$
REP = 20 ft
DEP = 10 ft

⇒ $PD_1 = 0.2467$

Note that for the same inputs, the choice of damage function produces different estimates of PD_1 for the rectangular cookie cutter, Carlton damage function, and lethal area matrix, respectively, as shown in Table 9.1.

This table produces three conclusions.

1. Although the different representations of the damage function all have the same lethality (MAE_F), they result in different values of PD_1.
2. The highest fidelity representation—the lethal area matrix—would be expected to produce the most "accurate" result, followed by the Carlton, and then the cookie cutter.
3. There is considerable variability in the result depending on which damage function is used.

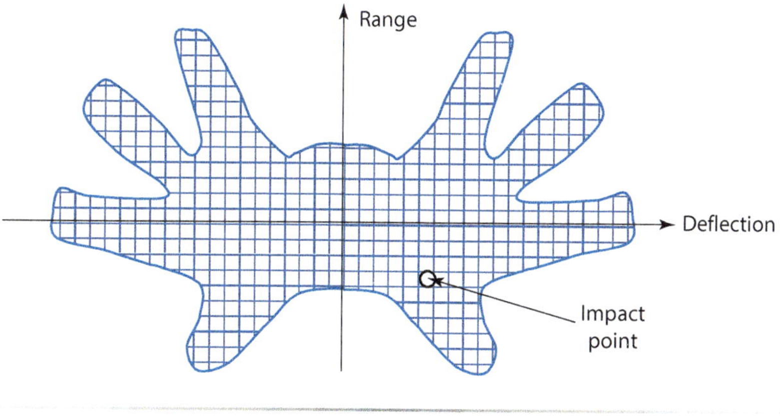

Fig. 9.4 Lethal area matrix.

Table 9.1 Effect of Damage Function on Calculated PD_1

Damage Function	PD_1
Rectangular cookie cutter	0.4655
Carlton	0.3901
Lethal area matrix	0.2467

9.7 Blast Damage Function

For completeness, we compute a value of PD_1 when the EI is an MAE_B using a Monte Carlo simulation. The process is basically the same as for an MAE_F, except the rectangular cookie cutter is square and independent of the impact angle. Reproducing Eq. (8.64) here:

$$L_{ET} = W_{ET} = \sqrt{MAE_B} \tag{9.13}$$

we can use this equation to determine L_{ET} and W_{ET} by modifying the Monte Carlo code shown previously to include the MAE_B, which produces the following result:

$MAE_F = 1000$ ft^2
I = N/A
REP = 20 ft
DEP = 10 ft

→

$PD_1 = 0.2904$

9.8 Multiple Weapon Attacks

Examples of multiple weapon attacks would be a stick of unguided bombs, a battery of artillery cannons firing a single volley at a target, or a single aircraft making repeated attacks on a target, releasing a single bomb on each run. In the previous sections only single weapon attacks were considered; therefore, the dependent and independent errors could be combined. However, it was pointed out in Chapter 4, Section 4.2 that when multiple weapons are considered, the dependent error will have to be kept separate from the independent error when calculating effectiveness; otherwise, incorrect results will be obtained.

The next question to address is: If multiple weapons are used in an attack, how is the effectiveness calculated? The answer to this question is that it depends on whether the attacks are dependent or independent; that is, does the impact point of the second and subsequent weapons depend on the impact point of the first? If the answer is yes, then we are dealing with dependent attacks; if the answer is no, then it is a case of independent attacks. Each has a different methodology to calculate effectiveness. Consider the following examples.

9.8.1 Multiple, Independent Attacks

Let's say an aircraft flies over the target and drops a bomb. It then circles around and drops another bomb, or another aircraft attacks the same target with the same weapon. The impact point of the second bomb is independent of the first, because each was independently aimed.

As another example, a moving tank shoots at an enemy tank, drives on, reaims the cannon, and fires another shot at the same target. Each round is independently aimed.

For independent attacks, the methodology for calculating the value of PD is straightforward. We calculate the probability of damage due to a single attack with a single weapon, PD_1, and use the powering up formula to calculate PD, the damage caused by n independent, identical attacks.

$$\text{PD} = 1 - (1 - PD_1)^n \tag{9.14}$$

Recall this formula is derived from the binomial distribution, which is only valid for independent Bernoulli trials.

9.8.2 Multiple, Dependent Attacks

For this example, consider an artillery cannon that fires multiple rounds at a target without reaiming between rounds. Each round has the same aiming error (denoted by the MPI), so the impact point is therefore not independent for each round.

As another example, an aircraft drops several GPS/INS guided weapons at a set of coordinates believed to be that of the intended target. Because each weapon is flying to the same set of coordinates, the target location error (TLE) is the same for each weapon, and each weapon delivery is therefore not independent.

To illustrate the effect of treating these errors separately, or otherwise, let us review the scenario shown in Fig. 9.5.

Here we have an automatic weapon firing a three-round burst at a target. After the first burst, the soldier examines where the rounds landed, reaims the weapon, fires another three rounds, and so on. The appropriate EI for this case would be the vulnerable area A_V in the normal plane, and we associate this with the target. Impacts in the normal plane for this scenario are shown in Fig. 9.6.

Again in this representation, the lethal area is associated with the target and each weapon is a dart (thrown three at one time), where the target is killed if one or more rounds lands inside it. Suppose the target is represented by a 10 ft × 10 ft vulnerable area and is fired on by a gun with a three-round burst. Assuming the MPI error has a standard deviation of 5 ft in range and deflection, and the precision error is 2 ft, also assumed equal in range and deflection, what is the expected probability of damaging the target?

Fig. 9.5 Automatic weapon firing at a target.

It is important to note that the MPI error is common to all rounds fired on a single occasion, and therefore is the dependent error, whereas the precision error is truly independent because it is different for each round for the same occasion (burst).

Conceptually, in order to reproduce the pattern of shot shown in Fig. 9.6 we would have a Monte Carlo simulation where we take a single random draw from the MPI distribution to determine its location in the normal plane and then three random draws from the precision error distribution to determine the locations of each round in the burst. We then determine if one or more rounds lands inside the vulnerable area, and if that occurs, the target is killed and the P_K counter incremented. Note that if more than one round does land inside the lethal area, the iteration results in a

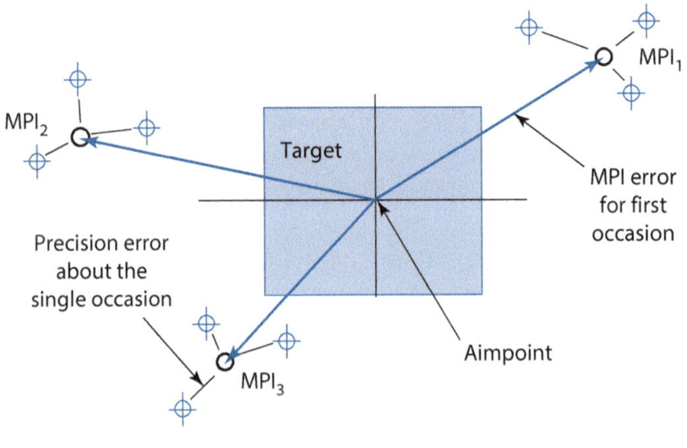

Fig. 9.6 Monte carlo simulation example.

single kill, because the target cannot be killed more than once in any burst; hence, the break statement after count is incremented to report a kill.

To calculate the probability of damage we implement the double loop Monte Carlo program illustrated functionally in Fig. 9.7.

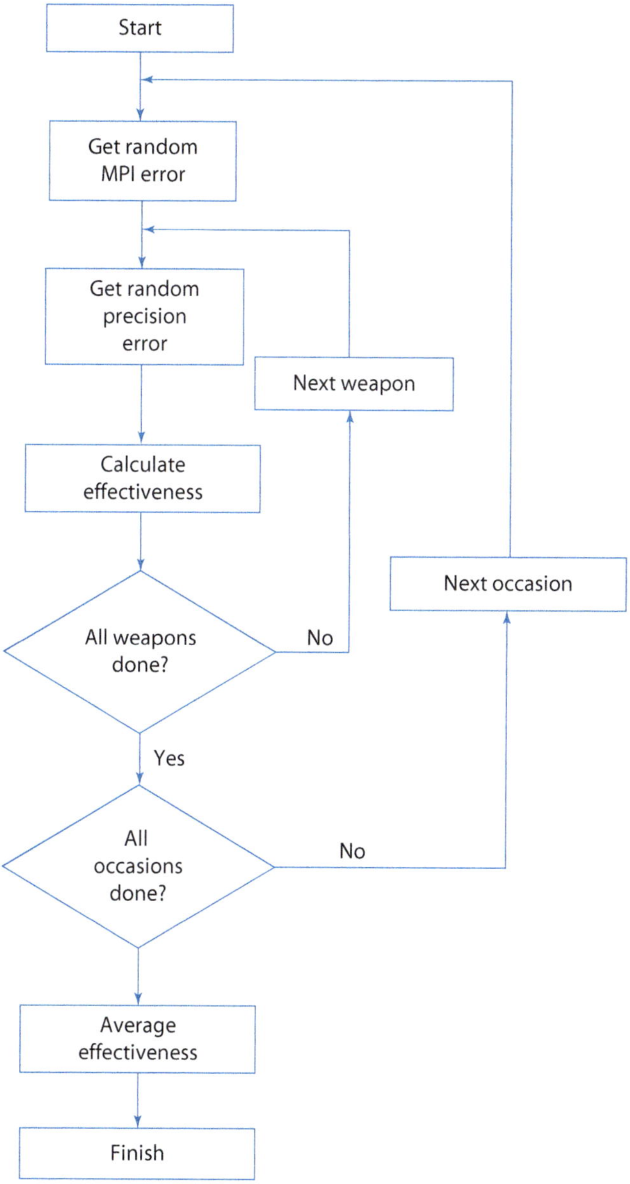

Fig. 9.7 Double loop code for Monte Carlo simulation.

This algorithm is easily implemented as a MATLAB M-file, a listing of which follows:

```
% Probability of Kill - Dependent fire
clear all;
LET=10.0;               % lethal area length
WET=10.0;               % lethal area width
MPI_sigmax=5.0;         % separate MPI and precision errors
MPI_sigmay=5.0;
Prec_sigmax=2.0;
Prec_sigmay=2.0;
m=5000000;              % number of iterations
n_burst=3;              % rounds/burst
count=0;                % initialize counter
%--------- Dependent fire --------------------------------------
for iter_MPI=1:m;                   % MPI loop
  x_MPI=MPI_sigmax*randn;           % MPI impact range
  y_MPI=MPI_sigmay*randn;           % MPI impact deflection
  for iter_round=1:n_burst;         % precision loop
    x_round=Prec_sigmax*randn;      % precision impact range
    y_round=Prec_sigmay*randn;      % precision impact deflection
    x_impact=x_MPI+x_round;         % total impact point range
    y_impact=y_MPI+y_round;         % total impact point deflection
    if ((abs(x_impact)<=LET/2)&(abs(y_impact)<=WET/2));
      count=count+1;                % impact inside rectangle
      break                         % exit loop if target killed
    end
  end
end
p=count/(m);                        % final prob. damage
t=['Pk for dependent fire Monte Carlo=',num2str(p)];
disp(t)
```

This program has a double loop—one for the MPI error and one for the precision error. Note the *break* statement, which exits to an inner loop if the target is hit, because subsequent hits cannot kill it again. Running the program for the specified 5×10^7 iterations produces the result PD = 0.6116.

If we make the common mistake and assume both the MPI and precision errors are independent, we would combine them by root sum squared (RSS). If this is done for the previous example, we get a single, combined error term $\sigma_x = \sigma_y = 5.385$ ft. Can we calculate the probability of damage to the target using this erroneously combined error? This is essentially the same program shown in Section 9.3 resulting in

$$\mathrm{PD}_1 = \mathrm{PD}_{1(R)} \times \mathrm{PD}_{1(D)} = 0.4194 \tag{9.15}$$

This is the probability that a single shot hits the lethal area, but to ensure compatibility with the burst fire case of three rounds we have to consider the other two rounds fired at the target. However, because these rounds are being considered (incorrectly) independent of each other, the PD may be calculated using the powering up formula.

$$PD = 1 - (1 - PD_1)^n = 1 - (1 - 0.4194)^3 = 0.8043 \tag{9.16}$$

Clearly, by comparing the two resulting P_K values, the effect of assuming independence of error sources introduces considerable error into the calculation of PD = 0.8043 vs PD = 0.6116. The reason there is a difference may be explained by considering another hypothetical case of using five weapons against a unitary target. These weapons are employed in two ways against a rectangular lethal area:

1. Five independent weapons aimed and fired one at a time
2. One independent volley (occasions) of five weapons

This comparison is shown in Fig. 9.8 where for the case of shooting one weapon at a time—independent fire (left)—each weapon may or may not kill the target. If it did, the next weapon sees a fresh target, which it too may or may not kill independently of the first weapon; therefore, three kills out of five would be recorded.

Now consider the dependent fire case on the right. Suppose that the first weapon to arrive does kill the target. If the next weapon to arrive (almost at the same time) would also have killed the target, this second kill is not counted because the target is already killed by the first weapon and therefore cannot be killed again. Therefore, only one kill would be recorded.

It is expected, therefore, that dependent fire will produce a lower probability of kill compared to the same number of weapons used independently, as confirmed by the numerical example discussed earlier.

It may be seen that instead of using the *break* statement on the precision error loop if any one of the rounds landed inside the lethal area, we could also calculate the probability of survival due to each round landing

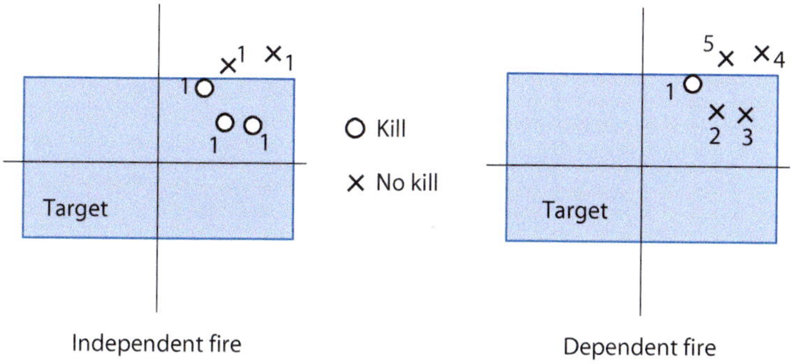

Fig. 9.8 Comparison of single round and salvo fire.

simultaneously as

$$P_S = (1 - P_{K1})(1 - P_{K2})(1 - P_{K3})(1 - P_{K4})(1 - P_{K5}) \tag{9.17}$$

where any of the $P_{Ki} = 1$ if the ith round lands inside the area. From this it is obvious that if one or more than one round lands inside the lethal area, the target doesn't survive. This may seem overly complicated; however, it is a useful concept to determine what happens if the damage is not represented by a rectangular cookie cutter but rather by a Carlton damage function or lethal area matrix.

In the case of a Carlton damage function, the P_{Ki} values in Eq. (9.17) are not simply a one or zero, but are given by

$$P_{Ki} = \exp - \left[\frac{x_i^2}{\mathrm{WR}_r^2} + \frac{y_i^2}{\mathrm{WR}_d^2}\right] \tag{9.18}$$

where (x_i, y_i) is the impact point of the ith round. Equation (9.17) may still be used to compute the probability of survival from all five rounds, with the final P_K being calculated from

$$P_K = 1 - P_S \tag{9.19}$$

For the case of using a lethal area matrix, the P_{Ki} will be the value inside the cell where the ith round lands, and Eqs. (9.17) and (9.19) are used in the same way.

An interesting and surprising point arises if we are concerned only with accuracy metrics such as CEP, as is typically the case in weapon testing. Suppose we revisit the case just discussed using a Monte Carlo approach to compare independent fire of, say, 50 weapons and dependent fire of 10 volleys of 5 weapons/occasion. The expected fall of shot (impact point data) would look like that shown in Fig. 9.9 for the two cases.

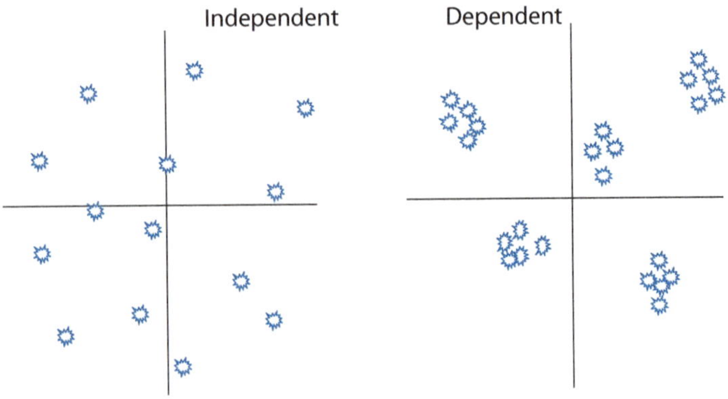

Fig. 9.9 Impact points for independent and dependent fire.

Remarkably, whether we deliver the weapon in 50 independent attacks or 10 volleys of 5 weapons each, the resultant CEP is the same even though the P_K will be different.

The conclusion, therefore, is that even though the MPI error is a dependent error, it may be RSS with the precision error for the purposes of determining accuracy, but it cannot be RSS with the precision error in order to determine effectiveness. Remember, this problem occurs only when there are multiple rounds/occasion. Setting the variable n_burst = 1 in the simulation shown in Fig. 9.8 gives PD = 0.4194, the same result predicted analytically for the single shot case.

Given that for a particular weapon delivery system we need to keep the MPI and precision error components separate, at least for the purpose of computing effectiveness, this requirement is not easily met in practice. In the weapon procurement process, the contractor usually has to meet some accuracy requirement such as CEP no greater than 20 ft, and this will be verified by acceptance testing of the prototype. However, the tests usually consist of single rounds independently fired over several occasions. Because each impact is for a single weapon, the MPI and precision errors are combined, and it is impossible to split them out of the single data point for each independent shot. This process works well for meeting the required accuracy specification, but the inability to determine the MPI and precision errors from the test data makes it impossible to predict the effectiveness of the weapon if more than one is used in an attack. This is an ongoing problem for analysts.

9.9 Required Number of Monte Carlo Iterations

In the simulations considered so far we have arbitrarily selected a large number of Monte Carlo iterations in order to achieve an accurate solution. This raises the general question of how many Monte Carlo iterations are needed to converge to a solution.

This was addressed in Chapter 2, Sections 2.14 and 2.18, where two alternatives were presented. Section 2.18 described an automated method for putting the Monte Carlo simulation in "free run" mode and collecting statistics on the resulting values of P_K from which a required number of iterations meeting a confidence level can be predicted. The simpler approach in Section 2.14 is expanded here by reconsidering the earlier cookie cutter example where $MAE_F = 2270$ ft^2, impact angle = 65 deg, REP = 20 ft, and DEP = 10 ft, resulting in $PD_1 = 0.4655$ for 10^7 iterations. We rerun the problem and vary the number of iterations, as summarized in Table 9.2, where we run the same simulation three times for the same number of iterations without changing anything.

One observation from the table shows that if three consecutive runs of the simulation produce widely varying results, then it implies more iterations

Table 9.2 Consecutive Runs of MC Simulation vs Iterations

Number of Iterations	Mean Values of the Damage Function (Runs 1, 2, and 3)
10^2	0.399, 0.456, 0.418
10^3	0.431, 0.484, 0.429
10^4	0.456, 0.461, 0.473
10^6	0.462, 0.466, 0.463
10^7	0.463, 0.463, 0.463

are needed. Notice how the bounds of the individual results narrow as the number of iterations increases.

The other observation is that at least 10^7 iterations are needed to get a consistent result to three decimal places. This is the only way to determine if sufficient iterations are being used, because the result will vary from simulation to simulation and may vary for a particular simulation when different input data are used.

9.10 Methodology Summary

It is worth summarizing some of the important points that the reader should take from this chapter.

- In determining effectiveness, the weapon lethal area may be considered a feature of the weapon and the target represented by a point, or the lethal area a feature of the target and the weapon represented by a point. Either may be used in an effectiveness calculation and are mathematically identical.
- Effectiveness calculation may be made using a Monte Carlo simulation using any of the damage functions we have developed—lethal area matrix, Carlton damage function, or a rectangular cookie cutter. It was seen that even if these different damage functions have the same lethality (MAE), the resulting PD_1 values computed are very different.
- If multiple weapons are used, it has to be determined whether they are used in a dependent or independent manner, because this will affect the resulting estimate of effectiveness. Also, the two principle error terms, dependent and independent errors, may be RSS together if only accuracy is required, but have to be kept separate if effectiveness calculations for multiple weapons are to be performed.

We have seen that effectiveness by one or more weapons, delivered independently or otherwise, may be calculated using a Monte Carlo approach. Although capable of providing benchmark measures of PD, these methods can be slow in producing the final result, depending on the complexity of the target and the number of iterations needed to converge to a solution.

Particularly in the past, when Monte Carlo simulation was not a widely available tool to the warfighter, requirements for a rapid calculation of effectiveness drove the methods of solution to closed form, analytical methods. Formulating these methods can present considerable challenges to analysts in their search for a rapid, accurate solution for calculating a PD value, and the methods used to achieve this goal are presented in the following chapters. Before starting, however, we try to categorize all target types we might encounter in order to structure our methods in a more logical manner.

9.11 Four Target Types

In demonstrating the general strategy of computing damage estimates, our focus has been on one or more weapons directed against a point or unitary target; however, this is but the first of four target types we will encounter.

1. *Unitary target:* A single target element (e.g., one tank, one artillery cannon, one personnel target, etc.). The damage estimate or metric is probability of damage PD for multiple weapons or PD_1 for a single weapon.
2. *Area of target elements:* A rectangular area containing an unspecified number of identical unitary target elements (e.g., a tank assembly area, a Petroleum, oil and lubricants facility, an area of troops in the open, etc.). The damage metric is fractional damage FD for multiple weapon attacks or FD_1 for single weapon attacks.
3. *Complex target:* A collection of different unitary targets grouped to perform a function (e.g., a SAM site, an electrical power plant, a ship, etc.). The damage metric is specific to the target, but typically a functional or mission kill.
4. *Special target:* These types of targets (e.g., bridges, tunnels, bunkers, buildings, etc.) require a custom methodology. The damage metric is also specific to the target (e.g., dropping a span for a bridge, blocking tunnel access, or a percentage of structural damage for a building).

The first two target types may be analyzed by deterministic methods (closed form) and are dealt with in Volume 1; the other two can only be analyzed by Monte Carlo methods and are covered in Volume 2.

Chapter 10 Calculating Damage Probability: Deterministic Methods

10.1 Introduction

In this chapter we attempt to reproduce the results obtained in the previous chapter by deterministic methods, that is, a closed form analytical process that is much faster than using a Monte Carlo simulation. Specifically, the objective is to obtain the same, or similar, results to those in Chapter 9, Sections 9.3, 9.5, and 9.6, employing the rectangular cookie cutter, Carlton damage function, and lethal area matrix, respectively.

We begin as before by considering the damage due to a single weapon delivery where the dependent error (mean point of impact [MPI]) and independent error (precision) may be root sum squared (RSS) to produce a single delivery error, and to single weapons that have a Gaussian delivery error. These generally include both unguided and guided rounds; however, recall that some guided weapons may have non-Gaussian delivery errors. These will be dealt with later in the chapter by modifying what is developed initially. Later in the chapter we also address the issue of multiple round deliveries where the MPI and precision errors must be kept separate.

10.2 Rectangular Damage Function

Recall Section 9.3 where the rectangular lethal area is associated with the target, the weapon is represented by a single impact point, and the Monte Carlo approach is to generate random samples of the impact point using the delivery error statistics and investigate if the weapon landed inside the lethal area. Over a sufficiently large population of samples, PD_1 is defined as the proportion that lands inside the lethal area, thereby killing the target. This is shown graphically in Fig. 10.1.

Considering just the range direction and defining the dimensions of the cookie cutter to be L_{ET} and W_{ET}, the impact point distribution appears as shown in Fig. 10.2.

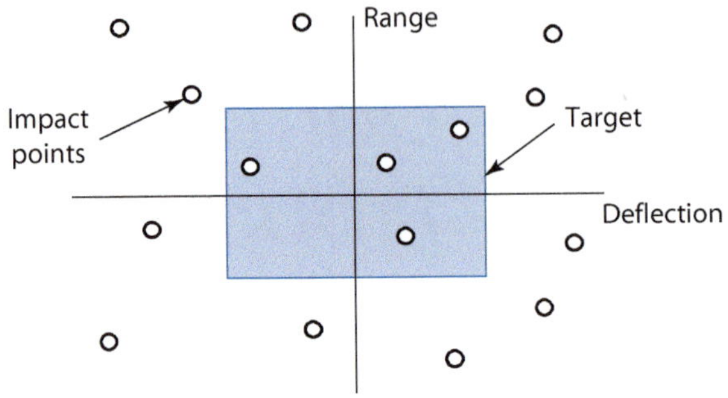

Fig. 10.1 Impact points.

In calculating the PD_1 in the range direction, the question we are asking is: What is the probability that a random sample of x lies inside the cookie cutter. It may be seen that this is given by

$$PD_{1x} = \int_{-L_{ET}/2}^{L_{ET}/2} \frac{1}{\sigma_x\sqrt{2\pi}} \exp\left[\frac{-x^2}{2\sigma_x^2}\right] dx \tag{10.1}$$

This is just the integral of the delivery error probability density function (PDF) over the extent of the target shown in Fig. 10.3.

Another way of looking at this is to recognize the damage function is unity between the limits shown, and we are sampling the random variable. For each sample we are calculating the value of the damage function, which will be either zero or one. These samples are then averaged. This is another way of saying we are calculating the average or expected value of the damage function, which may be calculated directly from the expected

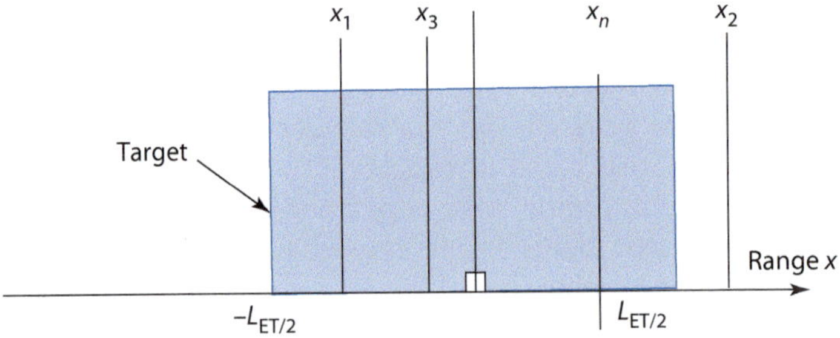

Fig. 10.2 Range direction impact points.

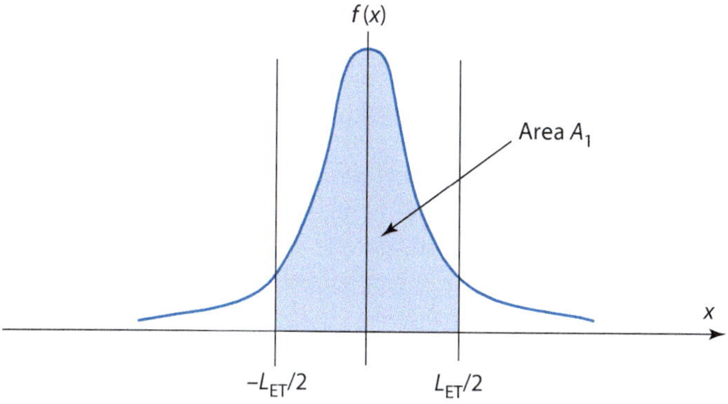

Fig. 10.3 Extent of target over accuracy PDF.

value theorem.

$$PD_{1x} = E(c(x)) = \int_{-\infty}^{+\infty} c(x)g(x)dx \tag{10.2}$$

Using the expected value theorem to calculate PD_1 for this damage function, the miss distance PDF $g(x)$ is given by

$$g(x) = \frac{1}{\sigma_x\sqrt{2\pi}}\exp\left[-\frac{x^2}{2\sigma_x^2}\right] \tag{10.3}$$

Using the damage function shown in Fig. 10.2, we can write

$$c(x) = P_{CD} = 1 \quad \text{for} \left(-\frac{L_{ET}}{2} < x < \frac{L_{ET}}{2}\right), \qquad c(x) = 0 \quad \text{otherwise} \tag{10.4}$$

Substituting Eqs. (10.3) and (10.4) into Eq. (10.13), we can write

$$PD_{1x} = \int_{-\infty}^{\infty} \frac{P_D}{\sigma_x\sqrt{2\pi}}\exp\left[\frac{-x^2}{2\sigma_x^2}\right]dx = \int_{-L_{ET}/2}^{L_{ET}/2} \frac{1}{\sigma_x\sqrt{2\pi}}\exp\left[\frac{-x^2}{2\sigma_x^2}\right]dx \tag{10.5}$$

which is the same result as Eq. (10.1). A similar calculation may be applied to the deflection direction, yielding

$$PD_{1y} = \int_{-\infty}^{\infty} \frac{P_D}{\sigma_y\sqrt{2\pi}}\exp\left[\frac{-y^2}{2\sigma_y^2}\right]dy = \int_{-W_{ET}/2}^{W_{ET}/2} \frac{1}{\sigma_y\sqrt{2\pi}}\exp\left[\frac{-y^2}{2\sigma_y^2}\right]dy \tag{10.6}$$

The overall PD_1 is obtained by multiplying the individual range and deflection values to get

$$PD_1 = PD_{1x} \times PD_{1y} \tag{10.7}$$

We now implement this solution and check with the Monte Carlo simulation result obtained previously. It is possible to evaluate the integral as the difference between two CDFs and use the statistical tables at the back of the Appendix A, volume 1; however, this is labor intensive. Instead we use an Excel spreadsheet and the worksheet function NORMDIST. This takes the form of

```
Result=NORMDIST(x, mean, sigma, FLAG)
```

In the function call, x is the point value on the PDF, mean and sigma define the normal distribution, and the flag determines what value the result takes. If FLAG = 1, the integral of the PDF up to the value of X is determined [i.e., $F(X)$], but if FLAG is zero, the point value on the PDF is returned [i.e., $f(X)$]. The integrals in Eqs. (10.5) and (10.6) would be calculated as follows:

$$PD_{1x} = (ND(L_{ET}/2, 0, \sigma_x, 1) - ND(-L_{ET}/2, 0, \sigma_x, 1)) \tag{10.8}$$

$$PD_{1y} = (ND(W_{ET}/2, 0, \sigma_y, 1) - ND(-W_{ET}/2, 0, \sigma_y, 1)) \tag{10.9}$$

where we use the abbreviation ND to represent the NORMDIST function. The process to calculate a PD_1 is as follows. Assumed inputs are REP, DEP, MAE_F, and the impact angle I. From these inputs, the following computations are made:

1. The lethal area aspect ratio parameter a is computed from Eq. (9.5).

$$a = MAX[1 - 0.8\cos(I), 0.3] \tag{10.10}$$

2. The value of L_{ET} is computed from Eq. (9.6).

$$L_{ET} = \sqrt{MAE_F \times a} \tag{10.11}$$

3. The value of W_{ET} is computed from Eq. (9.7).

$$W_{ET} = \frac{L_{ET}}{a} \tag{10.12}$$

4. PD_1 is calculated from Eqs. (10.8), (10.9), and (10.7).

Implementing these equations into an Excel spreadsheet using the data from Chapter 9, Section 9.3 produces the results shown in Table 10.1.

This compares with the value of 0.4655 obtained in Section 9.3 for a Monte Carlo simulation of 5×10^5 iterations.

Table 10.1 Calculating PD_1 for a Rectangular Damage Function

PD_1 calculator using cookie cutter damage function	
INPUTS	
MAEF	2270
Impact angle	65
REP	20
DEP	10
DERIVED	
a	0.6619
Let	38.7624
Wet	58.5619
sigma_x	29.6516
sigma_y	14.8258
RESULTS	
PD1-range	0.4867
PD1-deflection	0.9517
PD1-total	0.4632

10.3 Carlton Damage Function

By reconsidering the Monte Carlo approach as in the previous section, replacing the rectangular cookie cutter with the Carlton damage function transforms Fig. 10.2 into Fig. 10.4.

We follow the process in the previous section by noting that PD_1 can be calculated as the expected value of the damage function, which in this case takes the Carlton form. First focusing on the range direction, we get

$$PD_{1x} = E(c(x)) = \int_{-\infty}^{+\infty} c(x)g(x)\mathrm{d}x \tag{10.13}$$

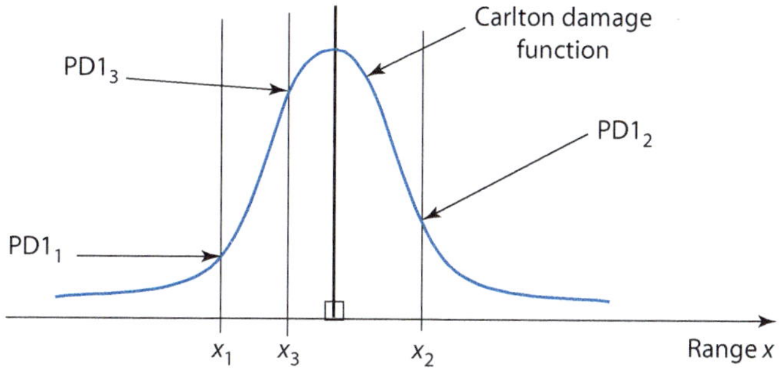

Fig. 10.4 Random samples of the Carlton damage function.

Again the Gaussian delivery error is represented by

$$g(x) = \frac{1}{\sigma_x\sqrt{2\pi}}\exp\left[-\frac{x^2}{2\sigma_x^2}\right] \tag{10.14}$$

The damage function is assumed to take the one-dimensional form of the Carlton damage function obtained from Eq. (8.50).

$$c(x) = \exp\left[-\frac{4x^2}{{L'_{\mathrm{ET}}}^2}\right] \tag{10.15}$$

Substituting into Eq. (10.13) gives

$$\begin{aligned} \mathrm{PD}_{1x} &= \frac{1}{\sigma_x\sqrt{2\pi}}\int_{x=-\infty}^{x=\infty}\exp\left[-\frac{4x^2}{{L'_{\mathrm{ET}}}^2}\right]\times\exp\left[-\frac{x^2}{2\sigma_x^2}\right]dx \\ &= \frac{1}{\sigma_x\sqrt{2\pi}}\int_{-\infty}^{+\infty}\exp\left[-x^2\left(\frac{4}{{L'_{\mathrm{ET}}}^2}+\frac{1}{2\sigma_x^2}\right)\right]dx \\ &= \frac{1}{\sigma_x\sqrt{2\pi}}\int_{-\infty}^{+\infty}\exp\left[-\frac{ax^2}{2\sigma_x^2}\right]dx \end{aligned} \tag{10.16}$$

where

$$a = \frac{8\sigma_x^2 + {L'_{\mathrm{ET}}}^2}{{L'_{\mathrm{ET}}}^2} \tag{10.17}$$

By setting

$$\sigma_u^2 = \frac{\sigma_x^2}{a} \tag{10.18}$$

we may write Eq. (10.16) in the form

$$\mathrm{PD}_{1x} = \frac{1}{\sigma_x\sqrt{2\pi}}\int_{-\infty}^{+\infty}\exp\left[-\frac{x^2}{2\sigma_u^2}\right]dx = \frac{\sigma_u}{\sigma_x}\left\{\frac{1}{\sigma_u\sqrt{2\pi}}\int_{-\infty}^{+\infty}\exp\left[-\frac{x^2}{2\sigma_u^2}\right]dx\right\} \tag{10.19}$$

The term in the braces may be recognized as the CDF of a normally distributed random variable, which is unity, hence

$$\mathrm{PD}_{1x} = \frac{\sigma_u}{\sigma_x} = \frac{1}{\sqrt{a}} = \frac{L'_{\mathrm{ET}}}{\sqrt{8\sigma_x^2 + {L'_{\mathrm{ET}}}^2}} \tag{10.20}$$

Recalling from the delivery accuracy section that $\sigma_x = \mathrm{REP}/0.6745$, Eq. (10.20) may be written in the form

$$\mathrm{PD}_{1x} = \frac{L'_{\mathrm{ET}}}{\sqrt{17.6\,\mathrm{REP}^2 + {L'_{\mathrm{ET}}}^2}} \tag{10.21}$$

Similarly, for the deflection direction

$$\mathrm{PD}_{1y} = \frac{W'_{\mathrm{ET}}}{\sqrt{17.6\,\mathrm{DEP}^2 + {W'_{\mathrm{ET}}}^2}} \tag{10.22}$$

The overall PD_1 is obtained by multiplying the individual range and deflection values to get

$$\mathrm{PD}_1 = \mathrm{PD}_{1x} \times \mathrm{PD}_{1y} = \frac{L'_{\mathrm{ET}} \times W'_{\mathrm{ET}}}{\sqrt{\left(17.6\,\mathrm{REP}^2 + {L'_{\mathrm{ET}}}^2\right)\left(17.6\,\mathrm{DEP}^2 + {W'_{\mathrm{ET}}}^2\right)}} \tag{10.23}$$

This equation indicates that for damage functions represented by the Carlton form, PD_1 may be calculated if L'_{ET}, W'_{ET}, REP, and DEP are known. It may surprise some readers that the higher fidelity Carlton damage function has a closed form solution in Eq. (10.23) whereas the simpler rectangular cookie cutter requires the NORMDIST function for its evaluation, as in Eqs. (10.8) and (10.9).

At this point we want to compare our analytical model solution with that of the Monte Carlo solution, so we review the case discussed in Chapter 9, Section 9.5. Assumed inputs are REP, DEP, MAE_F, and the impact angle I. From these inputs, the following computations are made:

1. The lethal area aspect ratio parameter a is computed from Eq. (9.9).

$$a = \mathrm{MAX}[1 - 0.8\cos(I),\ 0.3] \tag{10.24}$$

2. The value of L'_{ET} is computed from Eq. (9.10).

$$L'_{\mathrm{ET}} = 1.128\sqrt{\mathrm{MAE}_F \times a} \tag{10.25}$$

3. The value of W'_{ET} is computed from Eq. (9.11).

$$W'_{\mathrm{ET}} = \frac{L'_{\mathrm{ET}}}{a} \tag{10.26}$$

4. PD_1 is calculated from Eq. (10.23).

$$\mathrm{PD}_1 = \frac{L'_{\mathrm{ET}} \times W'_{\mathrm{ET}}}{\sqrt{\left(17.6\,\mathrm{REP}^2 + {L'_{\mathrm{ET}}}^2\right)\left(17.6\,\mathrm{DEP}^2 + {W'_{\mathrm{ET}}}^2\right)}} \tag{10.27}$$

Table 10.2 Computation of PD_1 Using Carlton Damage Function

PD_1 calculator using Carlton damage function	
INPUTS	
MAEF	2270
Impact angle	65
REP	20
DEP	10
DERIVED	
a	0.6619
L'et	43.7240
W'et	66.0578
RESULTS	
PD1-range	0.4621
PD1-deflection	0.8441
PD1-total	0.3901

Implementing these equations into an Excel spreadsheet using the data from Section 9.5 produces the results shown in Table 10.2.

This is close to the value of 0.3915 obtained using Monte Carlo simulations in Section 9.5.

10.4 Lethal Area Matrix Damage Function

It is possible to use the lethal area matrix to determine the total probability of damage PD_1 that includes the effect of delivery accuracy. Consider a single cell of the lethal area matrix shown in Fig. 10.5.

For the cell shown, Eq. (9.1) may be interpreted as follows:

$$PD_1(x, y) = P_K(x, y) \times \text{Prob}\{\text{hitting cell}(x, y)\} \tag{10.28}$$

The probability of survival associated with each cell is

$$PS_1(x, y) = 1 - PD_1(x, y) \tag{10.29}$$

The total probability of damage is obtained from

$$PD_1 = 1 - \prod_{x,y} PS_1(x, y) \tag{10.30}$$

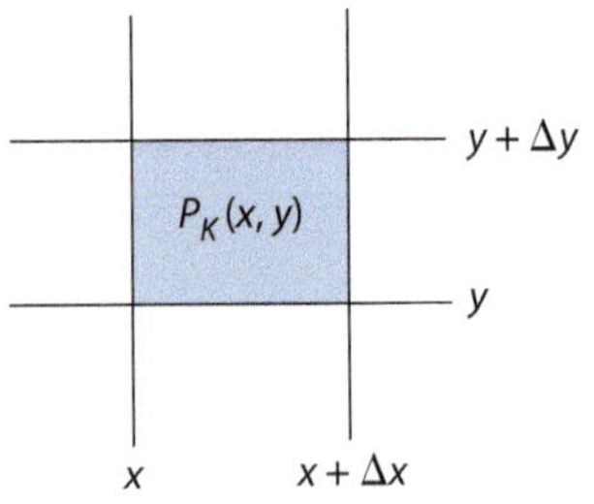

Fig. 10.5 Single cell in lethal area matrix.

The probability of hitting inside the cell may be obtained from the bivariate probability

Table 10.3 P_K Matrix

P_K	37.9	75.8	113.7	151.6	Deflection →
−42.9	0.0028	0	0	0	
−28.6	0.0064	0.0001	0.0006	0.0029	
−14.3	0.1402	0.0059	0.0099	0.0042	
0	0.5571	0.0459	0.0127	0.0045	
14.3	0.6794	0.0891	0.0156	0.0045	
28.6	0.1741	0.0927	0.0325	0.0116	
42.9	0.0060	0.0186	0.0258	0.0128	
57.2	0.0007	0.0050	0.0105	0.0072	

density function for delivery accuracy, which may be simplified assuming independence of range and deflection.

$$P_{\mathrm{HIT}}(x, y) = \int_{y}^{y+\Delta y}\int_{x}^{x+\Delta X} g(x, y)\, dxdy = \int_{x}^{x+\Delta x} g_1(x)dx \times \int_{y}^{y+\Delta y} g_2(y)dy \quad (10.31)$$

For a single cell the P_K inside it is independent of the probability of hitting in the cell, so Eq. (10.31) may be used to construct another matrix, the hit matrix, with the same cell dimensions and cell count as the lethal area matrix except that in each cell is the probability a weapon would impact inside it. From the lethal area matrix generated in Table 8.10, a small sample is shown in Table 10.3 in the region close to ground zero.

Using the example data of REP = 20 ft, DEP = 10 ft, we use Eq. (10.31) to produce the hit matrix shown in Table 10.4 from Eq. (10.31).

As for the lethal area matrix, the hit matrix is symmetrical about the range axis, so only half is shown here. Multiplying the two matrices together in

Table 10.4 P_{HIT} Matrix

P_{HIT}	37.9	75.8	113.7	151.6	Deflection →
−42.9	0.0007	0	0	0	
−28.6	0.0100	0.0024	0.0001	0	
−14.3	0.0561	0.0134	0.0007	0	
0	0.1328	0.0317	0.0017	0	
14.3	0.1328	0.0317	0.0017	0	
28.6	0.0561	0.0134	0.0007	0	
42.9	0.0100	0.0024	0.0001	0	
57.2	0.0007	0	0	0	

Table 10.5 Product of Lethal Area and Hit Matrices

$P_K \times P_{HIT}$	37.9	75.8	113.7	151.6	Deflection →
−42.9	0	0	0	0	
−28.6	0.0001	0	0	0	
−14.3	0.0079	0.0001	0	0	
0	0.0903	0.0028	0	0	
14.3	0.0098	0.0012	0	0	
28.6	0.0001	0	0	0	
42.9	0	0	0	0	
57.2	0	0	0	0	

Table 10.6 Survivability Matrix

PS	37.9	75.8	113.7	151.6
−42.9	1	1	1	1
−28.6	0.9999	1	1	1
−14.3	0.9921	0.9999	1	1
0	0.9097	0.9972	1	1
14.3	0.9902	0.9988	1	1
28.6	0.9999	1	1	1
42.9	1	1	1	1
57.2	1	1	1	1

accordance with Eq. (10.28) on a cell-by-cell basis gives the matrix shown in Table 10.5.

Finally, the PS_1 matrix is calculated using Eq. (10.29)—see Table 10.6.

Equation (10.30) is evaluated from the product of all cells in Table 10.6 subtracted from unity, giving $PD_1 = 0.1103$. Because we used only half the symmetrical LAM, the final result is twice this number, that is, $PD_1 = 0.2206$. This is somewhat smaller than the value 0.2467 obtained using the Monte Carlo simulation in Section 9.6 due to the fact that in Table 10.3 we used a truncated version of the full LAM.

10.5 PD_1 for Blast Damage Function

For targets that are sensitive to blast, the EI is an MAE_B lethal area and is represented by a square in the ground plane, as shown in Fig. 10.6.

Here the assumed inputs to the PD_1 calculation are REP, DEP, and MAE_B. The damage function used for this warhead will be the square cookie cutter shown in Fig. 10.6. From these inputs, the following calculations are made:

1. From Eq. (8.64) for blast weapons we have

$$L_{\text{ET}} = W_{\text{ET}} = \sqrt{\text{MAE}_B} \tag{10.32}$$

2. The standard deviations in the range and deflection directions are calculated from Eqs. (4.4) and (4.5).

$$\sigma_x = \frac{\text{REP}}{0.6745} \qquad \sigma_y = \frac{\text{DEP}}{0.6745} \tag{10.33}$$

3. Equation (10.5) is implemented using the NORMDIST worksheet function.

$$\text{PD}_{1x} = (\text{ND}(L_{\text{ET}}/2, 0, \sigma_x, 1) - \text{ND}(-L_{\text{ET}}/2, 0, \sigma_x, 1)) \tag{10.34}$$

4. Equation (10.6) is also implemented using the NORMDIST worksheet function.

$$\text{PD}_{1y} = (\text{ND}(W_{\text{ET}}/2, 0, \sigma_y, 1) - \text{ND}(-W_{\text{ET}}/2, 0, \sigma_y, 1)) \tag{10.35}$$

5. The final PD_1 is obtained from Eq. (10.7).

$$\text{PD}_1 = \text{PD}_{1x} \times \text{PD}_{1y} \tag{10.36}$$

Again, a simple spreadsheet implementation for Eqs. (10.32) to (10.36) is shown in Table 10.7.

This compares with the value 0.2904 obtained by Monte Carlo simulation in Section 9.7.

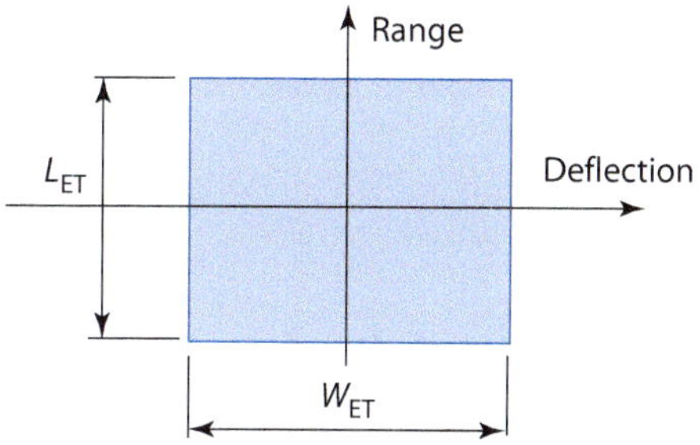

Fig. 10.6 Blast cookie cutter damage function.

Table 10.7 Computation of PD_1 for Blast Damage Function

PD_1 calculator for MAE_B	
INPUTS	
MAEB	1000
REP	20
DEP	10
DERIVED	
sigma_x	29.6516
sigma_y	14.8258
Let	31.6228
Wet	31.6228
RESULTS	
SSPD-range	0.4061
SSPD-deflection	0.7138
SSPD-total	0.2899

10.6 Some Computational Considerations

In this chapter so far we have calculated PD_1 in a spreadsheet using the Excel NORMDIST function to evaluate the area under a Gaussian PDF up to some given value. Just for completeness we restate how the NORMDIST function would be used.

$$P(a < x < b) = \int_a^b \frac{1}{\sigma_x\sqrt{2\pi}} \exp\left[\frac{-(x-\mu_x)^2}{2\sigma_x^2}\right] dx = F(b) - F(a) \tag{10.37}$$

$$= \mathrm{ND}(b, \mu_x, \sigma_x, 1) - \mathrm{ND}(a, \mu_x, \sigma_x, 1)$$

where the abbreviation ND is used for NORMDIST. An example is:

$$\texttt{NORMDIST}(0.7, 1, 2, 1) = 0.4404 \tag{10.38}$$

For applications that are not implemented in Excel, the user should determine if an equivalent function exists in their chosen computational environment. Some examples follow.

MATLAB uses the function normcdf to compute a cumulative density function with arguments normcdf(x, mean, sigma). For example

$$\texttt{normcdf}(0.7, 1, 2) = 0.4404 \tag{10.39}$$

If the programming environment is a procedural language such as FORTRAN or C, then $F(X)$ is not computed directly, but expressions for the standard normal variable CDF $F(Z)$ can be computed from the error function erf(x). This function is related to $F(Z)$ by the following equation:

$$F(z) = \frac{1}{2}\left[1 + \text{erf}\left(\frac{z}{\sqrt{2}}\right)\right] = \int_{x=-\infty}^{x=z} \frac{1}{\sigma_x\sqrt{2\pi}} \exp\left[\frac{-x^2}{2\sigma_x^2}\right] dx \qquad (10.40)$$

If no intrinsic functions are available, we can compute the CDF from a curve fit function.

$$F(z) = 0.5\left(1 - \sqrt{1 + \frac{1}{[1 + Az^2 + Bz^4 + Cz^6]^4}}\right) = \int_{x=0}^{x=z} \frac{1}{\sigma_x\sqrt{2\pi}} \exp\left[\frac{-x^2}{2\sigma_x^2}\right] dx \qquad (10.41)$$

where $A = 0.1591881356$, $B = 0.010062609$, and $C = 0.0005406609$. Whatever the programming environment, users should check reference material to determine the availability of intrinsic functions to compute statistical functions before developing their own.

Recall also from Chapter 2, Section 2.4 that for symmetrical integral limits about zero, which is the case for Eqs. (10.5) and (10.6), the Polya-Williams approximation may be used.

$$\frac{1}{\sqrt{2\pi}} \int_{-x}^{+x} \exp\left(-\frac{t^2}{2}\right) dt \leq \left[1 - \exp\left(\frac{2x^2}{\pi}\right)\right]^{1/2} \qquad (10.42)$$

10.7 Force Estimation

Suppose that the results of the computations for PD_1 produce the result $PD_1 = 0.4$, the probability of damaging the target with a single weapon. It may be necessary to use more than one weapon in order to achieve a probability of damage greater than 0.4, say, 0.7. The question is, how many weapons will this require? Recall the powering up formula [Eq. (2.65)] using the binomial distribution. This may be written in a more general form as follows:

$$P(\text{success from } n \text{ trials}) = 1 - \frac{n!}{1 \times n!}(1-p)^n = 1 - (1-p)^n \qquad (10.43)$$

where p is the probability of success of a single trial. In terms of total damage from n independently released weapons, this equation may be written as

$$PD = 1 - (1 - PD_1)^n \qquad (10.44)$$

In this case, PD is the resulting damage level caused by n weapons, and p is equal to PD_1. Equation (10.44) has to be solved for n, given PD and p, so we can write

$$1 - PD = (1 - PD_1)^n \qquad (10.45)$$

$$\log_e(1 - \mathrm{PD}) = n \log_e(1 - \mathrm{PD}_1) \tag{10.46}$$

resulting in

$$n = \frac{\log_e(1 - \mathrm{PD})}{\log_e(1 - \mathrm{PD}_1)} \tag{10.47}$$

Another common quantity to compute in force estimation studies is the probability of damage if only q weapons are available. This may be obtained directly from Eq. (10.44) as

$$\mathrm{PD} = 1 - (1 - \mathrm{PD}_1)^q \tag{10.48}$$

For the data just considered, the result would be

$$n = \frac{\log_e(1 - 0.7)}{\log_e(1 - 0.4)} = 2.36 \tag{10.49}$$

Clearly we can only have an integer number of weapons, so we round up, observing that three weapons would give more damage that that required (i.e., PD = 0.784).

A simple improvement to the spreadsheets using an MAE_F might be the addition of a trajectory program that would provide impact data given the release conditions of the weapon rather than asking the user to supply them. This and other improvements are discussed in the next section.

10.8 Template for Implementing Weaponeering Solutions

Although the simple spreadsheets given previously will provide PD_1 for most cases, a standard template will now be described that will cover these and other weaponeering solutions considered in later chapters. For simplicity, the focus of this spreadsheet is on air-delivered weapons, but the reader will see it is easily adapted to surface-launched weapons. In describing the template, each section will be covered in detail showing how the contents of that section contribute to the final calculation of PD_1 and PD. The complete spreadsheet will be divided into the following sections:

1. Constants used in the calculations
2. Target description
3. Weapon characteristics
4. Aircraft release parameters
5. Trajectory results
6. Delivery accuracy data
7. Weapon effectiveness
8. Effectiveness calculation
9. Force requirements

Where possible, the data given in each section and the calculations performed will be kept the same for all weaponeering solutions; for example, weapon trajectory will be the same for blast and fragment warheads, but the PD_1 calculations will be different. Section 8 in the previous list is where there will be the most significant differences between the methodologies. In order to illustrate this approach, a generalized spreadsheet will be presented for the first case, that of a single weapon against a unitary target, where each section of the spreadsheet will be dealt with separately. In the spreadsheet segments presented in the following sections, the shaded cells outlined in bold indicate where the user must supply data.

10.8.1 Constants Used in the Calculation

The only constants used are the acceleration due to gravity and the conversion factor from degrees to radians, as shown here:

CONSTANTS		
Gravitational acceleration, ft/s	g	32.200
Convert degrees to radians	deg2rad	0.017

10.8.2 Target Description

In many cases, the target parameters of interest will be the length and width in feet. For this case, however, where we have a unitary (point) target, these dimensions are not entered because they do not appear in the methodology.

TARGET PARAMETERS		
Target Dimensions LA, ft		N/A
Target Dimensions WA, ft		N/A

10.8.3 Weapon Characteristics

This section contains information about the drag properties of the weapon used by the trajectory program, precision error (ballistic dispersion), and detonation reliability. To keep the data unclassified, nominal values for a low drag and high drag weapon may be selected.

WEAPON PARAMETERS		
Drag: L = low, H = high		L
Ballistic dispersion, mils	Sigma-*b*	5.00
Reliability	*R*	0.94

A new factor is introduced here, the weapon detonation reliability *R*. This is obtained from JMEM data and is used to multiply the final PD_1 calculation to allow for the possibility that the weapon does not detonate correctly. The role of the reliability in the calculation of effects will be discussed shortly.

10.8.4 Aircraft Release Parameters

This section provides data for the inputs to the weapon trajectory program.

AIRCRAFT RELEASE DATA					
Airspeed, kt	*V*	450.00		760	ft/s
Dive angle, deg	theta	30.00		0.524	rad
Release altitude, ft	*yi*	3000			

The numbers on the right are the speed in ft/s and the dive angle in radians.

10.8.5 Trajectory Results

This is the output of the trajectory program. The model used is the point mass, linear drag model described in Chapter 3, Section 3.6.

TRAJECTORY RESULTS				
Vertical velocity, ft/s	*Vv*	–379.80		
Horizontal velocity, ft/s	*Vh*	657.83		
Drag constant, Cd/M	c0	0.01		
constant *a* used in trajectory	*a*	6060.20		
		6.32		**Dragless**
time of flight, s	*t_flight*	6.32		6.25
Horizontal velocity at impact, ft/s	*Vhi*	637.37		657.83
Vertical velocity at impact, ft/s	*Vvi*	568.35		580.90
Impact angle, deg	*I*	41.72		41.45
Slant range, ft	*Sri*	5075		5087
Ground range, ft	*Gri*	4093		4108

This routine gives data on impact conditions used in the PD_1 calculation, the impact angle in this case. The column on the right gives the results from the zero-drag case for comparative purposes.

10.8.6 Delivery Accuracy Data

This basically allows the user to input CEP in the ground plane or normal plane (feet or mils), or as independently specified REP and DEP. Only one column must be supplied, and the rightmost column represents the values used. If CEP is input, REP and DEP are calculated assuming a circular normal distribution resulting in equal values.

DELIVERY ACCURACY						
		Ground ft	Normal ft	Mils	Specify	REP/DEP
CEP, ft ground, normal, mils, spec	CEP	100.00	0	0		
Deflection error probable, ft	DEP	57.30	0	0	0.000	57.300
Range error probable, ft	REP	57.30	0	0	0.000	57.300
Ballistic error, normal plane, mils	sigma_b	5.00				
Ballistic error, deflection, ft	sigma-*bd*	17.10				
Ballistic error range, ft	sigma-*br*	25.70				
Adjusted DEP, ft	DEP'	59.80				
Adjusted REP, ft	REP'	62.80				

Note that ballistic error is included at this point to produce an effective REP and DEP. Because we are considering only single weapon release, the ballistic error can be combined with the REP and DEP using the root sum square method as indicated in Eq. (4.3).

10.8.7 Weapon Effectiveness

The purpose of this section is to compute parameters derived from the EI to be used in the PD_1 computation. It will, of course, depend on the type of damage mechanism, so both MAE_F and MAE_B options are available.

WEAPON EFFECTIVENESS					
Effectiveness index type			MAE_B	MAE_F	
Effectiveness value	*EI*		0	10,000	
Length/width ratio (frag only)	*a*	0.403			WET/LET
Effective target width, ft	W'_{ET}		0.000	177.707	177.707
Effective target length, ft	L'_{ET}		0.000	71.600	71.600
Effective target area, ft^2	A_{ET}		0.000	10000	10000

Again only one entry is to be used, and the spreadsheet detects which one and derives the appropriate dimensions of the damage function. The parameter a is computed from the impact angle for fragmentation warheads, Eq. (10.10).

10.8.8 PD_1 Calculation

This is the unique part of the methodology. For single weapons against unitary targets, it is the set of calculations performed in the simple spreadsheets of Tables 10.2 and 10.7. These, in turn, are derived from Eq. (10.23) for fragment warheads and Eq. (10.36) for blast warheads. They are, however, modified to take into account the weapon reliability R in the following manner:

$$\text{Frag:} \quad \text{PD}_1 = \frac{L'_{\text{ET}} \times W'_{\text{ET}} \times R}{\sqrt{\left(17.6(\text{REP}')^2 + L'^{2}_{\text{ET}}\right)\left(17.6(\text{DEP}')^2 + W'^{2}_{\text{ET}}\right)}} \tag{10.50}$$

$$\text{Blast:} \quad \text{PD}_1 = \text{PD}_{1x} \times \text{PD}_{1y} \times R \tag{10.51}$$

PD1 CALCULATION				
	Range		Deflection	
Sigma, ft	93.104		88.655	
PD1	0.262		0.578	
Combined PD1	**0.143**			

10.8.9 Force Requirements

This section deals with the issue of how many weapons (hence sorties) are required to achieve a particular probability of damage. The methodology used here is discussed in Section 10.7.

FORCE REQUIREMENTS	
Desired PD	0.700
# of sorties for this PD	7.831
# sorties available NA	4.000
Expected PD from NA sorties	0.459

The user is expected to supply the desired PD and the number of sorties available to compute the results. The complete spreadsheet is shown in Table 10.8.

10.9 Calculating PD_1 for Guided Weapons

The basic methodology outlined so far may also be used to compute the single weapon probability of damage PD_1 for guided weapons launched against unitary targets, subject to some modifications. These modifications will be discussed relative to the unguided method developed in previous sections.

10.9.1 Weapon Characteristics

For air-launched guided missiles it is no longer necessary to define the weapon as low drag or high drag, because there is no ballistic trajectory calculation for powered, guided weapons. The precision error also is not a factor

Table 10.8 Spreadsheet for Single Unguided Weapons Against Unitary Targets

METHOD 1-U SINGLE UNGUIDED WEAPON AGAINST UNITARY TARGETS						
CONSTANTS						
Gravitational acceleration, ft/s	g	32.200				
Convert degrees to radians	deg2rad	0.017				
TARGET PARAMETERS						
Target dimensions LA, ft		N/A				
Target dimensions WA, ft		N/A				
WEAPON PARAMETERS						
Drag: L = low, H = high		L				
Ballistic dispersion, mils	Sigma_b	5.00				
Reliability	*R*	0.94				
AIRCRAFT RELEASE DATA						
Airspeed, kt	*V*	450.00		760	ft/s	
Dive angle, deg	theta	30.00		0.524	rad	
Release altitude, ft	*yi*	3000				
TRAJECTORY RESULTS						
Vertical velocity, ft/s	*Vv*	–379.80				
Horizontal velocity, ft/s	*Vh*	657.83				
Drag constant, Cd/M	c0	0.01				
constant *a* used in trajectory	*a*	6060.20				
		6.32		**Dragless**		
time of flight, s	*t_flight*	6.32		6.25		
Horizontal velocity at impact, ft/s	*Vhi*	637.37		657.83		
Vertical velocity at impact, ft/s	*Vvi*	568.35		580.90		
Impact angle, deg	*I*	41.72		41.45		
Slant range, ft	*Sri*	5075		5087		
Ground range, ft	*Gri*	4093		4108		
DELIVERY ACCURACY						
		Ground ft	Normal ft	Mils	Spec.	REP/DEP
CEP, ft, ground, normal, mils, spec	CEP	100.00	0	0		
Deflection error probable, ft	DEP	57.30	0	0	0.000	57.300
Range error probable, ft	REP	57.30	0	0	0.000	57.300
Ballistic error, normal plane, mils	sigma_*b*	5.00				
Ballistic error, deflection, ft	sigma-*bd*	17.10				
Ballistic error range, ft	sigma-*br*	25.70				
Adjusted DEP, ft	DEP'	59.80				
Adjusted REP, ft	REP'	62.80				
WEAPON EFFECTIVENESS						
Effectiveness index type			MAE_*b*	MAE_ *f*		
Effectiveness value	*EI*		0	10000		
Length/width ratio (frag only)	*a*	0.403				WET/LET
Effective target width, ft	W'_{ET}		0.000	177.707		177.707
Effective target length, ft	L'_{ET}		0.000	71.600		71.600
Effective target area, ft^2	A_{ET}		0.000	10000		10000

(Continued)

Table 10.8 Spreadsheet for Single Unguided Weapons Against Unitary Targets (*Continued*)

PD1 CALCULATION						
	Range		Deflection			
Sigma, ft	93.104		88.655			
PD1	0.262		0.578			
Combined PD1	**0.143**					

FORCE REQUIREMENTS						
Desired PD	0.700					
# of sorties for this PD	7.831					
# sorties available NA	4.000					
Expected PD from NA sorties	0.459					

for any unpowered, guided weapons, and is therefore deleted from the computation and spreadsheet.

10.9.2 Trajectory

This topic was discussed in Chapter 3, but will be summarized here. For air-to-surface missiles the trajectory is usually taken as the harp angle, following from the assumption that the weapon moves from the aircraft in a straight line to the target on release. Some weapons have the ability for trajectory shaping, allowing a user-selected impact angle specifically for hardened targets where such angles maximize penetration by the weapon.

For laser-guided bombs, it may be desirable to release them from the ballistic point as for an unguided bomb and use the guidance system for minor corrections during the last few seconds of flight. The methodology therefore includes an unguided ballistic trajectory routine based on a low drag weapon to allow for this in the computation of impact angles. The spreadsheet allows for all of these options.

10.9.3 Delivery Accuracy

Recall Chapter 4, Section 4.14 dealing with specifying delivery accuracy for guided weapons, and in particular Fig. 4.29, reproduced in Fig. 10.7. As a result of this unique distribution, guided weapons need the values of P_{NM} and P_{HIT}, as well as REP and DEP to be specified in the delivery accuracy section of the spreadsheet.

10.9.4 Calculating PD_1

The calculation of PD_1 is a little different than for unguided weapons. Consider Fig. 10.7 again. Suppose $P_{HIT} = 1$ and $P_{NM} = 0$. This would indicate that the weapon always hits the target, and there is no Gaussian distribution of miss distances. As long as the weapon is well matched against

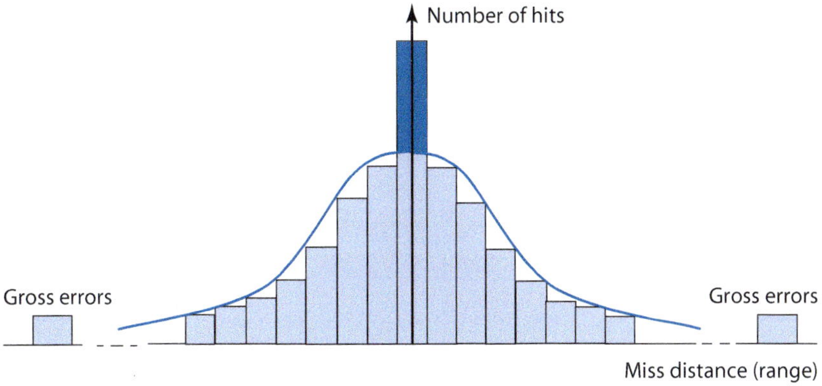

Fig. 10.7 Non-Gaussian distribution of guided weapons.

the target, this would always give $PD_1 = 1$. Now consider $P_{HIT} = 0$ and $P_{NM} = 1$. This implies the weapon behaves with a Gaussian distribution of miss distances, and the unguided methodology* just described would work just fine in these circumstances. For intermediate values of P_{HIT} and P_{NM}, we need an estimate of PD_1 that takes both factors into account. To achieve this, P_{HIT} and P_{NM} are used as weighting factors to combine the direct hit and Gaussian miss distribution values of PD_1. The weighted summation function takes the following form:

$$PD_1 = [PD1_1 \times P_{NM} + PD1_2 \times P_{HIT}] \tag{10.52}$$

In this equation, $PD1_1$ is the value for PD_1 computed using the unguided methodology described in the first part of this chapter. $PD1_2$ is the same as $PD1_1$ except that a value of CEP = 0 is used (or REP = DEP = 0) to simulate a weapon that always hits the target. It is seen that Eq. (10.52) is indeed a weighted combination of the pure Gaussian and direct hit PD_1 values.

Equation (10.52) gives the single weapon PD_1, but if more than one guided weapon is used, the resulting PD is obtained by powering up the single weapon PD_1 in the usual manner

$$PD = 1 - [1 - PD_1]^n \tag{10.53}$$

where n is the number of weapons launched. Note that an implicit assumption here is that each weapon launched is individually aimed at the target. All of these modifications are incorporated into the spreadsheet shown in Table 10.9, which covers the case of one or more guided weapons used against a unitary target.

*Although the methodology dealing with a weapon having a Gaussian distribution of miss distances is referred to as "unguided," guided weapons may also have essentially Gaussian distributions.

Table 10.9 Spreadsheet for Guided Weapons Against a Unitary Target

METHOD 1-G GUIDED WEAPONS AGAINST UNITARY TARGETS						
Example 1-4 Basic Manual						
CONSTANTS						
Gravitational acceleration ft/s	*g*	32.200				
Convert degrees to radians	deg2rad	0.017				
TARGET PARAMETERS						
Target dimensions LA, ft		N/A				
Target dimensions WA, ft		N/A				
WEAPON PARAMETERS						
Reliability	*R*	0.950				
AIRCRAFT RELEASE DATA						
Airspeed-KTAS, ft/s	*V*	500		844	ft/s	
Number of weapons salvoed	*n*	1.000				
Dive angle, deg	theta	30.000		0.524	rad	
Ground range at release, ft	*Gri*	25981		4.287	Nm	
Release altitude, ft	*yi*	15000				
TRAJECTORY						
Vertical velocity, ft/s	*Vv*	–422.00				
Horizontal velocity, ft/s	*Vh*	730.93				
Dragless trajectory for LGBs	*V*_imp	1069.62				
	TOF	20.11				
	Grd range	14,700.65				
	imp angle	55.653		**Dive angle**	**Input**	**LGB**
Impact angle, deg	I	55.653		0	0	55.65
DELIVERY ACCURACY						
		CEP ft ground	**CEP ft n**	**CEP mils**	**Spec**	**Total**
CEP, ft ground, normal or mils	CEP	0.000	0	0		
Deflection error probable, ft	DEP	0.000	0	0	50.000	50.000
Range error probable, ft	REP	0.000	0	0	50.000	50.000
Probability of near miss	*P*nm	0.900				
Probability of hit	*P*hit	0.050				
WEAPON EFFECTIVENESS						
Effectiveness index type			MAE_*b*	MAE_ *f*		
Effectiveness value	*EI*		0	20000		
Length/width ratio (frag only)	*a*	0.549				
Effective target element, ft	W'_{ET}	215.368	0.000	215.368		
	L'_{ET}	118.159	0.000	118.159		
	A_{ET}	20,000	0	20,000		

(Continued)

Table 10.9 Spreadsheet for Guided Weapons Against a Unitary Target (*Continued*)

PD1 CALCULATION						
	Range		**Deflection**			
Sigma, ft	74.129		74.129			
$PD1_1$ range/deflection	0.491		0.716			
$PD1_1$	0.352					
$PD1_2$ range/deflection	1.000		1.000			
$PD1_2$	1.000					
PD1	0.348					
PD	**0.348**					

FORCE REQUIREMENTS						
Desired PD	0.700					
# of sorties for this PD	2.814					
# sorties available NA	4.000					
Expected PD from NA sorties	0.819					

10.10 Multiple Weapon Attacks

We saw in Chapter 4 that for multiple weapon attacks we have to keep the MPI and precision errors separate because the former is the same for all weapons whereas the latter is different for each weapon. In the previous chapter this was handled in a straightforward manner by constructing the Monte Carlo simulation with a double loop—an outer loop for the MPI random draw, sampled once per attack, and an inner loop for the precision error, sampled once for each weapon in the attack.

Developing a closed form or deterministic equivalent to this MC solution proves challenging, particularly if a fast running solution is required; therefore, the JMEM methodology takes a different approach. This methodology attempts to reduce the number of error terms from two to one, and has the following objectives:

- Combine the precision error with the damage function, thereby eliminating it from the analysis.
- Because this leaves only one error term, the MPI, the methods described earlier in this chapter are applicable.

Consider a single weapon such as an artillery shell fired from a fixed cannon, and the damage function represented by a simple rectangle. If the cannon fires multiple rounds without reaiming, we will expect some variability in the location of the rectangular lethal areas on the ground plane around the DPI due to precision errors alone, as shown in Fig. 10.8.

Consider the following question: Can we draw a rectangle that encloses all of the weapon lethal areas? If we can, then we argue this enlarged lethal area may be considered to include all variability associated with precision error. Essentially we have enlarged the single weapon lethal area to

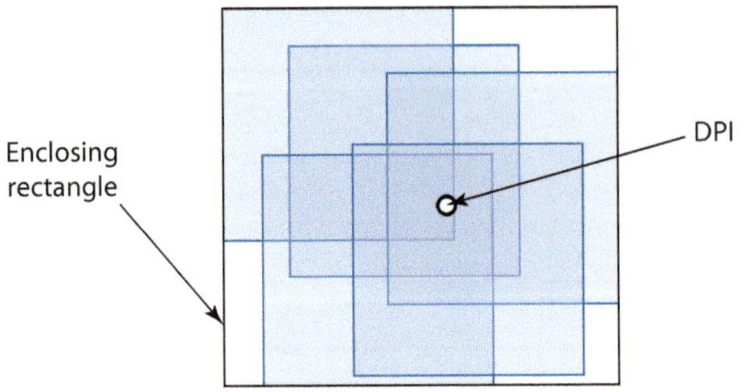

Fig. 10.8 Enclosing lethal areas for precision error.

account for the precision error, and in doing so this error is removed from further consideration.

Unfortunately, the answer to the question is no, because if we draw such a rectangle around 1000 randomly placed lethal areas and then continue the process we will eventually find one that extends beyond the enclosing rectangle we have drawn. Assuming that precision error is normally distributed in range and deflection, we see there is a small but finite probability the lethal area could land a considerable distance from the aim point.

The best we can do is to define such an enclosing rectangle that will contain a certain proportion of lethal areas. For example, suppose σ_{br} is the standard deviation of the precision error in range, L_{ET} is the length of the rectangular damage function, σ_{bd} is the standard deviation of the precision error in deflection, and W_{ET} is the width of the rectangular damage function. In the range direction the interval $L_{\mathrm{ET}} \pm \sigma_{br}$ will contain 68% of such rectangles, and in the deflection direction the interval $W_{\mathrm{ET}} \pm \sigma_{bd}$ will also contain 68% of the rectangles. The following statements therefore apply:

- A rectangle $(L_{\mathrm{ET}} \pm \sigma_{br}) \times (W_{\mathrm{ET}} \pm \sigma_{bd})$ will enclose about 46% (68% × 68%) of the lethal areas.
- A rectangle $(L_{\mathrm{ET}} \pm 2\sigma_{br}) \times (W_{\mathrm{ET}} \pm 2\sigma_{bd})$ will enclose about 90% of the lethal areas.
- A rectangle $(L_{\mathrm{ET}} \pm 3\sigma_{br}) \times (W_{\mathrm{ET}} \pm 3\sigma_{bd})$ will enclose about 98% of the lethal areas.

Whichever one we choose, the process is to expand the single weapon rectangular lethal area by an amount that depends on the precision error, as shown in Fig. 10.9.

The size of the enlarged cookie cutter is given by the following:

$$L_B = L_{\mathrm{ET}} + k \times \sigma_{br} \tag{10.54}$$

$$W_B = W_{\mathrm{ET}} + k \times \sigma_{bd} \tag{10.55}$$

where $k = 1, 2$, or 3 depending on the model or analyst's choice. Note that in doing this enlargement, we cannot leave the damage probability inside the cookie cutter at 1.0; otherwise, we have a weapon with increased lethality compared to the original one. The new damage probability inside the expanded lethal area is defined as the conditional probability of damage P_{CD}, and its value is determined as the ratio of lethal areas. This ensures the lethality of both rectangles are the same.

$$P_{CD1} = P_{CD}\frac{A_{ET}}{A_B} = P_{CD}\frac{L_{ET}W_{ET}}{L_B W_B} \tag{10.56}$$

The rectangle of area A_B now becomes the new weapon lethal area, and precision error has been removed from the effectiveness calculation. This leaves only the single, random MPI error, which may be used directly in a deterministic solution.

If we apply this approach to the case considered previously, where the effectiveness was calculated using a Monte Carlo simulation keeping the error terms separate, what will the result be if we use the approach just described? Given the data for this case, we expand the lethal area using $k = 4$ and the precision error of 2 ft to get

$$L_B = 10 + 4 \times 2 = 18 \tag{10.57}$$

$$W_B = 10 + 4 \times 2 = 18 \tag{10.58}$$

Equation (10.56) gives the probability of damage inside the enlarged lethal area as

$$P_{CD1} = P_{CD}\frac{10 \times 10}{18 \times 18} = 0.3086 \tag{10.59}$$

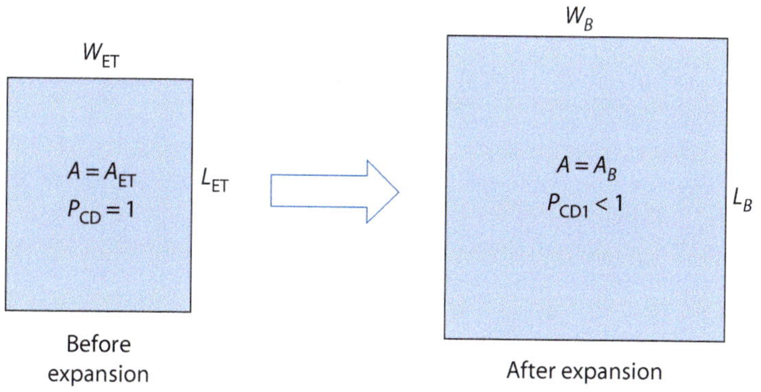

Fig. 10.9 Expansion of single weapon A_{ET} to account for precision error.

Using just the MPI error, Eq. (10.5) becomes

$$\begin{aligned} \mathrm{PD}_{1x} &= \frac{1}{\sqrt{2\pi}\sigma}\int_{-L_B/2}^{+L_B/2} \exp\left(\frac{x^2}{2\sigma^2}\right)dx \\ &= \left[\frac{1}{\sqrt{2\pi}\sigma}\int_{-9}^{9} \exp\left(\frac{x^2}{2\times 5^2}\right)dx\right] \\ &= [F(1.8) - F(-1.8)] = 0.9282 \end{aligned} \tag{10.60}$$

In calculating PD_1, we have to modify Eq. (10.7) to allow for the reduced conditional probability of damage inside the enlarged lethal area.

$$\mathrm{PD}_1 = P_{\mathrm{CD}} \times \mathrm{PD}_{1x} \times \mathrm{PD}_{1y} = 0.2659 \tag{10.61}$$

Because we have eliminated the precision error, we have only the MPI error left as the single random error, so we use the powering up formula for the three rounds in the burst.

$$\mathrm{PD} = 1 - (1 - \mathrm{PD}_1)^n = 1 - (1 - 0.2659)^3 = 0.6043 \tag{10.62}$$

Clearly this is closer to the Monte Carlo value of 0.6116 than the result where we incorrectly RSS the MPI and precision errors, and it illustrates the utility of this approach; however, variations from this result may occur for different values of k.

Although this method has been developed using a rectangular damage function, the same approach may be used for other forms as well. Consider the Gaussian or Carlton form of the damage function, which, for simplicity, will be considered only in the range direction.

$$P(x) = \exp - \left[\frac{x^2}{WR_r^2}\right] \tag{10.63}$$

Now consider the precision error to be also Gaussian in the range direction, given by the following PDF:

$$G(x) = \frac{1}{\sigma_{br}\sqrt{2\pi}} \exp - \left[\frac{x^2}{2\sigma_{br}^2}\right] \tag{10.64}$$

where as before, σ_{br} is the standard deviation of the precision error in the range direction. Multiplying these two Gaussian functions will result in a more diffuse damage function that takes the following form:

$$
\begin{aligned}
P'(x) = P(x)G(x) &= \frac{1}{\sigma_{br}\sqrt{2\pi}} \exp - \left[\frac{x^2}{WR_r^2}\right] \exp - \left[\frac{x^2}{2\sigma_{br}^2}\right] \\
&= \frac{1}{\sigma_{br}\sqrt{2\pi}} \exp - \left[\frac{x^2}{WR_r^2 + 2\sigma_{br}^2}\right] \\
&= \frac{1}{\sigma_{br}\sqrt{2\pi}} \exp - \left[\frac{4x^2}{(L'_{\mathrm{ET}})^2 + 8\sigma_{br}^2}\right]
\end{aligned}
\tag{10.65}
$$

Notice how the maximum damage at $x = 0$ is no longer unity, and the damage function extends over a larger area (diffused). The exponential falloff is increased, thereby preserving lethality. Figure 10.10 shows this effect for $L'_{\mathrm{ET}} = 10$ ft and $\sigma_{br} = 5$ ft.

This is similar to the effect seen in the rectangular cookie cutter where the size is enlarged and the effectiveness decreased. Note that in the previous equations, σ_{br} and σ_{bd} are the precision error standard deviations in feet in the ground plane and are usually obtained from the specification of σ_{b} in mils by using Eqs. (4.47) and (4.48) without the factor 0.6745.

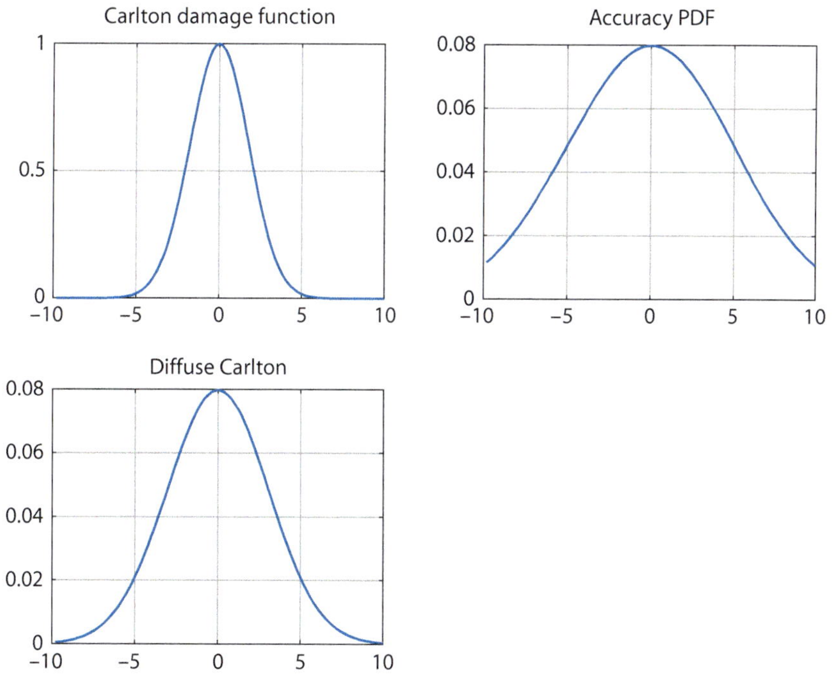

Fig. 10.10 Expansion of Carlton damage function.

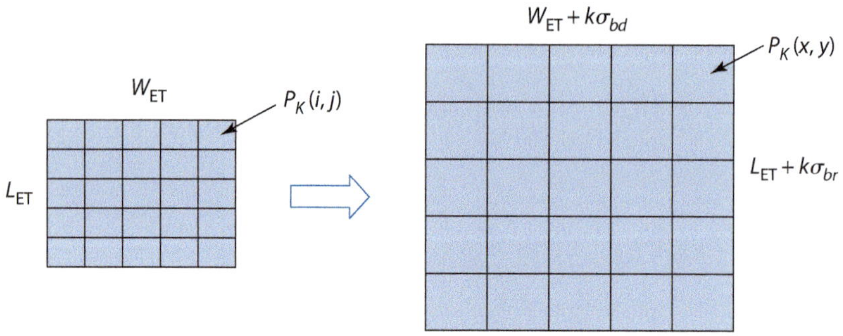

Fig. 10.11 Expansion of lethal area matrix.

If a model uses the lethal area matrix directly, it may also be expanded to allow for precision error, as shown schematically in Fig. 10.11.

Essentially, each cell in the LAM is treated as an individual rectangular cookie cutter and is expanded accordingly, implying the LAM as a whole is enlarged by the same amount. As before, the area ratio scaling shown in Eq. (10.59) may be used to determine the expanded matrix cell (dx, dy) P_K values.

$$P_K(x, y) = P_K(i, j) \times \frac{\mathrm{d}x\mathrm{d}y}{(\mathrm{d}x + k\sigma_{br})(\mathrm{d}y + k\sigma_{bd})} \tag{10.66}$$

Because we have described several means to describe damage functions—lethal area matrix, Carlton damage function, and cookie cutters—the question often arises: Which one should be used for a particular application? The answer is to use the highest fidelity model for which a mathematically tractable solution can be obtained in a short amount of time. It is important to keep in mind that the JMEM tools support the warfighter, and in most cases missions have to be planned, evaluated, and executed quickly. Sophisticated methodologies that may provide a 99% solution in hours is a luxury that can rarely be afforded compared to an approximate method giving the 90% solution in minutes.

10.11 Comparing Circular and Rectangular Cookie Cutter Damage Functions

For some analyses such as those involving blast warheads, it is convenient to represent the lethal area by a circle rather than a rectangular or square cookie cutter. For completeness, we now compare the effectiveness obtained by considering the weapon/target lethal to be different shapes. For example, suppose an MAE_B is 3000 ft^2 and the weapon CEP is 20 ft circular normal. What is the probability of damage if the target is represented by a cookie cutter damage function that is (1) circular or (2) square? This probability of damage is computed as the probability of hitting the lethal area when represented by either of the two shapes.

1. *Circular damage function:* The damage function is a circle of radius 30.9 ft, so the probability a single round hits this circle with $\sigma = 20/1.1774 = 17$ ft, obtained from the Rayleigh CDF.

$$P_{\mathrm{HIT}} = P(r < 17) = 1 - \exp\left[\frac{-30^2}{2 \times 17^2}\right] = 0.7892 \tag{10.67}$$

2. *Square damage function:* The damage function is a square of side 54.78 ft. The probability a single round hits this square in the range direction with $\sigma = 20/1.1774 = 17$ ft is given by the integral of the normal PDF.

$$\begin{aligned} P(-27.39 < x < 27.3) &= P(-1.61 < z < 1.61) \\ &= F(1.61) - F(-1.61) \\ &= F(1.61) - 1 + F(1.61) \\ &= 0.8926 \end{aligned} \tag{10.68}$$

Because we are dealing with a square, the probability of a hit in the deflection direction is the same; therefore,

$$P_{\mathrm{HIT}} = 0.8926^2 = 0.7967 \tag{10.69}$$

For this case, at least, there is less than 1% difference in the damage based on the shape of the lethal area.

10.12 Bomb Burial

When a single weapon impacts a yielding material such as soil or sand, it will generally become partially buried before it detonates, resulting in a degraded value for the PD_1. Based on test data and high fidelity computer models of the resulting blast and fragment patterns, the following empirical equations were found to represent the degradation of PD_1 for 10%, 25%, and 50% burial in terms of the unburied PD_1 value:

$$PD1_{10} = 0.3897\, PD_1^4 - 0.1047\, PD_1^3 - 0.2210\, PD_1^2 + 0.8443\, PD_1 \tag{10.70}$$

$$PD1_{25} = 0.2034\, PD_1^4 - 0.1051\, PD_1^3 - 0.1606\, PD_1^2 + 0.6321\, PD_1 \tag{10.71}$$

$$PD1_{50} = 0.3148\, PD_1^4 - 0.2279\, PD_1^3 - 0.3190\, PD_1^2 + 0.4013\, PD_1 \tag{10.72}$$

These equations are an empirical fit to high fidelity computer simulations of GP bombs penetrating a resistive material, and formed the legacy JMEM methodology for accounting for mitigated blast and fragment effects. These effects are shown in Fig. 10.12.

Some care is necessary when deciding whether bomb burial is possible for a particular combination of surface material and arrival conditions of the munition, because this effect may be counterintuitive. Surprisingly, the

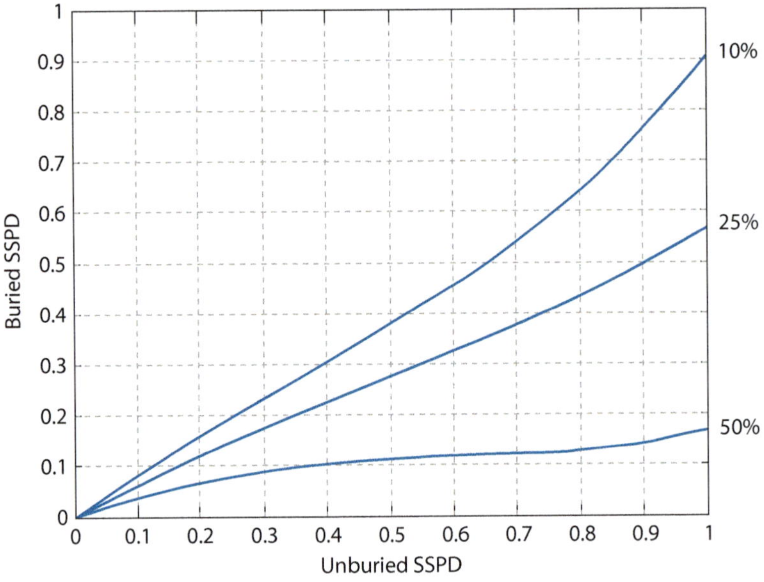

Fig. 10.12 Effect of bomb burial on PD_1.

slower the weapon impact speed, the more burial will occur in a soft material. If you slap a water surface with the palm of your hand, a large resistive force will be felt preventing submersion, but push slowly on the surface and the water quickly displaces and the hand easily moves below the surface. A high drag, slow moving bomb therefore may achieve more burial than a low drag, high impact speed bomb.

Better, more complex models exist for determining the penetration of warheads into a variety of materials. These are dealt with in Volume 2, Chapter 9.

Chapter 11 Single Weapons Directed Against an Area of Targets

11.1 Introduction

This chapter will extend the previous one to include targets that contain multiple target elements inside a defined area, corresponding to target type 2 described in Chapter 9, Section 9.11. Although an area of targets is sometimes referred to as simply an area target, there is an important point to keep in mind. A target such as a large storage tank may have significant dimensions and may be considered an area target; however, because there is only one storage tank it is a unitary target, and the methods developed in the previous chapter will apply. On the other hand, an area of targets may contain multiple, identical target elements, perhaps a collection of storage tanks.

In analyzing targets of this type, the following assumptions are made.

1. All target elements are unitary and identical.
2. They are contained in a minimum rectangular area of dimensions L_A and W_A.
3. The target elements are uniformly distributed within this rectangle.

Figures 11.1 and 11.2 show how real-world target scenarios are stylized into compliance with the definition of an area of target elements.

11.2 Measurement of Damage

We will begin with the same general expression for the PD_1 used before, namely the expected value of effectiveness using the one-dimensional Carlton damage function except we now integrate over the target extent, corresponding to Eq. 10.13.

$$PD_{1x} = \int_{-\infty}^{\infty} c(x)g(x)\,dx \tag{11.1}$$

Fig. 11.1 Area of vehicle elements of dimensions L_A by W_A.

where the accuracy function $g(x)$ and the Carlton damage function $c(x)$ are given by Eqs (10.14) and (10.15) respectively. Once more, we will consider the range and deflection directions independent of each other, and initially only consider the former. For a single weapon launched against the target, Fig. 10.4 may be modified to include the target size, which in this case is L_A in the range direction. The modified figure is shown below as Fig. 11.3.

The problem now is that the damage function varies over the target, causing more damage to the elements on the right side and less to the elements on the left. How can we calculate the "average" damage done to the aggregate of target elements? Two methods for dealing with this situation will now be described. In the first, a simplifying assumption will be made to enable a closed form solution to be made and is representative of the JMEM air-to-surface approach to areas of target elements. This assumption is to replace the Carlton damage function with an equivalent rectangular, cookie-cutter damage function. In the second method, the Carlton damage function

Fig. 11.2 Area of personnel elements also L_A by W_A.

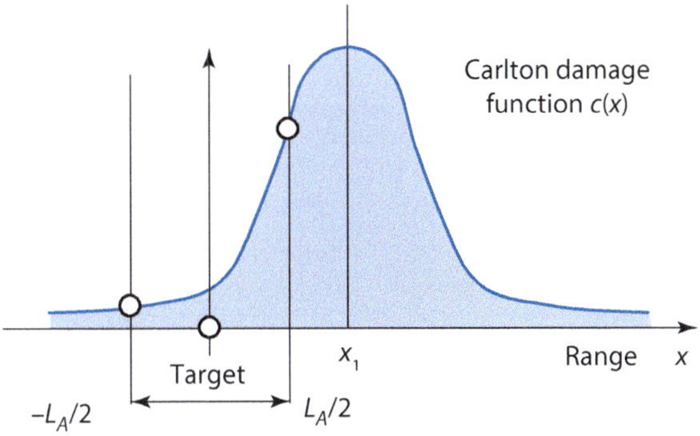

Fig. 11.3 Single weapon against an area target.

is preserved, but leads to a Monte Carlo solution; this is more indicative of the JMEM surface-to-surface approach.

In dealing with areas of targets, we do not describe the damage done by a single weapon in terms of a PD_1, as was done for a unitary target. Instead the damage metric is the expected fractional damage, or FD_1. This is because the target area contains a fixed number of target elements and of primary interest is the fraction of those elements damaged. Consider the situation shown in Fig. 11.4, where an area target is 50% covered by a rectangular weapon lethal area.

For this particular case, if there were 100 target elements, 50 would be affected by the weapon. If the probability of damage inside the weapon lethal area was 1.0, those 50 elements would be damaged to the level for which the lethal area is defined. Suppose we consider the general case where perhaps the probability of damage inside the lethal area is 0.8, then 40 elements would be damaged. We therefore can write

$$\mathrm{FD}_1 = F_C \times P_{\mathrm{CD}} \tag{11.2}$$

where FD_1 is the fractional damage due to a single weapon, F_C is the fractional coverage (50%), and P_{CD} is the conditional damage probability inside the rectangular lethal area. Usually $P_{CD} = 1$ for a single weapon cookie cutter lethal area but we will see later how this may not always be the case. However, because P_{CD} is fixed, the problem of finding FD_1 is the same as finding the fractional coverage F_C.

Of course we have assumed that the weapon lands in the location shown such that it covers exactly half the target. Because the impact point will be random, we should now ask the following: In a large number of trials, what is the average, or expected, fractional coverage of the target by the weapon lethal area? This situation is shown in Fig. 11.5.

This expected fractional coverage may be obtained from Eq. (11.2) to produce the following generalized result:

$$E(\mathrm{FD}_1) = E(F_C) \times P_{\mathrm{CD}} \tag{11.3}$$

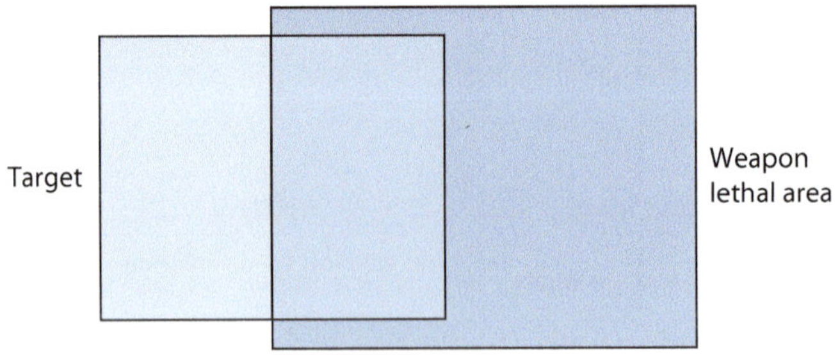

Fig. 11.4 Target half covered by weapon lethal area.

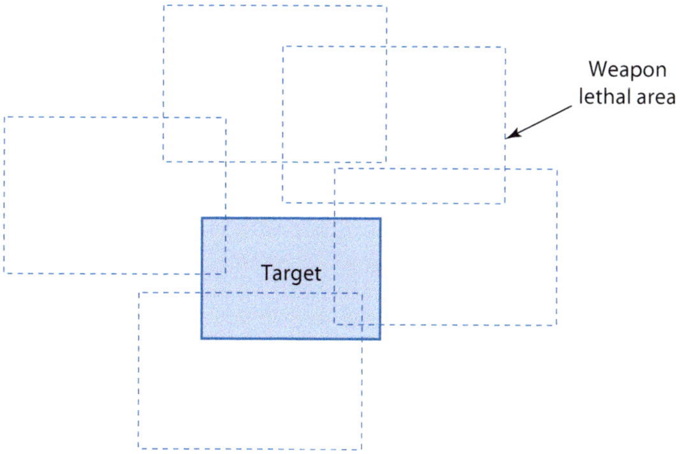

Fig. 11.5 Fractional coverage of area of targets.

In this equation, $E(\mathrm{FD}_1)$ is the expected fractional damage, and $E(F_C)$ is the expected fractional coverage. [Note that we usually write $E(\mathrm{FD}_1)$ simply as FD_1]. The problem, therefore, decomposes into one of finding the expected fractional coverage. This will be examined for the range and deflection directions separately, denoted by $E(F_R)$ and $E(F_D)$, respectively.

11.3 Weapon Represented by Rectangular Damage Function

We now examine in detail the case where the damage function is represented by a rectangular equivalent to the Carlton damage function, as described in Fig. 8.30 and shown again in Fig. 11.6. This damage function has a conditional probability of damage inside it of unity.

We can now replace the Carlton damage function with this rectangular form and investigate its interaction with the accuracy function over the

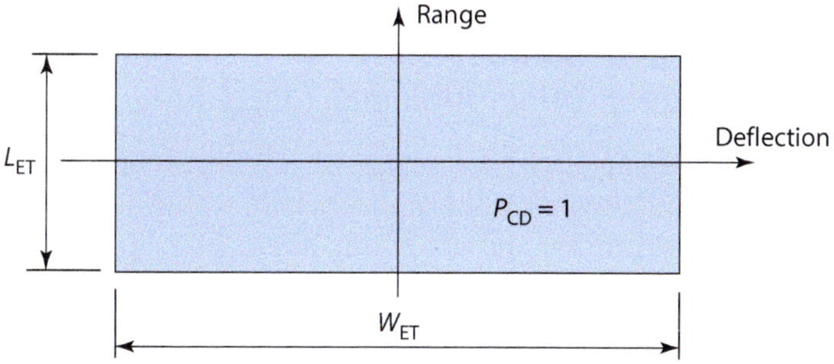

Fig. 11.6 Rectangular cookie cutter damage function.

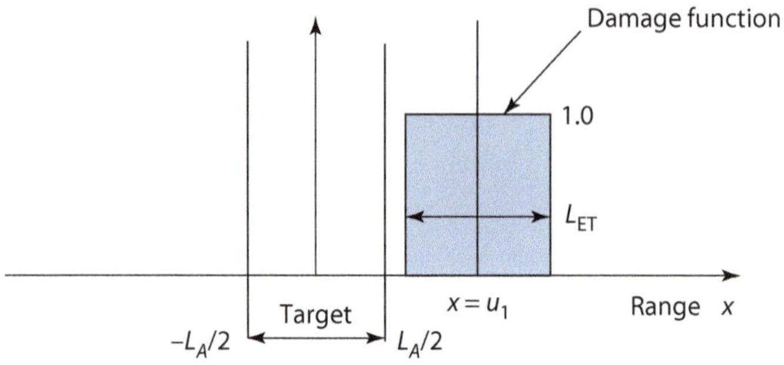

Fig. 11.7 Modified damage function and area target.

target, as shown in Fig. 11.7 for a random impact location in the range direction only.

In this figure we are preserving the same Monte Carlo approach used in the previous chapter. It shows a sample iteration where the weapon lands at a position relative to the target $x = u_1$.

For the particular situation shown in Fig. 11.7, no damage to the target would be done, because the weapon lethal area does not overlap any part of the target ($F_C = 0$); however, for another impact point, the lethal area may partially cover the target area ($F_C < 1$) or cover it completely ($F_C = 1$). It becomes necessary, therefore, to analyze in detail the fractional coverage of the target by the weapon lethal area for a random placement of the lethal area relative to the target.

Before doing this however, we investigate the issue of the relative size of the weapon lethal area compared to the area of target elements it is covering. This has two consequences:

1. It affects the aiming strategy used when multiple weapons are directed against an area of targets.
2. It can modify the mathematics of the fractional damage calculations

11.4 Relative Size of Lethal Area and Target Area

If the single weapon lethal area is larger than the area target, as shown in Fig. 11.8, then it would seem that only one weapon would be needed to completely cover the target area, in which case the center of the target is the desired point of impact, DPI. If multiple weapons were used however then each one would be aimed at the same point – the center of the area of targets.

However, when the target is larger than the weapon lethal area, as shown in Fig. 11.9, there is something of a dilemma regarding the selection of aimpoints for multiple weapon attacks in order get optimum coverage.

Fig. 11.8 Lethal area larger than target area.

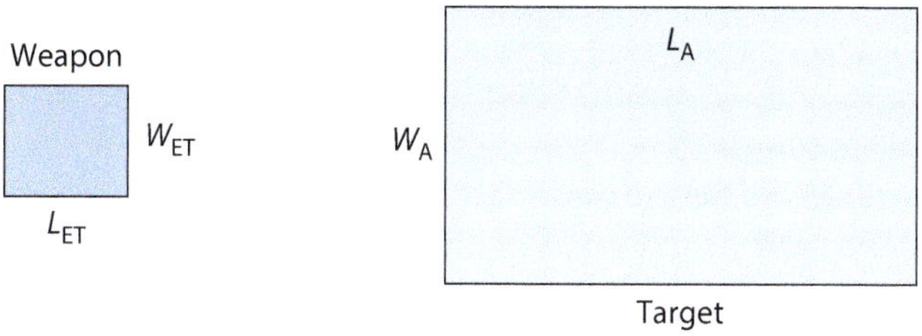

Fig. 11.9 Target area larger than lethal area.

Remember, this is not a stick delivery, but an attack involving multiple weapons directed at individual aimpoints.

Suppose four weapons are used against the target and that the MPI and precision accuracy of the weapon are both very good. If all weapons are aimed at the centre of the target as for the previous case, the situation shown in Fig. 11.10 might result.

In this case considerable damage is done to the center target elements, but the elements at the extremeties of the target are not damaged. Alternaitvely the four weapons could be aimed so as to produce more coverage of the target elements as shown in Fig. 11.11.

Clearly the fractional damage when the weapons are spread out would be larger than when they are aimed at a single DMPI. Since the aircrew would attempt to place weapons to get maximum target coverage, we need a way to account for this process which is independent of the number of weapons directed at the target and thereby not having to specify individual weapon aimpoints. Note that this process is only needed for the case where the target area is larger than the single weapon lethal area, otherwise the DMPI is the center of the area target.

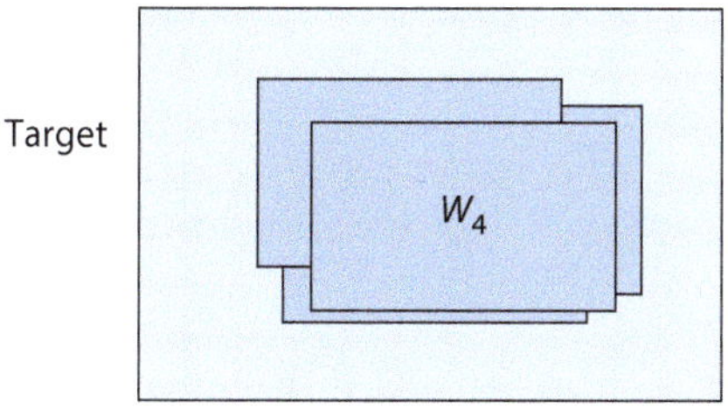

Fig. 11.10 Four weapons aimed at the target center.

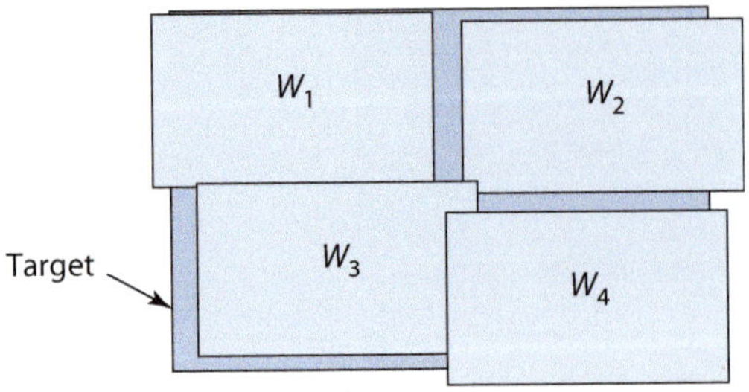

Fig. 11.11 Four weapons distributed over the target.

Consider the case shown in Fig. 11.11. The JMEM methodology assumes that however many weapons are used against the target, the aimpoints are randomly located inside the target area with a uniform distribution in both range and deflection, as indicated in Fig. 11.11. The process involves the following steps.

1. The single weapon lethal area is expanded, independently in range and direction, to the same size as the target. Mathematically this may be written as follows.

$$\text{IF}(L_{\text{ET}} < L_A) \quad \text{THEN} \quad (L_{\text{ET}} = L_A) \tag{11.4}$$

$$\text{IF}(W_{\text{ET}} < W_A) \quad \text{THEN} \quad (W_{\text{ET}} = W_A) \tag{11.5}$$

In this process however the JMEM uses a different notation for the expanded lethal area dimensions: L_{EP} and W_{EP}, for the equivalent pattern length and width respectively. Additionally we can allow for the case when the weapon is larger than the target. Equations (11.4) and (11.5) may be modified as

$$\text{IF}(L_{\text{ET}} < L_A) \quad \text{THEN} \quad (L_{\text{EP}} = L_A) \quad \text{ELSE} \quad (L_{\text{EP}} = L_{\text{ET}}) \tag{11.6}$$

$$\text{IF}(W_{\text{ET}} < W_A) \quad \text{THEN} \quad (W_{\text{EP}} = W_A) \quad \text{ELSE} \quad (W_{\text{EP}} = W_{\text{ET}}) \tag{11.7}$$

Physically, if both size conditions require expansion, the lethal area is transformed as shown below.

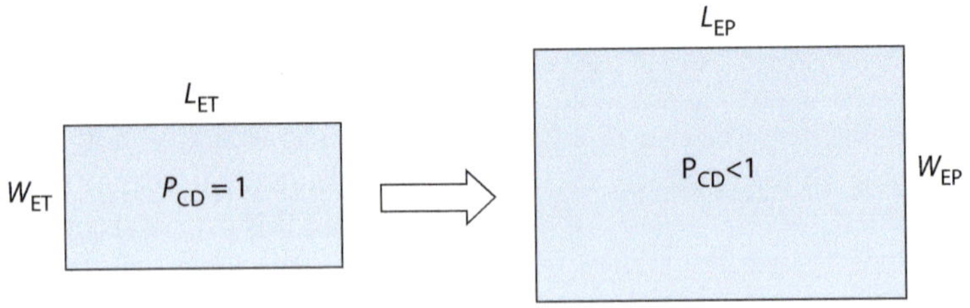

Fig. 11.12 Expansion of weapon lethal area.

2. To preserve lethality in the expanded lethal area, the conditional probability of damage has to be reduced by the ratio of the lethal areas.

$$P_{CD} = \frac{L_{ET} \times W_{ET}}{L_{EP} \times W_{EP}} = \frac{A_{ET}}{A_{EP}} \tag{11.8}$$

3. The expected fraction damage from a single weapon is given by

$$FD_1 = E(F_C) \times P_{CD} \tag{11.9}$$

4. Now that the new weapon lethal area is as large as the target area, all weapons are directed to the center of the target area, as would be the case if the lethal area was initially larger than the target area and expansion would be unnecessary.

This process may be summarized in Fig. 11.13 where the four uniformly distributed weapons are equivalent to four expanded weapon lethal areas lying directly on top of each other.

Equation (11.9) is the same expression that would apply to the case where the weapon lethal area is greater than the target, except that for that case $P_{CD} = 1$. To provide one computational method for dealing with both cases we may use the following procedure.

$$L_{EP} = \max(L_{ET}, L_A) \tag{11.10}$$

$$W_{EP} = \max(W_{ET}, W_A) \tag{11.11}$$

and that

$$FD_1 = E(F_C) \times P_{CD} \times R = E(F_C) \times \left[\frac{A_{ET}}{L_{EP} \times W_{EP}}\right] \times R \tag{11.12}$$

Equations (11.10)–(11.12) now apply whether the target area is larger or smaller than the weapon lethal area.

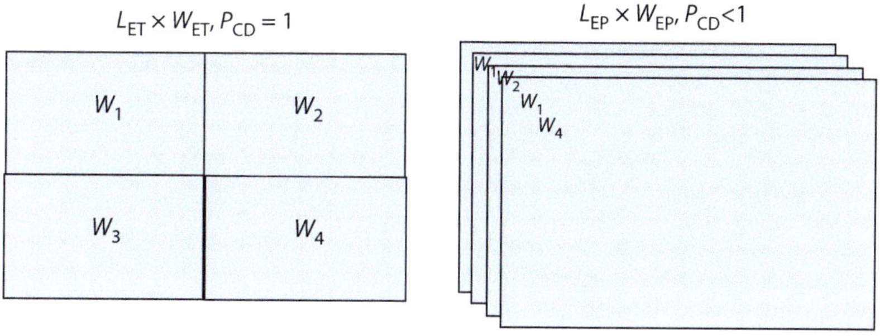

Fig. 11.13 Equivalence of lethal areas.

Note:

1. If the weapon is larger than the target $L_{EP} = L_{ET}$, $W_{EP} = W_{ET}$, no expansion of the weapon lethal area occurs and the conditional probability of damage is unity, as indicated by the square bracket in Eq. (11.12).
2. If the weapon is smaller than the target the dimensions of the lethal area are expanded, are given by Eqs (11.10), (11.11) and the conditional probability of damage is less than unity, again indicated by the square bracket in Eq. (11.12).
3. This method also covers cases where the weapon lethal area may be larger in range but not in deflection, and vice versa.
4. The method may be applied for single as well as multiple weapon attacks.

By using this process, we no longer need to have one aiming strategy for relatively large weapon lethal areas and a different one for weapons with relatively small lethal areas. As a result of this process the weapon will now be represented by a rectangle of dimensions L_{EP} by W_{EP} and an associated conditional probability of damage inside P_{CD}. This rectangular lethal area is at least as large as the target area and it is this rectangle that is used to calculate the fractional coverage of the target.

11.5 Fractional Coverage

Although the fractional coverage may be thought of in terms of a random placement of the weapon lethal area relative to the area of target elements, we apply a more methodical approach here. Consider what happens when the target area is fixed at the origin and the weapon lethal area is far to the left of it, as shown in Fig. 11.14. Here the position of the weapon is defined as the value of x from the origin to the center of the lethal area.

For this case we may simply write from observation that

$$F_R = 0 \tag{11.13}$$

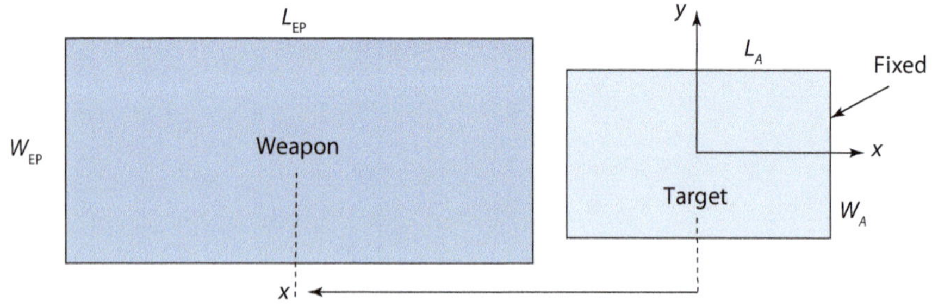

Fig. 11.14 Weapon misses the target to the left.

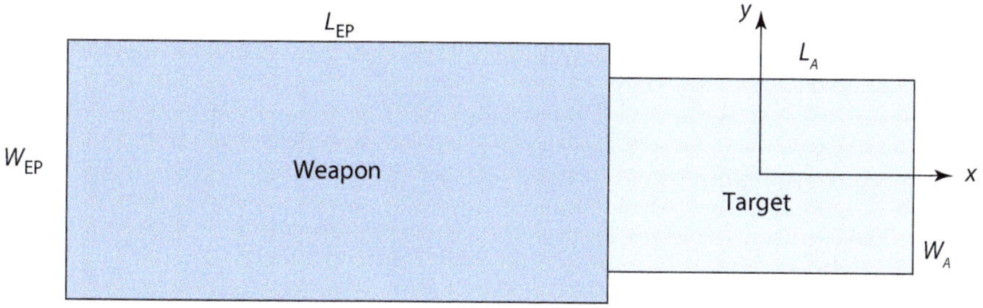

Fig. 11.15 Weapon abuts the target on the left.

Now consider the weapon moving to the right i.e. x increasing. Equation (11.13) will still apply until the condition shown in Fig. 11.15 occurs.

Further movement of the weapon area to the right will cause a partial coverage of the left side of the target area, as shown in Fig. 11.16.

We can see that the value of F_R is simply the fraction of the lethal area that overlaps the target area, which will vary from zero to unity depending on x; hence, the fractional coverage in range is defined as

$$F_R = \frac{\beta}{L_A} \tag{11.14}$$

From the geometry in Fig. 11.16, we can write the following

$$-\frac{L_A}{2} + \beta = x + \frac{L_{EP}}{2} \tag{11.15}$$

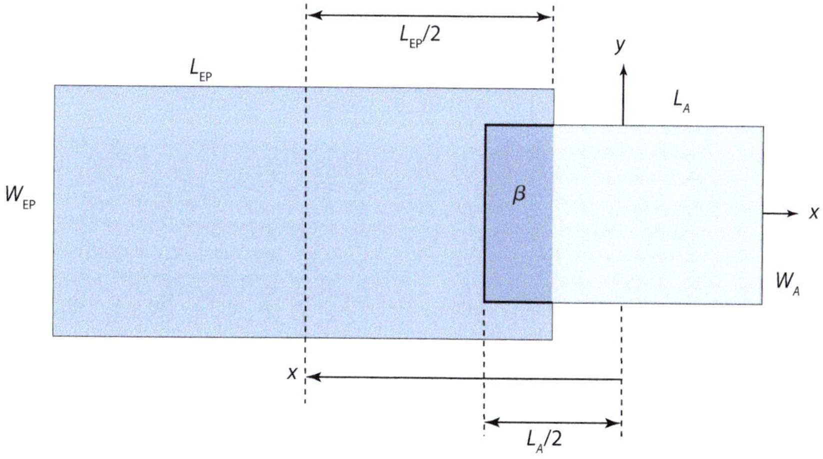

Fig. 11.16 Partial left overlap of target by weapon.

Substituting Eq. (11.15) into Eq. (11.14) yields

$$F_R = \frac{L_A + L_{EP}}{2L_A} + \frac{x}{L_A} \tag{11.16}$$

Setting $F_R = 0$ yields

$$x_{\max} = -\frac{L_{EP} + L_A}{2} \tag{11.17}$$

whereas setting $F_R = 1$ gives

$$x_{\min} = -\frac{L_{EP} - L_A}{2} \tag{11.18}$$

yielding the result that Eq. (11.16) is valid in the partial overlap region for

$$-\frac{L_{EP} - L_A}{2} < x < -\frac{L_{EP} + L_A}{2} \tag{11.19}$$

Once the weapon area completely covers the target area in which case

$$F_R = \frac{L_{EP}}{L_A} \tag{11.20}$$

The fractional coverage remains at this value until the weapon area is on the point of uncovering the target area on the left side, as shown in Fig. 11.17.

Moving the weapon lethal area further to the right begins to uncover the target area on the left side, as shown in Fig. 11.18.

Here the fractional coverage in the range direction γ is

$$F_R = \frac{\gamma}{L_A} \tag{11.21}$$

From the geometry in Fig. 11.18, we can write the following:

$$\frac{L_A}{2} - \gamma = x - \frac{L_{EP}}{2} \tag{11.22}$$

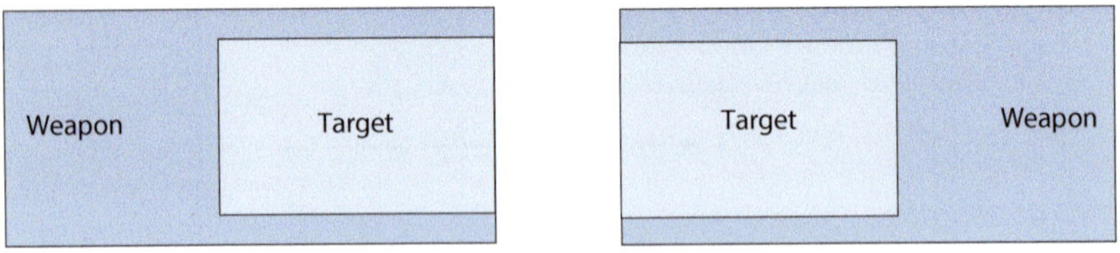

Fig. 11.17 Target area lying within weapon lethal area.

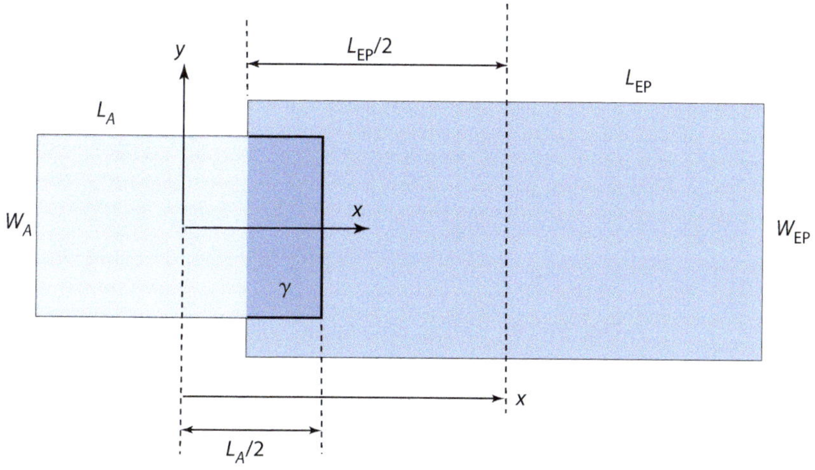

Fig. 11.18 Partial right overlap of target by weapon.

Substituting Eq. (11.22) into Eq. (11.21) yields

$$F_R = \frac{L_{EP} + L_A}{2L_A} - \frac{x}{L_A} \tag{11.23}$$

Again, setting $F_R = 0$ yields

$$x_{max} = \frac{L_{EP} + L_A}{2} \tag{11.24}$$

whereas setting $F_R = 1$ gives

$$x_{min} = \frac{L_{EP} - L_A}{2} \tag{11.25}$$

Note that Eq. (11.23) is valid in the partial overlap region for

$$\frac{L_{EP} - L_A}{2} < x < \frac{L_{EP} + L_A}{2} \tag{11.26}$$

Now use the shorthand notation

$$s = \frac{L_{EP} + L_A}{2} t = \frac{L_{EP} - L_A}{2} \tag{11.27}$$

The fractional coverage becomes zero when the geometry is as shown in Fig. 11.19.

Further movement of the weapon lethal area to the right results in $F_R = 0$. Fig. 11.20 summarizes fractional coverage as a function of x, the coordinate in the range direction measured from the center of the fixed

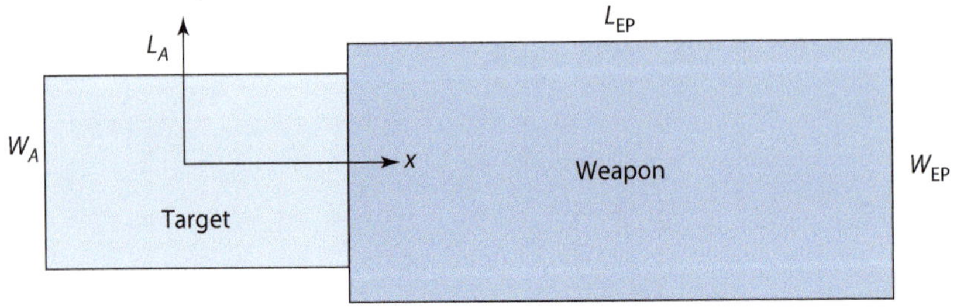

Fig. 11.19 Weapon abuts target on the right.

target to the center of the weapon lethal area. This fractional coverage function is a piecewise continuous trapezoidal function.

Again we reconsidering Eq. (11.3) and its solution using a Monte Carlo approach where for each iteration, the location in the range direction of the center of the weapon lethal area is a random draw from a zero mean Gaussian distribution. For a particular impact point, a distance x from the origin (DPI), the fractional coverage may be obtained from Fig. 11.20 above as shown in Fig. 11.7. Averaging over all Monte Carlo iterations, what we are calculating is the expected value of the fractional coverage function shown in Fig. 11.20, which can be obtained directly from the expected value theorem.

$$E(F_R) = \int_{x=-\infty}^{x=\infty} F_R(x)\,g(x)\mathrm{d}x \tag{11.28}$$

This is evaluated by substituting the piecewise functions of F_R in each region of x from Fig. 11.20, together with the probability density function of the

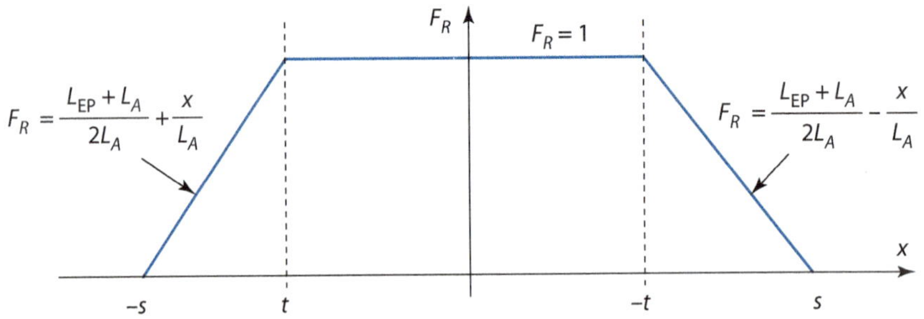

Fig. 11.20 Fractional coverage function in range.

aiming error $g(x)$, which has a PDF given by

$$g(x) = \frac{1}{\sigma_x\sqrt{2\pi}}\exp\left[\frac{-x^2}{2\sigma_x^2}\right] \tag{11.29}$$

Substituting into Eq. (11.28) gives the following:

$$E(F_R) = \int_{-\infty}^{+\infty} F_R(x)g(x)\mathrm{d}x = \frac{1}{\sigma_x\sqrt{2\pi}}\left\{\begin{array}{l} \int_t^{-t} \exp\left(\frac{-x^2}{2\sigma_x^2}\right)\mathrm{d}x \\ + \int_{-s}^{t}\left[\frac{L_{\mathrm{EP}}+L_A}{2L_A}+\frac{x}{L_A}\right]\exp\left(\frac{-x^2}{2\sigma_x^2}\right)\mathrm{d}x \\ + \int_{-t}^{s}\left[\frac{L_{\mathrm{EP}}+L_A}{2L_A}-\frac{x}{L_A}\right]\exp\left(\frac{-x^2}{2\sigma_x^2}\right)\mathrm{d}x \end{array}\right\} \tag{11.30}$$

The term in the braces corresponds to the positive slope, plateau, and negative slopes of Fig. 11.20, respectively. Expanding this equation gives the following:

$$E(F_R) = \frac{1}{\sigma_x\sqrt{2\pi}}\left\{\begin{array}{l} \int_t^{-t} \exp\left(\frac{-x^2}{2\sigma_x^2}\right)\mathrm{d}x + \left[\frac{L_{\mathrm{EP}}+L_A}{2L_A}\right]\int_{-s}^{t}\exp\left(\frac{-x^2}{2\sigma_x^2}\right)\mathrm{d}x \\ + \left[\frac{L_{\mathrm{EP}}+L_A}{2L_A}\right]\int_{-t}^{s}\exp\left(\frac{-x^2}{2\sigma_x^2}\right)\mathrm{d}x + \int_{-s}^{t}\frac{x}{L_A}\exp\left(\frac{-x^2}{2\sigma_x^2}\right)\mathrm{d}x \\ - \int_{-t}^{s}\frac{x}{L_A}\exp\left(\frac{-x^2}{2\sigma_x^2}\right)\mathrm{d}x \end{array}\right\} \tag{11.31}$$

Using shorthand notation, Eq. (11.31) may be written as the summation of five integrals.

$$E(F_R) = \{I_1 + I_2 + I_3 + I_4 - I_5\} \tag{11.32}$$

Now consider the normal cumulative distribution function with zero mean

$$F(x) = \frac{1}{\sigma_x\sqrt{2\pi}}\int_{-\infty}^{x}\exp\left[\frac{-x^2}{2\sigma_x^2}\right]\mathrm{d}x \tag{11.33}$$

If we solve these integrals in a spreadsheet, we can use the NORMDIST function (ND) of Excel eq. (11.33) may be written for the general case

$$\mathrm{ND}(p, 0, \sigma_x, 1) = \mathrm{ND}(p) = \frac{1}{\sigma_x\sqrt{2\pi}}\int_{-\infty}^{p}\exp\left[\frac{-x^2}{2\sigma_x^2}\right]\mathrm{d}x \tag{11.34}$$

Then, taking the $1/(\sigma_x\sqrt{2\pi})$ inside the braces in Eq. (11.31), and recognizing the distributions all have the same zero mean and variance, the first three integrals may be computed as

$$I_1 = [\mathrm{ND}(-t) - \mathrm{ND}(t)] \tag{11.35}$$

$$I_2 = \frac{L_{\mathrm{EP}} + L_A}{2L_A}[\mathrm{ND}(t) - \mathrm{ND}(-s)] \tag{11.36}$$

$$I_3 = \frac{L_{\mathrm{EP}} + L_A}{2L_A}[\mathrm{ND}(s) - \mathrm{ND}(-t)] \tag{11.37}$$

Integrals I_4 and I_5 are more difficult to evaluate due to the extra "x" inside the integral. These may be written generally as

$$I' = \frac{1}{\sigma_x\sqrt{2\pi}}\int_a^b x\exp\left[\frac{-x^2}{2\sigma_x^2}\right]\mathrm{d}x \tag{11.38}$$

Suppose this integral evaluates to some function $v(x)$, then the following must be true.

$$\frac{\mathrm{d}v}{\mathrm{d}x} = x\exp\left(\frac{-x^2}{2\sigma_x^2}\right) \tag{11.39}$$

Trying the function

$$v = -\sigma_x^2\exp\left[\frac{-x^2}{2\sigma_x^2}\right] \tag{11.40}$$

by differentiating to test Eq. (11.39) we get

$$\frac{\mathrm{d}v}{\mathrm{d}x} = (-\sigma_x^2)\left(-\frac{2x}{2\sigma_x^2}\right)\exp\left[\frac{-x^2}{2\sigma_x^2}\right] = x\exp\left[\frac{-x^2}{2\sigma_x^2}\right] \tag{11.41}$$

which is correct. Now considering the definite integral limits we get

$$\begin{aligned} I' &= \frac{1}{\sigma_x\sqrt{2\pi}}\int_a^b x\exp\left[\frac{-x^2}{2\sigma_x^2}\right]\mathrm{d}x \\ &= \frac{1}{\sigma_x\sqrt{2\pi}}\left\{-\sigma_x^2\exp\left[\frac{-x^2}{2\sigma_x^2}\right]\right\}_a^b \\ &= \frac{-\sigma_x}{\sqrt{2\pi}}\left\{\exp\left[\frac{-x^2}{2\sigma_x^2}\right]\right\}_a^b \end{aligned} \tag{11.42}$$

Therefore, integrals I_4 and I_5 may be written as follows:

$$I_4 = \frac{-\sigma_x^2}{L_A\sqrt{2\pi}}\left[\exp\left(\frac{-x^2}{2\sigma_x^2}\right)\right]_{-s}^{t} = \frac{-\sigma_x}{L_A\sqrt{2\pi}}\left[\exp\left(\frac{-t^2}{2\sigma_x^2}\right) - \exp\left(\frac{-(-s)^2}{2\sigma_x^2}\right)\right] \tag{11.43}$$

$$I_5 = \frac{-\sigma_x^2}{L_A\sqrt{2\pi}}\left[\exp\left(\frac{-x^2}{2\sigma_x^2}\right)\right]_{-t}^{s} = \frac{-\sigma_x}{L_A\sqrt{2\pi}}\left[\exp\left(\frac{-s^2}{2\sigma_x^2}\right) - \exp\left(\frac{-(-t)^2}{2\sigma_x^2}\right)\right] \tag{11.44}$$

Equations (11.35) through (11.37), (11.43), and (11.44) may be used to evaluate the fractional coverage in the range direction as defined by Eq. (11.32). The same approach may be taken for the corresponding deflection direction where the expected fractional coverage is

$$E(F_D) = \int_{y=-\infty}^{y=\infty} F_D(y)\,g(y)\mathrm{d}y \tag{11.45}$$

The applicable equations are

$$F_D = \frac{y}{W_A} \tag{11.46}$$

$$E(F_D) = \{I_1 + I_2 + I_3 + I_4 - I_5\} \tag{11.47}$$

$$I_1 = [\mathrm{ND}(-t) - \mathrm{ND}(t)] \tag{11.48}$$

$$I_2 = \frac{W_{\mathrm{EP}} + W_A}{2W_A}[\mathrm{ND}(t) - \mathrm{ND}(-s)] \tag{11.49}$$

$$I_3 = \frac{W_{\mathrm{EP}} + W_A}{2W_A}[\mathrm{ND}(s) - \mathrm{ND}(-t)] \tag{11.50}$$

$$I_4 = \frac{-\sigma_y}{W_A\sqrt{2\pi}}\left[\exp\left(\frac{-t^2}{2\sigma_y^2}\right) - \exp\left(\frac{-(-s)^2}{2\sigma_y^2}\right)\right] \tag{11.51}$$

$$I_5 = \frac{-\sigma_y}{W_A\sqrt{2\pi}}\left[\exp\left(\frac{-s^2}{2\sigma_y^2}\right) - \exp\left(\frac{-(-t)^2}{2\sigma_y^2}\right)\right] \tag{11.52}$$

where here

$$s = \frac{W_{\mathrm{EP}} + W_A}{2}, \qquad t = \frac{W_{\mathrm{EP}} - W_A}{2} \tag{11.53}$$

The total expected fractional coverage is obtained by multiplying the range and deflection values to get

$$E(F_C) = E(F_R) \times E(F_D) \tag{11.54}$$

The expected fractional damage representing the proportion of target elements damaged is obtained from Eq. (11.3)

$$\mathrm{FD}_1 = E(F_C) \times P_{\mathrm{CD}} \times R \tag{11.55}$$

where R is the weapon reliability and P_{CD} is the conditional probability of damage inside the rectangular damage function. Given FD_1, if the number of target elements in the area target is m, the number of elements damaged N_D is given by

$$N_D = \mathrm{FD}_1 \times m \tag{11.56}$$

Equation (11.55) gives the expected fractional damage due to a single sortie. If there are n independent sorties, the total fractional damage is given by

$$\mathrm{FD} = 1 - (1 - \mathrm{FD}_1)^n \tag{11.57}$$

Once again, the number of target elements damaged in n sorties is given by

$$N_D = \mathrm{FD} \times m \tag{11.58}$$

All the preceding analysis assumes the weapon lethal area is smaller than the area of targets. However, recall from section 11.4 that if the weapon is smaller than the target, it gets expanded to the target size and a reduced P_{CD} inside. This means that the preceding analysis is applicable whatever the relative size of weapon and target areas.

11.6 Spreadsheet to Compute FD_1

Table 11.1 is a spreadsheet for calculating FD_1 for an area of targets. It takes as input the area target dimensions, the lethal area dimensions and the weapon accuracy in terms of REP and DEP. The lethal area dimensions correspond to an $\mathrm{MAE}_F = 2000\ \mathrm{ft}^2$ and an impact angle of 65 degrees. It then computes L_{EP}, W_{EP}, P_{CD}, and the expected fractional coverage of the target by the lethal area, and as discussed previously is independent of their relative size. It shows the inputs needed for the calculation, and the final result computed is the product of the expected fractional coverage in range and deflection $E(F_R)$ $E(F_D)$ used in Eq. (11.55) above, assuming weapon reliability R is unity.

11.7 Weaponeering Spreadsheet

Now that the methodology for dealing with single unguided weapons and multiple guided weapons directed against an area of targets has been dealt

Table 11.1 Computation of FD_1.

FD₁ CALCULATOR			
RANGE		***DEFLECTION***	
INPUT PARAMETERS		***INPUT PARAMETERS***	
Target length LA (ft)	100.000	Target width WA (ft)	150.000
Weapon length LET(ft)	54.800	Weapon width WET(ft)	91.300
REP	10.000	DEP	0.010
LEP	54.800	WEP	91.300
Sigma range (ft)	14.826	Sigma deflection (ft)	0.015
INTEGRAL LIMITS		***INTEGRAL LIMITS***	
s	77.400	s	120.650
t	-22.600	t	-29.350
INTEGRALS		***INTEGRALS***	
integral #1	0.478	integral #1	0.609
integral #2	0.049	integral #2	0.000
integral #3	0.049	integral #3	0.000
integral #4	-0.018	integral #4	0.000
integral #5	0.018	integral #5	0.000
RESULTS			
Fractional coverage	0.540		0.609
Total fractional coverage	0.329		
FD₁	**0.3286**		

with, a simple spreadsheet will be introduced to enable rapid computation of FD_1. First however we summarize how the process used in the spreadsheet to compute FD_1 is implemented for fragmenting and blast warheads.

The analysis described so far represents the weapon lethal area as a rectangle, so it may be applied to fragment warhead weapons using a rectangular cookie cutter damage function weapons and to a blast warhead weapons using a square cookie cutter damage function. Each type of weapon will be dealt with separately, using the basic equations developed in this, and previous sections.

Fragmentation Weapons

The assumed inputs are: REP, DEP, MAE_F, L_A, W_A, and impact angle I. From these inputs, the following computations are made:

1. The parameter a is computed from Eq. (9.5)

$$a = \text{MAX}[1 - 0.8\cos(I),\ 0.3] \tag{11.59}$$

2. The value of L_{ET} is computed from Eq. (9.6)

$$L_{\mathrm{ET}} = \sqrt{\mathrm{MAE}_F \times a} \tag{11.60}$$

3. The value of W_{ET} is also computed from Eq. (9.7)

$$W_{\mathrm{ET}} = \frac{L_{\mathrm{ET}}}{a} \tag{11.61}$$

4. The expected fractional coverage in range and deflection is computed from Eqs. (11.32) and (11.47).
5. FD_1 is computed from Eq. (11.55)

Blast Weapons

The assumed inputs are: REP, DEP, MAE_B, L_A, and W_A. From these inputs, the following calculations are made:

1. L_{ET} and W_{ET} are calculated from Eq. (8.64)

$$L_{\mathrm{ET}} = W_{\mathrm{ET}} = \sqrt{\mathrm{MAE}_B} \tag{11.62}$$

2. The standard deviations in the range and deflection directions are calculated from

$$\sigma_x = \frac{\mathrm{REP}}{0.6745} \qquad \sigma_y = \frac{\mathrm{DEP}}{0.6745} \tag{11.63}$$

3. The expected fractional coverage in range and deflection is computed from Eqs. (11.32) and (11.47).
4. FD_1 is computed from Eq. (11.55)

The spreadsheet shown in Table 11.2 is the implementation of the weaponeering methodology for unitary unguided weapons against area targets, and follows the template described in the previous chapter. Most of the sections are similar to those in the previous chapter, but the FD_1 calculation is unique to this scenario (as shown in Table 11.1), including the ability to input a blast MAE. As before, the shaded cells indicate where user input is required.

11.8 Calculating FD_1 For Guided Weapons

The calculation of FD_1 for guided weapons is similar to that described in the previous chapter for unitary targets, that is, using Eq. (10.52) adapted for area targets.

$$\mathrm{FD}_1 = [\mathrm{FD1}_1 \times P_{\mathrm{NM}} + \mathrm{FD1}_2 \times P_{\mathrm{HIT}}] \tag{11.64}$$

In this equation, $\mathrm{FD1}_1$ is the value for FD_1 computed using the unguided methodology described in the first part of this chapter, that is, assuming a Gaussian miss distance characterized by REP and DEP, or CEP. $\mathrm{FD1}_2$ is

Table 11.2 Weaponeering Spreadsheet for Single Weapons Against an Area of Targets.

METHOD 2-U SINGLE WEAPON AGAINST AREA TARGET						
CONSTANTS						
Gravitational acceleration ft/s	32.200					
Convert deg to rad	0.017					
TARGET PARAMETERS						
Target area dimensions	LA	100.000				
	WA	150.000				
WEAPON PARAMETERS						
Drag: L = low, H = high		l				
Ballistic dispersion (mils)	Sigma-*b*	0.000				
Reliability	*R*	1.000				
AIRCRAFT RELEASE DATA						
Airspeed KTAS, ft/s	*V*	450.000		759.600		
Dive angle, deg	theta	0.000		0.000		
Release altitude, ft	*yi*	20000.000				
TRAJECTORY RESULTS						
Vertical velocity, ft/s	*Vv*	0.000				
Horizontal velocity, ft/s	*Vh*	759.600				
Drag constant, Cd/M	*c*0	0.005				
constant *a* used in trajectory	*a*	6440.000				
		36.312		**Dragless**		
time of flight, s	*t_flight*	36.312		35.25		
Horizontal velocity at impact	*Vhi*	633.483		759.60		
Vertical velocity at impact	*Vvi*	1069.236		1134.90		
Impact angle, deg	*I*	59.355		56.21		
Slant range, first weapon, ft	*Sri*	32190		33418		
Ground range, first weapon, ft	*Gri*	25223		26772		
DELIVERY ACCURACY						
		Ground ft	**Normal ft**	**Mils**	**Spec.**	**REP/DEP**
CEP, ft, ground, normal, or mils	CEP	0.000		0		
Deflection error probable, ft	DEP	0.00	0.00	0.00	30.00	30.00
Range error probable, ft	REP	0.00	0.00	0.00	20.00	20.00
Offset aimpoint, range, ft	*R*0	0.000				
Offset aimpoint, deflection, ft	*D*0	0.000				
Ballistic error, deflection, ft	sigma-*bd*	0.000		0		
Ballistic error range, ft	sigma-*br*	0.000				
Adjusted DEP	DEP_dash	30.00				
Adjusted REP	REP_dash	20.00				
WEAPON EFFECTIVENESS						
Effectiveness index type			**MAE_b**	**MAE_f**		
Effectiveness value	*EI*		0.000	5000		
Length/width ratio (frag only)	*a*	0.592				WET/LET
Effective target width, ft	WET		0.00	91.88		91.88
Effective target length, ft	LET		0.00	54.42		54.42
Effective target area, ft^2	AET		0	5000		5000
Cond. probability of damage	PCD	1.000				

(*Continued*)

Table 11.2 Weaponeering Spreadsheet for Single Weapons Against an Area of Targets (*Continued*).

***FD*₁ CALCULATION**						
	Range			**Deflection**		
Sigma range, ft	29.652			44.477		
LEP	54.416		WEP	91.884		
INTEGRAL LIMITS						
s	77.208		*s*	120.942		
t	–22.792		*t*	–29.058		
INTEGRALS						
integral #1	0.304		integral #1	0.298		
integral #2	0.167		integral #2	0.204		
integral #3	0.598		integral #3	0.204		
integral #4	–0.084		integral #4	–0.093		
Integral #5	0.084		Integral #5	0.093		
RESULTS						
Fractional coverage	0.900			0.522		
Combined fractional coverage	0.470					
FD₁	**0.4696**					
FORCE REQUIREMENTS						
Desired FD	0.500					
# of sorties for this FD	1.093					
# sorties available NA	4.000					
Expected FD from NA sorties	0.921					

the same as FD1$_1$ except that a value of CEP $= 0$ is used (or REP $=$ DEP $= 0$) to simulate a weapon that always hits the target. Again it is seen that Eq. (11.64) is indeed a weighted combination of the pure Gaussian and direct hit FD$_1$ values.

11.9 Weapon Represented by Carlton Damage Function

Recall that early in this chapter, mention was made of an alternative way to calculate FD$_1$ values that preserves the Carlton damage function representing the weapon. This process is described now. Figure 11.3 showed a single round aimed at the center of the target ($x = 0$) but impacting the ground plane at $x = u$ from this point, as illustrated in Fig. 11.23 for the range direction.

The figure shows the Carlton damage function superimposed on the location where the single round lands. The question is: What damage is done to the target? For a weapon landing at the center of the target, the one dimensional Carlton damage function $c(x)$ can be written as

$$c(x) = D_0 \exp\left[-D_0 \frac{x^2}{R_x^2}\right] \tag{11.65}$$

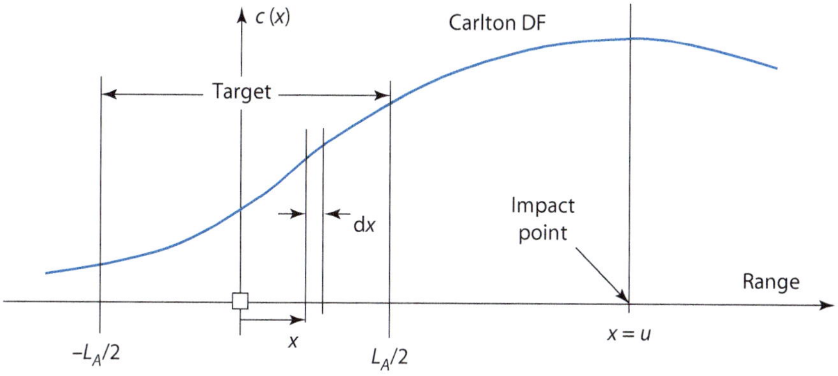

Fig. 11.21 Damage over the target extent for one round.

This is similar to the damage function discussed before* except for the additional term D_0 (≤ 1), which can be called the shaping factor. When $D_0 = 1$ we have the familiar form of the Carlton damage function, but for smaller values the damage function has a smaller peak value and is more spread out. The quantity R_X is the weapon radius previously denoted by WR_r in Eq. (8.38).

We may also write that the damage done to a target element at any value of x for a weapon landing a distance u from the center of the target can be determined from

$$c(x-u) = D_0 \exp\left[-D_0 \frac{(x-u)^2}{R_x^2}\right] \tag{11.66}$$

Consider a target element a distance x from the origin, width dx. We want to calculate a mean value of the damage function $\bar{c}$ over the whole target for this location of the round. This is obtained from the following equation.

$$\bar{c}(u) \times L_A = \int_{x=-L_A/2}^{x=L_A/2} c(x-u)\,\mathrm{d}x \tag{11.67}$$

For a large number of independent rounds, the mean value of the damage done to the target would be the expectation of $\bar{c}$. This requires knowledge of where the round is likely to fall, that is, what are the statistics of u? Assuming u to be normally distributed with a PDF $= g(u)$ in the range direction, we

*The Carlton damage function used in Eq. (8.38) had $D_0 = 1$, typical of AS methods, whereas SS methods have historically required different values of D_0 depending on the munition and weapon. For more information see Chapter 17.

obtain the following result.

$$\mathrm{EFD}_x = E[c(u)] = \int_{u=-\infty}^{u=+\infty} \bar{c}(u)g(u)\,du = \frac{1}{L_A}\int_{u=-\infty}^{u=+\infty}\left[\int_{x=-L_A/2}^{x=L_A/2} c(x-u)\mathrm{d}x\right] g(u)\,du \tag{11.68}$$

It is known that the expected value of PD_{1x} was confirmed by performing a Monte Carlo simulation. We will use the same association to solve Eq. (11.68). This process consists of the following:

1. Generating a value of u from a normal distribution characterized by zero mean and a standard deviation comprising the aiming error.
2. Given the target size, calculating $\bar{c}$ from Eq. (11.67).
3. Averaging all of the $\bar{c}$ values for the specified number of trials to obtain the expected value, corresponding to $\mathrm{FD1}_X$ from Eq. (11.68).

The second step may be simplified by substituting the Carlton damage function to obtain

$$\bar{c} = \frac{1}{L_A}\int_{x=-L_A/2}^{x=L_A/2} D_0 \exp\left[-D_0\frac{(x-u)^2}{R_x^2}\right]\mathrm{d}x \tag{11.69}$$

This may be written in the form

$$\bar{c} = \frac{D_0}{L_A}\int_{x=a}^{x=b} \exp\left[\frac{(x-\mu)^2}{2\sigma^2}\right]\mathrm{d}x \tag{11.70}$$

where

$$\mu = u \quad \text{and} \quad \sigma = \frac{R_x}{\sqrt{2D_0}} \tag{11.71}$$

Using the cumulative distribution function $F(z)$ of the normal distribution, enables Eq. (11.70) to be written as

$$\begin{aligned}\bar{c} = \frac{D_0}{L_A}\int_{x=a}^{x=b} \exp\left[-\frac{(x-\mu)^2}{2\sigma^2}\right]\mathrm{d}x &= \frac{D_0\sigma\sqrt{2\pi}}{L_A}\left[\frac{1}{\sigma\sqrt{2\pi}}\int_{x=a}^{x=b}\exp\left[-\frac{(x-\mu)^2}{2\sigma^2}\right]\mathrm{d}x\right] \\ &= \frac{D_0\sigma\sqrt{2\pi}}{L_A}[F(b)-F(a)] \\ &= \frac{R_x}{L_A}\sqrt{\frac{\pi}{D_0}}[F(b)-F(a)]\end{aligned} \tag{11.72}$$

Example 11.1

We will consider the case where $L_{ET} = 15$ ft, REP $= 20$ ft and L_A is a variable. We convert from the AS notation to the SS Carlton damage function notation using

$$R_x = \mathrm{WR}_R = L_{ET}/2 \tag{11.73}$$

and assume the shaping factor $D_0 = 1$. Coding the Monte Carlo simulation in MATLAB and running for 10^6 samples of u gives the results shown in Table 11.3 for different values of L_A.

In this table, the column labeled "FD_{1x} Carlton" is the result obtained from the analysis just discussed. In order to provide a check, the integral shown in Eq. (11.68) was numerically integrated using MATLAB, and resulted in the same values. Note that as the target size gets small (i.e., approaches a point), the FD_1 approaches the PD_1 value. Included for comparison is the column "FD_{1x} Rectangle," where the Carlton damage function is replaced by a rectangular MAE and Eq. (11.55) is used to calculate FD_1, and indicates the effect of simplifying the damage function.

It is also possible (ref [1]) to evaluate Eq. (11.68) using an Excel spreadsheet rather than numerical integration [1]. Changing the order of integration gives the following:

$$\begin{aligned}\mathrm{EFD}_x &= \frac{1}{L_A}\int_{u=-\infty}^{u=+\infty}\left[\int_{x=-L_A/2}^{x=L_A/2} c(x-u)\mathrm{d}x\right] g(u)\,\mathrm{d}u\\ &= \frac{1}{L_A}\int_{x=-L_A/2}^{x=L_A/2}\left[\int_{u=-\infty}^{u=\infty} c(x-u)g(u)\mathrm{d}u\right]\mathrm{d}x\end{aligned} \tag{11.74}$$

Table 11.3 Effectiveness Calculations.

L_A (feet)	FD_{1x} Rectangle	FD_{1x} Carlton
200	0.066	0.075
100	0.120	0.136
50	0.157	0.179
20	0.171	0.196
10	0.175	0.199
5	0.176	0.200
1	0.176	0.200
0.1	0.176	0.200

(Continued)

Example 11.1 (Continued)

The integral in the square bracket may be written as

$$\int_{u=-\infty}^{u=\infty} c(x-u)g(u)du = \frac{D_0}{\sigma_x\sqrt{2\pi}}\int_{-\infty}^{\infty}\exp\left[-D_0\frac{(x-\mu_x)^2}{R_x^2}\right]\exp\left[-\frac{u^2}{2\sigma_x^2}\right]du$$

$$= \frac{D_0}{\sigma_x\sqrt{2\pi}}\int_{-\infty}^{\infty}\exp\left[-\frac{\gamma^2}{2\beta^2\sigma_x^2}\left(u-\frac{\sigma_x^2}{\gamma^2}x\right)^2-\frac{x^2}{2\gamma^2}\right]du \tag{11.75}$$

where the constants β and γ are given by the following:

$$\beta^2 = \frac{R_x^2}{2D_0} \tag{11.76}$$

$$\gamma^2 = \sigma_x^2 + \beta^2 \tag{11.77}$$

Because the second term of the exponent is independent of u, it can be removed from the integration which then allows the integration, to be calculated to give the following:

$$\int_{u=-\infty}^{u=\infty} c(x-u)g(u)du = \frac{D_0}{\sigma_x\sqrt{2\pi}}\int_{-\infty}^{\infty}\exp\left[-\frac{\gamma^2}{2\beta^2\sigma_x^2}\left(u-\frac{\sigma_x^2}{\gamma^2}x\right)^2-\frac{x^2}{2\gamma^2}\right]du$$

$$= \frac{D_0}{\sigma_x\sqrt{2\pi}}\exp\left[\frac{x^2}{2\gamma^2}\right]\int_{-\infty}^{\infty}\exp\left[-\frac{\gamma^2}{2\beta^2\sigma_x^2}\left(u-\frac{\sigma_x^2}{\gamma^2}x\right)^2\right]du$$

$$= \frac{D_0}{\sigma_x\sqrt{2\pi}}\exp\left[\frac{x^2}{2\gamma^2}\right]\frac{\beta\sigma_x}{\gamma}\sqrt{2\pi} \tag{11.78}$$

Substituting this result and Eq. (11.76) into Eq. (11.74) gives

$$\text{EFD}_x = \frac{R_x\sqrt{\pi D_0}}{L_A}\times\frac{1}{\gamma\sqrt{2\pi}}\int_{-L_A/2}^{L_A/2}\exp\left[-\frac{x^2}{2\gamma^2}\right]dx \tag{11.79}$$

In this equation, the second term is just the integral of a Gaussian distribution that can be evaluated using the NORMDIST (ND) function of Excel as follows:

$$\text{EFD}_x = \frac{R_x\sqrt{\pi D_0}}{L_A}\left[\text{ND}\left(\frac{L_A}{2},0,\gamma,1\right)-\text{ND}\left(\frac{-L_A}{2},0,\gamma,1\right)\right] \tag{11.80}$$

It can be shown that Eq. (11.80) produces the same results as those shown in the last column of Table 11.3.

Reference

[1] Moon, S.-H., private communication, Korean Agency for Defense Development, July 2010.

Chapter 12 Stick Deliveries

12.1 Introduction

In many cases, aircraft will not deliver a single unguided weapon on the target, but will release multiple weapons in the form of a stick with the intent to have a larger lethal area on the ground in which damage can occur. An example is shown in Fig. 12.1.

A stick release involves more than one weapon being released—not at the same instant, but over a short period of time with a specified delay between them. The area on the ground covered by the weapons is called the *pattern*, and because sticks are usually used against area targets, the effectiveness measure is fractional damage FD. Examples of stick patterns are shown in Fig. 12.2, where the shaded areas are individual weapon lethal areas.

The basic approach in dealing with stick deliveries is as follows:

1. Assume the pattern on the ground is rectangular in shape.
2. Determine the dimensions of the rectangle from the release conditions and weapon trajectory.
3. Calculate the probability of damage, averaged within the pattern based on the density of weapons.

Fig. 12.1 Stick delivery.

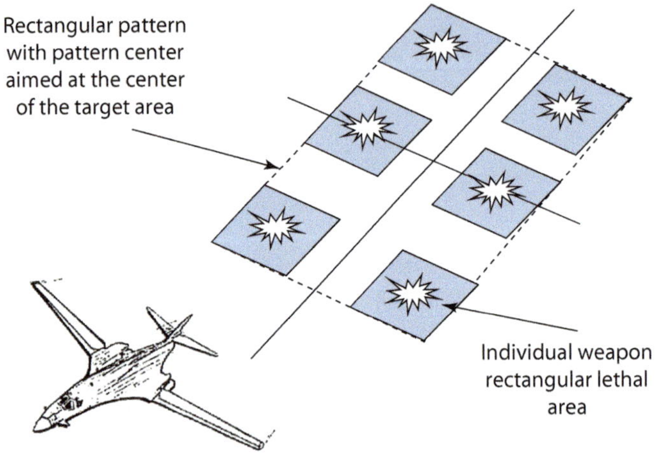

Fig. 12.2 Individual weapon impacts.

4. For an area of targets, calculate FD with Eq. (11.55) by considering the stick to be equivalent to a single effectiveness area with the characteristics defined previously.

Individual weapon impacts on the ground are determined by the number of weapons dropped simultaneously and the number of simultaneous events, as shown in Fig. 12.2.

This collection of individual weapon lethal areas is then aggregated into a single rectangle with a probability of damage P_{CDS} inside it, as in Fig. 12.3. The stick is then treated as a single weapon represented by a rectangular lethal area of known dimensions and a probability of damage inside it. The

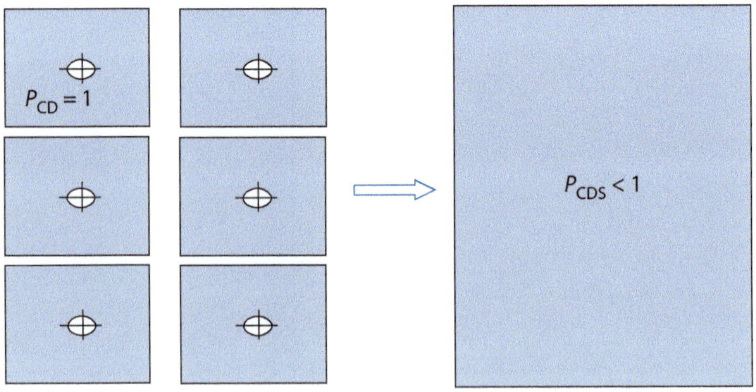

Fig. 12.3 Combining individual lethal areas into a pattern.

main task, therefore, is determining the pattern size and its associated conditional probability of damage P_{CDS}.

12.2 Determining the Pattern Dimensions

In the range direction, the pattern dimension is determined by the intervalometer setting. An intervalometer is a timer that sends electrical pulses to the bomb racks to release the weapons. For each pulse, one or more weapons may be released. The pilot is able to set the intervalometer to fixed values; for example, the A-10 intervalometer may be set between 0.010 and 0.999 s in 0.01-s steps.

Aircraft with more advanced avionics allow the pilot to enter a desired stick length on the ground, and sensors provide data regarding the aircraft state in order to calculate the correct intervalometer delays. The number of intervalometer pulses is denoted by the variable n_r, and the number of weapons released for each pulse is n_p. Figure 12.4 illustrates some weapon lethal area patterns for various n_r and n_p values.

If in Fig. 12.4 the shaded rectangles represent the lethal area of individual weapons, then the bounding rectangle represents the pattern, defined as the smallest rectangle to enclose all the weapons in a stick. Note that individual weapons may or may not overlap in range or deflection, or in one but not the other. This effect will have to be considered in determining the effect of overlapping or a sparse weapon distribution in the pattern. At this point we differentiate between the pattern and stick dimensions as shown in Fig. 12.5.

The stick dimensions L_s, W_s are measured from the centers of the extreme individual weapon effects areas, whereas the pattern is the rectangle L_P, W_P that encloses all the weapons in a stick. The first step in determining the pattern dimensions, therefore, is to determine the stick dimensions L_S and W_S.

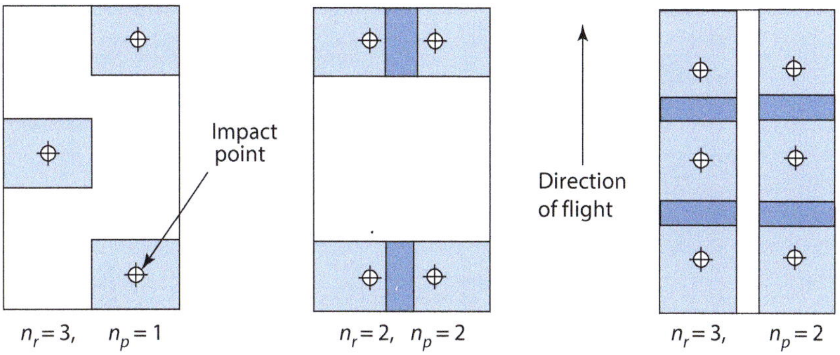

Fig. 12.4 Sample stick patterns for different n_r and n_p.

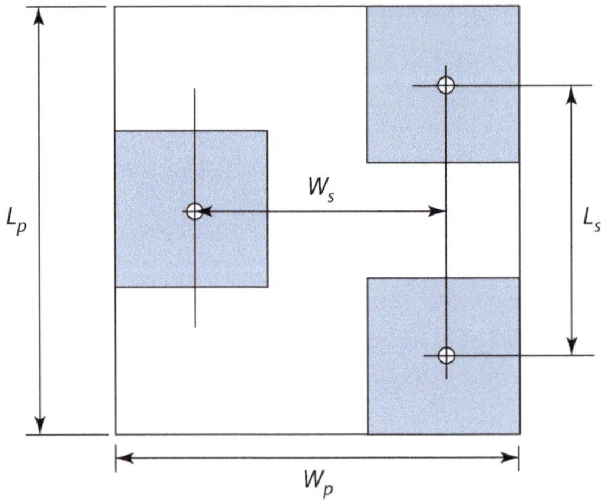

Fig. 12.5 Stick and pattern dimensions.

12.3 Calculating the Stick Width

Bombs may be carried on hard points on the aircraft wings and fuselage, or internally for larger aircraft. In addition, bomb racks capable of holding more than one weapon can be attached to the hard points to increase the payload of the aircraft. Figure 12.6 shows a multiple ejection rack (MER) bomb rack capable of holding six weapons on a single aircraft hard point.

Fig. 12.6 Multiple ejection rack (MER).

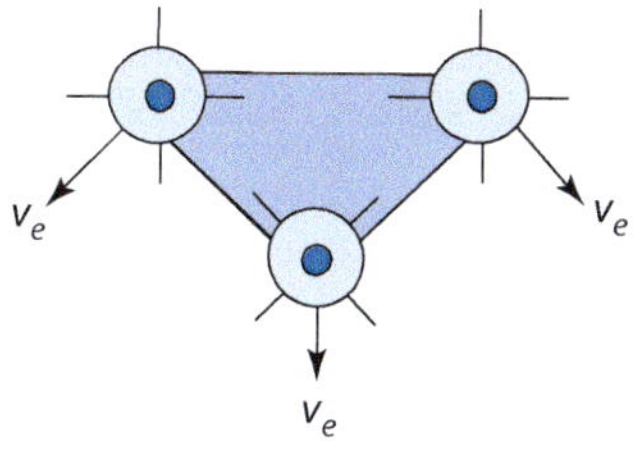

Fig. 12.7 Bomb ejection from a MER/TER.

A triple ejection rack (TER) carries three weapons at a single station, whereas a MER consists of two TER in tandem. Each weapon is ejected in the manner indicated in Fig. 12.7.

This figure shows that the weapons will spread in the deflection direction when ejected, and will therefore influence the pattern width on the ground. A main rack release is defined as one where the weapons drop vertically from the aircraft attachment points; therefore, the stick width is set to the intersta-tion distance on the aircraft. As shown in Fig. 12.8, this may vary considerably for each aircraft.

Because the stick width depends on each airframe, data for particular air-craft must be obtained, as indicated in Table 12.1.

For MER releases where a lateral velocity is present, the stick width is computed by determining the value of this velocity, as shown in Fig. 12.9.

In Fig. 12.9, μ is the horizontal angle between the aircraft and weapon vel-ocities, v_e is the ejection velocity, and v_a is the aircraft velocity. The angle φ is the angle between the fuselage and ejection velocity vector (usually 45 deg), and θ is the dive angle. From Fig. 12.9 it may be seen that

$$\tan \mu = \frac{v_e \sin \varphi}{v_a \cos \theta} \tag{12.1}$$

Figure 12.9 also shows the relationship between the ground range of the weapon R_B and the distance d equal to half the stick width. This figure

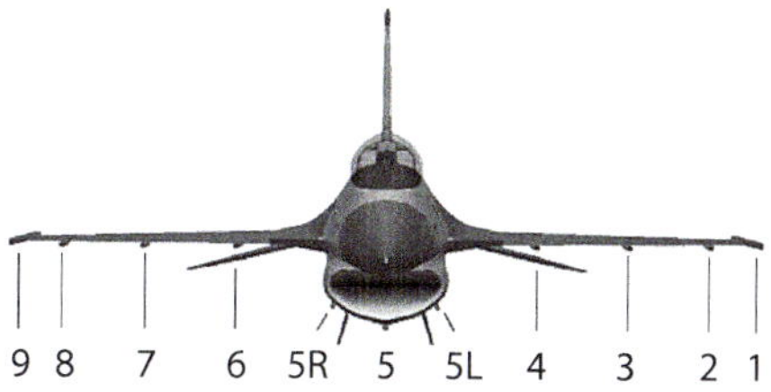

Fig. 12.8 Weapon station locations.

Table 12.1 Distance Between Stations (ft)

Aircraft	Centerline	Inboard	Intermediate	Outboard
A-10	0, 3.8	12	None	20
AV-8B	0	7.2	9.3	11.1
F-14	0, 4	None	None	None
F-15	0	None	None	20
F-16	0	12	None	20
B-52	None	None	None	36

shows that

$$d = R_B \tan \mu = \frac{R_B v_e \sin \varphi}{v_a \cos \theta} \tag{12.2}$$

The stick width therefore is given by

$$W_s = 2 \times d \tag{12.3}$$

Equations (12.2) and (12.3) are based on a zero drag approximation to the weapon ballistics in the deflection direction. In most cases we may assume that the ejection angle is 45 deg, resulting in

$$W_s = \frac{1.414 R_B \, v_e}{v_a \cos \theta} \tag{12.4}$$

The bomb fall range R_B is obtained from the weapon ballistics. Alternatively, we could use a zero drag ballistic model allowing the use of the time of flight, giving the stick width as

$$W_s = 2d = 2v_e \sin \varphi \times \text{TOF} \tag{12.5}$$

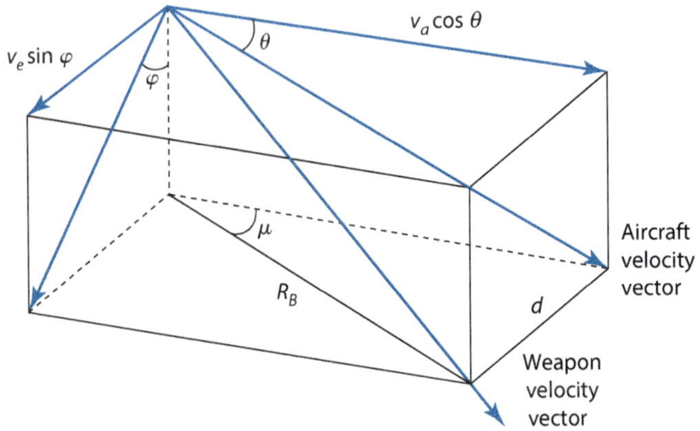

Fig. 12.9 Variables needed to compute stick width.

12.4 Calculating Stick Length

Whereas the stick width (and hence the pattern width) may be estimated from the locations of the hard points on the airframe or by Eq. (12.4), the stick length (and hence the pattern length) is calculated from the aircraft release conditions and the intervalometer settings.

For level deliveries, the problem becomes somewhat simplified because the stick length between the last release pulse and the first one is the same distance the aircraft travels horizontally, hence

$$L_s = v_a(n_r - 1)\Delta t \tag{12.6}$$

In this equation, v_a is the aircraft speed in the horizontal direction, and Δt is the intervalometer setting. This equation is based on the assumption that all weapon trajectories are identical, resulting in a uniform distribution of impact points in the ground plane.

For the dive case, however, these trajectories will not be identical, and a complete solution would involve computing the trajectory of all weapons released to obtain not just the stick length, but also the distribution of weapon impact points within the stick. The approach taken here is one in which the trajectory of the first and last weapons is calculated, and the stick length is calculated as shown in Fig. 12.10, which shows a typical dive release of a stick of bombs.

Recall Chapter 3, which discussed that in order to implement the trajectory calculation, we need to know the velocity, altitude, and dive angle of the aircraft at release. If the altitude of the first weapon released is y_i, the altitude

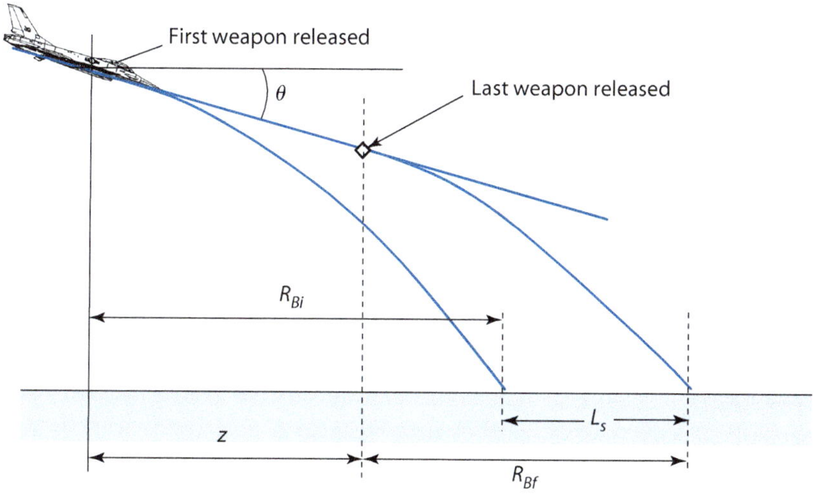

Fig. 12.10 Geometry of stick release in the range direction.

of the final weapon in the stick at release is given by

$$y_f = y_i - v_a(n_r - 1)\Delta t \sin\theta \tag{12.7}$$

where θ is the dive angle. Assuming the weapon velocity at release is the same as the aircraft, this enables the trajectory to be calculated, giving the horizontal ranges from the points of release (R_{Bi} and R_{Bf} in Fig. 12.10). It can be seen from Fig. 12.10 that

$$R_{Bi} + L_s = z + R_{Bf} \tag{12.8}$$

so

$$z = v_a(n_r - 1)\Delta t \cos\theta \tag{12.9}$$

We then get for the stick length

$$L_s = z + R_{Bf} - R_{Bi} = v_a(n_r - 1)\Delta t \cos\theta + R_{Bf} - R_{Bi} \tag{12.10}$$

Even if the intervalometer generates pulses at fixed intervals, the impact points of weapons in the range direction will not be equally spaced in the ground plane unless the aircraft is flying horizontally. Nevertheless, impact points in the ground plane are assumed to be uniformly distributed between the first and last weapon impact point for all dive angles.

This concludes the computations for the stick length and width. It remains to compute the size of the individual weapon lethal areas (shaded areas) in Fig. 12.5.

12.5 Precision Errors for Stick Deliveries

Recall Fig. 4.5 in Chapter 4, which showed there were two delivery errors, aiming and precision. This is now explained further by considering the process the pilot executes for a stick delivery. The intervalometer is set for the correct weapon spacing, and the pilot maneuvers towards the target. After selecting the mechanization (Continuously computed impact point (CCIP) or contimuously computed release point (CCRP)), the weapons are released at the ballistic point where the center of the stick is aimed at the desired mean point of impact (DMPI). Any aiming errors present at the point of weapon release apply to the stick as a whole, not to the individual weapons in the stick, because their spatial relationship is governed only by the intervalometer setting and stick width. Any aiming error moves the whole pattern relative to the aimpoint; however, the actual dimensions of the pattern will change slightly between successive deliveries (occasions) due to the precision error, as shown in Fig. 12.11.

Recall also Chapter 9, Section 9.8, where it was argued that the dependent errors and independent errors have to be kept separate; otherwise, incorrect PD will result. Because we use the rectangular damage function to represent individual weapons in the stick, the methods outlined in Chapter 10,

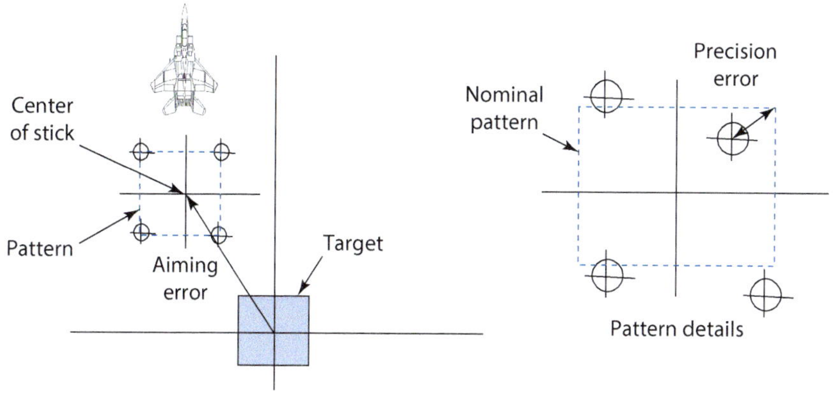

Fig. 12.11 Aiming error and precision error for sticks.

Section 10.10 will apply. This method expands the individual weapon lethal area by a multiple of the precision error standard deviation, as shown in Fig. 12.12.

As far as the value of P_{CD1} is concerned, we can preserve the lethality of the weapon by writing

$$P_{\mathrm{CD1}} = P_{\mathrm{CD}} \frac{L_{\mathrm{ET}} W_{\mathrm{ET}}}{L_B W_B} \tag{12.11}$$

Regarding the size of the enlarged lethal area, Section 10.10 suggested the following equations:

$$L_B = L_{\mathrm{ET}} + k \times \sigma_{br} \tag{12.12}$$

$$W_B = W_{\mathrm{ET}} + k \times \sigma_{bd} \tag{12.13}$$

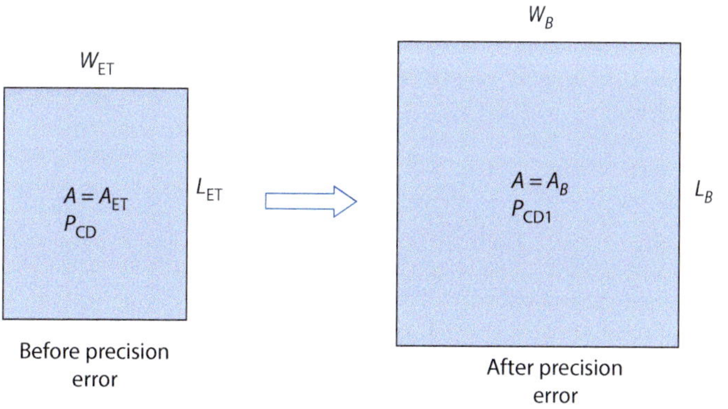

Fig. 12.12 Expanding the lethal area for precision error.

In the JMEM methodology for stick deliveries, however, a different approach is taken for determining the rectangle size, based on the Carlton damage function, even though we are enlarging a cookie cutter. We consider the Gaussian form of the damage function, which, for simplicity will be considered only in the range direction.

$$P(x) = \exp - \left[\frac{x^2}{WR_r^2} \right] \tag{12.14}$$

Similarly, considering the precision error to be also Gaussian in the range direction gives the following PDF:

$$G(x) = \frac{1}{\sigma_{br}\sqrt{2\pi}} \exp - \left[\frac{x^2}{2\sigma_{br}^2} \right] \tag{12.15}$$

where σ_{br} is the standard deviation of the precision error in the range direction. Combining these two Gaussian elements will result in a more diffuse damage function that takes the following form:

$$\begin{aligned} P(x, y) &= \frac{1}{\sigma_{br}\sqrt{2\pi}} \exp - \left[\frac{x^2}{\sqrt{WR_r^2 + 2\sigma_{br}^2}} \right] \\ &= \frac{1}{\sigma_{br}\sqrt{2\pi}} \exp - \left[\frac{4x^2}{\sqrt{(L'_{\mathrm{ET}})^2 + 8\sigma_{br}^2}} \right] \end{aligned} \tag{12.16}$$

If the lethal area is now considered to be rectangular rather than Gaussian, a similar expansion may be taken to determine the size of the enclosing rectangle shown in Fig. 10.8.

$$L_B = \sqrt{L_{\mathrm{ET}}^2 + 8\sigma_{br}^2} \tag{12.17}$$

$$W_B = \sqrt{W_{\mathrm{ET}}^2 + 8\sigma_{bd}^2} \tag{12.18}$$

Recall that L_{ET} and W_{ET} are obtained from the lethal area from the following equations:

$$L_{\mathrm{ET}} = \sqrt{\mathrm{MAE}_F \times a} \tag{12.19}$$

$$W_{\mathrm{ET}} = \frac{L_{\mathrm{ET}}}{a} \tag{12.20}$$

Again, in these equations, σ_{br} and σ_{bd} are the precision error standard deviations in feet in the ground plane; these are usually obtained from the specification of the normal plane precision error σ_b in mils by using Eqs. (4.47) and (4.48) without the factor 0.6745. In these equations, the slant range used is from the center of the stick to the aircraft at the average release altitude. Thus, the dimensions of the modified lethal area and the conditional probability of damage for a single weapon in the stick shown in Fig. 12.12 have been determined, and the effect of precision error is included.

12.6 Pattern Dimensions

Figure 12.5 defines the pattern dimensions in terms of the stick dimensions and what was referred to as the individual weapon lethal area. It may have been thought that this area was the effective target area A_{ET}, but it is not. It is, in fact, the A_{ET} modified by the precision error, as outlined in the previous section, the dimensions of which are L_B and W_B. Figure 12.13 modifies Fig. 12.5 by adding these dimensions.

From this figure, the pattern dimensions become

$$L_P = L_S + L_B \tag{12.21}$$

$$W_P = W_S + W_B \tag{12.22}$$

12.7 Weapon Sparsity and Overlap in the Pattern

We now turn our attention to calculating a uniform probability of damage inside the pattern defined in Fig. 12.13. Observe from Fig. 12.4 that the weapons in the stick may overlap, making the damage function of the stick more than the damage function from a single weapon, or the weapons may be sparsely distributed, causing the damage function to be less than the combined damage functions for the individual weapons within the pattern.

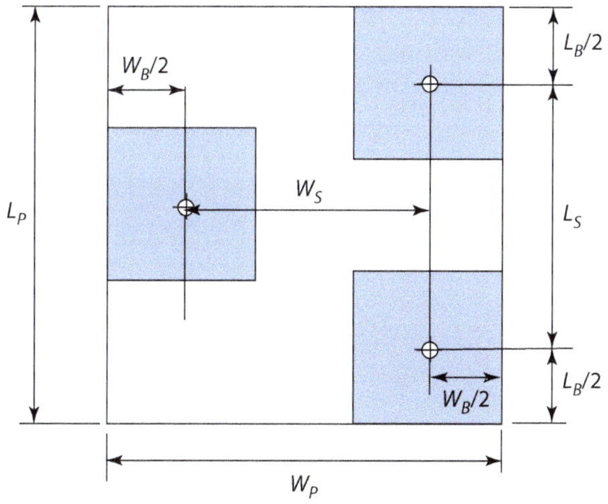

Fig. 12.13 Pattern dimensions from stick dimensions.

A method has to be developed to allow for the overlap or sparsity of the weapons in the pattern and therefore arrive at a uniform lethality over the whole of the pattern. This method will now be outlined by dealing with the deflection direction and range directions separately, because overlap may or may not occur in each of these directions independently. Remember that the pattern dimensions are fixed, and we have to fit individual weapon lethal areas (expanded to allow for precision error) into it.

12.7.1 Possible Overlap in Deflection Direction

Consider the case of $n_p = 2$, where two weapons are released for each release pulse. Figure 12.14 shows how the weapons may or may not overlap in the stick. Whether this happens or not depends only on the width of the expanded lethal area and the pattern width.

For the overlapping case, we define the degree of overlap in the deflection direction n_{od} to be

$$n_{od} = \frac{n_p W_B}{W_P} \tag{12.23}$$

which may be thought of as how much overlap of the n_p weapon lethal areas has to occur in order to fit into the fixed pattern width. The conditional probability of damage in the deflection direction $P_{CD/d}$ is obtained by powering up the single weapon damage function P_{CD} by the degree of overlap.

$$P_{CD/d} = 1 - (1 - P_{CD1})^{n_{od}} \tag{12.24}$$

A noninteger powering up index may seem counterintuitive; however, consider the case where the two weapon lethal area are just touching, as shown in Fig. 12.15.

From Eq. (12.23) we see that $n_{od} = 1$ and that Eq. (12.24) gives $P_{CD/d} = P_{CD1}$, which seems intuitive. Consider now the two weapon lethal areas lying exactly on top of each other, as shown in Fig. 12.16.

Now from Eq. (12.23) we see that $n_{od} = 2$ and that Eq. (12.24) gives a value of $P_{CD/d}$ from the powering up formula. Both of these cases have an integer value of the degree of overlap, so it now seems reasonable that for

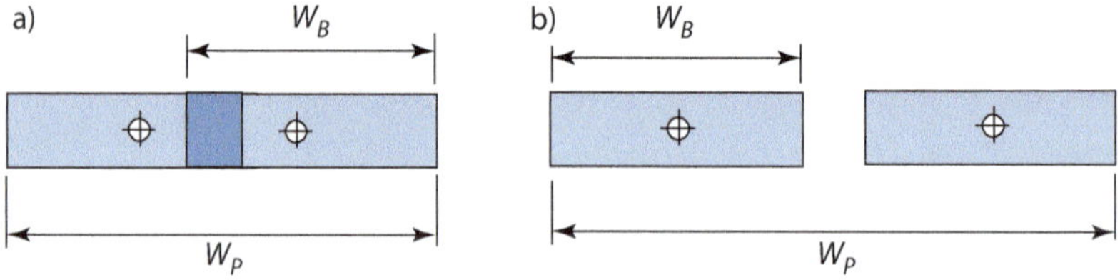

Fig. 12.14 Weapons overlapping (a) and not overlapping (b) in the deflection direction.

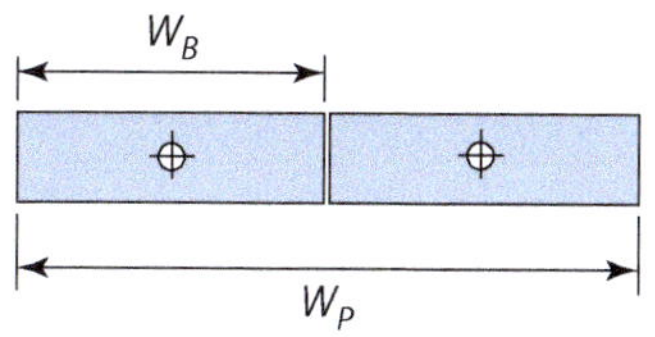

Fig. 12.15 Weapon lethal areas abutting.

a partial overlap, the powering up index should be between 1 and 2, as indicated in Eq. (12.24).

If the individual weapons do not overlap at all ($n_{od} < 1$), then we can write

$$P_{\mathrm{CD}/d} = n_p P_{\mathrm{CD1}} \frac{W_B}{W_P} \tag{12.25}$$

Here we are spreading out the lethality of the two weapons, resulting in $P_{\mathrm{CD}/d} < P_{\mathrm{CD1}}$.

In either case, we now have a single rectangle (weapon effective area) in the deflection direction with known dimensions and a damage probability inside it given by either Eq. (12.24) or Eq. (12.25), as shown in Fig. 12.17.

In the range direction there will be n_r such rectangles, one for each release pulse generated by the intervalometer.

12.7.2 Possible Overlap in the Range Direction

The same arguments may be applied to the range direction that were used in deflection, where overlap may or may not occur, as shown in Fig. 12.18.

Proceeding as before, for the overlapping case we define the degree of overlap in the range direction n_{or} to be

$$n_{or} = \frac{n_r L_B}{L_P} \tag{12.26}$$

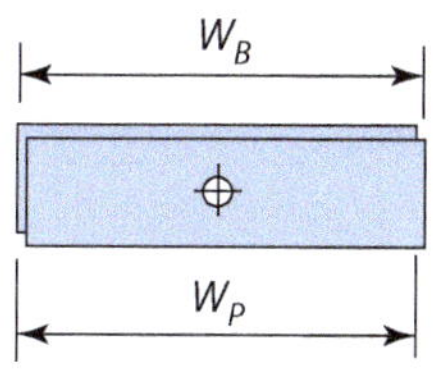

Fig. 12.16 Weapon lethal areas on top of each other.

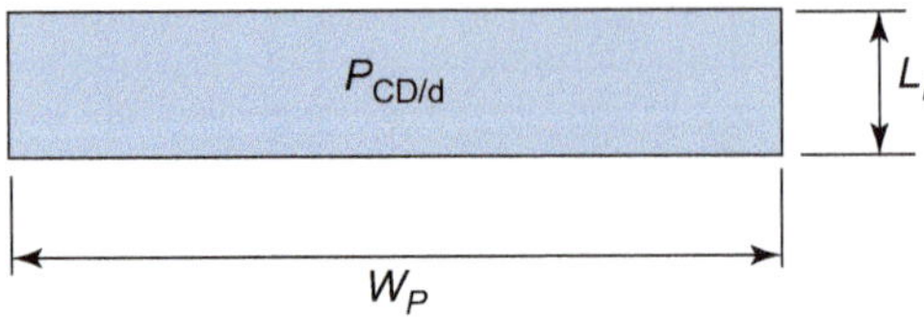

Fig. 12.17 Combining weapons in the deflection direction only.

which may be thought of as how much of the rectangles shown in Fig. 12.17 have to overlap in order to fit into the pattern length. The conditional probability of damage for the whole pattern P_{CDS} is obtained by powering up $P_{CD/d}$ by the degree of overlap in the range direction.

$$P_{CDS} = 1 - (1 - P_{CD/d})^{n_{or}} \tag{12.27}$$

If the individual rectangles do not overlap at all ($n_{or} < 1$), then we can write

$$P_{CDS} = n_r P_{CD/d} \frac{L_B}{L_P} \tag{12.28}$$

Equation (12.27) or (12.28) now defines a single rectangular pattern with known damage function and dimensions, as shown in Fig. 12.19.

Reviewing the procedure given in Chapter 11 for a single weapon against an area target, the stick as represented in Fig. 12.19 is now considered a single weapon lethal area. All of the fractional coverage analysis would proceed as before, beginning with modified equations for the effective pattern by comparing the stick lethal area dimensions to the target size,

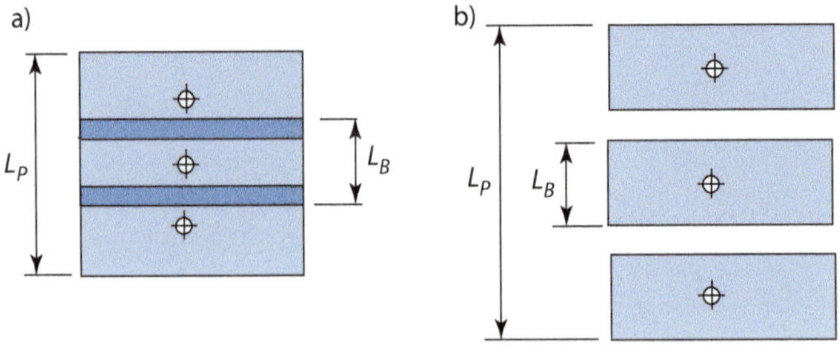

Fig. 12.18 Weapons overlapping (a) and not overlapping (b) in the range direction.

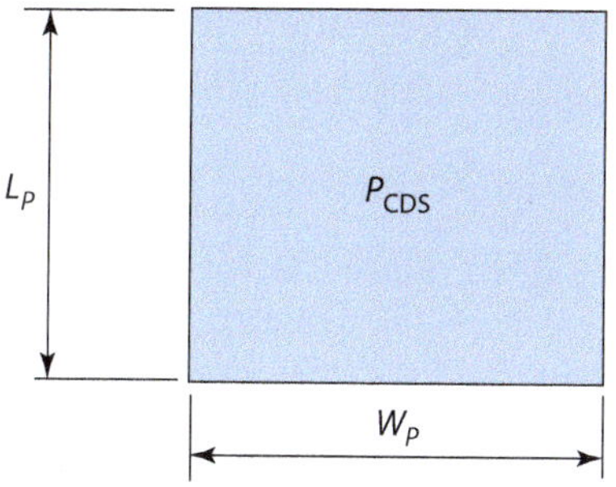

Fig. 12.19 Final pattern dimensions and damage probability.

except this time we compare the pattern size with the target area rather than the single weapon lethal area. The fractional damage is calculated in the usual manner.

$$FD_1 = E(F_C) \times P_{CDS} \times R \tag{12.29}$$

12.8 Summary of Computing Fractional Damage for Sticks

In summary, the FD_1 calculation as applied to sticks takes the following form:

1. The trajectory program is run to determine the downrange travel of the first and last weapons released. This enables the stick length to be calculated.
2. The stick width is assumed to be the same as the weapon stations on the aircraft dropping the weapons if it is a main rack release, or computed from Eq. (12.4) for MER or TER releases.
3. The slant range is calculated to the center of the stick on the ground.
4. Using the slant range, the ballistic error (mils) is converted to errors (ft) in the ground plane.
5. From the EI and impact angle, the single weapon effective area dimensions L_{ET} and W_{ET} are calculated.
6. The precision error is combined with the effective target length and width to produce L_B and W_B.
7. The pattern dimensions L_P and W_P are determined from Eqs. (12.21) and (12.22).
8. By considering the possibility of overlapping single weapons in the pattern, the conditional probability of damage P_{CDS} is computed.
9. The pattern is reduced to a single rectangle L_P by W_P with a cookie cutter damage function $P_{CD} = P_{CDS}$.

10. Given the aiming error of the pattern is the same as that of the single weapon in the pattern, the expected fractional coverage of the target is computed.
11. The final FD_1 is computed from Eq. (12.29).

The value of FD_1 that is computed is for a stick of weapons against an area target. What if we used a stick to attack a unitary target? This is dealt with by using the same analysis but assigning small dimensions to the target, such as $L_A = W_A = 0.01$ ft. Effects from multiple independently aimed sticks may be obtained from the usual powering up formula.

12.9 Spreadsheet Implementation

Table 12.2 shows an Excel implementation of the methodology for sticks against area of target elements. As mentioned before, this spreadsheet is easily adapted for unitary targets.

12.10 Advanced Method for Calculating Fractional Damage

A more accurate means to calculate FD values has been developed that is more computationally demanding but has been implemented in the JWS program. The difference in this newer method is that the individual weapons are not aggregated into a single rectangular lethal area. Instead, computing speed has increased sufficiently to allow the lethal area of individual weapons to be superimposed into a nonrectangular pattern. The calculation of fractional damage proceeds as follows:

1. The stick dimensions are determined in the manner previously discussed.
2. From the impact angle and MAE, the rectangular cookie cutter damage functions are determined.
3. These rectangles are expanded to allow for precision error and then centered on the individual weapon impact points. This pattern is centered on the target as shown in Fig. 12.20.
4. The shading indicates how many weapons, if any, lie on top of that part of the target.
5. The target will be covered by N rectangles, and each will be considered independently. If the ith rectangle of the target has n overlapping weapons ($0 \rightarrow 3$ in this case), the probability of damage inside that rectangle is obtained from the survivor rule.

$$P_{CDi} = 1 - (1 - P_{CD1})^n \tag{12.30}$$

Table 12.2 Weaponeering Spreadsheet for Stick Deliveries

METHOD 3 - STICK DELIVERY AGAINST AN AREA OF TARGETS						
Example problem 3-3 Basic Manual						
CONSTANTS						
Gravitational acceleration, ft/s	*g*	32.200				
Convert degrees to radians	deg2rad	0.017				
TARGET PARAMETERS						
Target length	LA	1000.00				
Target width	WA	1300.00				
WEAPON PARAMETERS						
Weapon: L = low drag, H = high drag		H				
Ballistic dispersion (mils)	Sigma-*b*	10.000				
Reliability	*R*	0.950				
AIRCRAFT RELEASE DATA						
Airspeed KTAS, ft/s	*V*	500.000		844.000	ft/s	
Dive angle, deg	theta	10.000		0.175	rad	
# of weapons, external racks	*ne*	0.000				
# of weapons, internal racks	*ni*	0.000				
# of release pulses	*nr*	6.000				
# weapons/release pulse	*np*	2.000				
Intervalometer setting, s	*dt*	0.130				
Release alt. first weapon, ft	*yi*	1132.00				
Release alt. last weapon, ft	*yn*	1036.73				
Average release altitude, ft	*ybar*	1084.36				
TRAJECTORY RESULTS						
Vertical velocity, ft/s	*Vv*	–146.559				
Horizontal velocity, ft/s	*Vh*	831.178				
Drag constant, Cd/M	*c0*	0.200				
constant *a* used in trajectory	*a*	14.441				
		1st wpn	**Last wpn**			
		7.377	6.772			
time of flight, s	*t_flight*	7.377	6.772			
Horizontal velocity at impact	*Vhi*	190.08	214.53			
Vertical velocity at impact	*Vvi*	157.70	157.27			
ground range from release, ft		3205	3083			
Impact angle, deg	*I*	39.680	36.246			
Stick length, ft	*Ls*	418.0				
Stick width, ft	*Ws*	40.000				
Slant range, first weapon, ft	*Sri*	3399				
Ground range to pattern center	*Gr*	3415				
Slant range to pattern center	*SR*	3583				
DELIVERY ACCURACY						
		Ground ft	**Normal ft**	**Mils**	**REP/DEP**	**REP/DEP**
CEP, ft ground, normal or mils	CEP	0.000		5		
Deflection error probable, ft	DEP	0.000	0	10.26	0	10.264
Range error probable, ft	REP	0.000	0	32.48	0	32.484
Ballistic error, deflection, ft	sigma-*bd*	35.826		10		
Ballistic error range, ft	sigma-*br*	56.109				

(Continued)

Table 12.2 Weaponeering Spreadsheet for Stick Deliveries (*Continued*)

WEAPON EFFECTIVENESS						
Effectiveness index type			MAE_*b*	MAE_*f*		
Effectiveness value	EI		0.000	23847.0		
Length/width ratio (frag only)	*a*	0.369				WET/LET
Effective target width, ft	W_{ET}		0.000	254.1	254.1	254.1
Effective target length, ft	L_{ET}		0.000	93.8	93.8	93.8
Effective target area, ft^2	A_{ET}		0.000	23,847.0		23847.0
Single weapon effective dim, ft	WB	273.6				
	LB	184.4				
	AB	50439				
Effective stick pattern, ft	WP	313.6				
	LP	602.4				
	AP	188904				
Conditional prob of damage	PCD_1	0.449				
Degree of overlap: deflection	No_d	1.745				
Cond prob damage: deflection	PCDd	0.647				
Degree of overlap: range	No_r	1.836				
Prob damage within pattern	PCDS	0.852				
***FD$_1$* CALCULATION**						
	Range		**Deflection**			
Sigma, ft	48.160		15.217			
Effective pattern size LEP/WEP	1000.000		1300.000			
INTEGRAL LIMITS						
s	1000.000		1300.00			
t	0.000		0.000			
a	14.683		–60.408			
b	0.000		0.000			
INTEGRALS						
integral #1	0.000		0.000			
integral #2	0.500		0.500			
integral #3	0.500		0.500			
integral #4 & integral #5	–0.038		–0.009			
Fractional coverage Fr/Fd	0.962		0.991			
FD$_1$	**0.112**					
FORCE REQUIREMENTS						
Desired PD	0.700					
# of sorties for this PD	10.132					
# sorties available NA	2					
Expected PD from NA sorties	0.212					

6. The expected fractional damage of this rectangle over the whole of the target area will be

$$\mathrm{FD}_i = F_{Ci} \times P_{\mathrm{CD}i} \tag{12.31}$$

where F_{Ci} is the expected fractional coverage of the whole target by the ith rectangular lethal area.

7. The fractional damage of the whole stick on the target will be summed for all rectangular subareas of the pattern covering the target.

$$\mathrm{FD} = \sum_{i=1}^{i=N} \mathrm{FD}_i \tag{12.32}$$

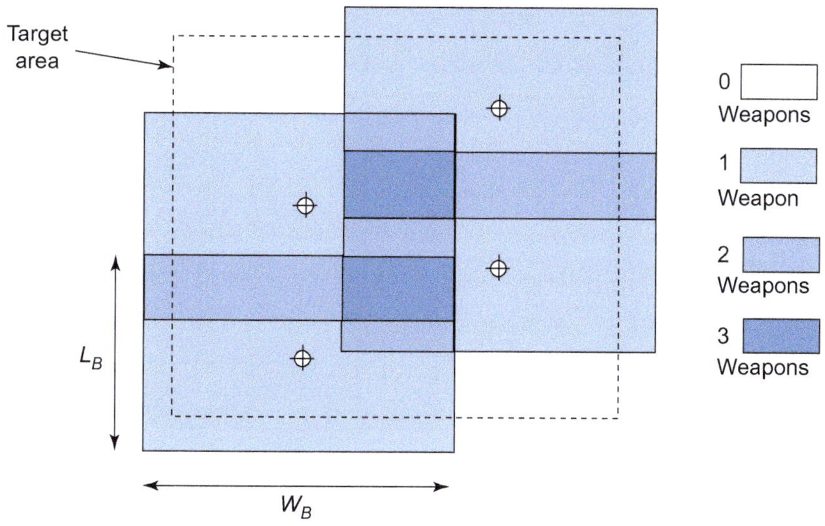

Fig. 12.20 Pattern overlapping an area target.

Example 12.1

Consider the scenario shown in Fig. 12.21 where a 10 ft × 10 ft target is attacked by a stick of two weapons side by side with 6 ft overlap. Each weapon lethal area measures 6 ft × 10 ft and has $P_{CD} = 0.8$. For simplicity, it is assumed that all aiming errors are zero so that the fractional coverage shown in the figure is the same as the expected fractional coverage.

From the figure it can be seen that we only need to consider overlap in the deflection direction. First we will aggregate the two weapons into a single

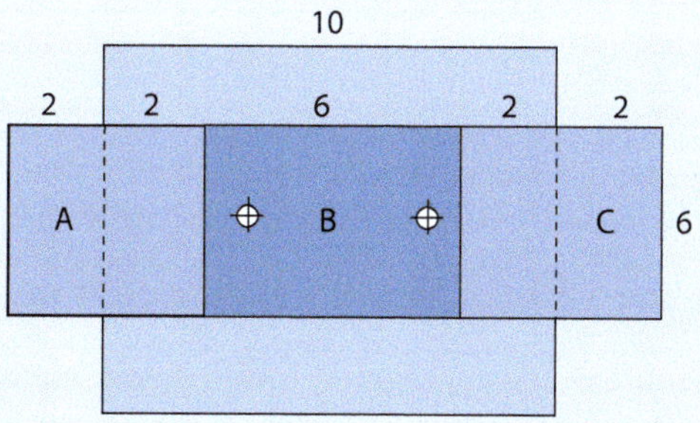

Fig. 12.21 Two weapons overlapping each other and the target.

(Continued)

Example 12.1 *(Continued)*

rectangular lethal area and constant $P_{CD/d}$, as was done earlier. From Eq. (12.23) we get

$$n_{od} = \frac{n_P W_B}{W_P} = \frac{2 \times 10}{14} = 1.429 \tag{12.33}$$

Equation (12.24) gives

$$P_{CD/d} = P_{CDS} = 1 - (1 - P_{CD1})^{n_{od}} = 0.90 \tag{12.34}$$

Assuming $R = 1$, Eq. (12.29) gives

$$FD = 0.9 \times \frac{10 \times 6}{10 \times 10} = 0.5400 \tag{12.35}$$

If we use the alternative method outlined earlier, it is seen that areas A and C have one weapon overlapping the target, area B has two, and all other areas of the target have no weapons overlapping them. Equation (12.31) gives

$$FD_A = FD_C = \frac{2 \times 6}{10 \times 10} \times 0.8 = 0.096 \tag{12.36}$$

For area B, two weapons overlap, so we get the P_{CD} from Eq. (12.30).

$$P_{CDB} = (1 - (1 - P_{CD})^n) = (1 - (1 - 0.8)^2) = 0.96 \tag{12.37}$$

The fractional damage of this rectangle over the target is

$$FD_B = 0.96 \times \frac{6 \times 6}{10 \times 10} = 0.3456 \tag{12.38}$$

The final FD is obtained from Eq. (12.32).

$$FD = 0.096 + 0.3456 + 0.096 = 0.5376 \tag{12.39}$$

Equations (12.35) and (12.39) give almost the same result in this case; however, when aiming errors are included, differences will be seen.

12.11 Effect on FD of Different Multiple Weapon Representations

We have seen how sticks of weapons may be dealt with in two ways:

1. Individual lethal areas are aggregated into a single rectangular area with known probability of damage.
2. The individual lethal area of each weapon is preserved.

Although the example considered earlier comparing the FD using both approaches yielded essentially the same result, these different

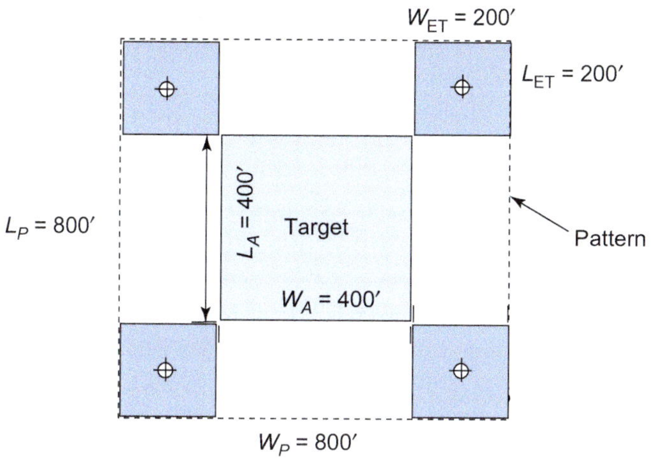

Fig. 12.22 Individual and aggregated lethal areas.

representations may lead to significant differences in the computed fractional damage. In order to illustrate the point, consider Fig. 12.22, which shows a 400 ft × 400 ft target attacked with four weapons, each having a lethal area of dimensions 200 ft × 200 ft. Consider also the special case of zero precision error and zero aiming error; that is, the center of the stick falls directly on the center of the target as shown in Fig. 12.22.

If the individual weapon lethal areas are considered, it is seen that the FD is zero, because all weapons have zero fractional coverage (i.e., they all miss).

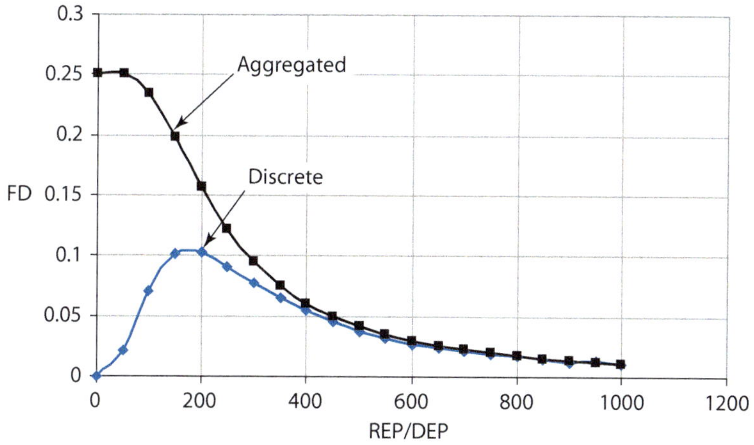

Fig. 12.23 FD from individual and aggregated weapon lethal areas.

If the individual weapon lethal areas are grouped into a pattern, however, this pattern will completely cover the target, although the probability of damage in the pattern will be reduced to the following:

$$P_{\mathrm{CDS}} = \frac{4 \times 200 \times 200}{800 \times 800} = 0.25 \qquad (12.40)$$

The fractional coverage will be unity (complete coverage), giving FD = 0.25. It might be expected that this difference will vary once aiming error is introduced, and this proves to be correct, as shown in Fig. 12.23.

As the aiming error becomes large, the difference between the two methods becomes smaller; however, the effect may again become significant if the weapon spacing is large compared to the size of the target area. This issue will be addressed further in Chapter 15, Section 15.8.

Chapter 13 Cluster Munitions

13.1 Introduction

To attack large area targets, large warhead unitary weapons or sticks of smaller unguided weapons may be used; however, better coverage is often obtained with the use of cluster weapons. A cluster weapon comprises a single canister that is released like a unitary weapon, but at a particular point in the trajectory it opens to release a large number of smaller submunitions, which then disperse before detonation, as indicated in Fig. 13.1. An air-dropped dispenser is often referred to as a *tactical munitions dispenser (TMD)*, whereas an artillery round carrying smaller mines is called a *cargo round*. An example of an air-launched weapon at the dispense point is shown in Fig. 13.1.

The submunitions may project small fragments designed to incapacitate personnel, or larger fragments more suited to lightly armored materiel targets such as trucks. Many cluster munitions carry a combination of such submunitions. Some cluster munitions have shaped charge warheads intended specifically for armored vehicles. The following sections describe

Fig. 13.1 Dispenser releasing cluster munitions.

the cluster munitions currently in the inventory. A single weapon may contain a few dozen large, sophisticated submunitions or several hundred grenade-size antipersonnel types.

In this chapter, the focus is on air-launched munitions; those fired from ground systems (i.e., artillery projectiles, rockets) will be discussed in detail in Chapter 16.

The basic approach in dealing with cluster munition deliveries is the same as for sticks and is shown in Fig. 13.2.

1. Assume the pattern on the ground is rectangular in shape and there is a uniform distribution of submunitions within it.
2. Determine the dimensions of the rectangle from the release conditions, weapon trajectory, and formulae for the submunition "spread."
3. Calculate the probability of damage, averaged within the pattern based on the density of weapons.
4. For an area of targets, calculate fractional damage with Eq. (11.45) by considering the stick to be equivalent to a single effectiveness area with the characteristics defined previously.

This collection of individual weapon lethal areas is then aggregated into a single rectangle with a probability of damage P_{CDS} inside it, as in Fig. 13.3.

Typically accuracy is poor, because the dispenser and submunitions are unguided, except for the INS-guided wind-corrected munitions dispenser (WCMD). The submunitions in particular may be influenced by unknown, local wind conditions if descent is controlled by a parachute. Like stick deliveries, they produce a large area of effects on the ground and are typically used against areas of target elements.

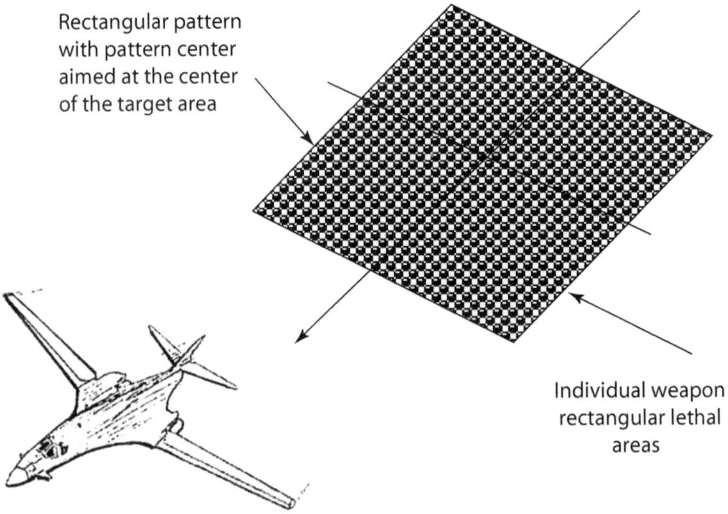

Fig. 13.2 Individual submunition impacts in the pattern.

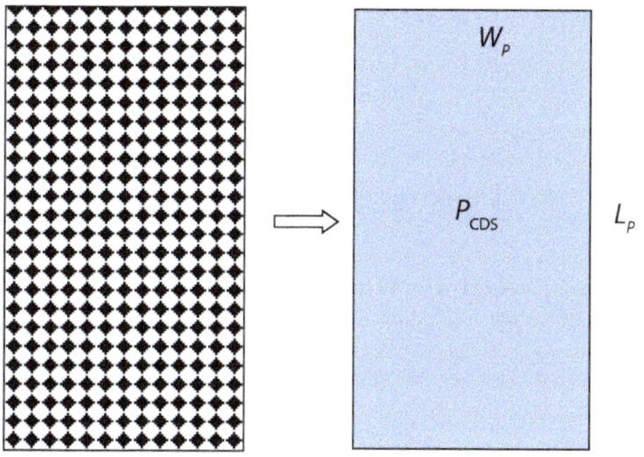

Fig. 13.3 Combining individual lethal areas into a pattern.

13.2 Types of Cluster Munitions

Before determining this effectiveness rectangle, it is worth looking at the different types of cluster munitions and the submunitions they carry.

13.2.1 CBU-99/CBU-100 (MK-20) Rockeye

The MK-20 Rockeye, shown in Fig. 13.4, is a free-fall, unguided cluster weapon designed to damage tanks and other armored vehicles. The system consists of a clamshell dispenser, a mechanical MK-339 timed fuze, and 247 dual-purpose fragmentation and armor-piercing shaped-charge bomblets. The bomblet weighs 1.32 lb and has a 0.4-lb shaped-charge warhead of high explosives, which produces up to 250,000 psi at the point of impact, penetrating approximately 7.5 in. of armor.

Rockeye is most efficiently used against an area of targets requiring penetration to kill. Fielded in 1968, the Rockeye dispenser is also used in the Gator air-delivered mine system. During Desert Storm, U.S. Marines used the weapon extensively, dropping 15,828 of the 27,987 total deployed

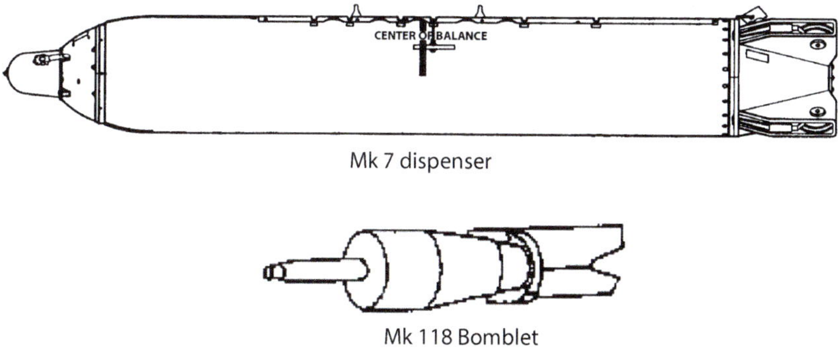

Fig. 13.4 Rockeye (Mk-20) dispenser and submunition.

Rockeyes against armor, artillery, and antipersonnel targets. The remainder were dropped by Air Force (5345) and Navy (6814) aircraft.

13.2.2 CBU-78/CBU-89/B Gator Mine Munition

The Gator mine system provides a means to place minefields on the ground rapidly using high-speed tactical aircraft. The minefields are used for area denial, diversion of moving ground forces, or immobilization of targets that may then be attacked using other direct attack weapons. This weapon and its submunitions are shown in Fig. 13.5.

The CBU-78 Gator is a tri-service weapon featuring antivehicle and antipersonnel landmines used adjacent to enemy forces to disrupt or deny use of selected areas. The 500-lb CBU-78 contains 45 antitank and 15 antipersonnel mines. These mines can be detonated by target sensors (magnetic field for antitank and trip line for antipersonnel) or by a disturbance/antidisturbance device. They also have a backup self-destruct time set before aircraft launch. During Desert Storm the Navy and the Marine Corps dropped 209 CBU-78s.

The Air Force CBU-89/B is a 1000-lb-class cluster weapon using the SUU-64/B tactical munitions dispenser (TMD). The TMD is the same general configuration used for the CBU-87/B combined effects munition. This commonality allows for high-rate, low-cost production of the dispenser. The Navy CBU-78/B is a 500-lb-class cluster weapon that uses the Mk-7 Rockeye dispenser. Rockeye has been in high-rate production for many years; the Mk7 dispenser is also a low-cost item. Both systems contain a mix of BLU-91/B antitank (AT) and BLU-92/B antipersonnel (AP)

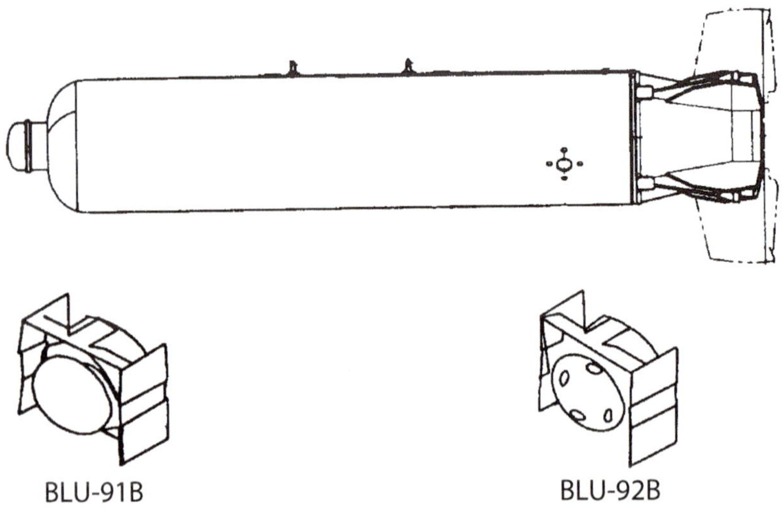

Fig. 13.5 CBU-89/B Gator dispenser and submunitions.

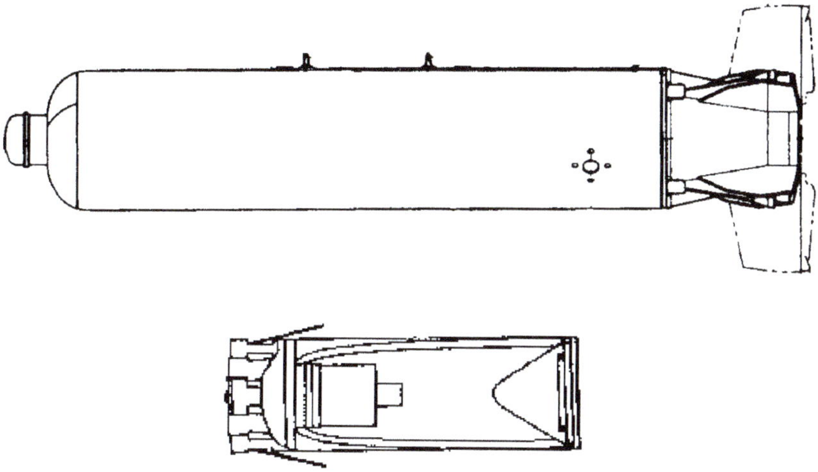

Fig. 13.6 CBU-87/B dispenser and BLU-97/B bomblet.

mines—72 AT and 22 AP for the CBU-89/B; 45 AT and 15 AP for the CBU-78/B. Again, commonality of mines for both systems also contributes to high-rate, low-cost production.

The BLU-91/B AT mine is the heart of the Gator system. Microelectronics in each mine detect the target, discriminate among armored vehicles, and detonate the mine when the target reaches the most vulnerable approach point. A Misznay-Schardin explosive charge defeats the belly armor of most vehicles. The BLU-92/B AP mine serves to discourage minefield clearing. Upon activation, the AP mine explosion sends high-velocity fragments in a horizontal plane over a wide area.

13.2.3 CBU-87 Combined Effect Munition

The CBU-87 is a 1000-lb, combined effects munition (CEM) for attacking soft target areas with detonating bomblets. It is shown in Fig. 13.6.

The CBU-87 CEM is an all-purpose, air-delivered cluster weapons system consisting of an SW-65 TMD with an optional FZU-39 proximity sensor. The BLU-97/B combined effects bomb (CEB) is effective against armor, personnel, and material and contains a shaped charge, scored steel casing, and a zirconium ring for antiarmor, fragmentation, and incendiary capability. The bomblet case is made of scored steel designed to break into approximately 300 preformed ingrain fragments for defeating light armor and personnel. A total of 202 of these bomblets are loaded in each dispenser, enabling a single payload attack against a variety of targets and achieving wide area coverage. During Desert Storm, the U.S. Air Force dropped 10,035 CBU-87s. During Allied Force, the United States dropped about 1100 cluster bombs, and most of these were CBU-87s. The failure to detonate rate for a standard cluster was approximately 5%.

13.2.4 CBU-97/B Sensor Fuzed Weapon

The CBU-97 is a 1000-lb-class weapon containing sensor-fuzed submunitions for attacking armor. The sensor-fuzed weapon (SFW) is an unpowered, top attack, wide area cluster munition designed to achieve multiple kills per aircraft pass against enemy armor and support vehicles. After release, the TMD opens and dispenses the 10 submunitions, which are parachute stabilized. Each of the 10 BLU-108/B submunitions contains four armor-penetrating projectiles with infrared sensors to detect armored targets.

At a preset altitude sensed by a radar altimeter, a rocket motor fires to spin the submunition and initiate an ascent. The submunition then releases its four projectiles, which are lofted over the target area. The projectile's sensor detects a vehicle's infrared signature, and an explosively formed penetrator (EFP) fires at the heat source from altitude. If no target is detected after a period of time, the projectiles automatically detonate after a preset time interval, causing damage to materiel and personnel. The various deployment phases of this weapon are shown in Fig. 13.7 and details of the warheads in Fig. 13.8.

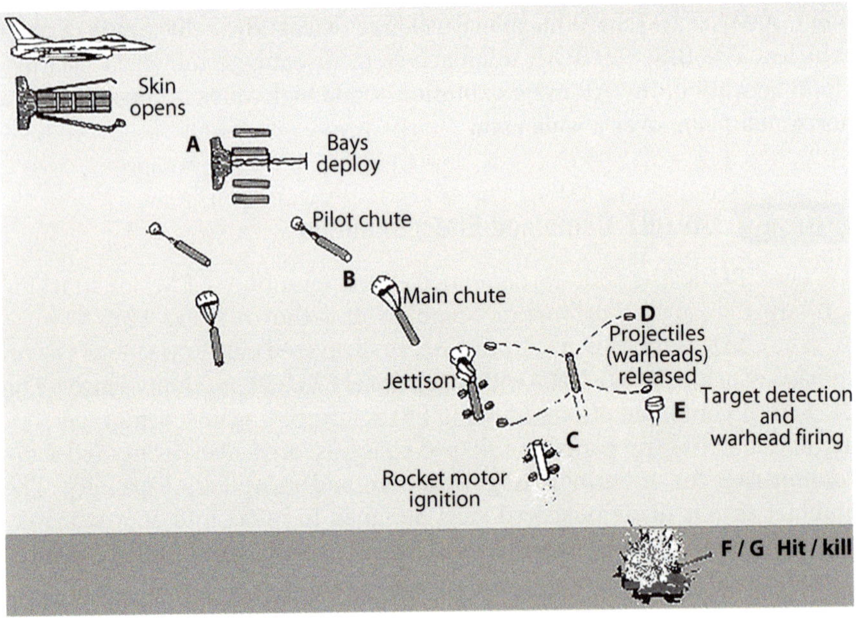

Fig. 13.7 Sensor-fuzed weapon operation

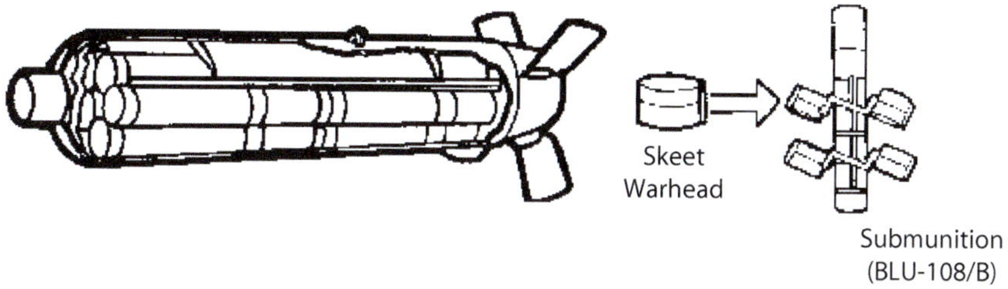

Fig. 13.8 SFW dispenser and BLU-108/B submunition

Figure 13.9 shows the elliptical track of the sensing axis for the individual weapons, together with the pattern obtained from a single dispenser containing 40 submunitions.

13.2.5 Wind-Corrected Munitions Dispenser (WCMD)

Low-altitude, high-speed lay-down deliveries were the preferred tactic to be used against heavily defended target sets such as the robust Soviet/Warsaw Pact threat envisioned in Central Europe. However, low-altitude deliveries were not the preferred option during Desert Storm, where the U.S. Air Force used medium/high-altitude weapons to provide fighter and bomber aircraft defense against short-range surface-to-air missiles and anti-aircraft artillery fire. Unfortunately, all the cluster munitions discussed so far are of the high drag, unguided type, which, if released from high altitude, have substantial delivery accuracy errors.

The wind-compensated munitions dispenser (WCMD, also known as Wick-Mid) remedies this current shortfall in TMD munitions, such as the CBU-87 CEM, CBU-89 GATOR, and CBU-97 SFW. The weapon uses inertial guidance only (no GPS). The WCMD program developed a tail kit for these inventory dispenser weapons. These weapons are now capable of accurate delivery from medium to high altitude when equipped with a WCMD kit, as shown in Fig. 13.10.

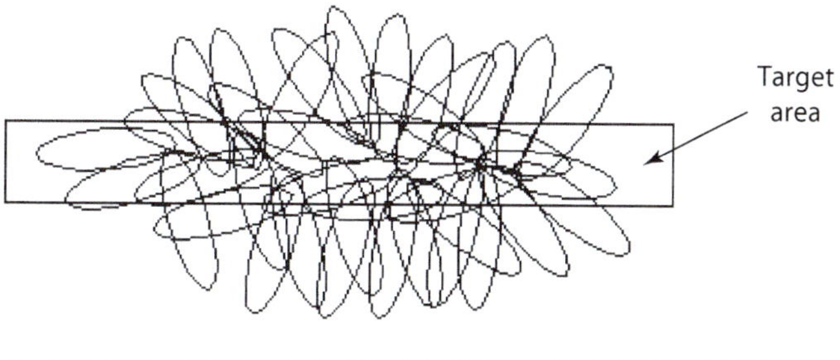

Fig. 13.9 SFW pattern geometry.

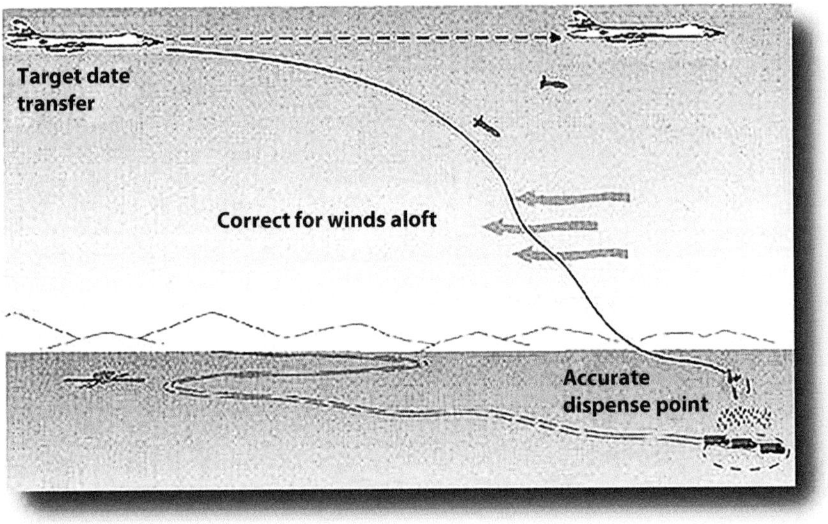

Fig. 13.10 WCMD operational scenario.

The WCMD weapon corrects for wind effects and errors during the weapon's ballistic fall by sensing the wind profile as it descends and guiding to a release point such that the released submunitions will drift onto the target area. Currently, the dispenser is achieving an accuracy of within 30 ft. Both fighter and bomber aircraft are able to employ WCMD from a wide range of altitudes, in adverse weather, using various tactics such as level, dive, and toss bombing, and bombing on coordinates.

Figure 13.11 shows how WCMD patterns are much tighter than other dispenser munitions.

Because WCMD is a kit to modify CBU-87 and CBU-89 conventional dispenser weapons, it also carries different payloads comprising antipersonnel and antimateriel submunitions.

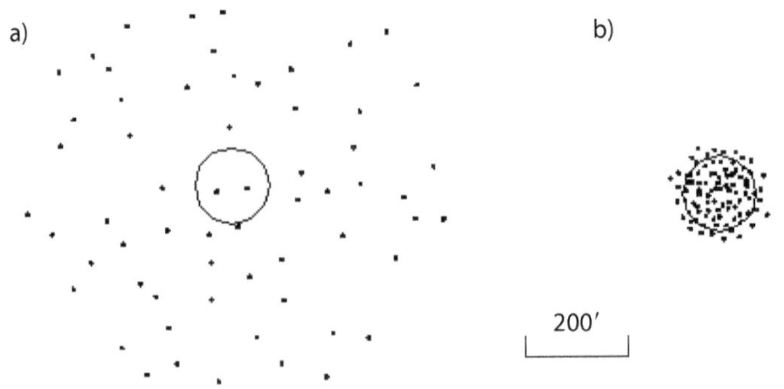

Fig. 13.11 Conventional (a) and WCMD (b) CBU patterns.

13.3 General Analytical Treatment of Cluster Munitions

The operation of cluster munitions is somewhat complex and varies from one type to another, although these weapons are almost always used against an area target. The computation of the fractional damage is also more complicated than for unitary weapons, and some assumptions have to be made to ensure a tractable solution that can be computed in a short period of time. In general, the process may be considered in two parts, corresponding to the two stages of the weapon trajectory.

1. The trajectory of the dispenser from the weapon release point to the dispense point is computed.
2. The trajectory of a single "mean" submunition from the dispense point to ground impact is computed. This locates the center of the pattern formed by the submunitions for the purposes of computing slant range and precision error effects. The size of the pattern is determined separately.

The goal is the same as for single weapons and stick deliveries, namely to determine the dimensions of an area of effectiveness on the ground corresponding to a lethal area for a unitary weapon, and a probability of damage within it. If this can be achieved, the cluster munition may be analyzed using techniques applicable to single weapons, and fractional damage values calculated in the usual manner. This is the same approach that was used for sticks. Figure 13.12 shows the two phases of cluster munition weapon delivery.

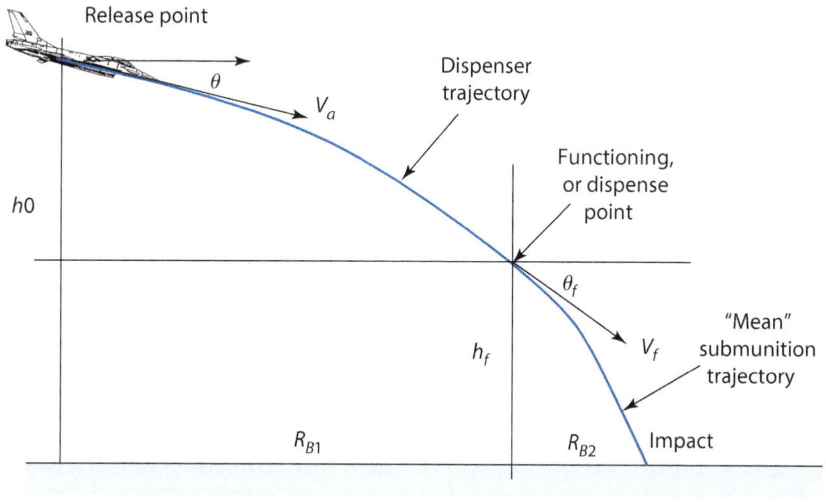

Fig. 13.12 Cluster munition and submunition trajectory.

The distinguishing operational features of the different cluster munitions may be summarized as follows:

- Some weapons dispense at a given functioning altitude (typically a few hundred feet), whereas others dispense at a functioning time after release from the launch aircraft, depending on the weapon fuzing. Altitude fuzing is usually preferred in order to generate more predictable coverage on the ground plane.
- Some weapons are spinning at the release point, producing a dispense pattern that is approximately rectangular in the ground plane. Other weapons do not spin, yet produce weapon patterns that are approximately circular in the normal plane. We are obviously interested in the pattern size in the ground plane as a means to obtain a lethal area, as was done for a unitary weapon.
- Because the number of submunitions in each dispenser is fixed, the pattern in the ground plane may contain overlapping areas of effectiveness from individual submunitions, or not, depending on the altitude at the dispense point and the resulting pattern size.

Our objective, therefore, is to allow for all these possibilities in computing FD values for cluster munition deliveries. In addition, we wish to allow for both point and area targets, and also for tactics that require the delivery of sticks of cluster munitions. The same basic spreadsheet template used for all previous weaponeering scenarios will be used in modified form for cluster munitions as well. The discussion that follows details those differences that distinguish this methodology from those previously encountered.

13.4 Trajectory Computations

Figure 13.12 indicates the basic trajectory geometry for both dispenser and submunition. The dispenser may be treated as an unguided weapon with given release conditions, as dealt with in Chapter 3. Two dispense conditions may apply—either a functioning time or a functioning altitude. The same linear drag model is used for dispensers as has been used for unitary weapons, although different drag coefficients are used for the dispenser and submunition. The objective is to determine effective "release conditions" for the submunitions at the dispense or functioning point, namely the dispenser velocity v_f, altitude h_f, and dispenser functioning angle θ_f. These parameters will be used to calculate the pattern size.

13.4.1 Dispenser Functioning Time t_f

This is the simpler case, because we do not have to compute a time of flight; it is simply the functioning time of the dispenser. We can compute the horizontal and vertical components of the dispenser velocity at the functioning time t_f from the linear drag model described in Chapter 3,

Section 3.6, as follows:

$$v_{hd} = v_{0h} \exp(-c_0 t_f) \tag{13.1}$$

$$v_{vd} = \left[v_{0v} - \frac{g}{c_0} \right] \exp(-c_0 t_f) + \frac{g}{c_0} \tag{13.2}$$

From these equations, the dispenser velocity and functioning angle are computed from

$$\theta_f = \tan^{-1} \left[\frac{v_{vd}}{v_{hd}} \right] \tag{13.3}$$

and

$$v_f = \sqrt{v_{hd}^2 + v_{vd}^2} \tag{13.4}$$

13.4.2 Dispenser Functioning Altitude h_f

This is a little more complicated, but essentially we solve the equation for the functioning altitude, Eq. (3.39), for h_f rather than the zero altitude case.

$$h_0 - \frac{1}{c_0} \left[v_{0v} - \frac{g}{c_0} \right] [1 - \exp(-c_0 t)] - \frac{gt}{c_0} = h_f \tag{13.5}$$

This gives the iterative relation $t = f(t)$ in Eq. (3.40) as

$$t = a[1 - \exp(-c_0 t)] + c_0(h_0 - h_f)/g \tag{13.6}$$

The constant a is given as before

$$a = \frac{1}{c_0} - \frac{v_{0v}}{g} \tag{13.7}$$

Equation (13.6) is solved iteratively for the corresponding functioning time t_f in a spreadsheet format as outlined before. This functioning time (corresponding to the input functioning altitude) may then be used in Eqs. (13.1) to (13.4) in order to obtain the submunition release conditions. For some methods, the ground range from the release point to dispense point is needed (Fig. 13.12), and this may be obtained as follows:

$$R_{B1} = \frac{v_{0h}}{c_0} \left[1 - \exp(-c_0 t_f) \right] \tag{13.8}$$

This equation applies to both time-functioning and altitude-functioning weapons.

13.4.3 Submunition Trajectory

To compute the single "mean" submunition trajectory, the functioning conditions may be used as initial release conditions, and the trajectory calculated down to impact on the ground. Parameters of interest here are the impact angle, impact velocity, time of flight of dispenser and submunition, and ground range traversed between dispense and impact R_{B2} (Fig. 13.12). The submunition trajectory defines the center of the pattern on the ground and coincides with the target (unitary) or center of the target (area) for zero delivery errors.

Table 13.1 shows a spreadsheet implementation of the trajectory methods, noting that different drag coefficients are used for different submunitions.

The impact angle of the mean submunition is used to determine the dimensions of a single submunition lethal area.

Table 13.1 Cluster Munition Trajectory

CLUSTER WEAPON TRAJECTORY				
Release Conditions				
initial altitude, ft	9.32E+02		**Zero Drag Solution**	
aircraft speed, kt/ft/s	4.50E+02	7.60E+02	v_impact	2.78E+02
Dive angle	1.00E+01	1.75E–01	Impact time	4.54E+00
functioning time, s	1.20E+00			
functioning altitude, ft	2.00E+02			
initial vertical velocity, ft/s	1.32E+02			
initial horizontal velocity, ft/s	7.48E+02			
drag constant dispenser	2.00E-01			
drag constant submunition	1.50E-01			
gravitational constant, ft/s^2	3.22E+01		**Altitude Function**	
			4.04E+00	*t* (TOF)
At the Dispense Point	**Time Function**		4.04E+00	*f*(*t*)
coefficient *a*_disp	9.04E-01			
horizontal velocity, ft/s	5.88E+02		3.33E+02	
vertical velocity, ft/se	1.38E+02		1.48E+02	
altitude, ft	7.70E+02		200	
downrange, ft	7.98E+02		2.07E+03	
dispenser velocity, ft/s	6.04E+02		3.65E+02	
dispenser angle, rad/deg	2.31E–01	1.32E+01	4.18E–01	2.39E+01
At Ground Impact (Submunition)				
coefficient *a*_submunition	2.38E+00		2.07E+00	
t (time of flight)	3.62E+00		9.41E–01	
f(*t*) also time of flight, s	3.62E+00		9.41E–01	
horizontal velocity, ft/s	3.42E+02		2.90E+02	
vertical velocity, ft/s	1.70E+02		1.57E+02	
altitude, ft	0.00E+00		0.00E+00	
downrange, ft	1.64E+03		2.93E+02	
submunition impact velocity, ft/s	3.82E+02		3.29E+02	
submunition impact angle, rad/deg	4.62E–01	2.65E+01	4.96E–01	2.84E+01
Summary Results				
total range: dispenser + sub, ft	2.44E+03		2.37E+03	
Impact angle: submunition, deg	2.65E+01		2.84E+01	
slant range release to impact, ft	2.64E+03		2.57E+03	
total time of flight, s	4.82E+00		4.98E+00	

13.5 Submunition Pattern Dimensions

For cluster weapons, the single pattern length and width of submunitions (L_{ET} and W_{ET}, in feet) and the single pattern effective area (A_{ET}, in square feet) are computed differently depending upon whether the pattern shape is rectangular or circular. For all cluster weapons, it is assumed that the submunition impacts are uniform randomly distributed within the pattern; therefore, the damage probability for a target element located in the pattern is uniform. Because there is a finite probability that a submunition will land at the edge of the cluster weapon pattern, it is possible to damage a target element that is located outside but near the edge of the cluster weapon pattern. This damage is generally referred to as *pattern edge effects*. To account for edge effects, an adjustment is added to the pattern dimensions to specify a lethal area that defines a cookie cutter damage function within which the probability of damage is constant and beyond which the probability of damage is zero.

To determine the pattern edge effect algorithm, pattern damage contours were computed for various numbers of submunitions and submunition EI values. These contours were based on the assumption that the submunition damage function can be represented by a Carlton function (i.e., a Gaussian damage function with a maximum probability of damage given a hit of $D_0 = 0.5$). This assumption of probability of damage was based on examination of edge effect predictions computed using a mainframe Monte Carlo computer program, and inspection of the data indicated that an exponential decay function was appropriate. The data were then fitted utilizing least squares techniques, and a cookie cutter edge effect adjustment that maintains the same total pattern effectiveness was derived.

Two equations are used to determine the outer pattern edge effects for all cluster weapons depending upon whether the pattern is rectangular in the ground plane or circular about the dispenser axis at the dispense point. These two cases will be considered separately.

13.6 Rectangular Patterns in the Ground Plane

In this section, we determine the pattern size and then calculate the uniform probability of damage inside the pattern; however, first we determine the edge effect distance that needs to be added to the pattern of submunitions.

13.6.1 Pattern Edge Effects

The edge effect equation is given by

$$R_A = 2\sqrt{A'_{\mathrm{ET}}/\pi}[1 - \exp(-0.18\, n_x)] \tag{13.9}$$

R_A is the edge effect adjustment (in feet), and n_x is the expected number of submunitions that can damage the target given target coverage by the pattern. This equation seems reasonable, because the first part is simply reducing A_{ET} to a circle, and the modifying factor in brackets is a scale factor to match mainframe simulations. Note that A'_{ET} is the single submunition effective target area. The value of n_x is given by

$$n_x = \frac{n_b \times R_S \times A'_{ET}}{L \times W} \tag{13.10}$$

where n_b is the number of submunitions in the dispenser, R_S is the submunition reliability, L is the length of the pattern (in feet), and W is the width of the pattern (in feet). Mainframe computer simulations were run for a wide range of release conditions, and curve fitting techniques were used to determine the coefficients for the simplified analysis that follows, which was found to reproduce the high fidelity simulations from which it was derived.

13.6.2 Pattern Length for Time Fuzes

The functioning altitude and functioning velocity are adjusted to allow for the functioning delay and are given by

$$h_1 = h_f - 38 \tag{13.11}$$

$$v_1 = v_f - 84 \tag{13.12}$$

Then two intermediate variables are then computed.

$$v_t = 0.041 \times R_{spin} + 13 \tag{13.13}$$

where R_{spin} is the dispenser spin rate in rev/min. The second intermediate variable is

$$T = \frac{v_t}{\sqrt{v_t^2 + v_1^2}} \tag{13.14}$$

If

$$v_t \cos \theta_f > v_1 \sin \theta_f \tag{13.15}$$

then the pattern length is computed using

$$L = 553 + 0.2h_1 \sin \theta_f - 10\sqrt{h_1/T} + 1.4v_t \cos \theta_f \tag{13.16}$$

otherwise the pattern length is given by

$$L = 581 + 0.5h_1 \sin \theta_f - 6.5\sqrt{h_1/T} - 583\sqrt{T} + 1.8v_t \cos \theta_f \tag{13.17}$$

If either value of L is less than 180 ft, it is set to 180 ft.

13.6.3 Pattern Width for Time Fuzes

The pattern width depends on the adjusted values calculated in Eqs. (13.11) and (13.12). If $h_1 < 525$ ft, two intermediate variables f and p are first computed as

$$f = \frac{v_1 \sin\theta_f}{h_1} \tag{13.18}$$

and

$$p = \sqrt{f^2 + 32/h_1} \tag{13.19}$$

Then, if $h_1 < 175$, W is computed using

$$W = 74 + 0.085R_{\text{spin}} - 0.001R_{\text{spin}}f - 0.015R_{\text{spin}}p - 39p \tag{13.20}$$

If $175 < h_1 < 525$, then W is computed from

$$W = 72 + 0.15R_{\text{spin}} + 0.13R_{\text{spin}}f - 0.21R_{\text{spin}}p - 36p \tag{13.21}$$

Finally, if $h_1 > 525$, the pattern width is

$$W = 45 + 1331T + 0.04h_1 \tag{13.22}$$

13.6.4 Pattern Length and Width for Altitude Fuzes

For altitude fuzing, the adjusted functioning altitude and velocity are

$$h_1 = h_f + 33 \tag{13.23}$$

$$v_1 = v_f - 92 \tag{13.24}$$

The intermediate variables v_t and T are computed as before using Eqs. (13.13) and (13.14). Then, if $h_1 < 640$ ft, L and W are computed from

$$L = -33.4 + 0.67h_1T\sin\theta_f + 1445T\cos\theta_f + 0.12h_1 \tag{13.25}$$

$$W = 24 + 2.6h_1T - 0.06h_1 \tag{13.26}$$

Otherwise, if $h_1 \geq 640$ ft, then L and W are computed from

$$L = 37.1 + 1430\,T\sin\theta_f + 362\;T\cos\theta_f \tag{13.27}$$

$$W = 12 + 1592\,T + 0.038\,h_1 \tag{13.28}$$

Now that the pattern dimensions have been determined, the dimensions of the pattern area A_p in the ground plane may be calculated by combining L and W with the edge effect parameter R_A, as shown in Fig. 13.13.

It follows that

$$L_{\text{ET}} = L + 2R_A \tag{13.29}$$

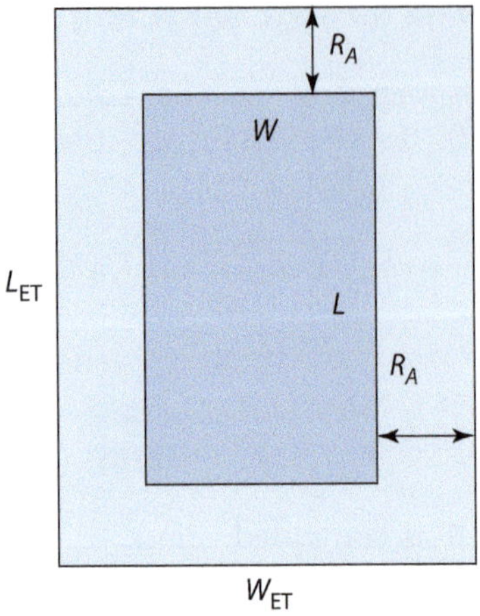

Fig. 13.13 Computing L_{ET} and W_{ET}.

$$W_{\text{ET}} = W + 2R_A \tag{13.30}$$

and the pattern area is given by

$$A_p = L_{\text{ET}} \times W_{\text{ET}} \tag{13.31}$$

13.6.5 Probability of Damage in the Pattern

At this point we consider the coverage within the pattern by the submunitions. The probability of damage to a target element within the pattern due to a single submunition of lethal area A'_{ET} is given by

$$P_S = \frac{A'_{\text{ET}} R_S}{L \times W} \tag{13.32}$$

Essentially we are diffusing the submunition lethal area over the whole of the pattern. Assuming the submunition impacts are uniformly distributed within the pattern and there are n_b submunitions, the probability of damage to the target is obtained by powering up P_s given in Eq. (13.32).

$$P_{\text{CD}} = 1 - (1 - P_S)^{n_b} \tag{13.33}$$

This equation may be approximated by the Morse-Kimball formula

$$P_{\text{CD}} \approx 1 - \exp(-n_b P_S) = 1 - \exp\left(-\frac{n_b R_S A'_{\text{ET}}}{L \times W}\right) = 1 - \exp(-n_x) \tag{13.34}$$

where

$$n_x = \frac{n_b R_S A'_{\text{ET}}}{L \times W} \tag{13.35}$$

13.7 Circular Patterns in the Normal Plane

Proceeding as before, the edge effect equation in the normal plane is

$$R_A = 2\sqrt{A'_{ET}\sin I/\pi[1-\exp(-0.18n_x)]} \tag{13.36}$$

Again, R_A is the edge effect adjustment (in feet), I is the submunition impact angle, n_x is the expected number of submunitions that can damage the target given target coverage by the pattern, and A'_{ET} is the single submunition effective target area in the ground plane. The value of n_x in this case is given by

$$n_x = \frac{n_b \times R_S \times A'_{ET} \times \sin I}{\pi R_p^2} \tag{13.37}$$

where R_p is the pattern radius in feet. The equations for computing the pattern radii for weapons were derived from actual impact data. It is assumed that the submunition acquires a spread velocity v_s in the normal plane at the dispense point, as shown in Fig. 13.14.

The spread velocity is computed using the empirical relationship

$$v_s = 5 + \frac{v_f}{168.8} + \left[\frac{v_f}{168.8}\right]^2 \tag{13.38}$$

If the time taken for the submunition to fall from the dispense point to the ground is dt, obtainable from the submunition trajectory calculation, the radius of the pattern in the normal plane at ground impact is

$$R_p = \left[5 + \frac{v_f}{168.8} + \left[\frac{v_f}{168.8}\right]^2\right]\mathrm{d}t \tag{13.39}$$

From the dispense point, therefore, the circular pattern gets larger and rotates so that it is perpendicular to the trajectory. The radius R_p has the

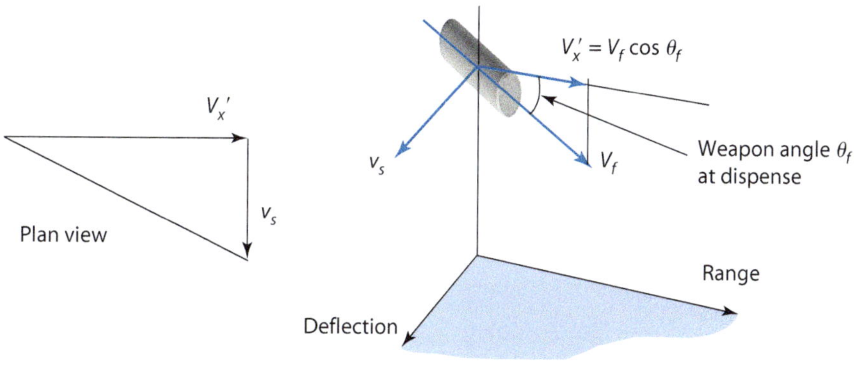

Fig. 13.14 Velocity vectors at the dispense point.

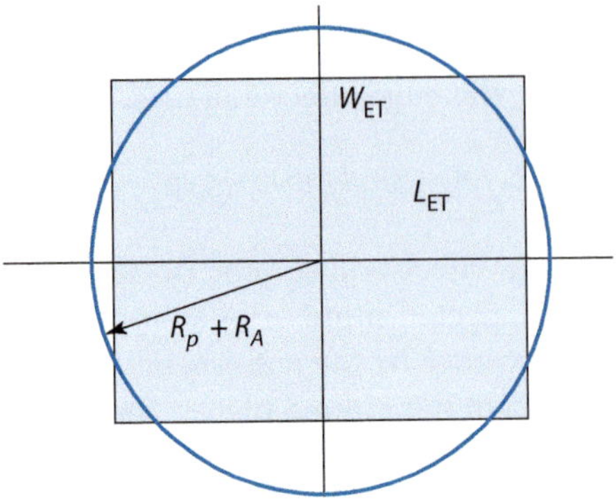

Fig. 13.15 Pattern lethal area in the normal plane.

edge effect R_A added to it and defines a circle in the normal plane. This is then represented by a square of equal area also in the normal plane, as shown in Fig. 13.15.

It follows that projecting the square shown in Fig. 13.15 onto the ground plane at impact produces a rectangle, the dimensions of which are

$$W_{\mathrm{ET}} = \sqrt{\pi}(R_p + R_A) \tag{13.40}$$

$$L_{\mathrm{ET}} = \frac{W_{\mathrm{ET}}}{\sin I} \tag{13.41}$$

Again, the pattern area is

$$A_p = L_{\mathrm{ET}} \times W_{\mathrm{ET}} \tag{13.42}$$

In this case, the probability of damage to a target element within the pattern due to a single submunition is given by

$$P_S = \frac{A'_{\mathrm{ET}} R_S}{\pi R_p^2} \tag{13.43}$$

Once more, the probability of damage to the target is obtained by powering up P_s given in Eq. (13.43)

$$P_{\mathrm{CD}} = 1 - (1 - P_S)^{n_b} \tag{13.44}$$

This equation may again be approximated by

$$P_{\mathrm{CD}} \approx 1 - \exp(-n_b P_S) = 1 - \exp\left(-\frac{n_b R_S A'_{\mathrm{ET}}}{\pi R_p^2}\right) = 1 - \exp(-n_x) \tag{13.45}$$

Table 13.2 shows the complete spreadsheet for trajectory and pattern, which now includes the calculation for different weapon pattern dimensions and probability of damage within them, the latter including both time and altitude fuzing.

Table 13.2 Trajectory and Pattern Dimensions for All Cases

CLUSTER WEAPON TRAJECTORY				
Release conditions				
initial altitude, ft	9.32E+02		**Zero drag solution**	
aircraft speed, kt/ft/s	4.50E+02	7.60E+02	v_impact	2.78E+02
Dive angle	1.00E+01	1.75E–01	impact time	4.54E+00
functioning time, s	1.20E+00			
functioning altitude, ft	2.00E+02			
initial vertical velocity, ft/s	1.32E+02			
initial horizontal velocity, ft/s	7.48E+02			
drag constant dispenser	2.00E–01			
drag constant submunition	1.50E–01			
gravitational constant, ft/s^2	3.22E+01		**Altitude Function**	
			4.04E+00	*t* (TOF)
At the dispense point (dispenser)	**Time Function**		4.04E+00	*f*(*t*)
coefficient a_disp	9.04E–01			
horizontal velocity, ft/s	5.88E+02		3.33E+02	
vertical velocity, ft/s	1.38E+02		1.48E+02	
altitude, ft	7.70E+02		200	
downrange, ft	7.98E+02		2.07E+03	
dispenser velocity, ft/s	6.04E+02		3.65E+02	
dispenser angle, rad/deg	2.31E–01	1.32E+01	4.18E–01	2.39E+01
At ground impact (submunition)				
coefficient *a*_submunition	2.38E+00		2.07E+00	
t (time of flight)	3.62E+00		9.41E–01	
f(*t*) (also time of flight), s	3.62E+00		9.41E–01	
horizontal velocity, ft/s	3.42E+02		2.90E+02	
vertical velocity, ft/s	1.70E+02		1.57E+02	
altitude, ft	0.00E+00		0.00E+00	
downrange, ft	1.64E+03		2.93E+02	
submunition impact velocity, ft/s	3.82E+02		3.29E+02	
submunition impact angle, rad/deg	4.62E–01	2.65E+01	4.96E–01	2.84E+01
Summary results				
total range, dispenser + sub, ft	2.44E+03		2.37E+03	
impact angle: submunition, deg	2.65E+01		2.84E+01	
slant range release to impact, ft	2.64E+03		2.57E+03	
total time of flight, s	4.82E+00		4.98E+00	
Pattern Dimensions	**Weapon #1 (time)**	**Weapon #2 (time)**	**Weapon #2 (Altitude)**	
pattern radius, ft	7.74E+01			
constant *H*_1		7.32E+02	2.33E+02	
constant *V*_1		5.20E+02	2.73E+02	
constant *V*_*T*		1.30E+01	1.30E+01	
constant *T*		2.50E-02	4.76E-02	
constant *F*		1.62E–01		
constant *P*		2.65E–01		
pattern length, ft	1.74E+02	1.80E+02	6.94E+01	
pattern width, ft	7.74E+01	1.08E+02	7.26E+01	
selection factor	0	0	1	
pattern length, ft	6.94E+01			
pattern width, ft	7.26E+01			
impact angle, deg	2.84E+01			
slant range, ft	2.57E+03			

13.8 Effect of Precision Error

The effective target length L_{ET} and width W_{ET} may be considered equivalent to the unitary weapon we considered in Chapter 12, Section 12.5 dealing with stick deliveries. We can now add in the precision error as we did for sticks to obtain

$$L_B = \sqrt{(L_{ET})^2 + 8\left[\frac{\text{SR} \times \sigma_b}{1000 \sin I}\right]^2} \tag{13.46}$$

$$W_B = \sqrt{(W_{ET})^2 + 8\left[\frac{\text{SR} \times \sigma_b}{1000}\right]^2} \tag{13.47}$$

These equations are similar to Eqs. (12.17) and (12.18) except that the precision error here (σ_b) is in mils rather than feet in the ground plane. The slant range SR is obtained from Fig. 13.12.

$$\text{SR} = \sqrt{h_0^2 + (R_{B1} + R_{B2})^2} \tag{13.48}$$

Note that because the slant range is from release through dispense to impact, the precision error represents an aggregation of both the dispenser and submunition individual precision errors.

13.9 Spreadsheet Implementation

At this point, a cluster weapon can be regarded as a single unitary warhead weapon with a rectangular lethal area, allowing us to also consider stick deliveries of cluster munitions. Table 13.3 shows the spreadsheet template for weaponeering this type of munition against an area of targets.

The following points regarding the spreadsheet should be observed:

- Unitary targets are distinguished from area targets by simply describing them with small values of L_A and W_A as for stick deliveries.
- Either functioning time or altitude may be specified, as well as the weapon spin speed and the particular weapon characteristic selected in the trajectory worksheet—see the selection factor. The appropriate pattern dimensions are returned to the main spreadsheet in the Trajectory Results section.
- The computation of W_B and L_B is performed in the Weapon Effectiveness section, and from that point the analysis is the same as for stick deliveries.
- Sticks of cluster munitions may seem complicated, but they are a simple extension of the methodology for unitary warheads.

Table 13.3 Weaponeering Spreadsheet for Cluster Munitions

METHOD 5-CLUSTER WEAPON AGAINST AN AREA OF TARGETS					
Comments for this worksheet	*Basic Manual example problem 5.2*				
CONSTANTS					
Gravitational acceleration, ft/s	*g*	32.200			
Convert degrees to radians	deg2rad	0.017			
TARGET PARAMETERS					
Target length, ft	LA	2000			
Target width, ft	WA	500			
WEAPON PARAMETERS					
Drag: L = low, H = high, D = dispenser		D			
Number of submunitions/dispenser	*n_b*	202			
Ballistic dispersion, mils	Sigma-*b*	10.200			
Functioning time, s	*T_f*	1.600			
Functioning altitude, ft	*H_f*	0.000			
Single dispenser pattern radius, ft	*R_p*	75.000			
Single dispenser pattern length, ft	*L*	180.000			
Single dispenser pattern width, ft	*W*	272.000			
Submunition reliability	*Rs*	0.930			
Dispenser reliability	*R*	0.800			
AIRCRAFT RELEASE DATA					
Airspeed: KTAS	*V*	550.000		928.400	ft/sec
Dive angle, deg	theta	0.000		0.000	radians
# of release pulses	*nr*	8.000			
# weapons/release pulse	*np*	1.000			
Intervalometer setting, s	*dt*	0.200			
Release alt. first weapon, ft	*yi*	1000.00			
Release alt. last weapon, ft	*yn*	1000.00			
Average release altitude, ft	*ybar*	1000.00			
TRAJECTORY RESULTS					
Vertical velocity, ft/s	*Vv*	0.000			
Horizontal velocity, ft/s	*Vh*	928.400			
Drag constant, Cd/M	*c0*	0.375			
constant *a* used in trajectory	*a*	85.867			
		14.300			
time of flight, s	*t_flight*	14.300			
Horizontal velocity at impact	*Vhi*	4.353			
Vertical velocity at impact	*Vvi*	85.464			
Impact angle, deg	*I*	87.084			
Weapon spacing on ground, ft	*Sb*	185.680			
Stick length, ft	*Ls*	1299.76			
Stick width, ft	*Ws*	18.000	See Data Sheet		
Slant range, first weapon, ft	*Sri*	2659.30			
Ground range, first weapon, ft	*Gri*	2464.12			
Ground range, ft	*Gr*	3114.00			
Slant range, ft	*SR*	3270.63			

Table 13.3 Weaponeering Spreadsheet for Cluster Munitions (Continued)

DELIVERY ACCURACY					
			CEP mils	**CEP ground**	**REP/DEP**
CEP, ft ground, normal or mils	CEP		0	0	
Deflection error probable, ft	DEP	150.000	0	0	150.000
Range error probable, ft	REP	170.000	0	0	170.000
Ballistic error, deflection, ft	sigma-*bd*	33.360		10.2	
Ballistic error range, ft	sigma-*br*	33.404			

WEAPON EFFECTIVENESS					
Effectiveness index type			MAE_*b*	MAE_*f*	AET
Effectiveness value: submunition	*EI*		0.000	4666.00	4666.000
# submunitions causing damage	*nx*	17.903			
Edge effect adjustment, ft	*RA*	74.006			
Single cluster weapon					
Single dispenser width with RA	WET	420.0			
Single dispenser length with RA	LET	328.0			
Effective target area with RA	AET	137769			
Pd one submunition	Ps	0.089			
Pd all submunitions	Pcd_wep	1.000			
Single weapon effective width, ft	WB	430.5			
Single weapon effective length ft	LB	341.3			
Stick calculation					
Effective stick pattern, ft	WP	448.5			
	LP	1641.1			
	AP	736005			
Conditional prob of damage	PCD1	0.938			
Degree of overlap: deflection	No_*d*	0.960			
Cond prob damage: deflection	PCD*d*	0.900			
Degree of overlap: range	No_*r*	1.664			
Prob damage within pattern	Pcds	0.978			

FD_1 CALCULATION					
	Range		**Deflection**		
Sigma, ft	252.039		222.38		
Effective pattern size LEP/WEP	2000.000		500.00		
Integral Limits					
s	2000.000		500.00		
t	0.000		0.000		
a	–5.611		–1.590		
b	0.000		0.000		
Integrals					
integral #1	0.0		0.0		
integral #2	0.5		0.5		
integral #3	0.5		0.5		
integral #4 & integral #5	–0.1		–0.3		
Fractional coverage Fr/Fd	0.899		0.649		
FD_1	**0.336**				

FORCE REQUIREMENTS					
Desired FD	0.700				
# of sorties for this FD	2.938				
# sorties available NA	4.000				
Expected FD from NA sorties	0.806				

Chapter 14 Projectiles

14.1 Introduction

In this section we consider the effect of aircraft-mounted guns and unguided rockets on ground targets. Most fighter/attack aircraft have forward-firing guns intended for air-to-air combat, although they may also be used for ground attack. The USAF A-10 aircraft is one of the few platforms where the gun is considered a primary weapon against ground targets; it consists of the GAU-8 seven-barreled 30-mm cannon, as shown in Fig. 14.1.

The fire control system for the gun compensates for the effects of wind and aircraft maneuver, and places an aiming cue on the heads-up display (HUD) that allows the pilot to select the target in a manner similar to Continuously computed impact point (CCIP*) bombing. In some aircraft, air-to-air radar will lock on to the target to give accurate range information. The gun fires a mix of both high-explosive incendiary (HEI) and armor-piercing incendiary (API) ammunition. The PGU-13/B HEI round employs

Fig. 14.1 A-10 aircraft with GAU-8 30-mm cannon.

*See Volume 2, Chapter 4.

a standard M505 fuze and explosive mixture with a body of naturally fragmenting material that is effective against lighter vehicle and materiel targets. The PGU-14/B API round has a lightweight body that contains a subcaliber, high density penetrator of depleted uranium (DU). In addition to its penetrating capability, DU is a natural pyrophoric material, which enhances the incendiary effects. The firing rate is approximately 3900 rounds/min.

Another example of an aircraft gun system is the AC-130 series of gunships, which have a variety of weapon systems pointing out of the port side of the aircraft, as shown in Fig. 14.2. Depending on the variant, weapons consist of a mix of 7.62-mm projectile, 20-mm cannon, 40-mm cannon, and a single 105-mm adapted artillery cannon.

This aircraft will fly in a low altitude profile at constant bank angle and speed (approximately circular) around the target while the gun crew executes a ranging process similar to the observer-adjusted field artillery process described in Volume 2, Chapter 5, Section 5.3. Miss distances are input to the fire control system, which then estimates and implements corrections to the gun pointing angle compensating for Meteorological (MET) variations

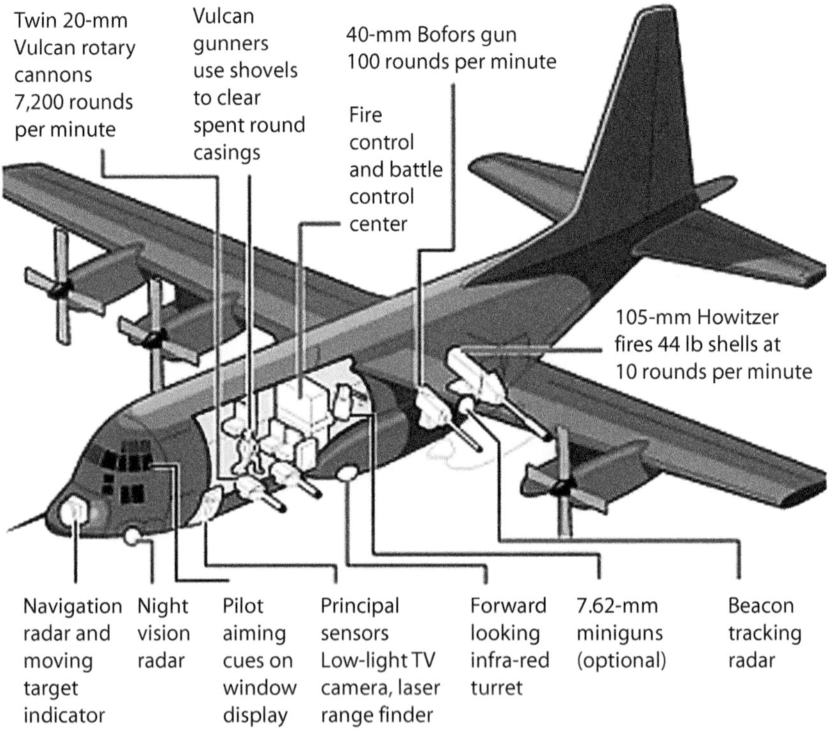

Fig. 14.2 AC-130 H gunship.

Fig. 14.3 Launcher options for 2.75-in. rockets.

and particularly the wind component. After at least one circuit of the target where about four or five ranging shots have been made, adjustments are complete and the fire for effect phase takes place.

In addition to guns, other weapon systems covered in this chapter include unguided rockets. Figure 14.3 shows the two configurations of launcher for 2.75-in. unguided rockets. One contains 19 rounds and the other contains 7; either may be fitted to a wide range of fixed wing aircraft and helicopters.

Rocket warheads vary; Fig. 14.4 shows those fitted to the 2.75-in. Hydra 70 series.

The LAU-10 rocket pod can be armed with up to four Zuni 5-in. rockets with varying warheads, typically a 50-lb high explosive round. These larger rockets are carried on either fixed wing or rotary wing aircraft.

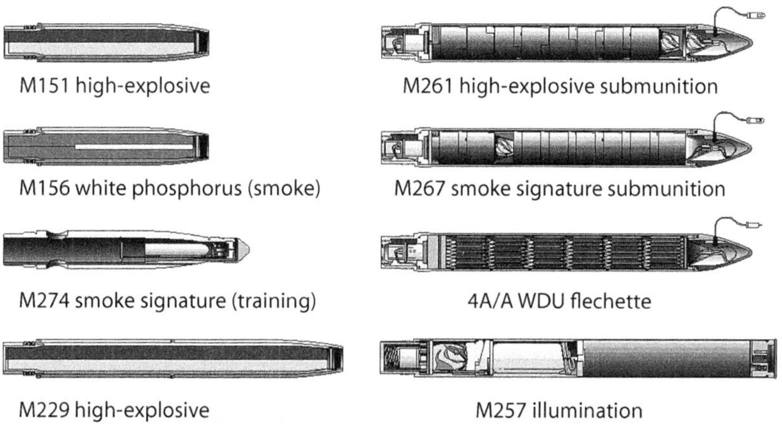

Fig. 14.4 Warheads for 2.75-in. rockets.

The models described in this chapter for projectiles begin with the simplest possible, where the trajectory of the round is assumed to be a straight line between the aircraft and the target.

14.2 Simplified Projectile Methodology

Figure 14.5 shows the geometry of an aircraft attacking a ground target with unguided rockets or guns. Although this view shows a strafing attack with forward firing guns, the same geometry would apply to the gunship weapons, except of course that the projectile trajectory is normal to the aircraft axis.

The following assumptions will be made regarding the attack:

- All rounds are fired rapidly, meaning that the position of the aircraft does not change during the firing. This firing is called a *salvo* for rockets and a *burst* for a gun. This assumption implies a constant accuracy.
- No trajectory of the weapon is considered, implying the round moves in a straight line from the aircraft to the target at high speed. At firing ranges of 5000 ft or less, this assumption is reasonable; however, the A-10 may attack targets at ranges much greater than this, in which case the projectile trajectory has to be included.
- All rounds fired are uncorrelated with other rounds fired during the same burst or salvo. In practice the shooter will be able to see where the rounds are striking because the bullet mix includes tracers every four or five rounds, so the shooter can "walk" the bullet stream onto the target. Although addressed later in the chapter, this effect is ignored for now.
- If more than one round is required to kill the target, the damage caused by a particular hit is uncorrelated with the damage caused by other hits. For some specific gun/target combinations, test results have estimated an average number of rounds needed to successfully destroy the target. An example is the Mk-15 Phalanx CIWS ship-mounted gun system where more than one hit is required to defeat an attacking missile.

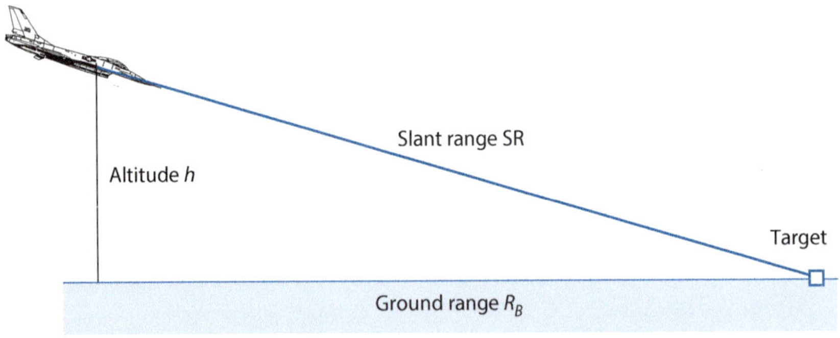

Fig. 14.5 Attack geometry for rockets and guns.

14.3 Accuracy Considerations

In calculating effectiveness, all calculations will be performed in the normal plane and will assume that all accuracy measures are normally distributed in this plane also. The first stage is to combine the aiming error for projectiles with the precision error. Data are usually given in the normal plane (mils), but the aiming error is an error probable measure (50%), whereas the precision error σ_b is a standard deviation (67%). Considering the range direction, first we convert the precision error to an error probable using Eqs. (4.4) and (4.5), which gives

$$\mathrm{REP}_b = 0.6745\sigma_b \tag{14.1}$$

In some cases, where the pilot is attacking an area of target elements on the ground plane, a strafing box is defined by the strafing length L_S and width W_S, as shown in Fig. 14.6.

Because this box defines an area in which the pilot will randomly aim the gun and move the aimpoint, it is considered a component of precision error. The total precision error in the presence of a strafing box is computed by combining the round-to-round dispersion in the normal plane with the projection of the strafing box by the round impact angle as follows:

$$\sigma_{br} = \sqrt{(\sigma_b \times \mathrm{SR}/1000)^2 + (L_S \times \sin I/2)^2} \tag{14.2}$$

$$\sigma_{bd} = \sqrt{(\sigma_b \times \mathrm{SR}/1000)^2 + (W_S/2)^2} \tag{14.3}$$

If the pilot aims at a unitary target, the strafing box dimensions are set to zero. Because we are going to calculate the effectiveness as the probability a single round will hit the target, the method assumes that a combined range error probable (REP) value (in mils) in the normal plane is obtained by root sum squaring (RSS) the error probable values of the aiming and ballistic

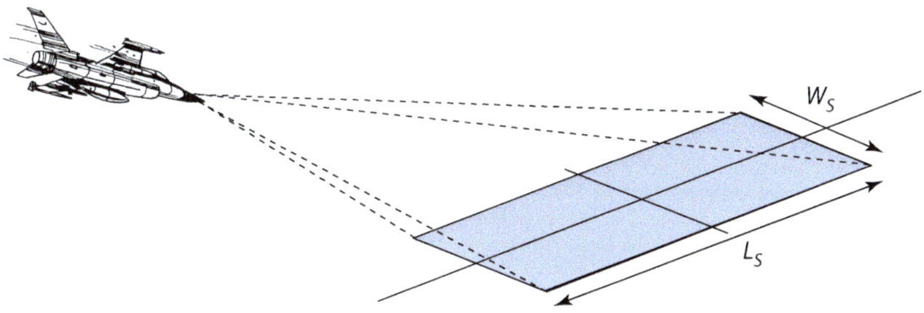

Fig. 14.6 Ground plane strafing box.

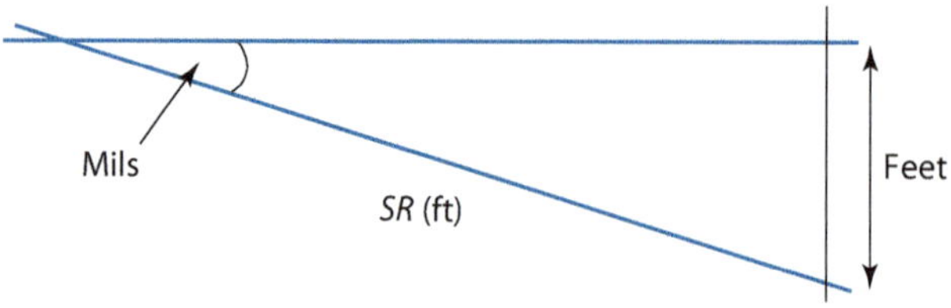

Fig. 14.7 Converting REP in mils to REP in feet.

errors as follows[†]:

$$\mathrm{REP}'_N = \sqrt{\mathrm{REP}_N{}^2 + (0.6745\sigma_{br})^2} \tag{14.4}$$

$$\mathrm{DEP}'_N = \sqrt{\mathrm{DEP}_N{}^2 + (0.6745\sigma_{bd})^2} \tag{14.5}$$

These equations describe a bivariate normal distribution in the normal plane including both aiming errors, precision errors, and the variability of the strafing dimensions, if present. Because both REP and deflection error probable (DEP) are measured in mils, we obtain REP and DEP in feet in the normal plane at the slant range SR from Fig. 14.7 as

$$\mathrm{REP}\,(ft) = \frac{\mathrm{SR} \times \mathrm{REP}'_N}{1000} \tag{14.6}$$

$$\mathrm{DEP}\,(ft) = \frac{\mathrm{SR} \times \mathrm{DEP}'_N}{1000} \tag{14.7}$$

Finally, the standard deviation in feet at the slant range SR is given by

$$\sigma'_{\mathrm{NR}} = \frac{\mathrm{REP}_N}{0.6745} \tag{14.8}$$

A similar expression is derived for the deflection direction

$$\sigma'_{\mathrm{ND}} = \frac{\mathrm{DEP}_N}{0.6745} \tag{14.9}$$

This completes the normal plane accuracy analysis; attention is now directed to the damage function.

14.4 Damage Functions

Damage functions for guns and rockets are usually expressed either as a mean area of effectiveness ($\mathrm{MAE}_B/\mathrm{MAE}_F$) or as a vulnerable area A_V

[†]We know these error sources have to be applied separately, so this assumption is a shortcoming of the method.

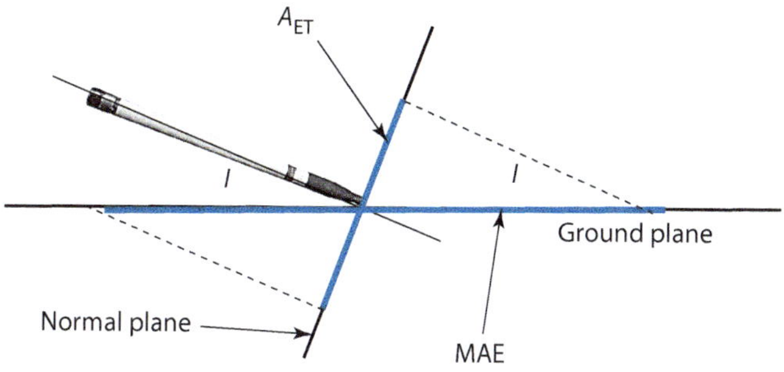

Fig. 14.8 Rotating from the ground to the normal plane.

effectiveness index (EI). Gun ammunition may be penetrating (ball round) or explosive, but rockets are always explosive. The methods for penetrating round munitions may be considered similar to those used for fragments, where the round is simply a single large fragment; therefore, vulnerable area is the appropriate EI. All computations will be dealt with in the normal plane, so the vulnerable area presents no problem, because it is always specified in this manner.[‡] Hence, we can write for the effective target area in the normal plane

$$A_{ET} = A_v \tag{14.10}$$

For an MAE that is always specified in the ground plane, however, the equivalent A_{ET} in the normal plane has to be computed by projecting from the ground plane into the normal plane by the impact angle, as seen in Fig. 14.8.

If the ground plane MAE dimensions are L_{ET-G} and W_{ET-G}, then the projection into the normal plane gives the following:

$$L_{ET} = L_{ET-G} \times \sin I \tag{14.11}$$

$$W_{ET} = W_{ET-G} \tag{14.12}$$

In this equation, I is the impact angle of the projectile, which, ignoring any trajectory considerations, is the dive angle of the aircraft at the firing point. This result applies if the effectiveness index is an MAE_B or an MAE_F.

14.5 Effectiveness Calculations

The probability of damage is considered to be the probability a single round hits the target area in the normal plane. The first step to calculating this is to assume that if we are using a vulnerable area EI then the A_v

[‡]For example, an output file similar to that shown in Table 7.4 would provide the EI, albeit averaged around the target azimuth.

is represented as a square of dimensions L_{ET} by W_{ET} such that

$$L_{\mathrm{ET}} = W_{\mathrm{ET}} = \sqrt{A_{\mathrm{ET}}} = \sqrt{A_V} \tag{14.13}$$

If the EI is an MAE, then the target area has dimensions given by Eqs. (14.11) and (14.12).

Either way, this situation may be represented in the normal plane by Fig. 14.9, which also shows a rectangle of sides σ'_{NR} and σ'_{ND}, representing the combined weapon accuracy.

The probability that one round will fall in the range direction within the $\pm L_T/2$ value is PD_{1x} and is equal to the area A beneath the Gaussian PDF between these limits, as shown in Fig. 14.10.

We can write

$$\mathrm{PD}_{1x} = \int_{x=-L_{\mathrm{ET}}/2}^{x=L_{\mathrm{ET}}/2} f(x)\mathrm{d}x \tag{14.14}$$

Using the NORMDIST function of Excel, this can be written as

$$\mathrm{PD}_{1x} = \{\mathrm{ND}[L_{\mathrm{ET}}/2,\ 0,\ \sigma'_{\mathrm{NR}},\ 1]\} - \{\mathrm{ND}[-L_{\mathrm{ET}}/2,\ 0,\ \sigma'_{\mathrm{NR}},\ 1]\} \tag{14.15}$$

From Fig. 14.10 we can also write for the deflection direction

$$\mathrm{PD}_{1y} = \{\mathrm{ND}[W_{\mathrm{ET}}/2,\ 0,\ \sigma'_{\mathrm{ND}},\ 1]\} - \{\mathrm{ND}[-W_{\mathrm{ET}}/2,\ 0,\ \sigma'_{\mathrm{ND}},\ 1]\} \tag{14.16}$$

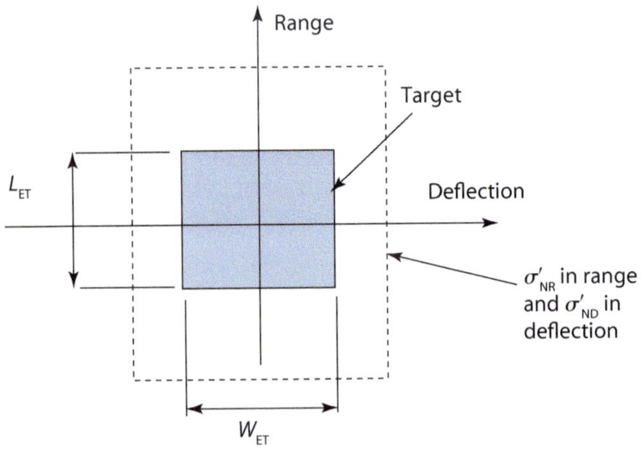

Fig. 14.9 A_{ET} in the normal plane.

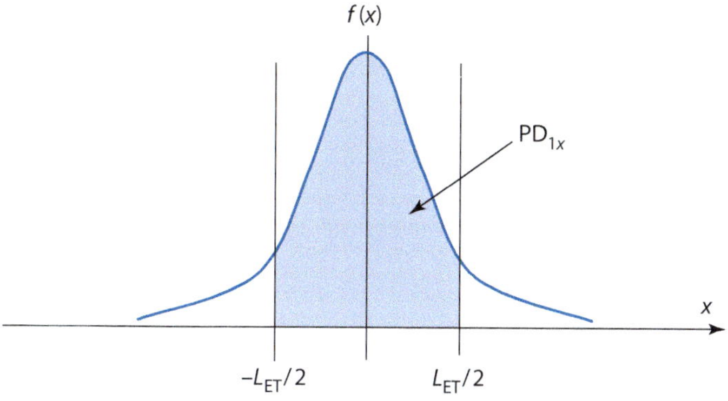

Fig. 14.10 Probability one round hits square target.

So, the probability that if only one round is fired it will hit the target is given by

$$PD_1 = PD_{1x} \times PD_{1y} \tag{14.17}$$

If we fire a total of n uncorrelated rounds, the probability that at least one of n will hit the target is obtained by powering up the single hit probability given by Eq. (14.17).

$$PD = 1 - (1 - PD_1)^n \tag{14.18}$$

Note that in using Eq. (14.18) to power up for multiple rounds we are implying that each is independent; however, recall back in Eq. (14.4) we RSS the aiming and precision errors. Clearly this can only be done for accuracy purposes; that is, Eq. (14.4) is correct, however many rounds are fired, but the effectiveness calculation in Eq. (14.18) is not correct and will give a value higher than that achieved. This is a shortcoming of this simple method; the correct approach would be to use the error terms separately in a Monte Carlo simulation, as shown in Fig. 9.7.

14.6 PD When More Than One Round is Required for a Kill

In the previous analysis, PD was computed assuming that only one hit was required to achieve the required kill level for which the EI was calculated. For some targets, experiments have shown that more than one round is required to produce the necessary damage. The methodology has to be modified to accommodate this feature, and to do this the binomial distribution is utilized.

Chapter 2, Section 2.9 described the binomial distribution in detail and illustrated its relationship to the powering up technique discussed earlier. Excel has a useful binomial distribution function that avoids having to reproduce the data shown in Appendix A, Table A.2. The BINOMDIST function is described here:

```
Returns the individual term binomial distribution
probability.

Syntax: BINOMDIST(number_s,trials,probability_s,cumulative)
        Number_s is the number of successes in trials.
        Trials is the number of independent trials.
        Probability_s is the probability of success on
        each trial.

        Cumulative is a logical value that determines the form
        of the function. If cumulative is TRUE, then BINOMDIST
        returns the cumulative distribution function, which
        is the probability that there are at most number_s suc-
        cesses; if FALSE, it returns the probability mass
        function, which is the probability that there are
        number_s successes.
```

Example 14.1

The flip of a coin can only result in heads or tails. The probability of the first flip being heads is 0.5, and the probability of exactly 6 of 10 flips being heads is

BINOMDIST(6, 10, 0.5, FALSE) equals 0.205078

We can use the function instead of powering up in the following manner:

1. The value of PD_1 is computed as before from Eq. (14.17). Recall that this is the probability that one round will hit (and kill) the target.
2. The probability that at least one out of n rounds fired hits the target is computed from

$$PD = 1 - \text{BINOMDIST}(0, \text{n}, PD_1, 1) \tag{14.19}$$

   The term BINOMDIST(0, n, PD_1, 1) returns the value that at most zero rounds hit the target, so subtracting this value from unity gives the probability at least one round does. This will give the same result as Eq. (14.18)
3. If n_{rK} is the required number of hits to obtain the kill level of interest, then the probability of this happening is PD_{rK}, where

$$PD_{rK} = 1 - \text{BINOMDIST}[(\text{n}_{\text{rK}} - 1), \text{n}, PD_1, 1] \tag{14.20}$$

   This is obtained from Eq. (2.60), which determines the probability that at least x events out of n are successful.

Suppose we fire $n = 100$ rounds, and at least 10 rounds have to hit the target to kill it. The BINOMDIST function in Eq. (14.20) returns the probability that at most nine rounds miss the target. Again, subtracting this value from unity gives the probability that at least 10 rounds strike the target.

14.7 Spreadsheet Implementation

A complete spreadsheet implementation of this methodology for projectiles is given in Table 14.1. Most of the steps should be apparent from the analysis in this chapter, but the following points should be noted:

- The reliability (of firing) is accounted for in the Aircraft Release Data block.
- The impact angle of the weapon is set to the aircraft dive angle.
- The Delivery Accuracy data block combines the precision errors with the aiming error with no strafing box.
- The value of A_{ET} in the normal plane accounts for the impact angle for the MAE_F value if given, but is not used for the vulnerable area.
- The binomial distribution function accounts for the number of rounds needed to kill the target, specified in the Weapon Effectiveness data block.

14.8 Higher Fidelity Model

The problem with the simplified methodology just described is that the dependent and independent errors are combined into a single accuracy, thereby allowing an analytical solution to be obtained. We know this leads to errors in the damage calculation, so it is valid to ask if a higher fidelity approach is possible.

The answer is that it is, but it depends on the availability of target vulnerability data and will result in a Monte Carlo–based approach; therefore, it may take longer to compute. Recall in Chapter 7, Section 7.7 where we performed shotline analysis on a target, as reproduced in Fig. 14.11.

We computed the shotline $P_{k/sh}$ multiplied by the presented area of the cell containing the shotline to get the cell vulnerable area, and then summed over the target to get an A_V for that azimuth and elevation. Instead of doing this, we could just associate the shotline P_K with the cell for which it is calculated and present these data overlaying the target. This produces what is known as a cell-by-cell ($\mathrm{C} \times \mathrm{C}$) data file; an equivalent image is shown in Fig. 14.12, where the P_K values are color coded however in this image we see only gray shades.

If we know the direction of attack in terms of azimuth and elevation and if there is a $\mathrm{C} \times \mathrm{C}$ vulnerability file available, then we can establish a Monte Carlo simulation keeping the dependent and independent errors separate for burst fire weapons, as shown in Fig. 9.7. The process would be:

1. Select an aimpoint (e.g., the centroid of the target presented area).
2. Take a random draw from the dependent error to determine the MPI.
3. For each round in the burst, take a random draw from the independent error to determine the cell each round impacts.

Table 14.1 Spreadsheet to Calculate PD for Projectiles

METHOD 4 - GUNS AND ROCKETS				
COMMENTS: Example 4-1 Basic Manual(PD = 0.22)				
CONSTANTS				
Gravitational acceleration, ft/s	*g*	32.200		
Convert degrees to radians	deg2rad	0.017		
WEAPON PARAMETERS				
Ballistic dispersion normal, mils	sigma_*b*	9.000		
Reliability	R	0.830		
AIRCRAFT RELEASE DATA				
Airspeed: KTAS, ft/s	*V*	450		759.600
Dive angle, deg	theta	30		0.524
Number of rounds fired	*n*	38		
Number of rounds actually fired	*nf*	32		
Slant range, ft	SR	5000		
TRAJECTORY RESULTS				
Impact angle, deg	*I*	30		
Ground range, ft	*Gr*	4330		
Altitude, ft	*y*_bar	2500		
DELIVERY ACCURACY				
		REP	**DEP**	
REP/DEP normal, mils (no BD)		12.00	9.00	
CEP, normal, mils (no BD)		0.00	0.00	0.00
REP/DEP, normal, mils (with BD)		13.448	10.856	
WEAPON EFFECTIVENESS				
Effectiveness type		MAE (blast or frag)	Av_n	
Effectiveness value: enter one only		500.00	0.000	
Effectiveness in the normal plane		250.000	0.000	
Number of hits to kill the target (> 0)	*n_rk*	1		
Target area in normal plane	AET	250.000	0.000	
SINGLE SORTIE EFFECTIVENESS				
		Range	**Deflection**	
Target dimension, ft	*T_d*	15.811	15.811	
Sigma at the slant range, ft	sigma_*sl*	99.689	80.474	
PD_1		0.063	0.078	
Prob. exactly 1 round hits target	PD1	0.00495		
Prob. at least 1 of nf hits target	P_{HIT}	0.142		
Prob. target killed	PD	0.142		

(Continued)

Table 14.1 Spreadsheet to Calculate PD for Projectiles (*Continued*)

FORCE REQUIREMENTS				
Desired PD	0.700			
# of sorties for this PD	7.832			
# sorties available NA	15.000			
Expected PD from NA sorties	0.900			

4. From the C × C file, extract the (i, j) cell P_K for the qth round and use the survivor rule to calculate the P_K for the burst.

$$\overline{P}_K = 1 - \prod_{q=1}^{q=nrb} (1 - P_K(i, j, q)) \tag{14.21}$$

This process would be repeated for as many iterations as needed to converge to a solution.

14.9 Gunnery Model with Correlation: Aimpoint Wander

The models described previously assumed that each impact point in range or deflection was independent of all other impact points. Considering, for example, the impact point relative to the Desired mean point of impact (DMPI) in the range direction, each impact point in the burst is obtained as a random draw from a normal distribution with zero mean and a standard deviation obtained from REP.

Experience has shown that for high rate of fire guns, this assumption may not be valid, and a phenomenon known as aimpoint wander, or just aim wander, is present. Aim wander is the process that occurs when an automatic weapon is fired at a fixed target in which the aimpoint will move in the normal plane mainly due to vibration of the gun barrel tube.

This effect has been observed in tests and combat for both aerial gunnery and ground-based direct fire situations, and has been analyzed since the 1940s [1]; this is explained further by Dawson [2].

Figure 14.13 shows that if a single round is fired from a gun, it excites a lateral vibration once the bullet leaves the tube.

Clearly if a burst of rounds is fired, subsequent bullets will be affected by this vibration if the firing rate is close to or exceeds the same frequency of vibration.

This effect, which occurs for all rapid fire weapon systems, implies that each impact point is not independent of other impact points but is correlated to them. Specifically, the impact point of the next round depends on the impact point of the current round and is therefore correlated to it, as shown in Fig. 14.14.

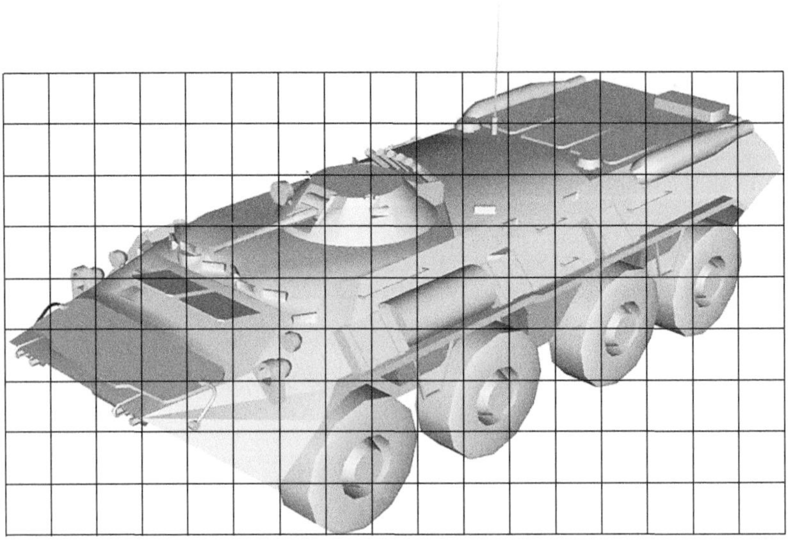

Fig. 14.11 Shotline analysis of a target.

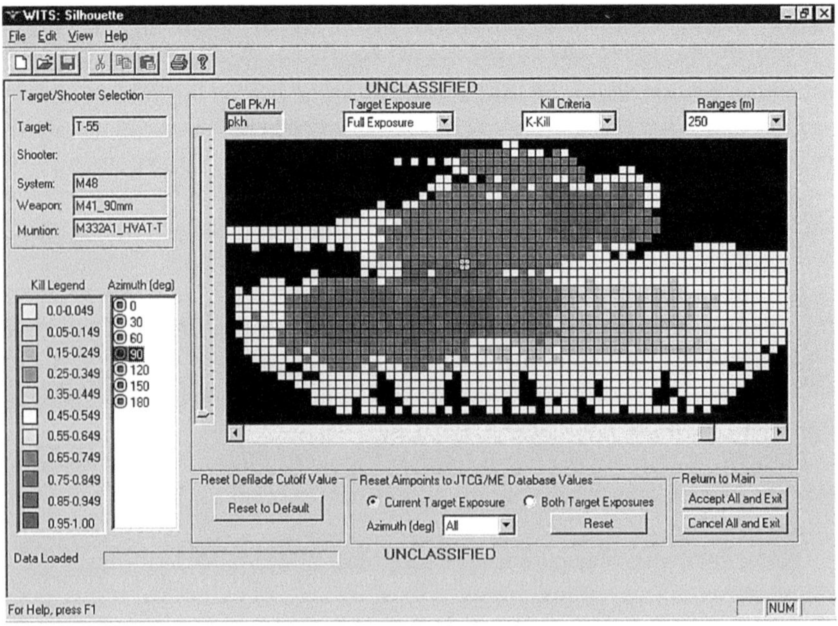

Fig. 14.12 Cell-by-cell data file.

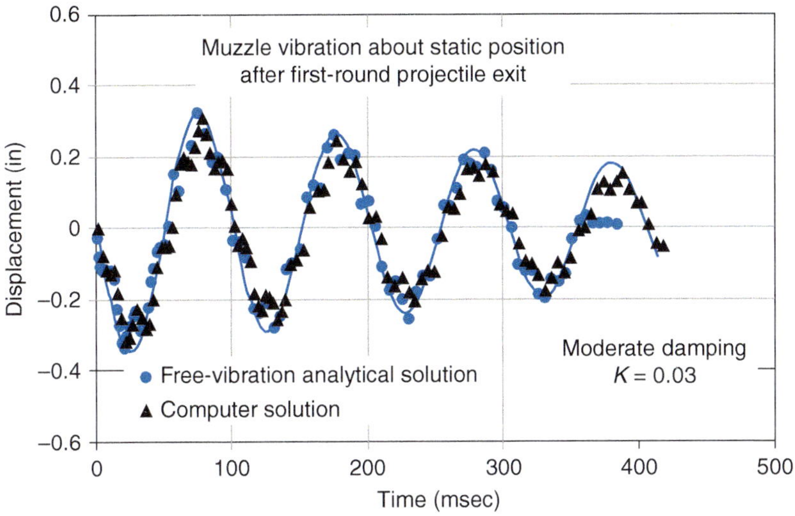

Fig. 14.13 Lateral vibration of gun barrel for single shot.

This figure requires some additional explanation. Recall Chapter 4, where the occasion-to-occasion and round-to-round errors were discussed. Figure 14.14 looks similar to Fig. 4.3 where the individual impact points are grouped around the MPI. In Fig. 14.14, however, we are assuming that the precision error is zero, and the dispersion we see is due only to the aim-point itself moving in the normal plane.

In the absence of precision error, uncorrelated rounds would be grouped around the DMPI with a dispersion determined by the aiming error, whereas for fully correlated rounds the MPI is fixed at the first random draw for the aiming error, and all rounds would pass through this point. For an

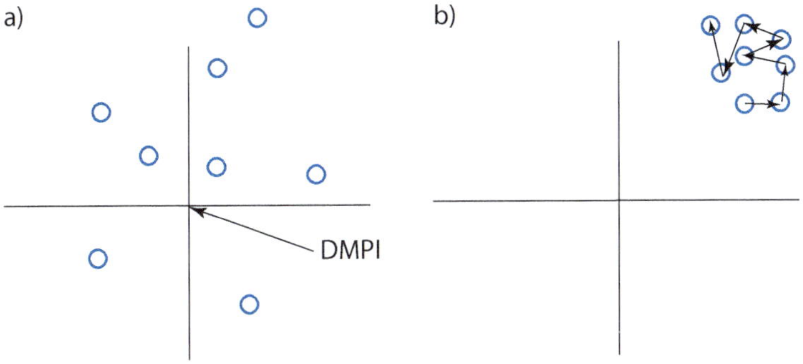

Fig. 14.14 Uncorrelated (left) and correlated (right) impacts.

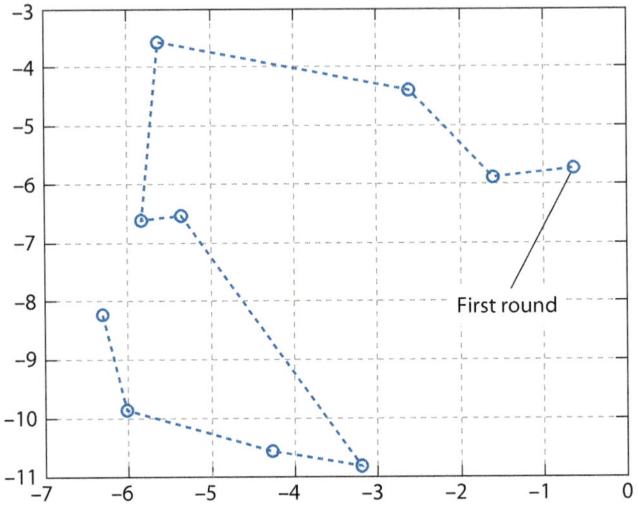

Fig. 14.15 Aimpoint wander in the normal plane.

intermediate value of correlation, the aimpoint will move in the normal plane. The dispersion we see in Fig. 14.14, therefore, is due to aimpoint wander and not precision error, which would be added to the aimpoint generated for each round. A more detailed example of this process is shown in Fig. 14.15. It will be shown in the next section that the amount of correlation is related to the firing rate.

The figure shows the impact point of the first round (−0.6, −5.7) drawn from a normal distribution of the aiming error, and the impact points for the subsequent, correlated rounds.

14.10 Round-to-Round Correlation

To simplify the process as much as possible, the effects of precision error will be ignored. In order to study the process mathematically, some additional statistics need to be explained. We begin by considering a sequence of random data points such as

$$x(i) = 3, 6, 9, 8, 7, 5, 4 \tag{14.22}$$

The correlation between these points may be determined by calculating the correlation coefficient ρ from the following formula:

$$\rho_k = \frac{\sum_{i=1}^{i=n-k} (x_i - \overline{x})(x_{i+k} - \overline{x})}{\sum\limits_{i=1}^{i=k} (x_i - \overline{x})^2} \tag{14.23}$$

This defines the kth order autocorrelation coefficient for a set of n data points. The order represents the correlation between one element of the data set with the kth element ahead of it. We will consider only the lag coefficient of order 1, meaning the correlation of adjacent elements in the data. For this case, Eq. (14.23) becomes

$$\rho_1 = \frac{\sum_{i=1}^{i=n-1}(x_i - \overline{x})(x_{i+1} - \overline{x})}{\sum_{i=1}^{i=n}(x_i - \overline{x})^2} \tag{14.24}$$

For the data set given in Eq. (14.22), the numerator of Eq. (14.24) becomes (3 – 6)(6 – 6) + (6 – 6)(9 – 6) . . . (5 – 6)(4 – 6) = 9 (i.e., six terms). The denominator is $(3 - 6)^2 + (6 - 6)^2 \ldots + (4 - 6)^2 = 28$ (i.e., seven terms). The correlation coefficient is therefore $\rho = 9/28 = 0.32$. A sequence of uncorrelated random numbers would generate $\rho = 0$, whereas a sequence of identical numbers would generate $\rho = 1$.

Now consider a set of 10 samples of miss distances in the range direction where the impact points are correlated. Five such sets might be tabulated, as shown in Table 14.2. Note that sample 5 is the one plotted in Fig. 14.15.

The following assumptions are made regarding the data:

- Clearly x_1 is normally distributed because that is how it was generated; however, we further assume that x_i for a given i (a column of data) is also normally distributed.
- Each sequence of impact points (a row of data) is stationary; that is, their statistical properties do not change with time.
- The correlation coefficient ρ for a sequence of data (a row) is of the form

$$\rho = \exp(-a \times \Delta t) \tag{14.25}$$

where Δt is some time interval and a is a positive constant to be determined.

Consider Eq. (14.25) in relationship to the rate of fire R_F of the weapon. If R_F is high, then all rounds will be grouped closely together and close to the impact point of the first round, because the shooter will have little time to

Table 14.2 Five Samples of 10 Impact Points

Sample #	Impact Point x_i in the Range Direction (ft)									
1	5.7	6.6	7.3	7.6	7.0	4.9	3.8	3.9	4.0	1.8
2	6.6	8.1	9.2	9.4	7.5	5.0	5.9	2.2	1.7	3.6
3	−5.3	−3.3	−6.4	−3.1	−2.6	−1.5	−0.6	−0.4	−1.1	−2.5
4	11.8	9.9	9.8	10.6	11.7	11.9	10.4	10.7	10.8	10.1
5	−0.6	−1.5	−2.6	−5.6	−5.8	−5.3	−3.1	−4.2	−6.0	−6.3

adjust the position of the weapon before all rounds are discharged. If R_F is low, however, then the weapon may be retrained at the DMPI between rounds. It seems logical, therefore, that high rate of fire is associated with highly correlated impact points and low rate of fire is associated with low correlation between rounds. Equation (14.25) is therefore written in the form

$$\rho = \exp\left(-\frac{a}{R_F}\right) \tag{14.26}$$

It may be seen that large R_F gives ρ about unity and small R_F yields ρ close to zero, matching the expected relationship to the rate of fire. The first extreme case ($\rho = 1$) is usually referred to as the salvo case, whereas the other extreme ($\rho = 0$) is the uncorrelated or independent fire case represented by the model in the first part of this chapter.

Returning to the three assumptions just made, it can be shown that each sample of impact points may be represented by a lag 1 Markov process where each data point depends only on the previous one. In Chapter 2 we defined the probability density function for a normally distributed random variable x as follows:

$$f(x) = \frac{1}{\sigma\sqrt{2\pi}} \exp\left[-(x-\mu)^2/2\sigma^2\right] \tag{14.27}$$

We are now going to consider the probability density function of the condition variable $f(x_{i+1}|x_i)$. This reads as the probability density function of the next impact point given the current impact point. For example, consider sample 3, $i = 4$ in Table 14.2. The data point is -3.1. Examine the column for $i = 5$ corresponding to the next impact point. This column represents a data set that is normally distributed (assumption 1) and will therefore have a mean and standard deviation relative to the value 3.1. It may be shown that the probability density function of the conditional variable for a lag one Markov process is given by

$$f(x_{i+1}|x_i) = \frac{1}{\sigma(1-\rho^2)\sqrt{2\pi}} \exp\left[-\frac{x_{i+1} - \rho x_i}{2\sigma(1-\rho^2)}\right] \tag{14.28}$$

Comparing Eqs. (14.27) and (14.28), it can be seen that the PDF for the conditional variable is Gaussian with a mean value and standard deviation given by

$$\mu_{x(i+1)} = \rho x_i \tag{14.29}$$

and

$$S_{x(i+1)} = \sigma\left(1-\rho^2\right) \tag{14.30}$$

In these equations, ρ is the correlation coefficient and σ is the standard deviation of the population.

14.11 Generating Correlated Impact Points

How would we generate a sequence of impact points where there is correlation between the rounds? Consider first how this would be accomplished if correlation were zero (i.e., each round is independent of the others). If z_i is a sequence of normally distributed random numbers with zero mean and standard deviation of one, then the physical domain variable x with mean μ and standard deviation σ is obtained from the familiar mapping

$$x = \mu + \sigma z \tag{14.31}$$

Considering Eqs. (14.28) through (14.30) for the case where correlation is present, we may infer that the conditional variable may be obtained in the same manner because it will be normally distributed also with, of course, a different mean and standard deviation. Hence we may write

$$x_{k+1} = \rho x_k + \sigma\sqrt{1-\rho^2}z \tag{14.32}$$

In this equation, z is the normal standard random number as before, and σ is the standard deviation associated with the aiming error. Equation (14.32) provides a simple recursive means to generate a sequence of random numbers that are correlated from the current value and a random number drawn from a zero mean, unity variance normal distribution.

Because the sequence depends on the current value, the question is what is the start point for the sequence? The answer is a random number generated by Eq. (14.31) for the initial aimpoint. The sequence therefore becomes

$$\begin{aligned}
x_1 &= \sigma z_1 \\
x_2 &= \rho x_1 + \sigma\sqrt{1-\rho^2}z_2 \\
x_3 &= \rho x_2 + \sigma\sqrt{1-\rho^2}z_3 \\
&\vdots \\
x_k &= \rho x_{k-1} + \sigma\sqrt{1-\rho^2}z_k
\end{aligned} \tag{14.33}$$

Extending this concept to the deflection direction, it is now possible to generate impact points in the normal plane if the aiming errors and correlation coefficient are given. It is from such a pair of sequences that Fig. 14.15 was generated. Note that it is assumed here that the correlation coefficient is the same in range and deflection, although it is simple to modify the sequence for different values.

14.12 Correlation Coefficients

Equation (14.26) gave the general form used for the correlation coefficients, but the JMEM uses a slightly different form based on observed data. The forms used are different in range and deflection, and are given by

$$\rho_R = \mathrm{MAX}\{\exp(-1.5T) - 3T^2, 0\} \tag{14.34}$$

$$\rho_D = \mathrm{MAX}\{\exp\left(-0.25(T)^{0.4}\right) - 5.35T^2, 0\} \tag{14.35}$$

where the firing period T (seconds) is given in terms of the firing rate R_F

$$T = \frac{60}{R_F} \tag{14.36}$$

14.13 Effectiveness Calculations with Correlation

The approach to calculating effectiveness will be to use a Monte Carlo simulation in which a sequence of impact points is generated based on the analysis of the previous sections for correlated fire. The effectiveness calculation will then depend on the damage mechanism chosen to represent the weapon/target interaction.

14.14 Effectiveness for Penetrating Munitions with Correlation

For this case, we will represent the target as a vulnerable area in the normal plane. If this vulnerable area is given, the target may be represented as a square in the normal plane, whereas if the vulnerable area in the ground plane is given, it is first represented as a square on the ground plane and then projected as a rectangle in the normal plane by rotating it through the impact angle I.

A Monte Carlo simulation will generate the impact points relative to the DMPI, which is the center of the normal plane vulnerable area. A sample iteration is shown in Fig. 14.16.

If the target requires one hit to kill it, then each round impact point is investigated to determine if it is inside the vulnerable area. If it is, the target is killed; otherwise, it is not. A record of the number of times the target is killed N_K is maintained over the total number of iterations N_{ITER} conducted. The resultant probability of kill is then given by the following equation:

$$P_K = \frac{N_K}{N_{\mathrm{ITER}}} \tag{14.37}$$

This method could also be adapted to the case where more than one round is required to kill the target. The following MATLAB code implements the algorithm, noting how precision error is added to the impact point.

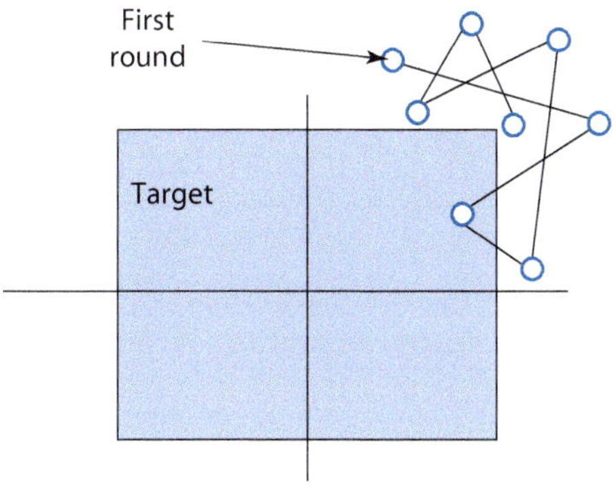

Fig. 14.16 Burst fire against a vulnerable area.

```
% Monte Carlo Gun Method with Correlation
clear all
sr=1000;          % slant range (ft)
rf=4000;          % firing rate in rounds per minute
rep=5;            % aiming error in mils, normal plane - range
dep=5;            % aiming error in mils, normal plane - deflection
rep_b=0;          % ballistic error mils, normal plane - range
dep_b=0;          % ballistic error mils, normal plane - deflection
n=10;             % number of rounds fired
van=100;          % vulnerable area in the normal plane (ft^2)
hit_max=1;        % number of hits to kill
let=sqrt(van);    % target dimension, range
wet=let;          % target dimension, deflection
iter_max=10000; % number of Monte Carlo iterations

sigma_r=sr*rep/(0.6745*1000);      % SD aiming in feet normal
sigma_d=sr*dep/(0.6745*1000);      % SD aiming in feet normal
sigma_br=sr*rep_b/(0.6745*1000); % SD ballistic in feet normal
sigma_bd=sr*dep_b/(0.6745*1000); % SD ballistic in feet normal
t=60/rf;
rho_r=max((exp(-1.5*t) -3*t*t),0);          % corr. coeff.,range
rho_d=max((exp(-0.25*t^0.4)-5.35*t*t),0);  % corr. coef. defl
pk_count=0;

for j=1:iter_max               % start of Monte Carlo loop
  count=0;                     % hits/burst, for 1 iteration
  x(1)=sigma_r*randn+sigma_br*randn; % get initial impact point
  y(1)=sigma_d*randn+sigma_bd*randn;

  for i=2:n                      % get remaining impact points
x(i)=(rho_r*x(i-1)+sqrt(1-rho_r*rho_r)*sigma_r*randn);
y(i)=(rho_d*y(i-1)+sqrt(1-rho_d*rho_d)*sigma_d*randn
x(i)=x(i)+sigma_br*randn;  % precision error
y(i)=y(i)+sigma_bd*randn;  % precision error
  end
  for i=1:n                        % is impact inside target
    if(abs(x(i))<let/2)          % hits in range
```

```
        if(abs(y(i))<wet/2)         % hits in deflection
          count=count+1;            % round hits target
        end
      end
    end                             % end checking hit loop
     if(count>(hit_max-1))           % target killed for this burst?
       pk_count=pk_count+1;         % yes, increment counter
     end
  end                               % end of Monte Carlo loop
  x(1)
  y(1)
  plot(y,x,'o-');                   % plot last impact points
  grid; xlabel('range');
  ylabel('deflection')
  pk_mean=pk_count/iter_max         % final Pk calculation
```

A few observations comparing the correlated round model with the independent round model are in order.

- For a single round, both models give the same result, as expected.
- If the correlation coefficient is set to unity, again both models give the same result. This is because for the correlated model all rounds pass through the same impact point; therefore, one round is as effective as any number more than one. The impact point distribution therefore is normal about the DMPI.
- For the data shown in the code, the correlation coefficients become 0.977 and 0.953 in range and deflection, respectively. The P_K values are 0.538 for the correlated model and 0.944 for the independent model. This might be expected because the correlation effect diffuses the impact points about the first round, rather than the DMPI, leading to a larger dispersion of rounds, hence a smaller P_K.

It is beyond the scope of this text to develop effectiveness models for all other damage functions other than the one described; however, it is not difficult to see how this would be done. For example,

- For high explosive/fragmenting rounds against unitary targets where the EI might be MAE_F or MAE_B, the target may be represented as a rectangle whose dimensions are L_{ET} and W_{ET}, respectively. These can be calculated using the methods described in Chapter 10, Section 10.2. The effectiveness would then be calculated in the same way as for a vulnerable area.
- Area targets can be analyzed by representing the damage function again as a rectangle, superimposing this lethal area onto the round impact points and using the methods developed for stick deliveries to compute an EFD.
- If the damage function is effective miss distance (EMD) (discussed in Chapter 15), this can also be represented as a rectangle on the ground plane. This would then be projected into the normal plane and analyzed as described in the previous sections.

- Higher fidelity models may be developed to represent the damage function in the form of a Carlton function or a lethal area matrix directly. These would be superimposed on the round impact points and the effect on a unitary or area target analyzed using the ARTQUICK or Matrix Evaluator methods (described in Chapter 16).

References

[1] Cunningham, L. B. C., and Hynd, W. R. B., "Random Processes in Problems of Air Warfare," Supplement to the *Journal of the Royal Statistical Society*, Vol. 8, No. 1, (1946), pp. 62–97.

[2] Dawson, T., "Gun Barrel Vibrations of Rapid-Fire Medium Caliber Guns," 43rd Annual Armament Systems: Gun and Missile Conference, U.S. Naval Academy, 23 April 2008.

Chapter 15 Weaponeering for Specific Targets Using Effectiveness Indices

15.1 Introduction

Up to this point, the weaponeering methodologies dealt with have related to particular types of weapons (e.g., guided and unguided bombs, guns, rockets, sticks, and cluster munitions). The combined effect of a particular weapon on a particular target has been accounted for in the value of the effectiveness index (EI), which has been specified by MAE_F, MAE_B, or vulnerable area A_V. The EI values, usually derived from programs such as GFSM, are then used to compute L_{ET} and W_{ET}, which are then used in the effectiveness calculation. Many air-to-surface attacks are directed against the types of targets that are not well analyzed by the generalized methods dealt with so far (e.g., GFSM analysis). Examples of these targets include bridges, buildings, vertical targets, and ships. For these classes of targets, other methods have been used, often using EI values derived from the dimensions of the target. This section deals with these techniques and outlines the extension of the methodologies developed so far in analyzing them.

Effectiveness indices discussed in this section include

- Bridge effectiveness index (BEI)
- Effective miss distance (EMD)
- Mean area of effectiveness for buildings (MAE_{BLDG})

It may be questioned why the EI approach for targets such as those listed here should be discussed at all, because JWS now has non-EI-based, higher fidelity methods for dealing with some of them. The answer is that these methods are still fast running and may be preferred by some users. Cataloging the methodology for archival purposes also has some merit.

A distinguishing feature of the targets in this section is that they have dimensions, whereas for MAE_F and MAE_B, the target was considered a point. We introduce the variables L_T and W_T to represent the target dimensions, and these values are used in determining the lethal area with dimensions L_{ET} and W_{ET}.

In dealing with such targets, we will use the basic effectiveness calculation defined by Eq. (9.1), reproduced here:

$$\text{Prob}\{\text{damage}\} = \text{Prob}\{\text{hit}\} \times \text{Prob}\{\text{damage if hit}\} \tag{15.1}$$

Also recall Chapter 9, Section 9.3, where the duality of representing the weapon/target by a lethal area was discussed. In this chapter it will be the target that is represented by a finite size lethal area.

The probability of damage given a hit is denoted as P_{HD} and is not a separate EI, but is involved in the effectiveness calculations. In most cases considered here, P_{HD} is set to unity; the exception is in analyzing bridges. The problem therefore reduces to one of calculating P_{HIT}.

The basic approach in dealing with specific targets such as those mentioned earlier is to compute the dimensions in the ground plane of the effective target area, namely L_{ET} and W_{ET}. Having done this, the methods for computing PD_1 or FD_1 already developed may be applied directly to the target under study. The process, however, is dependent on whether

1. We are dealing with a unitary target or an area of targets.
2. The attack is with a single weapon, or individually aimed multiple weapons, or a stick of unguided weapons.

15.1.1 Single Weapon vs Unitary Target

For this case we will derive L_{ET} and W_{ET} differently for each target type and use these dimensions to represent the target; therefore, it is appropriate to use the representation shown in Fig. 9.2b (in Chapter 9) where the damage function is a rectangular cookie cutter. The effectiveness PD_1 for a single weapon is the probability of the weapon hitting within the lethal area representing the target, given the accuracy of delivery, as shown in Fig. 15.1.

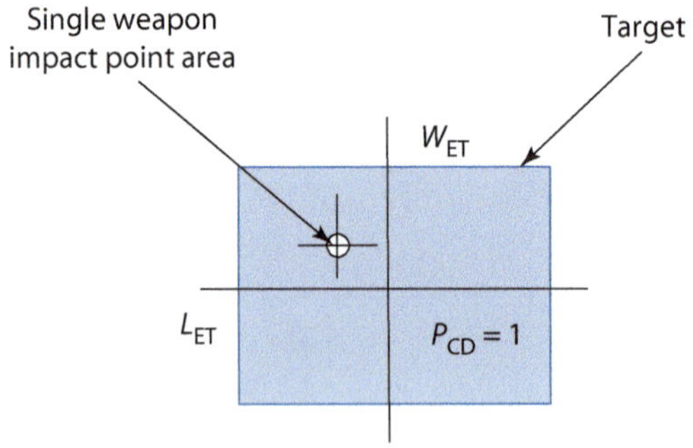

Fig. 15.1 Unitary targets with single weapon.

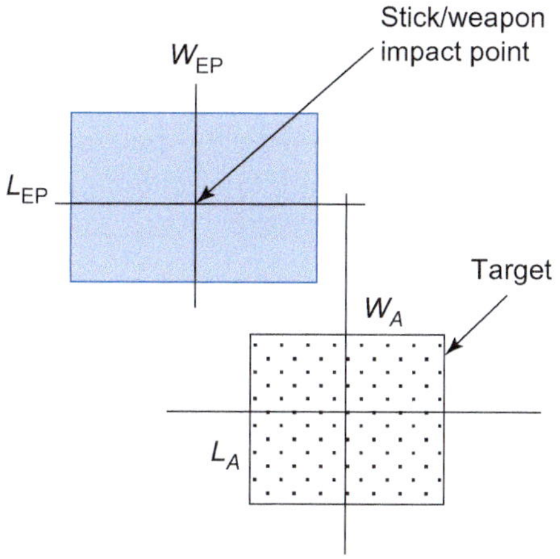

Fig. 15.2 Area target with a stick or single weapon.

If the attack uses single unguided or guided weapons, the effectiveness becomes the probability of a single weapon landing inside the rectangle of dimensions L_{ET} by W_{ET}. The probability of hitting inside the rectangle in range and deflection, respectively, becomes

$$P_{HIT-X} = \int_{-\infty}^{\infty} \frac{P_D}{\sigma_x\sqrt{2\pi}} \exp\left[\frac{-x^2}{2\sigma_x^2}\right] dx = \int_{-L_{ET}/2}^{L_{ET}/2} \frac{1}{\sigma_x\sqrt{2\pi}} \exp\left[\frac{-x^2}{2\sigma_x^2}\right] dx \quad (15.2)$$

$$P_{HIT-Y} = \int_{-\infty}^{\infty} \frac{P_D}{\sigma_y\sqrt{2\pi}} \exp\left[\frac{-y^2}{2\sigma_y^2}\right] dy = \int_{-W_{ET}/2}^{W_{ET}/2} \frac{1}{\sigma_y\sqrt{2\pi}} \exp\left[\frac{-y^2}{2\sigma_y^2}\right] dy \quad (15.3)$$

Combining these equations, the probability of hit may be written as follows:

$$P_{HIT} = P_{HIT-X} \times P_{HIT-Y} \quad (15.4)$$

Hence the probability of damage resulting from a single weapon, including the reliability R, is given by the following:

$$PD_1 = P_{HD} \times P_{HIT} \times R \quad (15.5)$$

15.1.2 Single Weapon or Stick vs an Area of Targets

For bridge and building targets, an area of uniformly distributed, identical targets is unlikely to occur; however, it is possible to have an area of target elements defined by an EMD effectiveness index. Again there are two methods for dealing with an area of targets, depending upon whether a single weapon or a stick of weapons is used, as indicated in Fig. 15.2. In either case, the lethal area is considered rectangular.

If the attack uses sticks of unguided weapons, then the methods described in Chapter 12 would apply to calculate the fractional damage. The single weapon lethal areas with dimensions L_{ET} and W_{ET} would be enlarged to account for the precision error as given by Eqs. (12.17) and (12.18).

$$L_B = \sqrt{L_{\mathrm{ET}}^2 + 8x_{br}^2} \tag{15.6}$$

$$W_B = \sqrt{W_{\mathrm{ET}}^2 + 8x_{bd}^2} \tag{15.7}$$

Then the pattern dimensions would be determined from the stick dimensions.

$$L_P = L_S + L_B \tag{15.8}$$

$$W_P = W_S + W_B \tag{15.9}$$

The effective pattern dimensions L_{EP} and W_{EP} would be calculated next.

$$L_{\mathrm{EP}} = L_P \tag{15.10}$$

$$W_{\mathrm{EP}} = W_P \tag{15.11}$$

If a single weapon is used, then the methods of Chapter 11 would apply, leading to

$$L_{\mathrm{EP}} = L_{\mathrm{ET}} \tag{15.12}$$

$$W_{\mathrm{EP}} = W_{\mathrm{ET}} \tag{15.13}$$

Finally, the effectiveness is calculated from

$$\mathrm{FD}_1 = E(F_C) \times R \times P_{\mathrm{HD}} \times \left[\frac{A_{\mathrm{ET}}}{L_{\mathrm{EP}} \times W_{\mathrm{EP}}} \right] \tag{15.14}$$

In both cases, the fractional coverage calculation would be of the pattern (L_{EP}, W_{EP}) over the area of target elements (L_A, W_A), and P_{HD} is set to unity.

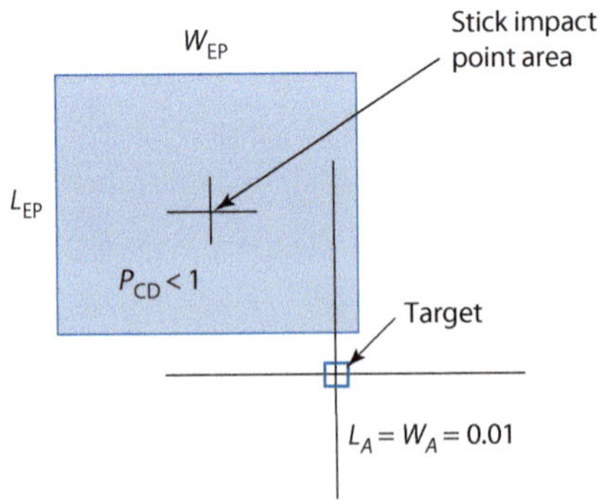

Fig. 15.3 Unitary targets with a stick of weapons.

15.1.3 Stick vs Unitary Target

If a stick of weapons is used, then the effectiveness measure is the fractional damage. A fractional coverage calculation is made where L_{ET} and

Table 15.1 Summary of Basic Methodologies

	Unitary Target	Area of Target Elements
Single weapon ($L_{ET} \times W_{ET}$)	Eq. (15.5)	Eqs. (15.12), (15.13), and (15.14)
Stick ($L_{EP} \times W_{EP}$)	Eq. (15.14) with $L_A = W_A = 0.01$ ft	Eqs. (15.10), (15.11), and (15.14)

W_{ET} are used in determining the pattern size L_{EP} and W_{EP}, but because we are dealing with a unitary target, we set the target area to a small value such as $L_A = W_A = 0.01$ ft. This is shown in Fig. 15.3.

One might ask: Why not treat the pattern rectangle as the target and calculate the probability of hitting it, as shown in Fig. 15.1, the unitary weapon case? The JMEM, however, performs a fractional coverage calculation on a very small target instead. The damage metric is denoted as a PD_1, even though a fractional coverage calculation is performed. The reasoning is that we are still dealing with a unitary target, hence we may write

$$PD_1 = FD_1 = E(F_C) \times R \times P_{HD} \times \left[\frac{A_{ET}}{L_{EP} \times W_{EP}}\right] \tag{15.15}$$

These methodologies are summarized in Table 15.1.

15.2 Bridges

Bridges are constructed in many different forms and materials, and their vulnerability to air-launched weapons depends on many factors. Some examples of construction methods and features are shown in Fig. 15.4.

Clearly it will be difficult to generalize an overall damage function for all bridges, but data obtained from bomb damage assessment (BDA) studies (WWII era) have provided some guidance on the issue.

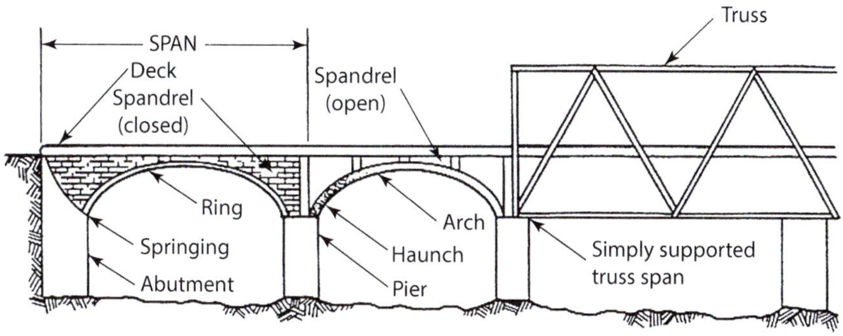

Fig. 15.4 Bridge construction elements.

The bridge will have a physical length L_T and width W_T, but the effective target length L_{ET} and width W_{ET} are defined relative to the attack direction of the aircraft, as shown in the Fig. 15.5.

For the BEI method, the weapon lethal area dimensions are set equal to the bridge dimensions, so for Fig. 15.5a we have

$$L_{ET} = W_T \tag{15.16}$$

$$W_{ET} = L_T \tag{15.17}$$

whereas for Fig. 15.5b

$$L_{ET} = L_T \tag{15.18}$$

$$W_{ET} = W_T \tag{15.19}$$

If the attack uses single unguided weapons or multiple independently aimed guided weapons, the single weapon effectiveness is given from Table 15.1 as

$$PD_1 = P_{HD} \times P_{HIT} \times R \tag{15.20}$$

For bridges, values of P_{HD} can be defined in terms of a classified EI known as the bridge effectiveness index (BEI), which is a function of bridge type and weapon, and has units of feet. For example, suppose a 2000-lb unguided bomb is used to attack a simply supported, reinforced concrete slab bridge where the damage criterion is to drop a span, which is typical for bridges. The BEI might have a value of 20 ft, which represents the width of bridge that may be damaged.

So, if we have a 20-ft-wide bridge and the bomb hits the bridge on its centerline, the damage width and bridge width are the same and the span will fall, hence $P_{HD} = 1$ for this case. However, the bomb may still hit the bridge but not at the centerline, which may or may not collapse the span, resulting in $P_{HD} < 1$; see Fig. 15.6.

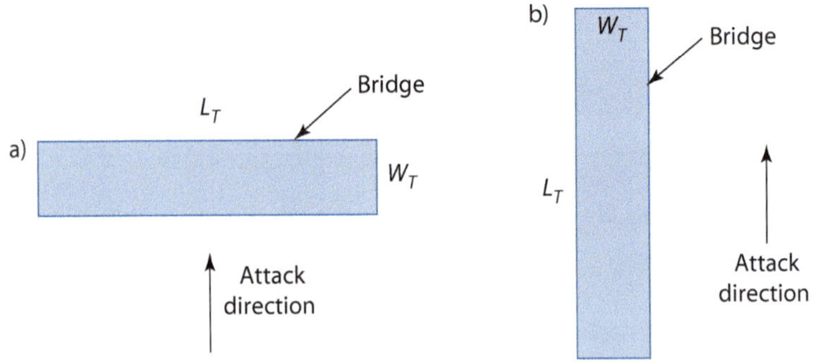

Fig. 15.5 Definition of L_{ET} and W_{ET} for bridge target.

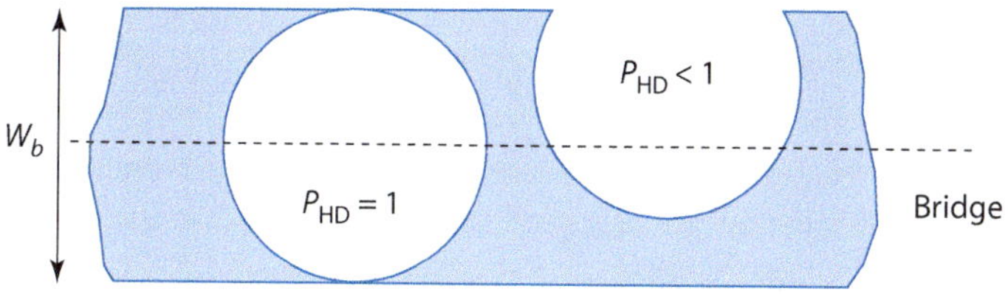

Fig. 15.6 Impacts on bridge deck.

It was found, therefore, that for random impacts on the bridge, the following expression fits the limited data available:

$$P_{HD} = 1 - \exp\left[-\frac{BEI}{W_b}\right] \tag{15.21}$$

where W_b is the bridge width (in feet) and is defined by

$$W_b = \mathrm{MIN}(L_T, W_T) \tag{15.22}$$

The use of Eq. (15.21) extends a limited amount of effectiveness data for specific bridges to other bridges with different widths but similar construction or strength. P_{HD} as a function of bridge width and a specific BEI is shown in Fig. 15.7.

A bridge is considered a unitary target, so Table 15.1 is used to determine the correct equations to verify effectiveness.

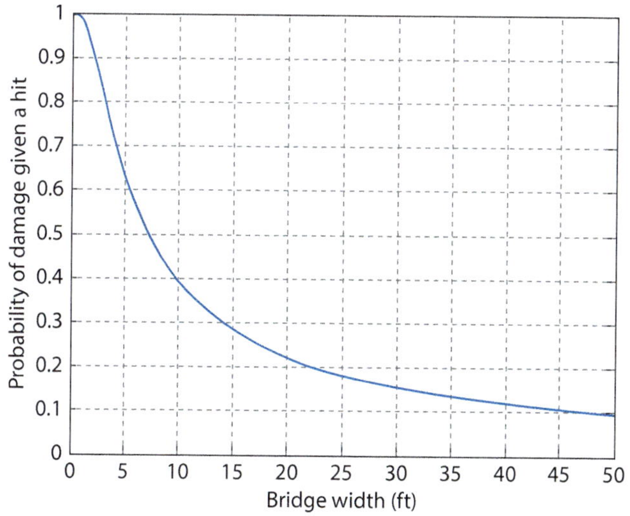

Fig. 15.7 P_{HD} as a function of bridge width for BEI = 5 ft.

Because the bridge assessment routine assumes random hits on the top view of the bridge, it will not provide accurate estimates for guided bombs that are aimed at specific critical bridge components, such as beams or pilings. This method can also be inaccurate for determining the effectiveness of weapons that are delivered with an accuracy of 10 mils or less. For these cases, more advanced methods must be used, as discussed in Volume 2, Chapter 15.

Example 15.1

A bridge has dimensions 30 ft wide by 500 ft long and is attacked by a single unguided weapon aimed at the center. The aircraft attacks in a direction parallel to the 500-ft length of the bridge. Calculate the probability of dropping the span of the bridge, given that for this type of construction and the weapon used, the value of the effectiveness index is BEI = 5 ft. First we calculate P_{HD} from

$$P_{HD} = 1 - \exp\left[-\frac{BEI}{W_b}\right] = 1 - \exp\left[-\frac{5}{30}\right] = 0.154 \tag{15.23}$$

Then the values of L_{ET} and W_{ET} form the bridge dimensions, taking into account the direction of attack we have.

$$L_{ET} = 500, \quad W_{ET} = 30 \tag{15.24}$$

Table 15.2 shows a segment of a spreadsheet that relates to the prior analysis, where the values of L_{ET} and W_{ET} are calculated for specific BEI.

Because the scenario is a single unguided weapon against a unitary target, Eq. (15.5) applies, and Table 15.3 shows a modified version of the spreadsheet used in Chapter 10, Section 10.8 to illustrate the weaponeering solution for this target.

This table is representative of bridge attacks using unguided weapons and indicates the difficulty of a successful single mission, or the large number of missions needed for a reasonable PD. It was results such as these that prompted the development of laser-guided bombs, which were originally used against bridges in Vietnam.

Table 15.2 Weapon Effectiveness for Bridges

WEAPON EFFECTIVENESS			
Effectiveness value	BEI		5.000
Probability of damage if hit	P_{HD}	0.154	
Length/width ratio (frag only)	a	0.337	
Effective target length, ft	WET		30.00
Effective target width, ft	LET		500.00
Effective target area,^{2} ft	AET		2303

(Continued)

Example 15.1 *(Continued)*

Table 15.3 Weaponeering Spreadsheet for Bridge Attacks

METHOD 1 Bridge: SINGLE UNGUIDED WEAPON VS BRIDGES				
CONSTANTS				
Gravitational acceleration, ft/s	*g*	32.200		
Convert degrees to radians	deg2rad	0.017		
TARGET PARAMETERS				
Target dimensions LT, ft		500.000		
Target dimensions WT, ft		30.000		
WEAPON PARAMETERS				
Drag: L = low, H = high, D = dispenser		L		
Ballistic dispersion (mils)	Sigma-*b*	2.000		
Reliability	*R*	0.950		
AIRCRAFT RELEASE DATA				
Airspeed KTAS, ft/s	*V*	550.000	928.400	
Dive angle, deg	theta	30.000	0.524	
Release altitude, ft	*yi*	1200.000		
TRAJECTORY RESULTS				
Vertical velocity, ft/s	*Vv*	–464.200		
Horizontal velocity, ft/s	*Vh*	804.018		
Drag constant, Cd/M	*c*0	0.005		
constant *a* used in trajectory	*a*	5975.800		
		2.400		
time of flight, s	*t_*flight	2.400		
Horizontalvelocity at impact	Vhi	794.426		
Vertical velocity at impact	Vvi	535.493		
Impact angle, deg	*I*	33.982		
Slant range, ft	Sri	2262.820		
Ground range, ft	Gri	1918.425		

(Continued)

Example 15.1 *(Continued)*

Table 15.3 Weaponeering Spreadsheet for Bridge Attacks (*Continued*)

DELIVERY ACCURACY				
		CEP ground	Normal	REP/DEP
CEP, ftground, normal or mils	CEP	0.000	36	
Deflection error probable, ft	DEP	0.000	20.628	20.628
Range error probable, ft	REP	0.000	36.91	36.906
Ballistic error, deflection, ft	sigma-*bd*	3.050		
Ballistic error range, ft	sigma-*br*	5.457		
Adjusted DEP	DEP_dash	20.852		
Adjusted REP	REP_dash	37.307		
WEAPON EFFECTIVENESS				
Effectiveness value	BEI		5.000	
Probability of damage if hit	Phd	0.154		
Length/width ratio (frag only)	*a*	0.337		
Effective target length, ft	WET		30.00	
Effective target width, ft	LET		500.00	
Effective target area, ^{2}ft	AET		2303	
PD_1 CALCULATION				
	Range		Deflection	
Sigma, ft	55.311		30.915	
PD_1 *x* and *y*	1.000		0.372	
PD_1	**0.054**			
FORCE REQUIREMENTS				
Desired PD	0.700			
# of sorties for this PD	21.557			
# sorties available NA	4.000			
Expected PD from NA sorties	0.200			

15.3 Effective Miss Distance (EMD) Damage Function

If we consider a typical MAE_F for a particular weapon and target combination, such as an Mk.84 against an armored vehicle, a value such as 500 ft^2 might be appropriate. Obviously the size of the vehicle is fixed, so there is little ambiguity in defining a single EI.

There are targets, however, that can exist in many different sizes, for example, cylindrical liquid petroleum, oil, and lubricant (POL) storage tanks, which can vary from a few feet to several hundred feet in diameter. These storage tanks might have the same construction method (e.g., 1/4-in. rolled steel plate welded along the edges). Consider that the damage criterion for attacking such a tank is to puncture it so that it drains within 10 min—see Appendix B.

For a given weapon, say an Mk-82 unguided bomb, if we hit the tank we would clearly achieve the damage required. Suppose we miss it by 5 ft—is the target damaged sufficiently? The answer is yes, surely. What if we miss by 10 ft? Again the answer is probably yes. As we increase the miss distance we will get to a point where the required damage is not met. This distance is a new EI called the effective miss distance (EMD) and depends on the target type, weapon, and damage criterion. It is not dependent on the target size; therefore, we do not need to compute different EI values for each possible POL tank size.

For the POL tank example, the EMD might be 100 ft, which allows us to define a value of A_{ET} for any given physical size tank. The area surrounding the target and enclosed by the EMD is treated exactly as if it is part of the target, as shown in Fig. 15.8.

For this rectangular target shown, L_{ET} and W_{ET} are computed using

$$L_{\text{ET}} = L_T + 2 \times \text{EMD} \tag{15.25}$$

$$W_{\text{ET}} = W_T + 2 \times \text{EMD} \tag{15.26}$$

where, as usual, L_T is the dimension of the target parallel to the aircraft flight path (in feet), and W_T is the dimension of the target element perpendicular to the aircraft flight path (in feet). For this case, A_{ET} is again computed using Eq. (15.27).

$$A_{\text{ET}} = L_{\text{ET}} \times W_{\text{ET}} \tag{15.27}$$

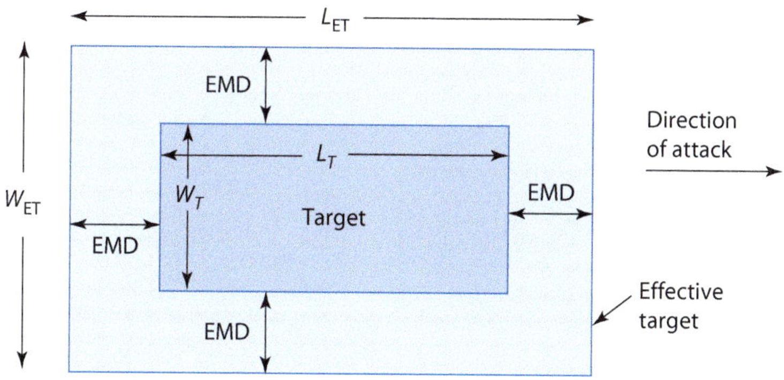

Fig. 15.8 EMD for two-dimensional rectangular target in the ground plane.

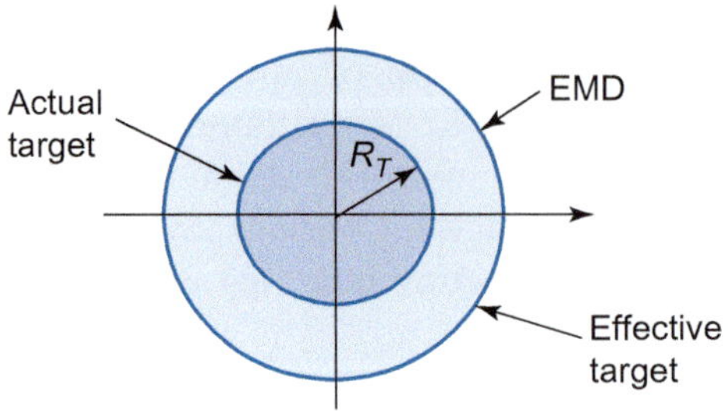

Fig. 15.9 Circular target.

The two-dimensional circular target is shown in Fig. 15.9.

It is approximated by a square of equal area having dimensions L_{ET} and W_{ET} in the ground plane, where L_{ET} and W_{ET} are defined by

$$L_{ET} = W_{ET} = \sqrt{\pi}\,(R_T + \text{EMD}) \tag{15.28}$$

Here R_T is the radius of the target (in feet).

We can extend the concept of EMD to targets having three dimensions, as shown in Fig. 15.10. From this figure we see that the target may be hit on the roof while the projected ground plane impact point relative to the building footprint would suggest that the building was not hit.

What we have to do is generate an equivalent rectangle in the ground plane that, if hit, would correspond to hitting the building, taking into account the height. This is done by defining the building "shadow."

For three-dimensional targets such as the one shown in Fig. 15.10, the shadow length (L_{SH}, in feet) is defined from Fig. 15.11, where the right-hand upper corner of the target is projected down to the ground plane by the impact angle of the weapon to the point P. The shadow length L_S is defined as the distance beyond the downrange EMD to point P, and is

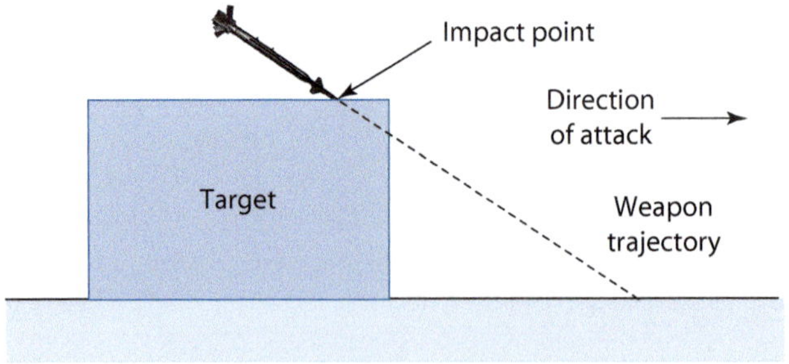

Fig. 15.10 Weapon striking the top of a three-dimensional target.

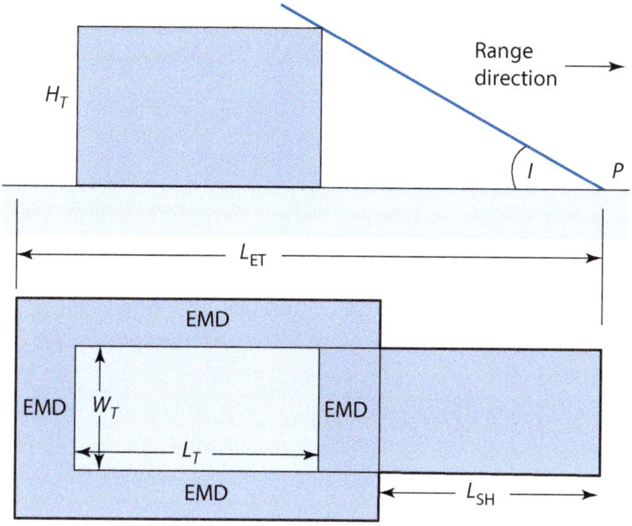

Fig. 15.11 Three-dimensional target with EMD.

computed using the expression

$$L_{SH} = \frac{H_T}{\tan(I)} - \text{EMD} \tag{15.29}$$

where H_T is the target element height (in feet) and I is the impact angle of the weapon. If L_{SH} is less than or equal to zero (i.e., the shadow does not extend beyond the EMD), L_{ET} and W_{ET} are computed using Eqs. (15.25) and (15.26), respectively, for rectangular targets, or Eq. (15.28) for circular targets, and A_{ET} is computed using Eq. (15.27). If L_{SH} is greater than zero, as indicated in Fig. 15.11, L_{ET} is computed using

$$L_{ET} = L_T + 2 \times \text{EMD} + L_{SH} \tag{15.30}$$

for rectangular targets, or

$$L_{ET} = \sqrt{\pi}\,(R_T + \text{EMD}) + L_{SH} \tag{15.31}$$

for cylindrical targets. The effective target area in the ground plane (A_T, in square feet) is then computed using

$$A_T = [L_T + 2\text{EMD}][W_T + 2\text{EMD}] + L_{SH} W_T \tag{15.32}$$

for rectangular targets, or

$$A_T = \pi(R_T + \text{EMD})^2 + 2R_T L_{SH} \tag{15.33}$$

for cylindrical targets. W_{ET} is then adjusted so that the same effective target area in the ground plane will be produced, that is,

$$W_{\text{ET}} = \frac{A_T}{L_{\text{ET}}} \tag{15.34}$$

A_{ET} can be computed using Eq. (15.27), but is equal to A_T.

Vertical targets such as tunnel entrances, caves, and shelter doors can be weaponeered by entering a target length L_T of zero, the door width as the target width W_T, and the door height as the target height H_T. Typical damage definitions for these targets include the production of rubble at the entrance to prevent passage of materiel or personnel. As such, rubble may be produced either by placing the weapon inside the tunnel or by hitting an area around the portal or the apron in front of the portal. Therefore, as indicated in Fig. 15.12, there will be an EMD in the vertical plane and in front of the target.

Because the target length is zero, the effective target length is the length of the shadow plus the EMD in front of the target, or

$$L_{\text{ET}} = \frac{H_T + \text{EMD}}{\tan(I)} + \text{EMD} \tag{15.35}$$

This may be seen from the side view of the tunnel entrance shown in Fig. 15.13. W_{ET} is computed using Eq. (15.26), and A_{ET} is from Eq. (15.27).

Once L_{ET} and W_{ET} have been computed, the effectiveness is calculated as shown in Table 15.1 for a unitary target. If we have an area of identical target elements, such as a POL tank farm, we would consider this an area of target elements, and a different set of equations from Table 15.1 would be used to calculate effectiveness.

Table 15.4 shows how target data are input for the spreadsheet implementation of the EMD damage function, and allows for either a

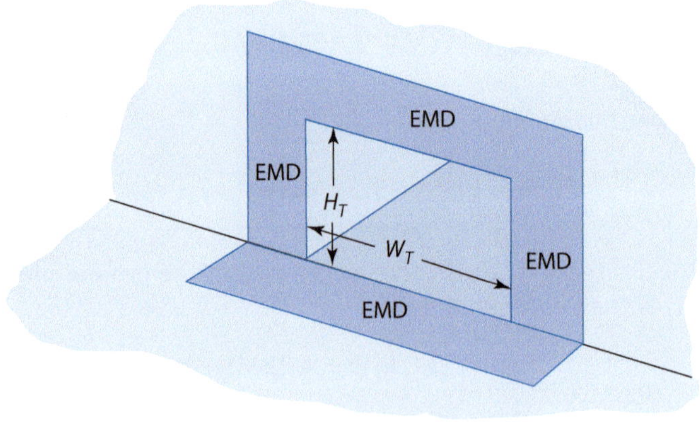

Fig. 15.12 Vertical target with EMD.

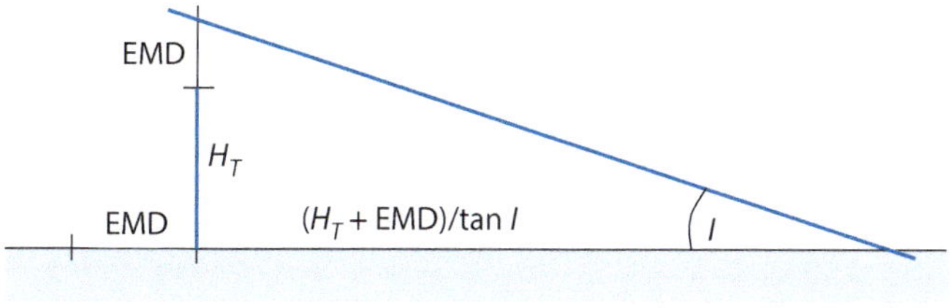

Fig. 15.13 Side view of tunnel entrance.

Table 15.4 Target Specification of the EMD Damage Function

TARGET PARAMETERS		
Target length LT, ft		30.000
Target width WT, ft		20.000
Target radius RT, ft		0.000
Target height HT, ft		40.000

rectangular or cylindrical target. The target used in the example is a single 30-ft × 20-ft rectangle, 40 ft high.

Table 15.5 shows how the weapon lethal area dimensions are calculated. This is done for both target geometry cases, and examination of the input data determines which one is selected under the Resultant column. In order to get the shadow length, the trajectory program has to be run for the release under consideration to obtain the weapon impact angle. For the case shown, the impact angle used is 36 deg.

Once the values of L_{ET} and W_{ET} have been found, the effectiveness calculation will depend on whether a single weapon is used or a stick, and on whether we are dealing with a unitary target or an area of EMD target elements, as specified in Table 15.1.

Table 15.6 is a complete weaponeering spreadsheet implementation for the EMD damage function and deals with a single unguided bomb dropped against a single cylindrical target 40 ft high and 20 ft in diameter.

Table 15.5 Weapon Effectiveness for EMD Damage Functions

WEAPON EFFECTIVENESS				
Effectiveness value	EMD	5.000		
Target geometry		**Rectangle**	**Cylindrical**	**Resultant**
Shadow length, ft	LSH	49.606	49.606	
Effective target length, ft	LET	89.606	58.468	89.606
Target area	AT	2192.115	78.540	
Effective target width. ft	WET	24.464	1.343	24.464
Effective target area	AET	2192.115	78.540	

Table 15.6 Weaponeering Spreadsheet for EMD Damage Functions

METHOD 1-EMD: SINGLE UNGUIDED WEAPON/EMD DAMAGE FUNCTION					
CONSTANTS					
Gravitational acceleration, ft/s	*g*	32.200			
Convert degrees to radians	deg2rad	0.017			
TARGET PARAMETERS					
Target length LT, ft		0.000			
Target width WT, ft		0.000			
Target radius RT, ft		10.000			
Target height HT, ft		40.000			
WEAPON PARAMETERS					
Drag: L = low, H = high, D = dispenser		L			
Ballistic dispersion, mils	Sigma-*b*	2.000			
Reliability	*R*	0.950			
AIRCRAFT RELEASE DATA					
Airspeed KTAS, ft/s	*V*	550.000		928.400	ft/s
Dive angle, deg	theta	30.000		0.524	rad
Release altitude, ft	*yi*	2000.000			
TRAJECTORY RESULTS					
Vertical velocity, ft/s	*Vv*	-464			
Horizontal velocity, ft/s	*Vh*	804			
Drag constant, Cd/M	*c0*	0.005			
constanta used in trajectory	*a*	5976			
		3.838			
time of flight, s	*t*_flight	3.838			
Horizontal velocity at impact	Vhi	789			
Vertical velocity at impact	Vvi	578			
Impact angle, deg	*I*	36.224			
Slant range, ft	Sri	3652			
Ground range, ft	Gri	3056			
DELIVERY ACCURACY					
		CEP ft ground	**CEP ft n**	**Spec**	**REP/DEP**
CEP, ft ground, normal or mils	CEP	0.000	13		
Deflection error probable, ft	DEP	0.000	7.449	0.000	7.449
Range error probable, ft	REP	0.000	12.605	0.000	12.605
Ballistic error, deflection, ft	sigma-*bd*	4.923			
Ballistic error range, ft	sigma-*br*	8.331			
Adjusted DEP	DEP_dash	4.923			
Adjusted REP	REP_dash	8.331			
WEAPON EFFECTIVENESS					
Effectiveness value	EMD		10.000		
Target geometry			**Rectangle**	**Cylinder**	**Resultant**
Shadow length, ft	LSH		44.606	44.606	
Effective target length, ft	LET		64.606	80.055	80.055
Target area	AT		400.000	2148.751	
Effective target width, ft	WET		6.191	26.841	26.841
Effective target area	AET		400.000	2148.751	

(*Continued*)

Table 15.6 Weaponeering Spreadsheet for EMD Damage Functions (*Continued*)

PD CALCULATION					
	Range			**Deflection**	
Sigma, ft	12.352			7.299	
PD1_*r*/PD1_*d*	0.916			0.793	
PD$_1$	**0.690**				
FORCE REQUIREMENTS					
Desired PD	0.700				
# of sorties for this PD	1.028				
# sorties available NA	4.000				
Expected PD from NA sorties	0.991				

Notice again how the target dimensions are used to calculate L_{ET} and W_{ET}, but the methods shown in Fig. 15.3a are used because the target is considered unitary.

15.4 Aboveground Buildings

As for bridges, building construction materials and methods vary considerably, and subtle changes in construction techniques can influence considerably the effect of a weapon on such a target. The early JMEM approach essentially correlates building damage to building type for a specific weapon based on battle damage assessment. This correlation is in the form of an effectiveness index called the MAE-Building and is given the symbol MAE_{BLDG}. It extends sparse data for a particular building type to a range of building sizes and weapons. In this sense, the method is similar to that described earlier for bridges. The process begins by classifying buildings according to their building construction code, as indicated in Table 15.7.

Examples of some building types are shown in Fig. 15.14.

Early methodologies treated buildings as two-dimensional targets and were adequate for weapon deliveries that produced high impact angles; however, these did not account for the increase in effectiveness that can be realized by considering target height when low impact angles occur, as shown in Fig. 15.10. Therefore, in the MAE_{BLDG} method, buildings are treated as three-dimensional targets with an EMD. One of the requirements

Table 15.7 Classification of Sample Building Types

Building Type	JMEM Code
Multistory masonry load bearing wall	1c
Multistory framed—steel or concrete reinforced	1a
Single story load-bearing masonry wall	4b
Single story framed, column and slab construction	3b

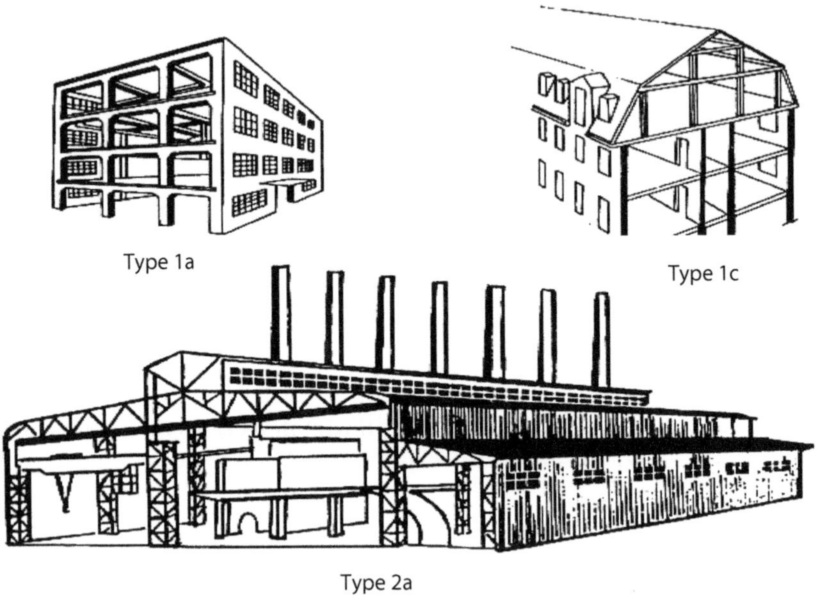

Fig. 15.14 Sample JMEM building types.

of the methodology was that the results be consistent with those for the two-dimensional area target method for high-altitude deliveries when target height was not considered.

Because buildings are damaged primarily by blast, the methodology for evaluating damage using MAE_{BLDG} is a modification of the MAE_B routines that provide for entering the target height H_T. With these routines, the building dimensions are input as L_T, W_T, and H_T, and the MAE_{BLDG} values are the MAE_B values given in classified JMEM manuals. If H_T is set to zero, these routines give the same results as those given when MAE_B is the EI type and when the building length and width are input as the target area length and width (L_T and W_T).

The effectiveness methodology is somewhat different for buildings compared to other target types dealt with in this chapter in that, although the building is a unitary target, a fractional coverage calculation is made on the building footprint.

MAE_{BLDG} is used to describe the damage function for buildings and represents the total floor area damaged, averaged over all the floors. For this cookie cutter damage function, P_{HD} is unity out to a specific distance from the impact point and drops to zero beyond that distance. Again, the numerical values of MAE_{BLDG} are classified and are obtained from the JMEM Basic Manual in tabular form (in square feet), listed as a function of the building type and weapon. It is assumed the weapon is fuzed such that it penetrates the building before detonation.

In the JMEM/AS open-ended methods presented here, all MAE_{BLDG} values are assumed to be circular in the ground plane and are approximated by squares, regardless of the impact angle.

The first stage of analysis is to represent the three-dimensional building as a two-dimensional rectangle in the ground plane by using the EMD approach of the previous section. The method begins by comparing the MAE_{BLDG} EI value to the physical size of the building by calculating an initial value of L_{ET} from the following equation:

$$L_{ET} = \sqrt{MAE_{BLDG}} \tag{15.36}$$

Then, if L_{ET} is greater than the length of the building (L_T in feet), the equivalent miss distance for the building (EMD in feet) is computed using

$$EMD = \frac{L_{ET} - L_T}{2} \tag{15.37}$$

as shown in Fig. 15.8; otherwise, EMD is set to zero. If the impact angle is 90 deg, no target shadow of the building height exists, and the shadow length and width (L_{SH} and W_{SH}, in feet) are set to zero. For all other impact angles, L_{SH} is computed as before using

$$L_{SH} = \frac{H_T}{\tan(I)} - EMD \tag{15.38}$$

Proceeding as for the EMD damage function, the new effective target length is now calculated from

$$L_{ET} = L_T + 2 \times EMD + L_{SH} \tag{15.39}$$

The effective target area in the ground plane (A_T, in square feet) is then computed using

$$A_{ET} = [L_T + 2EMD][W_T + 2EMD] + L_{SH} W_T \tag{15.40}$$

and W_{ET} is computed from

$$W_{ET} = \frac{A_{ET}}{L_{ET}} \tag{15.41}$$

Before calculating effectiveness from L_{ET} and W_{ET}, we demonstrate with an example.

The methodology for calculating building damage is somewhat different from that summarized in Table 15.1 and will be a fractional damage calculation of a pattern area over the actual target footprint, where the pattern area depends on the weapons used.

Example 15.2

Consider the building target shown in Fig. 15.15, which has dimensions of 100 ft (L_T) × 100 ft (W_T) and is 50 ft (H_T) high. Calculate L_{ET} and W_{ET} for the building if MAE_{BLDG} is 25,000 ft^2 and a weapon impact angle of 45 deg is specified. An initial calculation reveals that

$$L_{ET} = \sqrt{MAE_{bldg}} = \sqrt{25{,}000} = 158.1' \tag{15.42}$$

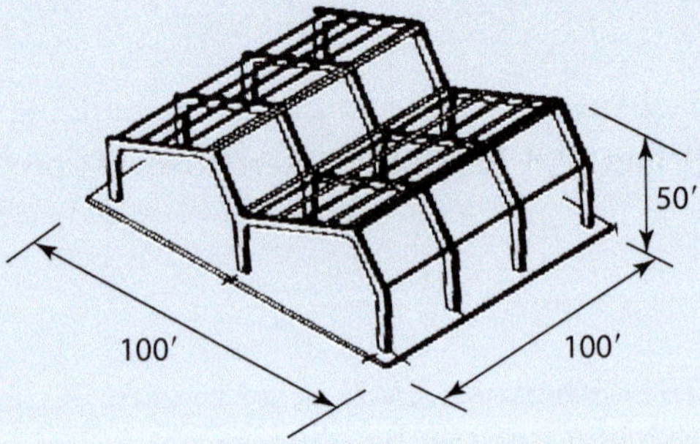

Fig. 15.15 Building target.

The EMD value is

$$EMD = \frac{L_{ET} - L_T}{2} = \frac{158.1 - 100}{2} = 29.1' \tag{15.43}$$

Calculating the shadow length as

$$L_{SH} = \frac{H_T}{\tan(I)} - EMD = \frac{50}{\tan 45} - 29.1 = 20.9' \tag{15.44}$$

the effective target length is recomputed as

$$L_{ET} = L_T + 2 \times EMD + L_{SH} = 100 + 2 \times 29.1 + 20.9 = 179.1' \tag{15.45}$$

This makes the effective target area

$$A_{ET} = [L_T + 2EMD][W_T + 2EMD] + L_{SH}W_T = 27{,}094\ \text{ft}^2 \tag{15.46}$$

from which we obtain

$$W_{ET} = \frac{A_{ET}}{L_{ET}} = \frac{27{,}094}{179.1} = 151.3' \tag{15.47}$$

Table 15.8 shows how the methodology for this type of damage function is implemented in a spreadsheet to calculate L_{ET} and W_{ET}.

Table 15.8 Implementing the MAE_{BLDG} Damage Function

WEAPON EFFECTIVENESS				
Effectiveness index type				MAE_bldg
Effectiveness value				25,000.0
Initial LET/WET	LET/WET			158.1
Equivalent EMD	EMD			29.1
Shadow length, ft	LSH			20.9
Effective target length, ft	LET			179.1
Target area	AT			27,094.3
Effective target width, ft	WET			151.3

Given that L_{ET} and W_{ET} have been determined, the process of calculating damage is somewhat obscure from [1]; however, what follows is a methodology that provides results compatible with the OEM program (JMEM A/S Open-Ended Methods) for a wide variety of weapon and target relative sizes.

1. The damage criterion is fractional damage FD, which is interpreted as structural fractional damage to the building. Although this is a subjective term when observing a damaged building, some guidelines as to its interpretation will be provided in Section 15.5.
2. For simplicity we will consider single weapon delivery only. If multiple, independently aimed weapons are used, the single fractional damage FD_1 will be powered up for the number of weapons used.

$$FD = 1 - (1 - FD_1)^n \tag{15.48}$$

3. If stick deliveries are of interest, the methods of Chapter 12 are used to produce an equivalent rectangular pattern of known dimensions and a conditional probability inside it given by P_{CDS}. For single weapon attacks, we set $P_{CDS} = 1$.
4. The process begins by calculating the expected fractional coverage $E(F_C)$ of the lethal area rectangle $L_{ET} \times W_{ET}$ over the target footprint $L_T \times W_T$.
5. The single weapon fractional damage is then given by the following:

$$FD_1 = E(F_C) \times R \times \text{MIN}\left(1, \frac{\text{MAE}_{\text{BLDG}}}{L_T \times W_T}\right) \tag{15.49}$$

6. The conditional probability of damage is calculated from

$$P_{CD} = \frac{A_{ET}}{L_{EP} \times W_{EP}} \tag{15.50}$$

which for a single weapon is always unity.

Table 15.9 Single Weapons Against Buildings

METHOD 2-BLDG: ONE WEAPON AGAINST BUILDING TARGET						
CONSTANTS						
Gravity accel, ft/s	32.200					
Convert deg to rad	0.017					
TARGET PARAMETERS						
Target area dimensions	LT	100.000				
	WT	100.000				
	HT	50.000				
WEAPON PARAMETERS						
Reliability	*R*	1.000				
TRAJECTORY RESULTS						
Impact angle, deg	*I*	45				
DELIVERY ACCURACY						
		Ground ft	**Normal ft**		**REP/DEP**	**REP/DEP**
CEP, ft ground, or mils	CEP	0.000				
DEP, ft	DEP	0.00	0.00		10.00	10.00
REP, ft	REP	0.00	0.00		20.00	20.00
Phit	Phit	0.00				
Pnm	Pnm	1.00				
Offset aimpoint, range	Ro	0.000				
Offset aimpoint, defl.	Do	0.000				
WEAPON EFFECTIVENESS						
EI type				MAE_bldg		
Effectiveness value				25000.0		
Initial LET/WET	LET/WET			158.1		
Equivalent EMD	EMD			29.1		
Shadow length, ft	LSH			20.9		
Effective target length	LET			179.1		
Target area	AT			27,094.3		
Effective target width	WET			151.3		
Target length	LT			100.0		
Target width	WT			100.0		
Conditional probability damage	Pcd			1.00	1.00	
$FD_1 - P_{nm}$						
	Range			**Deflection**		
Sigma range, ft	29.652			14.826		
LEP	179.057		WEP	151.317		
INTEGRAL LIMITS						
s	139.528		*s*	125.658		
t	39.528		*t*	25.658		
INTEGRALS						
integral #1	−1.464		int #1	−1.387		
integral #2	1.268		int #2	1.204		
integral #3	1.268		int #3	1.204		
integral #4	−0.049		int #4	−0.013		
integral #5	0.049		int #5	0.013		

(*Continued*)

Table 15.9 Single Weapons Against Buildings (*Continued*)

RESULTS						
Fractional coverage	0.975			0.995		
Combined fractional coverage	0.970					
FD_1	**0.9699**					
$FD_1 - P_{hit}$						
	RANGE			DEFLECTION		
Sigma range, ft	0.010			0.010		
LEP	179.057		WEP	151.317		
INTEGRAL LIMITS						
s	139.528		*s*	125.658		
t	39.528		*t*	25.658		
INTEGRALS						
integral #1	–1.791		int #1	–1.513		
integral #2	1.395		int #2	1.257		
integral #3	1.395		int #3	1.257		
integral #4	0.000		int #4	0.000		
integral #5	0.000		int #5	0.000		
RESULTS						
Fractional coverage	1.000			1.000		
Combined fractional coverage	1.000					
FD_1	**1.0000**					
Combined $E(FC_1)$	**0.9699**					
PCD	**1.0000**					
EFD	**0.9699**					

7. Finally, the expected fractional damage is

$$\mathrm{EFD} = \mathrm{FD}_1 \times P_{\mathrm{CD}} \tag{15.51}$$

8. Note that if we are considering a guided weapon where P_{HIT} is not zero, these calculations are performed twice for two different accuracies, one for the P_{NM} value and one where CEP $= 0$, and then combined in the usual way.

$$\mathrm{EFD} = P_{\mathrm{NM}} \times \mathrm{EFD}_{\mathrm{PNM}} + P_{\mathrm{HIT}} \times \mathrm{EFD}_{\mathrm{PHIT}} \tag{15.52}$$

Note that even though we are talking about a unitary target, the effectiveness measure is written as a fractional damage and corresponds to the proportion of volumetric structural damage.

Table 15.9 shows a complete weaponeering implementation of the building methodology for this case of a single guided weapon against a building target; it is similar to the methodology discussed in Chapter 11, Section 11.7, except for the inclusion of the $\mathrm{MAE}_{\mathrm{BLDG}}$ damage function and the methodology for dealing with both guided and unguided weapons. Unguided weapons may be used by setting $P_{\mathrm{HIT}} = 0$ and $P_{\mathrm{NM}} = 1$.

Intuitively this high value seems appropriate because we have a building with a 10,000-ft^2 footprint and a weapon with 25,000 ft^2 with a reasonable accuracy of 30 ft CEP. For brevity, the spreadsheets for independently aimed weapons and sticks of weapons are not presented here.

15.5 Interpretation of Building Damage

Building damage is a subjective measure, so it is difficult to assign it a value. Clearly the value of 0.95 generated in the previous section would correspond to severe damage, probably involving collapse. But what does 50% structural damage mean? Does the building collapse? Does whatever function being conducted in the building cease? There are very few data points to link quantitative damage estimates to real buildings; however, the following examples might provide some guidance.

- The Alfred P. Murrah federal building in Oklahoma City, shown in Fig. 15.16, was estimated to have sustained 30% structural damage and as a result had to be demolished. Clearly this level of damage ended any functionality in the building, though most of the structure still stood.
- Sometimes the commander's intent is to totally destroy a building and prevent any remaining structure to provide cover for combat troops. This is customarily associated with over 70% structural damage and basically reduces the building to a large area of flat rubble.

Volume 2, Chapter 14 looks at more detailed building damage estimation models and allows functional as well as structural damage to be calculated.

Fig. 15.16 Building damage in Oklahoma City.

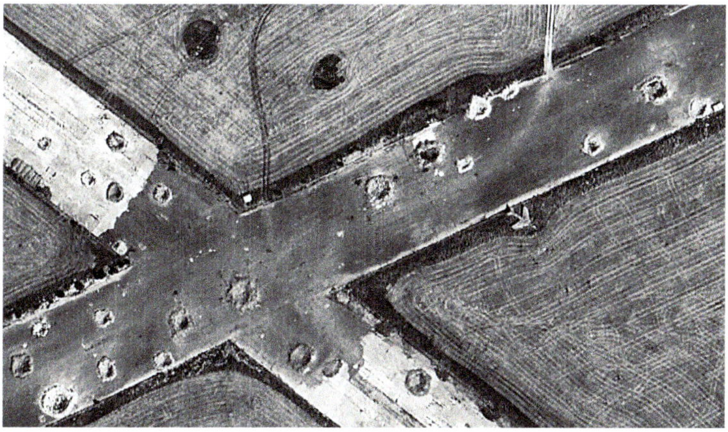

Fig. 15.17 Bomb crater damage to runways.

15.6 Linear Target Method

In this section we investigate the chances of a single weapon represented by a circular damage function partially or totally cutting a linear, rectangular target. Typical applications of this scenario involve a weapon creating a circular crater of diameter D_C impacting a long and narrow rectangular target of interest such as an aircraft runway, road, bridge, or railway.

Figure 15.17 shows the effect of multiple bomb crater damage to a runway; although such a target is clearly unusable, it should be kept in mind that some damage may be repaired more quickly than others. For example, roads and runways may be temporarily repaired by filling in the holes with rubble and placing a metal plate over the patch. Other linear targets such as bridges or railway lines may take longer to bring back into use.

It should also be noted that a crater does not necessarily have to damage the whole width of the target in order to make it inoperable. The damage by a single crater in Fig. 15.17 may prevent any aircraft using the runway for take-offs or landings.

We limit the analysis to a single weapon aimed at the center of a linear target, which is assumed long enough to ignore damage to its ends; therefore, only damage to the target width is of interest. Early JMEM methods attempted to derive an analytical solution to this problem, but they made many assumptions in the process; therefore, the solution presented here is based on a Monte Carlo simulation and is sufficiently simple that it runs in less than a second on an ordinary personal computer.

Consider the scenario in Fig. 15.18 showing a random impact point relative to the linear target. It is assumed that the attack direction is at an angle θ to the longest direction and that the impact point in weapon coordinates (x_w, y_w) is a random draw from the range and deflection accuracies REP and DEP (or σ_x, σ_y). The impact point in target coordinates (x, y) is given by

$$x_i = x_w \cos\theta - y_w \sin\theta \tag{15.53}$$

$$y_i = x_w \sin\theta + y_w \cos\theta \tag{15.54}$$

If the crater radius is r_c, we can calculate the effect of the crater on the target.

The maximum and minimum distances of the crater in the y direction are given by the following equations:

$$y_{\max} = y_i + r \tag{15.55}$$

$$y_{\min} = y_i - r \tag{15.56}$$

Note that if $y_{\min} > w/2$ or $y_{\max} \leq w/2$, the crater does not intersect the target and no damage results. If the crater does intersect the target, then one of the three situations shown in Fig. 15.19 may occur.

Consider the damage width above ($dw1$) and below ($dw2$) the impact point, defined as the distance in the y direction where the crater intersects the target, as shown in Fig. 15.19. We see the following relations hold:

$$\begin{aligned} &\text{IF } y_i > w/2 \text{ AND } y_{\min} < w/2 \text{ THEN} \\ &\quad dw1 = 0, \\ &\quad dw2 = \text{abs}(\min(y_{\min}, w/2) - y_i) \end{aligned} \tag{15.57}$$

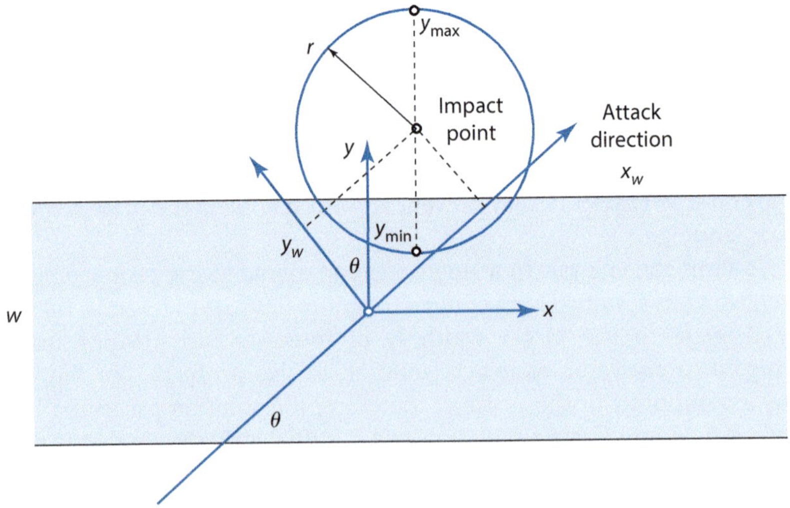

Fig. 15.18 Geometry of impact point and target.

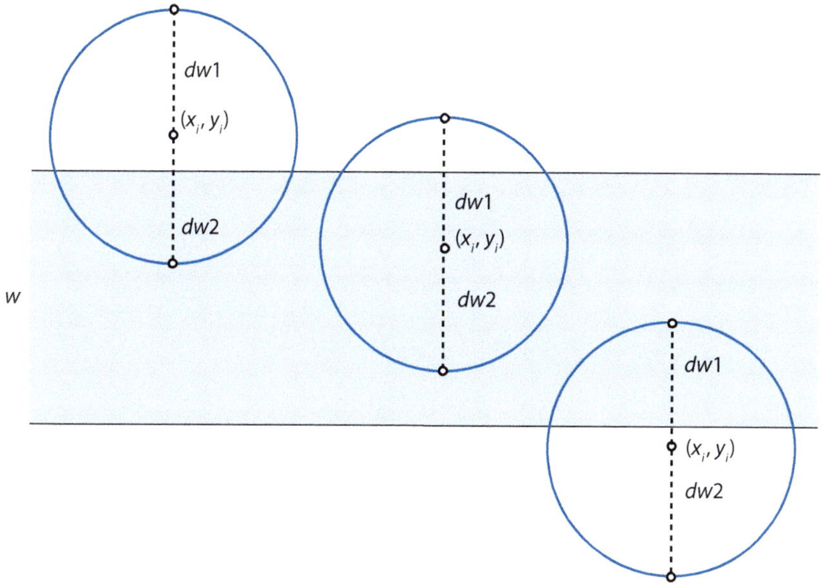

Fig. 15.19 Possible interactions of crater and target.

$$\begin{aligned}&\text{IF}\, y_i < w/2 \text{ AND } y_i > -w/2 \text{ THEN}\\&\quad dw1 = \text{abs}(\max(y_{\max},\ w/2) - y_i),\\&\quad dw2 = \text{abs}(\min(y_{\min},\ w/2) - y_i)\end{aligned} \tag{15.58}$$

$$\begin{aligned}&\text{IF}\, y_i < -w/2 \text{ AND } y_{\max} > -w/2 \text{ THEN}\\&\quad dw1 = \text{abs}(\max(y_{\max},\ w/2) - y_i),\\&\quad dw2 = 0\end{aligned} \tag{15.59}$$

The total damage width is then

$$\text{damage width} = dw1 + dw2 \tag{15.60}$$

and the proportion of target width damaged is $dw1 + dw2$. This algorithm has been implemented in a MATLAB script as shown here:

```
% Linear target method - single weapon
clear;clc
w=50;              % target width
r=20;              % crater radius
theta=45;          % attack angle
sigma_xw=50;       % accuracy in range (weapon axis)
sigma_yw=30;       % accuracy in deflection (weapon axis)
imax=50000;        % number of Monte Carlo iterations
```

```
for i=1:imax          % start of MC loop
    xw=normrnd(0,sigma_xw);                  % x impact on weapon axis
    yw=normrnd(0,sigma_yw);                  % y impact on weapon axis
    xi=xw*cosd(theta)-yw*sind(theta);        % x impact on target axis
    yi=xw*sind(theta)+yw*cosd(theta);        % y impact on weapon axis
    ymax=yi+r;                               % top of crater
    ymin=yi-r;                               % bottom of crater
    if ymin > w/2                            % crater above target
        cond=1;
        dw1=0;
        dw2=0;
    elseif ymax < -w/2                       % crater below target
        cond=2;
        dw1=0;
        dw2=0;
    elseif yi > w/2 && ymin < w/2     % some crater segment on target
        cond=3;
        dw1=0;
        dw2=abs(max(ymin,-w/2)-yi);
    elseif yi < w/2 && yi > -w/2
        cond=4;
        dw1=abs(min(ymax,w/2)-yi);
        dw2=abs(min(ymin,w/2)-yi);
    elseif yi < -w/2 && ymax > -w/2
        cond=5;
        dw1=abs(min(ymax,w/2)-yi);
        dw2=0;
    end
    damage_width=dw1+dw2;                 % width of crater on target
    prop_damage(i)=damage_width/w;        % proportion of DW this time
end                                       % end of MC loop
damage=mean(prop_damage);                 % average over all iterations
fprintf('Condition = % i, Proportion of damage = % .4f\n', cond,
damage)
```

Running the program produces the following observations:

- If the accuracy in range and direction is the same, the damage to the target is independent of the attack angle.
- If the accuracy in range is different from that in deflection, maximum damage is achieved with an attack angle that puts the smallest dispersion across the target width. This means that if REP < DEP, theta should be set to 90 deg; however, if DEP < REP, theta should be zero.

A further refinement would be to consider a stick of bombs; however, this case is beyond the scope of this book.

15.7 Indirect Aim Points

The methods developed in earlier chapters for calculating target damage can be adapted to assess the effects on the surrounding structures. Before this

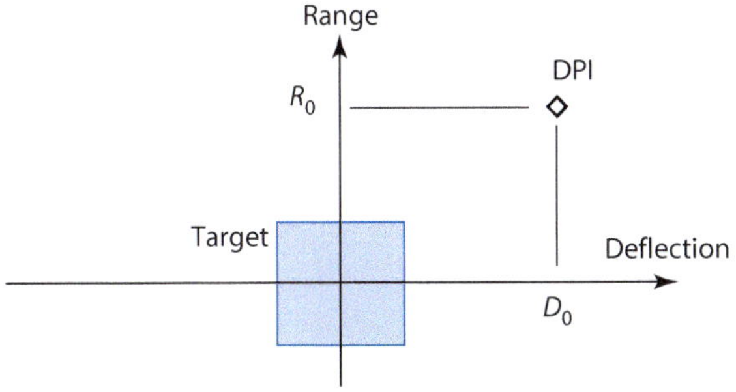

Fig. 15.20 Coordinates of offset aim point.

is looked at in detail, however, we will investigate the effect of damage sustained by a target when the weapon is not aimed directly at it. In all the previous weaponeering methodologies developed so far, it has been assumed that weapons were aimed at the center of the target because it seems that this will result in maximum damage.

We shall see soon that there are good reasons for studying the case where the weapons are directed to some other point away from the target, as shown in Fig. 15.20. The figure shows a target, which may be a unitary or area type located at the origin, but an indirect aim point (DPI) is located at coordinates (R_o, D_o) relative to the center of the target.

We will rework the weaponeering methodologies of Chapters 10 and 11 to look at this case, beginning with single weapons directed against unitary targets.

15.7.1 Single Fragmentation Weapons Against Unitary Targets

Recall the general expression for PD_1 in the range direction as in Eq. (10.2).

$$\mathrm{PD}_{1x} = \int_{-\infty}^{+\infty} c(x)g(x)\mathrm{d}x \tag{15.61}$$

The miss distance function $g(x)$ is assumed to be normally distributed, allowing it to be expressed in the form

$$g(x) = \frac{1}{\sigma_x\sqrt{2\pi}}\exp\left[-\frac{(x-\mu_x)^2}{2\sigma_x^2}\right] \tag{15.62}$$

The damage function for fragmentation warheads is assumed to take the one-dimensional form of the Carlton damage function.

$$c(x) = \exp\left[-\frac{4x^2}{L_{\mathrm{ET}}'^{\,2}}\right] \tag{15.63}$$

In Chapter 10, it was assumed that the mean value of the miss distance function μ_x was zero, and the PD_1 was derived accordingly. Now, however, the offset in the range direction R_o may be thought of as the mean value of the desired miss distance, allowing us to write for the miss distance function

$$g(x) = \frac{1}{\sigma_x\sqrt{2\pi}}\exp\left[-\frac{(x-R_o)^2}{2\sigma_x^2}\right] \tag{15.64}$$

Substituting this and the one-dimensional Carlton damage function into the expression for PD_1 in the range direction gives

$$\begin{aligned}
PD_{1x} &= \frac{1}{\sigma_x\sqrt{2\pi}}\int_{x=-\infty}^{x=\infty} \exp\left[-\frac{4x^2}{L'^{2}_{ET}}\right] \times \exp\left[-\frac{(x-R_o)^2}{2\sigma_x^2}\right] dx \\
&= \frac{1}{\sigma_x\sqrt{2\pi}}\int_{-\infty}^{+\infty} \exp-\left[\frac{4x^2}{L'^{2}_{ET}} + \frac{x^2}{2\sigma_x^2} - \frac{2xR_o}{2\sigma_x^2} + \frac{R_o^2}{2\sigma_x^2}\right] dx \\
&= \frac{1}{\sigma_x\sqrt{2\pi}}\int_{-\infty}^{+\infty} \exp(-p)dx
\end{aligned} \tag{15.65}$$

Consider the simplification of the exponent p.

$$p = \frac{8x^2\sigma_x^2 + x^2 L'^{2}_{ET} - 2xR_o L'^{2}_{ET} + R_o L'^{2}_{ET}}{2L'^{2}_{ET}\sigma_x^2} \tag{15.66}$$

After some manipulation this may be written as

$$\begin{aligned}
p = &\left[\left(x^2 - \frac{R_o L'^{2}_{ET}}{8\sigma_x^2 + L'^{2}_{ET}}\right)^2 - \frac{R_o^2 L'^{4}_{ET}}{\left(8\sigma_x^2 + L'^{2}_{ET}\right)^2} + \frac{R_o^2 L'^{2}_{ET}}{\left(8\sigma_x^2 + L'^{2}_{ET}\right)}\right] \\
&\times \left[\frac{8\sigma_x^2 + L'^{2}_{ET}}{2\sigma_x^2 L'^{2}_{ET}}\right]
\end{aligned} \tag{15.67}$$

Further simplification results in

$$p = \left[\left(x^2 - \frac{R_o L'^{2}_{ET}}{8\sigma_x^2 + L'^{2}_{ET}}\right)^2 \left[\frac{8\sigma_x^2 + L'^{2}_{ET}}{2\sigma_x^2 L'^{2}_{ET}}\right] + \frac{4R_o^2}{\left(8\sigma_x^2 + L'^{2}_{ET}\right)}\right] \tag{15.68}$$

Substituting back into Eq. (15.65) for PD1 gives

$$\mathrm{PD}_{1x} = \frac{1}{\sigma_x\sqrt{2\pi}} \int_{x=-\infty}^{x=\infty} \exp - \left\{ \left(x^2 - \frac{R_o {L'_{\mathrm{ET}}}^2}{8\sigma_x^2 + {L'_{\mathrm{ET}}}^2} \right)^2 \left[\frac{8\sigma_x^2 + {L'_{\mathrm{ET}}}^2}{2\sigma_x^2 {L'_{\mathrm{ET}}}^2} \right] \right\} \times \exp - \left[\frac{4R_o^2}{\left(8\sigma_x^2 + {L'_{\mathrm{ET}}}^2\right)} \right] \mathrm{d}x \tag{15.69}$$

or

$$\mathrm{PD}_{1x} = \frac{1}{\sigma_x\sqrt{2\pi}} \exp - \left[\frac{4R_o^2}{\left(8\sigma_x^2 + {L'_{\mathrm{ET}}}^2\right)} \right] \times \int_{x=-\infty}^{x=\infty} \exp - \left\{ \left(x^2 - \frac{R_o {L'_{\mathrm{ET}}}^2}{8\sigma_x^2 + {L'_{\mathrm{ET}}}^2} \right)^2 \left[\frac{8\sigma_x^2 + {L'_{\mathrm{ET}}}^2}{2\sigma_x^2 {L'_{\mathrm{ET}}}^2} \right] \right\} \mathrm{d}x \tag{15.70}$$

Consider the integral part of Eq. (15.70), denoted as I_1.

$$I_1 = \frac{1}{\sigma_x\sqrt{2\pi}} \int_{x=-\infty}^{x=\infty} \exp - \left[\frac{\left(x^2 - \mu\right)^2}{2\sigma_x^2} a \right] \mathrm{d}x \tag{15.71}$$

where

$$a = \left[\frac{8\sigma_x^2 + {L'_{\mathrm{ET}}}^2}{{L'_{\mathrm{ET}}}^2} \right] \tag{15.72}$$

We know that

$$\frac{1}{\sigma_u\sqrt{2\pi}} \int_{u=-\infty}^{u=+\infty} \exp\left[-\frac{(u-\mu_u)^2}{2\sigma_u^2} \right] \mathrm{d}u = 1 \tag{15.73}$$

so we can write as before

$$\int_{u=-\infty}^{u=+\infty} \exp\left[-\frac{(u-\mu_u)^2}{2\sigma_u^2} \right] \mathrm{d}u = \sigma_u\sqrt{2\pi} \tag{15.74}$$

Comparing Eqs. (15.71) and (15.73), we can write that

$$I_1 = \frac{1}{\sigma_x\sqrt{2\pi}} \sigma_u\sqrt{2\pi} = \frac{1}{\sigma_x\sqrt{2\pi}} \frac{\sigma_x}{\sqrt{a}} \sqrt{2\pi} = \frac{1}{\sqrt{a}} = \sqrt{\frac{{L'_{\mathrm{ET}}}^2}{8\sigma_x^2 + {L'_{\mathrm{ET}}}^2}} \tag{15.75}$$

Finally, from Eq. (15.70), PD_1 in the range direction is given by

$$PD_{1x} = \frac{L'_{ET}}{\sqrt{8\sigma_x^2 + {L'_{ET}}^2}} \times \exp - \left[\frac{4R_o^2}{\left(8\sigma_x^2 + {L'_{ET}}^2\right)} \right] \tag{15.76}$$

Recalling from the delivery accuracy section that $\sigma_x = \text{REP}/0.6745$, Eq. (15.76) may be written in the form

$$PD_{1x} = \frac{L'_{ET}}{\sqrt{17.6\text{REP}^2 + {L'_{ET}}^2}} \times \exp - \left[\frac{4R_o^2}{\left(17.6\text{REP}^2 + {L'_{ET}}^2\right)} \right] \tag{15.77}$$

Similarly for the deflection direction

$$PD_{1y} = \frac{W'_{ET}}{\sqrt{17.6\text{DEP}^2 + {W'_{ET}}^2}} \times \exp - \left[\frac{4D_o^2}{\left(17.6\text{DEP}^2 + {W'_{ET}}^2\right)} \right] \tag{15.78}$$

As before,

$$PD_1 = PD_{1x} \times PD_{1y} \tag{15.79}$$

It is seen that these results are the same as those obtained previously in Eqs. (10.21) and (10.22), except for the exponential terms in PD_{1x} and $PD1_{1y}$. The spreadsheet derived in Chapter 10 for unitary weapons/point targets can be modified to include indirect aimpoints using Eqs. (15.77) and (15.78).

15.7.2 Single Blast Weapons Against Unitary Targets

Referring back to Chapter 10, Section 10.5, the PD_1 values in range and deflection directions can be obtained from Eqs. (10.34) and (10.35) by including R_o and D_o as the nonzero mean values of the delivery error distributions.

$$\begin{aligned} PD_{1x} &= \int_{\frac{-L_{ET}}{2}}^{\frac{L_{ET}}{2}} \frac{1}{\sigma_x\sqrt{2\pi}} \exp\left[\frac{-(x-R_o)^2}{2\sigma_x^2}\right] dx \\ &= \text{ND}\left[\frac{L_{ET}}{2}, R_o, \sigma_x\right] - \text{ND}\left[-\frac{L_{ET}}{2}, R_o, \sigma_x\right] \end{aligned} \tag{15.80}$$

Table 15.10 Spreadsheet Changes to Include Indirect Aimpoints

DELIVERY ACCURACY					
		CEP ft ground	CEP ft n	CEP mils	REP/DEP
CEP, ft ground, normal or mils	CEP	100.000		0	
Delivery reliability	RD	1.000			
Deflection error probable, ft	DEP	57.300	0	0	57.300
Range error probable, ft	REP	57.300	0	0	57.300
Offset aimpoint range, ft	Ro	50.000			
Offset aimpoint deflection, ft	Do	50.000			
Ballistic error, deflection, ft	sigma-*bd*	17.103		5	
Ballistic error range, ft	sigma-*br*	25.698			
Adjusted DEP	DEP_dash	59.798			
Adjusted REP	REP_dash	62.799			

$$PD_{1y} = \int_{-\frac{W_{ET}}{2}}^{\frac{W_{ET}}{2}} \frac{1}{\sigma_y\sqrt{2\pi}} \exp\left[\frac{-(y-D_o)^2}{2\sigma_y^2}\right] dy$$

$$= ND\left[\frac{W_{ET}}{2}, D_o, \sigma_y\right] - ND\left[-\frac{W_{ET}}{2}, D_o, \sigma_y\right] \tag{15.81}$$

Again, the overall PD_1 is calculated from Eq. (15.79). The values of the offsets R_o and D_o are usually assigned in the delivery accuracy data block of the spreadsheet. Table 15.10 shows this modification for the spreadsheet described in Chapter 10, Section 10.8.6.

15.7.3 Single Rectangular Lethal Area Against an Area of Targets

The main difference here is that the finite target size means that the target may be partially covered by the weapon effective area, and the expected fractional coverage has to be calculated as in Chapter 11. We will consider the same situation described by Chapter 11, Section 11.5, but this time an indirect aim point (R_o, D_o) is included. Assume as in the previous section that the accuracy function takes the following form:

$$g(x) = \frac{1}{\sigma_x\sqrt{2\pi}} \exp\left[-\frac{(x-R_o)^2}{2\sigma_x^2}\right] \tag{15.82}$$

The expected fractional coverage in the range direction is given by

$$E(F_R) = \int_{x=-\infty}^{x=\infty} F_R(x)\, g(x)\, dx \tag{15.83}$$

substituting Eq. (15.82) and $F_R(x)$ from Fig. 11.20 into Eq. (15.83), we get

$$E(F_R) = \frac{1}{\sigma_x\sqrt{2\pi}}$$

$$\left\{ \begin{aligned} &\frac{L_{EP}}{L_A}\int_t^{-t}\exp\left(\frac{-(x-R_o)^2}{2\sigma_x^2}\right)dx + \left[\frac{L_{EP}+L_A}{2L_A}\right]\int_{-s}^{t}\exp\left(\frac{-(x-R_o)^2}{2\sigma_x^2}\right)dx \\ &+\left[\frac{L_{EP}+L_A}{2L_A}\right]\int_{-t}^{s}\exp\left(\frac{-(x-R_o)^2}{2\sigma_x^2}\right)dx + \int_{-s}^{t}\frac{x}{L_A}\exp\left(\frac{-(x-R_o)^2}{2\sigma_x^2}\right)dx \\ &-\int_{-t}^{s}\frac{x}{L_A}\exp\left(\frac{-(x-R_o)^2}{2\sigma_x^2}\right)dx \end{aligned} \right\} \tag{15.84}$$

The integral limits are

$$s = \frac{L_{EP}+L_A}{2} \qquad t = \frac{L_{EP}-L_A}{2} \tag{15.85}$$

Using shorthand notation, Eq. (15.84) may be written as

$$E(F_R) = [I_1 + I_2 + I_3 + I_4 - I_5] \tag{15.86}$$

Recalling the NORMDIST (ND) function from Excel, and considering R_o to be a mean value for the aiming error distribution, the first three integrals may be written as before, except we include the offset as follows:

$$I_1 = \frac{L_{EP}}{L_A}[\mathrm{ND}(-t, R_o, \sigma_x) - \mathrm{ND}(t, R_o, \sigma_x)] \tag{15.87}$$

$$I_2 = \frac{L_{EP}+L_A}{2L_A}[\mathrm{ND}(t, R_o, \sigma_x) - \mathrm{ND}(-s, R_o, \sigma_x)] \tag{15.88}$$

$$I_3 = \frac{L_{EP}+L_A}{2L_A}[\mathrm{ND}(s, R_o, \sigma_x) - \mathrm{ND}(-t, R_o, \sigma_x)] \tag{15.89}$$

Integrals I_4 and I_5 may be evaluated by putting them in the general form I', where

$$\begin{aligned} I' &= \frac{1}{\sigma_x\sqrt{2\pi}}\int_a^b x.\exp\left[-\frac{(x-R_o)^2}{2\sigma_x^2}\right]dx = \frac{1}{\sigma_x\sqrt{2\pi}}\int_a^b (x-R_o).\exp\left[-\frac{(x-R_o)^2}{2\sigma_x^2}\right]dx \\ &+\frac{1}{\sigma_x\sqrt{2\pi}}\int_a^b R_o\exp\left[-\frac{(x-R_o)^2}{2\sigma_x^2}\right]dx \end{aligned} \tag{15.90}$$

Setting $u = x - R_0$ and, therefore $dx = du$, the first integral on the right-hand side of Eq. (15.90) can be written as

$$\frac{1}{\sigma_x\sqrt{2\pi}}\int_a^b (x - R_o).\exp\left[-\frac{(x - R_o)^2}{2\sigma_x^2}\right]dx$$

$$= \frac{1}{\sigma_x\sqrt{2\pi}}\int_a^b u.\exp\left[-\frac{u^{\ 2}}{2\sigma_x^2}\right]du \tag{15.91}$$

Comparing Eq. (15.91) with Eq. (11.42), we can write

$$\frac{1}{\sigma_x\sqrt{2\pi}}\int_a^b u.\exp\left[-\frac{u^2}{2\sigma_x^2}\right]du = \frac{-\sigma_x}{\sqrt{2\pi}}\left\{-\exp\left[\frac{-u^2}{2\sigma_x^2}\right]\right\}_a^b$$
$$= \frac{-\sigma_x}{\sqrt{2\pi}}\left\{-\exp\left[\frac{-(b - R_0)^2}{2\sigma_x^2}\right] - \exp\left[\frac{-(a - R_0)^2}{2\sigma_x^2}\right]\right\} \tag{15.92}$$

The second term in Eq. (15.90) is evaluated using the NORMDIST function as

$$\frac{1}{\sigma_x\sqrt{2\pi}}\int_a^b R_o\exp\left[-\frac{(x - R_o)^2}{2\sigma_x^2}\right]dx = R_0[\mathrm{ND}(b, \sigma_x, R_0, 1) - \mathrm{ND}(a, \sigma_x, R_0, 1)] \tag{15.93}$$

Combining Eqs. (15.92) and (15.93), we can evaluate the remaining two integrals as

$$I_4 = \int_{-s}^{t} \frac{x}{L_A}\exp\left(\frac{-(x - R_o)^2}{2\sigma_x^2}\right)dx = \frac{-\sigma_x}{L_A\sqrt{2\pi}}\left\{-\exp\left[\frac{-(t - R_0)^2}{2\sigma_x^2}\right]\right.$$
$$\left. - \exp\left[\frac{-((-s) - R_0)^2}{2\sigma_x^2}\right]\right\} + \frac{R_o}{L_A}[\mathrm{ND}(t, R_o, \sigma_x) - \mathrm{ND}(-s, R_o, \sigma_x)] \tag{15.94}$$

$$I_5 = \int_{-t}^{s} \frac{-x}{L_A}\exp\left(\frac{-(x - R_o)^2}{2\sigma_x^2}\right)dx = \frac{-\sigma_x}{L_A\sqrt{2\pi}}\left\{-\exp\left[\frac{-(s - R_0)^2}{2\sigma_x^2}\right]\right.$$
$$\left. - \exp\left[\frac{-((-t) - R_0)^2}{2\sigma_x^2}\right]\right\} + \frac{R_o}{L_A}[\mathrm{ND}(s, R_o, \sigma_x) - \mathrm{ND}(-t, R_o, \sigma_x)] \tag{15.95}$$

Table 15.11 Fractional Damage with Offset Aimpoints

FD₁ CALCULATOR WITH OFFSETS			
RANGE		***DEFLECTION***	
INPUT PARAMETERS		***INPUT PARAMETERS***	
Target length LA (ft)	100.000	Target width WA (ft)	80.000
Weapon length LET(ft)	36.400	Weapon width WET(ft)	54.950
REP	20.000	DEP	10.000
LEP	100.000	WEP	80.000
Sigma range (ft)	29.652	Sigma deflection (ft)	14.826
Offset aimpoint R0	10.000	Offset aimpoint D0	10.000
INTEGRAL LIMITS		***INTEGRAL LIMITS***	
s	100.000	s	80.000
t	0.000	t	0.000
INTEGRALS		***INTEGRALS***	
integral #1	0.000	integral #1	0.000
integral #2	0.368	integral #2	0.250
integral #3	0.631	integral #3	0.750
integral #4	-0.075	integral #4	-0.028
integrall #5	0.174	integrall #5	0.153
RESULTS			
Fractional coverage	0.750		0.820
Total fractional coverage	0.615		
FD₁	**0.1538**		

This analysis applies to the range direction, but similar equations may be derived for the deflection direction by substituting W_{EP} for L_{EP} and W_A for L_A to calculate $E(F_D)$. Finally, we get the following equation for the total fractional coverage:

$$\text{FD}_1 = E(F_R) \times E(F_D) \times P_{CD} \times R \qquad (15.96)$$

Again, the spreadsheet described in Chapter 11 can be modified to include the indirect aimpoint, and the FD_1 computations modified as described by the previous equations. An example of a modified spreadsheet is shown in Table 15.11.

15.8 Improved Stick/Volley Methodology

We saw in Chapter 12, Section 12.11 how misleading results might be obtained if a stick or volley of weapons aggregates the individual lethal areas into a single rectangle that is then overlaid onto an area of target elements, as indicated in Fig. 12.22 and reproduced here in Fig. 15.21. In the case given, the differences are more obvious if the stick may be considered "sparse" compared to the target area, and the delivery errors are small.

This methodology applies a pattern of lethal areas against a rectangular area of target elements; it is therefore quite general and may be used to represent a stick of air-delivered bombs or a volley of surface-launched artillery projectiles.

Given the method outlined in Section 15.7.3, we can improve on the previous method where the individual weapon lethal areas are aggregated, by treating them individually. If we accept the run time consequences of using a Monte Carlo implementation of sampling the errors, then we can implement a flow chart similar to that shown in Fig. 9.7, adapted to an area of targets.

Assuming we can determine the fractional damage for a single weapon given an offset from the center of the target area, that is, the origin (Table 15.11), the question is: What contributes to this total offset? The components are

- The desired aimpoint of the individual weapon in the stick
- The common dependent error for the whole stick, relative to the origin
- The independent error for each weapon in the stick relative to the dependent error

For example, consider the top left weapon in Fig. 15.21. If the origin is located at the center of the target with range pointing up, the coordinates of this aimpoint are $(x_i, y_i) = (300, 300)$. If we make a random draw from the MPI error $(x_{i-\text{MPI}}, y_{i-\text{MPI}})$ then the center of the ith weapon lethal area is

$$x_I = x_i + x_{i+\text{MPI}}, \quad y_I = y_i + y_{i+\text{MPI}} \tag{15.97}$$

We could now add a random draw from the precision error in x and y for each weapon to determine the location of the impact point, and then determine the fractional coverage of this rectangular lethal area over the target

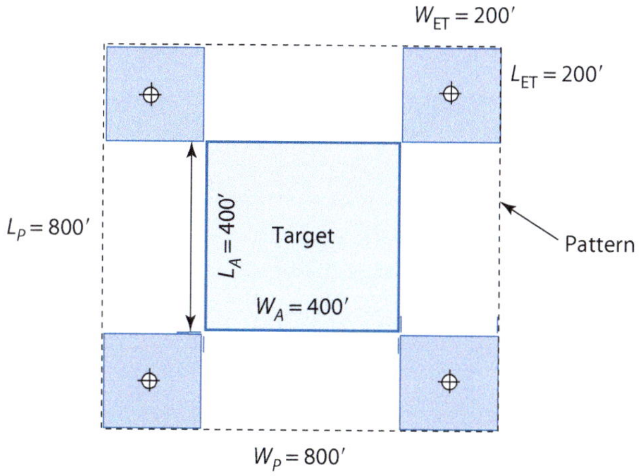

Fig. 15.21 Fractional coverage of sparse stick.

```
Dim FC_weap(20) As Double
sx = REP_MPI / 0.6745                                              ' convert aiming error to sigma x
sy = DEP_MPI / 0.6745                                              ' convert aiming error to sigma y
sigma_bd = REP_Prec / 0.6745                                       ' convert to deflection value
sigma_br = DEP_Prec / 0.6745                                       ' convert to range value
    For iter_MPI = 1 To iter_max                                   ' MPI loop sampling
        x_MPI = Application.NormInv(Rnd(), 0, sx)                  ' sample MPI error range
        y_MPI = Application.NormInv(Rnd(), 0, sy)                  ' sample MPI error deflection
        For iter_round = 1 To N_weaps                              ' loop through each weapon in the volley
            x_impact = x_MPI + x_aim(iter_round)                   ' constant offset for each weapon x
            y_impact = y_MPI + y_aim(iter_round)                   ' constant offset for each weapon y
            FC_weap(iter_round) = FC_offset(LA, WA, L_et, W_et, sigma_br, sigma_bd, x_impact, y_impact)
        Next iter_round                                            ' end precision loop
    Next iter_MPI                                                  ' end MPI loop

    FC_tot = 0                                                     ' initial total FC
    For j = 1 To N_weaps                                           ' loop through all weapons in stick
        FC_tot = FC_tot + FC_weap(j)                               ' update total FC
    Next j                                                         ' end of weapon loop
    FC_MC = FC_tot                                                 ' final FC
    Lb = L_et
    Wb = W_et
```

Fig. 15.22 Code segment for improved stick method.

rectangle. However, because this uniquely determines where the weapon lands, the fractional coverage could be determined from Fig. 11.20. Therefore, we use the expected value theorem as implemented in Chapter 11, Section 11.5, where the accuracy used is the precision error. This way we only need one loop, the one for the MPI error.

The expected fractional coverage is calculated for a fixed MPI error for each weapon in turn to give, in the example under consideration, four values of FD_1, which are then summed to get the total FD.

For the example shown in Fig. 15.21, the following data would apply: $L_S = W_S = 600$ ft, $L_A = W_A = 400$ ft. If we set $MAE_F = 40{,}000\ \text{ft}^2$ and the impact angle to 90 deg, we get the square lethal area measuring 200 ft × 200 ft. Implementing the methodology just described and setting all delivery errors to zero, the weapon lethal areas land in the positions indicated, all FD_1 values become zero, and therefore their sum is FD = 0. A code segment implementing this is shown in Fig. 15.22.

Changing the delivery errors REP and DEP MPI and precision to, say, 20 ft, some possible overlap of one or more weapon lethal areas in the stick will occur; therefore, there will be a resultant FD. The result from running the program is FD = 0.021.

Alternatively, using the AS sticks method, or the SQ2 method for artillery, we would aggregate the lethal areas into a pattern 800 ft × 800 ft with a conditional probability of damage inside of

$$P_{CDS} = \frac{4 \times (200 \times 200) \times 1}{800 \times 800} = 0.25 \qquad (15.98)$$

Again setting the delivery errors to zero, this rectangle will totally cover the target area, giving a fractional coverage of unity but a fractional damage of $FD = FC \times P_{CDS} = 0.25$. Compare this to the case using discrete weapon lethal areas of FD = 0, and the advantage of the improved methodology becomes obvious.

Table 15.12 Calculating L_{ET} and W_{ET} for Various EI Values

EI	L_{ET}	W_{ET}	A_{ET}	P_{HD}
MAE_F	$\sqrt{MAE_F \times a}$	L_{ET}/a	MAE_F	1.0
MAE_B	$\sqrt{MAE_B}$	L_{ET}	MAE_B	1.0
A_V(n)	$W_{ET}/\sin I$	$\sqrt{A_V}$	$A_V/\sin I$	See note 4
A_V(gnd)	W_{ET}	$\sqrt{A_V}$	A_V	See note 4
BEI	L_T	W_T	$L_{ET} \times W_{ET}$	$1 - \exp(-BEI/W_B)$
EMD 2D rect	$L_T + 2 \times EMD$	$W_T + 2 \times EMD$	$L_{ET} \times W_{ET}$	1.0
EMD 3D rect	$L_T + 2 \times EMD + L_{SH}$	see note 2	$L_{ET} \times W_{ET}$	See note 4
EMD 3D cyl	$\sqrt{\pi}(R_T + EMD) + L_{SH}$	see note 3	$L_{ET} \times W_{ET}$	See note 4
EMD vertical	$(H_T + EMD)/\tan I + EMD$	$W_T + 2\,EMD$	$L_{ET} \times W_{ET}$	See note 4
MAE_{BLDG}	$L_T + 2 \times EMD + L_{SH}$	see note 2	$L_{ET} \times W_{ET}$	1.0

$L_{SH} = H_T/\tan I - EMD$. If $H_T = 0$ or if $L_{SH} < 0$, set $L_{SH} = 0$.

$W_{ET} = [(L_T + 2\,EMD)(W_T + 2\,EMD) + L_{SH}\,W_T]/L_{ET}$.

$W_{ET} = \left[\pi(R_T + EMD)^2 + 2\,R_T\,L_{SH}\right]/L_{ET}$

Usually $P_{HD} = 1$, but users sometimes input the presented area together with a value of $P_{HD} < 1$.

Table 15.13 Methodology Locator for Different EI Types

EI ►	MAE_F		MAE_B		A_{VN}	
Weapon ▼	**Unitary**	**Area**	**Unitary**	**Area**	**Unitary**	**Area**
Single	Ch. 10 when $R_0 = D_0 = 0$, Ch. 15 otherwise	Ch. 11 when $R_0 = D_0 = 0$, Ch. 15 otherwise	Ch. 10, Sect. 10.5 when $R_0 = D_0 = 0$, Ch. 15 otherwise	Ch. 11 when $R_0 = D_0 = 0$, Ch. 15 otherwise	Ch. 14	N/A
Stick	Ch. 13 with $L_A = W_A = 0.01$	Ch. 13	Ch. 13 with $L_A = W_A = 0.01$	Ch. 13	Ch. 12 with $L_A = W_A = 0.01$	N/A
EI ►	**BEI**		**EMD**		MAE_{BLDG}	D_C
Weapon ▼	**Unitary**	**Area**	**Unitary**	**Area**	**Unitary**	**Unitary**
Single	Sect. 15.2	N/A	Sect. 15.3	Sect. 15.3	Sect. 15.4	Sect. 15.6
Stick	Sect. 15.2	N/A	Sect. 15.3	Sect. 15.3	N/A	N/A

15.9 Summary of Weaponeering Methodologies and Chapter Locator

Considering the methodologies discussed in Chapters 10–15, it may be seen that there is a commonality of approach that, simply stated, involves the following:

1. Determine L_{ET} and W_{ET} for the appropriate EI.
2. Determine if the delivery is a single weapon or a stick.
3. Identify the target as unitary or an area of targets.
4. Consider if offset aimpoints are used.
5. Select the appropriate methodology to compute PD_1 or FD_1.

For all the EI types discussed, Table 15.12 summarizes how to calculate the appropriate L_{ET} and W_{ET}.

Table 15.13 summarizes where to locate the corresponding methodologies for computing effectiveness.

Note also the following:

1. The Carlton damage function is used only for the unitary target/single weapon combination; otherwise, a rectangular cookie cutter damage function is used.
2. N/A means methods may or may not exist, but an operational scenario of this type is unlikely.

Reference

[1] "Derivation of JMEM/AS Open-End Methods," JTCG/ME Publication 61, JTCG/ME-3-7, Revision 3, 2004 Jan. 21.

Chapter 16 Surface-to-Surface Weaponeering: Indirect Fire

Surface-to-surface (SS) applications may be divided into two broad areas:

1. *Indirect fire* means the shooter does not have direct, visual contact with the target. The position of the shooter is known, perhaps with a GPS, and the position of the target is estimated. From these data and knowledge of the weapon ballistics, a firing solution is computed. Typical systems include artillery, mortars, and larger caliber ship-mounted guns. Examples are shown in Fig. 16.1.
2. *Direct fire* means the shooter does have visual contact with the target, and aim is assisted by placing an aiming cue on the target. Examples include small caliber weapons and a tank main gun—see Fig. 16.2.

The subject of indirect fire is dealt with in this chapter; direct fire follows in the next chapter.

16.1 Introduction

This section will begin with an introduction to the main differences between terminology and definitions used in previous air-to-surface (AS) methodologies and those used for surface-to-surface (SS) techniques. Then

Fig. 16.1 Examples of indirect fire.

Fig. 16.2 Examples of direct fire.

the SS methodologies will be described by comparing them to the AS versions already developed. Finally, the implementation of these methodologies into several SS weaponeering tools will be described. The accuracy section of the effectiveness calculation for these systems was developed in Chapter 4, so that part will not be discussed extensively here.

Although not traditionally considered indirect fire systems, air-launched coordinate seeking weapons (GPS guided) may also be considered within this group. The weapon is not guided by anything that distinguishes the target such as a radar image, just the location on the Earth where the target is thought to be. Therefore, these weapons share many characteristics with unguided artillery rounds.

16.2 Terminology and Combat Scenarios

A typical combat scenario for indirect fire might consist of a number of individual, but identical, weapons firing several rounds at a target location given as a grid coordinate set on a map. The number of weapons is referred to as a *battery*, and is typically four or six. The number of rounds fired on a particular occasion is termed a *volley* or a *salvo*, and is typically less than 10. Establishing the battery and laying each gun for a fire support mission is a complicated process [1], so only an overview will be provided here.

Because indirect fire unguided systems are essentially shooting "blind," they can be somewhat inaccurate and as such are traditionally used against an area of targets. At the very least, precise known coordinates of the battery and target, together with a good trajectory model for the rounds being fired, are all required in order to put rounds on target. Feedback from a forward observer who can see where rounds land relative to the target allows substantial improvement in delivery accuracy. Within the battery is a reference gun (see Fig. 16.3), which in conjunction with the location of the center of the target area forms the prime shooter – target geometry. The position of each gun relative to this reference gun and the aimpoints within the target area generate individual firing solutions.

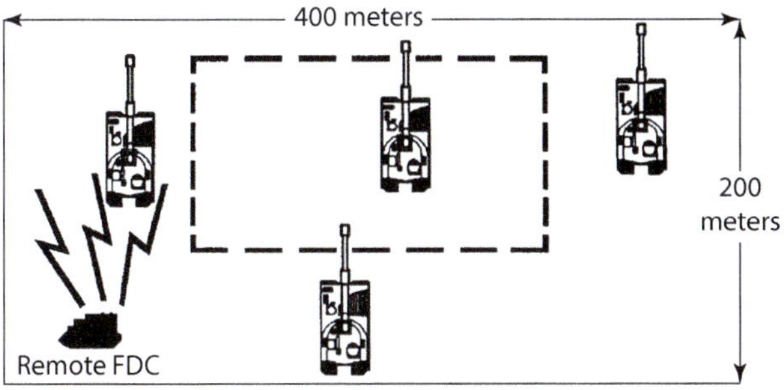

Fig. 16.3 Battery elements relative to reference gun.

Enemy counterfire radars are able to track the trajectory of incoming artillery rounds and therefore determine the point from which they were fired. It is usual, therefore, in a high threat environment for artillery to fire only a few volleys and then move to another location to avoid being attacked by weapons directed by such radars.

Rounds fired by artillery may be high explosive (HE) unitary rounds or rounds that dispense submunitions. Usually the submunition rounds are referred to as improved conventional munitions (ICMs). Although artillery fire using unguided munitions can be inaccurate, the use of guided munitions does allow single round shots against point targets.

Within the area target, each weapon in the battery will have its own aimpoint, where the selection of aimpoints is determined by the fire control system (FCS), which is currently the Advanced Field Artillery Tactical Data System (AFATDS). This computes the aimpoints for each gun, and is transmitted electronically directly to a weapon display. An exception is the use of longer range unguided rockets, which being less accurate than conventional artillery will have a single aimpoint. This allows the natural precision error of rounds to spread them over the target area.

We are dealing with an area of targets only, so the weapon effectiveness is usually described in terms of the fractional damage discussed in Chapter 11. The term *expected fractional damage (EFD)* is usually reserved for materiel targets, whereas the term *expected fractional casualties (EFC)* is used for personnel targets. They are calculated in a similar though not identical manner to the EFD in Chapter 11.

16.3 Aimpoint Selection

For howitzers, the legacy aimpoint policy is to spread the rounds from individual weapons uniformly over the target area. Figure 16.4 shows how this is done for a four-gun battery shooting against rectangular targets.

The aimpoints for n artillery weapons distributed uniformly within a rectangular target area having n_x and n_y aimpoints in the range and deflection

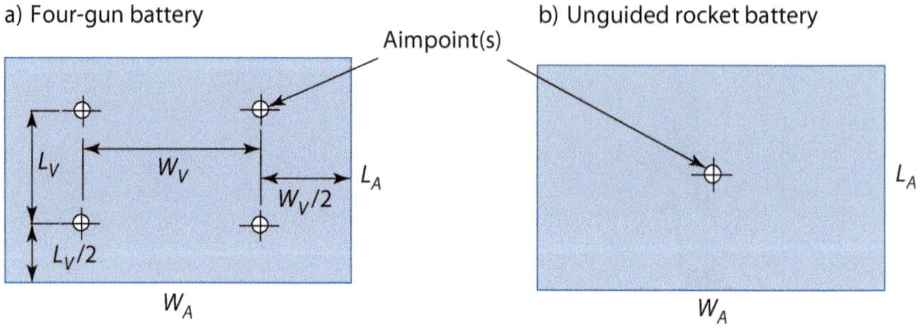

Fig. 16.4 Aimpoints for rectangular targets.

directions, respectively, will be given by

$$x_i = \frac{-L_A}{2} + \frac{L_A}{2n_x} + \frac{(i-1)L_A}{n_x} \qquad i = 1,\ 2\ldots n_x \tag{16.1}$$

$$y_i = \frac{-W_A}{2} + \frac{W_A}{2n_y} + \frac{(i-1)W_A}{n} \qquad i = 1,\ 2\ldots n_y \tag{16.2}$$

where

$$n_x \times n_y = n \tag{16.3}$$

The volley length and width are then given by

$$L_V = \frac{(n_x - 1)L_A}{n_x} \tag{16.4}$$

$$W_V = \frac{(n_y - 1)W_A}{n_y} \tag{16.5}$$

Aimpoints inside circular targets are shown in Fig. 16.5 for four- and six-gun batteries.

For the four-gun case, the volley pattern length and width are given by

$$L_V = W_V = R \tag{16.6}$$

whereas for the six-gun case they are

$$L_V = R \tag{16.7}$$

$$W_V = R\cos(30) \tag{16.8}$$

For the general case of n weapons, the volley length will be the same as in Eq. (16.7); the width is given by the following:

$$W_V = R\cos\left(90 - \frac{360}{n}\right) \tag{16.9}$$

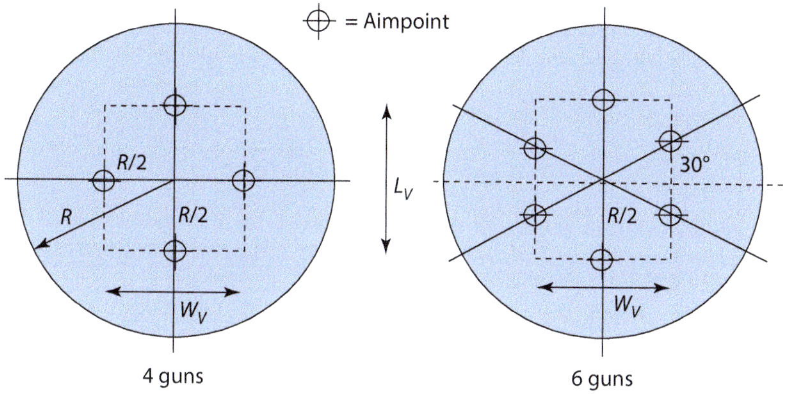

Fig. 16.5 Aimpoints for circular targets.

The equivalence of the fall of shot in a single volley of cannons to the stick of unguided bombs dropped from an aircraft will be noted. This similarity will now be extended to calculating the effectiveness of indirect fire by relating the analysis to that described in previous chapters for air-launched weapons.

More recently, improved aimpoint determination algorithms have been developed that take into account not only the size of the target area, but also the lethality of the warhead and delivery accuracy to provide greater efficiency of target coverage. In addition, newer systems may have aimpoint algorithms specifically tailored to that weapon.

16.4 Weapon Lethal Area

Methods for representing the lethal area of munitions are different for high-explosive (HE) unitary and improved conventional munition (ICM) warheads; however, the same basic representations are the same as for air-launched weapons. Each will be dealt with separately.

16.4.1 Unitary Warheads

The effect of a unitary warhead on a ground target may be represented by the lethal area matrix discussed previously in Chapter 8. A simplified example is shown in Table 16.1. The cell value represents the probability of damaging the target to the prescribed level if the target is located in that particular cell. The small circle indicates the weapon detonation point, and the shaded elements represent the nonzero cell P_K values. In this case, a representative cell size is about* 15 m × 8 m.

*Recall that in the JMEM AS quantities are imperial, whereas SS quantities are metric.

Table 16.1 Simplified PK, Damage, or Lethal Area Matrix

0	0	0	0	0	0	0	0	0	0	0	0	0	0
0	0	0	0	0	0.001	0	0	0.001	0	0	0	0	0
0	0	0	0	0.001	0.002	0.003	0.003	0.002	0.001	0	0	0	0
0	0.001	0.002	0.003	0.004	0.007	0.014	0.014	0.007	0.004	0.003	0.002	0.001	0
0	0	0	0.002	0.008	0.04	0.227	0.227	0.04	0.008	0.002	0	0	0
0	0	0	0	0.001	0.002	0.049	0.049	0.002	0.001	0	0	0	0
0	0	0	0	0	0	0.002	0.002	0	0	0	0	0	0
0	0	0	0	0	0	0	0	0	0	0	0	0	0

From a damage matrix such as this, a lethal area designated as the MAE_F may be calculated from

$$A_L = MAE_F = \sum_{i=1}^{n} P_K(i) \times A_{cell} \tag{16.10}$$

where there are n nonzero cell P_K values. A Carlton damage function may also be used to represent a damage matrix in the following form:

$$P(x, y) = D_0 \exp - \left\{ D_0 \left[\frac{x^2}{R_x^2} + \frac{y^2}{R_y^2} \right] \right\} \tag{16.11}$$

The parameters R_x and R_y are the weapon radii introduced in Chapter 8, Section 8.15,[†] and the shaping parameter D_0 was introduced in Chapter 11, Section 11.9. Once these parameters have been determined correctly, conservation of lethality implies the following expression will be true:

$$\int_{-\infty}^{\infty} \int_{-\infty}^{\infty} P(x, y)\, dx\, dy = A_L = MAE_F \tag{16.12}$$

The analysis for determining the weapon radii is similar to that described in Chapter 8, Section 8.15. Usually the value of MAE_F is given as an input together with the impact angle of the munition. Representing the lethal area as an ellipse, we can write

$$MAE_F = \pi R_x R_y \tag{16.13}$$

The aspect ratio of the ellipse is defined by

$$a = \frac{R_x}{R_y} \tag{16.14}$$

where it has been found that for most cases

$$a = 1 - 0.8 \cos(I) \tag{16.15}$$

This expression is different from the AS Eq. (8.43) in order to allow for shallower impact angles that may result from direct fire situations. Given the MAE_F and impact angle I, Eqs. (16.13) through (16.15) may be solved for R_x and R_y.

$$R_x = \sqrt{\frac{MAE_F \times a}{\pi}} \tag{16.16}$$

$$R_y = \frac{R_x}{a} \tag{16.17}$$

[†]In the AS methods, the weapon radii are given the notation WR_R and WR_D.

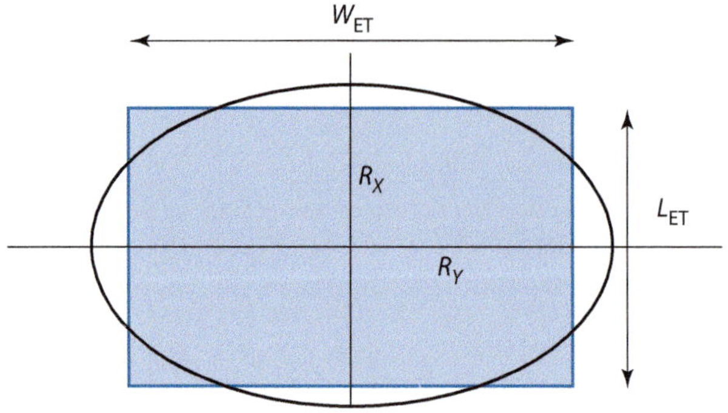

Fig. 16.6 Rectangular equivalent to elliptical lethal area.

This provides all the parameters needed to specify the Carlton damage function.

In many cases, when the damage function is given in the form of an elliptical cookie cutter, an equivalent L_{ET} and W_{ET} for a rectangular cookie cutter are needed. Figure 16.6 shows this conversion.

By equating the areas in Fig. 16.6 we obtain

$$\pi R_X R_Y = L_{ET} W_{ET} \tag{16.18}$$

giving

$$L_{ET} = \sqrt{\pi} R_X, \qquad W_{ET} = \sqrt{\pi} R_Y \tag{16.19}$$

This may be written in the following form:

$$L_{ET} = \sqrt{\text{MAE}_F \times a}, \qquad W_{ET} = L_{ET}/a \tag{16.20}$$

16.4.2 ICM Warheads

As for air-launched dispenser weapons, ICM warheads burst at a prescribed altitude or time from firing and dispense a fixed number of submunitions over an area on the ground. Different warheads contain different numbers of submunitions, and different mixes of antimateriel and antipersonnel types. A single ICM will produce a pattern on the ground that is assumed to be elliptical in shape and have dimensions that are functions of the range between the howitzer and the target. Table 16.2 shows typical precomputed pattern sizes for an artillery round.

Note a difference here between AS and SS treatment of ICM pattern dimensions. Air-delivered cluster munition pattern dimensions are computed from the functioning parameters, as described in Chapter 13, Section 13.5. For SS ICM, however, the pattern sizes are precomputed and stored in tabular form.

Table 16.2 Elliptical ICM Pattern Dimensions

Range (km)	Average Height of Burst (m)	Ellipse Axis (range, m)	Ellipse Axis (deflection, m)
4	290	57	55
7	340	67	59
12	425	103	60
14	425	111	73

The effect of an individual submunition on a target is represented by the same type of damage matrix used for HE rounds, although the derived lethal area will be much smaller. For example, a grenade-type submunition against personnel targets may have a lethal area of around 30 m^2 (i.e., approximately a circle of radius 3 m).

It is usual to assume that individual submunitions are distributed in the pattern in a uniformly distributed random manner, meaning that each submunition has an equal probability of landing anywhere within it. Suppose the single submunition lethal area is a_L and the pattern area is A_P. The uniform probability of damage for a single submunition within the pattern is

$$P_S = \frac{a_L}{A_P} \tag{16.21}$$

Essentially, here we are spreading out the submunition lethal area over the whole of the pattern. The pattern probability of damage is obtained by powering up Eq. (16.21) for N_S submunitions of reliability R_S to give

$$P_{\mathrm{DP}} = 1 - (1 - P_S)^{N_S R_S} \tag{16.22}$$

Because there is usually a large number of submunitions, the Morse-Kimball approximation to Eq. (16.22) may be used to obtain

$$P_{\mathrm{DP}} = 1 - \exp\left(-\frac{N_S R_S a_L}{A_P}\right) \tag{16.23}$$

The elliptical pattern is converted to an equivalent rectangular lethal area, as shown in Fig. 16.6, and together with the damage probability given by Eq. (16.23), the weapon is now treated as a unitary warhead.

16.5 Delivery Accuracy

For indirect fire weapons, there are four principle types of delivery accuracy metrics to consider:

1. Fixed bias—mean of MPI errors is not zero.
2. Mean point of impact (MPI) error, equivalent to aiming errors for air-launched weapons.

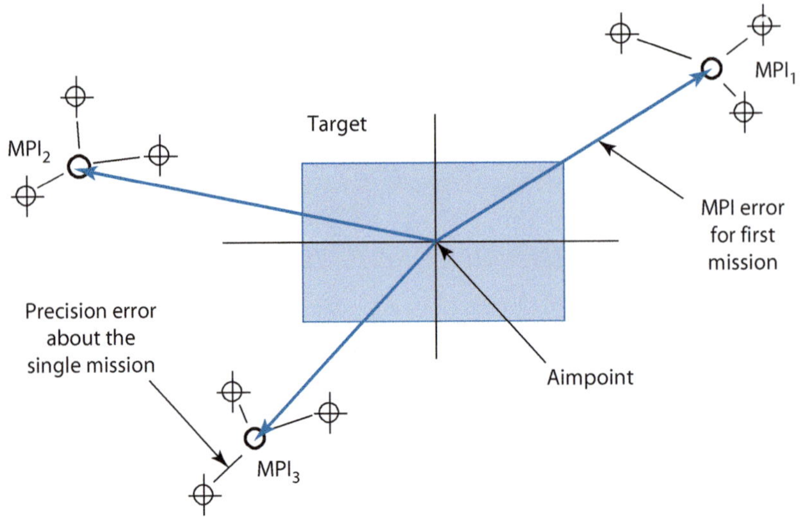

Fig. 16.7 MPI and precision errors.

3. Precision error, same as for air-launched weapons.
4. Gun and target location errors (TLEs).

All accuracy measures are usually specified in meters in the ground plane, although the precision error may be given in mils in the normal plane. A fixed bias error was discussed in Chapter 4, Section 4.2, and the derivation of artillery MPI and precision errors was also dealt with in Chapter 4.

One of the main contributors to these errors is the weather, and winds in particular. Considerable effort is made to obtain accurate, current meteorological data to be used in ballistic table calculations, which in the past involved the release of sensor-equipped balloons to measure wind speed and direction, air temperature, and humidity. Newer methods involving computer-predicted weather conditions are under development. Clearly these data may vary considerably from mission to mission. MPI errors are assumed to be normally distributed with zero mean (unless a fixed bias is present) and known standard deviations $\sigma_{\text{MPI-}x}$ and $\sigma_{\text{MPI-}y}$. The relationship between MPI and precision errors is shown again in Fig. 16.7.

16.5.1 Gun and Target Location Errors

Because we are dealing with indirect fire systems, the weapon system operators do not have direct visual confirmation of where their rounds land relative to the target. In the worst case, map grid coordinates are used to locate the gun and target. What is important here is the position (bearing and range) of the target relative to the gun; however, for indirect fire, each of these is usually determined independently. Hence there will be two separate errors, one in locating the gun with reference to a fixed coordinate system (latitude, longitude, elevation) and another in locating the target with respect to the same coordinated frame. From these data the position of the target relative to the gun may be determined. It is this relative error that is referred to as the target location error (TLE).

There may, however, be sensor systems that attempt to provide this information with varying degrees of accuracy; for example, a forward observer (FO) may relay information back to the weapon crew on how much in range and direction they are missing the target when single ranging shots are fired. Each system the observer uses has an associated accuracy with which the target location and fall of shot are estimated. These sensor-specific TLE values are usually measured and tabulated. Once more, it is assumed that these errors are normally distributed though uncorrelated in range and deflection with zero means and standard deviations S_X and S_Y.

Target location errors may occur due to inaccurate map grid references or GPS coordinates. TLEs also come from various target acquisition systems such as airborne or ground-based radars or optical sensors. Although observer-adjusted fire attempts to mitigate these, TLEs are usually present in some form. TLEs may be considered occasion-to-occasion type errors, as is the MPI error, and because they are both of the same type but independent, their variances (or error probable) may be root sum squared together in the following manner:

$$\mathrm{MPI}'_x = \sqrt{\mathrm{MPI}_x^2 + S_x^2} \tag{16.24}$$

$$\mathrm{MPI}'_y = \sqrt{\mathrm{MPI}_y^2 + S_y^2} \tag{16.25}$$

where S_x and S_y are the standard deviations of the TLE in the range and deflection directions, respectively.

16.5.2 Delivery Accuracy Data

As discussed in Chapter 4, Section 4.21, SS delivery accuracy data are tabulated as a function of range and delivery technique, an example of which is reproduced in Table 16.3. Future plans call for generation of the data "on the fly" for the specific firing solution requested by the user.

Table 16.3 Sample Delivery Accuracy Data

Impact Conditions				Probable Error (50%)					
				Precision Error (m)		MPI (m)			
						Observer Adjusted		Nonadjusted	
Range	Charge	θ	V_I	Range	Defl.	Range	Defl.	Range	Defl.
3000	4	13	246	13.5	1.3	22.3	7.4	29.7	10.1
5000	5	19	263	12.1	2.7	21.6	7.4	44.5	12.8
7000	6	26	269	11.5	1.3	21.6	6.7	70.1	20.2
9000	7	31	278	15.5	2.0	22.9	7.4	80.3	33.7
11,000	7	44	283	18.2	3.4	25.0	8.1	108.6	51.9
13,500	8	49	300	35.0	8.0	40.0	12.0	131.0	73.0

The problem with data in this format is that for range/charge combinations not listed in the table, interpolation is required.

16.6 Munition Trajectory

Details of the methods for computing projectile trajectories were dealt with in Chapter 3. In calculating effectiveness, it is possible to determine the trajectory of the round being fired, but at the present time this is not done. Impact conditions are computed for a variety of firing conditions and the results stored again in tabular form. These are then interpolated to obtain terminal conditions for the case under consideration. An example of such data is also shown in Table 16.3. Standardization efforts for SS munitions trajectories are more advanced than for air-delivered weapons. For example, the NATO Artillery Ballistics Kernel (NABK) represents a multinationally accepted trajectory program used by the United States and other countries to preserve commonality of trajectory results. It is a four degrees of freedom (4-DOF) program with an increasing complement of munitions that provides trajectory results for most artillery-fired projectiles. This model is described in detail in Volume 2, Chapter 2.

16.7 Naval Gunfire

Naval gunfire directed against shore targets (fire support) is similar to land-based artillery, including the type of ammunition (HE and ICM) and fire control systems used. There are, however, a few differences as noted here:

- Currently all Navy fire support guns are of 5-in. caliber. A second digit in their designation is the length of the barrel in calibers; for example, 5 in./54 and 5 in./62 represent two tubes of the same 5-in. caliber, but with barrel lengths of 54 × 5 in. and 62 × 5 in., respectively.
- Navy guns are also used to attack aircraft as well as ground targets; therefore, the rounds fired may have proximity fuzes.
- The aiming policy is different. For target areas with a radius smaller than 200 m, a single aimpoint at the target center is selected. For larger targets, it is divided into quadrants, and the center of the quadrant is taken as the aimpoint. In addition, unitary targets can also be attacked with ship-based weapons.
- Because targets may be located using ship-based radar, delivery accuracy is sometimes given as a function of offset distance from a radar significant reference point. This is similar to offset bombing for air-delivered weapons.

The same tool used for field artillery accuracy assessment (JWAM) is also used for Navy fire support accuracy assessment, but there are a few more error budget terms unique to a ship-board weapon system. These relate specifically to the aiming accuracy in azimuth and elevation and are decomposed into several independent sources as listed here:

Round-to-round errors:

- Gun jump
- Gun response
- Computer
- Stabilization
- Deck flexure
- Gyrocompass

Occasion-to-occasion errors:

- Gun jump
- Alignment
- Computer
- Stabilization

These independent error sources are root sum squared (RSS) in the usual manner to obtain the overall gun aiming accuracies.

16.8 Comparing Sequential Dependent and Independent FD Events

The issue of dependency is concerned with the occurrence of a sequence of events and whether each event is influenced by the previous ones. This topic, which has particular significance for indirect fire, was addressed in Chapter 9, Section 9.8 for point targets and will be expanded further here for area targets. Recall the differentiation between independent and dependent attacks using the following examples:

- An aircraft flies over an area of targets and drops a stick of unguided bombs, where the intent is to aim the center of the stick at the center of the target. Shortly after, another aircraft repeats the attack. If the second pilot aims the stick at the same location as the first, irrespective of the results of the first attack, each stick is considered to be independent.
- An artillery battery fires a volley of rounds at the same target where again the intent is to place the center of the volley on the center of the target area. The guns in the battery are then reloaded and a second volley is fired. Because the guns are not reaimed between volleys, each volley is considered dependent.

In both of these examples, we may constitute a rectangular pattern of known dimensions for the stick or volley with known P_{CD} inside it so lethal area on the ground plane may be considered emanating from a single weapon. This is achieved by means of the following process:

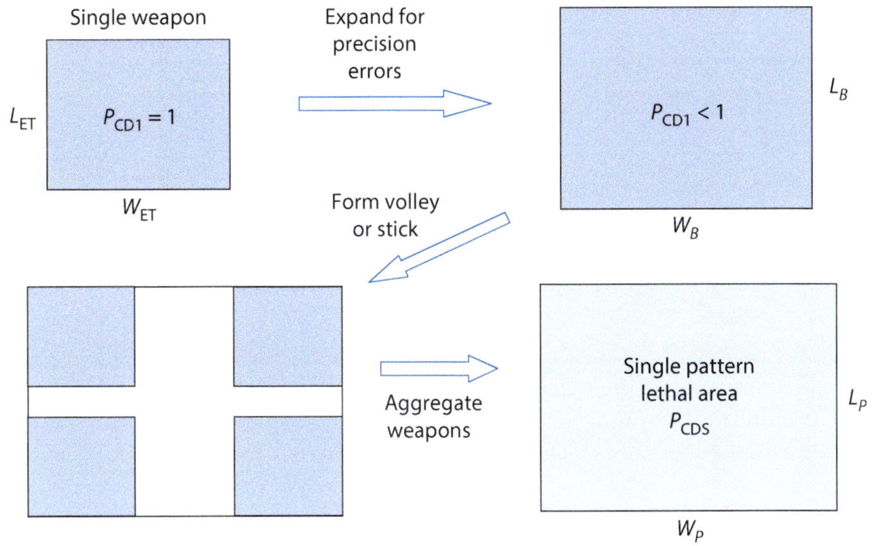

Fig. 16.8 Single weapon representation of a stick or volley.

1. Expand the single weapon lethal area to account for precision errors.
2. Aggregate the pattern from the number of weapons dropped simultaneously from the aircraft or fired from the battery artillery.
3. Evaluate the conditional probability of damage P_{CDS} within the pattern.

The process is shown schematically in Fig. 16.8.

The issue addressed next is how to power up the single event (stick or volley) estimation of damage to account for multiple events.

16.8.1 Independent Event Methodology

In order to derive the basic equation governing independent event effectiveness, consider an area of target of elements that is subjected to three attacks, as shown in Fig. 16.9.

The lethal area in this figure is the single pattern shown in Fig. 16.8. For this case, each event is considered a separate occasion; therefore, an appreciable amount of time elapses before the next event occurs. It is therefore an implicit assumption that after each event, the target elements become redistributed (uniformly) within the target area; hence, the density of dots in the target area indicates the relative density of target elements indicating how the density decreases with subsequent events.

We assume all weapons have the same P_{CD1} and that the effectiveness measure will be fractional damage; hence, for each attack we have the

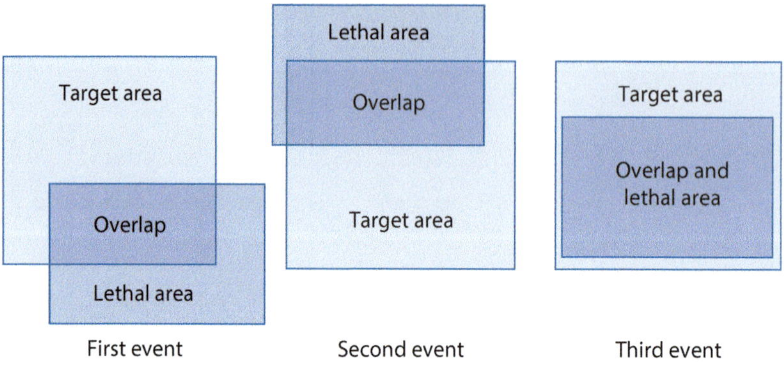

Fig. 16.9 Three independent events.

following equation:

$$\mathrm{FD}_1 = F_{C1}P_{\mathrm{CDS}} \tag{16.26}$$

The proportion of target elements that survive the first attack is given by

$$\mathrm{FS}_1 = 1 - \mathrm{FD}_1 = 1 - F_{C1}P_{\mathrm{CDS}} \tag{16.27}$$

The proportion of elements that survive the first attack and second attack is given by

$$\mathrm{FS}_2 = (1 - F_{C1}P_{\mathrm{CDS}})(1 - F_{C2}P_{\mathrm{CDS}}) \tag{16.28}$$

The proportion of elements that survive n attacks is given by

$$\mathrm{FS} = \prod_{i=1}^{i=n}(1 - F_{Ci}P_{\mathrm{CDS}}) \tag{16.29}$$

Hence the fractional damage is

$$\mathrm{FD} = 1 - \mathrm{FS} = 1 - \prod_{i=1}^{i=n}(1 - F_{Ci}P_{\mathrm{CDS}}) \tag{16.30}$$

It may be shown numerically that for a large number of trials for a given n, the expected fractional damage given by Eq. (16.30) is equal to

$$\mathrm{FD} = 1 - \prod_{i=1}^{n}(1 - E(F_C)P_{\mathrm{CDS}}) = 1 - (1 - E(F_C)P_{\mathrm{CDS}})^n \tag{16.31}$$

where $E(F_C)$ is the expected fractional coverage of the weapon lethal area over the target area. This equation is the same as that developed in Eq. (11.57) for air-launched weapons against area targets, where the quantity

$E(F_C)P_{CD}$ was referred to as the expected fractional damage FD_1 for a single sortie, that is,

$$FD_1 = E(F_C) \times P_{CDS} \tag{16.32}$$

We may therefore write Eq. (16.31) in the more familiar form of Eq. (11.57).

$$FD = 1 - (1 - FD_1)^n \tag{16.33}$$

16.8.2 Dependent Event Methodology

Dependent event methodology is different from the one just described because of the following:

- Because the precision error has been accounted for by enlarging the single weapon lethal area, and because the gun is not reaimed for dependent events, each lethal area falls in exactly the same place.
- Because dependent events are assumed to occur quickly compared to independent events, no redistribution of target elements within the bounding rectangle occurs, but redistribution does occur within the same overlap area of each event.

This implies that because the target elements outside the overlap area cannot be damaged regardless of the number of volleys, the fractional damage is limited by the fractional coverage (Fig. 16.10).

For all n events, all the lethal areas are lying exactly on top of each other, so this results in a single rectangular lethal area with a probability of damage inside it of

$$P_{CD} = 1 - (1 - P_{CDS})^n \tag{16.34}$$

The fractional damage for the example shown in Fig. 16.10 becomes

$$FD_1 = F_C P_{CD} = F_C[1 - (1 - P_{CDS})^n] \tag{16.35}$$

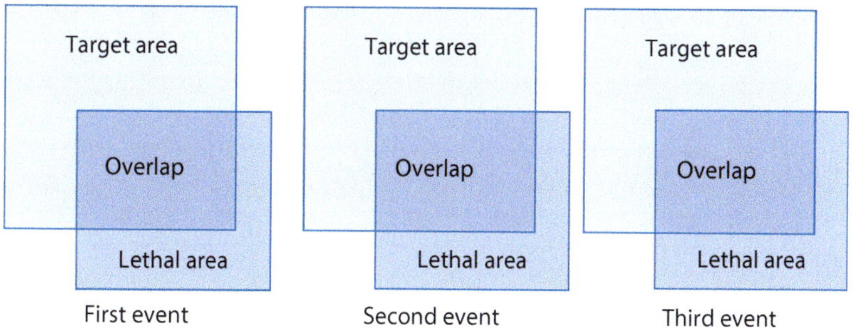

Fig. 16.10 Three dependent events.

where F_C is the fractional coverage of the lethal area over the target and is the same for all volleys. For the general case, the specific fractional coverage shown in the figure becomes the expected fractional coverage, hence we conclude that

$$\mathrm{FD} = E(F_C)[1 - (1 - P_{\mathrm{CDS}})^n] \tag{16.36}$$

16.8.3 Comparison of the Methodologies

The final results from the derivation of effectiveness measures are as follows:

Independent events

$$\mathrm{FD} = 1 - (1 - E(F_C)P_{\mathrm{CDS}})^n \tag{16.37}$$

Dependent events

$$\mathrm{FD} = E(F_C)(1 - (1 - P_{\mathrm{CDS}})^n). \tag{16.38}$$

When only one sortie or volley is delivered, Eqs. (16.37) and (16.38) both reduce to

$$\mathrm{FD} = E(F_C) \times P_{\mathrm{CD1}} \tag{16.39}$$

However, for $n > 1$ these equations are not equal. In the extreme case of a large number of sequential events where $n \gg 1$, the fractional damage for independent events approaches unity, whereas the fractional damage for dependent events approaches the expected fractional coverage, as seen in Fig. 16.11.

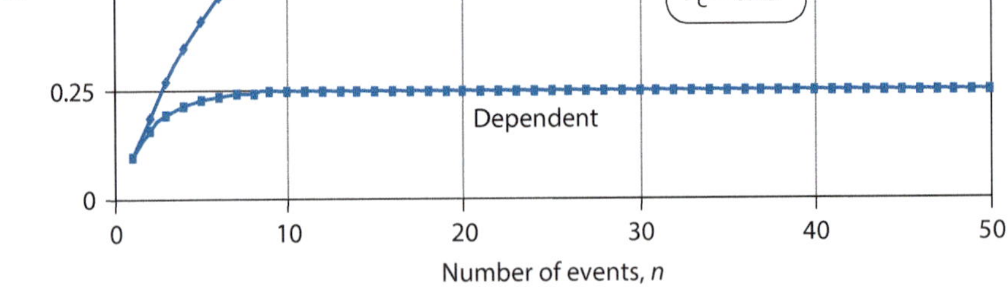

Fig. 16.11 Comparison of methods for multiple events.

Table 16.4 Models for Calculating FD

Model Fidelity	Model Name	Damage Function	Multiple Round Representation
Low	SuperQuickie2 SQ2	Rectangle	Rectangular pattern
Medium	ARTQUICK	Carlton (unitary), Rectangle (ICM)	Individual Carlton damage functions
High	Matrix Evaluator	P_K matrix	Individual LAM damage functions

Whenever we wish to calculate the effect of multiple weapons directed towards a target, it is necessary to determine whether these events are dependent or independent so that the correct equation [Eq. (16.37) or (16.38)] is used. The reason we get different results for dependent and independent attacks has been explained in Chapter 9, Section 9.8, at least for unitary targets.

16.9 Computational Tools for Surface-to-Surface Weaponeering

The generalized methodologies outlined so far have been implemented in several tools that provide various degrees of fidelity for surface-to-surface engagements. The main difference between them is the way that the damage function is represented and, as usual, increased fidelity or accuracy of results requires larger computer resources and longer computational times. A further difference between the models is in the way that multiple rounds are represented, either as an aggregated single rectangular pattern or individually. Table 16.4 lists the different models and how they represent individual and multiple weapon lethal areas.

Each model will be dealt with and the modifications to the preceding air-to-surface methodologies, if any, will be described. In the following sections, the high fidelity model Matrix Evaluator will be described before the ARTQUICK model. This is because, historically, this is the order in which these two models were developed, the latter being a faster running version of the former, but sharing many of its characteristics. The first model to be discussed, however, is simpler than both of them.

16.10 Superquickie 2, Unitary Warhead

SuperQuickie 2 (SQ2) is a model that is very similar to the AS model for stick deliveries (Chapter 12) against an area of targets and shares much of the methodology. The best way to explain how this simple model works is to examine a spreadsheet implementation of the method, beginning with an example of a unitary warhead.

SS- INDIRECT FIRE (SQ2) - Unitary		
Notes: Example 4.1 SS Basic manual		
# rounds/volley	nr	4
# of volleys fired	nv	6
# aimpoints in range direction	Nwr	2
# aimpoints in deflection direction	Nwd	2
Target radius (if specified)		0
Target length (if specified)	LA	250
Target width (if specified)	WA	250
Target length used	LT	250
Target width used	WT	250
Range from weapon to target (km)	Range	16
TLE	TLE	0
Precision error range	REP_p	48.6
Precision error deflection	DEP_p	7.4
MPI error range	REP_m	54.6
MPI error deflection	DEP_m	14.2
Total MPI error range (including TLE)	REP_tm	54.600
Total MPI error deflection (including TLE)	DEP_tm	14.200

This part of the table inputs the data relating to the number of weapons, aimpoints and how they are distributed in range and deflection, and the target size and range. The target may be rectangular or circular. Also, the three error sources are input from tabular data similar to those shown in Table 16.3. The total MPI errors are computed from Eqs. (16.24) and (16.25).

Round impact angle of fall	theta	40
Round reliability	R	0.993
Single round lethal area	AL	420

This section inputs the impact angle and munition reliability, both from tabular data, and the single round lethal area from target vulnerability data.

Volley length	LV	125
Volley width	WV	125
Lethal area aspect ratio	a	0.39
Single round pattern damage length	Let	12.751
Single round pattern damage width	Wet	32.938
Single round lethal area	Aet	420.000

In the preceding section, volley length and width are computed from the aimpoint placement Eqs. (16.4) and (16.5), because the target area is rectangular.

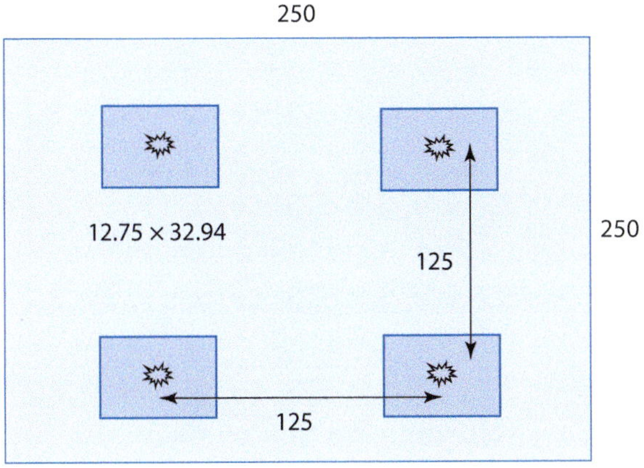

Fig. 16.12 Unadjusted pattern overlaying target.

Notice how the single-round weapon effects are represented as an equivalent rectangle, the aspect ratio of which is determined by the impact angle, or angle of fall. The lethal area aspect ratio is obtained from Eq. (16.15), and the single round lethal area dimensions from Eq. (16.20). This pattern is shown in Fig. 16.12 placed on top of the target area for comparison.

The next section enlarges the individual weapon lethal areas to account for precision error. The SS approach is different from the AS approach in Chapter 12, Section 12.5, instead using the method in Chapter 10, Section 10.10, specifically Eqs. (10.54) and (10.55). For convenience these are reproduced here:

$$L_B = L_{\mathrm{ET}} + k\,x_{\mathrm{br}} \tag{16.40}$$

$$W_B = W_{\mathrm{ET}} + k\,x_{\mathrm{bd}} \tag{16.41}$$

Recall that x_{br} and x_{bd} are the standard deviations of the precision error in feet on the ground plane. The quantity k is called the *pattern adjustment factor*, and clearly as the value of k increases there is a diminishing probability that the lethal area will fall outside the rectangle generated by Eqs.(16.40) and(16.41). The SQ2 method requires the user to input values for k that are considered to be munition dependent, the specific values being obtained from tabular data. For this example, we will use $k = 4$.

This enables the adjusted pattern dimensions to be calculated.

Single round adjusted pattern length	Lb	207.151
Single round adjusted pattern width	Wb	62.538
Single round adjusted pattern area	Ab	12955
Conditional probability of damage	PCD1	0.032

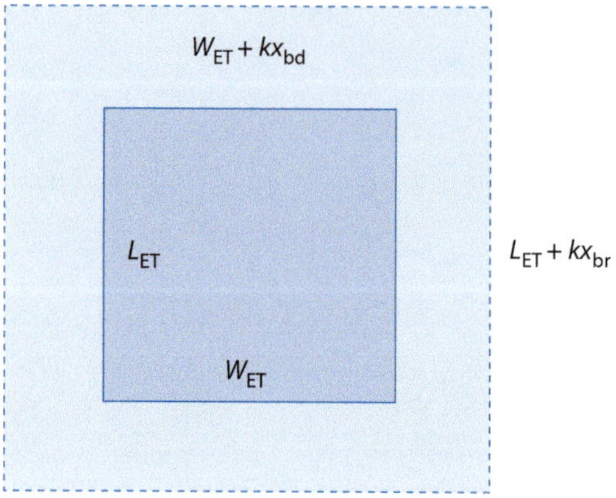

Fig. 16.13 Enlarging lethal area for precision error.

The conditional probability of damage in the enlarged single weapon lethal area (see Fig. 16.13) is given by

$$\mathrm{PCD}_1 = \frac{L_{\mathrm{ET}} W_{\mathrm{ET}}}{L_B W_B} = \frac{A_{\mathrm{ET}}}{A_B} \tag{16.42}$$

Once the adjusted lethal area dimensions are known, the pattern length and width for all weapons in the volley are calculated from

$$L_{VP} = L_V + L_B \tag{16.43}$$

and

$$W_{VP} = W_V + W_B \tag{16.44}$$

Volley damage pattern length	Lvp	332.15
Volley damage pattern width	Wvp	187.54
Volley damage pattern area	Avp	62291

The next step is to compute the conditional probability of damage within the pattern; this will depend on the amount of overlap or sparsity of the individual weapon lethal areas. Again, the method of computing this is the same as for air-delivered sticks (Chapter 12, Section 12.7), where the deflection direction is considered first, then range. For completeness, it is reproduced here.

16.10.1 Possible Overlap in the Deflection Direction

Consider the case of $n_p = 2$, where two weapons are released for each release pulse. Figure 16.14 shows how the areas may or may not overlap in

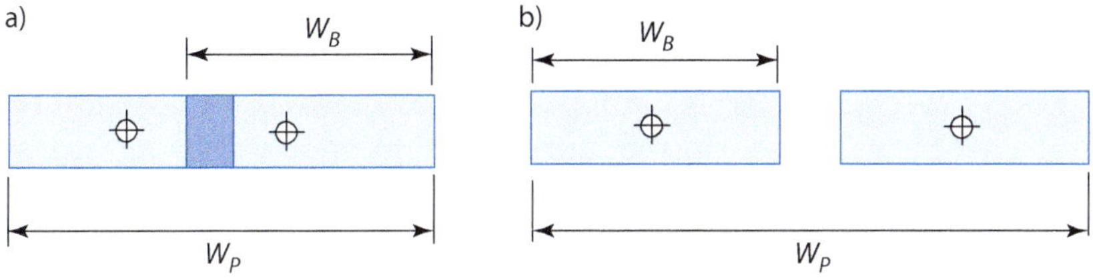

Fig. 16.14 Weapons overlapping (left) and not overlapping (right) in the deflection direction.

the pattern. Whether this happens or not depends only on the width of the expanded lethal area and the pattern width.

For the overlapping case, we define the degree of overlap in the deflection direction n_{od} to be

$$n_{\mathrm{od}} = \frac{n_P W_B}{W_P} \tag{16.45}$$

which may be thought of as how much overlap of the n_p weapon lethal areas has to occur in order to fit into the fixed pattern width. The conditional probability of damage in the deflection direction $P_{\mathrm{CD/d}}$ is obtained by powering up the single weapon damage function P_{CD} by the degree of overlap.

$$P_{\mathrm{CD/d}} = 1 - (1 - P_{\mathrm{CD1}})^{n_{\mathrm{od}}} \tag{16.46}$$

A noninteger powering up index may seem counterintuitive; however, consider the case where the two weapon lethal area are just touching, as shown in Fig. 16.15.

From Eq. (16.45) we see that $n_{\mathrm{od}} = 1$ and that Eq. (16.46) gives $P_{\mathrm{CD/d}} = P_{\mathrm{CD1}}$, which seems intuitive. Consider now the two weapon lethal areas lying exactly on top of each other, as shown in Fig. 16.16.

Now from Eq. (16.45) we see that $n_{\mathrm{od}} = 2$ and that Eq. (16.46) gives a value of $P_{\mathrm{CD/d}}$ from the powering up formula. Both of these cases have an integer value of the degree of overlap, so it now seems reasonable that for a partial overlap, the powering up index should be between 1 and 2, as indicated in Eq. (16.46). If the individual weapons do not overlap at all ($n_{\mathrm{od}} < 1$), then we can write

$$P_{\mathrm{CD/d}} = n_p P_{\mathrm{CD1}} \frac{W_B}{W_P} \tag{16.47}$$

Here we are spreading out the lethality of the two weapons, resulting in $P_{\mathrm{CD/d}} < P_{\mathrm{CD1}}$.

In either case we now have a single rectangle (weapon effective area) in the

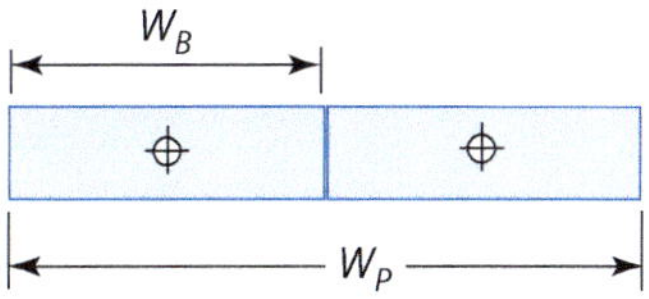

Fig. 16.15 Weapon lethal areas abutting.

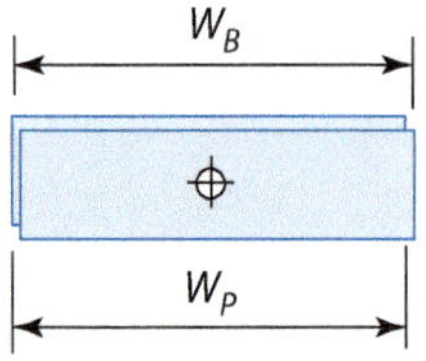

Fig. 16.16 Weapon lethal areas on top of each other.

deflection direction with known dimensions and a damage probability inside it given by either Eq. (16.46) or Eq. (16.47), as shown in Fig. 16.17.

In the range direction there will be n_r such rectangles, one for each release pulse generated by the intervalometer.

16.10.2 Possible Overlap in the Range Direction

The same arguments that were used in deflection may be applied to the range direction, where overlap may or may not occur, as shown in Fig. 16.18.

Proceeding as before, for the overlapping case we define the degree of overlap in the range direction n_{or} to be

$$n_{\text{or}} = \frac{n_r L_B}{L_P} \tag{16.48}$$

which may be thought of as how much of the rectangles shown in Fig. 16.18 have to overlap in order to fit into the pattern length. The conditional probability of damage for the whole pattern P_{CDS} is obtained by powering up

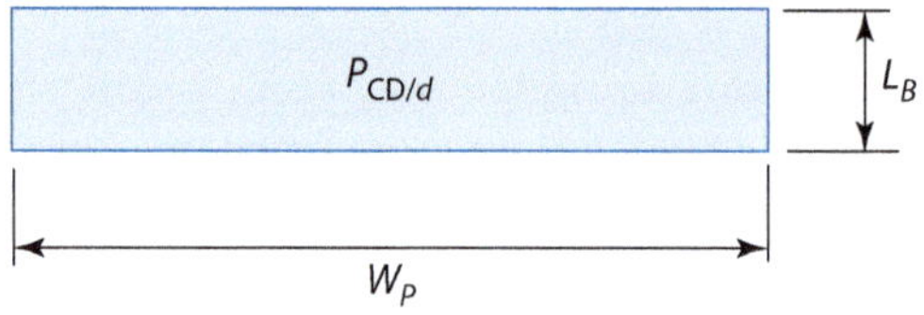

Fig. 16.17 Combining weapons in the deflection direction only.

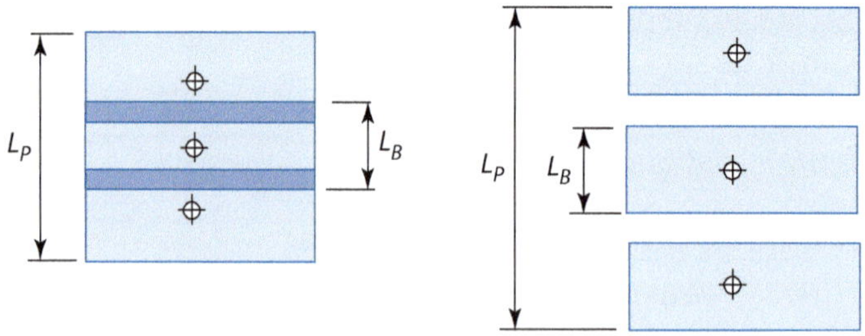

Fig. 16.18 Weapons overlapping (left) and not overlapping (right) in the range direction.

$P_{CD/d}$ by the degree of overlap in the range direction.

$$P_{CDS} = 1 - \left(1 - P_{CD/d}\right)^{n_{or}} \tag{16.49}$$

If the individual rectangles do not overlap at all ($n_{or} < 1$), then we can write

$$P_{CDS} = n_r P_{CD/d} \frac{L_B}{L_P} \tag{16.50}$$

Equations (16.49) or (16.50) now define a single rectangular pattern with known damage function and dimensions, as shown in Fig. 16.19.

The following spreadsheet shows the implementation of these calculations:

Overlap in deflection	no_d	0.667
Probdamage in deflection	PCDd	0.022
Overlap in range	no_r	1.247
Pattern probability of damage	PCD	0.027
Probability of damage for nv volleys		0.150

This rectangular damage function is now directed towards a rectangular target L_A by W_A with an accuracy governed by $\mathrm{REP_{TM}}$ and $\mathrm{DEP_{TM}}$, the aiming error combined with the TLE. The expected fractional coverage F_R and F_D is computed in the manner described in Chapter 11, Section 11.5.

Fractional Coverage Calculation	**Range**	**Deflection**
Sigma range	80.949	21.053
Integrallimits		
s	291.076	218.769
t	41.076	–31.231
a	2.543	7.348
b	0.359	–1.049
Integrals		
integral #1	–0.516	0.647
integral #2	0.808	0.060
integral #3	0.808	0.060
integral #4 & integral #5	–0.227	–0.022
Fractional coverage	0.873	0.745

This section of the spreadsheet computes the single volley fractional coverage of the target area by the volley area in both range and deflection.

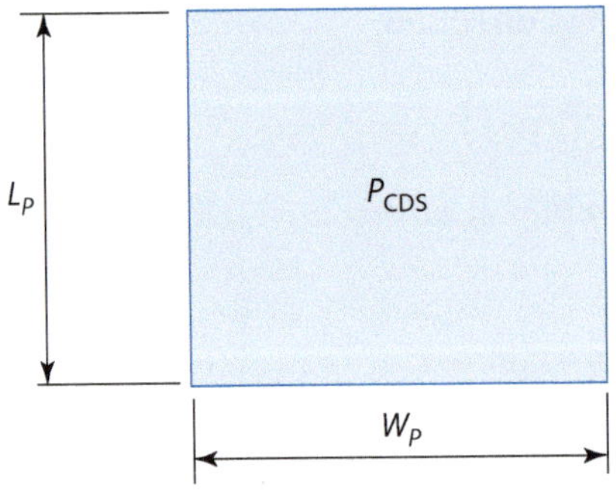

Fig. 16.19 Final pattern dimensions and damage probability.

16.11 Superquickie 2, Improved Conventional Munition

A similar spreadsheet can be developed for ICMs that has many of the same characteristics as the previous case, but some differences. The first two parts of the spreadsheet are similar to that for the unitary warhead.

SS- INDIRECT FIRE (SQ2) - ICM		
Notes: Example 4.2 SS Basic manual		
# rounds/volley	nr	6
# of volleys fired	nv	9
# aimpoints in range direction	Nwr	2
# aimpoints in deflection direction	Nwd	3
Target radius (ifspecified)		100
Target length (if specified)		0
Target width (if specified)		0
Target length used	LT	177.2
Target width used	WT	177.2
Range from weapon to target (km)	Range	16
TLE	TLE	0
Precision error range	REP_p	45.2
Precision errodeflection	DEP_p	7.4
MPI error range	REP_m	78.6
MPI error deflection	DEP_m	45.7
Total MPI error range (including TLE)	REP_tm	78.600
Total MPI error deflection (including TLE)	DEP_tm	45.700

This deals with the target size, aimpoints, and error parameters.

Round reliability	Rr	0.85
Submunition reliability	Rs	0.9
Number of submunitions/round	Ns	60
Single round pattern length	Lsp	182.52
Single round pattern width	Wsp	106.32
Single round submunition lethal area	AL	44

In this section, details of the submunition are entered, including their number, reliabilities (round and submunition), and the single round pattern size. The probability of damage due to one submunition in the pattern area is given by Eq. (16.21)

$$P_S = \frac{R_S a_L}{A_P} \tag{16.51}$$

where R_s is the submunition reliability. The probability of damage within the pattern due to all submunitions is obtained by powering up P_S by the number of submunitions and using the Morse-Kimball equation to simplify.

$$P_{DP} = 1 - (1 - P_S)^{N_S} \tag{16.52}$$

This simplifies to

$$P_{DP} = 1 - \exp(-N_S P_S) \tag{16.53}$$

and is shown in the following segment:

Pd due to one submunition in pattern	Ps	0.002
Pd due to all submunitions	Pdp	0.115

At this point, the single weapon cluster munition is represented by a single rectangular lethal area of known dimensions L_{SP}, W_{SP} and conditional probability of damage P_{DP}. The remaining methodology is therefore the same as for a unitary warhead and involves the following steps:

1. Expanding the single weapon lethal area to account for precision errors using the pattern adjustment factor k
2. Calculating the volley dimensions L_{VP} and W_{VP}
3. Determining the damage probability inside the rectangular volley by considering the degrees of overlap in deflection and range sequentially
4. Calculating the fractional coverage in range and deflection
5. Powering up for the number of volleys specified by the user
6. Computing the FD from Eq. (16.54)

The remaining portion of the spreadsheet is shown here:

Volley length	LV	100.00
Volley width	WV	86.60
Pattern adjustment factor	k	3.000
Single round adjusted pattern length	Lb	318.120
Single round adjusted pattern width	Wb	128.520
Single round adjusted pattern area	Ab	40885
Conditional probability of damage	PCD_1	0.055

Volley damage pattern length	Lvp	418.12
Volley damage pattern width	Wvp	215.12
Volley damage pattern area	Avp	89948
Overlap in deflection	no_d	1.792
Prob damage in deflection	PCDd	0.096
Overlap in range	no_r	1.522
Pattern probability of damage	PCD	0.142
Probability of damage for nv volleys		0.686
Fractional Coverage Calculation	**Range**	**Deflection**
Sigma range	116.531	67.754
Integral limits		
s	297.683	196.185
t	120.437	18.940
a	1.806	2.047
b	0.731	0.198
Integrals		
integral #1	–1.648	–0.267
integral #2	1.418	0.673
integral #3	1.418	0.673
integral #4 & integral #5	–0.287	–0.289
Fractional coverage	0.899	0.790
Overlap in deflection	no_d	1.792
Prob damage in deflection	PCDd	0.096
Overlap in range	no_r	1.522
Pattern probability of damage	PCD	0.142
Probability of damage for nv volleys		0.686
Expected fractional damage or casualties	Fd, Fc	**0.488**

For multiple volleys, the single probability of damage is powered up for n_v volleys, and Eq. (16.38) is used to compute the expected fractional damage.

$$\mathrm{FD} = F_R F_D (1 - (1 - R P_{\mathrm{CD}})^{n_v}) \tag{16.54}$$

where R is the round reliability.

16.12 High Fidelity Model: Matrix Evaluator (ME) Program

It has to be said that the Matrix Evaluator (ME) program is the legacy (1970s) methodology and computer tool that was the original way that JMEM effectiveness data were calculated. Its origins are that it was a U.S. Army model used to calculate point values of weapon–target engagements, but when the JMEM mission was established it was then employed to generate tabular data for sets of air-to-surface engagements that formed the basis of the original JMEMs. Clearly, given the computer resources of the time, results could only be computed over a long period of time and the results stored in books—the original "Manuals" in the title JMEM. These results are sometimes referred to as "precomputed" PD values because they have been determined offline for a selected set of engagement parameters.

If these parameters match those of interest to the operational user, the results are the most accurate estimates of damage available; however, if they don't match exactly, judicious interpolation is required. It is because of the limited set of scenarios available that the so-called "open-ended" methods were developed for any scenario the user specifies. These open-ended methods are based on the methodologies discussed in earlier chapters. After this, open-ended methods were developed for both AS and SS engagements, but Matrix Evaluator was used in the validation of the faster running codes before being incorporated into the JMEMs.

The Matrix Evaluator methodology has many of the same characteristics of the methods previously discussed, but its implementation is somewhat different, as described in the following sections.

16.12.1 Damage Function

In Matrix Evaluator, the weapon damage function is represented by a lethal area matrix (LAM), such as the one output from the GFSM model. A simplified example is shown in Table 16.5.

This matrix is sometimes referred to as a single weapon damage grid. Matrix Evaluator is the only effectiveness method that uses the lethal area matrix directly.

16.12.2 Fractional Coverage

Fractional coverage has usually been calculated as the expected overlap of a rectangular weapon lethal area over a rectangular target, as developed in Chapter 11. Here we shall simply consider the expected fractional overlap of two generalized rectangles, as shown in Fig. 16.20.

Here the first rectangle of dimensions L_1 by W_1 is located at the origin and fixed, whereas the center of the second rectangle measuring L_2 by W_2 has coordinates (x, y) relative to the fixed rectangle. It is assumed as before that x and y are uncorrelated normally distributed random variables with known standard deviations; however, we shall further assume that x and y may have nonzero mean values μ_x and μ_y. The computation of fractional coverage with nonzero mean values was described in Chapter 15, Section 15.7.3, but it should be realized that in order to compute FC, all that is needed are the dimensions of the rectangles—σ_x, μ_x and σ_y, μ_y—describing

Table 16.5 Lethal Area Matrix Used for the Damage Function

0	0	0	0	0	0	0	0	0	0	0	0	0	0
0	0	0	0	0	0.001	0	0	0.001	0	0	0	0	0
0	0	0	0	0.001	0.002	0.003	0.003	0.002	0.001	0	0	0	0
0	0.001	0.002	0.003	0.004	0.007	0.014	0.014	0.007	0.004	0.003	0.002	0.001	0
0	0	0	0.002	0.008	0.04	0.227	0.227	0.04	0.008	0.002	0	0	0
0	0	0	0	0.001	0.002	0.049	0.049	0.002	0.001	0	0	0	0
0	0	0	0	0	0	0.002	0.002	0	0	0	0	0	0
0	0	0	0	0	0	0	0	0	0	0	0	0	0

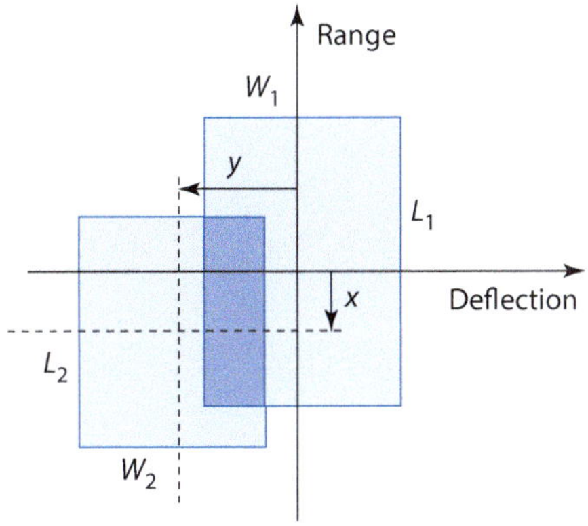

Fig. 16.20 Fractional coverage of two rectangles.

the bivariate distribution of the center of the second rectangle relative to the center of the first.

16.12.3 Precision Error

Precision error is allowed for by enlarging the weapon damage grid. This is done for the grid as a whole, enlarging each cell by the same amount. The expansion is done in the same manner as for the SQ2 method, except the pattern adjustment parameter k is not a user input but fixed at $k = 6$ (i.e., $\pm 3\sigma$).

$$L_B = L_{\text{ET}} + 6x_{\text{br}} \tag{16.55}$$

$$W_B = W_{\text{ET}} + 6x_{\text{bd}} \tag{16.56}$$

where x_{br} and x_{bd} are the standard deviations of precision error in range and deflection, respectively. This expansion is shown in Fig. 16.21.

This new grid is known as the ballistic grid. Clearly we cannot keep the same cell P_K values because this will result in a more lethal weapon, so they have to be adjusted to preserve lethality. The question is how to do this adjustment. A simple way, based again on Chapter 10, Section 10.10, Eq. (10.56), would be to scale the original P_K by the area ratio of the cells, which when applied to the whole matrix would give the following:

$$P'_K = P_K \frac{A(i, j)}{A(x, y)} \tag{16.57}$$

where the areas are defined in Fig. 16.21. This will certainly preserve lethality because

$$A_L = \text{MAE}_F = \sum_{i=1}^{n} \sum_{j=1}^{m} P_K(i, j) A(i, j) = \sum_{x=1}^{n} \sum_{y=1}^{m} P'_K(x, y) A(x, y) \tag{16.58}$$

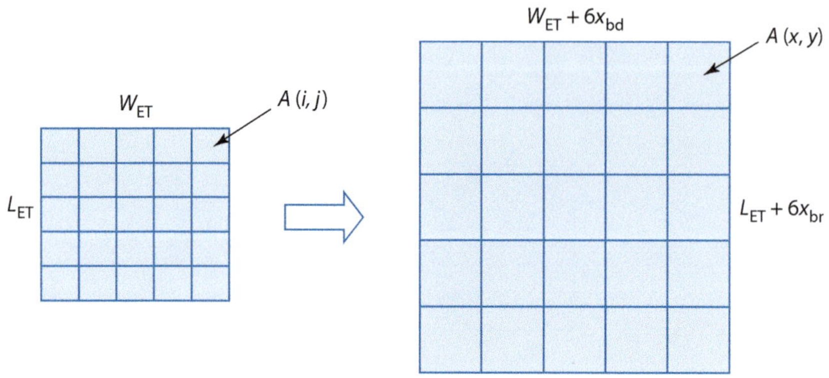

Fig. 16.21 Enlargement of weapon grid to form ballistic grid.

What the legacy Matrix Evaluator actually did was to pick a single cell (x, y) on the ballistic grid and compute the expected fractional coverage of each cell of the original damage matrix and sum them to obtain the new damage probability in that cell.

$$P'_K(x, y) = \sum_{i=1}^{n} \sum_{j=1}^{m} P_K(i, j)\, F_C(i, j, x, y) \tag{16.59}$$

Here $F_C(i, j, x, y)$ is the expected fractional coverage of ballistic cell (x, y) by weapon damage cell (i, j). This fractional coverage calculation uses the offset of the cell in question from the center of the weapon grid as a mean value in order to maintain the spatial relationship between adjacent cells, as described in the previous section. Intuitively using fractional coverage in this way is more reasonable than simple scaling because it reflects the probability, however small, that a cell in the bottom left of the ballistic grid could be covered by the top right cell of the damage grid. Considering the choice between Eq. (16.57) and Eq. (16.59), it may be seen that there are an infinite number of sets of $P_K{}'$ values that could preserve lethality, and Eqs. (16.57) and (16.59) represent just two of them.

16.12.4 Determining the Volley Damage Grid

Now the effect of multiple weapons will be taken into account, whether in terms of a stick or volley. In either case, the individual weapon aimpoints allow the shape of the volley to be determined, as shown in Fig. 16.22.

For AS applications, the individual damage grids are placed on the impact points predicted for the stick (Chapter 12, Section 12.6), whereas for artillery volleys they are placed on the aimpoints, the same as for SQ2. The outer dimensions of the volley damage grid are defined as the smallest rectangle containing the individual ballistic grids.

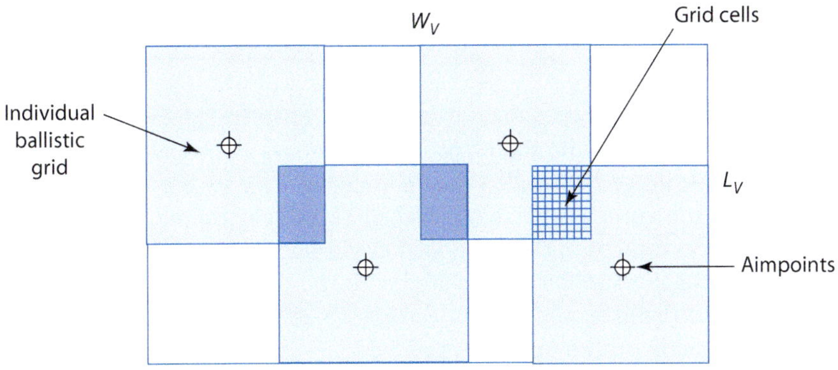

Fig. 16.22 Volley damage grid for four weapons.

Matrix Evaluator then establishes a grid within the volley damage grid defining a set of cells. For simplicity, assume the volley cells are the same size as the ballistic grid cells, and the latter are "snapped" to the volley grid. We can define the P_K of each volley cell, denoted by $S(i, j)$, by inspecting how many ballistic cells lie on top of it. One of three situations may arise:

1. The volley cell has no ballistic cell on it.
2. The volley cell has one ballistic cell on it.
3. The volley cell has more than one ballistic cell on it.

If a volley cell has no ballistic cell lying on it, the volley cell $S(i, j)$ is clearly zero. If the cell has one weapon on it, the corresponding $P_K(i, j)$ for that ballistic damage matrix cell is the value stored in the corresponding volley cell, that is,

$$S(i, j) = P'_K(i, j) \tag{16.60}$$

For those cells where there are two or more ballistic cells overlaying a single volley cell, the damage value is calculated by using the survivor rule.

$$S(i, j) = 1 - \prod_{q=1}^{q=m} \left[1 - (P'_K(i, j, q)\right] \tag{16.61}$$

Here

- $S(i, j)$ is the damage level in volley cell i, j.
- ${P_K}'(i, j, q)$ is the damage level in the (i, j) cell of the qth ballistic grid that lies on top of volley cell (i, j),
- m is the number of weapons overlapping the volley cell.

16.12.5 Allowing for Aiming Error

At this point we have a single volley damage grid with P_K values for each cell, and an area target, as shown in simplified form in Fig. 16.23.

We could allow for aiming error in the same way as we allowed for precision error—enlarge the single volley damage grid by six times the standard deviation of the aiming error and recalculate the corresponding cell P_K. If this were to be done, the resulting grid could be laid directly on top of the target, because all variability associated with the delivery of the weapon has been removed (ballistic and aiming errors); therefore, the center of the grid must land exactly on the target center.

This approach is not taken in the legacy version Matrix Evaluator.[‡] Instead we consider that a weapon may land anywhere relative to the target, but it is unlikely ($P < 0.5\%$) that it will land outside $\pm 3\sigma_{AIM}$ of its boundaries, where σ_{AIM} is the aiming error standard deviation. To implement this observation, the target is enlarged by this amount to form the apparent target area (ATA), as shown in Fig. 16.24. The ATA is then overlaid onto the volley damage grid as shown in Fig. 16.25.

As shown in Fig. 16.25, only certain cells of the damage grid overlap the apparent target area, and the following analysis is limited to only those cells. In this sense the ATA indicates which cells of the volley damage grid might possibly interact with the actual target. It may be seen that this process is essentially the same as the more logical one of moving the volley damage grid over the actual target with the randomness of the aiming error to determine which volley damage grid cells will interact with the actual target. In the legacy version of ME this also had the effect of limiting the number of cells involved in the following stages of computation.

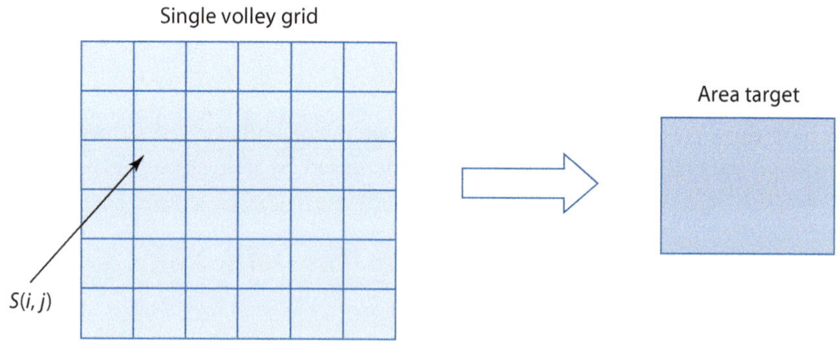

Fig. 16.23 Volley grid projected onto area target.

[‡]See Section 16.15 for the current implementation.

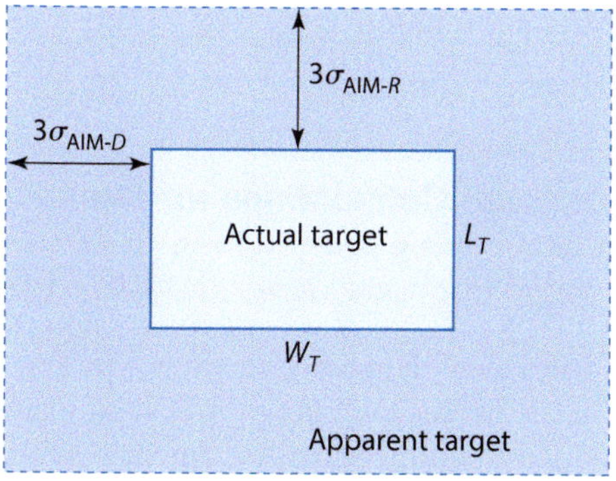

Fig. 16.24 Actual and apparent target geometries.

To calculate the expected fractional damage, each cell overlaying the ATA is taken in turn by:

1. Computing the expected fractional coverage $F_C(i, j)$ of each cell of the volley damage grid upon the actual (not apparent) target area, again using the offset of the cell in question from the center of the volley grid as a mean value in order to maintain the spatial relationship between adjacent cells, and then
2. Multiplying the fractional coverage by the damage probability $S(i, j)$ for that cell to obtain the cell fractional damage.

We finally sum all elements of the volley grid to give the expected fractional damage for a single volley.

$$\text{EFD}_1 = \sum_{i=1}^{i_{\text{MAX}}} \sum_{j=1}^{j_{\text{MAX}}} S(i, j) F_C(i, j) \tag{16.62}$$

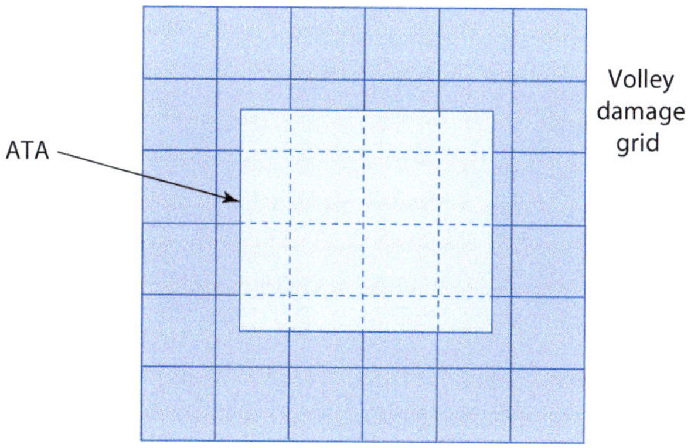

Fig. 16.25 ATA overlaying the volley damage grid.

remembering that the summation is limited to those cells overlapping the apparent target area.

16.13 Fractional Damage Calculations for Multiple Events

All the preceding analysis is for a single volley or stick, but when multiple events occur it is handled differently for dependent and independent fire. For dependent events, the volley damage grid shown in Fig. 16.22 is duplicated for each volley. What is more, each volley will land exactly on the previous one. This will produce a multiple volley damage grid that has the same overall size and the same number and sized grid cells all lying exactly on top of each other, as shown in Fig. 16.26.

The damage probability of each cell of the N-volley damage grid will be obtained using the survivor rule powered up for the N volleys.

$$S_N(i,j) = 1 - [1 - R \times S(i,j)]^N \tag{16.63}$$

The process from this point on for calculating FD is then the same as for the single volley case. For multiple independent events, however, we cannot assume that each sortie lands directly on top of the previous one. The method used is to assume that the mission comprises only one sortie and use Eq. (16.62) to calculate FD_1.

$$\mathrm{FD}_1 = \sum_{i=1}^{i_{\mathrm{MAX}}} \sum_{j=1}^{j_{\mathrm{MAX}}} S(i,j) F_C(i,j) \tag{16.64}$$

The probability of damage for multiple sorties is then obtained from the usual formula for N independent events.

$$\mathrm{FD} = 1 - (1 - \mathrm{FD}_1)^N \tag{16.65}$$

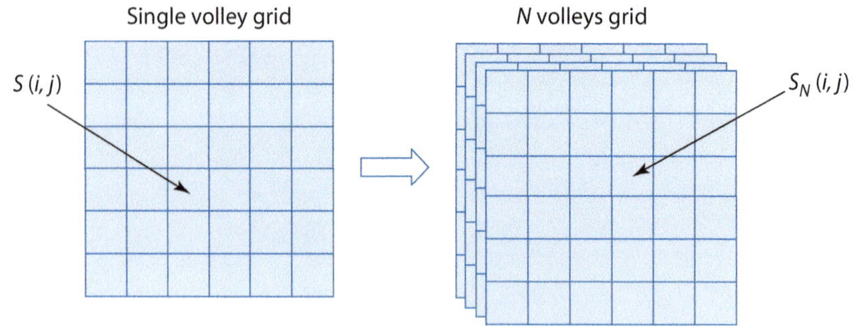

Fig. 16.26 Powering up single volley grid for N volleys.

16.14 Improved Conventional Munitions (ICM) in Matrix Evaluator

In dealing with ICM, Matrix Evaluator will generate a single weapon damage grid from the individual submunitions and then treat the effectiveness calculation the same way as for a unitary warhead.

In order to do this, the inputs to the program used are the pattern dimensions, which may specify an ellipse or rectangle depending on the weapon type, and the number of submunitions. Associated with a single submunition is a lethal area matrix generated by GFSM for the target in question. This would be obtained in the usual manner by locating a submunition at the origin of a ground plane grid (at a fixed altitude and angle) and moving the target to each grid cell in turn, calculating the probability of damage for that cell.

Because we are specifying the pattern size and number of submunitions independently, Matrix Evaluator uses an algorithm for placing individual submunition impact points within the defined pattern such that these points are distributed uniformly. Distributed uniformly is different from a statistically uniform distribution in that the submunitions are placed a constant distance apart (i.e., on a regular grid of points). A uniform distribution would result in

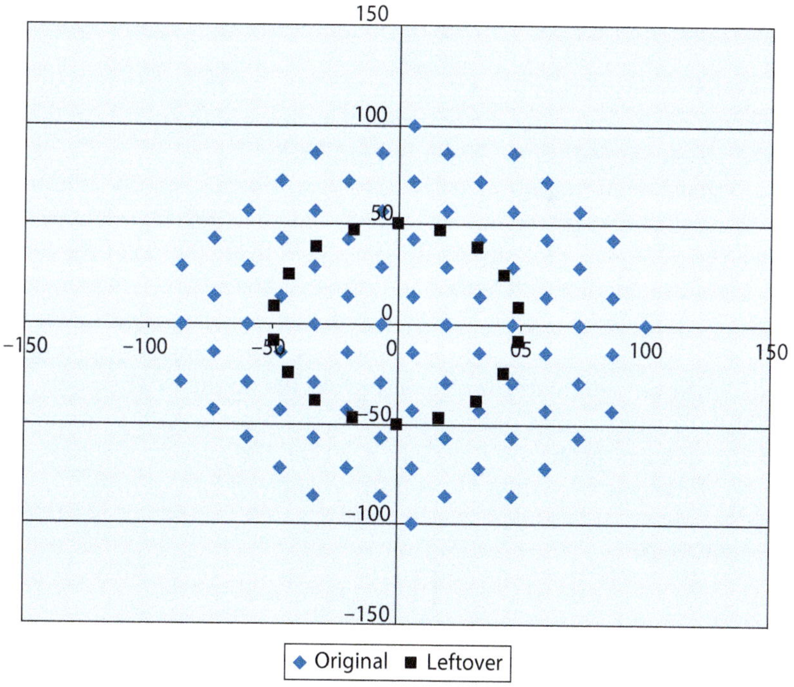

Fig. 16.27 Placement of submunitions within the pattern.

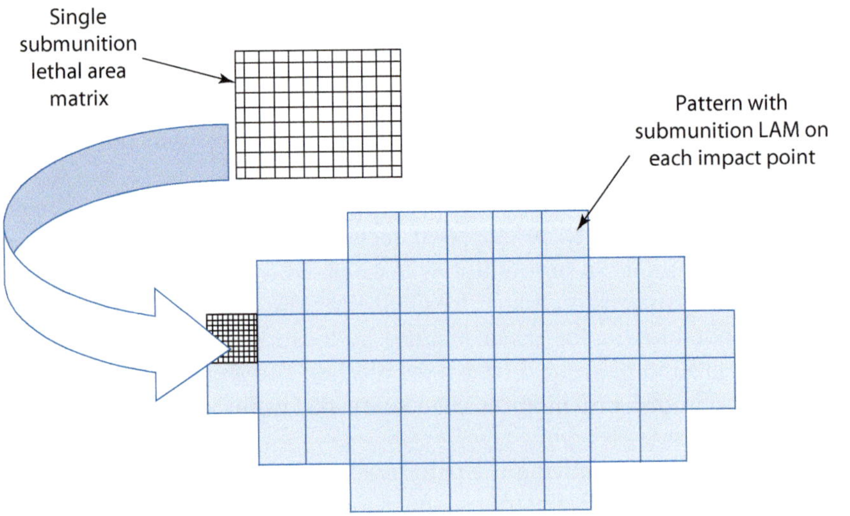

Fig. 16.28 ICM representation in Matrix Evaluator.

random placement of the submunitions. Usually in this process, there will be some submunitions left over, and the program will select locations within the pattern for these additional submunitions. Figure 16.27 shows an example of submunition placement for an elliptical pattern.

Once the submunition impact points are defined, the submunition lethal area matrix is laid down on each one with the center of the matrix coincident with the placement point, as shown in Fig. 16.28.

Note that the submunition lethal areas will not necessarily adjoin each other as might be suggested in Fig. 16.28, but may overlap or have space between them. Suppose for simplicity that the individual cells of the submunition LAM are "snapped" together (i.e., the cell boundaries are coincident), then the pattern shown in Fig. 16.28 will define a damage grid comprising overlapping submunition lethal area matrices. The damage probability in each cell may be obtained from the survivor rule applied to however many submunitions overlap that particular cell. If there are n such submunitions, where the damage probability in cell (p, q) for the ith weapon is $P_K(p, q, i)$, the total damage probability in cell (p, q) is given by

$$P_K(p, q) = 1 - \prod_{i=1}^{i=n} (1 - R \times P_K(p, q, i)) \tag{16.66}$$

If there are areas within the overall pattern that have no coverage by a submunition, the cell has a P_K of zero. This defines the single weapon damage grid, and this may be considered the same as a GFSM-generated lethal area matrix for a unitary HE warhead. The process of computing the value of FD may proceed as before:

1. Expanding the individual weapon damage grid to allow for precision error
2. Determining the volley damage grid
3. Powering up for multiple volleys if required
4. Projecting each cell of the N-volley damage grid against the area target

16.15 Matrix Evaluator V2.0

The original ME program dates from the 1970s and was developed at a time when computer memory, storage, and speed were very limited. Recently, the program has been updated to the 2.0 version in order to take advantage of improved computer capability. The methodology has also undergone some changes, and the interface and matrix structure have been modified in order to improve the fidelity of the model. These improvements will now be listed.

16.15.1 User-Defined Matrix Dimensions

The original GFSM program limited the lethal area matrix to a grid of 40 by 40 cells because this represented the maximum storage capacity of the time. This implied that the dimensions of each cell were determined by the overall effects size; large bombs had large cell sizes and small warheads had small cell sizes. This resulted in a loss of resolution close to the detonation point where the P_K was changing rapidly with distance. Because JMAE allows the user to specify the cell structure, the ability to logarithmically space the gridlines is particularly useful in improving resolution close to the detonation point, as shown in Fig. 16.29.

16.15.2 Fractional Coverage of Single Volley Damage Grid

A more consistent approach is now taken regarding the incorporation of delivery error. The method suggested in Section 16.12 is implemented, that is, the single volley damage grid is expanded by $\pm 3\sigma$ of the delivery error, where this error is the RSS of the MPI and TLE at the sigma level as shown here:

$$\sigma_{\mathrm{DE}} = \sqrt{\sigma_{\mathrm{MPI}}^2 + \sigma_{\mathrm{TLE}}^2} \tag{16.67}$$

The resulting grid is known as the single volley large damage grid (SVLDG) (Fig. 16.30).

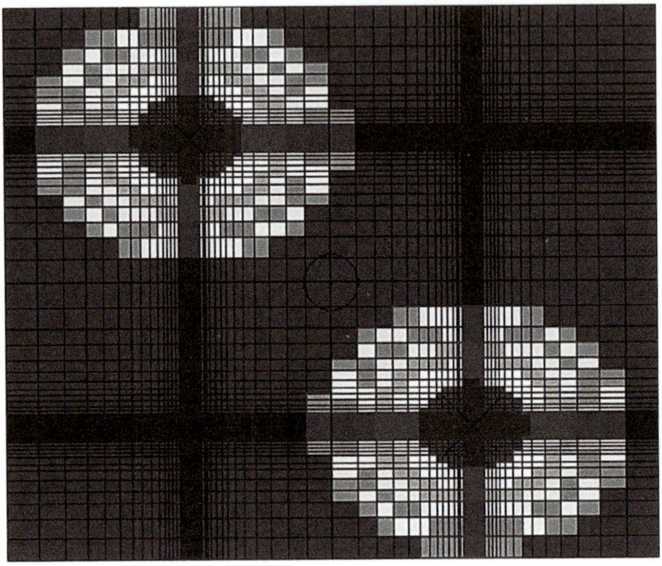

Fig. 16.29 Logarithmically spaced gridlines.

The new cell P_K is calculated as before by the summation over the expanded matrix of the expected fraction coverage of a single cell in the original matrix on a corresponding cell in the expanded matrix.

$$S'(x, y) = \sum_{i=1}^{n} \sum_{j=1}^{m} S(i, j) F_C(i, j, x, y) \tag{16.68}$$

Placing this expanded grid over the actual target area results in a situation where all variability on weapon accuracy has been removed; therefore, any cells outside the large damage grid may be ignored in the fractional damage calculation over the actual target area, as shown in Fig. 16.31.

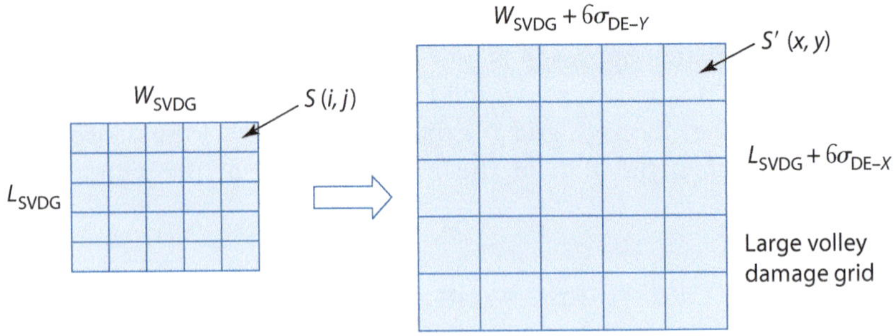

Fig. 16.30 Expansion of volley grid to include delivery error.

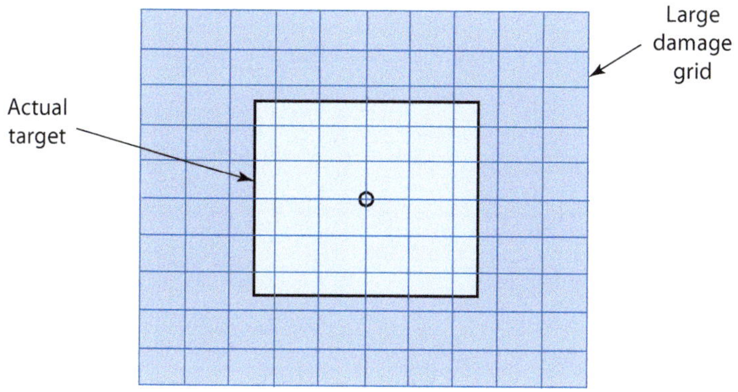

Fig. 16.31 Large damage grid covering the target.

Note that we now perform a fractional coverage calculation and not an expected fractional coverage calculation. We can write for the total damage

$$FD_1 = \sum_{i=1}^{i_{MAX}} \sum_{j=1}^{j_{MAX}} S'(i,j) F_C(i,j) \tag{16.69}$$

In this equation, $F_C(i,j)$ is the fractional coverage of cell (i,j) over the target. If the whole cell is inside the target boundary, then we may write

$$F_C = \frac{\text{cell area}}{\text{target area}} \tag{16.70}$$

For those cells partially covering the target, F_C will be less than this value. Due to this methodology difference, Matrix Evaluator 2.0 handles dependent and independent multiple events in a different manner than the legacy version.

16.15.3 Allowing for Multiple Dependent Events

For this case, the methodology is similar to the legacy version where N identical SVDGs are placed in top of each other to form the N volley damage grid (Fig. 16.32).

The P_K values for each cell are calculated using the survivor rule as before.

$$S_N(i,j) = 1 - [1 - R \times S(i,j)]^N \tag{16.71}$$

At this point, the NVDG is expanded to incorporate the delivery error to form the N volley large damage grid (Fig. 16.33).

Then the fractional coverage over the target area of the NVLDG is calculated in the same way as for the SVLDG.

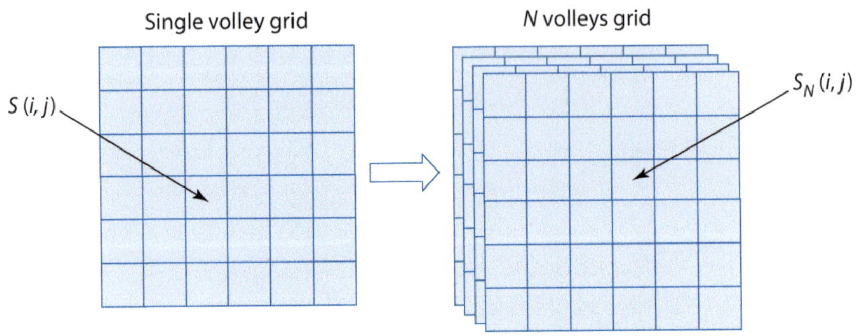

Fig. 16.32 Powering up the SVDG dependent events.

16.15.4 Allowing for Multiple Independent Events

Here, the SVDG is first expanded to allow for the delivery error to form the single volley large damage grid (SVLDG) (Fig. 16.34). Then, N identical SVLDGs are placed on top of each other to form the NVLDG (Fig. 16.35).

Again the P_K values for each cell are calculated using Eq. (16.71), and the fractional coverage over the target area calculation are performed as before. It may be seen that the difference in approaches is that for dependent events, the powering up of the SVDG is done first and then expanded by the delivery error, whereas for independent events the expansion is done first and then the powering up.

16.15.5 Treatment of Improved Conventional Munitions (ICMs)

Recall in the legacy model how each submunition was placed within the pattern by distributing them uniformly, and then aggregating each submunition lethal area matrix placed at each of these locations.

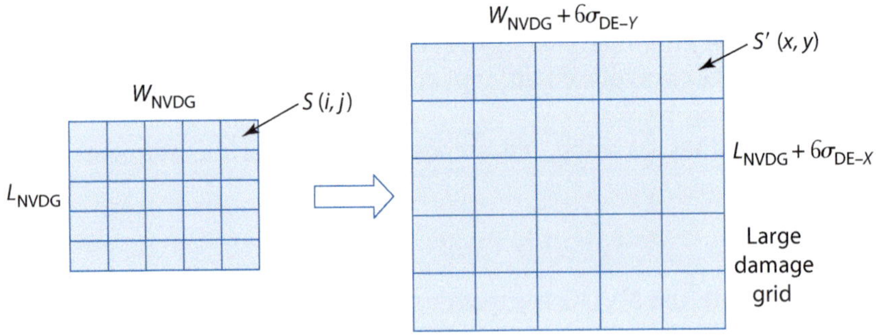

Fig. 16.33 Expansion of N volley grid to include delivery error.

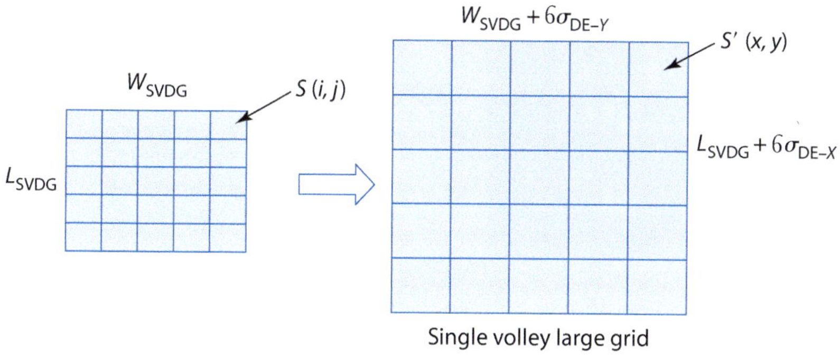

Fig. 16.34 Expansion of SVDG to include delivery error.

In v2.0 a different approach is taken. Although the pattern size is still fixed, the user specifies the number of cells inside this pattern. We now have the situation shown in Fig. 16.36, where on the left is a lethal area matrix for a single submunition, and on the right is the complete pattern with the user-defined cell structure. The objective is to populate this pattern damage matrix with P_K values in each cell.

Consider the first cell (1, 1) in the pattern damage matrix. We calculate the expected fraction coverage of each cell in the single submunition damage matrix on the target cell (1, 1) in the single submunition pattern matrix, multiply it by the P_K of the single submunition damage matrix, and sum for all single submunition damage matrix cells. That will result in the P_K in cell (1, 1) of the damage matrix. We can express this for the cell in question (1, 1) as follows:

$$P_{KP}(1, 1) = \sum_{i=1}^{n}\sum_{j=1}^{m} P_{KS}(i, j)\, F_C(i, j, 1, 1) \qquad (16.72)$$

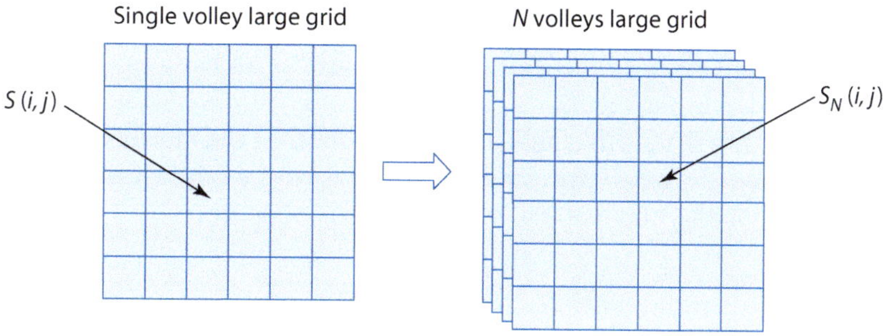

Fig. 16.35 Powering up the SVLDG for independent events.

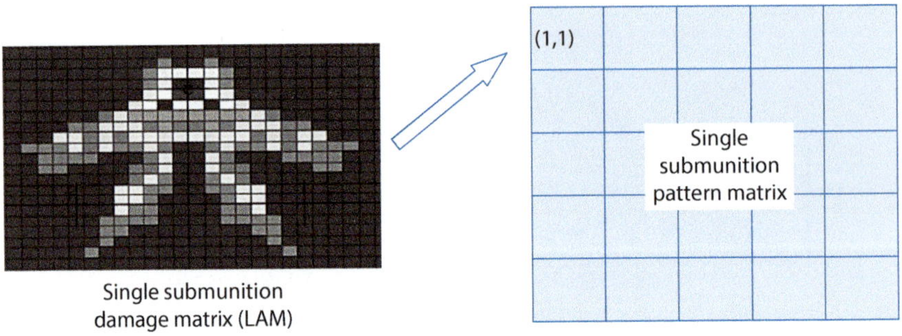

Fig. 16.36 EFC of single submunition damage matrix on pattern.

In this expression, P_{KP} is the P_K in the single submunition pattern matrix, P_{KS} is the P_K in the single submunition damage matrix, and F_C is the fractional coverage of each cell of the single submunition damage matrix on cell (1, 1) of the single submunition pattern matrix.

The significant difference here, however, is that when performing the expected fractional coverage calculation, we do not use a Gaussian distribution, but a uniform distribution, because it is assumed the submunitions will be uniformly distributed within the pattern. This process is repeated for each cell in the single submunition pattern damage matrix with the end result of distributing the single submunition damage matrix over the whole pattern, populating each cell with a P_K. This process is analogous to that described in Section 16.4, where the single submunition lethal area, represented by an area a_L, is spread out over the whole pattern area A_P. Mathematically, Eq. (16.72) is extended to the general cell (x, y) in the single submunition pattern matrix.

$$P_{KP}(x, y) = \sum_{i=1}^{n} \sum_{j=1}^{m} P_{KS}(i, j)\, F_C(i, j, x, y) \tag{16.73}$$

The effect of all n submunitions is now determined by considering n individual single submunition pattern damage matrices all lying on top of each other, as shown in Fig. 16.37.

We then power up this single ICM distribution by the number of submunitions on a cell by cell basis using the survivor rule, as shown in the figure. The cell P_K for the final pattern damage matrix becomes

$$P_K(i, j) = 1 - (1 - P_{K1}(i, j))^n \tag{16.74}$$

From this point on, the pattern matrix may be treated as an equivalent HE warhead in the effectiveness calculation.

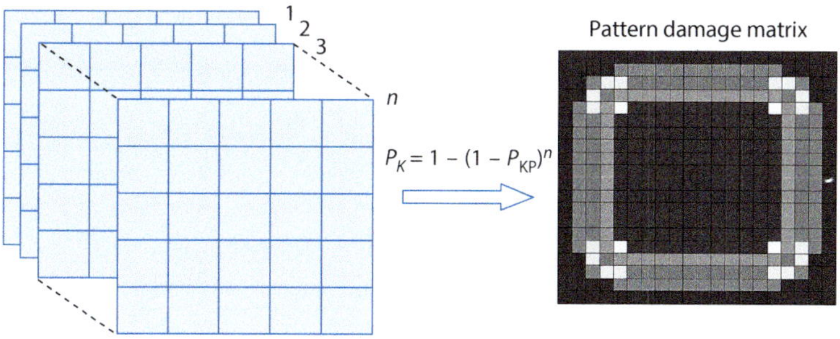

Fig. 16.37 Powering up the single submunition pattern matrices.

16.16 ARTQUIK Method

This is a quick-running artillery-based method of intermediate fidelity for calculating FD for both unitary weapons and ICM, and addresses only dependent event engagements. ARTQUIK was developed as a way to improve the execution speed of Matrix Evaluator by reducing the number of cell operations; hence, the main difference between this methodology and Matrix Evaluator is the representation of the damage function. Instead of the lethal area matrix, the Carlton damage function is used, shown for the range direction only in the following form:

$$c(x) = D_0 \exp\left[-D_0 \frac{x^2}{R_x^2}\right] \tag{16.75}$$

Recall the quantity D_0 was introduced earlier, and its effect on the Carlton damage function is to make the peak value smaller but spread out the rest of the surface over a larger area on the ground plane. Its purpose in SS methodologies, however, is to provide an adjustment parameter so that the ARTQUICK method provides results closer to those obtained from the Matrix Evaluator program. It has been found that values of D_0 are weapon system dependent; sample values are given in Table 16.6.

The effectiveness methodology computes the expected fractional damage probability based on the type of munition used (HE or ICM). The method uses the inputs listed in Table 16.7.

Table 16.6 D_0 Values for Carlton Damage Function

Weapon System	D_0
Artillery	0.2
Mortars	0.4
Navy 5-in. gun	0.5

Table 16.7 Summary of ARTQUIK Input Data

General	Unitary Warhead
Number of rounds/volley	Lethal area
Number of volleys	Angle of fall
Aimpoints	
Number of integration steps	**ICM Warhead**
Target size	Submunition lethal area
Round reliability	Number of submunitions
D_0 values	Submunition reliability
Precision error	Pattern dimensions
MPI error	
TLE	

The implementation of the methodology for unitary and ICM warheads will now be given.[§]

16.17 Damage Done by a Single Weapon with an Offset Aimpoint

In this section, multiple weapons comprising the volley are represented as individual Carlton damage functions, as shown in Fig. 16.38. In this figure, a four-gun volley is assumed and is superimposed on the target, where the ellipses represent the Carlton damage functions produced by each gun.

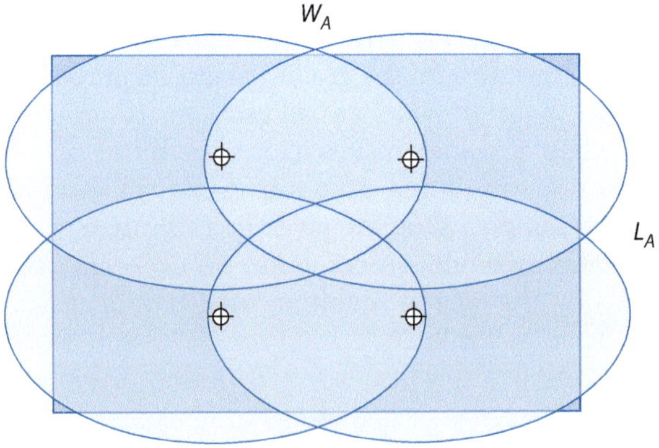

Fig. 16.38 Individual warhead Carlton damage functions overlaying the target area.

[§]Some variables in this section are the same as for the Matrix Evaluator model, but have a different notation. This is done to provide consistency with the respective user and analyst manuals.

The analysis will begin by considering the effect of a single weapon that is not aimed at the center of the target but at an offset aimpoint. Then, the effect of accumulating multiple weapons will be addressed.

Recall that for multiple weapons firing on an area target, each weapon will usually be assigned an aimpoint, as shown in Fig. 16.4. We have to account for the fact that each weapon in the battery will be aimed at a location other than the center of the target. Figure 16.39 shows one of n_r weapons firing at an area target, although once more only the range direction is currently under analysis.

Also shown in the figure is the ith aimpoint located at a_{xi} along the target length according to Fig. 16.4. Note that the miss distance is with respect to the aimpoint, and that the Carlton damage function has been superimposed on the weapon impact point.

If the weapon is aimed at the ith aimpoint with range coordinate a_{xi} and misses by a distance u in range, the damage done at a point x from the origin may be obtained from the Carlton damage function

$$P_R(x) = D_0 \exp\left\{-D_0 \frac{(x - a_{xi} - u)^2}{R_x^2}\right\}$$
$$= D_0 \exp\left\{-\frac{(x - a_{xi} - u)^2}{2\sigma_{Dx}^2}\right\} \tag{16.76}$$

where

$$\sigma_{Dx}^2 = \frac{R_x^2}{2D_0} \tag{16.77}$$

This is essentially rewriting the Carlton damage function to allow for an offset aimpoint (from the target center) and a miss distance from that aimpoint.

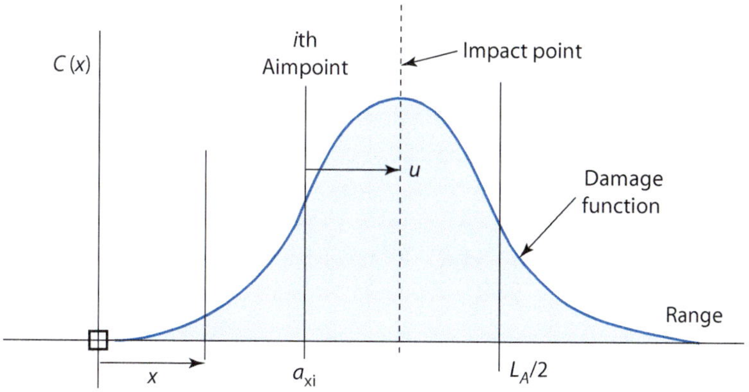

Fig. 16.39 Damage to area target for a single weapon.

16.18 Combining the Damage Function with Precision Error

As for SQ2 and Matrix Evaluator, the damage function is now combined with the precision error in a manner similar to that described in Chapter 10, Section 10.10, although the notation is a little different than Eq. (10.65) due to the different form of the Carlton damage function.

If the precision error distribution relative to the aimpoint may be considered Gaussian normal with zero mean, it may be expressed as

$$E_{\mathrm{PR}}(u) = \frac{1}{\sigma_x\sqrt{2\pi}}\exp\left\{-\frac{u^2}{2\sigma_x^2}\right\} \tag{16.78}$$

The damage function expressed relative to the aimpoint may also be written as

$$P_R(u) = D_0 \exp\left\{-D_0\frac{u^2}{R_x^2}\right\} \tag{16.79}$$

Combining the damage function with the precision error results in a modified damage function

$$P_{\mathrm{DR}}(u) = \frac{R_x}{\sqrt{2}Q_1}\exp\left\{-\frac{1}{2}\left[\frac{u}{Q_1}\right]^2\right\} \tag{16.80}$$

where the combined "standard deviation" Q_1 is given by the RSS of the standard deviations of the precision error and damage function, that is,

$$Q_1 = \sqrt{\frac{R_x^2}{2D_0} + \sigma_x^2} \tag{16.81}$$

If we now write the modified damage function to include the ith aimpoint and refer it to the origin in Fig. 16.39, that is, in the form of Eq. (16.76), we get

$$P_R(x) = P_{\mathrm{DR}}(i) = \frac{R_x}{\sqrt{2}Q_1}\exp\left\{-\frac{1}{2}\left[\frac{x - a_{xi} - u}{Q_1}\right]^2\right\} \tag{16.82}$$

Similarly for the deflection direction

$$P_{\mathrm{DD}}(i) = \frac{R_y}{\sqrt{2}Q_2}\exp\left\{-\frac{1}{2}\left[\frac{y - a_{yi} - v}{Q_2}\right]^2\right\} \tag{16.83}$$

where

$$Q_2 = \sqrt{\frac{R_y^2}{2D_0} + \sigma_y^2} \tag{16.84}$$

16.19 Defining the Volley Damage Grid

It was recognized that one of the factors slowing down Matrix Evaluator was the potentially large number of cells that had to be aggregated from each warhead lethal area. In ARTQUICK, this is avoided by the user defining a single grid for all warheads in the volley and determining the cell P_K by aggregating the effect from each Carlton damage function. The single volley damage grid is defined as shown in Fig. 16.40.

The user specifies the number of cells in range N_R and deflection N_D, and the physical size of the cells is calculated from the target dimensions and delivery accuracy as in the following equations:

$$dx = \frac{(4 + L_A/2\sigma_{\text{tar}})}{N_R} \times \sigma_{\text{tar}} \tag{16.85}$$

$$dy = \frac{(4 + W_A/2\sigma_{\text{tad}})}{N_D} \times \sigma_{\text{tad}} \tag{16.86}$$

Here, σ_{tar} and σ_{tad} are the combined MPI and TLE errors defined by

$$\sigma_{\text{tar/d}} = \sqrt{(\sigma_{\text{MPI}})^2 + (\sigma_{\text{TLE}})^2} \tag{16.87}$$

For example, if the user specifies a grid of 12 × 12 cells, the target is 30 m × 60 m, the MPI errors are 45 m (range) and 15 m (deflection), and TLE is 50 m, the grid size comes out to 21.9 m × 16.8 m. The volley grid size becomes 12 × 21.9 = 262.5 m in range and 12 × 16.8 = 201.6 m in deflection, considerably larger than the target area.

Note that so far, we have accounted for precision errors with Eqs. (16.81) and (16.84) and target location errors with Eq. (16.87).

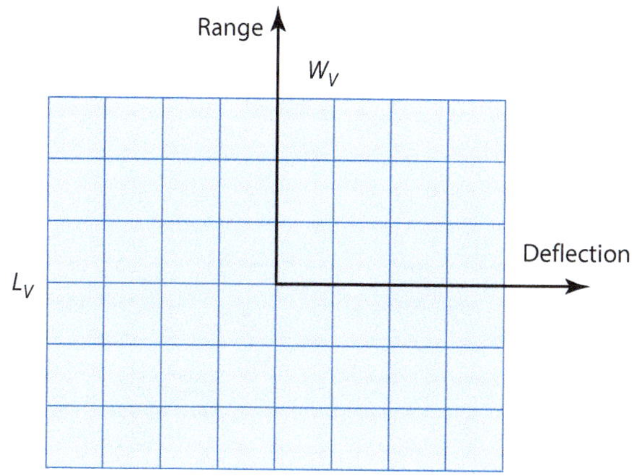

Fig. 16.40 Single volley damage grid.

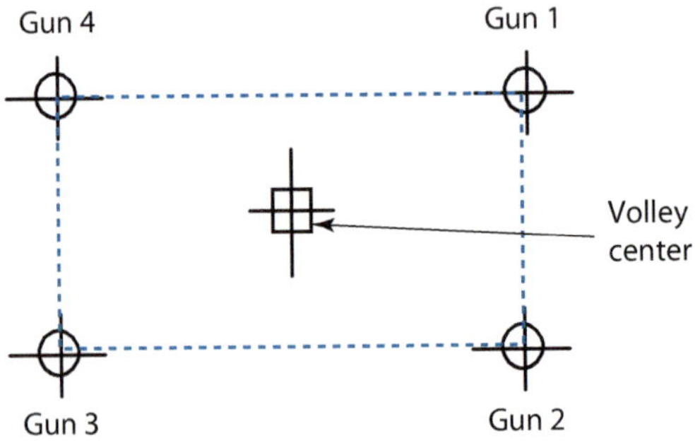

Fig. 16.41 Four-gun volley pattern.

The next step is to determine the P_K in each cell in the damage grid by placing the individual weapon impact points on top of the weapon damage grid. Assuming for convenience that we are dealing with a four-gun battery, the aimpoint or volley pattern might look like that shown in Fig. 16.41.

Because precision error has been accounted for, we can now place the volley pattern onto the weapon damage grid with the center of the grid coincident with the center of the volley, as shown in Fig. 16.42.

There will be a Carlton damage function centered on each of the individual round impact points, and each will contribute damage to every cell in the grid. The total damage in the range direction done to target elements inside cell (p, q) shown in the figure from each weapon in a single volley is given by Eq. (16.82), where x is the range coordinate of the center of the cell and $u = 0$. The accumulated damage done to this cell by all n weapons in the volley is obtained from the survivor rule as

$$P_{\mathrm{DR1}}(p, q) = 1 - \prod_{i=1}^{i=n} (1 - R \times P_{\mathrm{DR}}(i)) \tag{16.88}$$

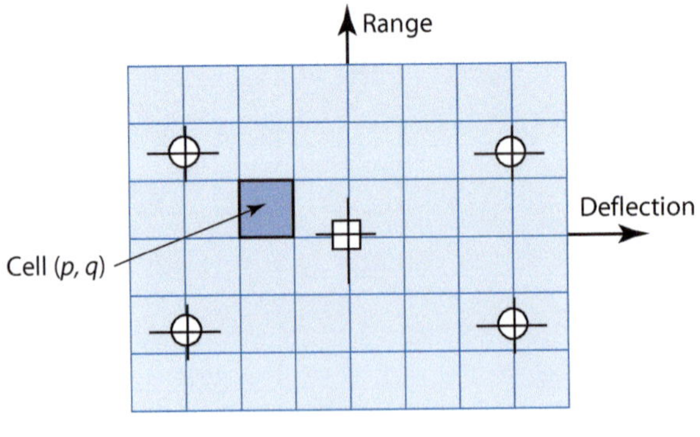

Fig. 16.42 Single volley damage grid.

where R is the weapon reliability and $P_{\mathrm{DR}}(i)$ is given by Eq. (16.82). Remember, this method is for dependent fire only; therefore, we can use the survivor rule to power up for the number of volleys n_V giving the N volley P_K for each cell.

$$P_{\mathrm{DRN}}(p, q) = 1 - \prod_{i=1}^{i=n_V} (1 - P_{\mathrm{DR1}}(p, q)) \tag{16.89}$$

Similarly, for the deflection direction we can write

$$P_{\mathrm{DDN}}(p, q) = 1 - \prod_{i=1}^{i=n_V} (1 - P_{\mathrm{DD1}}(p, q)) \tag{16.90}$$

Finally, the cell damage probability becomes

$$P_{\mathrm{DN}}(p, q) = P_{\mathrm{DRN}}(p, q) \times P_{\mathrm{DDN}}(p, q) \tag{16.91}$$

At this point we have the N-volley damage grid with the dimensions shown in Fig. 16.40, where each cell has a damage probability associated with it obtained from Eq. (16.91). The next step is to consider how this grid interacts with the area target as a function of the aiming error.

The final stage is to calculate the expected fractional damage on the target including the effect of aiming error. The situation is shown in Fig. 16.43 and is identical to that shown for the Matrix Evaluator program earlier.

The method used to calculate the FD is exactly the same as for the legacy version of Matrix Evaluator, namely:

1. Expanding the actual target to the apparent target size using the standard deviation of the aiming error
2. Overlaying the apparent target onto the volley damage grid to identify which cells damage the target

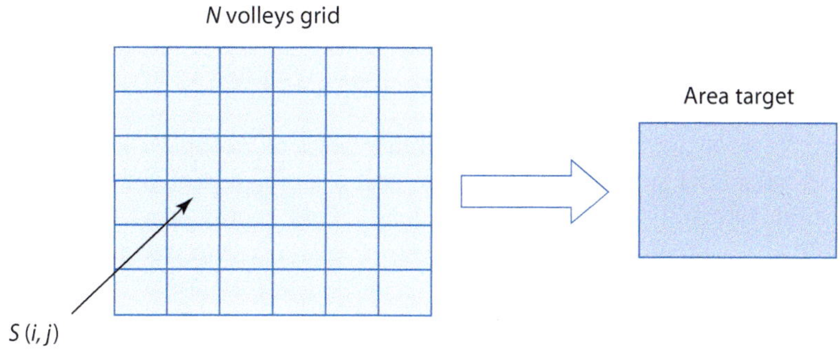

Fig. 16.43 Volley damage grid projected at area target.

3. Calculating the expected fractional coverage of each cell in the damage grid identified in step 2 over the actual target area, multiplying by the damage probability of the cell in question and summing over all cells

$$FD1 = \sum_{i=1}^{i_{MAX}} \sum_{j=1}^{j_{MAX}} S(i, j)\, F_C(i, j) \tag{16.92}$$

Note that in ME v2.0, the approach is modified in the manner described in Section 16.15.

16.20 ARTQUIK Model for Improved Conventional Munitions (ICMs)

In addressing ICM warheads, ARTQUIK does not use the Carlton damage function, but reverts back to a rectangular damage function as used in the low fidelity model. Given the nature and purpose of ICM weapons, this is not an unreasonable approximation to the lethal area. Inputs to the program include the elliptical dimensions of the pattern and number of submunitions.

The methodology for dealing with ICM is the same as for unitary weapons as far as accuracy is concerned, but there are differences when it comes to specifying the damage function. These differences are in both the magnitude of the damage function and the shape of the resulting lethal area.

The basic methodology outlined in Section 16.4 is followed, namely reducing the single submunition elliptical lethal area to a rectangle.

$$L_{ET} = \sqrt{\pi} R_X, \qquad W_{ET} = \sqrt{\pi} R_Y \tag{16.93}$$

The probability of damage due to a single submunition in the pattern is

$$P_S = \frac{a_L}{A_P} = \frac{a_L}{\pi R_X R_Y} = \frac{a_L}{L_{ET} W_{ET}} \tag{16.94}$$

Because it is assumed that the submunitions are uniformly distributed within the pattern,[¶] the probability of damage in the pattern due to all submunitions is

$$P_{DP} = R[1 - \exp(-R_{SUB} N_S P_S)] \tag{16.95}$$

where R_{SUB} is the submunition reliability. At this point we have a pattern with dimensions supplied by the user, and a damage probability within the resultant rectangle given by Eq. (16.95). This rectangular lethal area is expanded to account for precision errors in the same manner as Matrix Evaluator.

$$L_B = L_{ET} + 6x_{br} \tag{16.96}$$

[¶]Here the submunitions are statistically uniformly distributed, unlike distributed uniformly submunitions in the Matrix Evaluator approach.

$$W_B = W_{ET} + 6x_{bd} \tag{16.97}$$

$$A_B = L_B W_B \tag{16.98}$$

The probability of damage is consequently reduced to account for the precision error expansion

$$P'_{DP} = P_{DP} \frac{L \times W}{L_B \times W_B} \tag{16.99}$$

In this model, we do not group all rounds comprising a volley into a single pattern, but represent them individually, as described for unitary warheads (Fig. 16.44).

Notice in the figure that overlap of individual patterns does occur but may not depending on the relative size of the pattern and individual munition lethal areas. We are now able to determine the dimensions of the rectangular areas shown in Fig. 16.44, and by powering up the number of overlapping patterns we can determine the damage probability in each one. The process now proceeds as for unitary weapons, that is,

1. A weapon damage grid is established based on the target size, combined aiming and TLE errors, and the desired number of cells—Eqs. (16.85) and (16.86).
2. The damage grid is overlaid by the weapon patterns—Fig. 16.44.
3. The probability of damage in each cell of the damage grid is calculated using the survivor rule—Eq. (16.88).
4. These values are powered up to allow for multiple volleys—Eq. (16.91).
5. The fractional coverage of each cell of the damage grid is calculated and accumulated to produce the final FD—Eq. (16.92).

It may be noted that for ICMs where the weapon lethal area is rectangular with a constant P_D rather than Carlton, a closed form solution may be computed in a manner similar to that of sticks dealt with in Chapter 12. The

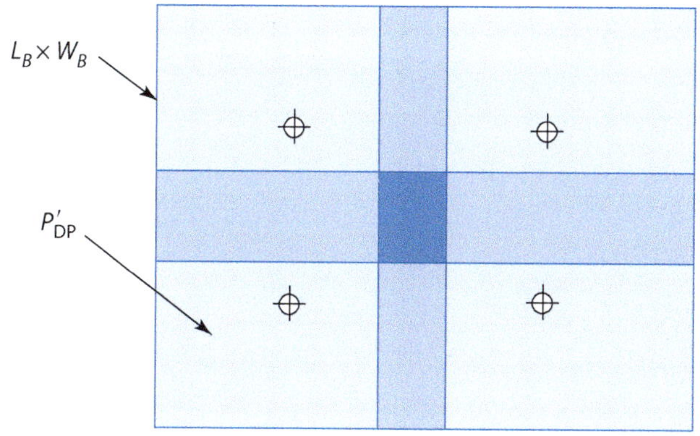

Fig. 16.44 Individual weapon lethal areas on the target.

ARTQUIK program does not do this, however, but uses the method outlined previously.

16.21 Stochastic Modeling of Indirect Fire

Now that computers are relatively fast, it is possible to develop a Monte Carlo based methodology to calculate fractional damage for indirect, battery fire on an area of target elements to calculate fractional damage, including the effect of multiple volleys. The basic approach follows the double loop Monte Carlo loop for a single volley is shown in Figure 9.7 where for each sample of MPI error defining the volley center there are as many samples of precision error as there are rounds in the volley. Based on these impact points, the fractional damage may be calculated for the fall of shot over the target area rectangle. In detail, the method consists of the following steps.

The area containing the identical, uniformly distributed target elements is placed at the origin. Suppose for simplicity it is assumed there are 10 elements along each side of this rectangle therefore there are 100 elements total. These elements are represented by a 2-D "kill matrix" K where if the element value is unity it is killed, otherwise it survives and has the value 0. The array is initialized to all zero values and is shown in Figure 16.45. We are interested in calculating the fractional damage for an n-volley fire mission.

For the first volley, a random draw is taken from the dependent error (MPI) distribution and placed on the ground as shown in Figure 16.46.

Suppose we have a four gun battery and the volley geometry is L_V by W_V. The volley rectangle is drawn centered on the MPI as shown in Figure 16.46 indicating the intended impact points of the four rounds at the corners.

Four samples are drawn from the precision error distribution to determine the actual impact points of each weapon lethal area. From the EI value and impact angle, the rectangular cookie cutter lethal areas dimensions

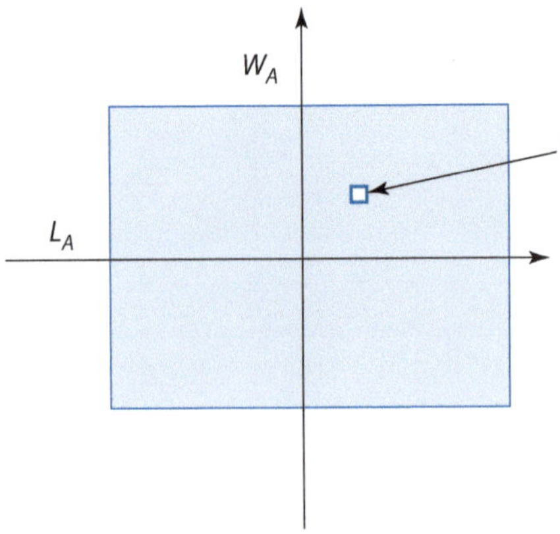

Figure 16.45 Area if target elements (10 × 10) placed at the origin

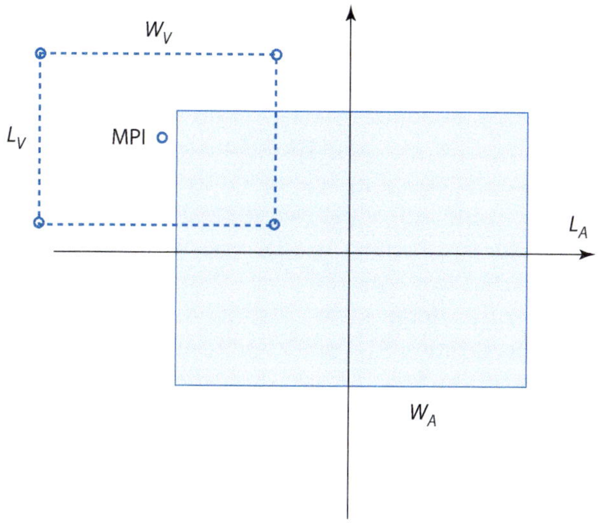

Figure 16.46 Volley centered on sample MPI

(L_{ET} by W_{ET}) are calculated and drawn centered on the actual impact points of each round, as shown in Figure 16.47.

Taking each target element in turn, we determine if it is enclosed by each of the round lethal areas comprising the volley, and if it is the value of that element in the kill matrix is set to unity. As seen in Figure 16.47, in the

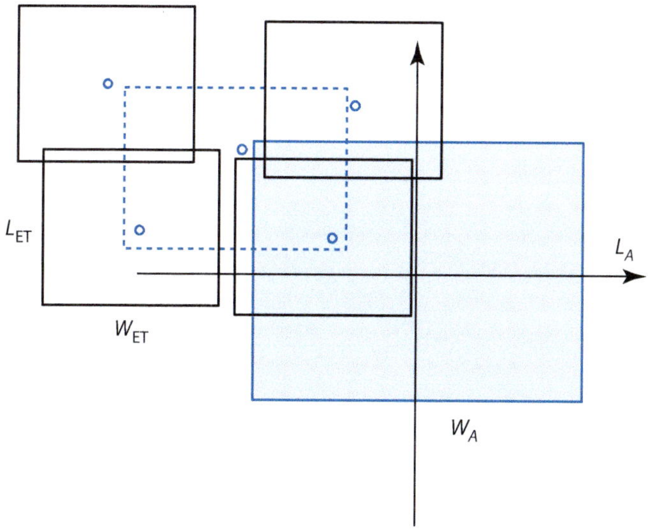

Figure 16.47 Lethal areas placed on actual impact points

top left of the target area some elements fall inside two weapon lethal areas, so if it is killed by the first weapon its value is already set to unity and since the second weapon cannot kill it again it remains set to unity. This determines which target elements are killed by each round in this volley.

Keeping the same MPI error and the kill matrix unchanged the process is repeated for the second volley by taking another four draws from the precision distribution so the individual round impact points will have different locations relative to the corners of the volley rectangle. The individual element coverage process is repeated to see if any additional target elements have been killed for this iteration and if so more elements of the kill matrix are set to unity. This is repeated n times and results in an n-volley kill matrix for a single iteration of the Monte Carlo simulation from which the fractional damage may be calculated by summing all the elements in the kill matrix and dividing by the total number of elements present, which for the example case of 100 elements is given by the following.

$$\mathrm{FD}_n = \frac{\sum_{i=1}^{i=10}\sum_{j=1}^{j=10} K(i,j)}{100} \tag{16.100}$$

The whole process is repeated for a user specified number of Monte Carlo iterations and the resulting values of fractional damage averaged. Sample results from this approach are shown in Figure 16.48.

This figure might be initially confusing to the reader, particularly following the discussion of dependent and independent fractional damage calculations in section 16.8 and Figure 16.11 in particular. There it was argued

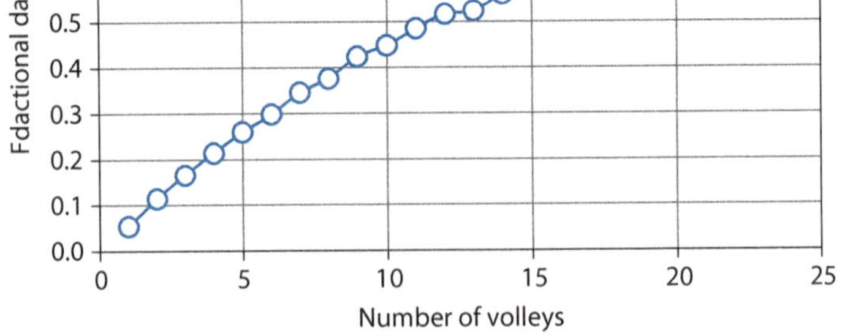

Figure 16.48 Fractional damage for n volleys

that for multiple weapon dependent fire, which artillery certainly is, the fractional damage approaches the fractional coverage as the number of volleys increases, while Figure 16.48 shows it slowly approaching 1.0. How is this apparent contradiction resolved?

It will be seen that in the analysis of section 16.8 it was assumed that the precision error was combined with the single weapon lethal area and that is what is displayed in Figure 16.9 and Figure 16.10. That being so, the only remaining delivery error is the MPI which for a single draw in a Monte Carlo simulation will place each lethal area directly on top of the one before. The FD will indeed approach the FC for this case.

For the Monte Carlo approach just described however, the precision error is not combined with the single weapon lethal area but kept separate. Still the single draw of the MPI will be common to all weapons but the precision error will provide some variability in determining the actual impact point. It is argued therefore that whatever the magnitude of the MPI and precision errors, given enough Monte Carlo iterations, it is possible for at least part of one lethal area to partially cover the target, and in the limit the fractional damage will approach unity.

It is concluded therefore is that for methodologies where the basic, single weapon lethal area is combined with the precision errors (e.g. SQ2 and Matrix Evaluator) the fractional damage will approach the fractional coverage as the number of volleys increases, and equation 16.54 will apply. For methodologies where this does not occur (Monte Carlo), the fractional damage will approach unity as the number of volleys increases.

16.22 Summary of Model Features

This chapter reviews the basis of surface-to-surface indirect methodology used primarily against an area of target elements. It shows a fundamental difference with the air-to-surface methods presented so far for the effectiveness of multiple events, namely a difference in the way single events are powered up. Three different models are described, each exhibiting a different level of fidelity and differing mainly in the representation of damage functions.

16.22.1 SUPERQUICKIE 2 (SQ2)

1. This is the lowest fidelity model, and it uses a rectangular representation for the individual weapon lethal area.
2. This lethal area is increased to account for precision error in a manner similar to AS methods, but the equation is different.
3. The aimpoints on the area target are selected in accordance with firing doctrine, and the volley pattern is laid down over the area target.
4. Individual weapon lethal areas are considered to overlap or not in a manner identical to AS stick deliveries.

5. The single volley is then represented as an equivalent single weapon lethal area of known dimensions and known conditional probability of damage inside it.
6. The effectiveness calculation proceeds in the usual manner of computing the expected fractional damage of this rectangle over the target area.
7. Multiple volleys are powered up using the SS indirect fire formula.
8. ICM (cluster) munitions are dealt with in the same way as for AS methods, except the powering up process is for multiple volleys.

16.22.2 ARTQUIK

1. This is a higher fidelity model than SQ2 and uses a Carlton damage function representation for the individual weapon lethal area. Unlike the AS methods that use Carlton, this implementation uses a shaping factor D_0.
2. This single weapon damage function is mathematically combined with precision error, spreading out the damage function in a manner similar to expanding the rectangular cookie cutter damage function of SQ2.
3. The aimpoints on the area target are selected in accordance with firing doctrine and the volley pattern laid down over the ground plane. A single volley damage grid comprising a defined number of cells is generated, and the damage accumulated in each cell is computed using the survivor rule and the damage contribution from each individual weapon Carlton functions.
4. The cell elements are powered up by the number of volleys N to obtain the N-volley damage grid. This N-volley damage grid is overlaid on top of the area target, and the FD calculated in the usual manner.
5. For ICMs, the Carlton damage function is not used, but a rectangular cookie cutter characteristic of this weapon system is used. The process is the same as for HE rounds described in steps 3 and 4.

16.22.3 MATRIX EVALUATOR (ME)

This is the highest fidelity model available, and so it is the slowest to run. It is applicable for both AS and SS scenarios.

1. This model uses a lethal area matrix (LAM) representation for the individual weapon lethal area derived directly from the GFSM or AJEM programs. Apart from this difference, the method is very similar to ARTQUIK. In this model, the LAM is called the single weapon damage grid.
2. This single weapon damage grid is enlarged cell by cell to account for the precision error in a manner similar to the AS process but having a different mathematical formulation. This produces the single weapon

ballistic grid, where the original cell P_K values have been adjusted after the enlarging process to preserve lethality.

3. The aimpoints on the area target are selected in accordance with firing doctrine and the volley pattern laid down over the ground plane. A single volley damage grid enclosing the single weapon ballistic grids is generated, and the damage accumulated in each cell is computed using the survivor rule and the damage from each individual weapon ballistic damage grid cell overlaying it.
4. Powering up for multiple events depends on whether those events are independent or dependent upon each other.
   - *For dependent multiple events:* The cell elements are powered up by the number of volleys N to obtain the N-volley damage grid. This N-volley damage grid is overlaid on top of the area target, and the FD calculated in the manner explained in the text.
   - *For independent multiple events:* The single volley damage grid is overlaid on top of the area target, and the FD for one occasion calculated in the manner explained in the text. Then the AS powering up formula is used to calculate the effect of multiple events.
5. For ICMs, Matrix Evaluator uses the LAM for a single submunition against the target element contained within the defined target area (e.g., troops, vehicles, etc.). The program then has an algorithm for placing each submunition within the specific pattern of the weapon being used. By using the survivor rule, these overlapping (or not) single submunitions define a single weapon lethal area in the ground plane. From this point on, the method proceeds as for a unitary warhead indicated in Steps 2–7 (Section 16.22.1).
6. Improvements to the legacy ME model have resulted in ME 2.0, which includes higher fidelity modeling of the lethal area matrix and more consistent fractional coverage calculations.

Reference

[1] "Tactics, Techniques and Procedures for the Field Artillery Cannon Battery," FM 6-50, U.S. Marine Corps, Dec. 1996.

Chapter 17 Direct Fire: Infantry, Vehicles, and Small Boats

17.1 Introduction

Direct fire engagements are considered to be those where the shooter has visual contact with the target and places an aiming cue on it. This contact may be directly with the unaided eye or with the assistance of a sensor, such as an image magnifier or night vision device. After an aimpoint on the target is selected, the weapon fires either a single round or a burst of rounds. Aircraft-mounted guns strafing a ground target would qualify as a direct fire scenario, but because this was dealt with in Chapter 14, the focus here is on surface-to-surface engagements.

In many cases, direct fire involves a penetration round such as a bullet or sabot, although occasionally a fragmenting round may be used. In this chapter we will study three common scenarios:

- Direct fire against dismounted personnel
- Direct fire against vehicular targets
- Naval direct fire methodology

It is important to remember that for surface-to-surface engagements in general and these three scenarios in particular, we are not performing a weaponeering calculation in the strictest sense of the definition; that is, no weapon selection is being made to increase effectiveness. What we are calculating is the effect of a selected weapon on a target. As an M1A1 tank commander stated, "When I see a T-72 tank in my gun sight, I don't consult the JMEM to decide what round to fire—it's a DU penetrator." The JMEM will, however, determine the outcome of a single engagement of this type.

The United States and other countries use many large scale force-on-force models that simulate a battle between large numbers of troops and vehicles that include maneuver. In running such a model, a tactical situation may occur, for example, where three Bradley AFVs will engage a company of enemy troops, or four Abrams tanks will attack a column of T-62 tanks. At the core of these multi-element engagements, the JMEM will supply the outcome of the one-on-one attack, even though no user weapon selection has taken place. That decision is preset in the

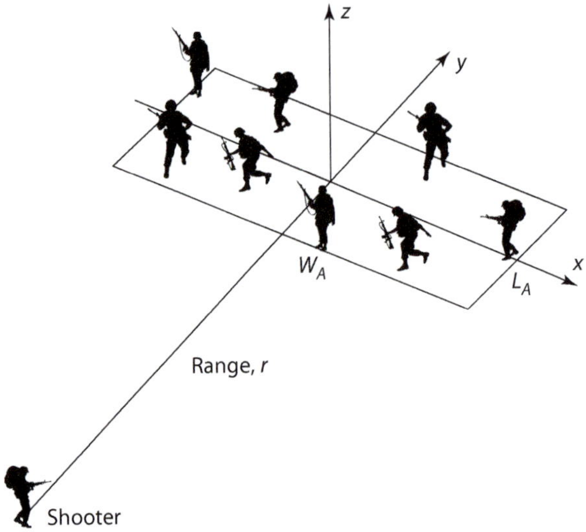

Fig. 17.1 Engagement geometry.

program and based on field tactics. This issue should be kept in mind when considering the applicability of these methodologies.

17.2 Direct Fire Against Personnel Targets: FBAR

This methodology simulates the engagement of burst firing weapons against dismounted infantry. The personnel targets are uniformly distributed within a rectangular target area centered on a Cartesian coordinate frame. Figure 17.1 shows the geometry of this scenario.

Inputs to the methodology are the height and width of an individual soldier target, the dimensions of the smallest rectangle enclosing all the individual targets, and the range to the shooter from the Cartesian origin. The targets may change posture, which affects their presented area to the shooter.

The method uses a Monte Carlo simulation and begins by randomly placing personnel targets within the specified target area using a uniform distribution. An initial aimpoint is selected, a burst is fired, another aimpoint is selected, another burst is fired, and so on. Traversing the target area firing at each aimpoint is known as a sweep, and several sweeps of the target area may be made within each iteration of the simulation. Errors associated with aimpoint accuracy and dispersion of individual rounds within a burst will determine if the target is hit. Rounds missing the intended target may still hit other personnel nearby or behind. The vulnerability of the target is defined by the probability of incapacitation given a hit $P_{I/H}$ and is a function of range and other factors such as whether the target is wearing body armor. For each scenario, the number of hits is recorded together with the number of personnel incapacitated.

Further iterations are then performed, and the mean effectiveness values are recomputed. Between iterations, the locations of the targets within the target area will be changed. Also the MPI and precision errors will change, leading to different numbers of targets hit. The different phases of the methods operation will now be described in the order in which they take place.

17.2.1 Target Placement

The personnel targets are placed at coordinates within the specified target area in a uniform random manner. A check is made to ensure that individual targets are greater than some minimum value (typically 5 m) from each other in order to avoid tactically unrealistic grouping of personnel. The coordinates of each target are stored and represent ground plane x, y values for each target.

17.2.2 Aimpoint Selection

Aimpoints are either uniformly distributed or placed on individual target elements. If the range is greater than some user-supplied value, the aimpoints are uniformly distributed across the x-axis, whereas for ranges shorter than this critical value, weapons are aimed directly at personnel. The aimpoints X_{AIM} and Y_{AIM} are also x, y coordinate pairs on the ground plane.

17.2.3 Centers of Burst

The nominal center of a burst is different from the aimpoint and is located directly above it at a distance equal to half the height of a personnel target. The weapon is actually aimed at the nominal center of burst. The actual center of burst is displaced from the nominal center of burst by the aiming error, as shown in Fig. 17.2. In FBAR terminology, the between-burst error is the aiming or MPI error.

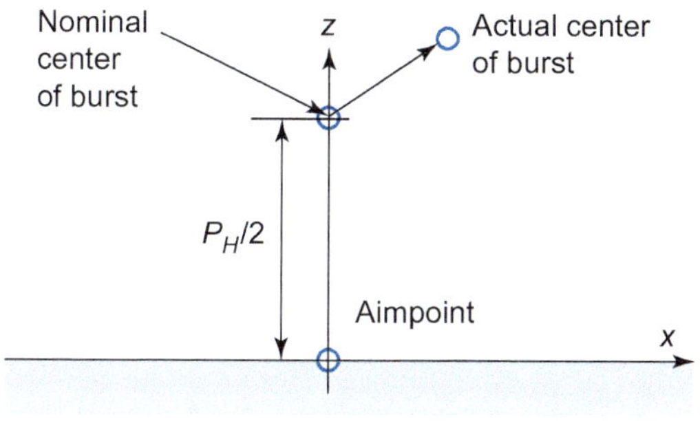

Fig. 17.2 Aimpoint and centers of burst.

The center of burst, or center of impact (CI), is related to the aimpoint by the following equations:

$$\mathrm{CI}_X = X_{\mathrm{AIM}} + \mathrm{BSE}_X + \sigma_{\mathrm{XBB}} \times \mathrm{RN} \tag{17.1}$$
$$\mathrm{CI}_Z = Z_{\mathrm{AIM}} + \mathrm{BSE}_Z + \sigma_{\mathrm{ZBB}} \times \mathrm{RN} \tag{17.2}$$

In these equations CI_X, CI_Z are the X, Z coordinates of the nominal center of the burst on the $x-z$ plane; BSE_X, BSE_Z are the constant bias errors due to weapon boresight misalignment, if known; σ_{XBB}, σ_{ZBB} are the standard deviations of the between-burst random errors; and RN is a random number drawn for each burst from a normally distributed set with zero mean and unity standard deviation. The last two terms in each of these equations comes from the mapping for the standard normal variable z.

$$x = \mu_x + \sigma_x \times z \tag{17.3}$$

If Z_{AIM} is set to half the height of a personnel target, Fig. 17.3 shows for distant targets the relationship between the aimpoint and center of burst coordinates.

The aimpoint, nominal, and actual center of burst all lie in the $x-z$ plane.

17.2.4 Round Impact Points

Next, the X and Z coordinates of each round impact point on a vertical plane containing the nominal center of burst are calculated.

$$\mathrm{XI} = \mathrm{CI}_X + \sigma_{\mathrm{XWB}} \times \mathrm{RN} \tag{17.4}$$
$$\mathrm{ZI} = \mathrm{CI}_Z + \sigma_{\mathrm{ZWB}} \times \mathrm{RN} \tag{17.5}$$

Here XI, ZI are the coordinates of the individual rounds within the burst, and σ_{XWB}, σ_{ZWB} are the standard deviations of the normally distributed errors of individual rounds within the burst and represent the precision errors. RN is a normally distributed random number as before, but drawn

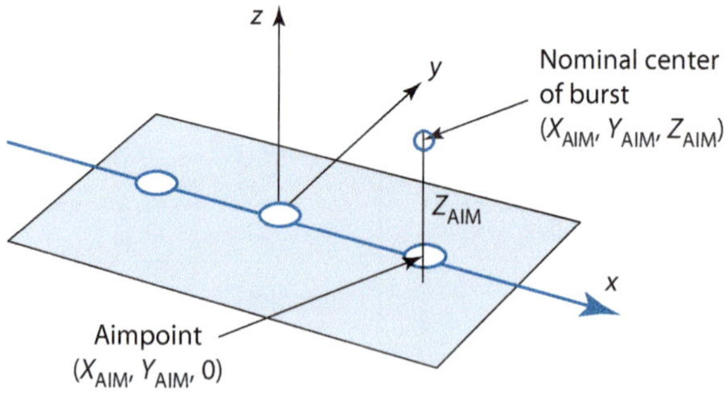

Fig. 17.3 Aimpoint and nominal center of burst coordinates.

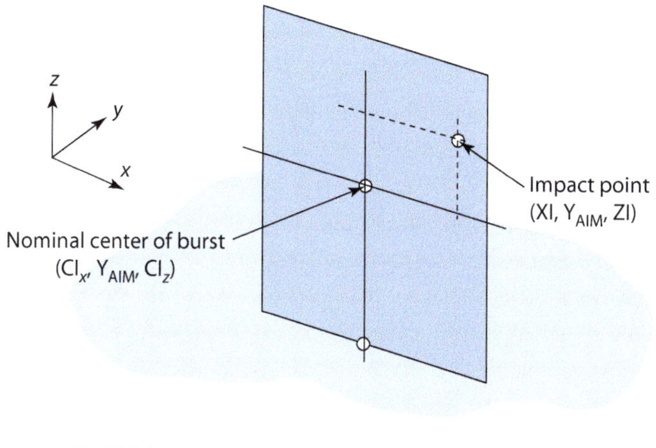

Fig. 17.4 Individual round impact point.

for each round in the burst. Figure 17.4 shows where an individual round impacts the vertical plane relative to the nominal center of burst.

Again in FBAR jargon, the within-burst error is otherwise known as the precision error. It is important to note that FBAR does not have a trajectory program for the rounds fired, but the terminal effects are considered. The methodology now modifies the ZI coordinate due to the effect of a variation in the muzzle velocity of the munition. This is done by first calculating the vertical component of the round velocity at the range from the shooter to the center of the target area.

$$v_Z = v_{\mathrm{REM}} \times \sin(I) \tag{17.6}$$

Here v_{REM} is the remaining velocity of the round at this range, and I is the impact angle, or angle of fall, of the round on the ground plane, both of which have to be input by the user. Assuming that the time of flight variation is normally distributed with standard deviation σ_{TOF}, the adjusted vertical coordinate of the round is

$$\mathrm{ZI} = \mathrm{ZI} + v_Z \times \sigma_{\mathrm{TOF}} \times \mathrm{RN} \tag{17.7}$$

The Y coordinate of the impact point of each round of the burst on the ground plane is computed with the aid of Fig. 17.5.

Hence we get

$$\mathrm{YI} = Y_{\mathrm{AIM}} + \frac{\mathrm{ZI}}{\tan(I)} \tag{17.8}$$

We now have impact points in the ground plane for all rounds fired in a sweep.

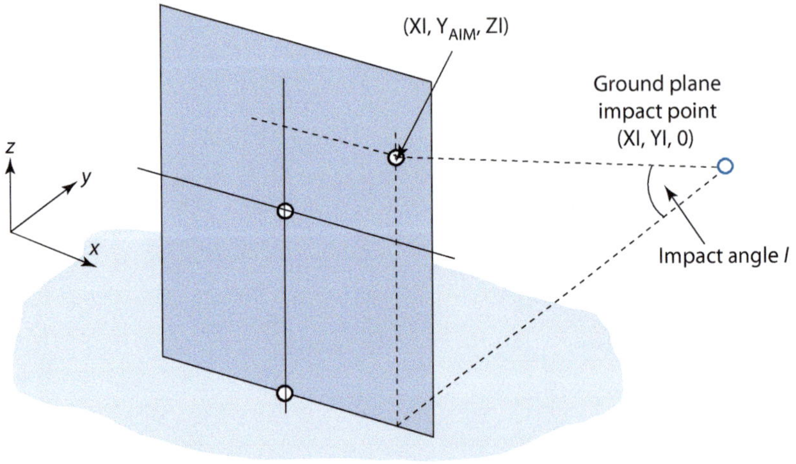

Fig. 17.5 Impact of round in the ground plane.

17.2.5 Effectiveness Calculation

The vertical rectangular representation of each personnel target is now checked to determine if a round has passed through it. This is done by projecting the target "shadow" onto the ground plane by the angle of fall *I*. If a round lies within the shadow, it must have passed through the target, and a hit has occurred. The shadow geometry is shown in Fig. 17.6.

If a target is hit, a determination has to be made if it is incapacitated. This is done by drawing a uniformly distributed random number and comparing it to the probability of incapacitation given a hit. If the number is less than this

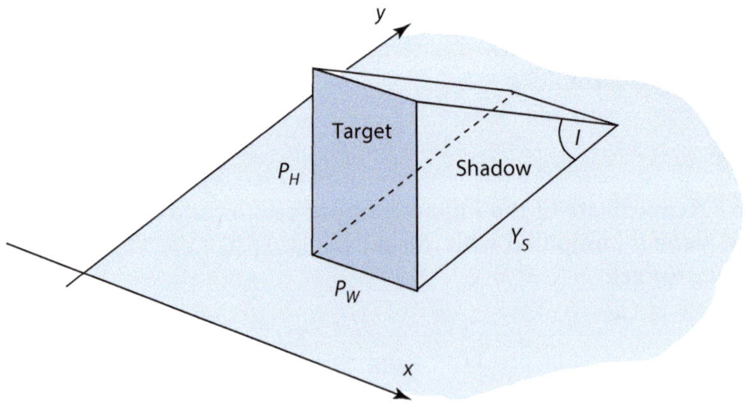

Fig. 17.6 Shadow projected by personnel target.

value, the target is incapacitated; otherwise, it is not. If the target is incapacitated, the counter keeping track of incapacitations N_{INC} is incremented. An incapacitated target is then given the x coordinate of 999 m, to remove it from the engagement and prevent it from being hit again.

For fragmentation rounds, the same process of calculating whether a bullet strikes an individual target is followed, calculating the number of hits and the number of personnel incapacitated. If a fragmenting round strikes a personnel target, its fragmentation effect is considered lost; however, if the round misses all targets and impacts the ground, the impact point is considered as the detonation point and a check is made for further indirect casualties due to fragments. This is done by assuming the probability of an indirect casualty decreases exponentially with distance from the target, decreasing the $P_{\mathrm{I/H}}$ in the following manner:

$$P_{\mathrm{CAS}} = P_{\mathrm{I/H}} \exp\left(-\frac{\pi r^2}{A_L}\right) \tag{17.9}$$

where r is the distance from the point of detonation to the individual personnel target being considered, and A_L is the lethal area of the fragmenting round. Taking a uniformly distributed random variable and comparing it to P_{CAS} in Eq. (17.9) determines if the target becomes a casualty, thereby increasing the casualty counter N_{CAS}. Preserving different casualty counts for N_{HIT} and N_{CAS} enables estimates to be made of direct and indirect casualties for fragmenting munitions.

17.2.6 Program Loop Structure

The structure shown in Fig. 17.7 will explain which parameters are varied and in which order as the Monte Carlo simulation is executed.

```
BEGIN Monte Carlo iteration loop
        Locate targets
        Locate aimpoints
            BEGIN aimpoints loop
                BEGIN bursts loop
                    BEGIN rounds/burst loop
                        BEGIN targets loop
                            Is target hit?
                            Is target incapacitated?
                        END targets loop
                    END rounds/burst loop
                END bursts loop
            END aimpoints loop
END Monte Carlo iteration loop
Calculate mean effectiveness values
```

Fig. 17.7 Program structure.

17.2.7 Implementation

A MATLAB program is presented next to duplicate the major features of the methodology and to enable simple effectiveness calculations of direct fire against personnel targets. The various sections will not be described in detail, but the part shown in Table 17.1 represents the input data.

Most of the inputs have been discussed and are explained by the comments part of the table. The program outputs include:

1. The proportion of personnel directly hit per sweep $N_{\mathrm{HIT}}/N_{\mathrm{MEN}}$
2. The proportion of personnel indirectly hit per sweep $N_{\mathrm{CAS}}/N_{\mathrm{MEN}}$

Table 17.1 Inputs to MATLAB Version of Methodology

n_run	=	100;	% number of Monte Carlo trials
wa	=	20;	% target area width
la	=	20;	% target area length
ph	=	2;	% personnel height
pw	=	0.5;	% personnel width
R	=	50;	% range
p_inc	=	0.49;	% probability of incapacitation
theta	=	0.05;	% angle of fall, mils
tof	=	0.05;	% time of flight
tof_sig	=	0;	% tof standard deviation, % 1 sigma
vr	=	1000;	% velocity remaining
al	=	500;	% lethal area
r_error	=	0;	% range estimation error, % 1 sigma
n_aim	=	5;	% number of aimpoints
sigma_1x	=	7.2;	% between burst sigma x-axis, mils
sigma_1z	=	7.2;	% between burst sigma z-axis, mils
sigma_2x	=	1;	% within burst sigma x-axis, mils
sigma_2z	=	1;	% within burst sigma z-axis, mils
b_errorv	=	0;	% vertical boresight error, mils
b_errorh	=	0;	% horizontal boresight error, mils
n_men	=	8;	% number of personnel targets
n_sweep	=	1;	% number of sweeps of target area
n_bursts	=	1;	% number of bursts/aimpoint
nrb	=	3;	% number of rounds/burst
d_min	=	5;	% minimum separation of targets
z_burst	=	1;	% height of burst
r_max	=	1000;	% maximum range to aim at personnel

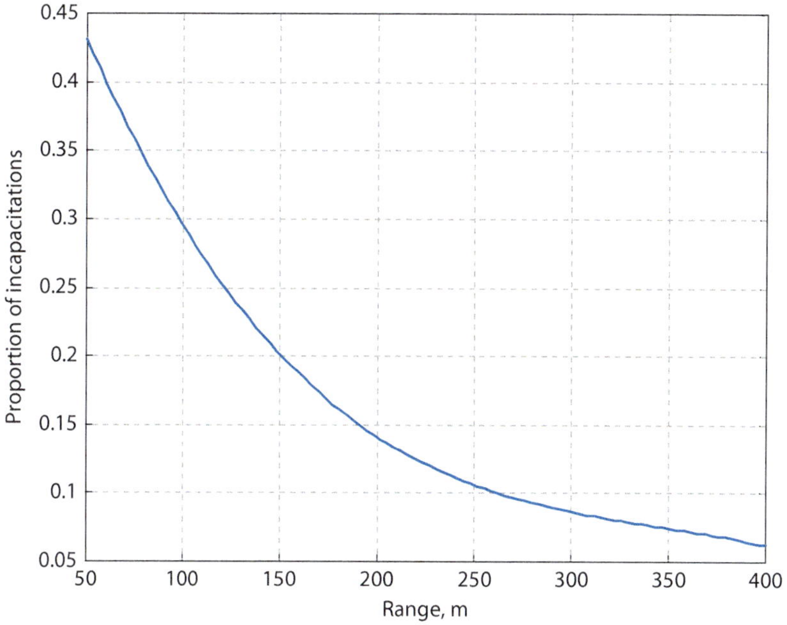

Fig. 17.8 Output from FBAR program.

17.2.8 Sample Results

We will consider the case shown in Table 17.1, where direct fire from a semiautomatic weapon is directed against an eight-man platoon. One three-round burst is aimed at each element of the platoon, and the effectiveness is determined as a function of range for nonfragmenting rounds. The results are shown in Fig. 17.8.

The results shown in Fig. 17.8 are as expected, with the effectiveness decreasing with range. As the range becomes smaller, the proportion of incapacitations approaches the probability of incapacitation given a hit.

17.3 Direct Fire Against Vehicles: Passive Vehicle Target Model (PVTM)

The method discussed in this section has been developed to simulate direct fire weapons in both burst and single-shot modes against stationary vehicles. The vehicle is regarded as passive in that it does not return fire. The model is stochastic like the personnel model, in that the location of the round(s) on the target are determined by a combination of random numbers representing the aiming accuracy, precision error, and so on. A Monte Carlo simulation over a large number of trials allows average effectiveness to be determined.

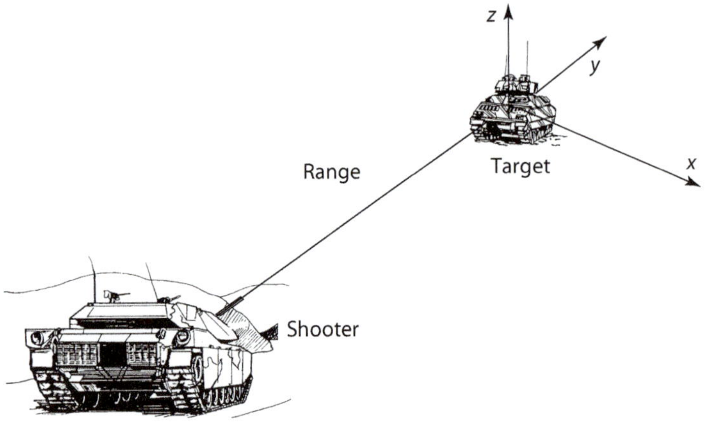

Fig. 17.9 Coordinate system for direct fire.

Although in practice the user may use weapons to "range-in" to the target, this is not a feature of the model, only the "fire-for-effect" phase of the engagement. In order to compute the measures of effectiveness (MOEs), weapon system characteristics and target vulnerability must be known. The outputs from the program include the probability of kill P_K for a given burst or single shot, or the number of bursts/shots required to achieve a specific P_K. The coordinate system used in the model is the same as that used for FBAR and is shown in Fig. 17.9.

17.3.1 Target Representation

Recall the shotline analysis described in Chapter 7, Section 7.7, and in particular Fig. 7.12 showing how the target is overlaid with a grid of cells down through which a fragment is considered to travel. In that section, the shotline P_K was calculated, multiplied by the cell presented area to give the cell vulnerable area and summed for all cells over the target. This process is common to many customers of vulnerability data, of which the JMEM is but one, and each has its own way of postprocessing the resulting data. For example, we described how COVART averaged the A_V around the azimuth, but other applications do not. PVTM is one such application that doesn't.

The target is divided into a number of cells (typically 4 in. × 4 in.) for a given aspect. For each cell, the P_K for a given weapon, range, and kill criterion is defined. The data file containing this information is usually referred to as a cell-by-cell or C × C file. A visualization of this file is shown in Fig. 17.10.

In practice, each cell would have a discrete value of P_K associated with it rather than the range of P_K shown in the legend on the left of the figure.

These C × C files are generated for different azimuth and elevation angles by a process similar to the shotline analysis described in Chapter 7, Section 7.7. The target vulnerability may be determined at a component level if data are available, or at a compartment level. For the latter, for example, a mobility kill would be achieved if the engine compartment were penetrated by a fragment of sufficient size. Note that the C × C file for Fig. 17.10 includes the range to target, kill definition, target exposure azimuth angle, weapon, munition, and target.

17.3.2 Aimpoint Selection

The cell at the center of the target in Fig. 17.10 is the default aimpoint, defined to be the centroid of the presented area, or the centroid of the exposed area for a target in defilade. The centroidal aimpoint coordinates are

$$X_{\mathrm{AIM}} = \frac{\sum x_{\mathrm{cell}}}{n_{\mathrm{cell}}} \tag{17.10}$$

$$Z_{\mathrm{AIM}} = \frac{\sum z_{\mathrm{cell}}}{n_{\mathrm{cell}}} \tag{17.11}$$

Here X_{AIM} and Z_{AIM} are the coordinates of the aimpoint for the current azimuth angle, x_{cell} and z_{cell} are the coordinates of the center of each cell,

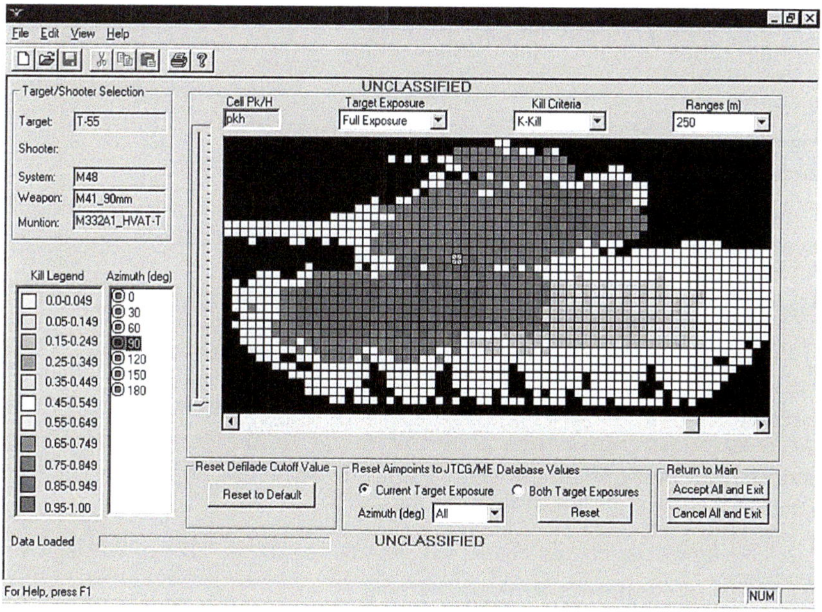

Fig. 17.10 Cell-by-cell P_K values for sample target.

and there are n_{cell} cells in the target for this aspect. The user may select an alternative aimpoint for the view in question, based on the observed areas of high vulnerability indicated in Fig. 17.10, corresponding to the magazine and crew compartment for a K-kill. A different kill definition, say an M-kill, would highlight other regions of the target presented area.

The general firing sequence consists of the shooter aiming at the aimpoint and firing a burst of rounds. The weapon is then retrained on the same aimpoint and another burst fired, corresponding to a different occasion. This may continue for a specified number of bursts. The accuracy of the rounds depends on both the burst-to-burst errors (aiming/MPI) and the within-burst errors (ballistic/precision). Single shots may be represented by a burst comprising one round.

17.3.3 Center of Burst

The center of a burst fired at the aimpoint is calculated in a similar manner to that in the personnel model from

$$\text{CI}_X = X_{\text{AIM}} + \text{BSE}_X + (\sigma_{\text{XBB}} \times \text{RN})R/1000 \tag{17.12}$$

$$\text{CI}_Z = Z_{\text{AIM}} + \text{BSE}_Z + (\sigma_{\text{ZBB}} \times \text{RN})R/1000 \tag{17.13}$$

In these equations, BSE_X and BSE_Z are the constant boresight errors, σ_{XBB} and σ_{ZBB} are the standard deviations (in mils) of the between burst random errors, R is the range to the target, and RN is a random number drawn for each burst from a normally distributed set with zero mean and unity standard deviation. The basis of these equations was discussed in the section on FBAR.

17.3.4 Round Impact Point

For each round in the burst, the impact point is given by the following equations:

$$\text{XI} = \text{CI}_X + (\sigma_{\text{XWB}} \times \text{RN})R/1000 \tag{17.14}$$

$$\text{ZI} = \text{CI}_Z + (\sigma_{\text{ZWB}} \times \text{RN})R/1000 \tag{17.15}$$

Again σ_{XWB} and σ_{ZWB} are the standard deviations (in mils) of the within-burst random errors, R is the range to the target, and RN is a random number drawn for each round from a normally distributed set with zero mean and unity standard deviation.

17.3.5 Calculating Probability of Kill

The cell where each round strikes the target is determined from Eqs. (17.14) and (17.15), and the corresponding P_K value is extracted from the

```
BEGIN Monte Carlo iteration loop
        Locate aimpoint
            BEGIN bursts loop
                BEGIN rounds/burst loop
                    Locate round impact point on target
                    Get P_K for that cell
                END rounds/burst loop
                Compute P_K for this burst
            END bursts loop
        Compute P_K for all bursts
END Monte Carlo iteration loop
Calculate mean effectiveness values
```

Fig. 17.11 Program structure.

C × C file. The probability the target is killed to the specified level for all rounds in the burst is obtained from the survivor rule.

$$\overline{P}_K = 1 - \prod_{q=1}^{q=nrb} (1 - P_K(i, j, q)) \tag{17.16}$$

where $P_K(i, j, q)$ is the P_K value for the cell located at coordinates (i, j) struck by the qth round of a burst comprising n_{rb} rounds.

17.3.6 Program Loop Structure

Figure 17.11 illustrates the order of the computations.

17.3.7 Sample Results

This is somewhat difficult to represent in a simple model due to the complexity of the C × C file for particular targets. However, an example will be presented based on an oversimplification of such a file, as shown in Fig. 17.12.

In this figure, which is the same target considered in Chapter 6, only nine cells will be considered, each of which if hit by a specific munition will precipitate a K-kill with the probability shown in the cell. Each cell is assumed to measure 3 m long by 1.5 m high. Note the shift in the origin between Fig. 17.9 and Fig. 17.12 to assist in the generation of random impact points.

The aimpoint is given by Eqs. (17.10) and (17.11) as $X_{\mathrm{AIM}} = 4.5$ m, $Z_{\mathrm{AIM}} = 2.25$ m, or the center of the middle cell. As before, a Monte Carlo simulation has been written in MATLAB for this case, where the inputs consist of

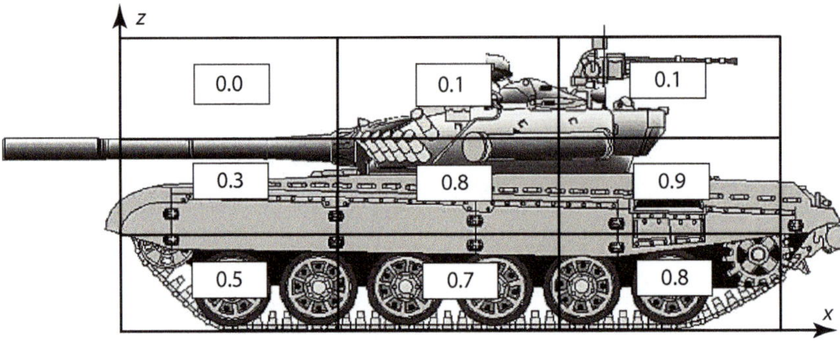

Fig. 17.12 Simplified C × C file for a sample target.

- Target C × C data (as shown in Fig. 17.12)
- Aiming accuracy and precision error data for the weapon system used
- Engagement scenario such as range, rounds/burst, etc.

Table 17.2 shows a sample data set for this simplified version of PVTM; it consists of a single three-round burst fired at this target at various ranges.

The results of the program for the data given in Table 17.2 are shown in Fig. 17.13.

Again the results are as expected, with the P_K decreasing quickly with range to the target.

Table 17.2 Inputs to MATLAB Effectiveness Program

n_run	=	2000;	% number of Monte Carlo trials
R	=	250;	% range
sigma_1x	=	5;	% between burst sigma *x*-axis, mils
sigma_1z	=	5;	% between burst sigma *y*-axis, mils
sigma_2x	=	3;	% within burst sigma *x*-axis, mils
sigma_2z	=	3;	% within burst sigma *y*-axis, mils
b_errorv	=	0;	% vertical boresight error, mils
b_errorh	=	0;	% horizontal boresight error, mils
n_bursts	=	1;	% number of bursts/aimpoint
nrb	=	3;	% number of rounds/burst
% .... Target cell by cell data			
tgt_data = [0.0 0.1 0.1;			
0.3 0.8 0.9;			
0.5 0.7 0.8];			
cell_x = 3.0;			
cell_z = 1.5;			

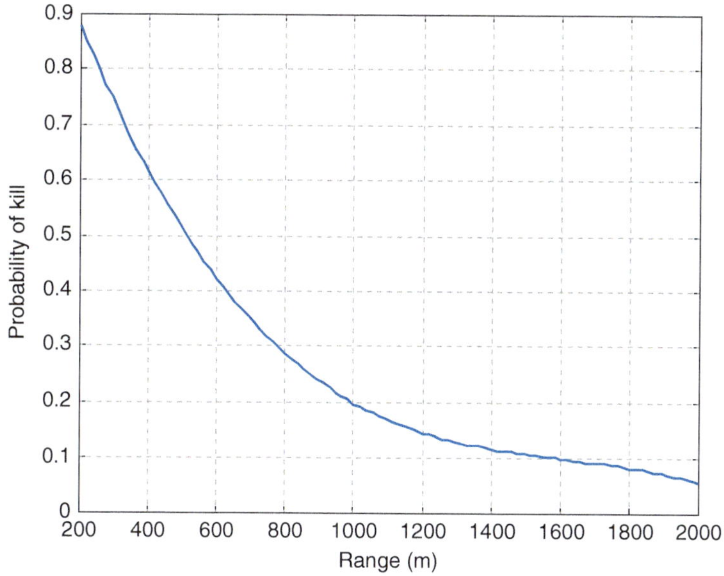

Fig. 17.13 Output as a function of range.

17.4 Naval Direct Fire Systems: JGEM

In recent years, considerable concern has grown due to the threat posed to naval surface ships by multiple, small, high speed boats. At close range, the most suitable weapons to engage such threats are rapid fire, small to medium caliber deck-mounted gun systems, as shown in Fig. 17.14, which typically range from 25-mm (1-in.) to 76-mm (3-in.) caliber. The material that follows is based on [1].

In order to assess the effectiveness of such guns against highly maneuverable small boats, the performance of the total weapon system has to be considered, by which we mean determining the effect on performance of the following factors:

- Manually or automatically (servo) trained
- Visual identification of the aimpoint or sensor driven (radar/thermal)
- Engagement range (100 yd to 10 n miles)
- Fire control system capability to provide "lead"
- Stabilized or unstabilized for ship motion

In order to perform such effectiveness assessments, a program known as the Joint Gun Engagement Model (JGEM) has been developed and accredited by the JTCG/ME, and will be described in the following sections.

JGEM is a Monte Carlo simulation of the engagement between a surface ship–mounted gun weapon system (GWS) and one or more surface targets. It may be described as an "end-to-end" model in that it includes the effects

Fig. 17.14 Small and medium caliber naval gun systems.

of target and shooter motions, accuracy of aiming the gun, munition trajectory, target vulnerability, and effectiveness of the warhead against the target in question. Some of its distinguishing features include the capability to study moving targets and a generic way to model target-tracking sensors such as radar.

JGEM is divided into two main software components: the Direct Fire Effectiveness (DFE) module and the graphical user interface (GUI). User inputs for the given scenario and engagement and range table information are accepted through the GUI into the DFE, where a Monte Carlo simulation predicts the effectiveness of a given small or medium caliber weapon in a given engagement scenario. Scenario results are output to the GUI.

17.4.1 Coordinate System

The absolute reference frame is Cartesian with the direction of motion of the ship, and hence the GWS in the y direction. The $x-y$ plane represents the flat earth, and the z direction represents altitude measured positive up. Two angular measures for gun pointing and projectiles are also appropriate: bearing B and elevation E. The former represents an angular rotation in the $x-y$ plane in the clockwise direction from the y-axis. The latter represents an angular rotation from the $x-y$ plane in clockwise direction towards the z-axis. A relative frame, which is also Cartesian, is used aboard the ship with the origin at the GWS location. Figure 17.15 shows the absolute reference frames of the model.

17.4.2 Gunship and Target State Estimation

The true gunship and target states (specifically, position and velocity) are input into the engagement data separately as vectors that represent the x, y, and z positions of each vessel throughout the engagement. Target tracking is achieved by modeling the estimates of both the gunship and target states and applying an alpha-beta (simplified Kalman) filter to determine their future positions.

A second-order ($n = 2$) Lagrangian interpolation formula is used to determine the true states of the target for any time t. Note that the exact same formulae given in the following equations are applied to determine the gunship states (simply interchange the subscript t with a subscript g). The input values for the gunship and target states are given by Eqs. (17.17) and (17.18).

$$X_g(t_i) = \{\, t_i \quad x_i \quad y_i \quad z_i \,\}_g\big|_{i=1,2,\ldots} \tag{17.17}$$

$$X_t(t_j) = \{\, t_j \quad x_j \quad y_j \quad z_j \,\}_t\big|_{j=1,2,\ldots} \tag{17.18}$$

The summation of $n = 2$ Lagrange polynomials yields the gunship or target position for time t. Equation (17.19) shows the formula for Lagrange polynomials, and Eqs. (17.20) and (17.21) show the determination of the target position and velocity, respectively.

$$L_k = X_k(t_k) \prod_{\substack{i=0 \\ i^1 k}}^{n} \frac{t - t_i}{t_k - t_i} \tag{17.19}$$

$$X_t(t) = \sum\nolimits_{k=0}^{2} L_k(t) \tag{17.20}$$

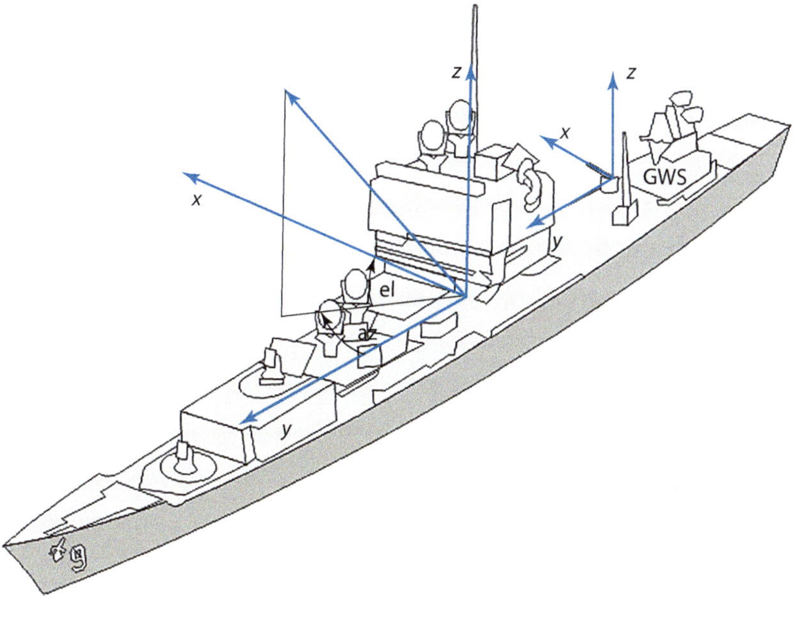

Fig. 17.15 JGEM coordinate frames.

$$\dot{X}_t(t) = \frac{\mathrm{d}}{\mathrm{d}t}\sum_{k=0}^{2} L_k(t) = \sum_{k=0}^{2}\frac{\mathrm{d}L_k(t)}{\mathrm{d}t} \tag{17.21}$$

This process yields the target and gunship positions and velocities for the engagement.

17.4.3 Target Tracking

During the fire-control (FC) cycle, the target position relative to the gunship is determined for three quantities: range, bearing and elevation. These are computed by using the input states, geometry, and trigonometry from Eqs. (17.22) through (17.24) to give the required state. Note that noise in these measurements has been added to represent random errors of the tracker.

$$R(t) = \sqrt{[x_t(t) - x_g(t)]^2 + [y_t(t) - y_g(t)]^2 + [z_t(t) - z_g(t)]^2} + \eta_R \tag{17.22}$$

$$B(t) = \tan^{-1}\left(\frac{x_t(t) - x_g(t)}{y_t(t) - y_g(t)}\right) + \eta_B \tag{17.23}$$

$$E(t) = \sin^{-1}\left(\frac{z_t(t) - z_g(t)}{R(t)}\right) + \eta_E \tag{17.24}$$

Once these quantities are determined, the measured target position relative to the gunship is transformed into Cartesian coordinates.

$$X_m(t) = \begin{bmatrix} x_m(t) \\ y_m(t) \\ z_m(t) \end{bmatrix} = \begin{bmatrix} R(t)\cos E(t)\sin B(t) \\ R(t)\cos E(t)\cos B(t) \\ R(t)\sin E(t) \end{bmatrix} \tag{17.25}$$

17.4.4 Alpha-Beta Filter for Target Tracking

The predictor-corrector alpha-beta filter is a fixed-gain Kalman filter used for target tracking. It does not take into account accelerations, and so is a simpler formulation than a conventional alpha-beta-gamma filter. There are two main parts of this filter: correction (or smoothing) and prediction. The former is accomplished by the following:

$$\begin{aligned} \hat{X}_h(t) &= \hat{X}_p(t) + \alpha_{\mathrm{DF}}(X_m(t) - \hat{X}_p(t)) \\ \hat{\dot{X}}_h(t) &= \dot{X}_p(t) + \frac{\beta_{\mathrm{DF}}}{\Delta t}(X_m(t) - \hat{X}_p(t)) \end{aligned} \tag{17.26}$$

The parameters α_{DF} and β_{DF} are functions of the gain index k, which is defined as the quotient of the FC track settling time and the FC cycle time Δt. Equations (17.27) and (17.28) show the determination of α_{DF} and β_{DF}, respectively.

$$\alpha_{DF} = \frac{2(2k-1)}{k(k+1)} \tag{17.27}$$

$$\beta_{DF} = \frac{6}{k(k-1)} \tag{17.28}$$

Prediction occurs by simply updating the estimate of the target position and velocity for the next time step, as shown in the following equation:

$$\begin{aligned} X_f(t+\Delta t) &= X_h(t) + \Delta t \hat{\dot{X}}_h(t) \\ \dot{X}_f(t+\Delta t) &= \hat{\dot{X}}_h(t) \end{aligned} \tag{17.29}$$

As a simplified example, take the case of a target moving with x and y velocities of 10 m/s from the origin to an x position of 150 m and a y position of 650 m. The time of simulation is $t = 15$ s. The implementation of the alpha-beta filter is accomplished for the single parameter of range from the gunship (origin) and velocity along the range direction by root sum squaring the position and velocity components separately and then implementing the filter.

Let the initial estimate of the target position and velocity be 450 m and 0 m/s, respectively. For this case, the gain index is $k = 20$, so $\alpha_{DF} = 0.1857$ and $\beta_{DF} = 0.01579$. The time step used is $\Delta t = 0.1$ s. Figure 17.16 shows the results of the alpha-beta filter in predicting the target position.

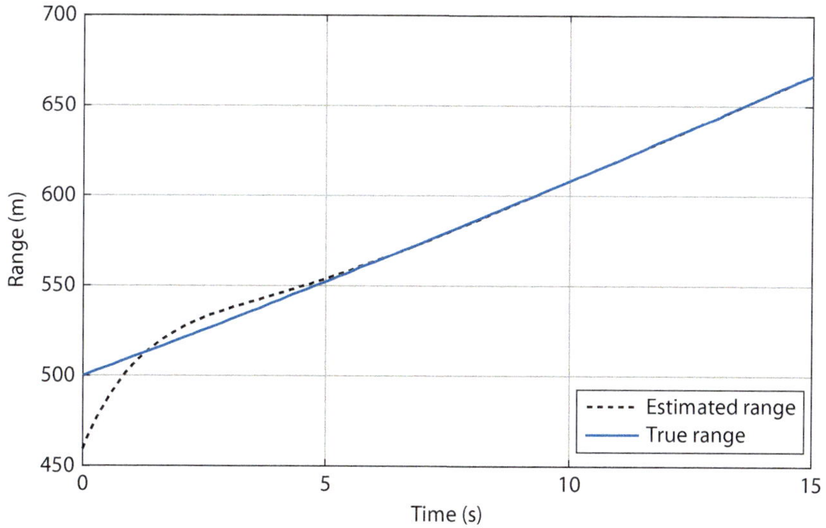

Fig. 17.16 Range estimate for alpha-beta filter.

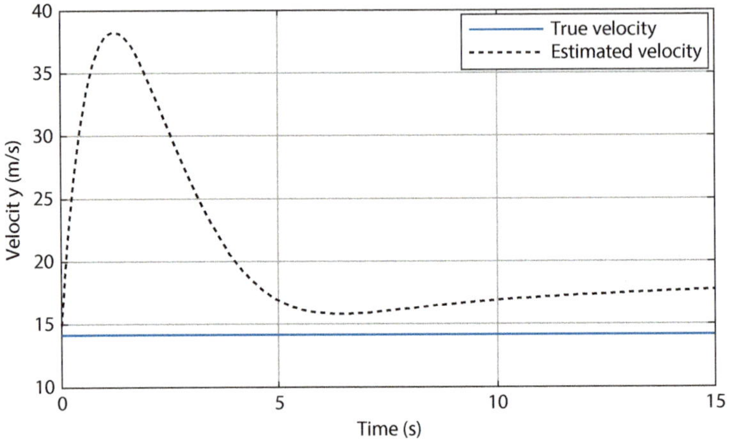

Fig. 17.17 Velocity estimate for alpha-beta filter.

Figure 17.17 shows the alpha-beta filter performance with respect to velocity prediction. It can be easily seen that the velocity prediction is not as good as range prediction. This is due in part to the initial estimates of position and velocity chosen for this example.

17.4.5 Determining the Bullet–Target Intercept Point

Iteration is required to determine the bullet–target intercept point. An initial estimate of the ballistic solution is drawn from the range table for a given time of flight t_2 based on the estimated range to the target from the gunship. The velocity of the intercept point, including an error representing the noise in the measurement of ship speed η_{SPEED} with respect to the global coordinate frame, is given by Eq. (17.30).

$$\dot{X}_a(t+t_2) = \hat{\dot{X}}_h(t) + \left(\dot{X}_g(t) + \eta_{\mathrm{speed}}\right) \tag{17.30}$$

The determination of the position of the intercept point is achieved by summing the estimate of the target position with the product of the estimate of the intercept position with the sum of the time of flight t_2 and the time between the last FC cycle update and the firing of the GWS δt, represented in Eq. (17.31).

$$X_a(t+t_2) = \hat{X}_h(t) + \dot{X}_a(t)(t_2 + \delta t) \tag{17.31}$$

Equations (17.22) through (17.24) are used to determine the range, bearing, and elevation of the intercept position relative to the gunship at time $t + t_2$. A bilinear interpolation scheme is employed with the range table to extract the slant range $R_p(t_2)$, terminal velocity $v_p(t_2)$, and angle of fall AOF(t_2). Using these values, the computed bullet position $X_p(t_2)$ is computed as in Eq. (17.25). The computed bullet velocity equation is as follows:

$$\dot{X}_p(t+t_2) = \begin{bmatrix} x_p(t+t_2) \\ y_p(t+t_2) \\ z_p(t+t_2) \end{bmatrix} = \begin{bmatrix} v_p(t_2)\cos\left[E_a(t+t_2)-AoF(t_2)\right]\sin B_a(t+t_2) \\ v_p(t_2)\cos\left[E_a(t+t_2)-AoF(t_2)\right]\cos B_a(t+t_2) \\ v_p(t_2)\sin\left[E_a(t+t_2)-AoF(t_2)\right] \end{bmatrix} \tag{17.32}$$

In order to compute the intercept position of the bullet and target, the time increment δ, where the target and bullet have the same slant range (but not necessarily the same position), relative to the gunship is found by setting the absolute value of the difference between the interpolated slant range $R_p(t+t_2+\delta)$ and the computed slant range $R_a(t+t_2+\delta)$ equal to zero. Equations (17.33) and (17.34) define these quantities, while Eq. (17.35) shows the difference in slant range.

$$R_p(t+t_2+\delta) = \left|X_p(t+t_2)+\delta\dot{X}_p(t+t_2)\right| \tag{17.33}$$

$$R_a(t+t_2+\delta) = \left|X_a(t+t_2)+\delta\dot{X}_a(t+t_2)\right| \tag{17.34}$$

$$\left|R_p(t+t_2+\delta)-R_a(t+t_2+\delta)\right| = 0 \tag{17.35}$$

Solving for δ yields a quadratic with complex coefficients, so the positive root of the quadratic formula yields $t_{2-\min}$. Updates to the time of flight are given as $t_2 = t_2 + \delta$. Iteration ceases when the closest point of approach (CPA) of the bullet and target is less than a predetermined value. Let two error quantities be defined as in the following equations:

$$\varepsilon(t_2) \equiv X_p(t+t_2) - X_a(t+t_2) \tag{17.36}$$

$$\dot{\varepsilon}(t_2) \equiv \dot{X}_p(t+t_2) - \dot{X}_a(t+t_2) \tag{17.37}$$

The squared directed distance between the bullet and target at some interval Δ from the estimated time of flight is given in Eq. (17.38).

$$D^2(t_2+\Delta) = \left|\left[\varepsilon(t_2)+\Delta\dot{\varepsilon}(t_2)\right]\cdot\left[\varepsilon(t_2)+\Delta\dot{\varepsilon}(t_2)\right]\right| \tag{17.38}$$

The derivative of this expression with respect to the time interval Δ is set to zero.

$$\frac{\mathrm{d}(D^2(t_2+\Delta))}{\mathrm{d}\Delta} = 0 \tag{17.39}$$

Now solving for Δ yields

$$\Delta = \frac{-\varepsilon(t_2)\times\dot{\varepsilon}(t_2)}{|\dot{\varepsilon}(t_2)\times\dot{\varepsilon}(t_2)|} \tag{17.40}$$

Substituting this into Eq. (17.38) yields the CPA squared.

$$D^2(t_2+\Delta) = \varepsilon(t_2)\cdot\varepsilon(t_2) - \frac{[\varepsilon(t_2)\cdot\dot{\varepsilon}(t_2)]^2}{|\dot{\varepsilon}(t_2)\cdot\dot{\varepsilon}(t_2)|} \tag{17.41}$$

To prevent an infinite loop, a maximum of 10 iterations within this loop are allowed. Upon convergence of the FC solution, the true bullet position is determined through scaling the perturbation effects in the range table data by error budget values. First gunship- and wind-speed components along and across the line of fire must be calculated, as in the following equations:

$$\mu_{\text{tailwind}} = \frac{X_a(t+t_2)\mu_{\text{windspeed}}}{R_a(t+t_2)} \tag{17.42}$$

$$\mu_{\text{crosswind}} = \frac{X_a(t+t_2)\mu_{\text{windspeed}}}{R_a(t+t_2)} \tag{17.43}$$

$$\mu_{\text{alongshipspeed}} = \frac{X_a(t+t_2)\eta_{\text{shipspeed}}}{R_a(t+t_2)} \tag{17.44}$$

$$\mu_{\text{crossshipspeed}} = \frac{X_a(t+t_2)\eta_{\text{shipspeed}}}{R_a(t+t_2)} \tag{17.45}$$

The total bullet position offsets—relative to the aim point in height, deflection, and horizontal range (respectively)—are then calculated as in Eqs. (17.46) through (17.48).

$$\varepsilon_h = \sum \eta_i \frac{\partial \text{height}}{\partial \eta_i} S_i + \sum \mu_j \frac{\partial \text{height}}{\partial \mu_j} S_j \tag{17.46}$$

$$\varepsilon_d = \sum \eta_i \frac{\partial \text{deflection}}{\partial \eta_i} S_i + \sum \mu_j \frac{\partial \text{deflection}}{\partial \mu_j} S_j + \mu_{\text{gunazimuth}} R_a(t+t_2) \tag{17.47}$$

$$\varepsilon_r = \sum \eta_i \frac{\partial \text{horizontalrange}}{\partial \eta_i} S_i + \sum \mu_j \frac{\partial \text{horizontalrange}}{\partial \mu_j} S_j \tag{17.48}$$

The ballistic partial derivatives in these equations correspond to the parameters listed in Table 17.3.

The final bullet position is then determined by performing a basic coordinate transformation as in Eq. (17.49).

$$X_p(t+t_2) = X_a(t+t_2) + \begin{bmatrix} \cos B_a(t+t_2) & \sin B_a(t+t_2) & 0 \\ -\sin B_a(t+t_2) & \cos B_a(t+t_2) & 0 \\ 0 & 0 & 1 \end{bmatrix} \cdot \begin{bmatrix} \varepsilon_d \\ \varepsilon_r \\ \varepsilon_h \end{bmatrix} \tag{17.49}$$

The actual target position is found at time $t + t_2$, and the square of the CPA is recomputed by Eq. (17.41), where the parameter $\varepsilon'(t_2)$ is defined in Eq. (17.50).

$$\varepsilon'(t_2) = X_p(t+t_2) - X_t(t+t_2) + X_g(t) \tag{17.50}$$

Table 17.3 Ballistic Perturbations

Perturbation	Random Event for Each	Direction of Effect	Range Table Scale Factor
Precision error	Round	Range/deflection/height	0.6745 probable error scale
Gun laying	Round	Range/deflection/height	0.1 deg
Muzzle velocity	Round	Range/height	+10 m/s
Air pressure	Replication	Range/height	+100 mbar
Air temperature	Replication	Range/height	−10°C
Air density	Replication	Range/height	−1% standard
Wind speed	Replication	Range/deflection/height	10 m/s
Gunship speed	Round	Range/deflection	1 m/s

17.4.6 Determination of Target Kill

After the ballistic solution is computed (i.e., the bullet has a CPA that is within a specified value of the target position), it must be determined whether the bullet hit the target, and if so, whether the vulnerable area within the target was also hit resulting in a target kill. Unlike materiel targets discussed earlier, this model does not have an elevation-dependent set of RCAS files. Instead it has five vulnerable areas for the target's four sides and top, from which an equivalent A_V may be computed for a specified bullet approach direction. Figure 17.18 shows the resolution of the target areas into three orthogonal directions.

The top, frontal, and side areas of the target and its vulnerable areas are projected onto the target's direction of motion at the intercept time by the following equation:

$$A(t+t_2) = \begin{bmatrix} \hat{v}_i(t+t_2)A_{\text{side}} \\ \hat{v}_j(t+t_2)A_{\text{front}} \\ \hat{v}_k(t+t_2)A_{\text{top}} \end{bmatrix} \tag{17.51}$$

The unit vectors from Eq. (17.51) are computed by first computing the unit vector in the direction of target motion $(\hat{v}_j)$; then the unit vector parallel to the horizontal plane and perpendicular to $\hat{v}_j$ $(\hat{v}_i)$ is computed; and finally the vertical unit vector $(\hat{v}_k)$ is computed, as shown in the following equations:

$$\hat{v}_j(t+t_2) = \frac{1}{|\dot{X}_t(t+t_2)|}\begin{bmatrix} \dot{x}(t+t_2) \\ \dot{y}(t+t_2) \\ \dot{z}(t+t_2) \end{bmatrix} \tag{17.52}$$

$$\hat{v}_i(t+t_2) = \hat{v}_j(t+t_2) \wedge \hat{k} = \hat{v}_j(t+t_2) \wedge [0 \quad 0 \quad 1]^T \tag{17.53}$$

$$\hat{v}_k = \hat{v}_i \wedge \hat{v}_j \tag{17.54}$$

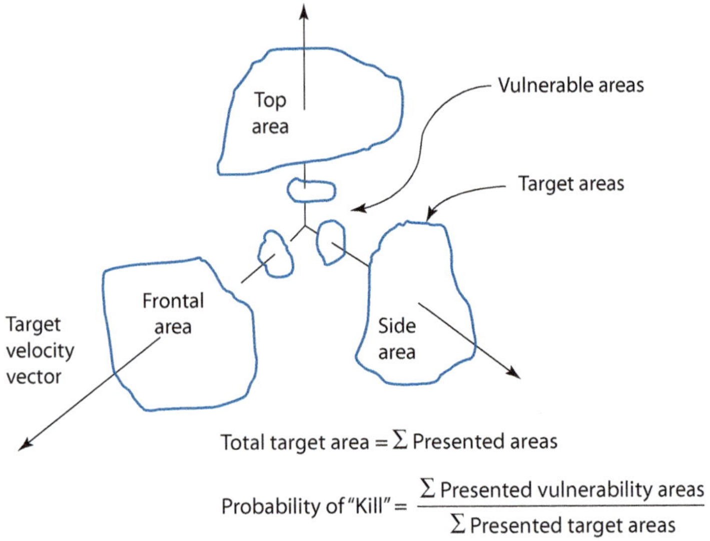

Fig. 17.18 Representation of presented and vulnerable areas.

The projection of the target area onto the normal plane $A_n(t+t_2)$ to the bullet–target velocity vector is given by

$$A_n(t+t_2) = \frac{\left|\dot{X}_p(t+t_2) \times A(t+t_2)\right|}{\left|\dot{X}_p(t+t_2)\right|} \tag{17.55}$$

Let $A_{nT}(t+t_2) = A_n(t+t_2)$ for the target area, and let $A_{nV}(t+t_2) = A_n(t+t_2)$ for the vulnerable area of the target. The radii of the presented areas of the target and its vulnerable components are computed as follows:

$$r_T(t+t_2) = \sqrt{\frac{A_{n_T}(t+t_2)}{\pi}} \tag{17.56}$$

$$r_V(t+t_2) = \sqrt{\frac{A_{n_V}(t+t_2)}{\pi}} \tag{17.57}$$

These radii are compared to the CPA of the bullet to determine if the target is hit. If so, a random draw from a uniform distribution is compared to the ratio A_V/A_T to determine if the vulnerable area is also hit; if it is, a target kill is recorded.

17.4.7 Program Output

The program accumulates the engagement outcome for each round fired, each salvo fired, and each target engaged. The number of rounds fired to each intercept range, in each salvo, and at each target is accumulated on a shot-by-shot basis. If a target hit was declared for the current shot, counters for hits to each intercept range, hits in each salvo, and hits on each target are incremented accordingly.

If a target kill was declared during a salvo, the corresponding counters are also incremented. Additionally, statistics for kill ranges and kill times are computed. Data corresponding to target kill can occur only once during a salvo. In other words, if all of the rounds of a salvo result in target kills, only the data for the last bullet of the salvo are entered into the statistics for target kills.

Average values are computed over the number of Monte Carlo replications in the simulation. For example, the average number of hits for the nth salvo is the total number of hits achieved by all nth salvos divided by the number of Monte Carlo replications. Note that some Monte Carlo replications may not have fired an nth salvo. Kill statistics are given by percentage of occurrence; that is, the nth target was killed x% of the time, or a target kill was realized between the indicated lower and upper ranges y% of the time.

At the end of each salvo, the fire control algorithms schedule the time for the next firing. The next firing commences after the kill assessment is completed if the target survived; otherwise, the next firing is delayed to account for the time needed to acquire and track the next target in the attack. Aiming corrections are applied only if the target survived. Targets are engaged sequentially; that is, the first target of the raid must be defeated before the gun system can fire upon the second target. The engagement will terminate if all the targets are defeated.

The program computes mean and standard deviations of intercept ranges and engagement times for measures of system performance. When the magnitude of data is relatively small, most simple averages and standard deviations can be computed by

$$\hat{x}_n = \frac{1}{n}\sum_{i=1}^{n} x_n \quad \text{and} \quad \sigma_n = \sqrt{\frac{1}{n-1}\sum_{i=1}^{n}(x_i - \hat{x}_n)^2} \tag{17.58}$$

However, significant roundoff and floating-point overflow errors can occur when there is a large amount of data of large magnitudes. Such is the case for large Monte Carlo trials and typical intercept ranges of gun systems. A much better way to calculate the means and standard deviations with floating point arithmetic is to use the following recurrence formula:

$$\begin{aligned} M_k &= M_{k-1} + \frac{x_k - M_{k-1}}{k} \\ S_k &= S_{k-1} + (x_k - M_{k-1}) \cdot (x_k - M_k) \end{aligned} \tag{17.59}$$

for $2 \leq k \leq n$ with $M_1 = x_1$ and $S_1 = 0$. M_K is the mean of the sequence x_i, $i = 1, 2 \ldots n$, and the standard deviation is given by

$$\sigma = \sqrt{S_n/(n-1)} \tag{17.60}$$

The final output of the Monte Carlo experiment is the expected number of kills; for example, 50% of the time the system will kill more targets than indicated, and 50% of the time it will kill fewer targets.

17.4.8 Airbursting Munitions

In addition to direct hit projectiles, JGEM is also able to determine the effect of fuzed blast/fragment warheads by using lethal area matrices (LAMs) that may be generated by either AJEM or GFSM [2]. In addition, three different fuze types may be selected, as described in Chapter 3, Section 3.15.

1. Impact or point detonating (PD)
2. Proximity (PX)
3. Time delay

Each fuze type allows for users to specify variability in its respective functional parameter (e.g., time, height of burst, PX distance, etc.). For each munition/target pairing, a set of LAM are precomputed as a function of range and height of burst (HOB). A program in JGEM called the Matrix Packaging Tool (MPT) stores and manages these LAM for use during the weapon fly-out phase of JGEM described earlier.

For each Monte Carlo iteration of the projectile flight, the appropriate fuze model, including its variability of function, is used to determine the detonation point in terms of range and altitude relative to the GWS. The MPT looks up the appropriate bracketing LAMs in range and altitude and performs the necessary interpolation to produce a single matrix for the detonation point. The P_K that is output for this iteration is determined by the location of the target center of vulnerability (COV) in this matrix.

17.4.9 Multiple-Round Effectiveness

The methodology described earlier uses the assumption that each round fired in a burst is independent of the other rounds, and although the powering up formula is not used explicitly, independence of rounds is implied. JGEM has a way to handle multiple shots in which damage may be accumulated from each contributing round by using a feature of AJEM not previously described.

Recall in Chapter 7, Section 7.12 we described AJEM as an end-to-end model, by which we mean it can calculate vulnerability (like COVART) and effectiveness (like GFSM). It is, however, JTCG/ME-approved only for

Fig. 17.19 Multiple-round shotline analysis.

the former, and as such AJEM is typically used to generate vulnerable area tables. JGEM used AJEM's effectiveness capability to handle multiple shotline analyses, thereby allowing for the destruction of multiple redundant components in a single burst. This can lead to a functional kill that would never occur with single-shot evaluations. This is illustrated in Fig. 17.19, where six rounds are fired at the target.

Given these randomly located impact points, and hence shotlines, an average P_K for this burst of six shots can be calculated using a Monte Carlo process that now includes killing more than one critical, redundant component. This results in a P_K vs number of hits graph of the type shown in Fig. 17.20, which is azimuth and elevation dependent. Note that in this simulation, it is assumed the round hits the target; whether this does, in fact, occur is handled in the JGEM fly-out method as described previously.

Again, a set of such curves has to be pregenerated to allow for variations in range and target aspect, and as such they become similar to the cell-by-cell

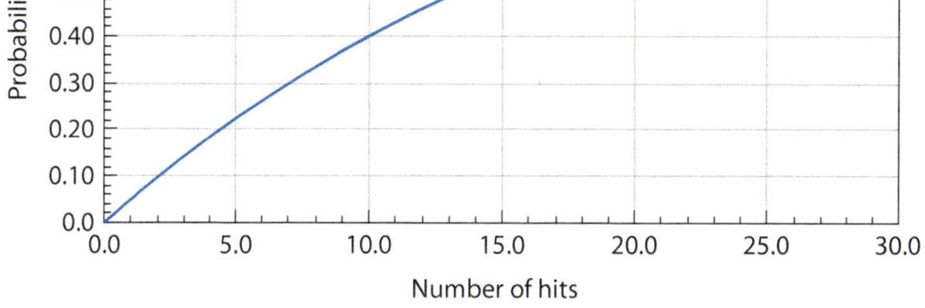

Fig. 17.20 Multiple hit P_K curve.

files used in the PVTM program described in Section 17.3 and shown in Fig. 17.10. At each target range, curves are generated for the front, port, starboard, and top of the target, and these will be used in a manner similar to the vulnerable areas shown in Fig. 17.18.

Because significant target motion is an important feature of JGEM, the target aspect and range may change appreciably while the target is maneuvering; this is tracked in the program to ensure the correct aspect for each round approaching the target is determined. This effect is significant for slower rate of fire GWS, which are usually larger caliber.

For single-shot hit-to-kill projectiles, JGEM determines for each hit if the target is killed by comparing the P_K to a random draw from a uniform distribution. A different approach is adopted for the multiple hit methodology, based on the observation that as the target maneuvers and receives more hits, the P_K will increase with time. This is used by taking a single random draw from a uniform distribution at the start of the engagement; when the P_K exceeds this value, the target is declared killed.

References

[1] McConkie, J., Hlywa, E., and Merrit, P., "Joint Gun Effectiveness Model Version 1.1, User and Analyst Manual for JWES 3.0," Naval Surface Warfare Center, Dahlgren Div., Dahlgren, VA, 30 July 2003.

[2] Rowles, S., "Joint Gun Effectiveness Model with Airbursting Methodology," Naval Surface Warfare Center, Dahlgren Div., Systems Concepts and Analysis Branch, Dahlgren, VA, April 2006.

Chapter 18 Computer Implementation of Weaponeering Methods

18.1 Introduction

This chapter begins by reviewing the historical development of various JMEM weaponeering tools. Following that, we develop an integrated computer tool capable of calculating damage for a wide set of targets in conjunction with a database of notional data using the effectiveness indices discussed previously. Because this will be based on methodologies developed in this book, it will allow users to run basic weaponeering examples to demonstrate what they have learned so far.

Keep in mind that the primary focus of these tools has been to support the warfighter; for example, pilots about to fly a strike mission may want to verify the probability of a successful outcome. Any automated process to provide this information should be relatively simple, fast running, and require a minimum of information to derive a result. Operational users often quote the requirement, "It is better to get the 70% answer in two minutes than the 99% answer in two hours."

There is, of course, another community that benefits from the work done in providing the warfighter with weaponeering tools, namely the analysts who are mainly responsible for their development in the first place. Analysts tend to operate in a more benign environment compared to the warfighter, typically have more time to run computer programs on up-to-date hardware, and are not usually subjected to life-threatening consequences if they obtain the wrong answer. It is fair to say, however, that the analytical community is well aware of who they are working for and strive to get it right the first time for their customers.

The platforms on which the various tools run have kept pace with computational advances over the last 50 years of development, ranging from manual computations through programmable calculators to the capable personal computers of today. Again, given the emphasis on warfighter utilization, porting onto alternative, perhaps less capable though more convenient devices such as tablets and cellular telephones has also received attention.

18.2 Historical Development of Warfighter Tools

Given the initial establishment of the JTCG/ME based on consolidating methods to provide effectiveness estimates for air-delivered weapons, the first step was to review the methodologies used by the various services to compute PD_1 and to agree on a single methodology. This resulted in several methods to compute weapon effects for a variety of air-launched weapon systems—unguided and guided bombs against unitary targets, against areas of target elements, sticks, cluster munitions, and projectiles. The basic tools used in the original JMEM methodologies were:

- COVART, which took target data (TGM and FALT) and generated vulnerable areas
- GFSM, which combined COVART data with Z-data to produce lethal area matrices
- Matrix Evaluator (ME), which incorporated delivery accuracy data to compute a P_K

18.2.1 Developments 1967–1980

The methods listed in the previous section were initially implemented on mainframe computers of 1960s vintage running the Matrix Evaluator program and attempted to provide estimates for all inventoried aircraft, weapons, and release conditions.

As might be imagined, this produced considerable amounts of data, which were recorded in orange-covered manuals—the Joint Munitions Effectiveness Manuals, the original JMEMs—and are termed the precalculated methods. They focused on air-launched weapons, and much of the original work involved keeping the data current. For example, if a new weapon was inventoried and its warhead characterized, it had to be evaluated against all targets and delivery conditions available, and the resulting data updated in the manuals.

Even though these manuals formed a library of considerable size, there was never the granularity the users desired, and the need arose for a method the individual warfighter or analyst could implement for the specific case of interest. This approach was termed the *open-ended method*, where the emphasis changed to ease of use and speed in getting an answer.

Clearly the legacy programs (COVART, GFSM, and ME) could not be used easily in such an environment, so approximations were made in order to generate a process that could be completed by hand, using graphical and tabular data for more complex calculations. It was acknowledged that the accuracy of the results could not equal those of the precomputed data, but the user was willing to sacrifice fidelity for speed to get a result that was "good enough." The analytical methods described in Chapters 10–15 form the basis of the methods developed.

The fundamental changes to the precalculated methodology were to retain COVART and GFSM to calculate a lethal area matrix, but instead of using ME, the matrix was reduced to an effectiveness index, the IE, which could be represented by either a Carlton damage function or a rectangular cookie cutter. This allowed the analytical methods described in Chapters 11–16 to determine effects for a single bomb or stick to be evaluated against unitary or areas of targets, as well as simple projectiles and cluster munitions. The methods were later extended to cover specific target types such as bridges, buildings, and linear targets as well as the effective miss distance EI.

Regarding the implementation platform, the first attempt at a user tool was called the *hand methods*. This approach used a multistep, user-completed form for each method (e.g., unguided single weapons against a unitary target, single weapons against an area of targets, sticks, cluster munitions, projectiles, etc.). The form would be completed by hand, step by step, with more difficult calculations being obtained from graphs or nomograms. An example of a nomogram to determine intervalometer settings for sticks is shown in Fig. 18.1, and a section of the hand method for clusters is shown in Fig. 18.2.

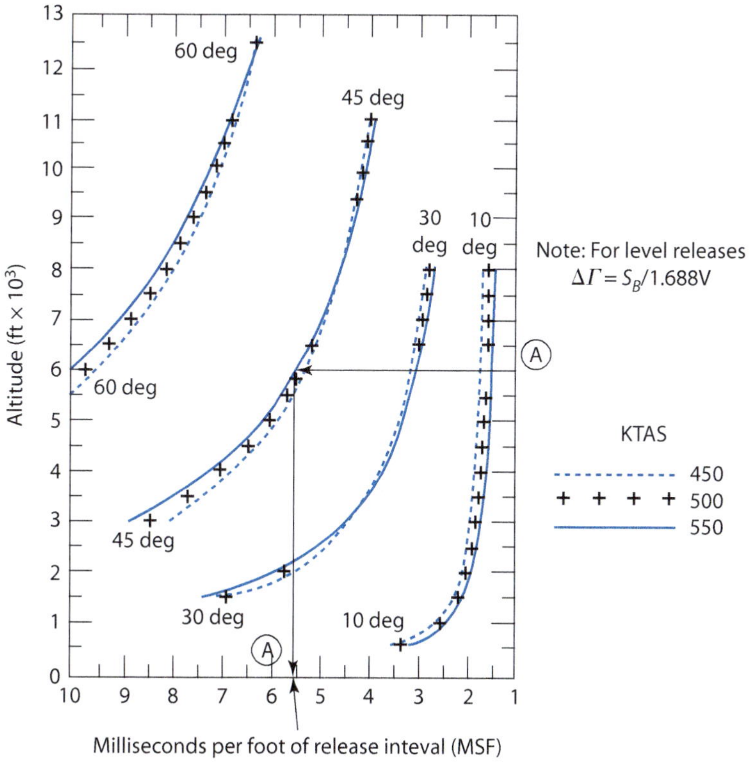

Fig. 18.1 Nomogram for stick spacing.

Both Fuze Types

11.	Functioning Angle, degrees	θ_f	—	—	—	Figure A-III-8 (Sheets 7 thru 12). Use CBU-52/B and CBU-58/B system graphs for low and high values.
12.	Functioning Velocity, ft/sec	V_f	—	—	—	
13.	Ground Range to Functioning, ft	GR_1	—	—	—	

III. SUBMUNITION BALLISTICS

14.	Submunition Velocity at Dispenser Functioning, ft/sec	V_D	—	—	—	Actual velocity equals dispenser functioning velocity (Step 12). Low and high values depend on graphs available for the submunition.
15.	Ground Range, ft	GR_2	—	—	—	For CBU-52/58: Interpolate on submunition velocity (Step 14). Figure A-III-8 (Sheets 13 thru 18). For Rockeye and APAM: Interpolate on airspeed (Step 4). Figure A-III-9 (Sheets 4 thru 9). For APAM: R_p = (1.15)(Rockeye R_p).
16.	Impact Angle, degrees	I	33	—	—	
17.	Time to Impact, sec	TF_B	—	—	—	
18.	Outer Pattern Radius, ft	R_p	80	—	—	
19.	Inner Pattern Radius, ft	R_{pi}	0	X	X	For CBU-52: $R_{pi} = R_p - 125$. For CBU-58: $R_{pi} = R_p - 175$. For Rockeye and APAM: $R_{pi} = 0$.

NOTE: For 95 percent of the BLU-61 or BLU-63 submunitions to function, TF_B must be greater than the value obtained from Figure A-III-8 (Sheet 19).

IV. TOTAL FLIGHT CONDITIONS

20.	Time of Flight, sec	T_F	6.0	—	—	For CBU-52/58: $T_F = T_D + TF_B$. For Rockeye and APAM: Figure A-III-9 (Sheets 4 thru 9).
21.	Ground Range, ft	GR	2,750	—	—	For CBU-52/58: $GR = GR_1 + GR_2$. For Rockeye and APAM: Figure A-III-9 (Sheets 4 thru 9).
22.	Slant Range, ft	SR	2,903	X	X	$SR = \sqrt{(GR)^2 + (Y_1)^2}$

UNCLASSIFIED

Fig. 18.2 Extract from original hand method.

Fig. 18.3 Magnetic card, programmable calculator.

The reader may conclude that some of the spreadsheet solutions presented in earlier chapters bear a close resemblance to the original hand methods, except of course the Excel program is able to perform all of the computations that originally had to be done by hand.

The first step to automate the process began in the 1970s when programmable, handheld calculators were used. The Texas Instruments TI-59 calculator shown in Fig. 18.3 used a programmable magnetic strip that could read programs into the device, and then the calculator could be used as an input/output device.

Because it had only 960 programming steps and 100 memory registers, simplifying the methodology from Matrix Evaluator to something that would work on this device was clearly a challenge, and this is where the methods described in Chapters 11–16 came into their own. Using deterministic methods, they could be accommodated within the programming limitations and produced results (the 70% solution) in seconds directly on a handheld device.

18.2.2 Developments 1980–1995

In the early 1980s, the widespread use of personal computers with a stable operating system and a common programming environment (FORTRAN, C) led to the first PC implementation of the methodologies. The IBM Personal Computer with the Disk Operating System (DOS) was chosen for its widespread acceptability within the Department of Defense (DoD), particularly within the operational community. With the PC's storage capability and execution speed, the methods could be implemented more accurately with fewer simplifying restrictions in order to obtain a better answer.

The original implementation in the early 1980s on a PC platform was called PC OEM and ran under the DOS operating system; however, when Microsoft introduced the Windows operating system in 1985, PC OEM was transitioned to WinJMEM around 1990. WinJMEM was incrementally refined and improved, resulting in the JMEM Air-to-surface Weaponeering System, JAWS 1.0, which was released in 1995. This was the first fully integrated weaponeering tool available to the operational user containing both methodologies and a classified database. For the first time, modified versions of JAWS 1.0 were released to coalition partners such as the Republic of Korea (ROK) and NATO.

Although the majority of the WinJMEM methods were deterministic, improving computational speed allowed some simple Monte Carlo methods to be incorporated into the suite of programs. An example was the Target Complex (TARCOM) program that modeled target type #3—a collection of dissimilar (nonhomogeneous) unitary targets (see Volume 2, Chapter 7).

Around the same time as JAWS development took place, the surface-to-surface community was included in the JMEM community with the release of the Joint Weapon Effectiveness System (JWES 1.0). This served the indirect fire community with tools such as SuperQuickie 2 and ARTQUIK (Chapter 16), and direct fire with FBAR and PVTM (Chapter 17). The Navy Joint Gun Engagement Model (JGEM) was added at a later date.

18.2.3 Developments 2000–present

These two parallel developmental streams (JAWS and JWES) continued into the early 2000s with two distinct products:

1. The air-to-surface Joint Air-to-surface Weaponeering Tool (JAWS). JAWS v1.0 was released in 1995.
2. The surface-to-surface program Joint Weapon Effectiveness System (JWES).

Although this approach worked well for the operational users, it did not suit more senior military officers who were more focused on the target rather than the weapon system. For example, if they wanted to attack a target 50 miles away, they could use air-delivered guided bombs or perhaps a salvo of guided multiple launched rockets. To compare the effect of each weapon, both JAWS and JWES had to be run separately, probably on two computers, and the results compared manually. These senior commanders wanted a single combined, target-centric program so that once a target was selected, all available weapons that might have an effect on it were available, irrespective of whether they were air or surface launched.

This requirement resulted in the current product—the JMEM Weaponeering System (JWS). Combining the two service-focused tools led to another round of consolidation of methodologies, which in the prior 20 years of development produced similar but not identical methods. For example, JAWS had a methodology for machine gun effects against ground targets whereas JWES had a different approach. JAWS had a vulnerability model of the T-72 tank and so did JWES—were they the same? The first step was to incorporate the two programs under a single wrapper, a trivial-sounding advance but one that was fraught with problems and resulted in JWS v1.0, which was released in 2005. The first fully integrated program, JWS v2.0, was released in 2009 with further refinements continuing to the present day.

Over the last decade, computational speed has increased to the point where the architecture used to generate the precomputed solutions may now be performed in near real time, although the basic models have been improved. This process is shown in Fig. 18.4.

It is perhaps ironic that as JWS has developed into an all-encompassing collection of weaponeering tools, giving the user a wide choice of weapon systems and targets to evaluate, its complexity in use is daunting to many operational users. As a result, there is now a requirement for table lookup solutions for common weaponeering cases that users encounter, saving the time needed to obtain them directly by running the JWS program. These weaponeering tables are reminiscent of the precomputed solutions of 50 years ago using the same (updated) tools!

The progression of JMEM platforms has matched computer development up to the present time so that the latest versions of the JWS program still run on Windows-based PCs. However, another operational constraint that has governed the level of fidelity of the models is the type of computer available to the operational user. The model developer (analyst) probably has the latest hardware and software available to them, but not so the operational user. These users have computer systems one or two developmental steps behind the analysts; therefore, the models have to run on relatively old systems.

18.2.4 JWS Releasability

A significant challenge to current JWS development is to make it releasable to coalition partners. For example, the Desert Storm/Desert Shield coalition included 39 different countries that were expected to fight in a coordinated way against a common enemy.

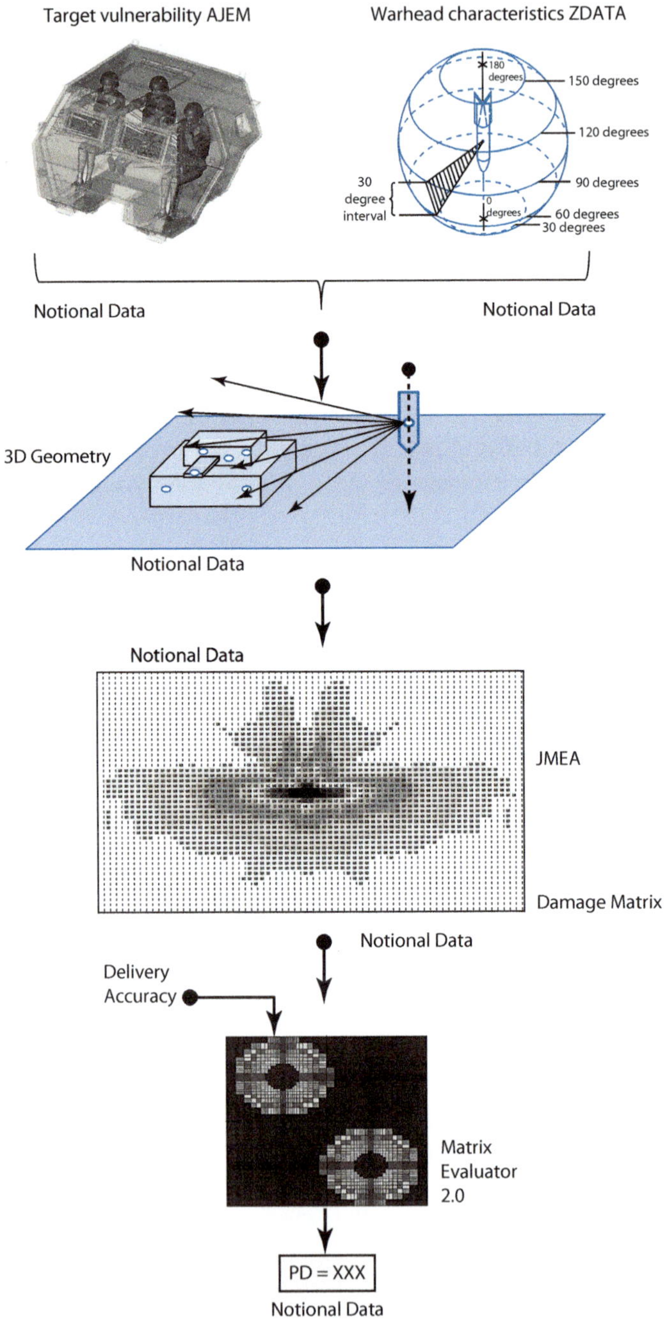

Fig. 18.4 Current process for computing damage.

The basic argument is: If we fight together we should train and mission plan together; otherwise, military operations cannot be effectively coordinated. The basic problem is: If our JWS weaponeering tool contains data classified to the secret level, not releasable to foreign individuals or agencies, how can JWS be used to assist coalition partners to mission plan with the United States? In previous conflicts, certain coalition countries had access to use U.S. weaponeering tools in a restricted area only under the supervision of U.S. personnel (i.e., visual use only). This solution is not ideal because those countries had a certain learning curve to overcome before effective use of the tools could be achieved. A better approach to solving this issue is to write the JWS program in a modular format so that parts may be added or removed for a custom version for a particular country.

The question is: What components should a country-specific version of JWS contain? This depends on the status of the foreign military sales (FMS) portfolio for that country. If a weapon system has been purchased from the United States, that weapon system is retained in the product; otherwise, it is removed. Although this approach is reasonable for weapons, it presents challenges regarding what target vulnerability data may be distributed.

18.2.5 Programming Language Considerations

The early programs such as Matrix Evaluator and GFSM were written in FORTRAN, and these were promulgated when the OEM running under DOS was implemented. Since then JWS has been written in various types of the C programming language (C, C++, and C#). Because JWS is made up of several computational modules, they are typically developed independently of JWS and incorporated once they have been through a Verification, Validation and Approval (VV&A) process and tri-service approved by the appropriate JTCG/ME Working Group.

In some cases, operational users and others in the weaponeering community request the stand-alone version of a module to be used outside of JWS with user-supplied data. The rationale for this is that JWS is a large and complex program requiring strict conditions for its installation and running. If a user is interested in just a small part of JWS, such as bomb penetration through layers of soil, rock, and concrete to breach a bunker, is there a simple stand-alone tool to accomplish this?

The answer in this case is yes, because the penetration module was developed separately and then integrated into the JWS program by the production contractor team. The question then arises regarding what programming language is used for initial module development and how easy is it to distribute to customers requiring it?

In the past, module development was usually done in C++ because this would provide easy integration into JWS; however, this is not a programming environment many of the operational users have available, which makes distribution difficult. Similarly, analysts may have used MATLAB for module development because it has extensive function support to perform complex tasks and excellent plotting capabilities to visualize any outputs from the methodology. Unfortunately, MATLAB is not widely available to the warfighter either because anything other than the student edition may be prohibitively expensive to purchase and challenging to install on DoD computer systems. In addition, if development used any high level MATLAB-specific functions, it proves difficult translating these into the native JWS language for later integration.

A suitable compromise has been to use the Visual Basic programming environment to develop modules that users might like to obtain as a stand-alone program. Visual Basic has the following attributes:

- It is a simple programming language to learn, and although it may not have the richness of functions available in MATLAB, it has good computational and graphical display capability. Not using intrinsic functions may involve more code but makes translation into JWS native code much easier.
- The main reason it is favored is that it runs under the Microsoft Excel program; therefore, it requires no additional software to be installed to execute. Modules developed with Visual Basic may be distributed to users as a simple spreadsheet that will run under Microsoft Office. Obviously there are large numbers of DoD computers running Microsoft Office; therefore, distribution of modules is a simple task.

Increasing threats to DoD computer systems from cyber-attack will further restrict what can and cannot be loaded onto them, which will make installation of more esoteric programming environments even more challenging in the future.

18.3 Integrated Weaponeering Tool

In this section we describe the development of a weaponeering tool having the following requirements:

- The air-to-surface (AS) tool should have similar capabilities as the OEM program described earlier in that it uses deterministic methods, as detailed in Chapters 10–15.
- It should handle single and multiple weapons utilizing the appropriate dependent and independent errors.
- A suitable graphical interface will be included similar to the View Points option in OEM so that stick/volley design may be achieved.
- An additional surface-to-surface (SS) component for indirect and direct fire weapon systems with a choice of models for the user to select from (chapters 16, 17).
- This SS component incorporates deterministic and Monte Carlo solutions (chapters 9, 10 and 16).

- A simplified database containing notional data will be provided to allow the user to generate unclassified though reasonable estimates of weapon effects.
- The program will be written in Visual Basic to allow for ease of distribution.
- The air-to-surface portion of the program will be validated using OEM 5.0 using identical weaponeering cases.

18.3.1 Air-to-Surface Program

The user interface for the weaponeering tool is shown in Fig. 18.5, and a comparative view of OEM 5.0 is shown in Fig. 18.6 for the same case. The following features and differences from OEM 5.0 regarding the interface will be noted:

- The Accuracy group includes the $P_{\mathrm{HIT}}/P_{\mathrm{NM}}$ methodology, allowing for Gaussian and non-Gaussian distributions.
- The ballistic error is now part of the accuracy group and is input as a sigma value.

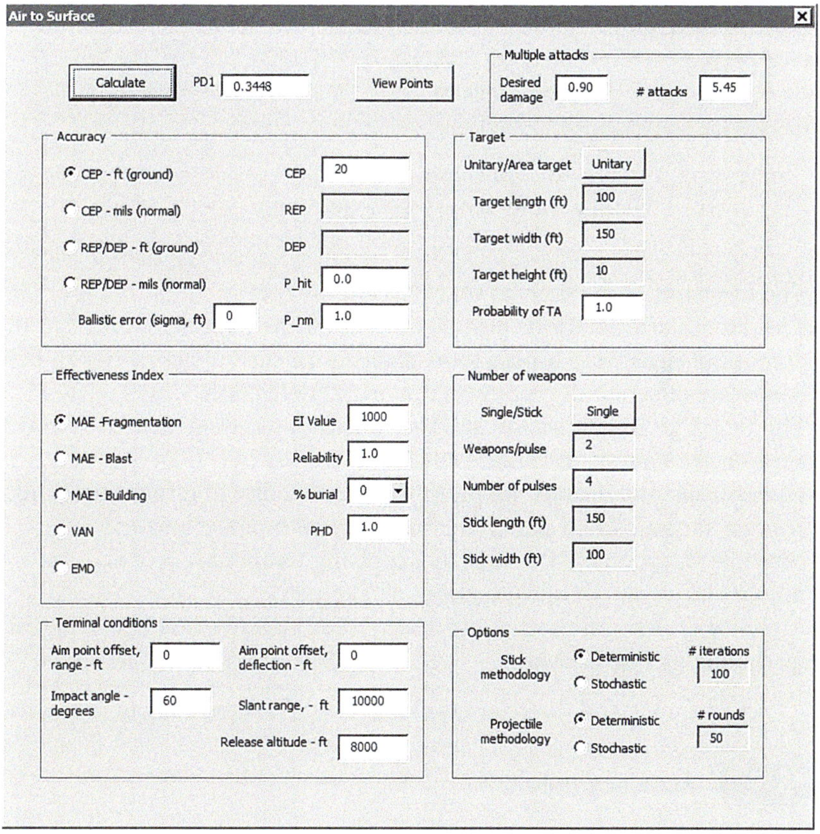

Fig. 18.5 Weaponeering tool: AS component.

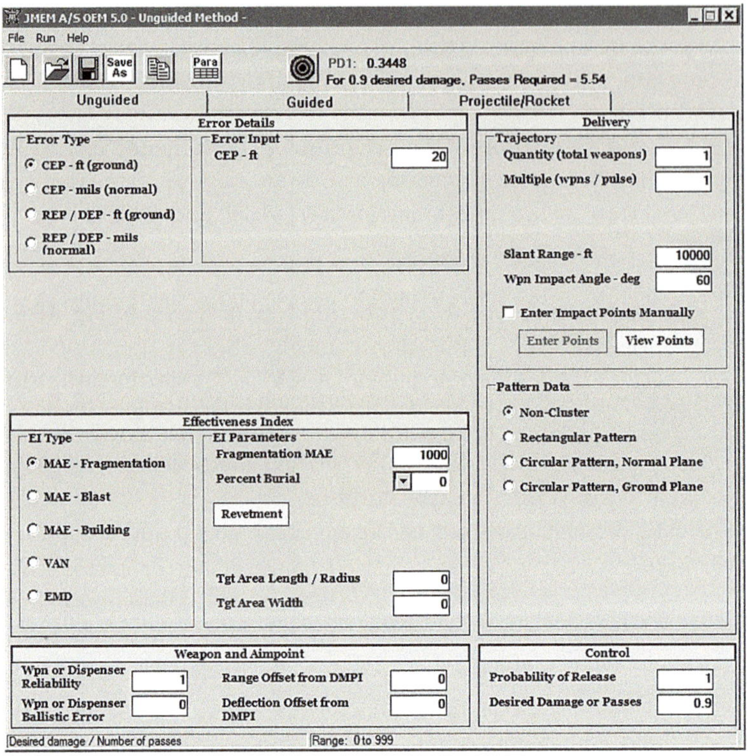

Fig. 18.6 OEM for same data as Fig. 18.5.

- The EI section is the same as OEM except the revetment option is omitted.
- The Terminal Conditions group contains all the parameters needed to determine lethal area aspect ratio, slant range scaling geometry, and any deliberate offset aiming point in range and deflection.
- The target group allows for unitary or area of target elements, with the appropriate data fields made available or grayed out. If an area target is selected, the dimensions are input. The probability of target acquisition is another target-related parameter and may be input, if known.
- Multiple weapons are handled by supplying weapons/pulse and the number of pulses along with the stick geometry.
- A choice of deterministic or stochastic stick effectiveness and projectile methodology is available.

These sections and the relevant chapters will be described in more detail.

18.3.2 Accuracy Inputs

These may be specified in either the ground plane or the normal plane, the former being in feet whereas the latter are in mils. The inputs may be in terms of a circular error probable (CEP) (50%) or range error probable (REP) and deflection error probable (DEP). The ballistic (independent) error is specified in mils in the normal plane and is a sigma value.

In order to convert normal plane mils to equivalent linear normal plane measures at the target, the slant range is required, and to convert to ground plane measures the release altitude is also input. These terms and conversions are described in Chapter 4, Sections 4.7, 4.10, and 4.12.

18.3.3 Effectiveness Indices

This follows the same set as OEM 5.0 and includes mean area of effectiveness MAE_F, MAE_B, and MAE_{BLDG}; normal plane vulnerable area V_{AN}; and effective miss distance (EMD). Once the EI type has been specified, its value is input into the EI Value box. Other effectiveness terms that may be used in the damage calculation are also specified including the weapon (fuze) reliability, the estimated percent burial of the warhead, and the probability of damage given a hit P_{HD}.

These EI have been explained in Chapter 6 (V_{AN}, MAE_F, and MAE_B) and Chapter 15 (EMD and MAE_{BLDG}). The burial methodology is described in Chapter 10, Section 10.12.

18.3.4 Terminal Conditions

The program does not include any trajectory calculations, so the user has to estimate these or obtain them from additional tools they may have access to. These variables include slant range, release altitude, and impact angle. These parameters were discussed in Chapter 3. In this area, an offset aim point in range and direction may be specified; its effect on the damage calculation is described in Chapter 15, Section 15.7.

18.3.5 Target Parameters

These are used in fractional damage calculations for areas of unitary targets and for when the EMD and MAE_{BLDG} types are specified. The target is defined by its length, width, and height. All targets are assumed orthogonal to the attack direction with the length aligned with the range axis. Target dimensions are discussed in Chapters 11 and 15.

Also in this section is the probability of target acquisition, which is usually set to unity; however, if sufficient details of the target and its environment are known, as well as the sensor used to detect it, the methods in Volume 2, Chapter 20 may be applied.

18.3.6 Number of Weapons

The choices are a single, unitary weapon or a stick of unguided bombs. If a stick is selected, it is defined in terms of the stick length, stick width,

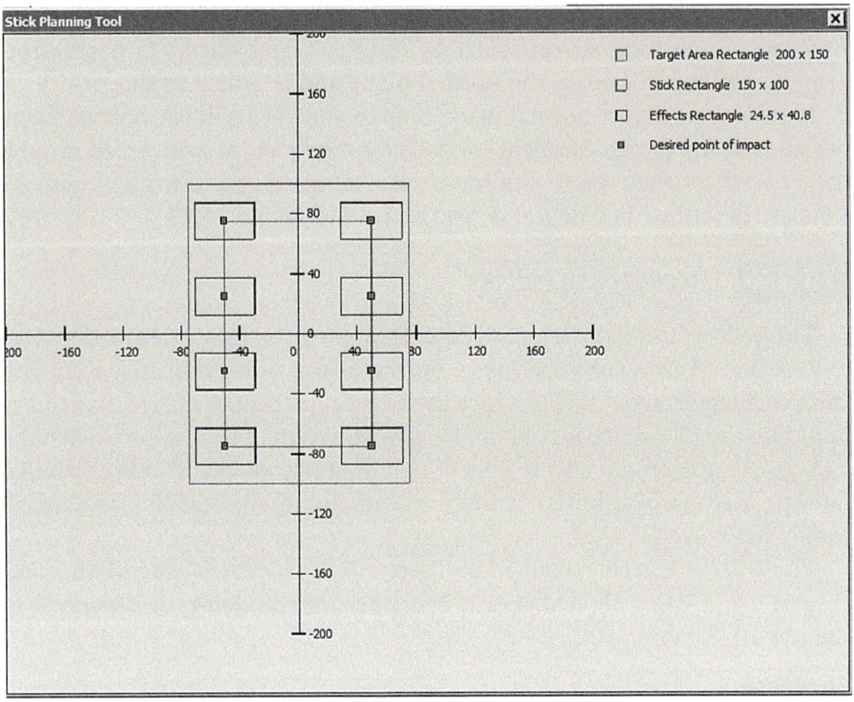

Fig. 18.7 View Points option showing stick and target.

number of weapons released for each intervalometer pulse, and number of pulses. Stick attacks are described in Chapter 12.

When weaponeering stick attacks against an area target, the View Points option allows the stick geometry and number of weapons to be adjusted to maximize coverage of the target. An example is shown in Fig. 18.7, where a stick of eight weapons (2×4) are dropped against an area of targets. The View Points option allows the user to see fractional coverage for given EI, impact angle, stick length, and stick width inputs. The graphic shows the pattern overlaying the target with no delivery errors.

Note that the View Points window has to be closed before further calculations can be performed.

18.3.7 Stick Methodology

The two options here are deterministic or stochastic (Monte Carlo). In the deterministic approach, the methods of Chapter 12 are used where the stick is aggregated into a single rectangular area of known dimensions with a conditional probability of damage inside it. This is then played against a rectangular area of target elements.

The stochastic approach follows that described in Chapter 15, Section 15.8, which uses the aimpoints defined by the number of weapons and the stick geometry, adds on the randomized aiming (dependent) error, and then performs a fractional coverage calculation using the ballistic dispersion (dependent) error. There are differences between the two approaches, and as indicated in Chapter 12, Section 12.11, these may be observed by the user.

18.3.8 Projectile Methodology

Again there are two options: deterministic and stochastic. In the former, the fast-running, simple AS methodology described in Chapter 14, Section 14.2 is used where the target is represented by a vulnerable area, the dependent and independent errors are (incorrectly) combined, and the probability of a single round hitting the vulnerable area is calculated. This is then powered up by the number of rounds in the burst.

The stochastic method keeps the dependent and independent errors separate but uses the same target representation of a single vulnerable area. The methodology is implemented by the double loop Monte Carlo simulation shown in Fig. 9.7 and described in Chapter 9, Section 9.8.

18.3.9 Multiple Attacks

The main spreadsheet calculates the effectiveness of a single attack and produces a value of PD_1 or FD_1. The Multiple Attacks section allows the user to input a required damage level, and the program calculates the number of attacks needed to achieve the specified damage level, as defined by Eqs. (10.47) and (11.57).

Note that this will result in a decimal value, and because the number of attacks has to be an integer, the result should be rounded up to obtain the number of attacks that will give at least the required damage.

18.3.10 Surface-to-Surface Program

This program addresses both indirect fire and direct fire and has the user interface shown in Figure 18.8. As for the Air-to-Surface program, the View Points option displays the target, volley and impact points geometry.

18.3.11 Indirect Fire Program

This models artillery battery fire against an area of target elements, where there is option for either a four or six gun battery. Unitary targets are not usually considered but may be modeled by making the target size small.

The damage function used is the rectangular cookie-cutter where the area is the EI and the aspect ratio determined from the weapon impact angle.

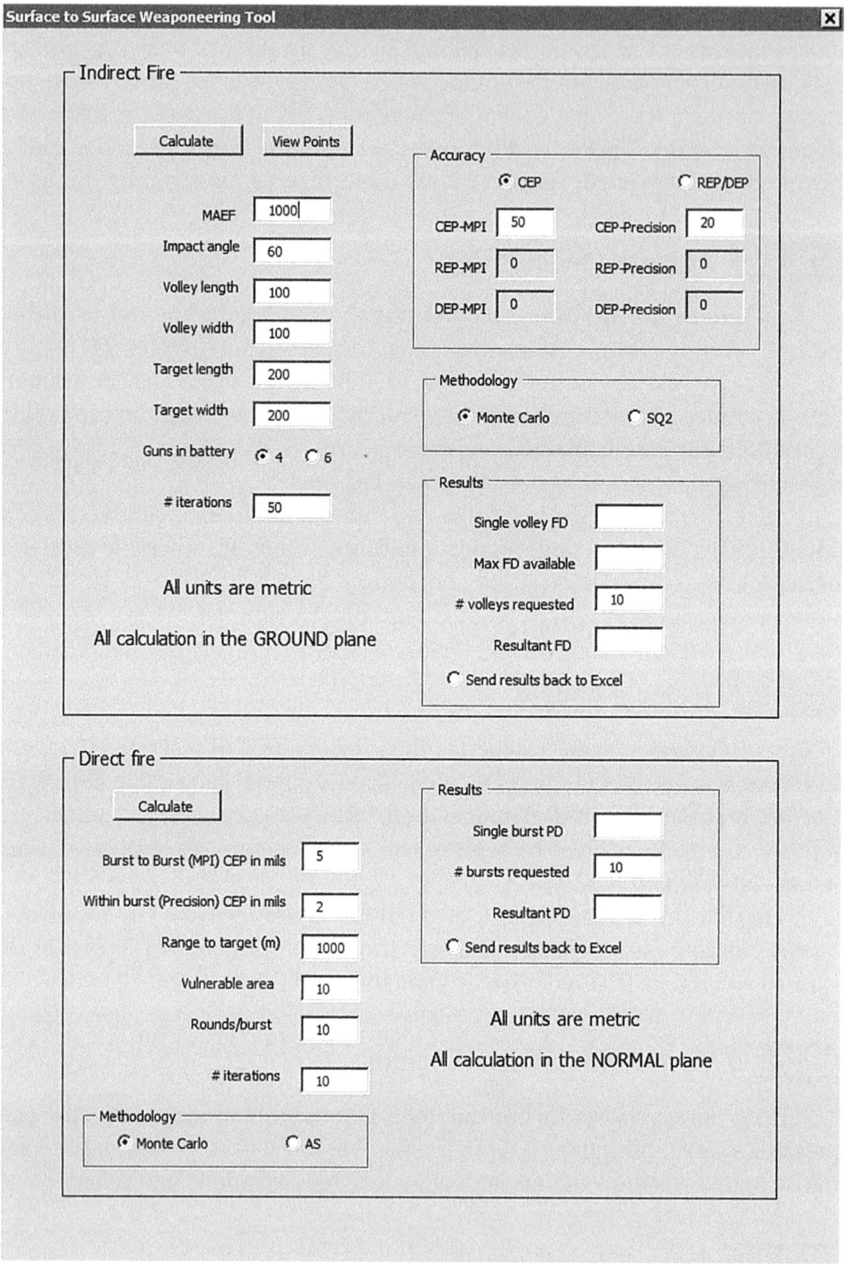

Fig. 18.8 Weaponeering tool SS component.

Most of the inputs are self-explanatory so only a brief description will be provided.

MAE_F	used for unguided artillery shells
Impact angle	used to determine damage function shape
Volley length	from aimpoint strategy – chapter 16.3
Volley width	from aimpoint strategy – chapter 16.3
Target length	area target – chapter 11
Target width	area target – chapter 11
Guns in battery	user selectable
Accuracy	REP/DEP for MPI and precision – chapter 4

The methodology section allows for both deterministic and stochastic solutions which will now be described.

18.3.12 Indirect fire Deterministic Methodology

The deterministic method used is the SuperQuickie process described in chapter 16.11, and is similar the sticks methodology used in the AS program.

The outcome of a single volley from all guns in the battery is shown in the Results section as the single volley fractional damage. Recall from chapter 16.8 the difference between dependent and independent sequential attacks. In general, air delivered sticks of bombs form a sequence of independent attacks so FD_1 may be powered up using the survivor rule. In the SQ2 method however multiple volleys from an artillery battery result in a sequence of dependent attacks and since the lethal area is combined with the precision error, the survivor rule cannot be used.

Further, as the number of volleys increases, the dependent fire FD approaches the expected fractional coverage, whereas the independent fire FD approaches unity as the number of attacks increases, as described in chapter 16.8.3. The fractional coverage is displayed in the Results section as the maximum fractional damage that may be achieved. The user may specify a given number of volleys and the independent event powering up formula (Eq. (16.36)) used to calculate the resultant fractional damage FD.

Figure 18.9 shows an example of a four gun volley overlaying a rectangle of target elements where each weapon lethal area is shown. Figure 18.10 show the SQ2 method of aggregating the lethal areas into a single rectangle overlaying the same target

18.3.13 Indirect Fire Stochastic Methodology

Here, the individual lethal areas shown in Figure 18.9 are preserved, as are the two errors – MPI and precision. The methodology follows that described in chapter 16.21 by sampling the MPI for a volley and precision error for each

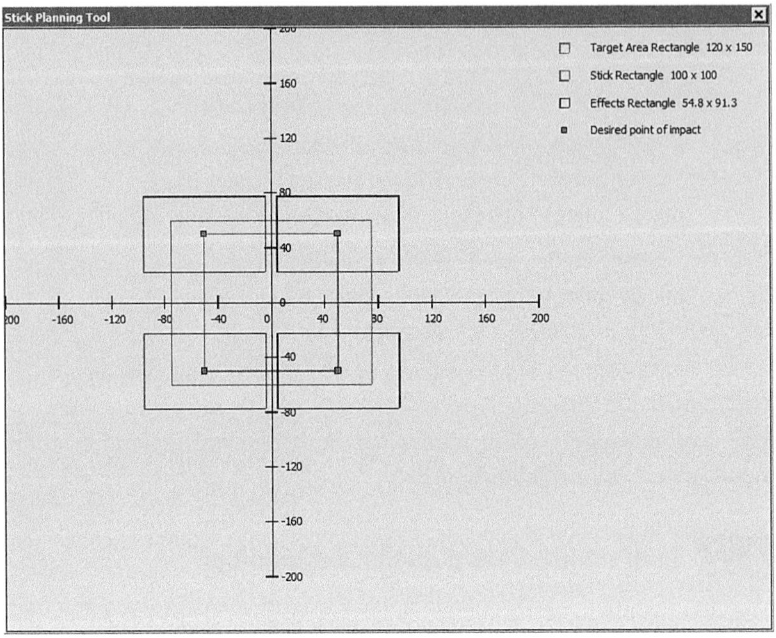

Fig. 18.9 Volley showing individual weapon lethal areas.

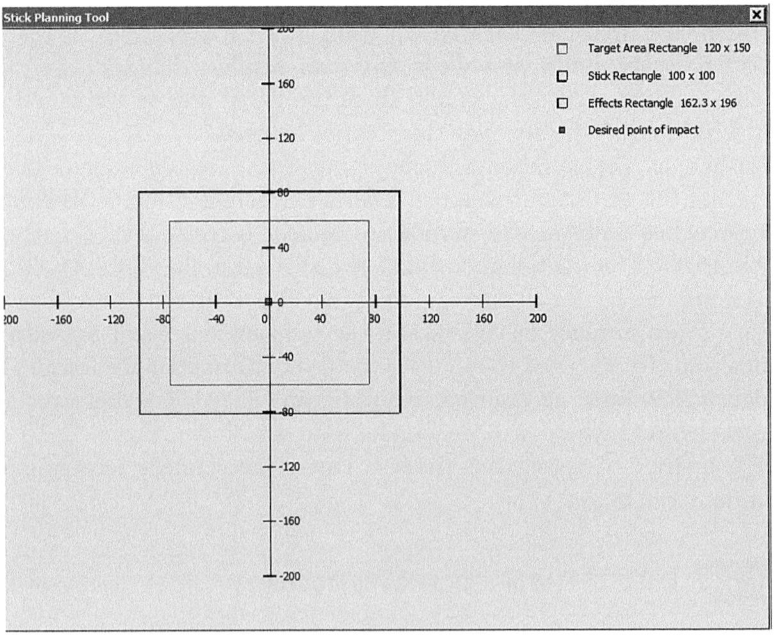

Fig. 18.10 Volley representation using SQ2 option.

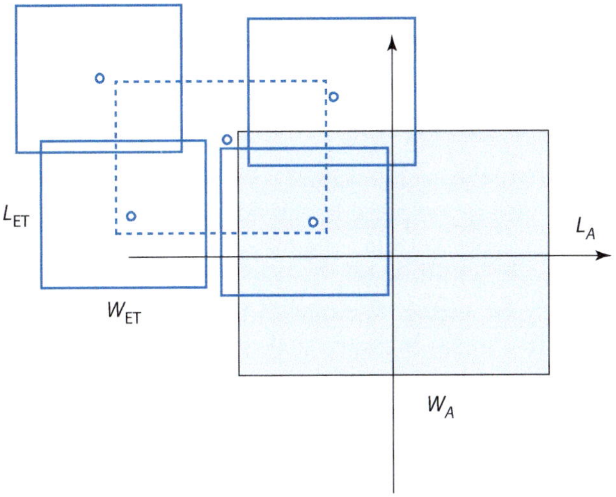

Fig. 18.11 Volley representation for stochastic method.

round resulting in a typical fall of shot shown in Figure 16.47 reproduced below.

The fractional damage is calculated for a single volley by inspecting how many of the discrete target elements comprising the target are damaged by this, and subsequent volleys. The fractional damage is calculated for one volley, a user specified number of volleys and a graph is plotted for up to 20 volleys. It is interesting to investigate the same scenario when evaluated by deterministic and stochastic methods, as shown in Figure 18.12 and Figure 18.13 respectively.

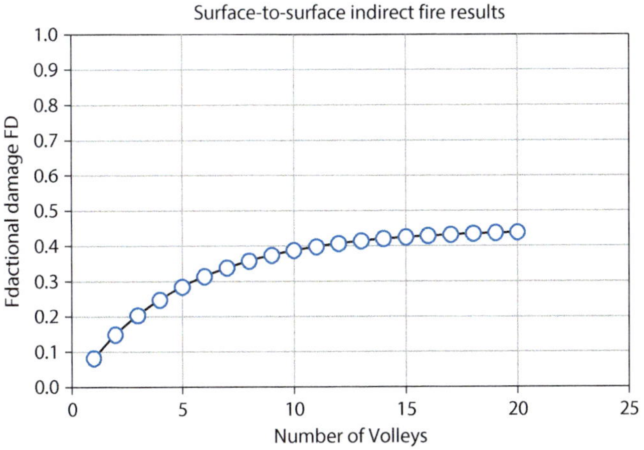

Fig. 18.12 Fractional damage vs. # volleys - deterministic.

The difference in the curves is explained in chapter 16.21 and is because of the two different approaches.

18.3.14 Direct Fire Program

For this program, the accuracies are specified in mils in the normal plane and the target is also represented in the normal plane by a vulnerable area. The dependent accuracy is called the burst-to-burst while the independent accuracy is termed within-burst, corresponding to MPI and precision errors respectively. Inputting the range to the target allows these accuracies to be converted to a linear measure at that range. Projectile trajectories are assumed to be straight lines. Since the target is represented by a vulnerable area, PD is defined as the probability that at least one round in the burst strikes this area. As for indirect fire, the user may specify either a deterministic or stochastic solution.

18.3.15 Direct Fire Deterministic Program

This is implemented as the simplified methodology described in chapter 14.2 and was attributed to the air-to-surface gunnery model. The target vulnerable area is represented by a square and the two errors are (incorrectly) combined. This approach will give the same results as the projectile methodology in the Air-to-Surface program described previously.

18.3.16 Direct Fire Stochastic Program

This approach is described in chapter 9.8 and uses a double loop Monte Carlo simulation shown in Figure 18.14. Here, one sample is drawn for the

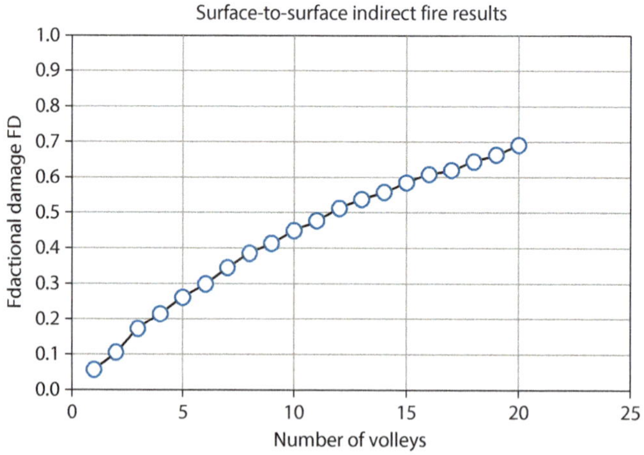

Fig. 18.13 Fractional damage vs. # volleys - stochastic.

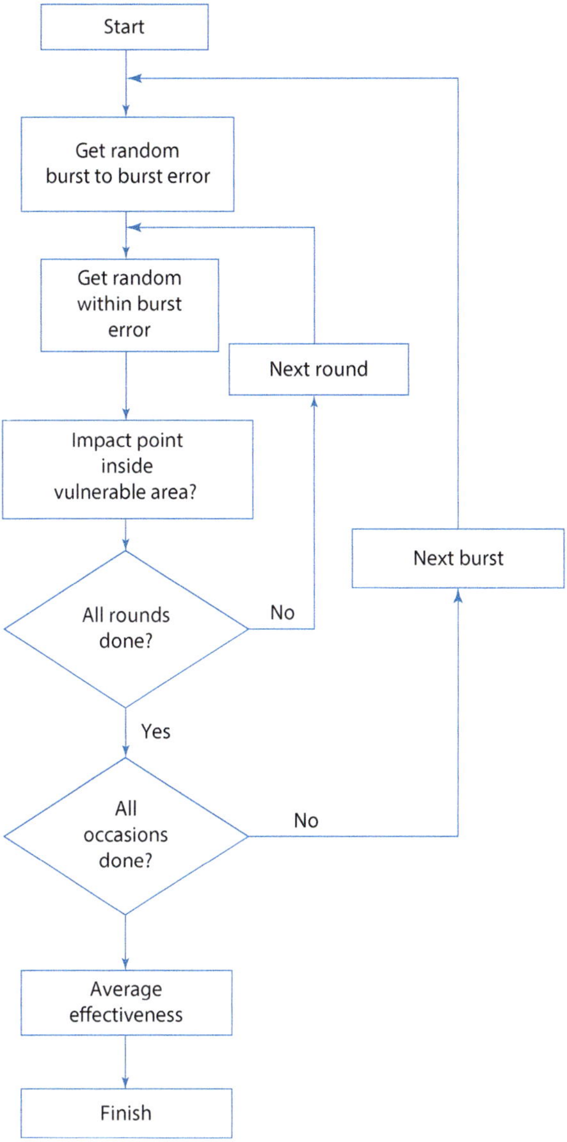

Fig. 18.14 Monte Carlo simulation for direct, burst fire weapons.

burst-to-burst error and then a sample is drawn for each round in the burst. If any rounds fall inside the vulnerable area then the target is killed, otherwise it survives. Results are averaged over a user specified number of Monte Carlo iterations.

Again results are plotted for up to twenty bursts and the results shown graphically in the main program window, as shown below for the same inputs.

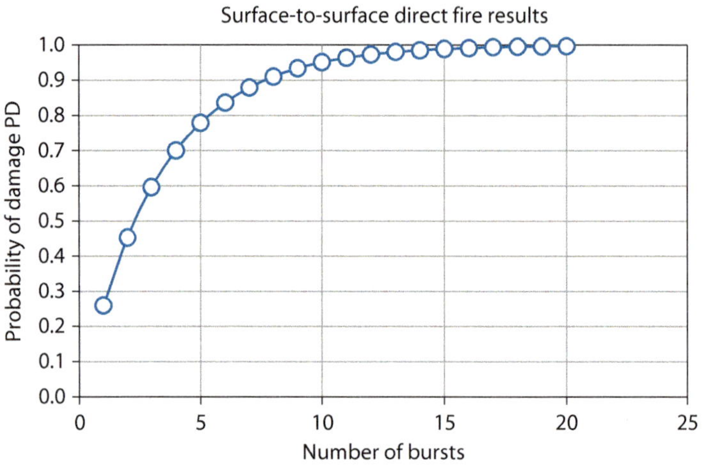

Fig. 18.15 Probability of damage vs. # bursts – deterministic.

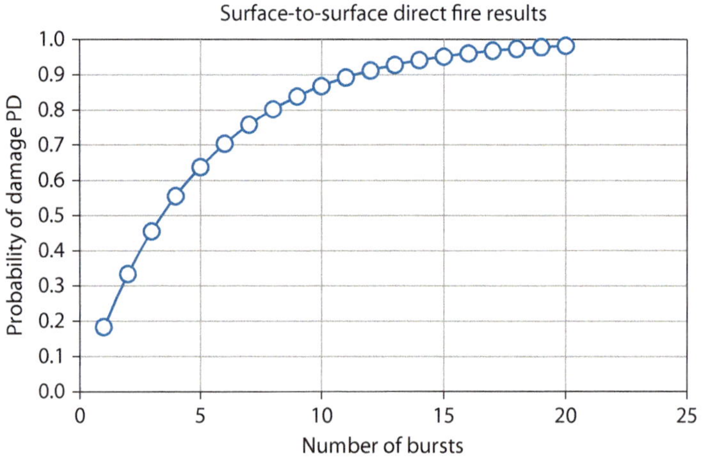

Fig. 18.16 Probability of damage vs. # bursts – stochastic.

Note that these programs reflect their status at the time of writing, and modifications and additions may occur over time.

18.4 Unclassified Database

As previously discussed, the basis of this book is to explain methods to calculate conventional weapon effects in terms of simple mathematical and engineering principles, resulting in techniques that are open source with

NOTIONAL DATA - FOR EDUCATIONAL PURPOSES ONLY

Target	Kill level	Combined effects bomblet A_V (ft^2), MAE_F	Small HE bomb warhead MAE_F (ft^2)	Medium HE bomb warhead MAE_F (ft^2)	Large HE bomb warhead MAE_F (ft^2)	Small high explosive (HE) rocket/missile warhead <10lb MAE_F (ft^2)	Small HEAT rocket/missile warhead <5" dia. A_V (ft^2), MAE_F	Large shaped charge >5 " dia. A_V (ft^2)
MB Tank	K	5 (A_V)	750	900	1000	0	5 (A_V)	30
APC	K	10(A_V)	1500	2000	5000	50	200	50
SP Howitzer	M	250	2500	4000	7000	200	400	50
MLRS	F	100	20000	30000	50000	300	200	N/A
AA artillery	F	20	5000	10000	15000	200	100	N/A
Parked Aircraft	PTO-4	600	40000	60000	90000	4000	2000	N/A
Truck	M	100	15000	30000	45000	1500	1000	N/A
Radar	K	0	2500	4500	10000	N/A	N/A	N/A
Bridge - single span girder	BEI		5	15	40			
Building SS Residential	MAE_{BLDG}		4000	7500	15000			
Building MS Military	MAE_{BLDG}		500	1000	2000			
Building MS Commercial	MAE_{BLDG}		2000	10000	20000			
Personnel standing 5 min assault	MAE_F		15000	30000	40000			
Personnel in foxhole 30 sec defense	MAE_F		3000	6000	12000			
Tunnel	EMD		10	15	25			
Buried POL Tank	EMD		25	50	75			

NOTIONAL DATA - FOR EDUCATIONAL PURPOSES ONLY

Fig. 18.17 Weapon–target database.

no classification restrictions. The difference between the classified JWS program and the one just described is that JWS contains classified data relating to weapon lethality, accuracy and target vulnerability. In many cases, entering the same data into both programs will produce the same results.

It follows that anyone wishing to employ the methods described or the Weaponeering tool outlined in this chapter has to supply their own data. In order to support promulgation of this subject, the JTCG/ME has kindly supplied an unclassified data base so that representative results may be obtained from the methods provided in the book. It must be understood however that all the data shown is notional and any association with any weapon or target, or their combination should not be inferred, and the data is to be used for educational purposes only.

The database is shown in Figure 18.17 and shows a variety of surface mobile targets as well as some infrastructure targets such as buildings, bridges, tunnels and a POL tank. Personnel targets are also addressed. Weapon types are focused on air delivered weapons and include guided and unguided bombs, rockets, shaped charge and cluster munitions. Note that not all possible weapon-target pairings have data, since some weapons are not appropriate for all targets, see appendix.

Appendix A

Standard Statistical Tables

Table A.1 CDF for Normal Distribution

	Second Digit of Z									
Z	0.00	0.01	0.02	0.03	0.04	0.05	0.06	0.07	0.08	0.09
0.0	0.5000	0.5040	0.5080	0.5120	0.5160	0.5199	0.5239	0.5279	0.5319	0.5359
0.1	0.5398	0.5438	0.5478	0.5517	0.5557	0.5596	0.5636	0.5675	0.5714	0.5753
0.2	0.5793	0.5832	0.5871	0.5910	0.5948	0.5987	0.6026	0.6064	0.6103	0.6141
0.3	0.6179	0.6217	0.6255	0.6293	0.6331	0.6368	0.6406	0.6443	0.6480	0.6517
0.4	0.6554	0.6591	0.6628	0.6664	0.6700	0.6736	0.6772	0.6808	0.6844	0.6879
0.5	0.6915	0.6950	0.6985	0.7019	0.7054	0.7088	0.7123	0.7157	0.7190	0.7224
0.6	0.7257	0.7291	0.7324	0.7357	0.7389	0.7422	0.7454	0.7486	0.7517	0.7549
0.7	0.7580	0.7611	0.7642	0.7673	0.7704	0.7734	0.7764	0.7794	0.7823	0.7852
0.8	0.7881	0.7910	0.7939	0.7967	0.7995	0.8023	0.8051	0.8078	0.8106	0.8133
0.9	0.8159	0.8186	0.8212	0.8238	0.8264	0.8289	0.8315	0.8340	0.8365	0.8389
1.0	0.8413	0.8438	0.8461	0.8485	0.8508	0.8531	0.8554	0.8577	0.8599	0.8621
1.1	0.8643	0.8665	0.8686	0.8708	0.8729	0.8749	0.8770	0.8790	0.8810	0.8830
1.2	0.8849	0.8869	0.8888	0.8907	0.8925	0.8944	0.8962	0.8980	0.8997	0.9015
1.3	0.9032	0.9049	0.9066	0.9082	0.9099	0.9115	0.9131	0.9147	0.9162	0.9177
1.4	0.9192	0.9207	0.9222	0.9236	0.9251	0.9265	0.9279	0.9292	0.9306	0.9319
1.5	0.9332	0.9345	0.9357	0.9370	0.9382	0.9394	0.9406	0.9418	0.9429	0.9441

1.6	0.9452	0.9463	0.9474	0.9484	0.9495	0.9505	0.9515	0.9525	0.9535	0.9545
1.7	0.9554	0.9564	0.9573	0.9582	0.9591	0.9599	0.9608	0.9616	0.9625	0.9633
1.8	0.9641	0.9649	0.9656	0.9664	0.9671	0.9678	0.9686	0.9693	0.9699	0.9706
1.9	0.9713	0.9719	0.9726	0.9732	0.9738	0.9744	0.9750	0.9756	0.9761	0.9767
2.0	0.9772	0.9778	0.9783	0.9788	0.9793	0.9798	0.9803	0.9808	0.9812	0.9817
2.1	0.9821	0.9826	0.9830	0.9834	0.9838	0.9842	0.9846	0.9850	0.9854	0.9857
2.2	0.9861	0.9864	0.9868	0.9871	0.9875	0.9878	0.9881	0.9884	0.9887	0.9890
2.3	0.9893	0.9896	0.9898	0.9901	0.9904	0.9906	0.9909	0.9911	0.9913	0.9916
2.4	0.9918	0.9920	0.9922	0.9925	0.9927	0.9929	0.9931	0.9932	0.9934	0.9936
2.5	0.9938	0.9940	0.9941	0.9943	0.9945	0.9946	0.9948	0.9949	0.9951	0.9952
2.6	0.9953	0.9955	0.9956	0.9957	0.9959	0.9960	0.9961	0.9962	0.9963	0.9964
2.7	0.9965	0.9966	0.9967	0.9968	0.9969	0.9970	0.9971	0.9972	0.9973	0.9974
2.8	0.9974	0.9975	0.9976	0.9977	0.9977	0.9978	0.9979	0.9979	0.9980	0.9981
2.9	0.9981	0.9982	0.9982	0.9983	0.9984	0.9984	0.9985	0.9985	0.9986	0.9986
3.0	0.9987	0.9987	0.9987	0.9988	0.9988	0.9989	0.9989	0.9989	0.9990	0.9990
3.1	0.9990	0.9991	0.9991	0.9991	0.9992	0.9992	0.9992	0.9992	0.9993	0.9993
3.2	0.9993	0.9993	0.9994	0.9994	0.9994	0.9994	0.9994	0.9995	0.9995	0.9995
3.3	0.9995	0.9995	0.9995	0.9996	0.9996	0.9996	0.9996	0.9996	0.9996	0.9997
3.4	0.9997	0.9997	0.9997	0.9997	0.9997	0.9997	0.9997	0.9997	0.9997	0.9998
3.5	0.9998	0.9998	0.9998	0.9998	0.9998	0.9998	0.9998	0.9998	0.9998	0.9998

Table A.2 CDF for Binomial Distribution

n	*x*	*p* = 0.1	0.2	0.3	0.4	0.5	0.6	0.7	0.8	0.9
4	0	0.65610	0.40960	0.24010	0.12960	0.06250	0.02560	0.00810	0.00160	0.00010
4	1	0.94770	0.81920	0.65170	0.47520	0.31250	0.17920	0.08370	0.02720	0.00370
4	2	0.99630	0.97280	0.91630	0.82080	0.68750	0.52480	0.34830	0.18080	0.05230
4	3	0.99999	0.99847	0.99190	0.97440	0.93750	0.87040	0.75990	0.59040	0.34390
4	4	1.00000	1.00000	1.00000	1.00000	1.00000	1.00000	1.00000	1.00000	1.00000
5	0	0.59049	0.32768	0.16807	0.07776	0.03125	0.01024	0.00243	0.00032	0.00001
5	1	0.91854	0.73728	0.52822	0.33696	0.18750	0.08704	0.03078	0.00672	0.00046
5	2	0.99144	0.94208	0.83692	0.68256	0.50000	0.31744	0.16308	0.05792	0.00856
5	3	0.99954	0.99328	0.96922	0.91296	0.81250	0.66304	0.47178	0.26272	0.08146
5	4	0.99999	0.99968	0.99757	0.98976	0.96875	0.92224	0.83193	0.67232	0.40951
5	5	1.00000	1.00000	1.00000	1.00000	1.00000	1.00000	1.00000	1.00000	1.00000
6	0	0.53144	0.26214	0.11765	0.04666	0.01562	0.00410	0.00073	0.00006	0.00000
6	1	0.88574	0.65536	0.42018	0.23328	0.10938	0.04096	0.01094	0.00160	0.00006
6	2	0.98415	0.90112	0.74431	0.54432	0.34375	0.17920	0.07047	0.01696	0.00127
6	3	0.99873	0.98304	0.92953	0.82080	0.65625	0.45568	0.25569	0.09888	0.01585
6	4	0.99994	0.99840	0.98906	0.95904	0.89062	0.76672	0.57982	0.34464	0.11426
6	5	1.00000	0.99994	0.99927	0.99590	0.98438	0.95334	0.88235	0.73786	0.46856
6	6	1.00000	1.00000	1.00000	1.00000	1.00000	1.00000	1.00000	1.00000	1.00000

7	0	0.47830	0.20972	0.08235	0.02799	0.00781	0.00164	0.00022	0.00001	0.00000
7	1	0.85031	0.57672	0.32942	0.15863	0.06250	0.01884	0.00379	0.00037	0.00001
7	2	0.97431	0.85197	0.64707	0.41990	0.22656	0.09626	0.02880	0.00467	0.00018
7	3	0.99727	0.96666	0.87396	0.71021	0.50000	0.28979	0.12604	0.03334	0.00273
7	4	0.99982	0.99533	0.97120	0.90374	0.77344	0.58010	0.35293	0.14803	0.02569
7	5	0.99999	0.99963	0.99621	0.98116	0.93750	0.84137	0.67058	0.42328	0.14969
7	6	1.00000	0.99999	0.99978	0.99836	0.99219	0.97201	0.91765	0.79028	0.52170
7	7	1.00000	1.00000	1.00000	1.00000	1.00000	1.00000	1.00000	1.00000	1.00000
8	0	0.43047	0.16777	0.05765	0.01680	0.00391	0.00066	0.00007	0.00000	0.00000
8	1	0.81310	0.50332	0.25530	0.10638	0.03516	0.00852	0.00129	0.00008	0.00000
8	2	0.96191	0.79692	0.55177	0.31539	0.14453	0.04981	0.01129	0.00123	0.00002
8	3	0.99498	0.94372	0.80590	0.59409	0.36328	0.17367	0.05797	0.01041	0.00043
8	4	0.99957	0.98959	0.94203	0.82633	0.63672	0.40591	0.19410	0.05628	0.00502
8	5	0.99998	0.99877	0.98871	0.95019	0.85547	0.68461	0.44823	0.20308	0.03809
8	6	1.00000	0.99992	0.99871	0.99148	0.96484	0.89362	0.74470	0.49668	0.18690
8	7	1.00000	1.00000	0.99993	0.99934	0.99609	0.98320	0.94235	0.83223	0.56953
8	8	1.00000	1.00000	1.00000	1.00000	1.00000	1.00000	1.00000	1.00000	1.00000
9	0	0.38742	0.13422	0.04035	0.01008	0.00195	0.00026	0.00002	0.00000	0.00000
9	1	0.77484	0.43621	0.19600	0.07054	0.01953	0.00380	0.00043	0.00002	0.00000

(Continued)

Table A.2 CDF for Binomial Distribution *(Continued)*

n	x	$p = 0.1$	0.2	0.3	0.4	0.5	0.6	0.7	0.8	0.9
9	2 |	0.94703	0.73820	0.46283	0.23179	0.08984	0.02503	0.00429	0.00031	0.00000
9	3 |	0.99167	0.91436	0.72966	0.48261	0.25391	0.09935	0.02529	0.00307	0.00006
9	4 |	0.99911	0.98042	0.90119	0.73343	0.50000	0.26657	0.09881	0.01958	0.00089
9	5 |	0.99994	0.99693	0.97471	0.90065	0.74609	0.51739	0.27034	0.08564	0.00833
9	6 |	1.00000	0.99969	0.99571	0.97497	0.91016	0.76821	0.53717	0.26180	0.05297
9	7 |	1.00000	0.99998	0.99957	0.99620	0.98047	0.92946	0.80400	0.56379	0.22516
9	8 |	1.00000	1.00000	0.99998	0.99974	0.99805	0.98992	0.95965	0.86578	0.61258
9	9 |	1.00000	1.00000	1.00000	1.00000	1.00000	1.00000	1.00000	1.00000	1.00000

Table A.3 Chi–Squared Tabular Data

DOF	Probability									
	0.995	0.99	0.975	0.95	0.9	0.1	0.05	0.025	0.01	0.005
1	0.000	0.000	0.001	0.004	0.016	2.706	3.841	5.024	6.635	7.879
2	0.010	0.020	0.051	0.103	0.211	4.605	5.991	7.378	9.210	10.597
3	0.072	0.115	0.216	0.352	0.584	6.251	7.815	9.348	11.345	12.838
4	0.207	0.297	0.484	0.711	1.064	7.779	9.488	11.143	13.277	14.860
5	0.412	0.554	0.831	1.145	1.610	9.236	11.070	12.832	15.086	16.750
6	0.676	0.872	1.237	1.635	2.204	10.645	12.592	14.449	16.812	18.548
7	0.989	1.239	1.690	2.167	2.833	12.017	14.067	16.013	18.475	20.278
8	1.344	1.647	2.180	2.733	3.490	13.362	15.507	17.535	20.090	21.955
9	1.735	2.088	2.700	3.325	4.168	14.684	16.919	19.023	21.666	23.589
10	2.156	2.558	3.247	3.940	4.865	15.987	18.307	20.483	23.209	25.188
11	2.603	3.053	3.816	4.575	5.578	17.275	19.675	21.920	24.725	26.757
12	3.074	3.571	4.404	5.226	6.304	18.549	21.026	23.337	26.217	28.300
13	3.565	4.107	5.009	5.892	7.041	19.812	22.362	24.736	27.688	29.819
14	4.075	4.660	5.629	6.571	7.790	21.064	23.685	26.119	29.141	31.319
15	4.601	5.229	6.262	7.261	8.547	22.307	24.996	27.488	30.578	32.801
16	5.142	5.812	6.908	7.962	9.312	23.542	26.296	28.845	32.000	34.267
17	5.697	6.408	7.564	8.672	10.085	24.769	27.587	30.191	33.409	35.718
18	6.265	7.015	8.231	9.390	10.865	25.989	28.869	31.526	34.805	37.156
19	6.844	7.633	8.907	10.117	11.651	27.204	30.144	32.852	36.191	38.582
20	7.434	8.260	9.591	10.851	12.443	28.412	31.410	34.170	37.566	39.997

Appendix B

Weapon Selection Based on Target and Damage Criteria

The following tables indicate the appropriate weapon selection for specific target types. Also indicated in the tables are definitions of damage criteria that relate to the particular target.

Table B.1 Target-damage criteria and appropriate weapon types

Target	Damage Criteria		Appropriate Weapon Type[1]
1. Personnel	Defense – 30 s Assault – 30 s Assault – 5 min Supply – 12 h	The term *casualty criteria* is used for personnel targets and is comparable to the term *damage criteria* for materiel targets. A casualty results when the capability of the individual soldier to perform military duties is reduced to a sufficient degree but the criteria depend upon the tactical situation. For the analysis of casualties, four tactical roles have been defined, and a time element has been assigned to each: defense: 30 s; assault: 30 s; assault: 5 min; and supply: 12 h. Tactical roles and variations of incapacitation are defined in JMEM *Target Vulnerability*. The time element is the period after which the wound will result in incapacitation to meet the criterion for each role. Chemical weapons and firebombs may have additional psychological effects on troops, causing an immediate lowering of unit combat effectiveness. Assault $\leq$ 30 s is representative of defense $\leq$ 5 min. Assault $\leq$ 5 min is representative of defense $\leq$ 12 h.	The appropriate weapon type depends upon the posture assumed by the target and the cover taken, such as lying prone behind appropriate shielding or under cover in foxholes. *For personnel in the open (standing or prone):* The appropriate weapon types are high-explosive (HE) gun projectiles, fragmentation submunitions, airburst and instantaneous-surface-burst general-purpose (GP) bombs, chemical weapons, and firebombs. *For personnel in foxholes:* The appropriate weapons are fragmentation submunitions in the foxholes, airburst GP bombs, chemical weapons, and fuel-air explosive (FAE) weapons with the fuel-air cloud over the foxholes. *For personnel under light cover:* Cluster weapons dispersing small shaped-charge submunitions (assuming the shaped-charge jet will penetrate the cover) should be effective, as well as rockets or gun projectiles with armor-piercing warheads that strike the covered foxholes.

Table B.1 Target-damage criteria and appropriate weapon types *(Continued)*

Target	Damage Criteria		Appropriate Weapon Type[1]
			Short-time-delay-fuzed GP bombs that hit the foxhole or strike close enough to crater the foxhole and kill the occupants should be appropriate. Guided weapons will be difficult to use because the target presents limited signature for directing a guided weapon. Chemical weapons should be effective, as well as FAE weapons where the fuel-air cloud is over the lightly covered foxhole.
			For personnel under heavy cover, such as concrete bunkers: Delay-fuzed bombs fuzed for maximum cratering, and guided weapons with large warheads and penetration capability are desirable.
2. Armored vehicles			
a. Tanks	Mobility, M0-kill	Damage that causes a vehicle to be incapable of executing controlled movement and is not repairable by the crew on the battlefield. The damage is assumed to occur instantaneously.	Shaped-charge weapons, armor-piercing projectiles, and contact-fuzed GP bombs. Guided bombs, such as electro-optically guided bombs (EOGBs) and laser-guided bombs (LGBs).
	Mobility, M40-kill	Same as M0, but mobility is lost within 40 min.	Same as for M0-kill.

(Continued)

Table B.1 Target-damage criteria and appropriate weapon types *(Continued)*

Target	Damage Criteria		Appropriate Weapon Type[1]
	Firepower, F-kill	Defeat of the main armament, either because the crew has been rendered incapable of operating it or because the armament or its associated equipment has been rendered inoperative and the gun cannot be effectively fired. It is also assumed that the equipment is not repairable by the crew on the battlefield. The damage is assumed to occur instantaneously.	Same as for M0-kill.
	Catastrophic, K-kill	Damage beyond repair or to the extent that repair is not economically feasible. The damage is assumed to occur instantaneously.	Same as for M0-kill.
b. Armored personnel carriers (APCs), reconnaissance and infantry combat vehicles, and scout cars	Mobility, M0-kill	Damage that causes a vehicle to be incapable of executing controlled movement and is not repairable by the crew on the battlefield. The damage is assumed to occur instantaneously.	Shaped-charge weapons, armor-piercing projectiles, and contact-fuzed GP bombs. Guided bombs, such as EOGBs and LGBs.
	Mobility, M40-kill	Same as M0, but mobility is lost within 40 min.	Same as for M0-kill.
	Firepower, F-kill	Damage to the primary armament, associated equipment, or crewmembers required to operate the armament, rendering the vehicle inoperative. It is assumed that armament and associated equipment are not repairable by the crew on the battlefield. Damage is assumed to occur instantaneously.	Same as for M0-kill.

Table B.1 Target-damage criteria and appropriate weapon types *(Continued)*

Target	Damage Criteria		Appropriate Weapon Type[1]
	Catastrophic, K-kill	Damage beyond repair or to the extent that repair is not economically feasible.	Same as for MO-kill.
	Passenger, P-kill	Incapacitation of transported personnel. The P-kill is a calculated ratio of the number of personnel incapacitated to the total number of personnel being transported in the passenger compartment of the APC. Incapacitation will be based on assault, 5-min wound ballistic criteria for personnel.	Shaped-charge weapons, rockets or gun projectiles with armor-piercing warheads, and contact-fuzed GP bombs. Also guided weapons as specified above.
c. Mobile assault guns	Same as tanks.	Same as tanks.	Same as tanks.
3. Field Artillery			
a. Towed and self-propelled	Firepower, F-kill	Damage to the gun that will prevent the accurate delivery of munitions on intended targets and that cannot be repaired in the field. Incapacitation of the crew is not included.	*Against guns in the open:* Contact-fuzed GP bombs or delay-fuzed GP bombs employed for shallow cratering and overturning of guns. If contact fuze is instantaneous and impact angle of GP bomb is carefully controlled, the combined blast-fragmentation can be quite effective against such hard targets. Dispenser weapons using conical-shaped-charge bomblets may be effective (but the vulnerable area of the gun is restricted). Little damage can be expected for holes in gun tube unless they are quite large or located in areas of maximum internal pressure. EOGBs and LGBs, if they can be

(Continued)

Table B.1 Target-damage criteria and appropriate weapon types *(Continued)*

Target	Damage Criteria		Appropriate Weapon Type[1]
			accurately guided to the target, and guided weapons employing large LSC warheads should be quite effective. Fragmentation bomblets and airburst GP bombs will be useful mainly against the personnel and crew associated with the field-artillery piece. *Against guns in revetments:* Contact-fuzed GP bombs or airburst GP bombs detonating at lower altitudes. High-altitude bursts (of the order of 50 ft and above the Mk 82) will be relatively ineffective because of velocity slowdown and low density of fragments. The use of GP bombs assumes that one can hit in the revetment to achieve damage, thus EOGBs and LGBs will be effective if target characteristics or signatures are such that the weapon can be guided to the target. Also, guided weapons employing largest LSC warheads should be effective. Cluster or dispenser weapons using conical-shape-charge bomblets impacting on the target can be effective against limited areas of the guns and especially effective against the crew of the revetted guns.

Table B.1 Target-damage criteria and appropriate weapon types *(Continued)*

Target	Damage Criteria		Appropriate Weapon Type[1]
			Against guns in covered positions: Delay fuzed bombs, GP bombs fuzed for maximum cratering, and guided weapons with large warheads and some penetration capability would be desirable.
b. Self-propelled	Mobility, M0-kill	Same as for tanks.	Same as for tanks.
	Mobility, M40-kill	Same as M0, but mobility is lost within 40 min.	Same as for M0-kill.
4. Antiaircraft artillery			
a. Automatic AA guns	Firepower, F-kill	Damage to the weapon that will prevent the accurate delivery of munitions on intended targets and that cannot be repaired in the field. Incapacitation of the crew is not included nor is any off-carriage fire-control equipment, such as a radar or fire-control director. Because various modes of operation from fully automatic to fully manual are prevalent, which can require different components to be damaged, each mode of operation may be evaluated separately. When this occurs, consult TH 61A1-3-1, *Target Vulnerability* (JMEM), for a complete discussion of the damage criteria for each mode of operation.	Same as for field artillery. *For damage to off-carriage fire-control equipment,* such as a radar or fire-control director, airburst GP bombs or instantaneous-contact-burst GP bombs would be quite effective. The same can be said of dispenser weapons with fragmentation or conical-shaped-charge bomblets. For light honeycomb radar vans, firebombs are effective given a direct hit on the target or a near miss in front of the target.
b. Self-propelled AA guns	M0-kill M40-kill	Same as for tanks. (See Armored Vehicles) Same as for tanks. (See Armored Vehicles)	Same as for tanks, except that the target has less armor. Some of the self-propelled AA guns (e.g., ZSU-57/2) are open on the top

(Continued)

Table B.1 Target-damage criteria and appropriate weapon types *(Continued)*

Target	Damage Criteria		Appropriate Weapon Type[1]
	F-kill	Same as for tanks. (See Armored Vehicles)	and will be susceptible to HEI projectiles and dispenser-delivered shaped-charge submunitions.
	K-kill	Same as for tanks. (See Armored Vehicles)	
5. Rocket Launchers			
a. Multiround launchers	Firepower, F-kill	Sufficient damage to all rocket launcher tubes and/or the launcher system that makes the weapon or system inoperable without major repair.	GP bombs with contact fuzing for direct hit or near miss or proximity-fuzed GP bombs. Conical-shaped-charge submunitions delivered in dispenser systems should be effective, because small conical-shaped charges striking the rocket warheads may cause high-order detonations. A high-order detonation of one warhead in a multiround launcher should sufficiently damage the remaining launcher tubes for an F-kill. There is also a possibility of other warheads detonating sympathetically. (The probability of this happening depends on the construction of the rocket-launcher tubes and size of the rocket warheads.) Guided weapons with large warheads should be effective.
b. Single-round launchers	Mobility, M0-kill	Damage that causes the vehicle to be incapable of executing controlled movement and is not repairable on the battlefield.	Same as for trucks. (See Land Transportation.)
	Mobility, M40-kill	Same as M0, but mobility is lost within 40 min.	Same as for M0-kill.

Table B.1 Target-damage criteria and appropriate weapon types *(Continued)*

Target	Damage Criteria		Appropriate Weapon Type[1]
	Firepower, F-kill	Damage to the rocket or launcher components that prevents successful launching of the rocket.	Same as for tanks. (See Armored Vehicles.)
			GP bombs with contact fuzing for direct hit or near miss or proximity-fuzed GP bombs. Conical-shaped-charge submunitions delivered in dispenser systems should be effective, because small conical-shaped charges striking the rocket warhead may cause high-order detonation. Guided weapons should be effective.
6. Missile Sites			
a. Surface-to-air missile (SAM) sites	Firepower, F-kill	Damage severe enough to prevent firing missiles effectively against airborne targets for at least 4 h. Accomplished by damage to all missiles/launchers, tracking radars, or fire control vans. Damage to acquisition radars will affect the site ability to acquire targets but not to fire missiles using external sources of target acquisition.	Proximity or contact-fuzed GP bombs delivered inside the revetments. Antiradiation missiles to temporarily disable the radar, followed by dispenser weapons using shaped-charge or fragmentation submunitions or retarded contact-fuzed bombs. Guided weapons with standoff capability.
b. Surface-to-surface missile (SSM) sites	Mobility, M0/M40-kill	Same as for tanks. (See Armored Vehicles.)	Same as for tanks. (See Armored Vehicles.)
	Firepower, F-kill	Damage that either prevents the missile from being launched or results in a decision not to launch the missile because of things such as loss of ability of the warhead to produce a nuclear detonation,	*Against liquid-fueled missiles:* Contact-fuzed GP bombs, proximity-fuzed bombs bursting at reasonably low altitudes, and dispenser or cluster weapons, followed by incendiary

(Continued)

Table B.1 Target-damage criteria and appropriate weapon types *(Continued)*

Target	Damage Criteria		Appropriate Weapon Type[1]
		loss of ability to guide to the target, or loss of proper rocket motor function.	munitions. Firebombs delivered after the target is perforated could also be effective. *Against solid-fueled missiles:* Contact-fuzed GP bombs. EOGBs or LGBs and small shaped-charge dispenser weapons.
7. Radar Installations			
a. Radars: fixed and mobile	Fire-control, F-kill	Sufficient damage to prevent the radar system from performing its intended function for at least 4 h. This applies to radars whose loss of ability to track targets or direct fire prevents missile or gun firings and to radars whose loss of function prevents early warning or ground control intercept.	Airburst or instantaneous-contact-burst GP bombs. Antiradiation missiles are effective if the target is radiating. Dispenser weapons with fragmentation or shaped-charge submunitions. For light honeycomb radar vans, firebombs are effective given a hit or near miss in front of the target. Projectiles (e.g., 20-mm rounds) are effective against radar vans, given a hit. EOGBs or LGBs with fragmentation or linear-shaped-charge (LSC) warheads.
	Mission, MSN-kill	Damage that prevents the jamming unit from completing its mission.	
	Mobility, MO/ M40-kill	Same as for tanks. (See Armored Vehicles.)	Same as for tanks. (See Armored Vehicles.)
	Catastrophic, K-kill	Damage beyond repair or repair is not economically feasible. Damage is assumed to occur instantaneously.	Airburst or instantaneous-contact-burst GP bombs. EOGBs or LGBs with fragmentation or LSC warheads.
b. Communications jammers	Mission, MSN kill	Damage that prevents the jamming unit from completing its mission.	Guided and unguided GP bombs. Guided missiles with fragmentation or LSC

Table B.1 Target-damage criteria and appropriate weapon types *(Continued)*

Target	Damage Criteria		Appropriate Weapon Type[1]
			warheads. Cluster or dispenser munitions. Rocket attacks.
8. Airfields			
a. Parked aircraft	Prevent takeoff, PTO-kill	Damage that requires repair other than normal preflight maintenance before level nonmaneuvering flight can be achieved and maintained for at least 5 min. Subcategories are identified by the number of hours of required repairs, as follows: PTO: Repairs requiring at least 5 min PTO4: Repairs requiring at least 4 h PTO24: Repairs requiring at least 24 h	For both damage criteria—PTO and K-kill: *Against aircraft in the open or in uncovered or light covered revetments:* Strafing attacks with guns or rockets; small fragmentation weapons; dispenser weapons with fragmentation or small shaped-charge bomblets; GP bombs fuzed for airburst over the target aircraft; GP bombs with instantaneous contact fuzes that hit within the revetment; EOGBs or LGBs with fragmentation or LSC warheads. Penetrating weapons to open fuel tanks, followed by incendiary munitions, should yield high probability of K-kill. Firebombs would be effective given a direct hit on each aircraft target.
	Catastrophic-on-ground, COG-kill	Nonrepairable damage to the aircraft so as to render it unfit for any purpose except cannibalization and scrap.	*Against aircraft in heavily covered revetments:* Guided bombs or missiles, assuming sufficient target signature for weapons to guide on, and with weapon capable of penetrating through roof and fuzed to detonate after perforation.
b. Runways	Interdiction, I-kill	Surface cratering sufficient to prevent aircraft from taking off or landing. No undamaged part of the	GP bombs with steel nose plug and short-time-delay tail fuze for runway penetration and

(Continued)

Table B.1 Target-damage criteria and appropriate weapon types *(Continued)*

Target	Damage Criteria		Appropriate Weapon Type[1]
		runway is long enough or wide enough for use as a takeoff strip.	cratering. *Note: Do not use nose and tail fuzing because nose fuze may cause explosive to react under impact force and deflagrate on the runway surface rather than penetrate on the basis of the time set on the tail delay fuze. In such case, damage will be minimal.*
c. Hangars	—	Damage is defined as structural damage of a specific level. Hangars that have sustained structural damage of 50% or more should be considered unusable. Aircraft and hangar shops will be severely damaged in the area under the collapsed hangar.	GP bombs fuzed for short delay striking the hangar or impacting close enough to severely damage load-bearing members of the hangar. EOGBs or LGBs with large warheads and short-delay fuzing should be effective.
d. Hardened aircraft shelters	Shelter breach, PTO-kill of aircraft	Damage to the aircraft by causing sufficient concrete spallation due to blast and/or earth shock to achieve the PTO damage level as stated in 8a above. PTO damage may be achieved by blocking aircraft egress from the hangarette because of spall or by jamming or rendering the front blast door inoperable in a closed or partially closed position or by cratering the area in front of the shelter to prevent aircraft egress. PTO damage will result where a large bomb penetrates the overburden to within the breaching radius of portions of the shelter arch wall yet does not penetrate into the interior of the shelter.	Large GP bombs, delay fuzed for maximum penetration. Guided weapons are appropriate against the shelter door provided target signature is adequate for guidance.

Table B.1 Target-damage criteria and appropriate weapon types *(Continued)*

Target	Damage Criteria		Appropriate Weapon Type[1]
		A detonation within the breach radius of any other point along the shelter will require further analysis to determine if a PTO-kill is produced on the aircraft by spall from the shelter walls or if a shelter destruct kill has been achieved.	
	Shelter penetration, K-kill of aircraft	Weapon penetrates shelter roof, walls, or doors and inflicts catastrophic damage on the aircraft so as to render it unfit for use and uneconomical to repair.	Any type of penetrating weapon that can place a damage mechanism with sufficient damage potential to kill the aircraft within the target. If the shelter doors are open, this can be achieved by accurate placement of the weapon in the closure. Guided missiles with large shaped charges attacking the doors can be effective in defeating the doors.
	Shelter destruct	Damage causing shelter collapse that results in an aircraft K-kill (defined in 8a above). Shelter destruct will result where a large bomb penetrates the arch wall and detonates in the interior of the shelter, collapsing the shelter and destroying the aircraft inside.	Shelter destruction can only be obtained with larger delay-fuzed GP bombs utilizing a nose plug and tail fuze.
9. Field Fortifications	Catastrophic, K-kill	Breaching the walls or overhead cover to cause breakdown of structural integrity and possible filling of the interior by dirt, scabbing, and debris.	GP bombs fuzed for optimum penetration and optimum cratering; guided weapons that can be guided on the field-fortification target and that impact on the target with explosive weights of 190 lb or greater (Mk 82 GP bomb or larger).

(Continued)

Table B.1 Target-damage criteria and appropriate weapon types *(Continued)*

Target	Damage Criteria		Appropriate Weapon Type[1]
10. Supply Depots and Dumps			
a. Stacked ammunition	Catastrophic, K-kill	Damage resulting in complete destruction of the stack. A high-order detonation of a projectile in a stack is assumed to destroy the entire stack.	Large blast and fragmentation weapons, direct hit by conical-shaped-charge and LSC weapons. (Small cluster fragmenting weapons are not appropriate.) Against wood-crated ammunition, fire and incendiary weapons are appropriate.
b. POL surface (aboveground) storage tanks	—	Destruction of the tank, ignition of the contents resulting in a sustained fire, or damage to the tank through blast or fragmentation resulting in drainage of the tank within 10 min or less after attack leaving less than 25% of the original contents.	Large blast and fragmentation weapons, direct hit by large shaped-charge or LSC weapons. Small cluster weapons alone are not appropriate.
c. POL storage drums	—	Destruction of drums or ignition of contents resulting in sustained fire to 50% of the drums within the unit.	Direct hit by LSC or conical-shaped-charge weapons, small fragmenting submunitions capable of igniting the contents, and all GP weapons resulting in destruction of the drums. Guns using HEI ammunition are effective resulting in sustained fires. (Primarily against gasoline.) FAE weapons are also effective.
d. POL underground storage	—	Destruction of the tanks making them unusable or ignition of the contents resulting in a sustained fire.	GP bombs fuzed for penetration and/or cratering.

Table B.1 Target-damage criteria and appropriate weapon types *(Continued)*

Target	Damage Criteria		Appropriate Weapon Type[1]
11. Land Transportation			
a. Roads		Road cratered sufficiently to prevent passage of vehicular traffic for a specified period of time.	Large GP bombs dropped singly or smaller bombs in either salvo or stick release, fuzed for penetration of reinforced concrete or graded earth.
b. Trucks	Mobility, M-kill	Damage that immobilizes a moving vehicle or renders it incapable of controlled movement within a given time. This type of kill is usually conditional based on provisions for repairing the damage.	GP bombs, fragmentation and incendiary submunitions, rockets, shaped-charge weapons, and strafing attacks are effective for inflicting M-, I-, and possible K-kills on light vehicles. Firebombs are effective against vehicles transporting flammable or explosive cargo. Combined attacks with fragmentation weapons that cause fuel to leak, followed by incendiary munitions are very effective. Dispenser systems containing fragmentation submunitions with high-velocity fragments or shaped-charge submunitions, and guided weapons that can be guided on the signatures emanating from a truck.
	Category A	Vehicle is immobilized within 5 mi.	
	Category B	Vehicle is immobilized within 20 min.	
	Category C	Vehicle is immobilized within 40 min.	
	Catastrophic, K-kill	Damage that renders the vehicle unfit for any purpose except salvage.	Same as for M-kill, except that weapons that will enhance fuel fires should be employed. Fragmentation weapons followed by incendiary weapons are effective. Direct hits and near misses with large weapons are needed for K-kill.

(Continued)

Table B.1 Target-damage criteria and appropriate weapon types *(Continued)*

Target	Damage Criteria		Appropriate Weapon Type[1]
c. Railways	Cut	Severe damage to, or undermining of, a section of track so that a train cannot pass until the damaged section of track is repaired or replaced.	Land mines and GP bombs fuzed to produce large craters.
d. Rolling stock	Mobility, M-kill	Overturn or other damage to the extent that a car is unable to move over the track network on its own truck assemblies without some repair or replacement of components.	Derailment may be caused by short-delay-fuzed GP bombs impacting quite near the car. Damage to trucks and undercarriage will be greater with instantaneous-fuzed GP bombs. Guided weapons with large warheads are effective. Against wooden cars and reactive cargo, use incendiary weapons.
e. Locomotives (1) Moving	S2	Train stops within 5 min and is not derailed; stops within 1 mile.	Instantaneous-fuzed GP bombs striking the locomotive; guided weapons with large fragmentation, LSC, or conical-shaped-charge warheads; cluster of shaped charges striking vital components of locomotive; armor-piercing projectiles of .50 caliber or larger and rockets are effective, given a hit.
	S3	Train stops within 60 min.	Instantaneous-fuzed GP bombs striking the locomotive; guided weapons with large fragmentation, LSC, or conical-shaped-charge warheads; armor-piercing projectiles of 7.62-mm or larger and rockets are effective, given a hit.

Table B.1 Target-damage criteria and appropriate weapon types *(Continued)*

Target	Damage Criteria		Appropriate Weapon Type[1]
	Q1 (quick kill)	Locomotive stops within 1 min.	Same as for S2.
(2) Stationary	R2	Requires 4 to 15 days to repair.	Large GP bombs with instantaneous firing; guided weapons with large HE and LSC warheads.
	R3	Requires more than 15 days to repair.	Same as for R2.
f. Tunnels	Block	Breaching of ceiling or walls causing complete breakdown of structural integrity and filling by dirt, water, or debris so as to render the tunnel impassable.	HE GP or penetration bombs fuzed for maximum penetration and cratering.
g. Bridges			
(1) Permanent	Drop span	Damage to the extent that it causes a span to drop.	Large GP bombs with instantaneous fuzing for maximum damage on the bridge superstructure, guided weapons with large HE or LSC warheads.
(2) Temporary	Interdiction, I-kill	Interdiction of specified load-capacity categories of traffic as follows:	*Against timber-deck bridges:* HE weapons. *Against cable bridges:* Weapons that produce earth cratering.
	Category 1	Up to 0.5 ton: includes human and animal foot traffic, bicycles, loaded hand carts, etc.	*Against wood or reed pontoon bridges:* Fragments, projectiles, and conical-shaped charges.
	Category 2	Up to 5 tons: includes light passenger vehicles and small cargo trucks.	*Against bamboo bridges:* Weapons delay-fuzed for underwater explosion.
	Category 3	Up to 15 tons: includes cargo trucks and light weapon carriers.	

(Continued)

Table B.1 Target-damage criteria and appropriate weapon types *(Continued)*

Target	Damage Criteria		Appropriate Weapon Type[1]
(3) Tactical floating	Interdiction	Rupture of adjacent pontoons.	Same as for permanent bridges.
h. Piers	Pier failure	Breaching of piers.	GP bombs with instantaneous fuzing for contact with pier faces above the water and delay fuzing for contact with pier faces under water or in earth.
i. Abutment	Collapse	Damage to the extent that it causes a span to drop.	GP bombs with delay fuzing delivered in the fill behind the abutment from the approach end of the bridge.
12. Water Transportation			
a. Inland canals	Block	Reduction of the water cross-sectional area below some critical value, so as to block the canal against vessel passage.	GP, demolition, and penetration bombs, delay-fuzed for maximum cratering in the canal bed.
b. Locks	Pin shear	Shearing of hinge pins. Deformation or other damage to the gates or caissons so they cannot be moved or cannot retain water in the lock. Pumps and gate-opening machinery damaged, rendering the lock useless. Breaching of levees, dikes, or dams.	GP bombs, delay-fuzed for underwater detonation, air-delivered torpedoes, and floating mines. *Against housed supporting equipment:* GP bombs and rockets with short-delay fuzes. *Against supporting equipment that is not housed:* instantaneous-fuzed GP bombs and rockets.
c. Port facilities	–	Blocking port-entrance channels and wharves by sinking vessels. Damage to the port rail- or road-clearance facilities. Damage or destruction of port warehouses and transit sheds or damage to loading cranes on wharves.	Appropriate weapons for attacking rails, bridges, ships, locks, and warehouses are presented in the pertinent sections of this table. GP bombs are appropriate for attacking wharves; when delay-fuzed for

Table B.1 Target-damage criteria and appropriate weapon types *(Continued)*

Target	Damage Criteria		Appropriate Weapon Type[1]
			cratering, they are suitable against breakwaters and jetties.
d. Commercial ships	SW kill, M-kill, Cargo kill	See Naval Vessels, Surface Combatants for definitions of kills.	Torpedoes, delay-fuzed GP bombs, and guided missiles with penetration capability, delay-fuzed for internal burst in ship.
e. Junks	Sink, K-kill	Sinking kill based on the buoyancy/cargo relationship. Requires structural damage in critical portion of the hull, uncontrolled leaks in all compartments, flooding within 20 min, loss of maneuverability, inability to reach shallow water, and loss by sinking if the junk is motorized or carrying high-density cargo.	Direct hits from large-caliber warheads (5.0 in. or larger). Bombs, delay-fuzed for underwater detonation, bursting near or under the keel of the junk. Hits near the waterline from 20-mm armor-piercing-incendiary (API) or HEI projectiles.
f. Sampans		Same as junks.	Same as junks.
g. Barges and small craft	Sink, K-kill	Sinking	GP bombs and guided missiles with short-time- delay or instantaneous fuzing striking the target or near the target; guns and rockets (if cargo is reactive).
h. Naval vessels			
(1) Surface combatants	Sink	Sinking kill based on the buoyancy/cargo relationship. Requires structural damage in critical portion of the hull, uncontrolled leaks in all compartments, flooding within 20 min, loss of maneuverability. Sinking occurs within 1 h.	Guided weapons with contact fuzes; cluster weapons containing conical-shaped-charge warheads; proximity- and contact-fuzed GP bombs; antiradiation missiles, delay-fuzed GP bombs.

(Continued)

Table B.1 Target-damage criteria and appropriate weapon types *(Continued)*

Target	Damage Criteria		Appropriate Weapon Type[1]
	Mobility, M-kill	Damage such that the ship cannot maintain steerage or all engines are stopped.	Same as sink.
	Firepower, F-kill	Damage to a ship such that its ability to use its offensive armament is lost. Firepower kill is separated into specific types of weapon systems on ships such as long-range missiles, short-range missiles, surface-to-surface missiles, guns, antisubmarine warfare, or antisurface warfare.	Same as sink.
(2) Submarines	SW kill	Damage that will result in pressure hull rupture and internal flooding of the submarine.	Same as M-kill weapon types.
			Data-link, electro-optical guided weapons with delay fuze; torpedoes; and delay-fuzed GP bombs.
13. Buildings	—	Structural damage of a specified level (percentage of roof or floor area). Structural damage is damage to principal load-carrying members (trusses, beams, columns, load-bearing walls) requiring replacement or special support during repair. Buildings that have sustained structural damage of 50% or more are considered unusable.	GP bombs striking the target with short-delay fuzing; guided weapons with LSC and large HE warheads; incendiary weapons (appropriate only for combustible contents).
14. Powerplants	—	Damage to the extent that the plant cannot generate or transmit the specified percentage of its output capacity for at least 30 days.	GP bombs with short-delay fuzing; guided weapons with LSC and large HE warheads.

Table B.1 Target-damage criteria and appropriate weapon types *(Continued)*

Target	Damage Criteria		Appropriate Weapon Type[1]
		Damage to the extent that the plant cannot generate or transmit for at least 6 mo.	Same as 30-day criterion.
15. Transformers	>1 day	Damage to the transformers so that transmission of electric power is not possible for at least 1 day.	Guided and unguided GP bombs; guided missiles with fragmentation and LSC warheads; cluster or dispenser munitions; rocket attacks.
	>1 mo	Damage to the transformers so that transmission of electric power is not possible for at least 1 mo.	Guided and unguided GP bombs; guided missiles with fragmentation and LSC warheads; cluster or dispenser munitions; 5-in.-rocket attacks.
	>6 mo	Damage to the transformers so that transmission of electric power is not possible for at least 6 mo.	Guided and unguided GP bombs.
16. Dams			
a. Earthen dams	—	Breaching of the dam so that erosion significantly reduces the storage capacity of the reservoir.	Largest available GP bombs with delay fuzing for maximum cratering effectiveness.
b. Concrete dams	—	Breaching or displacing the dam to release impounded water. No reliance on erosion.	Depth bombs with hydrostatic fuzing; large GP bombs with delay fuzing.
17. Industrial Targets			
a. Missile and aircraft factories	—	Damage to components so that the plant is no longer capable of the specified	GP bombs with short-delay fuzing; guided weapons with LSC or large HE warheads.

(Continued)

Table B.1 Target-damage criteria and appropriate weapon types *(Continued)*

Target	Damage Criteria		Appropriate Weapon Type[1]
		percentage of its output capacity for the times specified. <table><tr><th>Percent of Damage</th><th>Days to Full Recuperation</th><th>Production Loss (Days)</th></tr><tr><td>10</td><td>96</td><td>46</td></tr><tr><td>20</td><td>134</td><td>74</td></tr><tr><td>30</td><td>147</td><td>91</td></tr><tr><td>40</td><td>149</td><td>100</td></tr><tr><td>50</td><td>150</td><td>104</td></tr><tr><td>60</td><td>150</td><td>105</td></tr></table>	
b. Steel mills	—	Damage to components so that the plant is no longer capable of the specified percentage of its output capacity for the times specified. <table><tr><th>Percent of Damage</th><th>Months to Full Recuperation</th><th>Production Loss (Months)</th></tr><tr><td>10</td><td>9</td><td>4</td></tr><tr><td>20</td><td>18</td><td>11</td></tr><tr><td>30</td><td>23</td><td>18</td></tr><tr><td>50</td><td>24</td><td>23</td></tr><tr><td>70</td><td>25</td><td>24</td></tr></table>	Same as for missile and aircraft factories but with larger warheads, if available.
c. Aluminum plants	—	Undefined	Same as for steel mills.
d. Magnesium plants	—	Undefined	Same as for steel mills.
e. Other nonferrous	—	Undefined	Same as for steel mills.

Table B.1 Target-damage criteria and appropriate weapon types *(Continued)*

<table>
<tr><th>Target</th><th colspan="2">Damage Criteria</th><th>Appropriate Weapon Type[1]</th></tr>
<tr><td>f. Light industry</td><td>—</td><td>Undefined</td><td>Same as for aircraft and missile factories. Against combustible buildings or combustible building contents: incendiaries.</td></tr>
<tr><td>g. Manufacturing plants, machinery, and electrical equipment</td><td>—</td><td>
<table>
<tr><th>Percent of Damage</th><th>Days to Full Recuperation</th><th>Production Loss (Days)</th></tr>
<tr><td colspan="3">Tank and Heavy-Motor-Vehicle Manufacturing Plants</td></tr>
<tr><td>20</td><td>102</td><td>54</td></tr>
<tr><td>30</td><td>118</td><td>64</td></tr>
<tr><td>50</td><td>142</td><td>88</td></tr>
<tr><td>60</td><td>151</td><td>100</td></tr>
<tr><td>70</td><td>156</td><td>111</td></tr>
<tr><td colspan="3">Heavy Ordnance Manufacturing Plants</td></tr>
<tr><td>20</td><td>68</td><td>26</td></tr>
<tr><td>30</td><td>76</td><td>37</td></tr>
<tr><td>50</td><td>90</td><td>58</td></tr>
<tr><td>60</td><td>94</td><td>68</td></tr>
<tr><td>70</td><td>97</td><td>76</td></tr>
</table>
</td><td>Same as for missile and aircraft factories.</td></tr>
<tr><td>h. Explosives</td><td>—</td><td>Undefined</td><td>Against raw-material storage, manufacturing, and loading areas: Small incendiaries supported by GP bombs with instantaneous fuzing.</td></tr>
</table>

(Continued)

Table B.1 Target-damage criteria and appropriate weapon types *(Continued)*

<table>
<tr><th>Target</th><th colspan="2">Damage Criteria</th><th>Appropriate Weapon Type[1]</th></tr>
<tr><td></td><td></td><td></td><td>Against multistory nitration buildings: GP bombs with short-delay fuzing</td></tr>
<tr><td>i. Synthetic rubber plants</td><td>—</td><td>Undefined</td><td>Against massive buildings: GP bombs with short-delay fuzing.
Against areas with rubber and other chemicals: incendiaries.</td></tr>
<tr><td>j. Explosives</td><td>—</td><td></td><td></td></tr>
<tr><td>(1) Oil refinery</td><td></td><td>Damage to critical components to the extent that the plant is no longer capable of the specified percentage of its output capacity for the times specified.
<table>
<tr><th>Percent of Damage</th><th>Days to Full Recuperation</th><th>Production Loss (Days)</th></tr>
<tr><td>30</td><td>65</td><td>45</td></tr>
<tr><td>50</td><td>90</td><td>75</td></tr>
<tr><td>70</td><td>135</td><td>130</td></tr>
</table></td><td>GP bombs with instantaneous fuzing; guided weapons with LSC and large HE warheads.</td></tr>
<tr><td>(2) Other</td><td>—</td><td>Undefined</td><td>Against chlorine and caustic soda, nitrogen, hydrogen, and sulfuric acid plants: GP bombs with instantaneous fuzing.
Against synthetic ammonia plants: Rockets and bombs with instantaneous fuzing.</td></tr>
<tr><td>k. Abrasives factories</td><td>—</td><td>Undefined</td><td></td></tr>
</table>

Table B.1 Target-damage criteria and appropriate weapon types *(Continued)*

Target	Damage Criteria		Appropriate Weapon Type[1]
			GP bombs with short-delay fuzing to disable cranes and power source.
l. POL pumping stations		Damage that results in loss of 100% of the pumping capacity and requires at least 1 mo to repair.	GP bombs with instantaneous fuzing.
m. Nuclear-materials production	—	Undefined	*Against heavy buildings:* GP bombs. *Against light buildings:* Bombs with delay in nose and no delay in tail. *Against equipment:* Same as for light industry. *Against power source:* Same as for transformers.

[1]For additional information, refer to JMEM *Target Vulnerability*.

INDEX

Note: page numbers are followed by f or t (indicating figures or tables).

Supporting Materials

To download supplemental material files, please go to AIAA's electronic library, Aerospace Research Central (ARC), at arc.aiaa.org. Use the menu bar at the top to navigate to Books > AIAA Education Series; then, sort alphanumerically by title to navigate to the desired book's landing page.

A complete listing of titles in the AIAA Education Series is available from AIAA's electronic library, Aerospace Research Central (ARC), at arc.aiaa.org. Visit ARC frequently to stay abreast of product changes, corrections, special offers, and new publications.

AIAA is committed to devoting resources to the education of both practicing and future aerospace professionals. In 1996, the AIAA Foundation was founded. Its programs enhance scientific literacy and advance the arts and sciences of aerospace. For more information, please visit www.aiaafoundation.org.